# Experimental Design, ANOVA, and Regression

# Experimental Design, ANOVA, and Regression

**Richard A. Damon, Jr.**

*University of Massachusetts at Amherst*

**Walter R. Harvey**

*The Ohio State University*

**HARPER & ROW, PUBLISHERS, New York**

*Cambridge, Philadelphia, San Francisco, Washington,
London, Mexico City, São Paulo, Singapore, Sydney*

1817

Sponsoring Editor: Claudia M. Wilson
Project Editor: Thomas R. Farrell
Cover Design: Wanda Lubelska Design
Text Art: RDL Artset, Ltd.
Production: Kewal K. Sharma
Compositor: TAPSCO, Inc.
Printer and Binder: R. R. Donnelley & Sons Company

Experimental Design, ANOVA, and Regression

**Library of Congress Cataloging-in-Publication Data**

Damon, Richard A.
    Experimental design, ANOVA, and regression.

    Bibliography: p.
    Includes index.
    1. Biometry.  I. Harvey, Walter R. (Walter Robert),
1919–    . II. Title
QH323.5.D35  1987     574.072     86-11970
ISBN 0-06-041479-0

86  87  88  89  9  8  7  6  5  4  3  2  1

# Contents

## CHAPTER 7 SPLIT-PLOT DESIGNS   317

## CHAPTER 8 ANALYSIS OF COVARIANCE   389

## CHAPTER 9 MATRIX ALGEBRA   423

# Preface

We have each taught a second course in statistical methods for graduate students in the biological sciences for many years. In addition, one of us (W.R.H.) has taught an applied course in least-squares analysis and variance component estimation for 20 years. This text grew out of our experiences in teaching those courses and from our extensive consulting experience on research problems of faculty and graduate students.

Our goal was to develop a text that could be used for a second and perhaps third course in statistical methods for graduate students in the biological sciences. Therefore, it is assumed that students using this text will have taken one or perhaps two courses in introductory statistical methods. Introductory statistical concepts are reviewed very briefly in Chapter 1.

Currently, there are numerous computer packages available for completing statistical analyses. We believe it is important, however, that students understand the detailed computations involved in completing any analysis. Therefore, this text includes many example sets of data, and the computations involved in completing the analyses are usually given in considerable detail.

It is also important that students first obtain a good understanding of data analysis and interpretation under balanced designs before attempting to understand the appropriate analysis of data where disproportionate subclass numbers exist. Therefore, we have first presented, in most chapters, appropriate analyses for balanced designs with examples and interpretation of results before presenting analyses of data with disproportionate subclass frequencies.

For those who are interested only in the balanced designs, Sections 3.19–3.21, 4.10, 5.17, 5.18, 7.9, and 8.8 can be omitted from consideration. For those who are concerned with unbalanced designs, or analyses of data with disproportionate subclass frequencies, these sections would be of prime concern.

New methodology that does not currently appear in textbooks, to our knowledge, includes (i) the least-squares analysis of data under the mixed model when dispro-portionate frequencies exist (Sections 3.21 and 7.9), (ii) curve fitting to least-squares means (Section 5.17), and (iii) the fitting of individual class and subclass regressions with least squares (Section 8.9).

We have made no attempt in this text to consider the analysis of categorical data or nonparametric data analyses. Currently, there are several texts available that cover each of these subjects.

The topic of regression is discussed in considerable detail since this is one of the most powerful tools used in the analysis of research data. The use of regression in conjunction with the analysis of variance—the analysis of covariance—is stressed because its importance in controlling experimental error is frequently overlooked.

The text also includes a rather thorough consideration of split-plot and repeated measures designs. These designs, used concomitantly with most of the major designs, are of great utility in biological research. Several examples, drawn from different areas of the biological sciences, are presented and considered in depth.

We should like to express our appreciation to James R. Vilkitis, Dale O. Everson, Trina A. Hosmer, and Ann S. Zamzow, who were kind enough to review various portions of the manuscript and who offered many helpful suggestions. We should also like to thank Linda S. Parsons, Anne M. Murray, Karen M. Truehart, and Debbie Gallagher for their fine work in typing and preparing the manuscript.

*Richard A. Damon, Jr.*
*Walter R. Harvey*

# Experimental Design, ANOVA, and Regression

# Chapter 1

# Review of Statistical Concepts and Definitions

## 1.1 INTRODUCTION

In this text we assume that the reader has some prior knowledge of statistics, such as that gained in an introductory course, as we do not attempt to develop some of the introductory concepts. This is a text dealing with applied statistics at an intermediate level concerned primarily with presenting methods for research workers in the biological sciences for dealing with statistical problems which they encounter. There are applications in many areas other than biology, however, and examples have been selected from several disciplines.

## 1.2 OBJECTIVES

The objectives of this text are twofold. The first is to acquaint the student with some of the principles of experimental design. A knowledge of these principles has many advantages to the researcher, such as ease of analysis of results, increased value and precision of results, decreased cost of experimentation, and, frequently, more conclusive results. The second objective is to prepare the student or researcher to deal with a mass of data, not collected through a planned experiment, in a logical and statistically sound manner. It is not always possible for students or research workers to conduct an experiment of their own design, because of such problems as lack of facilities, expense of conducting the experiment, or length of time involved in collecting sufficient data. Instead, they must rely on data collected by others over a period of time, frequently with no particular design in mind. Examples of this situation are the analysis of census or weather data and long-term genetic studies. It is therefore important to have a good understanding of the general methods of analysis of data in such cases.

## 1.3  COMMON DEFINITIONS

Research data consist of recorded observations or measurements made on experimental units. Experimental units arise from innumerable sources such as plants, animals, humans, batches of steel, and schools. An experimental unit may consist of a group of plants, a group of animals, a pen of birds, an individual animal, or an individual person to which one level or application of a treatment is applied. It should be noted that more than one characteristic or property such as intelligence, age, height, or weight of an experimental unit can be observed or measured. Any such characteristic or property displaying variability is called a *variable*. A single measurement or observation in a series of measurements or observations is known as a *variate*.

The characteristics or properties of the variables can be recorded as either *quantitative* or *qualitative* observations or measurements. *Quantitative measurements* or observations are those recorded in numerical measurements for such traits as weights of plants, ages of animals, and heights of pine trees. *Qualitative observations* occur when units belong to different treatment groups, different species, different sexes, different periods of time, and similar mutually exclusive categories.

Whether variables are made up of quantitative or qualitative observations they are categorized as *dependent* or *independent variables*. The difference between these two types of variable can perhaps be understood more clearly in the context of an example. Let us suppose that 30 mice of the same sex are selected at random from a particular strain of mice and are assigned in equal numbers to each of three separate environmental conditions at random. If the investigation is concerned with the effects of the different environmental conditions on rate of growth, the amount of gain in weight for each mouse could be determined over some period of time, resulting in one observation or measurement for each mouse on each of the separate conditions. Since the amount of gain is expected to depend on the particular environmental condition to which a mouse was subjected, the gains in weight are said to make up the *dependent variable*. A *dependent variable* then, as the name implies, is one whose magnitudes are dependent on one or more sources of variation. In this example, the amounts of gain are expected to differ, depending on which environmental condition the mice are assigned. The environmental conditions are said to make up an independent variable. Thus, the *independent variable* is one on which a portion of the variation in the dependent variable depends.

Both dependent and independent variables can be categorized in one of two ways, *continuous* or *discontinuous*. A *continuous variable* is one that can have any value within a given or theoretical range. The more accurately or precisely a measurement can be made, the more values are possible. Any single value is an estimate of a more exact value that could be determined with a finer measuring instrument. Biological variables such as height, weight, length, and rate of growth are said to be continuous variables.

In contrast to continuous variables, *discontinuous* (or discrete) variables can have only specific, unique values, and intermediate values falling between these values are not possible. Examples of discontinuous, or discrete, variables are such things as kernels per ear of corn, number of petals on a flower, number of spectators at a football game, or number of animals displaying immunity to a disease.

Under the usual experimental conditions, we are dealing with a group of experimental units drawn at random from a parent population of such units, and we wish to draw inferences with regard to the larger population. The experimental units are selected under the assumption that they are representative of the parent population so that the results of a given experiment or analysis are considered to be applicable to the entire population. That is, interest does not normally focus on the performance of the few experimental units utilized but is concerned with the population represented by the sample of units. Thus, the average of a sample set of units is expected to be an estimate of the average of the entire population. Also, if some measure of the variation among the observations making up the sample is computed, this measure is expected to be an estimate of the variation in the entire population.

Values such as the mean, or average, and any measure of variation calculated from a sample are designated as *statistics*. Each of these statistics is an estimate of the true population value (normally not known) which is called a *parameter*. Thus, if we calculate the average of a sample of observations, we are calculating a sample statistic which is an estimate of the average of all individuals in the population—a population parameter.

One of the concepts that is used repeatedly in statistical analysis is that of the *degrees of freedom*. Since in any random sample of $n$ observations each observation is independent of all others, it is stated that the quantity $\Sigma\ Y^2$ has $n$ degrees of freedom. Furthermore, any statistic calculated from the observations, such as the mean, is said to have only one degree of freedom since its value for a specified number of observations is dependent on $\Sigma\ Y$. Thus, the quantity

$$\sum_i (Y_i - \bar{y})^2 = \sum_i Y_i^2 - \frac{Y_.^2}{n}$$

has $n - 1$ degrees of freedom, $n$ degrees of freedom for the quantity $\Sigma\ Y_i^2$ minus one degree of freedom for $Y_.^2/n$ which is a function of $\Sigma\ Y$. Therefore, the degrees of freedom associated with the sum of squares of deviations about the mean are one fewer than the number of observations.

Similarly, if we had the means of several samples drawn from the same population, the quantity

$$\sum_i (\bar{y}_i - \bar{y}.)^2 = \sum_i \bar{y}_i^2 - \frac{(\Sigma_i\ \bar{y}_i)^2}{n_g}$$

has $n_g - 1$ degrees of freedom, $n_g$ degrees of freedom for the quantity $\Sigma_i\ \bar{y}_i^2$ minus one degree of freedom for $(\Sigma_i\ \bar{y}_i)^2/n_g$, which is a function of $\Sigma\ \bar{y}_i$, where $n_g$ is the number of groups. Therefore, the degrees of freedom associated with the sum of squares of deviations of sample means about the overall mean is one fewer than the number of sample means. So the number of degrees of freedom of the difference between two quantities is equal to the difference between the two corresponding degrees of freedom.

We shall be dealing frequently with degrees of freedom throughout the text, partitioning the total degrees of freedom according to the mathematical model applied to the data.

## 1.4 MATHEMATICAL MODELS, SIGMA, AND DOT NOTATION

It would be a very lengthy task to describe in words all the mathematical operations carried out in statistical computations. It would likewise be awkward to define in words the different sets of data with which we deal. We therefore resort to what may be thought of as shorthand procedures for such purposes. Let us look at the data displayed in Table 1.1, which we shall use to discuss some of the descriptive shortcuts.

As a first step, we can describe this set of observations by means of what is termed a *mathematical model*, which would be

$$Y_i = \mu + e_i \qquad i = 1, \ldots, 4 \tag{1.1}$$

where $Y_i$ represents any one of the individual weights, $\mu$ (mu) represents the average or mean of the universe of all possible weaning weights of Charolais steers, and $e_i$ represents a random error (chance variation) to which each weaning weight is subject. The mathematical model states that the weaning weight of any individual steer $Y_i$ is equal to the overall mean of an entire population or universe of weaning weights of Charolais steers plus some random or chance variation or error. The mean or average that we calculate from our sample is an estimate of the parameter $\mu$ and the error that we estimate is an estimate of the deviation from $\mu$, $e_i = Y_i - \mu$. In most mathematical models throughout this text, the Greek letter $\mu$ will be used to represent the population mean and Latin letters used to represent random errors and other effects in the model.

The random errors represented by $e_i$ in model (1.1) are usually assumed to be *normally and independently distributed* with a mean of 0 and a variance of $\sigma^2$, generally written NID(0, $\sigma^2$). This means that the errors ($e_i = Y_i - \mu$) follow the normal distribution. For a discussion of this important distribution any introductory text may be consulted.

The random error $e_i$ is the amount by which an individual observation deviates from the mean of all possible values, $\mu$. Thus, each value is subject to a chance, or uncontrolled and generally unknown source of variation, causing an increase or decrease from the overall mean. Since the random errors sum to 0 one would expect to find approximately equal numbers of positive and negative deviations in a sample. By independence of errors, it is meant that these errors follow one another in a random sequence; it is not expected that one would have a large number of positive deviations in sequence or a large number of negative deviations in sequence. Should this occur, a closer inspection of the experimental conditions should be undertaken. If it appears

**TABLE 1.1**  WEANING WEIGHTS
OF FOUR CHAROLAIS
STEERS (IN POUNDS)

| Steer number | Weight |
|:---:|:---:|
| 1 | $420 = Y_1$ |
| 2 | $480 = Y_2$ |
| 3 | $430 = Y_3$ |
| 4 | $465 = Y_4$ |

Sum $= 1795 = \Sigma_i Y_i = Y.$

that independence of errors does not hold for a set of data, a "runs test," as shown by Sokal and Rohlf (1969), can be performed. If the results of a runs test indicate that the errors are not independently distributed, caution must be taken in interpreting the results of the experiment. The lack of independence of errors indicates either some weakness in the way an experiment was laid out or a problem with the instrument of measurement. Since the estimates of the random errors are measured as deviations from the sample mean they must sum to 0 and their variance is an estimate of the variance $\sigma^2$ in the population which they represent.

If we add up the observations presented in Table 1.1 we can make use of the "sigma notation" and express the total as $\Sigma_i Y_i$, where $\Sigma$ means "the sum of" and the subscript $i$ indicates that all values of $Y_i$ have been summed. This is sometimes written $\Sigma_{i=1}^n Y_i$, indicating that there were $n$ observations in the sample and that all values from 1 to $n$ have been summed. This result can be written even more succinctly by utilizing the "dot notation." To do this we omit the sigma over which the summation has taken place and replace the subscript $i$ on the $Y_i$ with a dot, giving $\Sigma_i Y_i = Y.$, which represents the sum of the observations, as shown in Table 1.1. The sigma notation and the dot notation will be expanded in Chapters 2 and 3.

The $i$ subscript on a $Y$ designates any one of the individual $Y$ values and if we wish to specify a particular value of $Y$ the $i$ is replaced by the appropriate subscript. Thus, $Y_2 = 480$ and $Y_4 = 465$.

## 1.5  DESCRIPTIVE STATISTICS AND PARAMETERS

### 1.5.1  Mean

One of the first items of interest in a set of observations is a measure of a central or representative overall value. While such values as the mode, median, and midpoint of the range of values are occasionally used, the most commonly used measure is the average, generally referred to as the arithmetic mean, or simply the mean. The mean is obtained by dividing the total of all observations by the number of observations as shown in formula (1.2),

$$\bar{y} = \frac{\Sigma_i Y_i}{n} = \frac{Y.}{n} \qquad (1.2)$$

where $\bar{y}$ is the common symbol for the mean, sometimes written $\bar{Y}$, and $n$ is the number of observations in the sample. When using $n$ we are designating the number of observations in the group and thus need no subscript on $n$. When dealing with two or more groups we would need a subscript on $n$ to distinguish between the number of observations in each group. The same reasoning applies regarding subscripts on the $\bar{y}$. Since there is only one group involved in our example (Table 1.1) no subscript is necessary for the $\bar{y}$ or the $n$. The mean for the data in this table is

$$\bar{y} = \frac{Y.}{n} = \frac{1795}{4} = 448.75$$

and is an estimate of the unknown population parameter $\mu$.

## 1.5.2 Variance

When dealing with biological data, one is continually confronted with variability among observations. While various measures of this variability are used, the one of primary interest in statistical analysis is known as the variance. *Variance* in a sample or group of observations is measured as the sum of the squares of the deviations from the sample mean divided by one fewer than the number of observations. It has been shown that division by the total number of observations leads to a biased estimate of the population variance, while division by one fewer than the total number of observations leads to an unbiased estimate. The formula for calculating the variance in any sample set of data can be written

$$s^2 = \frac{\Sigma_i(Y_i - \bar{y})^2}{n - 1} \tag{1.3}$$

where $s^2$ is the symbol used for the variance of the observations in the sample and $n$ is the number of observations. The value $s^2$ is an estimate of the population variance, a parameter designated $\sigma^2$. The deviation $Y_i - \bar{y}$ is often written $y_i$, leading to the frequent use of the symbol $\bar{Y}$ for the mean, since $\Sigma_i\, y_i = 0$.

To calculate the variance of the observations in Table 1.1 we can set up Table 1.2 which shows the operations involved in calculating the variance using formula (1.3). The variance of 806.25 is an estimate of the population variance $\sigma^2$. While calculating the variance using formula (1.3) is relatively easy with only a few observations, it is quite cumbersome when a large number of observations are involved. It can be shown algebraically that

$$\frac{\Sigma_i(Y_i - \bar{y})^2}{n - 1} = \frac{\Sigma_i\, Y_i^2 - Y_{\cdot}^2/n}{n - 1} \tag{1.4}$$

The right-hand side of the equation, known as the "working formula," is easier to use with a large number of observations. Using formula (1.4) for calculating the variance of the observations shown in Table 1.1 we have

$$s^2 = \frac{\Sigma_i\, Y_i^2 - Y_{\cdot}^2/n}{n - 1} = \frac{807,925 - (1795)^2/4}{4 - 1}$$

$$= \frac{807,925.00 - 805,506.25}{3} = \frac{2418.75}{3} = 806.25$$

**TABLE 1.2**   CALCULATION OF VARIANCE USING DEVIATIONS
FROM THE MEAN

| $Y_i$ | $Y_i - \bar{y} = y_i$ | $(Y_i - \bar{y})^2 = y_i^2$ |
|---|---|---|
| 420 | −28.75 | 826.5625 |
| 480 | 31.25 | 976.5625 |
| 430 | −18.75 | 351.5625 |
| 465 | 16.25 | 264.0625 |

$Y_{\cdot} = 1795$        $\Sigma_i(Y_i - \bar{y}) = 0.00$        $\Sigma_i(Y_i - \bar{y})^2 = 2418.7500 = \Sigma_i\, y_i^2$

$$\bar{y} = \frac{Y_{\cdot}}{n} = \frac{1795}{4} = 448.75 \qquad s^2 = \frac{\Sigma_i(Y_i - \bar{y})^2}{n - 1} = \frac{\Sigma\, y_i^2}{n - 1} = \frac{2418.7500}{3} = 806.25$$

$\Sigma\ Y_i^2$ is read as the "sum of the squared $Y$'s," meaning that each $Y$ value is squared individually with these squares being summed. $Y_.^2$ is read as "the square of the sum of the $Y$'s."

### 1.5.3 Standard Deviation

While it is convenient mathematically to use the squared deviations as a measure of dispersion or variation about the mean, it is normal to think of this variation in terms of the original values. We can merely take the square root of the variance to return to the original scale of measurement. The square root of the variance is known as the *standard deviation* and is expressed as

$$s = \sqrt{\frac{\Sigma_i\ Y_i^2 - Y_.^2/n}{n - 1}} \tag{1.5}$$

where $s$ is the symbol for the standard deviation of a sample from the population and is an estimate of the parameter $\sigma$. In the example under discussion we have

$$s = \sqrt{806.25} = 28.39$$

If our data follow the normal curve, the population of observations would be visually described by the normal curve shown in Figure 1.1. That is, if one were able to plot the values of all the individuals in a population, their frequencies would follow a bell-shaped curve, or distribution, with a clustering of observations not greatly different in size than the mean and a reduction in numbers of observations as the size of the deviation from the mean increases in either direction. Statistical theory tells us that the area under the curve included by $\mu \pm 1\sigma$ would include 68.26 percent of the variates or observations, $\mu \pm 2\sigma$ would include 95.46 percent of the observations, and $\mu \pm 3\sigma$ would include 99.73 percent of the values. It can also be stated that 50 percent of the values would fall between $\mu \pm 0.674\sigma$, 95 percent of the values would fall between $\mu \pm 1.965\sigma$, and 99 percent of the values would fall between $\mu \pm 2.576\sigma$. Assuming that the estimates of the population standard deviation and the mean from

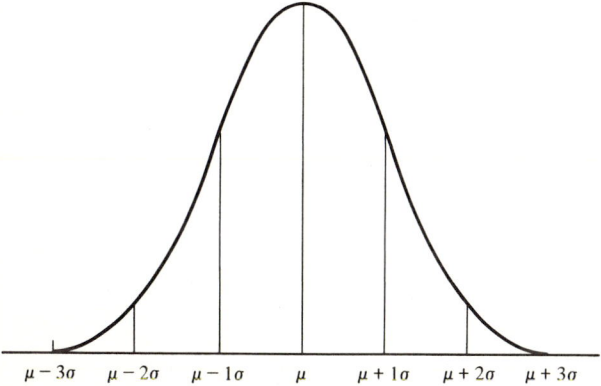

**Figure 1.1** Areas of the normal curve.

our sample are the population parameters (i.e., $\sigma = 28.38$ and $\mu = 448.75$), we would expect 50 percent of the weaning weights of Charolais steers to fall between 429.62 and 467.88 lb, 95 percent of the weaning weights to fall between 392.96 and 504.54 lb, and 99 percent to fall between 375.62 and 521.88 lb. It should be noted that four observations is an extremely small number from which to make such estimations.

### 1.5.4 Standard Error of the Mean

If a number of samples are drawn from a population, the mean of each of the samples could be calculated, and we would find that we had variability among these sample means just as we had variability among the individual observations in a single sample. If a large number of sample means were calculated, the mean of these sample means would be an estimate of the parameter $\mu$. The means would also be distributed in a normal curve similar to that of the original population. However, the variation among the means would be smaller than $\sigma_y$ since the mean of a sample will deviate less from the overall mean than will some members of the sample. The variance among a group of sample means would be measured as

$$s_{\bar{y}}^2 = \frac{\Sigma_i(\bar{y}_i - \bar{y}.)^2}{k - 1} \tag{1.6}$$

where $s_{\bar{y}}^2$ represents the variance of a group of sample means, $\bar{y}_i$ represents an individual sample mean, $\bar{y}.$ represents the mean of all sample means, and $k$ is the number of means included. The value $s_{\bar{y}}^2$ would be an estimate of the parameter $\sigma_{\bar{y}}^2$.

A common situation arises when we have only one sample mean and wish to estimate the variance to be expected in a distribution of several means. This variance is estimated as

$$s_{\bar{y}}^2 = \frac{s^2}{n} \tag{1.7}$$

where $s^2$ is the variance calculated among the observations within the sample and $n$ is the number of observations in the sample. The square root of the variance of the means, $s_{\bar{y}}$, would be the standard deviation of the mean. However, the standard deviation of a statistic such as a mean, whether calculated as the square root of Equation (1.6) or estimated as the square root of Equation (1.7), is normally referred to as the standard error of the statistic, in this case the *standard error of the mean*. The term standard deviation is usually reserved to describe the variation among individual observations. In our sample, the standard error of the mean would be

$$s_{\bar{y}} = \frac{s}{\sqrt{n}} = \frac{28.39}{\sqrt{4}} = 14.20$$

It can be seen that the standard error of the mean is inversely related to the square root of the sample size. As the sample size increases, the standard error of the mean decreases. The standard error of the mean serves the same purpose for the distribution of sample means as does the standard deviation for a distribution of individual observations. The standard error of the mean is an estimate of the parameter $\sigma_{\bar{y}}$ and a range of $\mu \pm 1\sigma_{\bar{y}}$ includes 68.26 percent of a population of means and a range of $\mu \pm 1.96\sigma_{\bar{y}}$ includes 95 percent of the means. The standard error of the mean is used frequently in statistics to indicate what amount of variation would be expected with continued sampling of a population of means.

### 1.5.5 Confidence Interval

The standard error of a statistic is used frequently to develop what is termed a confidence interval. A *confidence interval* is a range between upper and lower limits, which is expected to include the true population value of a parameter at a selected level of probability. The upper and lower limits are referred to as *confidence limits* and are a function of a $t$ value for a given level of probability and the standard error of the statistic.

The necessary $t$ values are found in Table A.2 and are derived from the $t$ distribution developed by William S. Gossett (Student, 1908) and perfected by R. A. Fisher (1926). W. S. Gossett (1876–1937) was a brewer and statistician who published under the name of Student. R. A. Fisher (1890–1962) was one of the pioneers in the field of statistics and made outstanding contributions in a great many areas of statistical theory and application. The statistic

$$t = \frac{\bar{y} - \mu}{s_{\bar{y}}} \tag{1.8}$$

is used repeatedly throughout the text. The curve for the $t$ distribution is symmetric, and as the degrees of freedom increase, the $t$ distribution comes closer to the normal curve in form. Values for degrees of freedom of $\infty$ are those of the normal distribution.

The $t$ distribution has found great utility in statistical procedures, and its application with regard to statistics other than the mean, and their standard errors, will be detailed in later sections of the text.

If we wished to develop the 95 percent confidence interval about the mean of a set of observations, we would calculate the confidence limits as

$$\text{Lower confidence limit} = -t_{0.05}s_{\bar{y}} \tag{1.9}$$

$$\text{Upper confidence limit} = +t_{0.05}s_{\bar{y}} \tag{1.10}$$

where the $t$ value is found in the table of the $t$ distribution (Table A.2) under the column headed 0.05 level of probability and in the row for $n - 1$ degrees of freedom, where $n$ is the number of observations in the sample. The confidence interval about the population mean can then be written

$$\bar{y} - t_{0.05}s_{\bar{y}} \le \mu \le \bar{y} + t_{0.05}s_{\bar{y}} \tag{1.11}$$

The 95 percent confidence interval for the mean of the four observations in Table 1.1 would be from

$$448.75 - (3.182)(14.20) \quad \text{to} \quad 448.75 + (3.182)(14.20)$$

or

$$403.57 \quad \text{to} \quad 493.93$$

Calculation of this interval leads to the statement that we feel 95 percent confident that the population mean $\mu$ lies between 403.57 and 493.93 lb. The interval is commonly written

$$\bar{y} \pm t_{0.05} s_{\bar{y}} \tag{1.12}$$

which in our example would be

$$448.75 \pm 45.18$$

### 1.5.6 Coefficient of Variation

While we can measure the variation among a set of observations in a sample by means of the standard deviation, there are occasions when we wish to compare the relative amounts of variation in two variables having different means, such as the birth weights and mature weights of mice. Since the mature weights would intuitively be expected to have a larger standard deviation because of the larger values involved, a comparison of the two standard deviations would not be very helpful. Again, we might be interested in a comparison of the relative amounts of variation between different characteristics or traits of the same animals, such as the weaning weights and white blood cell counts of rabbits. A statistic known as the coefficient of variation has been developed for use in such situations. The *coefficient of variation* is calculated as the standard deviation expressed as a percentage of the mean, the formula being

$$CV = \frac{(100)s}{\bar{y}} \tag{1.13}$$

It can be seen that this coefficient is independent of the unit of measurement and thus comparisons among coefficients of variation can be made to evaluate the relative amounts of variation. For the data displayed in Table 1.1 we would have

$$CV = \frac{(100)(28.39)}{448.75} = 6.33 \text{ percent}$$

Comparisons among coefficients of variation can be quite useful in designing experiments. Experiments involving those variables reflecting the greatest relative amount of variability would require larger numbers of experimental units to obtain precision equal to that of experiments with variables having smaller coefficients.

Researchers in particular areas are generally aware of the amount of variation to be expected in their experimental material and inspection of the coefficient of variation in a set of data may indicate problems in experimental techniques or methods of analysis.

### EXERCISES

**1.1.** Suppose that 10 apple trees from each of three different strains are to be studied for differences in production, the measure of productivity being bushels of apples produced by each tree. We would then have two variables, strains and bushels of apples per tree or yield. Categorize each of these two variables as to whether it is quantitative or qualitative, dependent or independent, continuous or discrete.

**1.2.** Consider an experiment in which four different groups of poultry, with 20 birds in each group, are maintained on rations containing different levels of protein. The response to the different levels of protein is measured in terms of number of eggs laid in a one-year period. Categorize each of the variables as to whether it is quantitative or qualitative, dependent or independent, continuous or discrete.

**1.3.** We are given the following times (in seconds) in which 10 students ran the 100-yard dash: 11.2, 9.4, 9.9, 10.6, 11.2, 11.6, 12.5, 10.1, 11.2, and 12.3. If the model for this set of data is $Y_i = \mu + e_i$ and $i$ goes from 1 to $n$, what is the value of $n$? Show the values for $Y_6$, $Y_8$, $Y_{i-7}$, and $Y_i - 1.0$ where $i = 9$. Calculate the value $\bar{y}$.

**1.4.** Suppose that six students were able to complete the following number of push-ups: 32, 24, 56, 48, 36, and 44. Calculate the mean number of push-ups for the group. Calculate $y_i = Y_i - \bar{y}$ for all observations and show that $\Sigma_i(Y_i - \bar{y}) = 0$. Calculate the variance of this set of observations using the formula $s^2 = \Sigma_i(Y_i - \bar{y})^2/(n - 1)$ and also by the formula $s^2 = (\Sigma_i Y_i^2 - Y_.^2/n)/(n - 1)$. Calculate the standard deviation $s$ and the standard error of the mean $s_{\bar{y}}$.

**1.5.** Show algebraically that $\Sigma_i(Y_i - \bar{y})^2$ equals $\Sigma_i Y_i^2 - Y_.^2/n$.

**1.6.** Suppose that the maximum speeds of fast balls thrown by eight different baseball pitchers were 87, 92, 91, 96, 84, 88, 92, and 90 mi/h. Calculate the mean, variance, standard deviation, standard error of the mean, 95 percent confidence interval about the mean, and the coefficient of variation for this set of data.

**1.7.** The grades for 12 students in a class of biometrics were 92, 74, 48, 93, 71, 97, 88, 67, 86, 94, 82, and 80. Calculate the mean, variance, standard deviation, standard error of the mean, 99 percent confidence interval about the mean, and the coefficient of variation for this set of data.

# Chapter 2

# One-Way Classification of Data

## 2.1 INTRODUCTION

Usually, observations resulting from a planned experiment are the result of measurements taken on experimental units such as plants, animals, or humans that have been subjected to one or more treatments which are expected to add to the variation normally present among random samples not subjected to any extraneous treatment. If two or more groups of experimental units are exposed to different treatments, it is logical to expect that these treatments will bring about a change in the amount of variation among the units. While the variation within the treatment groups would not be expected to change, since all units within a particular treatment group should give a similar response, the experimental units in one treatment group would be expected to vary from those in another treatment group in response to the applied treatments. Thus, the more effective the treatments are in causing differential responses among the groups, the greater will be the differences among the group or treatment means. If we find a greater variation among the means than would be expected due to chance alone, we conclude that the treatments are effective and that there are significant differences among the treatment means. The terms group means and treatment means are used synonymously.

## 2.2 ANALYSIS OF VARIANCE (ANOVA)

Since we know that chance variation alone will cause differences among the treatment means, it would be unsatisfactory to guess at whether or not there are real differences among the means. Some objective criterion is desired to remove guesswork or subjective evaluation from the decision process. The most powerful and useful statistical procedure utilized in the analysis of research data is called the *analysis of variance,* frequently referred to by the acronym ANOVA. Developed by R. A. Fisher, this technique is applied in many areas such as agronomy, genetics, psychology, zoology, entomology, geology, botany, physical education, and sociology. A great number of computer programs have been developed over the years to handle differing sets of data by means of analysis of variance.

The analysis of variance allows us to make what is known as a *test of significance* between or among treatment means, the results being given on the basis of the probability, due to chance alone, of finding a difference or differences as large as those obtained. The test of significance is cast in the form of accepting or rejecting what is termed a *null hypothesis,* the hypothesis of no difference. If we are concerned with the difference between two means, we can write the null hypothesis as

$$H_0: \mu_1 = \mu_2$$

meaning that we are hypothesizing that there is no difference between the population treatment means. If the null hypothesis is rejected, we have an *alternative hypothesis* $H_1$. This hypothesis can take the form

$$H_1: \mu_1 \neq \mu_2$$

the hypothesis that the population treatment means do differ. There are other null hypotheses and alternative hypotheses that might be considered, such as

$$H_0: \mu_1 \geq \mu_2 \qquad H_1: \mu_1 < \mu_2$$

$$H_0: \mu_1 \leq \mu_2 \qquad H_1: \mu_1 > \mu_2$$

In the analysis of variance, the null hypothesis is accepted or rejected on the basis of the test criterion

$$F = \frac{\text{Mean square for treatment means}}{\text{Mean square for error}} \tag{2.1}$$

in the most simple form. The meanings of the numerator and denominator will be shown in examples to follow. The original test of this nature was $Z = \log_e \sqrt{F}$ discovered by R. A. Fisher. G. W. Snedecor (1934) later put the distribution in the form of $F$ values, named in honor of Fisher. Table A.3, entered under degrees of freedom for numerator and degrees of freedom for denominator of the $F$ ratio, presents the $F$ values.

It is most common in statistical procedures to use the 5 or 1 percent levels shown in the $F$ table. If the value of $F$ found in the analysis is equal to or greater than the value found in the $F$ table at either the 5 or 1 percent level, we reject the null hypothesis and accept the alternative hypothesis. By rejecting the null hypothesis, we are guided by the fact that the probability of finding a difference as large as or larger than that obtained in the experiment is quite small ($P \leq 0.05$ or $P \leq 0.01$), and we conclude that there is a significant difference between the treatment means. The treatment means differ more than we would expect due to chance alone. If the null hypothesis is rejected at the 5 percent level, this means that there is less than 1 chance in 20 of finding a difference as large as or larger than that found just due to chance. Similarly, if the null hypothesis is rejected at the 1 percent level, we conclude that there is less than 1 chance in 100 of obtaining a difference as large as or larger than that found due to chance alone. If the null hypothesis is rejected at the 5 percent level, we state that there is a "significant" difference between the two means, and the difference is denoted by one asterisk on the $F$ value. If the null hypothesis is rejected at the 1 percent level, we state that there is a "highly significant" difference between the means and this is denoted by two asterisks on the $F$ value. While it is customary to use the 5 and 1 percent levels in tests of significance, findings of significance at higher levels such as 10, 15, and even 20 percent may be indicative of something other than chance causing differences.

When more than two treatments are involved the null hypothesis is that of no differences among population treatment means, with the alternative hypothesis that the population treatment means do differ.

## 2.3 ONE-WAY CLASSIFICATION: EQUAL NUMBERS

The simplest form of the analysis of variance is the one-way classification in which observations are classified only on the basis of one set of treatments or source of variation. An example of this is seen in Table 2.1, which presents the results of growing tree seedlings in two types of soil. The two groups are referred to as *categories, classes,* or *levels* of the *classification* soil type.

In going from a single sample or set of observations to two groups of observations, we have had to use an additional subscript on the $Y$'s. The first subscript $i$ tells us in which of the two groups an observation lies and the second subscript $j$ distinguishes among the observations within a particular group. Thus, $Y_{23} = 36$ tells us that 36 lies in the second type of soil and is the third observation in that soil type. If we add

**TABLE 2.1**  HEIGHT OF TREE SEEDLINGS IN CENTIMETERS GROWN ON TWO TYPES OF SOIL

| Soil type $i = 1$ | Soil type $i = 2$ |
|---|---|
| $54 = Y_{11}$ | $43 = Y_{21}$ |
| $57 = Y_{12}$ | $37 = Y_{22}$ |
| $64 = Y_{13}$ | $36 = Y_{23}$ |
| $68 = Y_{14}$ | $42 = Y_{24}$ |
| $50 = Y_{15}$ | $35 = Y_{25}$ |
| $293 = \Sigma_j Y_{1j} = Y_1.$ | $193 = \Sigma_j Y_{2j} = Y_2.$ |

$$Y_1. + Y_2. = \sum_i Y_i. = Y.. = 293 + 193 = 486$$

$$n_1 = n_2 = n_w = 5 \qquad n_g = 2 \qquad n_g n_w = N = 10$$

$$\sum_j Y_{1j}^2 = 17,385 \qquad \sum_j Y_{2j}^2 = 7503$$

$$\bar{y}_1 = \frac{Y_1.}{n_w} = \frac{293}{5} = 58.6 \qquad \bar{y}_2 = \frac{Y_2.}{n_w} = \frac{193}{5} = 38.6$$

$$\bar{y}. = \frac{Y..}{N} = \frac{486}{10} = 48.6$$

up or sum the observations in the first group, we will be adding up all the observations with different $j$ subscripts, the addition being said to be over the $j$ subscript, for $i = 1$. We then have $\Sigma_j Y_{1j} = 293$ and adding over the $j$ subscript in the second group we would have $\Sigma_j Y_{2j} = 193$. Putting each of these totals into the dot notation, we would omit the $\Sigma_j$ and substitute a dot for the $j$ subscript, indicating that we have summed over the $j$ subscript, giving $Y_1.$ and $Y_2..$ When we add the total for the first soil type to the total for the second soil type we are summing over both soil types. Since the soil types are specified by the $i$ subscript, we are therefore summing over $i$, giving $\Sigma_i Y_i. = Y..$, each of the group totals being a $Y_i..$

The figures at the foot of Table 2.1 show that, while two subscripts are necessary for the totals for the groups, only one subscript is necessary for the means. The symbol $n_g$ represents the number of groups and $n_w$ the number of observations within each group. Note that the product of these two yields the total number of observations $N$.

We can describe the set of data shown in Table 2.1 by the mathematical model

$$Y_{ij} = \mu + t_i + e_{ij} \qquad i = 1, 2 \qquad j = 1, \ldots, 5 \tag{2.2}$$

where $Y_{ij}$ is the $j$th observation in the $i$th level of soil type, $\mu$ is the overall mean of the population from which the plants were selected, $t_i$ is an effect due to the $i$th level of soil type, and $e_{ij}$ is the random error associated with the $j$th observation in the $i$th soil type.

As the term *analysis of variance* implies, we are going to analyze the variation that exists among the 10 observations in Table 2.1. We are then going to partition the total variation among the observations into appropriate sources as defined in our model. Our null hypothesis can be written

$$H_0: \mu_1 = \mu_2 \qquad H_1: \mu_1 \neq \mu_2 \tag{2.3}$$

which states our hypothesis that the two group means do not differ and the alternative hypothesis that the two groups do differ.

We shall accept or reject the null hypothesis depending on our computed $F$ value compared with $F$ values found in Table A.3. To carry out this $F$ test, we make the following calculations:

$$\text{Total uncorrected sum of squares} = \sum_i \sum_j Y_{ij}^2$$

$$= (54)^2 + (57)^2 + \cdots + (35)^2$$

$$= 24{,}888 \tag{2.4}$$

$$\text{Total sum} = \sum_i \sum_j Y_{ij} = Y..$$

$$= 54 + 57 + \cdots + 35 = 486$$

$$\text{Total corrected sum of squares} = \sum_i \sum_j (Y_{ij} - \bar{y}.)^2 = \sum_i \sum_j Y_{ij}^2 - \frac{Y_{..}^2}{N}$$

$$= 24{,}888 - \frac{(486)^2}{10} = 24{,}888.0 - 23{,}619.6$$

$$= 1268.4 \tag{2.5}$$

$$\text{Sum of squares between soil types} = n_w \sum_i (\bar{y}_i - \bar{y}.)^2 = \frac{\sum_i Y_{i.}^2}{n_w} - \frac{Y_{..}^2}{N}$$

$$= \frac{(293)^2 + (193)^2}{5} - \frac{(486)^2}{10}$$

$$= 24{,}619.6 - 23{,}619.6 = 1000.0 \qquad (2.6)$$

$$\text{Sum of squares within soil types} = \sum_i \sum_j (Y_{ij} - \bar{y}_i)^2$$

$$= \left( \sum_i \sum_j Y_{ij}^2 - \frac{Y_{..}^2}{N} \right) - \left( \frac{\sum_i Y_{i.}^2}{n_w} - \frac{Y_{..}^2}{N} \right)$$

$$= \sum_i \sum_j Y_{ij}^2 - \frac{\sum_i Y_{i.}^2}{n_w}$$

$$= 1268.4 - 1000.0 = 24{,}888.0 - 24{,}619.6$$

$$= 268.4$$

$$= \text{Total SS} - \text{Between soil type SS} \qquad (2.7)$$

The sum of squares for within treatments, which is seen to be the sum of squares of the deviations of the individual observations from their group means, is calculated either by the difference between the total corrected sum of squares and the corrected sum of squares between soil types [the first equality in Equation (2.7)] or the difference between the uncorrected total sum of squares and the uncorrected sum of squares between soil types [the second equality in Equation (2.7)]. We have added a second sigma in these formulas to indicate that we sum over the observations within the groups as well as over the groups. This is done for the individual observations as well as the squares of the individual observations. We have then partitioned the total sum of squares 1268.4 into its components: 1000.0, the sum of squares between soil types (treatments), and 268.4, the sum of squares within soil types. This can be seen algebraically as

$$\sum_i \sum_j (Y_{ij} - \bar{y}.)^2 = n_w \sum_i (\bar{y}_i - \bar{y}.)^2 + \sum_i \sum_j (Y_{ij} - \bar{y}_i)^2 \qquad (2.8)$$

or 
$$\sum_i \sum_j Y_{ij}^2 - \frac{Y_{..}^2}{N} = \left( \frac{\sum_i Y_{i.}^2}{n_w} - \frac{Y_{..}^2}{N} \right) + \left( \sum_i \sum_j Y_{ij}^2 - \frac{\sum_i Y_{i.}^2}{n_w} \right) \qquad (2.9)$$

This same procedure is followed repeatedly in the analysis of variance, the partitioning being more detailed as additional sources of variation are specified. It should be noted that when we speak of the total sum of squares, we are normally referring to the total *corrected* sum of squares (i.e., the correction term has been subtracted, the square of the overall total divided by the number of observations being known as the correction term).

The analysis of variance for this set of data is shown in Table 2.2, where df represents degrees of freedom.

Among the total of 10 observations we have nine degrees of freedom, after accounting for the mean, which are partitioned as was the total sum of squares. Since

**TABLE 2.2** ANALYSIS OF VARIANCE OF WEIGHTS OF VETCH PLANTS

| Source of variation | df | Sum of squares | Mean square | F |
|---|---|---|---|---|
| Total | 9 | 1268.4 | | |
| Between treatments | 1 | 1000.0 | 1000.00 | 29.81** |
| Within treatments | 8 | 268.4 | 33.55 | |

there are two levels, or classes, of soil type there is one degree of freedom *between* them. The five observations within the first level of soil type yield four degrees of freedom and the five observations within the second level of soil type yield an additional four degrees of freedom, giving a total of eight degrees of freedom *within* the treatments of soil types. This last line in an analysis of variance table is frequently designated as residual or error.

The column headed "Mean square" in Table 2.2 results from dividing the sum of squares by the corresponding degrees of freedom. The mean square for "within" treatments is an estimate of the variation among individual observations in the population. The mean square for "between" treatments, while being subject to this same variation, is subject to the additional variation brought about by the difference between the two treatment effects. Thus, the between treatment mean square would be expected to be larger than the within treatment mean square if the treatments or soil types cause any differences between the groups. The specific components of mean squares will be discussed in more detail in a later section.

Since we are interested in whether there is a significant difference between the two treatments or soil type means we proceed to a consideration of our null hypothesis. After calculating the $F$ ratio as described in formula (2.1), we enter a table of $F$ values (Table A.3) under the column headed 1 for the degrees of freedom for the numerator mean square and in the row for 8 degrees of freedom for the denominator mean square. If our calculated value of $F$ equals or exceeds that given for the 5 percent level, we reject the null hypothesis and state that there is a "significant" difference between the two treatment means. If our value equals or exceeds that at the 1 percent level we would reject the null hypothesis and conclude that there is a "highly significant" difference between the two means. A significant difference is denoted by an asterisk attached to the $F$ value and a highly significant difference is denoted by two asterisks. Our interpretation is that there is either less than 1 chance in 20 or less than 1 chance in 100 of obtaining a difference as large as or larger than the one obtained *due to chance alone.* Therefore, we conclude that the treatments have caused a significantly greater difference between the means than would have occurred at random.

While we have calculated the sum of squares between the soil types or treatments by use of formula (2.6) we could have calculated this sum of squares in terms of the mathematical model, Equation (2.2). The $t_i$ in the model are referred to as treatment effects and represent deviations from the overall mean. The $t_i$ are estimated by $\hat{t}_i$ that, with equal numbers in the groups, are

$$\hat{t}_i = \bar{y}_i - \bar{y}. \tag{2.10}$$

where $\bar{y}.$ is our estimate of $\mu$, sometimes designated $\hat{\mu}$. Since $\bar{y}. = 486/10$, we can calculate

$$\hat{t}_1 = \bar{y}_1 - \bar{y}.. = \frac{Y_1.}{n_1} - \frac{Y..}{N} = \frac{293}{5} - \frac{486}{10} = 10.0$$

$$\hat{t}_2 = \bar{y}_2 - \bar{y}.. = \frac{Y_2.}{n_2} - \frac{Y..}{N} = \frac{193}{5} - \frac{486}{10} = -10.0$$

Using these values, the sum of squares between, or due to, soil types can be calculated as

$$\sum_i n_i \hat{t}_i^2 = n_1 \hat{t}_1^2 + n_2 \hat{t}_2^2 \tag{2.11}$$

giving

$$(5)(10.0)^2 + (5)(-10.0)^2 = 1000.0$$

It would be quite cumbersome to calculate the sum of squares in this way with many levels of treatment, but we hope this does relate the calculations made in the analysis of variance to the mathematical model used to describe the data.

The random errors defined by the mathematical model can also be related to the computations in the analysis of variance. Using the symbol $\hat{e}_{ij}$ as our estimate of $e_{ij}$, we can substitute this symbol as well as the $\hat{t}_i$ and $\bar{y}.$ in the model, giving a sample model of

$$Y_{ij} = \bar{y}. + \hat{t}_i + \hat{e}_{ij} \tag{2.12}$$

Rearranging, we have

$$\hat{e}_{ij} = Y_{ij} - \bar{y}. - \hat{t}_i \tag{2.13}$$

and since $\hat{t}_i = \bar{y}_i - \bar{y}.$ this can be written

$$\hat{e}_{ij} = Y_{ij} - \bar{y}. - (\bar{y}_i - \bar{y}.) = Y_{ij} - \bar{y}_i \tag{2.14}$$

Therefore,

$$\sum_i \sum_j \hat{e}_{ij}^2 = \sum_i \sum_j (Y_{ij} - \bar{y}_i)^2 \tag{2.15}$$

as shown in formula (2.7). It can be seen from Equation (2.15) that the random errors measure the deviations of the individual observations from the treatment or group mean. The sum of the squares of these deviations, or random errors, yields the sum of squares within groups or soil types, previously calculated in a different manner.

As an example of the one-way classification, we have used only two groups. However, the number of groups possible is limited only by practical considerations and all the operations and discussion in this section apply equally to three or more levels or classes.

## 2.4 ONE-WAY CLASSIFICATION: UNEQUAL NUMBERS

The analysis of data with differing numbers of observations per treatment group proceeds in the same fashion as the analysis with equal numbers. The one modification to be made deals with the calculation of the uncorrected treatment sum of squares. With unequal numbers in the groups, each treatment total is squared and divided by the number of observations in the group. The results are summed up over treatments, yielding the uncorrected treatment sum of squares. This is in contrast to the situation with equal numbers where the treatment totals are squared and summed, then divided by the common $n$. The formula for calculating the treatment sum of squares with unequal numbers is

$$\sum_i \frac{Y_{i.}^2}{n_i} - \frac{Y_{..}^2}{N} \tag{2.16}$$

where $N$ is the total number of observations and $n_i$ is the number of observations in the $i$th treatment, in contrast to formula (2.6). The formula for the within treatment sum of squares is written

$$\sum_i \sum_j Y_{ij}^2 - \sum_i \frac{Y_{i.}^2}{n_i} \tag{2.17}$$

in contrast to formula (2.7).

As an example of the one-way classification with unequal numbers, let us consider the data displayed in Table 2.3. The mathematical model remains the same as for the case with equal numbers per treatment:

$$Y_{ij} = \mu + t_i + e_{ij} \qquad i = 1, 2 \qquad j = 1, \ldots, n_i \tag{2.18}$$

$$\text{Total sum of squares (corrected)} = \sum_i \sum_j Y_{ij}^2 - \frac{Y_{..}^2}{N}$$

$$= 106.89 - \frac{(37.3)^2}{14}$$

$$= 106.89 - 99.3779 = 7.5121 \tag{2.19}$$

$$\text{Sum of squares between treatments} = \sum_i \frac{Y_{i.}^2}{n_i} - \frac{Y_{..}^2}{N}$$

$$= \frac{(25.4)^2}{8} + \frac{(11.9)^2}{6} - \frac{(37.3)^2}{14}$$

$$= 80.6450 + 23.6017 - 99.3779$$

$$= 104.2467 - 99.3779 = 4.8688 \tag{2.20}$$

**TABLE 2.3** BIRTH WEIGHTS (IN POUNDS) OF POLAND CHINA PIGS IN TWO LITTERS

| Litter 1 | Litter 2 |
|----------|----------|
| 3.3 | 2.6 |
| 3.6 | 2.2 |
| 2.6 | 2.2 |
| 3.1 | 2.5 |
| 3.2 | 1.2 |
| 3.3 | 1.2 |
| 2.9 | |
| 3.4 | |
| $25.4 = Y_{1.}$ | $11.9 = Y_{2.}$ |

$$Y_{1.} + Y_{2.} = Y_{..} = 25.4 + 11.9 = 37.3$$

$$n_1 + n_2 = N = 8 + 6 = 14$$

$$\text{Sum of squares within treatments} = \sum_i \sum_j Y_{ij}^2 - \sum_i \frac{Y_{i.}^2}{n_i}$$

$$= 106.89 - 104.2467$$

$$= 2.6433 \tag{2.21}$$

The analysis is then completed as shown in Table 2.4.

Since we had a total of 14 observations, we have 13 degrees of freedom for the total with 1 degree of freedom between the two litters. Since the first litter had eight observations there are 7 degrees of freedom in that litter and 5 degrees of freedom among the six observations in the second litter, yielding 12 degrees of freedom within litters. The $F$ value of 22.10 with 1 and 12 degrees of freedom shows a highly significant difference between the average birth weights of the pigs in the two litters.

In the example of analysis of variance with equal numbers in the classes, or groups, in Section 2.3, we found that the sums of squares between groups and within groups could be obtained by calculations made in terms of the mathematical model. With unequal numbers in the classes some complications arise because the overall mean $\bar{y}. = Y../N$ is no longer the best estimate of the population mean. Instead, our estimate is

$$\hat{\mu} = \frac{\sum_i \bar{y}_i}{n_t} = \frac{25.4/8 + 11.9/6}{2} = \frac{3.1750 + 1.9833}{2} = 2.57915$$

an average of the two means. Our estimates of the treatment (or litter weight) effects are

$$\hat{t}_1 = \bar{y}_1 - \hat{\mu} = 3.1750 - 2.57915 = 0.5958$$

$$\hat{t}_2 = \bar{y}_2 - \hat{\mu} = 1.9833 - 2.57915 = -0.5958$$

We are not able to use the $\hat{t}$ values as we did previously to obtain the treatment sum of squares. Instead, we redefine $\bar{y}_i - \bar{y}.$ as

$$\hat{t}'_1 = \bar{y}_1 - \bar{y}. = \frac{25.4}{8} - \frac{37.3}{14} = 3.1750 - 2.6643 = 0.5107$$

$$\hat{t}'_2 = \bar{y}_2 - \bar{y}. = \frac{11.9}{6} - \frac{37.3}{14} = 1.9833 - 2.6643 = -0.6810$$

We can now calculate the treatment sum of squares as

$$\sum_i n_i \hat{t}'^2_i = (8)(0.5107)^2 + (6)(-0.6810)^2 = 2.0865 + 2.7826 = 4.8691$$

which is the value we obtained earlier (within rounding error).

**TABLE 2.4**   ANALYSIS OF VARIANCE OF BIRTH WEIGHTS OF
               POLAND CHINA PIGS

| Source of variation | df | Sum of squares | Mean square | $F$ |
|---|---|---|---|---|
| Total | 13 | 7.5121 | | |
| Between | 1 | 4.8688 | 4.8688 | 22.10** |
| Within | 12 | 2.6433 | 0.2203 | |

The sum of squares for within groups, however, can be calculated as before, as we can write

$$\hat{e}_{ij} = Y_{ij} - \hat{\mu} - \hat{t}_i$$

and since $\hat{t}_i = \bar{y}_i - \hat{\mu}$,

$$\hat{e}_{ij} = Y_{ij} - \hat{\mu} - \bar{y}_i + \hat{\mu} = Y_{ij} - \bar{y}_i$$

## 2.5 *t* TEST: EQUAL NUMBERS

The significance of the difference between the two treatment groups presented in Table 2.1 can also be tested by means of the *t* test with the same conclusion as with the analysis of variance. This test is based on the *t* distribution discovered by W. S. Gossett (Student, 1908) and perfected by R. A. Fisher (1926) as noted earlier. While the analysis of variance described heretofore can be applied to several treatment groups, the *t* test is limited to only two treatments. The *t* test is applied in many situations, most commonly when one is interested in testing for the significance of the difference between two statistics or between a statistic and some hypothesized value, such as whether a calculated correlation coefficient or regression coefficient differs from zero or some hypothesized value. The *t* test can be represented in a general form as

$$t = \frac{\text{Calculated value} - \text{Expected value}}{\text{Standard error of calculated value}} \quad (2.22)$$

where the calculated value can be a mean or other statistic and the expected value a parameter whose difference from the statistic is to be tested. If we wish to use the *t* test for our test of the significance of the difference between the two treatment means of the data in Table 2.1, our null hypothesis is that there is no difference between the two means; that is, we are sampling a population of mean differences whose mean is zero. We can then express formula (2.22) as

$$t = \frac{\bar{d} - \mu_0}{s_{\bar{d}}} \quad (2.23)$$

where $\bar{d}$ is the difference between the two means (the mean difference), $\mu_0$ is the hypothesized population mean difference, and $s_{\bar{d}}$ is the standard error of the mean difference.

Since our assumption is that $\mu_0 = 0$, the formula can be written

$$t = \frac{\bar{d}}{s_{\bar{d}}} = \frac{\bar{y}_1 - \bar{y}_2}{s_{\bar{y}_1 - \bar{y}_2}} \quad (2.24)$$

Since $s_{\bar{y}_1 - \bar{y}_2} = s\sqrt{1/n_1 + 1/n_2}$, we can first calculate $s$, which is an estimate of $\sigma$, by pooling the sums of squares from the two samples and dividing by the pooled degrees of freedom:

$$s = \sqrt{\frac{(\Sigma_j Y_{1j}^2 - Y_{1.}^2/n_1) + (\Sigma_j Y_{2j}^2 - Y_{2.}^2/n_2)}{(n_1 - 1) + (n_2 - 1)}}$$

$$= \sqrt{\frac{(n_1 - 1)s_1^2 + (n_2 - 1)s_2^2}{(n_1 - 1) + (n_2 - 1)}}. \quad (2.25)$$

This yields a weighted average of the sample variances. With equal numbers in the two groups, weighted and arithmetic averages of the variances are the same. The assumptions are made that the two populations are normally distributed and have a common variance. The general working formula for conducting the $t$ test is then

$$t = \frac{\bar{d}}{s_{\bar{d}}} = \frac{\bar{y}_1 - \bar{y}_2}{\sqrt{\dfrac{(\Sigma_j Y_{1j}^2 - Y_{1.}^2./n_1) + (\Sigma_j Y_{2j}^2 - Y_{2.}^2./n_2)}{(n_1 - 1) + (n_2 - 1)}} \sqrt{\dfrac{n_1 + n_2}{n_1 n_2}}} \tag{2.26}$$

In this form, the formula can be applied whether or not there are equal numbers of observations in the two groups. With equal numbers in the two groups, we can replace $\sqrt{(n_1 + n_2)/n_1 n_2}$ with $\sqrt{2/n}$, where $n$ is the number of observations in each of the two groups. As an example of the $t$ test with equal numbers let us refer back to the data presented in Table 2.1 where we found

$$Y_{1.} = 293 \qquad \bar{y}_1 = \frac{Y_{1.}}{n_1} = \frac{293}{5} = 58.6 \qquad \sum_j Y_{1j}^2 = 17,385$$

$$Y_{2.} = 193 \qquad \bar{y}_2 = \frac{Y_{2.}}{n_2} = \frac{193}{5} = 38.6 \qquad \sum_j Y_{2j}^2 = 7503$$

Now following formula (2.26) we have

$$t = \frac{\bar{d}}{s_{\bar{d}}} = \frac{293/5 - 193/5}{\sqrt{\dfrac{[17,385 - (293)^2/5] + [7503 - (193)^2/5]}{(5 - 1) + (5 - 1)}} \sqrt{\dfrac{2}{5}}}$$

$$= \frac{58.6 - 38.6}{\sqrt{\dfrac{(17,385 - 17,169.8) + (7503 - 7449.8)}{8}} \sqrt{\dfrac{2}{5}}}$$

$$= \frac{20}{\sqrt{268.4/8} \sqrt{0.40}} = \frac{20}{3.6633} = 5.46**$$

Since we are concerned only with the size of the difference between the two means, we can use the absolute value of $\bar{d}$ in the numerator of the $t$ ratio. The table of the $t$ distribution (Table A.2) is entered under the degrees of freedom of $(n_1 - 1) + (n_2 - 1)$ or 8 in the example, where it is found that the value of 5.46 exceeds the tabulated value of $t$ at both the 5 and 1 percent levels. We therefore conclude that there is a highly significant difference between the two treatment means. While we have not *proved* that the difference between the two means is certain, we declare the difference to be highly significant since the probability of obtaining a difference as large as or larger than this, *due to chance alone,* is less than 1 chance in 100.

It should be noted that when there are only two treatment groups, or in any situation where there is only one degree of freedom for the numerator and both a $t$ test and an $F$ test are carried out, $t^2 = F$. In our example $t^2 = 65.90 = F = 65.93$. (The difference between the two values is a result of rounding error.)

## 2.6 *t* TEST: UNEQUAL NUMBERS

Just as the data in Table 2.1 containing equal numbers in the two groups can be analyzed by the *t* test as well as by the analysis of variance using the *F* test, the data in Table 2.3 containing unequal numbers in the two groups can also be analyzed by either method. From the data in Table 2.3 we find

$$Y_{1.} = 25.4 \qquad \bar{y}_1 = \frac{Y_{1.}}{n_1} = \frac{25.4}{8} = 3.1750 \qquad \sum_j Y_{1j}^2 = 81.32$$

$$Y_{2.} = 11.9 \qquad \bar{y}_2 = \frac{Y_{2.}}{n_2} = \frac{11.9}{6} = 1.9833 \qquad \sum_j Y_{2j}^2 = 25.57$$

Following formula (2.26) we have

$$t = \frac{\bar{d}}{s_{\bar{d}}} = \frac{25.4/8 - 11.9/6}{\sqrt{\dfrac{[81.32 - (25.4)^2/8] + [25.57 - (11.9)^2/6]}{(8-1) + (6-1)}} \sqrt{\dfrac{8+6}{(8)(6)}}}$$

$$= \frac{3.1750 - 1.9833}{\sqrt{\dfrac{2.6433}{12}} \sqrt{\dfrac{14}{48}}} = \frac{.1917}{\sqrt{\dfrac{37.0062}{576}}} = \frac{1.1917}{0.2535} = 4.701^{**}$$

Again we can note that $t^2 = 22.10 = F = 22.10$. Entering the *t* table under the degrees of freedom of $(n_1 - 1) + (n_2 - 1) = 12$, we conclude that there is a highly significant difference between the two treatment groups since our value of 4.701 exceeds the value tabulated for the 1 percent level of 3.055 for 12 degrees of freedom.

The data displayed in Tables 2.1 and 2.3 consisted of only two categories or classes of experimental units. For this reason, both sets of data could be subjected to either a *t* test or an analysis of variance. However, should there be more than two categories of experimental units in a one-way classification, only an analysis of variance can be used. Thus, the *t* test is limited to the particular case where only two treatment classes exist, while the number of classes permissible with the analysis of variance is unlimited.

## 2.7 *t* TEST: UNEQUAL VARIANCES

The *t* tests we have discussed use the basic formula for the variance of the difference between two means,

$$\sigma_{\bar{y}_1 - \bar{y}_2}^2 = \frac{\sigma_1^2}{n_1} + \frac{\sigma_2^2}{n_2} \tag{2.27}$$

and in each case the sums of squares of deviations from treatment means in each sample were pooled, as were the degrees of freedom, to estimate the variance of the mean difference. The assumption was made that the two population variances were equal, in order to justify this pooled estimate. However, there are occasions when the variances in the two populations differ from one another, $\sigma_1^2 \neq \sigma_2^2$, and yet one wishes to carry out a $t$ test. This can be done by estimating $\sigma_1^2$ by $s_1^2$ and $\sigma_2^2$ by $s_2^2$ and using these estimates in the formula

$$t' = \frac{\bar{y}_1 - \bar{y}_2}{\sqrt{s_1^2/n_1 + s_2^2/n_2}} \tag{2.28}$$

Note that neither the sums of squares nor the degrees of freedom are pooled as was done previously.

The $t'$ quantity shown in Equation (2.28) does not follow exactly the $t$ distribution but is an approximation of it. Different forms of this distribution have been developed, each with accompanying tables. One is known as the Behrens–Fisher distribution and was presented by Fisher and Yates (1957). Another distribution was derived by Welch (1947) and tables for this distribution have been prepared by Aspin (1949). However, it is not necessary to refer to these tables because the regular $t$ distribution can be used by estimating the approximate degrees of freedom involved using a method suggested by Satterthwaite (1946). This approximation takes the form

$$df' = \frac{(s_1^2/n_1 + s_2^2/n_2)^2}{(s_1^2/n_1)^2/(n_1 - 1) + (s_2^2/n_2)^2/(n_2 - 1)} \tag{2.29}$$

This value is rounded down to the next smallest whole number to enter the $t$ table.

As an example of the computations involved in carrying out this test, we shall use some observations selected from an experiment conducted by Dr. Allen V. Barker of the Department of Plant and Soil Sciences at the University of Massachusetts. The data and the calculations leading to $t'$ and approximate degrees of freedom are shown in Table 2.5. The value of $t' = 5.05$ leads to the conclusion that there is a highly significant difference between the two treatment means.

It is rather obvious from merely looking at the two groups shown in Table 2.5 that there is a large difference between the variances in each of the groups. However, in many instances, a difference is only suspected and a statistical test of the difference between the two variances is desired. Such a test is discussed in Section 3.14.

Although the $t$ test is frequently completed using the form shown in Equation (2.28), it is more common to transform the data in such a way that the variances are homogeneous; that is, they do not differ significantly from one another. The transformation of data is discussed in Chapter 3.

## 2.8 NESTED CLASSIFICATION: BALANCED DESIGN

While the terms *hierarchal, hierarchical,* and *within-within* are also used to describe this new classification, the term *nested* will be used primarily throughout this text. The groupings specified in experimental data are designated as either *main* or *nested classifications.* In a *main classification* every level of classification can be identified independently of any other. This is the type of classification into which the observations dealt with in this chapter so far would fall. For example, in Table 2.1 all one needs to know to classify an observation properly with respect to soil type is whether it was grown in soil type 1 or soil type 2.

**TABLE 2.5**    NITRATE ACCUMULATION (IN GRAMS) IN SPINACH PLANTS GROWN AT TWO LEVELS OF NITROGEN

| Level 1 | Level 2 |
|---------|---------|
| 41.4 | 14.3 |
| 98.4 | 15.9 |
| 49.9 | 14.8 |
| 43.2 | 17.6 |
| 79.4 | 13.5 |
| 59.8 | — |

$$\Sigma\ Y_1 = \quad 372.1 \qquad\qquad \Sigma\ Y_2 = \quad 76.1$$
$$\Sigma\ Y_1^2 = 25{,}633.17 \qquad \Sigma\ Y_2^2 = 1168.35$$
$$\bar{y} = \quad 62.02 \qquad\qquad \bar{y} = \quad 15.22$$

$$\Sigma(Y_1 - \bar{y}_1)^2 = \Sigma\ Y_1^2 - \frac{(\Sigma\ Y_1)^2}{n_1} \qquad \Sigma(Y_2 - \bar{y}_2)^2 = \Sigma\ Y_2^2 - \frac{(\Sigma\ Y_2)^2}{n_2}$$

$$= 2556.77 \qquad\qquad = 10.11$$

$$s_1^2 = \ 511.35 \qquad\qquad s_2^2 = \ 2.53$$

$$\frac{s_1^2}{n_1} = \quad 85.22 \qquad\qquad \frac{s_2^2}{n_2} = \ 0.51$$

$$s_{\bar{y}_1 - \bar{y}_2} = \sqrt{\frac{s_1^2}{n_1} + \frac{s_2^2}{n_2}} = \sqrt{\frac{511.35}{6} + \frac{2.53}{5}} = \sqrt{85.73} = 9.26$$

$$t' = \frac{\bar{y}_1 - \bar{y}_2}{s_{\bar{y}_1 - \bar{y}_2}} = \frac{46.80}{9.26} = 5.05$$

$$df' = \frac{(85.22 + 0.51)^2}{(85.22)^2/5 + (0.51)^2/4} = 5$$

*Each* of the levels of a main classification can be subdivided into randomly selected subgroups, in which case the classification of groups is said to be *nested*. Each of the levels of the nested classification can also be subdivided into randomly selected subgroups, resulting in a second nesting or hierarchy, with this process being continued through any number of hierarchies. Nested classifications may also occur in subclasses made up of *combinations of levels* of main classifications.

The level of a nested classification is not completely identified until the level (or levels) of the classification (or classifications) within which it is nested is identified. Nested classifications then are subordinate to other classifications. It is important to note that the classes or groups making up the subordinate levels are usually randomly chosen. A nested classification is normally encountered in a sampling situation.

Let us consider an experiment in which it is desired to test the rate of weight gain of steers in a feedlot from three different breeds of cattle. One could select a certain number of steers representing the offspring from several different sires in each of the three breeds. If we had the offspring from four sires in each of the three breeds, in order to obtain the total for the first breed we would merely add up the gains of all steers belonging to the first breed. However, if we want the total for the first sire, we now have to distinguish which breed as well as which sire because the first sire in breed one is not the first sire in breed two since the sires have been selected at random within the breeds. Therefore, our totals are for *sire one in breed one, sire one in breed two*, and so on, giving us 12 totals instead of the 4 we would have had if the sires had been a main classification. So the levels of the sires are subordinate or dependent on the levels of the breeds. Furthermore, since the sires appearing in breed one cannot possibly appear in breed two, there can be no interaction between a nested classification and the main classification(s) within which it is nested, although it is possible for the nested classification to interact with some other main or nested classification should there be one. The subject of a statistical interaction will be discussed in a later section.

As an example of the nested classification let us look at the data in Table 2.6 supplied by Dr. Donald L. Anderson of the Department of Veterinary and Animal Sciences at the University of Massachusetts.

The model for this set of data is

$$Y_{ijk} = \mu + t_i + p_{ij} + e_{ijk} \qquad i = 1, \ldots, 4 \qquad j = 1, 2 \qquad k = 1, \ldots, 6 \quad (2.30)$$

where $Y_{ijk}$ is the $k$th observation in the $j$th level of $P$ (pen) and the $i$th level of $T$ (treatment), $\mu$ is the population mean, $t_i$ is the effect of the $i$th level of $T$, $p_{ij}$ is the effect of the $j$th level of $P$ in the $i$th level of $T$, and $e_{ijk}$ is the random error associated with the $k$th observation in the $j$th level of $P$ and the $i$th level of $T$. The model indicates that we are going to partition the total sum of squares into the sums of squares due to treatment effects, effects of pens within a treatment, and random variation within the pen × treatment subclasses.

The computations necessary to complete the analysis of variance follow:

$$\text{Total SS} = \sum_i \sum_j \sum_k Y_{ijk}^2 - \frac{Y^2 \cdots}{N}$$

$$= 25,515,538 - \frac{(34,190)^2}{48}$$

$$= 25,515,538 - 24,353,252 = 1,162,286 \quad (2.31)$$

**TABLE 2.6**  WEIGHT GAINS (IN GRAMS) FROM 10 TO 20 WEEKS OF CHICKENS
PLACED ON COMBINATIONS OF HIGH AND LOW
CALCIUM AND LYSINE

| Treatments | | | | | | | |
|---|---|---|---|---|---|---|---|
| LoCaLoL | | LoCaHiL | | HiCaLoL | | HiCaHiL | |
| Pen 1 | Pen 2 | Pen 1 | Pen 2 | Pen 1 | Pen 2 | Pen 1 | Pen 2 |
| 573 | 1041 | 618 | 943 | 731 | 416 | 518 | 416 |
| 636 | 814 | 926 | 640 | 845 | 729 | 782 | 729 |
| 883 | 498 | 717 | 373 | 866 | 590 | 938 | 590 |
| 550 | 890 | 677 | 907 | 729 | 552 | 755 | 552 |
| 613 | 636 | 659 | 734 | 770 | 776 | 672 | 776 |
| 901 | 685 | 817 | 1050 | 787 | 657 | 576 | 657 |
| $Y_{11.} =$ 4156 | $Y_{12.} =$ 4564 | $Y_{21.} =$ 4414 | $Y_{22.} =$ 4647 | $Y_{31.} =$ 4728 | $Y_{32.} =$ 3720 | $Y_{41.} =$ 4241 | $Y_{42.} =$ 3720 |

$$\text{Treatment SS} = \text{SS}_T = \frac{\Sigma_i \, Y_{i..}^2}{n_p n_w} - \frac{Y_{...}^2}{N}$$

$$= \frac{(8720)^2 + (9061)^2 + (8448)^2 + (7961)^2}{(2)(6)} - 24,353,252$$

$$= \frac{292,886,346}{12} - 24,353,252$$

$$= 24,407,196 - 24,353,252 = 53,944 \qquad (2.32)$$

$$\text{Pens within treatment SS} = \text{SS}_{P:T} = \frac{\Sigma_i \, \Sigma_j \, Y_{ij.}^2}{n_w} - \frac{\Sigma_i \, Y_{i..}^2}{n_p n_w}$$

$$= \frac{(4156)^2 + (4564)^2 + \cdots + (3720)^2}{6} - 24,407,196$$

$$= \frac{147,197,302}{6} - 24,407,196$$

$$= 24,532,884 - 24,407,196 = 125,688 \qquad (2.33)$$

$$\text{Error SS} = \sum_i \sum_j \sum_k Y_{ijk}^2 - \frac{\Sigma_i \, \Sigma_j \, Y_{ij.}^2}{n_w}$$

$$= 25,515,538 - 24,532,884 = 982,654 \qquad (2.34)$$

It can be seen that the letters $P{:}T$ represent the effect of the pens within treatments. This is the notation that will be used throughout the text, the nested effect appearing to the left of the colon and the effect(s) within which it is nested to the right of the colon.

With the nested classification we have one more subscript than we did with the one-way classification, necessary for the identification of the level of nested effect or classification. We now have $Y...$, which represents the overall total; $Y_{ij.}$, the total for a treatment $\times$ pen subclass, the dot indicating that we have added all observations within the subclass; and $Y_{i..}$, the total for a treatment, the two dots indicating that we have added all the observations in the two pens of each treatment. The $n_p$ is the number of levels or classes of pens which is the number of pens *in each treatment,* and $n_w$ is the number of observations in each pen $\times$ treatment subclass. $N$ is the product $n_t n_p n_w = (4)(2)(6) = 48$, where $n_t$ is the number of levels of treatments. Note that in the denominators of all sums to be squared there is an $n$ corresponding to each dot in the numerator, with the appropriate subscript. Each denominator is *always* composed of the number of observations in the sum(s) to be squared.

To complete the analysis of variance it is necessary to compute the appropriate degrees of freedom. These calculations are shown in Table 2.7. Note that in calculating the degrees of freedom for the nested effect, the $n$ corresponding to the nested effect together with a "$-1$" appears to the left of the colon and the $n$ corresponding to the effect(s) within which it is nested appears to the right of the colon. The completed analysis of variance for this set of data is shown in Table 2.8.

**TABLE 2.7** DEGREES OF FREEDOM FOR NESTED CLASSIFICATION: BALANCED DESIGN

| Source | Degrees of freedom | | |
|---|---|---|---|
| Total | $N - 1$ | $= 48 - 1$ | $= 47$ |
| $T$ | $n_t - 1$ | $= 4 - 1$ | $= 3$ |
| $P{:}T$ | $(n_p - 1)n_t = (2 - 1)(4)$ | | $= 4$ |
| $W$ | $\dfrac{n_w - 1}{n_w} N = \dfrac{6 - 1}{6} (48) = 40$ | | |

The $F$ ratio for testing the significance of treatment effects is $T/P{:}T$ and if the resulting $F$ value had been greater than 1, the $F$ table would be entered with three degrees of freedom for numerator mean square and four degrees of freedom for denominator mean square. The $F$ ratio for testing the significance of pen effects is $P{:}T/W$ and the corresponding degrees of freedom for entering the $F$ table are 4 and 40. In both cases, we conclude that there are no significant differences.

## 2.9 NESTED CLASSIFICATION: UNBALANCED DESIGN

The nested classification with unequal levels in the nested effects, or unequal numbers in the smallest subclasses, presents no problem in analysis other than slight modification of formulas for obtaining some of the sums of squares and degrees of freedom. As an example of this type of analysis, let us look at the data presented in Table 2.9. These data were supplied by Richard Larsen of the Department of Fisheries and Wildlife at the University of Massachusetts.

In this set of data we have two hierarchies of nesting, bricks nested within locations and slides nested within bricks. This calls for an additional classification in the mathematical model, which is

$$Y_{ijkl} = \mu + l_i + b_{ij} + s_{ijk} + e_{ijkl}$$

$$i = 1, 2 \qquad j = 1, \ldots, n_{b_i} \qquad k = 1, \ldots, n_{s_{ij}} \qquad (2.35)$$

$$n_{lb} = \text{number of } LB \text{ subclasses} \qquad n_{lbs} = \text{number of } LBS \text{ subclasses}$$

where $Y_{ijkl}$ is the $l$th observation on the $k$th slide on the $j$th brick in the $i$th location, $\mu$ is the population mean, $l_i$ is the effect of the $i$th location, $b_{ij}$ is the effect of the $j$th brick in the $i$th location, $s_{ijk}$ is the effect of the $k$th slide on the $j$th brick in the $i$th location, and $e_{ijkl}$ is the random error associated with the $Y_{ijkl}$ observation. The number

**TABLE 2.8** ANALYSIS OF VARIANCE FOR DATA PRESENTED IN TABLE 2.6

| Source | df | SS | MS | F |
|---|---|---|---|---|
| $T$ | 3 | 53,944 | 17,981 | 0.57 |
| $P{:}T$ | 4 | 125,688 | 31,422 | 1.28 |
| $W$ | 40 | 982,654 | 24,566 | |

**TABLE 2.9**  NUMBER OF DIATOMS PER SQUARE CENTIMETER COLONIZING
GLASS SLIDES ATTACHED TO BRICKS AT TWO LOCATIONS ON A
RIVER (NUMBERS HAVE BEEN CODED FOR EASE IN COMPUTATIONS)

| Location 1 | | | | |
|---|---|---|---|---|
| Brick 1 | | Brick 2 | | Brick 3 |
| Slide 1 | Slide 2 | Slide 1 | Slide 2 | Slide 1 |
| 102 | 500 | 142 | 119 | 243 |
| 111 | 480 | 125 | 114 | 189 |
| 400 | 112 | 221 | 461 | 362 |
| 380 | 103 | 464 | 382 | 264 |
| 210 | 225 | | 510 | |
| 245 | 361 | | 380 | |
| | 842 | | | |
| | 657 | | | |
| $Y_{111.} = 1448$ | $Y_{112.} = 3280$ | $Y_{121.} = 952$ | $Y_{122.} = 1966$ | $Y_{131.} = 1058$ |
| $Y_{11..} = 4728$ | | $Y_{12..} = 2918$ | | $Y_{13..} = 1058$ |
| | | $Y_{1...} = 8704$ | | |

| Location 2 | | | |
|---|---|---|---|
| Brick 1 | | Brick 2 | |
| Slide 1 | Slide 2 | Slide 1 | Slide 2 |
| 500 | 822 | 826 | 642 |
| 165 | 743 | 750 | 710 |
| 752 | 263 | 682 | 720 |
| 142 | 321 | 522 | 584 |
| 921 | 620 | 650 | 650 |
| 792 | 584 | 621 | |
| 871 | 841 | | |
| 900 | | | |
| $Y_{211.} = 5043$ | $Y_{212.} = 4194$ | $Y_{221.} = 4051$ | $Y_{222.} = 3306$ |
| $Y_{21..} = 9237$ | | $Y_{22..} = 7357$ | |
| | $Y_{2...} = 16{,}594$ | | |

of bricks in the $i$th location is designated as $n_{b_i}$ and the number of slides in the $ij$th
location $\times$ brick subclass is designated as $n_{s_{ij}}$. The sums needed for the analysis are
shown at the bottom of the table, the dots showing over what subscript(s) summation
has taken place. The computations necessary to complete the analysis of variance
follow, again using unequal number notation in the denominators:

$$\text{Total SS} = \sum_i \sum_j \sum_k \sum_l Y_{ijkl}^2 - \frac{Y_{....}^2}{n_{...}}$$

$$= 15{,}370{,}364 - \frac{(25{,}298)^2}{54}$$

$$= 15{,}370{,}364 - 11{,}851{,}645 = 3{,}518{,}719 \qquad (2.36)$$

where $n_{...} = \sum_i \sum_j \sum_k n_{ijk}$ and $n_{ijk}$ is the number of observations in the $ijk$th subclass.

$$\text{Location SS} = \text{SS}_L = \sum_i \frac{Y_{i...}^2}{n_{i..}} - \frac{Y_{....}^2}{n_{...}}$$

$$= \frac{(8704)^2}{28} + \frac{(16{,}594)^2}{26} - \frac{(25{,}298)^2}{54}$$

$$= 13{,}296{,}502 - 11{,}851{,}645 = 1{,}444{,}857 \qquad (2.37)$$

where $n_{i..}$ is the number of observations in the $i$th location.

$$\text{Brick SS} = \text{SS}_{B:L} = \sum_i \sum_j \frac{Y_{ij..}^2}{n_{ij.}} - \sum_i \frac{Y_{i...}^2}{n_{i..}}$$

$$= \frac{(4728)^2}{14} + \frac{(2918)^2}{10} + \frac{(1058)^2}{4} + \frac{(9237)^2}{15} + \frac{(7357)^2}{11} - 13{,}296{,}502$$

$$= 13{,}336{,}666 - 13{,}296{,}502 = 40{,}164 \qquad (2.38)$$

where $n_{ij.}$ is the number of observations in the $ij$th location $\times$ brick subclass.

$$\text{Slide SS} = \text{SS}_{S:BL} = \sum_i \sum_j \sum_k \frac{Y_{ijk.}^2}{n_{ijk}} - \sum_i \sum_j \frac{Y_{ij..}^2}{n_{ij.}}$$

$$= \frac{(1448)^2}{6} + \frac{(3280)^2}{8} + \cdots + \frac{(3306)^2}{5} - 13{,}336{,}666$$

$$= 13{,}457{,}674 - 13{,}336{,}666 = 121{,}008 \qquad (2.39)$$

where $n_{ijk}$ is the number of observations in the $ijk$th location $\times$ brick $\times$ slide subclass.

$$\text{Error SS} = \sum_i \sum_j \sum_k \sum_l Y_{ijkl}^2 - \sum_i \sum_j \sum_k \frac{Y_{ijk.}^2}{n_{ijk}}$$

$$= 15{,}370{,}364 - 13{,}457{,}674 = 1{,}912{,}690 \qquad (2.40)$$

Again, it is necessary to calculate the degrees of freedom before completing the analysis of variance. The necessary calculations are shown in Table 2.10.

The completed analysis of variance for this set of data is shown in Table 2.11.

The $F$ tests for locations and bricks within locations are both approximate for reasons which will be explained in the section dealing with the expectations of mean squares.

**TABLE 2.10**  DEGREES OF FREEDOM FOR NESTED CLASSIFICATION: UNBALANCED DESIGN

| Source | Degrees of freedom | | | |
|--------|---|---|---|---|
| Total | $N - 1$ | $= 54 - 1$ | $=$ | 53 |
| $L$ | $n_l - 1$ | $= 2 - 1$ | $=$ | 1 |
| $B:L$ | $n_{lb} - n_l$ | $= 5 - 2$ | $=$ | 3 |
| $S:BL$ | $n_{lbs} - n_{lb}$ | $= 9 - 5$ | $=$ | 4 |
| $W$ | $N - n_{lbs}$ | $= 54 - 9$ | $=$ | 45 |

**TABLE 2.11** ANALYSIS OF VARIANCE FOR DATA
PRESENTED IN TABLE 2.9

| Source | df | SS | MS | F |
|--------|-----|-----------|-----------|------------------------|
| L | 1 | 1,444,857 | 1,444,857 | $L/B{:}L = 107.92$** |
| B:L | 3 | 40,164 | 13,388 | $B{:}L/S{:}BL = {<}1$ |
| S:BL | 4 | 121,008 | 30,252 | $S{:}BL/W = {<}1$ |
| W | 45 | 1,912,690 | 42,504 | |

It is sometimes puzzling to students to find, in some instances, a different number of subscripts on the $Y$'s than on the $n$'s in the formulas. This arises from the fact that the $Y$'s, with whatever subscripts, are used to identify particular individuals or particular totals whereas the $n$'s are used to designate the *number* of individuals, starting with the smallest class or subclass and going on to the total. In Table 2.8, for example, $Y_{1124} = 103$ needs four subscripts to be able to identify it as the fourth observation on the second slide on the first brick in the first location. However, $n_{112} = 8$ needs only three subscripts to identify this as being the number of observations on the second slide on the first brick in the first location. The number of subscripts for any $Y$ is that necessary to identify an individual observation while the number of subscripts on any $n$ is that necessary to identify the number of observations in any class in the one-way classification, or the smallest subclass in a nested or multiway classification.

## 2.10 THE METHOD OF LEAST SQUARES

### 2.10.1 Introduction to Least-Squares Analysis

The usual methods of analyzing data that are classified in only one way are straightforward even though unequal numbers exist from class to class, as has been shown above. However, to understand how to use the method of least squares (sometimes referred to as the method of fitting constants) for the analysis of data that are classified in more than one way when unequal subclass frequencies exist, it is necessary to understand that the method may also be used for the analysis of data that are classified in only one way. It will be shown that exactly the same results are obtained with the method of least squares as with conventional methods, in this case.

### 2.10.2 One-Way Classification: Equal or Unequal Numbers

The general mathematical model underlying the analysis of data that are classified in only one way, as was given in Equation (2.18), is

$$Y_{ij} = \mu + a_i + e_{ij} \qquad i = 1, \ldots, n_a \qquad j = 1, \ldots, n_i \qquad (2.41)$$

where $Y_{ij}$ is the $j$th observation in the $i$th level of $A$, $\mu$ is the overall mean when equal frequencies exist, $a_i$ is the effect of the $i$th level of $A$ expressed as a deviation from the overall mean $\mu$, and $e_{ij}$ is the random error. Each of the terms on the right-hand side of the model, other than the random error, represents population parameters or "constants."

Application of the method of least squares requires the use of a set of simultaneous equations, referred to as the *least-squares normal equations.* The least-squares normal equations result from the use of differential calculus. However, it is not necessary to complete the differential calculus manipulations to write out the least-squares equations, since these equations always follow a systematic pattern. The first point to remember in the construction of this set of equations is that there must be one equation for each of the parameters (constants) to be estimated. The equations for the one-way classification are as follows:

$$\mu: \quad N\hat{\mu} + n_1\hat{a}_1 + n_2\hat{a}_2 + \cdots + n_{na}\hat{a}_{na} = Y.$$

$$a_1: \quad n_1\hat{\mu} + n_1\hat{a}_1 \qquad\qquad\qquad = Y_1.$$

$$a_2: \quad n_2\hat{\mu} \qquad\quad + n_2\hat{a}_2 \qquad\qquad = Y_2.$$

$$\vdots \quad \vdots \qquad\qquad\qquad\qquad\qquad \vdots$$

$$a_{na}: \quad n_{na}\hat{\mu} \qquad\qquad\qquad + n_{na}\hat{a}_{na} = Y_{na}.$$

With this notation, the letter and appropriate subscript on the left followed by a colon denotes the equation. The second subscript is omitted in the $n$'s since this subscript is always summed over. The column of values on the right-hand side of the equal sign is referred to as the right-hand member (RHM).

The least-squares equations may be represented in tabular form as follows:

|         | $\hat{\mu}$ | $\hat{a}_1$ | $\hat{a}_2$ | $\cdots$ | $\hat{a}_{na}$ | RHM |
|---------|-------------|-------------|-------------|----------|----------------|-----|
| $\mu:$     | $N$       | $n_1$     | $n_2$     | $\cdots$ | $n_{na}$     | $Y..$ |
| $a_1:$     | $n_1$     | $n_1$     | $0$       | $\cdots$ | $0$          | $Y_1.$ |
| $a_2:$     | $n_2$     | $0$       | $n_2$     | $\cdots$ | $0$          | $Y_2.$ |
| $\vdots$   | $\vdots$  | $\vdots$  | $\vdots$  | $\vdots$ | $\vdots$     | $\vdots$ |
| $a_{na}:$  | $n_{na}$  | $0$       | $0$       | $\cdots$ | $n_{na}$     | $Y_{na}.$ |

Four features of the least-squares equations should be noted from this table. First, the coefficients of the estimates of the parameters (the left-hand members) form a matrix that is symmetric about the main diagonal; that is, the elements on the right-hand side of the diagonal, which goes from upper left to lower right, are a mirror image of those on the left-hand side. Second, the coefficients of the $\hat{a}_i$ in the $\mu$ equation are the same as the coefficients on the main diagonal for the $a_i$ equations. Third, the off-diagonal elements in the section of rows and columns for the $a_i$ are all zero. Fourth, the sums of the coefficients by column in the $\hat{a}_i$ equations are equal to the coefficients in the $\mu$ equation. Likewise, the sum of the right-hand members for the $a_i$ equations is equal to the right-hand member for the $\mu$ equation.

The least-squares equations for the one-way classification may be shown more concisely as follows:

$$\mu: N\hat{\mu} + \sum_i n_i\hat{a}_i = Y..$$

$$\hat{a}_i: n_i\hat{\mu} + n_i\hat{a}_i = Y_i.$$

In tabular form, these are

|  | $\hat{\mu}$ | $\hat{a}_i$ | RHM |
|---|---|---|---|
| $\mu$: | $N$ | $n_i$ | $Y_{..}$ |
| $a_i$: | $n_i$ | $0$ | $Y_{i.}$ |
|  |  | $n_i$ |  |
|  |  | $0$ |  |

where the zeros in the $a_i$, $\hat{a}_i$ section indicate that the off-diagonal elements in this section are equal to zero.

Since the sums of the coefficients by column in the $a_i$ equations are equal to the coefficients in the $\mu$ equation, there is no unique solution to the least-squares set of equations. This means that the parameters $\mu$ and the $a_i$ cannot be estimated. Only linear functions of these parameters are estimable, that is, $\mu^*$ and $a_i^*$. Although there is an infinite number of sets of linear functions of the parameters which one could estimate, with each set containing $n_a$ different linear functions, only two of these are in common use and these will now be considered.

If one imposes the restriction (or constraint) on the least-squares equations that $\Sigma \hat{a}_i^* = 0$, then

$$E(\hat{a}_i^*) = a_i^* = a_i - \frac{\Sigma_i a_i}{n_a} \qquad (2.42)$$

and

$$E(\hat{\mu}^*) = \mu^* = \mu + \frac{\Sigma_i a_i}{n_a} \qquad (2.43)$$

where $E$ refers to "mathematical expectation." It is the value that is obtained for (  ) when averaged over all possible samples. To impose the restriction that $\Sigma_i \hat{a}_i^* = 0$ on the least-squares equations, the coefficients of one of the $\hat{a}_i$, say $\hat{a}_{n_a}$, are subtracted from the coefficients of the other $\hat{a}_i$. The resulting elements in the $a_{n_a}$ equation are then subtracted from the corresponding elements in the other $a_i$ equations to maintain symmetry in the equations. Likewise, the RHM for the $a_{n_a}$ equation is subtracted from the RHMs for the other $a_i$ equations. When these subtractions are completed the column and row for the $a_{n_a}$ equation is deleted and the resulting symmetric set of equations is solved to obtain estimates of $\mu^*$ and $a_i^*$. After subtraction, $n_a$ equations remain, that is, $n_a - 1$ for the $a_i^*$ and 1 for $\mu^*$ as seen below:

| | $\hat{\mu}^*$ | $\hat{a}_1^*$ | $\hat{a}_2^*$ | $\cdots$ | $\hat{a}_{n_a-1}^*$ | RHM |
|---|---|---|---|---|---|---|
| $\mu^*$: | $N$ | $n_1 - n_{n_a}$ | $n_2 - n_{n_a}$ | $\cdots$ | $n_{n_a-1} - n_{n_a}$ | $Y_{..}$ |
| $a_1^*$: | $n_1 - n_{n_a}$ | $n_1 + n_{n_a}$ | $n_{n_a}$ | $\cdots$ | $n_{n_a}$ | $Y_{1.} - Y_{n_a.}$ |
| $a_2^*$: | $n_2 - n_{n_a}$ | $n_{n_a}$ | $n_2 + n_{n_a}$ | $\cdots$ | $n_{n_a}$ | $Y_{2.} - Y_{n_a.}$ |
| $\vdots$ | $\vdots$ | $\vdots$ | $\vdots$ | | $\vdots$ | $\vdots$ |
| $a_{n_a-1}^*$: | $n_{n_a-1} - n_{n_a}$ | $n_{n_a}$ | $n_{n_a}$ | $\cdots$ | $n_{n_a-1} + n_{n_a}$ | $Y_{n_a-1.} - Y_{n_a.}$ |

$$(2.44)$$

One of the simplest restrictions (or constraints) that could be imposed to obtain a unique solution to the set of least-squares equations is simply to set one of the linear functions of the $a_i$ equal to zero. To do this one of the equations for the $a_i$ is deleted and the $a_i$ removed from every other equation. This is the procedure used by GLM for the analysis of general linear models in the statistical analysis system (SAS).

If the $a_{n_a}$ equation is deleted, the linear functions of the parameters that will then be estimated (say $\mu^0$ and $a_i^0$) are as follows:

$$E(\hat{\mu}^0) = \mu^0 = \mu + a_{n_a}$$

$$E(\hat{a}_1^0) = a_1^0 = a_1 - a_{n_a}$$

$$E(\hat{a}_2^0) = a_2^0 = a_2 - a_{n_a}$$

$$\vdots \qquad \vdots$$

$$E(\hat{a}_{n_a-1}^0) = a_{n_a-1}^0 = a_{n_a-1} - a_{n_a}$$

It should be noted that the difference between two $\hat{a}_i^*$ is identical with the difference between the same two $\hat{a}_i^0$; that is,

$$\hat{a}_1^* - \hat{a}_2^* = \hat{a}_1^0 - \hat{a}_2^0$$

since

$$E(\hat{a}_1^* - \hat{a}_2^*) = E(\hat{a}_1^0 - \hat{a}_2^0) = a_1 - a_2$$

Also, note that

$$\hat{a}_{n_a}^* = -\sum_{i=1}^{n_a-1} \hat{a}_i^*$$

$$= \frac{-\sum_{i=1}^{n_a-1} \hat{a}_i^0}{n_a} \tag{2.45}$$

Therefore, one can obtain $\hat{\mu}^*$ and the $\hat{a}^*$ from $\hat{\mu}^0$ and the $\hat{a}_i^0$, or vice versa, provided the linear functions of the parameters being estimated are known in each case.

The reduced set of least-squares equations (2.44) may be written in matrix notation as

$$\mathbf{C}\hat{\mathbf{B}} = \mathbf{Y} \tag{2.46}$$

where $\mathbf{C}$ is the square symmetric matrix consisting of the coefficients of the linear functions of the parameters that are to be estimated, $\hat{\mathbf{B}}$ is a column vector of the estimates, and $\mathbf{Y}$ is the reduced set of RHMs. Therefore, the solution of these equations is

$$\hat{\mathbf{B}} = \mathbf{C}^{-1}\mathbf{Y} \tag{2.47}$$

where $\mathbf{C}^{-1}$ is the inverse of the matrix $\mathbf{C}$. The reader will recall that the total uncorrected sum of squares for the one-way classification is $\sum_i \sum_j Y_{ij}^2$. This is the sum of squares accounted for by the full model, but the total reduction in sum of squares due to fitting the constants $\mu$ and $a_i$ is $R(\mu, a_i)$. In general, the total reduction in the sum of squares is

$$R(\mathbf{B}) = \hat{\mathbf{B}}'\mathbf{Y} \tag{2.48}$$

where $\hat{\mathbf{B}}'$ is a row vector of the solution to the set of least-squares equations. Therefore, for the one-way classification,

$$R(\mu, a_i) = \hat{\mu}^* Y.. + \hat{a}_1^*(Y_1. - Y_{na}.) + \hat{a}_2^*(Y_2. - Y_{na}.)$$

$$+ \cdots + \hat{a}_{na-1}^*(Y_{na-1}. - Y_{na}.) \tag{2.49}$$

The sum of squares for $a_i$ is the reduction in the sum of squares for $a_i$, adjusted for the mean $\mu$; that is,

$$R(a_i \mid \mu) = R(\mu, a_i) - R(\mu) \tag{2.50}$$

and

$$R(\mu) = \bar{y}. Y.. = \frac{Y^2..}{N} \tag{2.51}$$

the overall correction term, and

$$\bar{y} = \frac{Y..}{N} \tag{2.52}$$

the overall mean. The sum of squares for $a_i$ may also be computed by a "direct" procedure which makes use of the estimates of the linear functions of the $a_i$ and the elements of the inverse matrix $\mathbf{C}^{-1}$, which correspond by row and column to the $\hat{a}_i^*$. With this procedure

$$R(a_i \mid \mu) = \hat{\mathbf{B}}_A' \mathbf{Z}_A^{-1} \hat{\mathbf{B}}_A \tag{2.53}$$

where $\hat{\mathbf{B}}_A'$ is a row vector of the $\hat{a}_i^*$ from the solution vector, $\mathbf{Z}_A$ is the square segment of $\mathbf{C}^{-1}$ for the $\hat{a}_i^*$, and $\hat{\mathbf{B}}_A$ is a column vector of the $\hat{a}_i^*$ (Harvey, 1970).

The method of least-squares will now be applied to the analysis of the data given in Table 2.3. The specific model for this set of data is

$$Y_{ij} = \mu + t_i + e_{ij} \qquad i = 1, 2 \qquad j = 1, \ldots, n_i \tag{2.54}$$

where $t_i$ is the effect of the $i$th litter. The least-squares set of equations is

|        | $\hat{\mu}^*$ | $\hat{t}_1^*$ | $\hat{t}_2^*$ | RHM  |
|--------|------|------|------|------|
| $\mu$:   | 14   | 8    | 6    | 37.3 |
| $t_1$:   | 8    | 8    | 0    | 25.4 |
| $t_2$:   | 6    | 0    | 6    | 11.9 |

and the reduced set of equations, after imposing the restriction that $\Sigma_i \hat{t}_i^* = 0$, is

$$\begin{bmatrix} 14 & 2 \\ 2 & 14 \end{bmatrix} \begin{bmatrix} \hat{\mu}^* \\ \hat{t}_1^* \end{bmatrix} = \begin{bmatrix} 37.3 \\ 13.5 \end{bmatrix} \tag{2.55}$$

$$\quad \mathbf{C} \qquad\quad \hat{\mathbf{B}} \qquad\quad \mathbf{Y}$$

The solution of these equations is

$$\hat{\mathbf{B}} = \mathbf{C}^{-1}\mathbf{Y} = \begin{bmatrix} 14 & 2 \\ 2 & 14 \end{bmatrix}^{-1} \begin{bmatrix} 37.3 \\ 13.5 \end{bmatrix}$$

$$= \begin{bmatrix} 0.072917 & -0.010417 \\ -0.010417 & 0.072917 \end{bmatrix} \begin{bmatrix} 37.3 \\ 13.5 \end{bmatrix}$$

$$= \begin{bmatrix} 2.57917 \\ 0.59583 \end{bmatrix} \tag{2.56}$$

The total reduction in the sum of squares is

$$R(\mu, t_i) = R(\mathbf{B}) = \hat{\mathbf{B}}'\mathbf{Y}$$

$$= (2.57917 \quad 0.59583)\begin{bmatrix} 37.3 \\ 13.5 \end{bmatrix}$$

$$= 104.2467 \tag{2.57}$$

and the reduction due to fitting $\mu$ only is the overall correction term

$$\frac{Y^2_{..}}{N} = \frac{(37.3)^2}{14}$$

$$= 99.3779 \tag{2.58}$$

Therefore,

$$R(t_i \mid \mu) = R(\mu, t_i) - R(\mu)$$

$$= 104.2467 - 99.3779$$

$$= 4.8688 \tag{2.59}$$

Equation (2.53) may be used to compute the sum of squares for litters as follows:

$$R(t_i \mid \mu) = \hat{\mathbf{B}}'_T \mathbf{Z}_T^{-1} \hat{\mathbf{B}}_T$$

$$= (0.59583)(0.072917)^{-1}(0.59583)$$

$$= \frac{(0.59583)^2}{0.072917}$$

$$= 4.8687 \tag{2.60}$$

The reader should note that the sum of squares for litters is the same when computed by the method of least squares as when obtained by formula (2.16). We will see that simple formulas do not exist for obtaining the least-squares sum of squares in most cases when the underlying model contains two or more sets of cross-classified effects and unequal-subclass frequencies exist.

## EXERCISES

**2.1.** The data presented below are the fresh weights in grams of 24 plants, 8 plants on each of three treatments. Complete an analysis of variance for this set of data, which conforms to a one-way classification, and state your conclusions. After completion of the analysis calculate the constant (treatment effect) estimates $\hat{t}_i$ and find the sum of squares among treatments by the formula $\Sigma_i n_w \hat{t}_i^2$.

| Treatment 1 | | Treatment 2 | | Treatment 3 | |
|---|---|---|---|---|---|
| 41.4 | 52.5 | 35.8 | 63.1 | 59.8 | 77.8 |
| 49.9 | 53.4 | 26.2 | 51.3 | 98.4 | 87.6 |
| 43.2 | 51.7 | 45.8 | 54.4 | 79.4 | 83.4 |
| 40.1 | 52.2 | 49.0 | 56.1 | 80.0 | 84.7 |

**2.2.** A portion of the results of a study on fruit set in apple trees is reproduced here. The data presented below represent fruit set per cubic centimeter of limb circumference. Complete an analysis of variance for this set of data, which conforms to a one-way classification with unequal numbers, stating your conclusions. Show that the within-treatment sum of squares can be calculated by the formula $\Sigma_i \Sigma_j \hat{e}_{ij}^2$, where $e_{ij} = Y_{ij} - \bar{y}_i$.

| | |
|---|---|
| Treatment 1 | 4.79, 3.62, 3.42, 2.38, 2.15, 4.65, 3.33 |
| Treatment 2 | 0.05, 0.57, 0.05, 0.10, 0.63 |
| Treatment 3 | 3.61, 3.45, 2.17, 2.27 |

**2.3.** The weaning weights of 10 beef calves from Brangus dams in a cross-breeding experiment were 442, 410, 430, 451, 448, 460, 432, 462, 464, and 432. The weights of 10 calves from Hereford dams were 392, 410, 420, 381, 385, 415, 420, 397, 404, and 422. Test for the significance of the difference between the two groups of calves by means of the analysis of variance. Repeat the test, this time using the $t$ test, showing that $t^2 = F$.

**2.4.** The slaughter calf grades of eight calves sired by a Hereford bull were 14, 11, 15, 13, 11, 10, 14, and 12. The grades for six calves sired by a Brangus bull were 10, 14, 11, 10, 10, and 12. Test for the significance of the difference between the means of these two groups by both the analysis of variance and the $t$ test, showing that $t^2 = F$.

**2.5.** The weaning weights of beef calves sired by three Charolais sires and by three Angus sires are shown below. Carry out an analysis of variance for this set of data showing a test of significance between the two breeds of sires and a test among the sires within the breeds.

| Charolais | | | Angus | | |
|---|---|---|---|---|---|
| Sire 1 | Sire 2 | Sire 3 | Sire 1 | Sire 2 | Sire 3 |
| 480 | 430 | 473 | 354 | 410 | 423 |
| 440 | 420 | 438 | 341 | 413 | 414 |
| 465 | 461 | 451 | 362 | 390 | 398 |
| 450 | 433 | 432 | 379 | 401 | 407 |

**2.6.** In a study of the composition of soils from individual farms in different areas of Massachusetts, levels of phosphorus were found as presented below. Carry out an analysis of variance on this set of data including a test for the significance of the difference between the two areas and a test of the significance of the differences among the farms within the areas.

| Connecticut River valley | | | | Central area | | |
|---|---|---|---|---|---|---|
| Farm 1 | Farm 2 | Farm 3 | Farm 4 | Farm 1 | Farm 2 | Farm 3 |
| 109 | 130 | 147 | 85 | 12 | 21 | 8 |
| 105 | 135 | 126 | 97 | 17 | 17 | 9 |
| 115 | 122 | 131 | 92 | 14 | 13 | 7 |
| 119 | | | 83 | 21 | | 14 |
| | | | 74 | 27 | | 12 |
| | | | | | | 8 |

**2.7.** For the data given in Exercise 2.1, set up the least-squares equations, impose the restriction that $\Sigma_i \hat{t}_i = 0$, solve the equations for $\hat{\mu}$ and the $\hat{t}_i$, and compute the sums of squares for the analysis of variance using general least-squares procedures.

**2.8.** Using general least-squares procedures complete the analysis of variance for the data given in Exercise 2.2.

# Multiway Classification of Data

## 3.1 INTRODUCTION

It was pointed out in Chapter 2 that the analysis of variance is a method of partitioning the total sum of squares (the sum of squares of the deviations of the individual observations from the overall mean) into separate components. This partitioning is simply an algebraic manipulation conducted for the purpose of isolating the sources that are contributing to the variation among the individual observations. One of the purposes of this isolation of sources, which are defined by the mathematical model used, is to test the hypothesis that two or more treatment means are equal. Such a hypothesis is concerned with making a decision as to whether the means in the experiment or set of data have been drawn from the same population or whether they represent observations from populations with different means. The hypothesis is tested by what is referred to as a test of significance as seen in Chapter 2. As the text develops, hypotheses of many different types will be considered.

In Chapter 2 we first partitioned the total sum of squares into the sources of variation among or between treatment groups and within treatment groups. While it is common to talk in terms of differences among treatment groups, in fact, we are referring to differences among group means. After completing the partitioning and calculating the mean square for each source of variation, we completed a test of significance among the treatment means to accept or reject the null hypothesis, which is a hypothesis of no differences among the treatment means. By no differences among treatment means, we mean that there are no greater differences among these means than would be expected if they were a random group of means drawn from the same population. The test of significance was made by dividing the mean square for among treatments by the mean square for within treatments. Let us look at what is involved in this test.

Suppose that we consider a sample of observations from a given population. We can estimate the variance $\sigma^2$ in the population by

$$s^2 = \frac{\Sigma(Y - \bar{y})^2}{n_w - 1}$$

as shown previously.

Now let us select several samples of size $n_w$ from the same population. We can now estimate the variance in the population in two additional ways. First, using the information from the variation *within all the samples,* we can obtain a pooled estimate of the population variance $\sigma^2$ by

$$\frac{\Sigma_i \Sigma_j (Y_{ij} - \bar{y}_{i.})^2}{N - n_t} = \frac{\Sigma_i \Sigma_j Y_{ij}^2 - (\Sigma_i Y_{i.}^2)/n_w}{N - n_t} = \frac{\text{SS within samples}}{N - n_t} = s_p^2$$

where $N$ is the total number of observations and $n_t$ is the number of groups. To obtain a pooled estimate, the sums of squares are added up over all treatments.

Since the samples are all drawn from the same population, we can also estimate the variance in the population by using the variation *among the sample means.* We have previously seen that $s_{\bar{y}}^2 = s^2/n_w$, leading to the relationship $n_w s_{\bar{y}}^2 = s^2$, which is an estimate of the population variance $\sigma^2$. From the several samples of size $n_w$ we then have

$$\frac{n_w \Sigma_i (\bar{y}_i - \bar{y}.)^2}{n_t - 1} = \frac{n_w \Sigma_i \hat{t}_i^2}{n_t - 1} = \frac{(\Sigma_i Y_{i.}^2)/n_w - Y^2../N}{n_t - 1} = \frac{\text{SS among samples}}{n_t - 1} = n_w s_{\bar{y}}^2 = s^2$$

another estimate of the population variance.

It can be seen from the above that a ratio of the mean square for among samples and the mean square for within samples would be expected to be close to the value 1. The same would be true for the $F$ test of significance in a one-way analysis of variance where the treatment means really all came from the same population. However, the farther away from 1 the $F$ value becomes, the greater chance there is of the means coming from, or representing, different populations.

If indeed the treatment means do not represent or arise from the same population, the mean square for among treatments will contain an extra component resulting from the additional variation among the means. If there are no significant differences among the treatment means, then the mean square for among treatments and the mean square for within treatments each estimate the same thing, $\sigma^2$, the variance in the original or sampled population. If the treatment means do represent different populations then the variation among them is now representing the variance of a group of means arising from different populations. The variance of these population means is estimated by

$$s_t^2 = \frac{\Sigma \hat{t}^2}{n_t - 1}$$

It can be shown (see the detailed example presented by Jerome C. R. Li, 1957, p. 165) that the average of $s_{\bar{y}}^2$ of all possible sets of $n_t$ samples with $n_w$ observations is $\sigma^2/n_w + \sigma_t^2$. Therefore, the average among-sample mean square is expected to equal $n_w(\sigma^2/n_w + \sigma_t^2)$ or $\sigma^2 + n_w\sigma_t^2$. The variance component $\sigma_t^2$ is then the variance of the means, or treatment effects, drawn from different populations. The coefficient of $\sigma_t^2$ is necessary to put this quantity in terms of individual observations.

The $F$ test carried out in the one-way classification to test for the significance of treatment effects tests the hypothesis, in terms of variance components, that $\sigma^2 + n_w\sigma_t^2 = \sigma^2$, which is really the hypothesis that $\sigma_t^2 = 0$. The expected $F$ ratio is therefore $(\sigma^2 + n_w\sigma_t^2)/\sigma^2$. The null hypothesis, which is that there are no differences among the treatment means, is rejected if the value of $F$ equals or exceeds the value in the $F$ table for $n_t - 1$ and $N - n_t$ degrees of freedom for the numerator and denominator degrees of freedom, respectively.

The value $\sigma^2$ is known as the expected mean square for the within-sample mean square and the value $\sigma^2 + n_w\sigma_t^2$ is the expected value for among-treatments mean square. In Table 2.8 a test of the significance of differences among treatment means was made by a ratio of the mean square for among treatments over the mean square for among plants within treatments. In terms of variance components this ratio would be

$$F = \frac{\sigma^2 + 6\sigma_{p:t}^2 + 12\sigma_t^2}{\sigma^2 + 6\sigma_{p:t}^2}$$

The hypothesis is that $\sigma^2 + 6\sigma_{p:t}^2 + 12\sigma_t^2 = \sigma^2 + 6\sigma_{p:t}^2$ which is the hypothesis that $\sigma_t^2 = 0$.

In this chapter several additional models are considered, each having a different set of expected mean squares. Rules are presented for writing out the expectations of mean squares and the use of the expected mean squares in determining appropriate tests of significance is discussed. Several examples are presented to clarify their use. A thorough knowledge of this subject is vital to the correct analysis and interpretation of experimental data.

When conclusions are drawn from a statistical analysis of the data resulting from an experiment, they are based on tests of significance. However, these tests do not prove whether there are real differences existing among the treatment groups or not. Rather, the conclusions are based on the *probability* of real differences existing among the treatment groups or means. By common consent, over the years the 5 and 1 percent levels of significance have been used most. If it is found that the probability of the treatment means having been drawn from the same population is 5 percent or less, it is stated that there are "significant" differences among the means. This means that there is only 1 chance in 20 or less that the means would differ as much as they do if they had been drawn from the same population. Therefore, since the probability of such differences occurring due to chance alone is very slight, the conclusion is that real differences exist, although this has not been proved. If the probability is 1 percent or less then the statement is made that there are "highly significant" differences among the treatment means, although again this has not been proved.

Since conclusions about hypotheses are drawn on the basis of probability, it is then possible to make errors in the conclusions. Two types of error are involved, not too surprisingly referred to as Type I and Type II errors or errors of the first kind and errors of the second kind. If the hypothesis is true and it is accepted, then the correct conclusion has been reached and no error has been committed. If, however, the hypothesis is true and it is rejected by the test then a Type I error has been committed. If the hypothesis is false and it is accepted then a Type II error has been committed. Of course, if the hypothesis is false and it is rejected then no error has been committed. Thus, the two errors cannot be committed at the same time since acceptance of a false hypothesis leads to a Type II error and rejection of a true hypothesis leads to a Type I error.

The Type I error is the significance level utilized; that is, if the 5 percent level of significance is used there is a 5 percent chance that we will declare that significant differences exist among the treatment means when in fact they do not. Since this error will be committed 5 percent of the time, this is sometimes referred to as the error rate. Since lowering the Type I rate of error will make it more likely to accept the hypothesis that no differences exist among the treatment means when in fact they do, this means that lowering the Type I error increases the probability of committing a Type II error. Again, as the differences among the treatment means become smaller, the greater the probability becomes of committing a Type II error. The number of observations per treatment as well as the size of the within-group variance play important roles in the determination of the Type II error rate. Thus, when the real differences can be expected to be small, larger sample sizes are in order as well as in the situation where sample variances are large or expected to be large. For a more detailed discussion of Type I and Type II errors, see Steel and Torrie (1980, Chap. 5). It is believed that a remarkably large number of experiments report no significant differences where the probability of detecting such differences is indeed small owing to either small numbers of observations per treatment or large variances or both.

One of the important aspects in making statistical tests of significance is the decision as to whether the sources specified in the mathematical model are considered to be *random effects* or *fixed effects*. This distinction is vital since, as will be seen later in this section, this decision can make a great difference in the particular ratio of mean squares to be used in the $F$ tests and thus in the conclusions drawn from an analysis.

When we refer to the effects of a source of variation as being fixed, we mean that the parameters which they represent do not change from sample to sample or from one experiment to another. We consider them to be a finite group of effects and not a random sample from a population of such effects. Thus, if an experiment is concerned with testing the effects of three different levels of phosphorus on the yield of corn, these levels of phosphorus are considered to be fixed effects. They are not a random selection of all possible levels of phosphorus but rather the only levels in which the experimenter is interested. Furthermore, if the experiment is repeated several times, the experimenter can use the same levels each time, since these are the only levels of interest. When the fixed effect is a series of quantitative levels of a factor, that is, 0, 10, and 20 pounds of $N$, interpolation including intermediate values may be in order.

If the experiment with phosphorus is replicated, or repeated, several times, then we would have an additional source of variation which we can call replications, as well as the source of variation of phosphorus treatments. In the case of this second classification or source, we are not interested in the particular replications as such but want to consider them merely as a random selection from an infinite population of replications. Such a classification, one in which the levels are considered as a random selection representing an infinite population, is referred to as a random effect. There are, of course, situations where the replications of an experiment are considered for particular reasons to be fixed effects, but it is most common to consider them as random effects.

The difference between the two types of effect is reflected in types of population about which inferences are to be made. This difference is brought out by a difference in the way the variance components are written. Given an experiment in which we have the random effects of replications and the fixed effects of treatments, the expected mean square for the replication effects contains the variance component $\sigma_r^2$ which is the variance due to replications in the population of replications. The expected mean square for the treatment effects contains the component $\Sigma \, t^2/(n_t - 1)$ which is the variance due to the population of treatment levels included in the experiment. The treatments *are not* to be considered as a random sample from an infinite population. The difference between the two is that, in one case, we are interested in estimating the variance in a population from which samples have been drawn at random and, in the other case, we are interested in estimating the variance in a population represented only by the treatment levels included in the experiment. In this text, as will be noted later, an expression such as $\Sigma \, t^2/(n_t - 1)$ will be represented by $\sigma_t^2$ for convenience. For the random effect of replications, inferences are drawn about the entire population from which the replications are considered to be drawn. For the fixed effects, inferences are drawn only about the treatment levels included in the experiment.

When more than one observation exists in each of the replication by treatment subclasses, it is possible to test for the significance of an interaction between the two classifications or effects. The nature of a statistical interaction is defined and discussed later in this chapter. For the moment, we merely wish to point out that the interaction between two or more fixed effects is considered to be fixed. In the present example, the component for this effect, with its coefficient, could be written $n_w \, \Sigma_i \, \Sigma_j (rt)_{ij}^2/(n_r - 1)(n_t - 1)$, where $rt_{ij}$ is an interaction effect between the $i$th level of replication and $j$th level of treatment, $n_w$ is the number of observations in the $ij$ subclass, and $n_r$ and $n_t$ are the number of levels of replications and treatments, respectively.

The interaction between two random effects is considered to be random and it is most common to consider the interaction between a fixed effect and a random effect to be random. Unless all factors or classifications involved in an interaction are fixed effects, the interaction effect will be considered to be random. In our examples with replications and treatments, if one or both classifications are considered random then the variance component with its coefficient would be written $n_w \sigma_{rt}^2$.

It will be found that in writing out expectations of mean squares for different statistical models, all variance components for fixed interactions are deleted in the lines of the analysis of variance table denoting the fixed effects involved in the interactions. In addition, certain of the variance components for random interactions are also deleted. This occurs because we are concerned with making statements as to what results we expect due to the sources of variation in our model in the general situation or in a wide variety of situations, and not just what has happened in our particular experiment. In the example under discussion, we would like to make statements as to what results are expected from the different phosphorus levels under any random selection of replications, not just what has occurred with the few replicates involved in the experiment. We would also like to make a statement as to the magnitude of variation among replications expected when sampling a population and not just the variation involved when dealing with the particular levels of the fixed effect of phosphorus levels.

This chapter will deal also with the paired $t$ test, sometimes referred to as the paired comparison test. This test is sometimes considered as a special case of the analysis of variance since the paired $t$ test and the analysis of variance yield the same results. In an experiment dealing with paired comparisons, each member of a pair of experimental units is assigned one of two possible treatments. One of the treatments is frequently a control against which some treatment is to be tested or compared. A number of pairs, of course, must be included in the experiment. The difference between each of the members of the several pairs is calculated and the variance among these differences serves as a basis for a test of significance of the mean difference between the two groups. The mean difference between the two groups is the average of the differences or the difference between the means of the two groups.

The purpose of the pairing is this. The two members of a pair are expected to be quite similar or have a great deal in common, so that any difference between the two at the end of a trial or an experiment is expected to reflect in great measure the difference brought about by the treatments and not reflect initial differences. Initial differences among the pairs are expected to be greater than differences between members of a pair. Thus, the analysis is expected to remove as much of the variation among the pairs from the experimental error as is possible. The original pairing is vital to the success of this experimental approach. In a comparison of two teaching methods, the pairing criteria might include the pair of students being of the same age, the same sex, and approximately the same IQ.

Since twinning in sheep is quite common, pairs of twins are used frequently in feeding trials when only two treatments are involved. In swine, two littermates of the same sex and approximately the same weight might be employed. In an agronomic experiment one could select several plots of land and divide each block into two subplots, applying treatments A and B at random to the two subplots.

Given a set of differences between the units of several pairs, their standard deviation can be calculated as $s_d = \sqrt{\Sigma_i(D - \bar{d})^2/(n - 1)}$, where $D$ represents a difference, $\bar{d}$ is the mean difference, and $n$ is the number of pairs. From this a standard error of the mean difference can be calculated as $s_{\bar{d}} = s_d/\sqrt{n}$. We can now set up the null hypothesis of $H_0: \mu_a - \mu_b = 0$, with $H_1: \mu_a - \mu_b \neq 0$. The null hypothesis can now be tested by

$$t = \frac{\bar{d} - 0}{s_{\bar{d}}} \quad \text{or} \quad t = \frac{\bar{d}}{s_{\bar{d}}}$$

in the same manner as described in Chapter 2.

Several different models will be considered in this section in which two or more main classifications (cross-classified effects) are included. In addition, combinations of nested and main classifications will be considered. In all cases, the expectations of mean squares and appropriate $F$ ratios will play important roles in the completion of the analyses and statements of conclusions. When two or more classifications are involved in an analysis it is referred to as a multiway classification.

## 3.2 TWO-WAY CLASSIFICATION: SINGLE OBSERVATION PER SUBCLASS

Within any subclass (each combination of classifications $A$ and $B$ makes up a subclass) there may be one or more observations. We shall first consider the situation of one observation per subclass, using the data furnished by Dr. William J. Lord of the Plant and Soil Sciences Department of the University of Massachusetts. An experiment such as this is often referred to as the randomized block design. Only a portion of the complete set of data is included and is shown in Table 3.1.

These data can be described by the model

$$Y_{ij} = \mu + r_i + t_j + e_{ij} \qquad i = 1, \ldots, 8 \qquad j = 1, \ldots, 4 \qquad (3.1)$$

where $Y_{ij}$ is the observation in the $j$th class of $T$ and the $i$th class of $R$, $\mu$ is the effect of the population mean, $r_i$ is the effect of the $i$th level of $R$ (replication), $t_j$ is the effect of the $j$th level of $T$ (treatment), and $e_{ij}$ is the random error associated with the observation in the $ij$ subclass.

The calculations necessary to carry out the analysis of variance are quite similar to those of the one-way classification.

$$\text{Total SS} = \sum_i \sum_j Y_{ij}^2 - \frac{Y_{..}^2}{N}$$

$$= 42{,}314 - \frac{(1040)^2}{32} = 42{,}314 - 33{,}800 = 8514 \qquad (3.2)$$

where $N = n_r n_t n_w = (8)(4)(1) = 32$.

$$\text{Replication SS} = \text{SS}_R = \frac{\sum_i Y_{i.}^2}{n_t} - \frac{Y_{..}^2}{N}$$

$$= \frac{(144)^2 + (134)^2 + \cdots + (116)^2}{4} - \frac{(1040)^2}{32}$$

$$= \frac{136{,}084}{4} - 33{,}800 = 34{,}021 - 33{,}800 = 221 \qquad (3.3)$$

**TABLE 3.1**   WEIGHT INCREASE IN APPLE ROOT STOCKS (IN GRAMS)

| Treatment | Replications | | | | | | | | Treatment totals |
|---|---|---|---|---|---|---|---|---|---|
| | 1 | 2 | 3 | 4 | 5 | 6 | 7 | 8 | |
| 1 | 47 | 51 | 37 | 41 | 49 | 52 | 40 | 38 | $Y_{.1} = 355$ |
| 2 | 58 | 45 | 55 | 51 | 45 | 37 | 55 | 42 | $Y_{.2} = 388$ |
| 3 | 29 | 32 | 11 | 29 | 28 | 31 | 28 | 26 | $Y_{.3} = 214$ |
| 4 | 10 | 6 | 11 | 2 | 19 | 17 | 8 | 10 | $Y_{.4} = 83$ |
| Replication totals | $Y_{1.} =$ 144 | $Y_{2.} =$ 134 | $Y_{3.} =$ 114 | $Y_{4.} =$ 123 | $Y_{5.} =$ 141 | $Y_{6.} =$ 137 | $Y_{7.} =$ 131 | $Y_{8.} =$ 116 | $Y_{..} = 1040$ |

**TABLE 3.2**  CALCULATION OF THE DEGREES OF FREEDOM

| Source of variation | Degrees of freedom | | |
|---|---|---|---|
| Total | $N - 1 = n_r n_t n_w - 1 = (8)(4)(1) - 1$ | $= 31$ | |
| $R$ | $n_r - 1$ | $= 8 - 1$ | $= 7$ |
| $T$ | $n_t - 1$ | $= 4 - 1$ | $= 3$ |
| $RT$ | $(n_r - 1)(n_t - 1)$ | $= (7)(3)$ | $= 21$ |

$$\text{Treatment SS} = \text{SS}_T = \frac{\Sigma_j \, Y_{.j}^2}{n_r} - \frac{Y_{..}^2}{N}$$

$$= \frac{(355)^2 + (388)^2 + (214)^2 + (83)^2}{8} - \frac{(1040)^2}{32}$$

$$= \frac{329,254}{8} - 33,800 = 41,156.75 - 33,800 = 7356.75 \qquad (3.4)$$

$$\text{SS}_{RT} = \text{Total SS} - \text{SS}_R - \text{SS}_T$$

$$= \left( \sum_i \sum_j Y_{ij}^2 - \frac{Y_{..}^2}{N} \right) - \left( \frac{\Sigma_i \, Y_{i.}^2}{n_t} - \frac{Y_{..}^2}{N} \right) - \left( \frac{\Sigma_j \, Y_{.j}^2}{n_r} - \frac{Y_{..}^2}{N} \right)$$

$$= \sum_i \sum_j Y_{ij}^2 - \frac{\Sigma_i \, Y_{i.}^2}{n_t} - \frac{\Sigma_j \, Y_{.j}^2}{n_r} + \frac{Y_{..}^2}{N}$$

$$= 42,314 - 34,021 - 41,156.75 + 33,800 = 936.25 \qquad (3.5)$$

Again, $n_w$ is the number of observations in the smallest subclass.

Before the analysis can be completed, the degrees of freedom must be partitioned, as was the total sum of squares. This partitioning is shown in Table 3.2.

The analysis of variance then takes the form shown in Table 3.3.

The analysis of variance is completed by computing the $F$ value for treatments, the ratio between the mean square for treatments and the mean square for error or $RT$. Looking up the appropriate values in the $F$ table with 3 and 21 degrees of freedom, we find that our value of 55.01 not only exceeds the value of 3.07 at the 5 percent level but also exceeds the value of 4.87 at the 1 percent level. We therefore conclude that there are highly significant differences among the treatment means. The composition of the mean square on the bottom line in the analysis of variance table labeled $RT$ will be discussed at a later stage.

**TABLE 3.3**  ANALYSIS OF VARIANCE OF DATA
SHOWN IN TABLE 3.1

| Source | df | SS | MS | F |
|---|---|---|---|---|
| $R$ | 7 | 221.00 | 31.57 | |
| $T$ | 3 | 7356.75 | 2452.25 | 55.01** |
| $RT$ | 21 | 936.25 | 44.58 | |

Our estimates of the treatment effects, also referred to as main effects or constants, can be calculated as before by $\hat{t}_j = \bar{y}_{.j} - \bar{y}_{..}$ since we have equal numbers, and are estimates of the $t_j$ in the mathematical model. Since $\bar{y}_{..} = Y_{..}/N = 32.500$,

$$\hat{t}_1 = \bar{y}_{.1} - \bar{y}_{..} = 44.375 - 32.500 = 11.875$$

$$\hat{t}_2 = \bar{y}_{.2} - \bar{y}_{..} = 48.500 - 32.500 = 16.000$$

$$\hat{t}_3 = \bar{y}_{.3} - \bar{y}_{..} = 26.750 - 32.500 = -5.750$$

$$\hat{t}_4 = \bar{y}_{.4} - \bar{y}_{..} = 10.375 - 32.500 = -22.125$$

Because we are dealing with an equal number or balanced design, each treatment mean has the same standard error, which can be calculated as

$$s_{\bar{y}} = \sqrt{\frac{s^2}{n_r n_w}}$$

$$= \sqrt{\frac{44.58}{(8)(1)}} = \sqrt{5.5725} = 2.3606 \tag{3.6}$$

Because there is only one observation in each $RT$ subclass, $n_w$ is equal to 1, the number of observations in the smallest subclass. In formula (3.6), $s^2 = \mathrm{MS}_{RT}$, the mean square for $RT$.

## 3.3 PAIRED t TEST

Under particular conditions, a two-way classification can also be analyzed by the *paired t test*. These conditions are (1) that one of the classifications be made up of only two levels and (2) that we are not interested in a test of significance among the levels of the second classification. The classification containing two levels represents the two classes or levels of particular treatment whose difference we wish to test for significance. The levels or classes of the second classification are frequently made up of such items as identical twins, animals from the same litter, plots from the same block of ground, or leaves from the same plant. Thus, *in each level* of the second classification there is a pair of observations, and the interest lies in the difference between members of the pairs. The paired t test does not yield a test of significance *among* the pairs, but since the test is based on differences between units of the pairs, the variation due to differences among the pairs is, in effect, accounted for or eliminated. Therefore, a two-way classification amenable to a t test can be analyzed by either an analysis of variance or a paired t test, each yielding the same result.

A selection of a small portion of the data collected by Judith G. Wolcott of the Exercise Science Department of the University of Massachusetts will serve to illustrate this test, the data being shown in Table 3.4.

Let us first analyze this set of data by means of the analysis of variance, the results of this test to be compared later with the paired t test. For the analysis of variance we would have the model $Y_{ij} = \mu + s_i + t_j + e_{ij}$, where the $s_i$ represent the

**TABLE 3.4** MOVEMENT TIME (IN MILLISECONDS) USED IN ELBOW FLEXION OF MAXIMUM SPEED THROUGH 85°. TREATMENT $A$ CONSISTED OF NO WEIGHT ON THE ARM WHILE TREATMENT $B$ CONSISTED OF A LIGHT WEIGHT ON THE ARM

| Subject | Treatment $A$ | Treatment $B$ | Totals | Difference = $D$ |
|---|---|---|---|---|
| 1 | 151 | 167 | 318 | −16 |
| 2 | 152 | 176 | 328 | −24 |
| 3 | 152 | 159 | 311 | −7 |
| 4 | 168 | 189 | 357 | −21 |
| 5 | 180 | 178 | 358 | +2 |
| 6 | 192 | 224 | 416 | −32 |
| 7 | 139 | 143 | 282 | −4 |
| 8 | 162 | 180 | 342 | −18 |
| Totals | $Y_{.1} = 1296$ | $Y_{.2} = 1416$ | $Y_{..} = 2712$ | $\Sigma D = -120$ |
| Means | $\bar{y}_{.1} = 162$ | $\bar{y}_{.2} = 177$ | $\bar{y}_{..} = 169.5$ | |

subject effects and the $t_j$ represent the treatment effects. The computations would proceed as shown below:

$$\text{Total SS} = \sum_i \sum_j Y_{ij}^2 - \frac{Y_{..}^2}{N}$$

$$= 466{,}638 - \frac{(2712)^2}{16} = 466{,}638 - 459{,}684 = 6954 \qquad (3.7)$$

$$\text{Subject SS} = \text{SS}_S = \frac{\Sigma_i Y_{i.}^2}{n_t} - \frac{Y_{..}^2}{N}$$

$$= \frac{(318)^2 + (328)^2 + \cdots + (342)^2}{2} - \frac{(2712)^2}{16}$$

$$= \frac{930{,}586}{2} - 459{,}684 = 465{,}293 - 459{,}684 = 5609 \qquad (3.8)$$

$$\text{Treatment SS} = \text{SS}_T = \frac{\Sigma_j Y_{.j}^2}{n_s} - \frac{Y_{..}^2}{N}$$

$$= \frac{(1296)^2 + (1416)^2}{8} - \frac{(2712)^2}{16}$$

$$= \frac{3{,}684{,}672}{8} - 459{,}684 = 460{,}584 - 459{,}684 = 900 \qquad (3.9)$$

$$\text{SS}_{ST} = \sum_i \sum_j Y_{ij}^2 - \frac{\Sigma_i Y_{i.}^2}{n_t} - \frac{\Sigma_j Y_{.j}^2}{n_s} + \frac{Y_{..}^2}{N}$$

$$= 466{,}638 - 465{,}293 - 460{,}584 + 459{,}684 = 445 \qquad (3.10)$$

The analysis of variance can now be completed as shown in Table 3.5.

**TABLE 3.5**  ANALYSIS OF VARIANCE OF MOVEMENT TIME

| Source | df | SS | MS | F |
|--------|-----|------|--------|---------|
| Subjects ($S$) | 7 | 5609 | 801.29 | |
| Treatments ($T$) | 1 | 900 | 900.00 | 14.16** |
| $ST$ | 7 | 445 | 63.57 | |

The probability due to chance alone of obtaining a mean difference as large as that found between the two treatment groups is less than 1 percent as shown by the $F$ table. We therefore conclude that there is a highly significant difference between the two treatment means and indicate this with a double asterisk.

The analysis of these data by means of the paired $t$ test uses the formula

$$t = \frac{\bar{d}}{s_{\bar{d}}} = \frac{\Sigma\, D/n}{\sqrt{\dfrac{\Sigma\, D^2 - (\Sigma\, D)^2/n}{n(n-1)}}} \tag{3.11}$$

where $\bar{d}$ is the mean difference between the two treatments, $s_{\bar{d}}$ is the standard error of the mean difference, $D$ is an individual difference, and $n$ is the number of subjects.

If the first $n$ in the denominator of the standard error were omitted, we would have the standard deviation of the individual differences. Inclusion of this $n$ yields the standard error of the mean difference, which could be written

$$s_{\bar{d}} = \frac{s_d}{\sqrt{n}}$$

Applying formula (3.11) we obtain

$$t = \frac{120/8}{\sqrt{\dfrac{2690 - (120)^2/8}{(8)(8-1)}}} = \frac{15}{\sqrt{\dfrac{2690 - 1800}{56}}} = \frac{15}{\sqrt{\dfrac{890}{56}}}$$

$$= \frac{15}{3.9866} = 3.763**$$

$$t^2 = 14.16 = F$$

Since we are only concerned with the mean difference between the two treatments we can ignore the sign of $\bar{d}$ and $\Sigma\, D$ in the analysis. It can be seen that the same results are obtained with the paired $t$ test and the analysis of variance.

It is important to note the difference in results in this set of data should the source of variation of subjects not be taken into account. If this were the case it would result in our considering the experiment as consisting of a total of 16 subjects instead of 8, there being 8 *different* subjects on *each treatment,* dictating a one-way classification in an analysis of variance, or a regular $t$ test as opposed to a paired $t$ test. The analysis of these data by means of a one-way analysis, displayed in Table 3.6, shows the same sum of squares and mean square for treatments, but the sum of squares among subjects is now combined with the $ST$ sum of squares, with a similar combining of the corresponding degrees of freedom. The resulting $F$ test yields a value of 2.08, which, with

**TABLE 3.6** ONE-WAY ANALYSIS OF VARIANCE OF DATA CONTAINED IN TABLE 3.4

| Source | df | SS | MS | F |
|--------|----|----|----|----|
| $T$ | 1 | 900 | 900.00 | 2.08 |
| $W$ | 14 | 6054 | 432.43 | |

1 and 14 degrees of freedom, does not approach significance. Of course, a regular $t$ test would yield the same results. So it can be seen that when an extraneous source of variation can be identified and the corresponding sum of squares removed from the error term, greater precision can be obtained as well as a bias eliminated.

## 3.4 TWO-WAY CLASSIFICATION: MULTIPLE OBSERVATIONS PER SUBCLASS

While it is common to encounter data in which there is only one observation per subclass, it is perhaps more frequently the case that there is more than one observation per subclass; that is, more than one experimental unit is subjected to the same treatment combination or classified by the same levels of the sources of variation. Let us consider the data shown in Table 3.7, selected from a larger set of data supplied by Dr. Allen V. Barker of the Department of Plant and Soil Sciences at the University of Massachusetts. The model for this set of data is

$$Y_{ijk} = \mu + n_i + l_j + nl_{ij} + e_{ijk} \qquad i = 1, 2 \qquad j = 1, 2, 3 \tag{3.12}$$

**TABLE 3.7** NITRATE ACCUMULATION (IN GRAMS) IN SPINACH PLANTS RESULTING FROM TWO SOURCES OF NITROGEN, EACH APPLIED AT THREE LEVELS

| Levels | Sources of nitrogen | | Totals |
|--------|---------------------|---------|--------|
| | Ammonium sulfate | Nitrate | |
| 1 | $14.3 = Y_{111}$ | 17.6 | |
| | 15.9 | 24.0 | |
| | 14.8 | 13.5 | |
| | $45.0 = Y_{11.}$ | $55.1 = Y_{21.}$ | $100.1 = Y_{.1.}$ |
| 2 | 37.5 | 43.3 | |
| | 29.4 | 53.5 | |
| | 33.8 | 49.3 | |
| | $100.7 = Y_{12.}$ | $146.1 = Y_{22.}$ | $246.8 = Y_{.2.}$ |
| 3 | $41.4 = Y_{131}$ | 59.8 | |
| | 49.9 | 98.4 | |
| | 43.2 | 79.4 | |
| | $134.5 = Y_{13.}$ | $237.6 = Y_{23.}$ | $372.1 = Y_{.3.}$ |
| Totals | $280.2 = Y_{1..}$ | $438.8 = Y_{2..}$ | $719.0 = Y_{...}$ |

where $Y_{ijk}$ is the $k$th observation in the $j$th level of $L$ and the $i$th level of $N$, $\mu$ is the effect of the population mean, $n_i$ is the effect of the $i$th level of $N$ (source of nitrogen), $l_j$ is the effect of the $j$th level of $L$ (level of nitrogen), $nl_{ij}$ is the effect of the interaction between the $i$th level of $N$ and the $j$th level of $L$, and $e_{ijk}$ is the random error associated with the $k$th observation in the $i$th level of $N$ and the $j$th level of $L$. The new feature of this model is the interaction term which will be discussed in more detail shortly. First, we shall analyze the data by means of the analysis of variance in the usual fashion.

$$\text{Total SS} = \sum_i \sum_j \sum_k Y_{ijk}^2 - \frac{Y_{...}^2}{n_n n_l n_w = N}$$

$$= 37{,}958.2 - \frac{(719.0)^2}{(2)(3)(3) = 18}$$

$$= 37{,}958.2 - 28{,}720.0556 = 9238.1444 \tag{3.13}$$

$$\text{Nitrogen SS} = \text{SS}_N = \frac{\sum_i Y_{i..}^2}{n_l n_w} - \frac{Y_{...}^2}{N}$$

$$= \frac{(280.2)^2 + (438.8)^2}{(3)(3) = 9} - 28{,}720.0556$$

$$= 30{,}117.4978 - 28{,}720.0556 = 1397.4422 \tag{3.14}$$

$$\text{Level SS} = \text{SS}_L = \frac{\sum_j Y_{.j.}^2}{n_n n_w} - \frac{Y_{...}^2}{N}$$

$$= \frac{(100.1)^2 + (246.8)^2 + (372.1)^2}{(2)(3) = 6} - 28{,}720.0556$$

$$= 34{,}898.1100 - 28{,}720.0556 = 6178.0544 \tag{3.15}$$

$$\text{Interaction SS} = \text{SS}_{NL} = \left( \frac{\sum_i \sum_j Y_{ij.}^2}{n_w} - \frac{Y_{...}^2}{N} \right) - \left( \frac{\sum_i Y_{i..}^2}{n_l n_w} - \frac{Y_{...}^2}{N} \right)$$

$$- \left( \frac{\sum_j Y_{.j.}^2}{n_n n_w} - \frac{Y_{...}^2}{N} \right) \tag{3.16}$$

$$= \frac{\sum_i \sum_j Y_{ij.}^2}{n_w} - \frac{\sum_i Y_{i..}^2}{n_l n_w} - \frac{\sum_j Y_{.j.}^2}{n_n n_w} + \frac{Y_{...}^2}{N}$$

$$= \frac{(45.0)^2 + (100.7)^2 + \cdots + (237.6)^2}{3} - \frac{\sum_i Y_{i..}^2}{n_l n_w}$$

$$- \frac{\sum_j Y_{.j.}^2}{n_n n_w} + \frac{Y_{...}^2}{N}$$

$$= 37{,}030.2400 - 30{,}117.4978 - 34{,}898.1100$$

$$+ 28{,}720.0556 = 734.6878 \tag{3.17}$$

$$\text{Within SS} = SS_W = \left( \sum_i \sum_j \sum_k Y_{ijk}^2 - \frac{Y_{...}^2}{N} \right) - \left( \frac{\sum_i \sum_j Y_{ij.}^2}{n_w} - \frac{Y_{...}^2}{N} \right) \tag{3.18}$$

$$= \sum_i \sum_j \sum_k Y_{ijk}^2 - \frac{\sum_i \sum_j Y_{ij.}^2}{n_w}$$

$$= 37{,}958.2 - 37{,}030.24 = 927.96 \tag{3.19}$$

Since the calculation of the interaction sum of squares is new, it is a good idea to look at this in detail. Inspection of formula (3.16) shows that the sum of squares for the interaction is calculated by subtracting the corrected nitrogen sum of squares and the corrected level sum of squares from the corrected nitrogen $X$ level subclass sum of squares; that is,

$$SS_{NL} = SS_{\text{among } NL \text{ subclasses}} - SS_N - SS_L$$

Formula (3.17) shows that the interaction sum of squares can also be calculated using the uncorrected sums of squares and the overall correction term.

It is important for future statistical computation and understanding to keep clearly in mind the distinction between the sum of squares for an interaction and the sum of squares among the corresponding subclasses. Putting the above expression in a different form we have

$$SS_{\text{among } NL \text{ subclasses}} = SS_N + SS_L + SS_{NL}$$

where two or more subscripts on the SS will always refer to interactions and not subclasses.

The last calculation in the above analysis of variance, that for "within," refers to the sum of squares among observations within the subclasses. Expression (3.18) shows that if we remove the corrected sum of squares *among* subclasses, $\sum_i \sum_j Y_{ij.}^2 / n_w - Y_{...}^2/N$, from the total corrected sum of squares, $\sum_i \sum_j \sum_k Y_{ijk}^2 - Y_{...}^2/N$, we obtain the sum of squares *within* the subclasses. Expression (3.19) indicates that this sum of squares can also be obtained by using the uncorrected sums of squares. The within-subclass sum of squares is, in reality, a *pooled* sum of squares of the deviations of each observation within a subclass from its subclass mean. This sum of squares can be computed within each subclass and the individual subclass values added up over the subclasses as shown in Table 3.8. The values in the bottom

**TABLE 3.8**   CALCULATION OF WITHIN-SUBCLASS SUM OF SQUARES

| | 11 | 12 | 13 | 21 | 22 | 23 | Totals |
|---|---|---|---|---|---|---|---|
| | Subclass identification numbers | | | | | | |
| $\sum_k Y_{ijk}^2$ | 676.34 | 3,413.05 | 6,070.21 | 1,068.01 | 7,167.63 | 19,562.96 | $\sum_i \sum_j \sum_k Y_{ijk}^2 = 37{,}958.20$ |
| $\dfrac{Y_{ij.}^2}{n_w}$ | 675.00 | 3,380.16 | 6,030.08 | 1,012.00 | 7,115.07 | 18,817.92 | $\dfrac{\sum_i \sum_j Y_{ij.}^2}{n_w} = 37{,}030.23$ |
| $\sum_k Y_{ijk}^2 - \dfrac{Y_{ij.}^2}{n_w}$ | 1.34 | 32.89 | 40.13 | 56.01 | 52.56 | 745.04 | $SS_W$ = 927.97 |

**TABLE 3.9**   DEGREES OF FREEDOM FOR
                 DATA PRESENTED
                 IN TABLE 3.7

| Source | Degrees of freedom | | |
|--------|--------|--------|--------|
| Total | $N - 1$ | $= 18 - 1$ | $= 17$ |
| $N$ | $n_n - 1$ | $= 2 - 1$ | $= 1$ |
| $L$ | $n_l - 1$ | $= 3 - 1$ | $= 2$ |
| $NL$ | $(n_n - 1)(n_l - 1) = (1)(2)$ | | $= 2$ |
| $W$ | $\dfrac{n_w - 1}{n_w} N$ | $= \dfrac{3 - 1}{3}(18) = 12$ | |

line of the table are the sums of squares of the deviations of each individual observation from its subclass mean. If we add (pool) these values we have

$$\sum_i \sum_j \left( \sum_k Y_{ijk}^2 - \frac{Y_{ij.}^2}{n_w} \right) = \sum_i \sum_j \sum_k Y_{ijk}^2 - \frac{\Sigma_i \, \Sigma_j \, Y_{ij.}^2}{n_w} = \sum_i \sum_j \sum_k (Y_{ijk} - \bar{y}_{ij})^2$$

the within-subclass sum of squares.

Following the notation used previously, the appropriate degrees of freedom are calculated as shown in Table 3.9.

The degrees of freedom for the interaction are seen to be the product of the degrees of freedom for the two classifications involved in the interaction. Since there are three observations in each subclass, there are 2 degrees of freedom contributed by each of the six subclasses, resulting in 12 degrees of freedom for the within-subclass sum of squares. The completed analysis of variance is shown in Table 3.10.

If we consider the within mean square to be the appropriate denominator for testing the $N$, $L$, and $NL$ effects, the $F$ values would be as shown in Table 3.10. However, this is not always the case and more will be said later on the selection of appropriate denominators for testing the various effects.

## 3.5 INTERACTION

The analysis of variance contained in Table 3.10 shows a highly significant difference between the two sources of nitrogen and also a highly significant interaction between sources of nitrogen and levels of nitrogen. When two classifications are involved in an interaction it is referred to as a *two-way* or *first-order* interaction, and when three classifications are involved in an interaction it is referred to as a three-way or second-

**TABLE 3.10**   ANALYSIS OF VARIANCE OF DATA
                  SHOWN IN TABLE 3.7

| Source | df | SS | MS | F |
|--------|-----|--------|--------|--------|
| Nitrogen ($N$) | 1 | 1397.4422 | 1397.4422 | 18.07** |
| Levels ($L$) | 2 | 6178.0544 | 3089.0272 | 39.95** |
| $NL$ | 2 | 734.6878 | 367.3439 | 4.75** |
| $W$ | 12 | 927.9600 | 77.3300 | |

order interaction, and so on. The presence of a significant interaction in this, or any set of data, dictates that considerable caution must be exercised in drawing conclusions about the overall effects of the main classifications.

It is important to understand the nature of a statistical interaction because it plays an important role in the interpretation of the results of any experiment in which an interaction term is included in the model. In many experiments, the interaction is of primary interest with concern for the main effects assuming a lesser role.

If we are studying the response to two sets of main effects or classifications we shall look to see whether they act independently of one another or whether the responses to the levels of one set of effects are influenced by the levels of the second set of effects. Should we find that the responses to one set of treatments are influenced, or changed, by the levels of the second set of treatments, we consider that an interaction is present. If there is no interaction present, we anticipate that the response to given levels of one variable will be the same at all levels of the second variable. Since random error is expected to exert an influence on these responses, our decision as to whether a significant interaction exists is based on a statistical test of significance as shown in Table 3.10. Let us look at an example of an interaction shown graphically in Figure 3.1, adapted from Steel and Torrie (1980).

In both the $A$ and the $B$ diagrams of Figure 3.1 we have two levels of an $A$ effect and two levels of a $B$ effect. The first diagram shows that at the first level of $A$ we obtain an increase of 10 units as we go from the first level of $B$ to the second level of $B$ and that we also obtain the same difference of $+10$ units at the second level of $A$. The response to $B$ is seen to be independent of $A$ since we find the same difference between the two levels of $B$ at each level of $A$.

The second diagram shows quite a different situation since we now obtain an increase of 10 units when we go from the first level of $B$ to the second at the first level of $A$ but find that at the second level of $A$ the situation is reversed and we have a decrease as we go from the first level of $B$ to the second. Thus, we are not obtaining the same response to $B$ at the different levels of $A$; we conclude that we have an interaction between the two effects. In this discussion, $A$ and $B$ effects could be interchanged, so it is a matter of preference or logic as to which sequence the interaction is viewed.

When an interaction is suspected, or found to be significant, an inspection of the two-way table of subclass means is valuable in interpreting where and how the interaction is occurring. Let us look at the data in Table 3.11, made up of the means of the subclasses shown in Table 3.7.

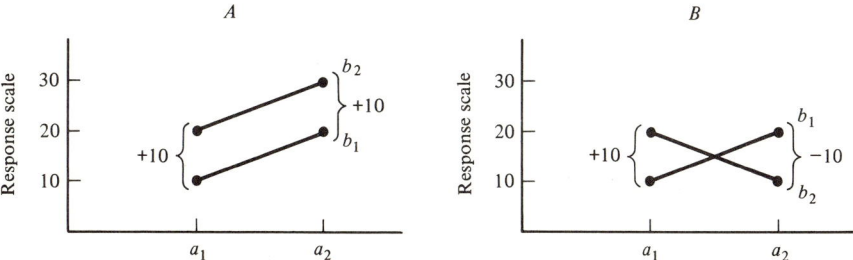

**Figure 3.1** Example of differential response.

**TABLE 3.11**　SUBCLASS MEANS FOR DATA DISPLAYED IN TABLE 3.7

| Source of nitrogen | Level of nitrogen | | |
| --- | --- | --- | --- |
| | 1 | 2 | 3 |
| Ammonium sulfate | $15.0000 = \bar{y}_{11}$ | $33.5667 = \bar{y}_{12}$ | $44.8333 = \bar{y}_{13}$ |
| Nitrate | $18.3667 = \bar{y}_{21}$ | $48.7000 = \bar{y}_{22}$ | $79.2000 = \bar{y}_{23}$ |

An inspection of these subclass means shows rather startling differences between the responses to the two sources of nitrogen at the different levels of application, the values of 3.6667, 15.1383, and 34.3667 showing clearly that the difference in response to the two sources of nitrogen is greatly dependent on the level at which they are applied. With no interaction present these differences are expected to vary only by an amount that could be accounted for due to chance or random error. It is also informative to look at the table in the opposite direction, that is, to examine the changes in response that take place from one level of application to another for each of the sources of nitrogen. It can be seen that there is a positive increase in response for both sources of nitrogen as the levels increase; however, the increase is far greater for the nitrate source than for the ammonium source. The highly significant interaction found in the analysis of variance is quite clearly explained by the differential responses we have noted.

When there are three or more classifications involved in an interaction, interpretation becomes steadily more difficult. One may still study two-way tables of subclass means, looking at the *AB* subclass means *at each level of C,* or *AC* subclass means *at each level of B,* and so on.

We have previously defined and discussed estimates of treatment effects, or constants, and shall now do the same for the interaction effects, or constants. Considering the model describing the data in Table 3.7 and substituting our estimates of the population values, we have the sample model

$$Y_{ijk} = \bar{y}.. + \hat{n}_i + \hat{l}_j + \hat{nl}_{ij} + \hat{e}_{ijk}$$

and rearranging we have

$$Y_{ijk} - \hat{e}_{ijk} = \bar{y}.. + \hat{n}_i + \hat{l}_j + \hat{nl}_{ij}$$

Since $\hat{e}_{ijk}$ is the deviation of the $k$th observation from its subclass mean, when this deviation is subtracted from the $k$th observation the result is the subclass mean. Therefore,

$$\bar{y}_{ij} = \bar{y}.. + \hat{n}_i + \hat{l}_j + \hat{nl}_{ij}$$

which is an important and frequently used statistical relationship. The interaction effect, or constant, is estimated as

$$\hat{nl}_{ij} = \bar{y}_{ij} - \bar{y}.. - \hat{n}_i - \hat{l}_j \tag{3.20}$$

In the model under consideration $\hat{n}_i$ is defined as $\bar{y}_i. - \bar{y}..$ and $\hat{l}_j$ is defined as $\bar{y}._j - \bar{y}..$ and these values are shown in Table 3.12.

Using the estimates of the main effects given in Table 3.12 together with the overall mean of 39.9444 and each individual subclass mean, we can substitute in Equation (3.20) and calculate estimates of the interaction effects, or constants, shown in Table 3.13.

**TABLE 3.12**  ESTIMATES OF MAIN EFFECTS

| $\hat{n}_i$ | $\hat{l}_j$ |
|---|---|
| $\hat{n}_1 = 31.1333 - 39.9444 = -8.8111$ | $\hat{l}_1 = 16.6833 - 39.9444 = -23.2611$ |
| $\hat{n}_2 = 48.7555 - 39.9444 = \phantom{-}8.8111$ | $\hat{l}_2 = 41.1333 - 39.9444 = \phantom{-}1.1889$ |
| | $\hat{l}_3 = 62.0166 - 39.9444 = \phantom{-}22.0722$ |

We have noted previously that estimates of the treatment effects, such as $\hat{n}_i$ and $\hat{l}$, sum to zero. It can be seen from Table 3.13 that the estimates of the interaction effects sum to zero by both row and column. Another way of expressing these relationships is

$$\sum_i \hat{n}_i = \sum_j \hat{l}_j = \sum_i \hat{nl}_{ij} = \sum_j \hat{nl}_{ij} = 0 \qquad (3.21)$$

This double restriction among the estimates of the interaction constants of summing to zero by row and column results in the $(n_n - 1)(n_l - 1)$ degrees of freedom among the interaction effects.

Just as we can calculate the sum of squares for main effects by using the estimated constants, we can also calculate the sum of squares for interaction, with equal subclass numbers, as

$$SS_{NL} = n_w \sum_i \sum_j \hat{nl}_{ij}^2 = (3)(244.8956) = 734.6868 \qquad (3.22)$$

## 3.6  SETTING UP THE ANALYSIS OF VARIANCE TABLE IN THE BALANCED DESIGN

The steps that we have taken so far can be summarized in a manner outlined by C. R. Henderson (1969)* and are shown in Sections 3.6.1–3.6.5. The balanced design is one in which the number of observations in the smallest classes or subclasses, $n_w$, are equal.

### 3.6.1  Notation

(a)  The source of variation for a main classification is denoted by a capital letter.

(b)  The source of variation for a nested classification is denoted by a capital letter followed by a colon and then the letter or letters denoting the class or classes within which it is nested. For example, $A{:}BC$ denotes $A$ nested in $B$ and $C$ classifications. This could be $A$ in $B$ in $C$ or $A$ in $C$ in $B$, or $A$ in $B \times C$ subclasses.

**TABLE 3.13**  ESTIMATES OF INTERACTION EFFECTS

| | | |
|---|---|---|
| $\hat{nl}_{11} = \phantom{-}7.1278$ | $\hat{nl}_{12} = \phantom{-}1.2444$ | $\hat{nl}_{13} = -8.3722$ |
| $\hat{nl}_{21} = -7.1278$ | $\hat{nl}_{22} = -1.2444$ | $\hat{nl}_{23} = \phantom{-}8.3722$ |

* Permission to use any part of this publication was granted by personal communication from the author.

(c) The source of variation for an interaction is denoted by a combination of letters identifying the interacting classifications, followed by a colon and the letter or letters of the classification within which the interaction is nested. For example, $AB:C$ denotes the interaction between $A$ and $B$ nested within $C$.

(d) The variation within the smallest subclasses is denoted by $W$. A smallest subclass is identified by a level of every classification in the design.

(e) The number of levels of a main classification is denoted by $n$ with a subscript that is the lowercase letter that identifies the source of variation. For example, if $A$ refers to a main classification, $n_a$ denotes the number of different levels of that classification.

(f) The number of levels of a nested classification is denoted by $n$ with a subscript that is the lowercase of the letter to the left of the colon in the identification for that course of variation. For example, in the nested classification $C:B$, the number of levels of $C$ in *each* level of $B$ is denoted by $n_c$.

### 3.6.2 Determining the Possible Lines in the Analysis of Variance Table

(a) One for within the smallest subclasses.

(b) One for each classification both main and nested.

(c) One for all possible interactions among classifications. To determine what interactions can exist, all possible pairs, trios, and so on of classification are formed by the following rules.

1. Write to the left of the colon in the symbol denoting the interaction the letters to the left of the colons in the classes being combined. (If no colon appears in a source of variation it is understood to be at the right of all letters.)
2. Write following the colon, but with no repetition of a letter, those letters to the right of the colons in the classifications being combined.
3. Delete any combination having a letter to the left of the colon that is repeated to the right of the colon.

To illustrate these rules for determining possible interactions suppose the sources of variation denoting classifications are $A$, $B$, $C:A$, and $D:AB$. The possible pairs of sources are $AB$, $AC:A$ which should be deleted because $A$ appears both to the left and to the right of the colon, $AD:AB$ which should be deleted, $BC:A$, $BD:AB$ which should be deleted, and $CD:AAB = CD:AB$. None of the possible trios can be used because $A$ and/or $B$ appear both to the left and to the right of the colon. The same is true for the combination of all four classifications.

### 3.6.3 Determining Degrees of Freedom

(a) The degrees of freedom for $W = [(n_w - 1)/n_w]N$, which of course reduces to zero when $n_w = 1$.

(b) The degrees of freedom for any other source of variation is the product of $(n - 1)$'s corresponding to letters to the left of the colon and $n$'s corresponding to letters to the right of the colon. Remember that if no colon appears it is understood to be to the right of all letters. To illustrate, the degrees of freedom for $AB$ are $(n_a - 1)(n_b - 1)$ and for $CD:AB$ they are $(n_c - 1)(n_d - 1)n_a n_b$.

As a check, the computed degrees of freedom should add to $N - 1$.

### 3.6.4 Computation of Sums of Squares in the Balanced Design

(a) Delete all lines of the table for which degrees of freedom are zero.

(b) Symbolic notation

1. A single lowercase letter denotes the uncorrected (for the mean) sum of squares of the corresponding main classification.
2. A combination of two or more lowercase letters denotes the uncorrected sum of squares for the corresponding subclass.
3. The correction factor, sometimes called the correction for the mean, is denoted by 1. The reason for this notation becomes apparent in Section 3.6.5.

**EXAMPLE**

Utilizing the data in Table 3.7 we would have

$$n = \frac{\Sigma_i Y_{i..}^2}{n_l n_w} = 30{,}117.4978 = \text{Uncorrected } SS_N$$

$$l = \frac{\Sigma_j Y_{.j.}^2}{n_n n_w} = 34{,}898.1100 = \text{Uncorrected } SS_L$$

$$nl = \frac{\Sigma_i \Sigma_j Y_{ij.}^2}{n_w} = 37{,}030.2400 = \text{Uncorrected } NL \text{ subclass SS}$$

where 1 denotes $Y_{...}^2/N = 28{,}720.0556$.

### 3.6.5 Rules for Computing a Mean Square

(a) Substitute for every letter to the left of the colon the corresponding lowercase letter minus one.

(b) Substitute for every letter to the right of the colon the corresponding lowercase letter.

(c) Expand algebraically and proceed according to the notation in Section 3.6.4(b).

(d) Divide by the degrees of freedom.

**EXAMPLE 1**

$$GH = \frac{1}{(n_g - 1)(n_h - 1)} (g - 1)(h - 1)$$

$$= \frac{1}{(n_g - 1)(n_h - 1)} (gh - g - h + 1)$$

$$= \frac{1}{(n_g - 1)(n_h - 1)} \text{(uncorrected } GH \text{ subclass sum of squares}$$

$$- \text{ uncorrected } G \text{ sum of squares}$$

$$- \text{ uncorrected } H \text{ sum of squares + correction factor)}$$

**EXAMPLE 2**

$$AB{:}C = \frac{1}{(n_a - 1)(n_b - 1)n_c} (a - 1)(b - 1)c$$

$$= \frac{1}{(n_a - 1)(n_b - 1)n_c} (abc - ac - bc + c)$$

$$= \frac{1}{(n_a - 1)(n_b - 1)n_c} \text{(uncorrected } ABC \text{ subclass sum of squares}$$

$$- \text{ uncorrected } AC \text{ subclass sum of squares}$$

$$- \text{ uncorrected } BC \text{ subclass sum of squares}$$

$$+ \text{ uncorrected } C \text{ sum of squares)}$$

(e) $W = 1/f_w$ times (uncorrected total sum of squares − uncorrected smallest subclass sum of squares). The uncorrected smallest subclass sum of squares is the one denoted by the letters of all classifications. $f_w$ symbolizes degrees of freedom.

(f) As a partial check the different sums of squares should add to the total sum of squares = uncorrected total sum of squares minus correction factor.

## 3.7 COMBINATION OF MAIN AND NESTED CLASSIFICATIONS

We have previously had illustrations of main classifications and hierarchical classifications in separate examples, but since we can have both classifications in the same set of data, let us look at such a situation. The data in Table 3.14 have been selected from a much larger group of figures resulting from a genetic experiment with poultry conducted by Dr. Thomas W. Fox of the Veterinary and Animal Sciences Department at the University of Massachusetts.

The first step in the analysis of variance is to decide on the sources of variation. In this case it can be seen that we have hatches, sires, dams, and observations as our sources. Next, we must decide into which classification each of these effects falls—main or nested. Considering the hatch effect first, we can see that any individual observation falls into the first or second hatch and so is completely identified as to the appropriate hatch by this single criterion, thus making hatch a main classification. It can also be seen that any individual observation falls into the first sire group or the second sire group with no further criterion of identification needed, making the sires a main classification. Looking next at the dams, we note that the first dam, number

**TABLE 3.14**   SIX-WEEK WEIGHTS OF CHICKS (IN GRAMS)

|  | Hatch | | | |
|---|---|---|---|---|
|  | 1 | | 2 | |
| Sires | Dam #807 | Dam #865 | Dam #807 | Dam #865 |
| #1630 | 550 | 630 | 470 | 540 |
|  | 660 | 590 | 540 | 570 |
| Subtotals | 1210 | 1220 | 1010 | 1110 |
|  | Dam #830 | Dam #918 | Dam #830 | Dam #918 |
| #1655 | 570 | 580 | 430 | 490 |
|  | 590 | 480 | 590 | 470 |
| Subtotals | 1160 | 1060 | 1020 | 960 |
| Totals | 2370 | 2280 | 2030 | 2070 |

807, appears only with the first sire, number 1630. Dam number 865 also appears only with the first sire while dams 830 and 918 appear only with the second sire. Therefore, to identify any dam we must specify not only her number but also the sire's number; that is, dam 807 is the first dam within the first sire, 830 is the first dam within the second sire, and 918 is the second dam within the second sire. Since the identification of any dam requires not only her own number but also the sire's number, the dams become a nested classification. Since both dams appear in both hatches they cannot be nested within hatches also and so are said to be cross-classified with respect to hatches. Note that there can be no interaction between dams and sires since the dams appear with only one sire. Finally, to identify completely any individual observation, we must state the identification of the hatch, the sire, and the dam. The observations become repeated observations within the smallest, or hatch $\times$ sire $\times$ dam, subclass. Our mathematical model for this set of data would be

$$Y_{ijkl} = \mu + h_i + s_j + hs_{ij} + d_{jk} + hd_{ijk} + e_{ijkl}$$

$$i = 1, 2 \quad j = 1, 2 \quad k = 1, 2 \quad l = 1, 2$$

where $h_i$ is the effect of the $i$th hatch, $s_j$ is the effect of the $j$th sire, and $d_{jk}$ is the effect of the $k$th dam in the $j$th sire.

Again, we shall start by calculating a total sum of squares and partition this total into component parts specified by the model. The necessary calculations follow, using the symbolic notation described in Section 3.6.4. In this notation, $N$ is the product of all lowercase $n$'s.

$$\text{Total SS} = \sum_i \sum_j \sum_k \sum_l Y_{ijkl}^2 - \frac{Y^2\cdots}{n_h n_s n_d n_w = N} = hsdw - (1)$$

$$= 4{,}845{,}900 - \frac{(8750)^2}{(2)(2)(2)(2) = 16} = 4{,}845{,}900 - 4{,}785{,}156.25$$

$$= 60{,}743.75 \tag{3.23}$$

$$\text{Hatch SS} = SS_H = \frac{\Sigma_i \, Y_{i\cdots}^2}{n_s n_d n_w} - \frac{Y_{\cdots}^2}{N} = h - (1)$$

$$= \frac{(4650)^2 + (4100)^2}{(2)(2)(2) = 8} - (1) = 4{,}804{,}062.50 - (1) = 18{,}906.25 \tag{3.24}$$

$$\text{Sire SS} = SS_S = \frac{\Sigma_j \, Y_{\cdot j \cdots}^2}{n_h n_d n_w} - \frac{Y_{\cdots}^2}{N} = s - (1)$$

$$= \frac{(4550)^2 + (4200)^2}{(2)(2)(2) = 8} - (1) = 4{,}792{,}812.50 - (1) = 7656.25 \tag{3.25}$$

$$SS_{HS} = \frac{\Sigma_i \Sigma_j \, Y_{ij\cdots}^2}{n_d n_w} - \frac{\Sigma_i \, Y_{i\cdots}^2}{n_s n_d n_w} - \frac{\Sigma_j \, Y_{\cdot j\cdots}^2}{n_h n_d n_w} + \frac{Y_{\cdots}^2}{N}$$

$$= hs - h - s + (1)$$

$$= \frac{(2340)^2 + (2220)^2 + (2120)^2 + (1980)^2}{(2)(2) = 4} - h - s + (1)$$

$$= 4{,}812{,}025.00 - h - s + (1) = 306.25 \tag{3.26}$$

$$SS_{D:S} = \frac{\Sigma_j \Sigma_k \, Y_{\cdot jk\cdot}^2}{n_h n_w} - \frac{\Sigma_j \, Y_{\cdot j\cdots}^2}{n_h n_d n_w} = (d-1)s = ds - s$$

$$= \frac{(2220)^2 + (2330)^2 + (2180)^2 + (2020)^2}{4} - s$$

$$= 4{,}797{,}525.00 - s = 4712.50 \tag{3.27}$$

$$SS_{HD:S} = \frac{\Sigma_i \Sigma_j \Sigma_k \, Y_{ijk\cdot}^2}{n_w} - \frac{\Sigma_i \Sigma_j \, Y_{ij\cdots}^2}{n_d n_w} - \frac{\Sigma_j \Sigma_k \, Y_{\cdot jk\cdot}^2}{n_h n_w} + \frac{\Sigma_j \, Y_{\cdot j\cdots}^2}{n_h n_d n_w}$$

$$= (h-1)(d-1)s = hds - hs - ds + s$$

$$= \frac{(1210)^2 + (1220)^2 + \cdots + (960)^2}{2} - hs - ds + s$$

$$= 4{,}817{,}950.00 - hs - ds + s = 1212.50 \tag{3.28}$$

**TABLE 3.15** ASSOCIATION OF TOTALS TO BE SQUARED, IN OBTAINING UNCORRECTED SUMS OF SQUARES, WITH ELEMENTS IN THE MODEL AND DETERMINATION OF DENOMINATOR

| Source of variation | Symbol in model | Summation | Totals to be squared | Number of totals[a] | | |
|---|---|---|---|---|---|---|
| $H$ | $h_i$ | $\Sigma_i$ | $Y_{i\cdots}$ | $n_h$ | $= 2$ | |
| $S$ | $s_j$ | $\Sigma_j$ | $Y_{\cdot j\cdots}$ | $n_s$ | $= 2$ | |
| $HS$ | $hs_{ij}$ | $\Sigma_i \Sigma_j$ | $Y_{ij\cdots}$ | $n_h n_s$ | $= (2)(2)$ | $= 4$ |
| $D:S$ | $d_{jk}$ | $\Sigma_j \Sigma_k$ | $Y_{\cdot jk\cdot}$ | $n_d n_s$ | $= (2)(2)$ | $= 4$ |
| $HD:S$ | $hds_{ijk}$ | $\Sigma_i \Sigma_j \Sigma_k$ | $Y_{ijk\cdot}$ | $n_h n_s n_d$ | $= (2)(2)(2)$ | $= 8$ |
| $W$ | $e_{ijkl}$ | $\Sigma_i \Sigma_j \Sigma_k \Sigma_l$ | $Y_{ijkl}$ | $n_h n_s n_d n_w$ | $= (2)(2)(2)(2)$ | $= 16$ |

[a] Division of $N$ by the value given in this column yields the appropriate denominators.

**TABLE 3.16**   DEGREES OF FREEDOM FOR DATA
                PRESENTED IN TABLE 3.14

| Source of variation | Degrees of freedom | | |
|---|---|---|---|
| Total | $N - 1$ | $= 16 - 1$ | $= 15$ |
| $H$ | $n_h - 1$ | $= 2 - 1$ | $= 1$ |
| $S$ | $n_s - 1$ | $= 2 - 1$ | $= 1$ |
| $HS$ | $(n_h - 1)(n_s - 1)$ | $= (2 - 1)(2 - 1)$ | $= 1$ |
| $D{:}S$ | $(n_d - 1)n_s$ | $= (2 - 1)(2)$ | $= 2$ |
| $HD{:}S$ | $(n_h - 1)(n_d - 1)n_s$ | $= (2 - 1)(2 - 1)(2)$ | $= 2$ |
| $W$ | $\dfrac{n_w - 1}{n_w} N$ | $= \dfrac{2 - 1}{2}(16)$ | $= 8$ |

$$\mathrm{SS}_W = \sum_i \sum_j \sum_k \sum_l Y_{ijkl}^2 - \frac{\Sigma_i \, \Sigma_j \, \Sigma_k \, Y_{ijk\cdot}^2}{n_w} = \sum_i \sum_j \sum_k \sum_l Y_{ijkl}^2 - hds$$

$$= 4{,}845{,}900.00 - hds = 27{,}950.00 \qquad\qquad (3.29)$$

It can be seen that the sums of squares obtained above for the various sources of variation sum to the total sum of squares. Before we complete the analysis of variance, note the relationship between the sources of variation specified in the mathematical model and the totals that are formed in calculating the appropriate sums of squares. A study of the symbols presented in Table 3.15 should be of help in determining which totals are required for each of the sources of variation, what classes or subclasses they are summed over, and also the appropriate denominators for each. While this table has been constructed for the particular set of data under consideration, a similar table can be developed for any model.

To proceed with the analysis of variance of these data we should next calculate the degrees of freedom attributed to each source of variation. The calculations are shown in Table 3.16.

The completed analysis of variance is shown in Table 3.17. *F* tests have not been calculated because further discussion of the composition of the mean squares is required to make decisions on the appropriate ratios to be used.

In Table 3.17 the bottom line of the sources of variation is labeled *W*, which exists only when the smallest subclass has more than one observation. This represents a within-subclass variation that, of course, cannot occur with only a single observation per subclass. The within-subclass sum of squares is really the sum of squares pooled

**TABLE 3.17**   ANALYSIS OF VARIANCE OF
                DATA PRESENTED IN TABLE 3.13

| Source | df | SS | MS |
|---|---|---|---|
| $H$ | 1 | 18,906.25 | 18,906.25 |
| $S$ | 1 | 7,656.25 | 7,656.25 |
| $HS$ | 1 | 306.25 | 306.25 |
| $D{:}S$ | 2 | 4,712.50 | 2,356.25 |
| $HD{:}S$ | 2 | 1,212.50 | 606.25 |
| $W$ | 8 | 27,950.00 | 3,493.75 |

**TABLE 3.18**   CALCULATION OF WITHIN-SUBCLASS SUM OF SQUARES

| | \multicolumn{8}{c}{Hatch × Sire × Dam Subclass Identification} | | |
|---|---|---|---|---|---|---|---|---|---|
| | 111 | 112 | 121 | 122 | 211 | 212 | 221 | 222 | Totals |
| $\sum_l Y_{ijkl}^2$ | 738,100 | 745,000 | 673,000 | 566,800 | 512,500 | 616,500 | 533,000 | 461,000 | $\sum_i \sum_j \sum_k \sum_l Y_{ijkl}^2$ = 4,845,900 |
| $\dfrac{Y_{ijk\cdot}^2}{n_w}$ | 732,050 | 744,200 | 672,800 | 561,800 | 510,050 | 616,050 | 520,200 | 460,800 | $\dfrac{\sum_i \sum_j \sum_k Y_{ijk\cdot}^2}{n_w}$ = 4,817,950 |
| $\sum_l Y_{ijkl}^2 - \dfrac{Y_{ijk\cdot}}{n_w}$ | 6,050 | 800 | 200 | 5,000 | 2,450 | 450 | 12,800 | 200 | $SS_w$ = 27,950 |

over all subclasses and can be calculated in that fashion. Although too cumbersome to be done in the usual analysis, Table 3.18 shows these calculations for our example. If we pool the sums of squares from within the subclasses we have

$$\sum_i \sum_j \sum_k \left( \sum_l Y_{ijkl}^2 - \frac{Y_{ijk\cdot}^2}{n_w} \right) = \sum_i \sum_j \sum_k \sum_l Y_{ijkl}^2 - \frac{\sum_i \sum_j \sum_k Y_{ijk\cdot}^2}{n_w}$$

$$= \sum_i \sum_j \sum_k \sum_l (Y_{ijkl} - \bar{y}_{ijk})^2$$

the within-subclass sum of squares. It can be seen that calculating the sums of squares of the deviations of the individual observations from the subclass means

$$\left( \sum_l Y_{ijkl} - \bar{y}_{ijk} \right)^2 = \sum_l Y_{ijkl}^2 - \frac{Y_{ijk\cdot}^2}{n_w}$$

and summing up or pooling over all subclasses yields the same results as using formula (3.29), which is really a shortcut method.

## 3.8  CROSS-CLASSIFIED EQUIVALENTS OF NESTED CLASSIFICATIONS

It can be shown that the sum of squares for $B{:}A$ in the nested classification contains the sums of squares that would have been calculated for $B$ and for the $AB$ interaction had the data been cross-classified instead of being nested. By cross-classification we mean that the second effect $B$ would also be considered as a main classification, rather than a nested one, and would interact with the $A$ effect. At times in statistical writings the term cross-classified effect is used synonymously with main effect.

The relationship between the sums of squares for nested effects and the sums of squares for main effects can be stated as follows:

*The sum of squares for a nested effect contains the sum of squares for the effect to the left of the colon plus the sums of squares for all possible interactions of the effect to the left of the colon with all effects (considered singly and in combination) to the right of the colon.*

**EXAMPLE 1** ─────────────────────────────────────────

$$SS_{D:S} = SS_D + SS_{DS}$$

Symbolically, $SS_{D:S} = (d - 1)s = ds - s$ which can be seen from

$$SS_D = d - 1$$

$$+ \ \underline{SS_{DS} = ds - d - s + 1}$$

Total $\quad SS_{D:S} = ds - s$

**EXAMPLE 2** ─────────────────────────────────────────

$$SS_{A:BC} = SS_A + SS_{AB} + SS_{AC} + SS_{ABC}$$

Symbolically, $SS_{A:BC} = (a - 1)bc = abc - bc$, as can be seen from

$$SS_A = a - 1$$

$$+ \ SS_{AB} = ab - a - b + 1$$

$$+ \ SS_{AC} = ac - a - c + 1$$

$$+ \ \underline{SS_{ABC} = abc - ab - ac - bc + a + b + c - 1}$$

Total $\quad SS_{A:BC} = abc - bc$

**EXAMPLE 3** ─────────────────────────────────────────

$$SS_{AC:BD} = SS_{AC} + SS_{ABC} + SS_{ACD} + SS_{ABCD}$$

Symbolically, $SS_{AC:BD} = (a - 1)(c - 1)bd = abcd - abd - bcd + bd$ as can be seen from

$$SS_{AC} = ac - a - c + 1$$

$$+ \ SS_{ABC} = abc - ab - ac - bc + a + b + c - 1$$

$$+ \ SS_{ACD} = acd - ac - ad - cd + a + c + d - 1$$

$$+ \ SS_{ABCD} = abcd - abc - abd - acd - bcd + ab + ac + ad + bc$$

$$\underline{+ \ bd + cd - a - b - c - d + 1}$$

Total $\quad SS_{AC:BD} = abcd - abd - bcd + bd$

## 3.9 FIXED AND RANDOM CLASSIFICATIONS

Until now we have ignored designating our classifications as being fixed or random, an essential decision before carrying out and interpreting an $F$ test in an analysis of variance. We shall first give some more formal definitions of fixed and random classifications or effects and further discussion of these effects. The decision as to whether a classification is fixed or random is a crucial one, making it necessary to have a clear understanding of their meanings. The most clear-cut definitions appear to be those presented by Henderson (1969).

**A.** Fixed classifications:
  (1) All levels of the classification are in the experiment, or
  (2) the only levels of interest to the experimenter are in the experiment, or
  (3) the levels in the experiment are from a normally distributed population but were not randomly chosen.
**B.** Random classification: The levels in the experiment are a random sample from a normally distributed population of levels.

The basic difference between the two types of classification is that a repetition of the experiment would call for the use of the same set of treatment levels for the fixed classification(s). However, where random classifications are involved, the repetition of the experiment would call for different levels drawn at random from the same population of levels. With fixed classifications we are concerned with drawing conclusions concerning only the levels which are included in the experiment and intermediate values and are not considering them as a random sample from all possible levels. Thus, we do not attempt to extrapolate beyond the limits of the specific levels included in the experiment. However, with a random classification we do draw conclusions concerning the entire population from which our levels are considered to be a random sample.

## 3.10 FIXED, RANDOM, AND MIXED MODELS

While we have written out mathematical models for each of the different examples which we have discussed, we have not described the models in terms of the classifications which the models contain. There are primarily three types of model although additional models have been suggested.

*Model I.* This is called the *fixed model* because all the classifications are fixed.

*Model II.* This is called the *random model,* or the *variance components* model, because all the classifications are random.

*Mixed Model.* When one or more fixed classifications and one or more random classifications appear in the same model it is referred to as a *mixed model.*

## 3.11 EXPECTATIONS OF MEAN SQUARES: BALANCED DESIGN

### 3.11.1 Method of Calculation

In the preceding analysis of variance tables we have calculated a mathematical value for the mean square of each line of the table. These values are equated to the variance components which the mean squares are expected to contain, together with appropriate coefficients. While several textbooks present rules by which the expectations of mean squares $E(MS)$ can be calculated, those presented by Henderson (1969) appear to be the clearest and most useful.

    **A.** All classifications in the model are random, Model II.

      (1) $\sigma^2$ values corresponding to each of the possible lines of the analysis of variance table appear in one or more of the expectations of mean squares. The subscripts on $\sigma^2$ are the same as the identification for the sources of variation. If the mean square cannot be computed because it has zero degrees of freedom, the corresponding $\sigma^2$ is retained nevertheless.

      (2) Regardless of whether there is a $W$ mean square, $\sigma_w^2$ appears with a coefficient of one (1) in the expectation of every mean square.

      (3) In addition to $\sigma_w^2$ the expectation of a mean square contains any $\sigma^2$ described in section A(1) above which has in its subscripts all the letter denoting that mean square.

      (4) Aside from $\sigma_w^2$, the coefficient of a particular $\sigma^2$ in a particular mean square is $N$ divided by the product of the $n$'s corresponding to the letters in the subscript of $\sigma^2$. For example, the coefficient of $\sigma_{ab:c}^2$ wherever it appears is $N/n_a n_b n_c$. This rule can also be stated as the product of all $n$'s excluding those corresponding to the letters in the subscript of $\sigma^2$.

    **B.** One or more classifications are fixed. The expectations are the same as in section A above except that certain $\sigma^2$ values vanish according to the following rule. Any $\sigma^2$ having to the left of the colon a letter denoting a fixed classification disappears from the expectations of all mean squares not containing this letter. Remember that if there is no colon in the subscript, the colon by definition is to the right of all letters. To illustrate, in a three-way classification with $C$ fixed $\sigma_{abc}^2$ vanishes from the expectations of $A$, $B$, and $AB$ since they do not contain the letter $C$. But $\sigma_{abc}^2$ does not vanish from $C$, $BC$, and $ABC$ all of which contain the letter $C$.

    The deletion rule given in 3.11.1B is that given by Henderson in the first edition (1959) of the cited publication. However, in the second edition (1969) he recommends that the variance components for interactions between fixed and random effects be retained. Thus, the variance components to be deleted would be those having to the left of the colon *only* letters denoting fixed classifications. They would not be deleted, of course, from the line pertaining to that $\sigma^2$. We prefer the earlier interpretation and have followed this procedure throughout the text.

    Many statistical texts reserve a symbol such as $\sigma^2$ for a random classification, using a different symbol for a fixed classification. Were $A$ a fixed classification, Steel and Torrie (1980) would express this component as $\sum \alpha_i^2/(a-1)$ and similarly, with $B$ a fixed effect, they would express $\sigma_{b:a}^2$ as $\sum_i \sum_j \beta_{ij}^2/(b-1)a$, where the lowercase letters refer to the number of levels. Snedecor and Cochran (1980) would express these two components as $\kappa_a$ and $\kappa_{b:a}$, $\kappa$ being the greek letter kappa. In this text the sum of squares of the true levels of a fixed classification divided by the appropriate degrees of freedom is defined as $\sigma^2$ with a subscript that is the lowercase of the letter or letters identifying the corresponding sum of squares. With $A$ and $B:A$ fixed classifications,

$$\frac{1}{n-1} \sum_i a_i^2 = \sigma_a^2 \quad \text{and} \quad \frac{1}{(n_b-1)n_a} \sum_i \sum_j b_{ij}^2 = \sigma_{b:a}^2$$

    The expectations of mean squares play a crucial role in the analysis of variance because they determine the appropriate ratios used in the $F$ test. The purpose of the

$F$ test is to develop a probability statement as to whether a particular source of variation causes a significant increase in the amount of variation in a set of data. For example, if 30 experimental units are allotted at random to three different levels of treatment $A$, we need a test which will tell us whether the three levels have increased the variation among the experimental units enough for us to believe that there are real differences among the three treatment levels. The variance due to treatment $A$ is designated by $\sigma_a^2$ and the $F$ ratio of mean squares must be such that the numerator containing $\sigma_a^2$ (together with its appropriate coefficient) must differ from the denominator by *only* the component $\sigma_a^2$. (This same procedure is followed in testing for any other $\sigma^2$.) Should there be no apparent denominator for the $F$ test, an approximate test devised by Satterthwaite (1946) may be used in some circumstances.

### 3.11.2  Examples of the Calculation of the Expectations of Mean Squares

**One-Way Classification**  Suppose we have three levels or classes of $A$ with eight observations in each level.

| Source | df | $E$(MS) |
|--------|-----|---------|
| $A$ | 2 | $\sigma_w^2 + 8\sigma_a^2$ |
| $W$ | 21 | $\sigma_w^2$ |

By rule A(1), we have $\sigma_a^2$ and $\sigma_w^2$ ($n_a = 3$, $n_w = 8$, $N = n_a n_w = 24$).
By rule A(2), $\sigma_w^2$ appears in each line.
By rule A(3), $\sigma_a^2$ appears in the line for $A$.
By rule A(4), $\sigma_a^2$ has a coefficient of $N/n_a = 8$.
By rule B, we do not delete any component in the $E$(MS).
Thus, the $F$ ratio for testing for the significance of $A$ is $(\sigma_w^2 + 8\sigma_a^2)/\sigma_w^2$ with 2 and 21 degrees of freedom.

**Two-Way Classification: Single Observation per Subclass**  Suppose we have eight levels of $A$, two levels of $B$, and one observation in each $AB$ subclass.

| Source | df | $E$(MS) | Source | df | $E$(MS) |
|--------|-----|---------|--------|-----|---------|
| $A$ | 7 | $\sigma_w^2 + \sigma_{ab}^2 + 2\sigma_a^2$ | $A$ | 7 | $\sigma_w^2 + \sigma_{ab}^2 + 2\sigma_a^2$ |
| $B$ | 1 | $\sigma_w^2 + \sigma_{ab}^2 + 8\sigma_b^2$ | $B$ | 1 | $\sigma_w^2 + \sigma_{ab}^2 + 8\sigma_b^2$ |
| $AB$ | 7 | $\sigma_w^2 + \sigma_{ab}^2$ | $AB$ | 7 | $\sigma_w^2 + \sigma_{ab}^2$ |
| $W$ | 0 | $\sigma_w^2$ | | | |

By rule A(1) we have $\sigma_a^2$, $\sigma_b^2$, $\sigma_{ab}^2$, and $\sigma_w^2$ ($n_a = 8$, $n_b = 2$, $n_w = 1$, and $N = n_a n_b n_w = 16$).
By rule A(2), $\sigma_w^2$ appears in each line.
By rule A(3), $\sigma_{ab}^2$ appears in the lines for $AB$, $B$, and $A$; $\sigma_b^2$ appears in the line for $B$; and $\sigma_a^2$ appears in the line for $A$.

By rule A(4), $\sigma_{ab}^2$ has a coefficient $N/n_a n_b = 1$; $\sigma_b^2$ has a coefficient of $N/n_b = 8$; and $\sigma_a^2$ has a coefficient of $N/n_a = 2$.

Since there are 0 degrees of freedom for $W$ this line drops out of the analysis of variance table so the table has been rewritten omitting this effect. It can be seen that there is no test for the $AB$ line of the table, which is sometimes labeled as an error. Both $A$ and $B$ are tested by $AB$. By rule B, if $B$ is considered to be a fixed classification, $\sigma_{ab}^2$ is deleted from the $A$ line and there would be no test for $A$. If both $A$ and $B$ are considered to be fixed classifications, $\sigma_{ab}^2$ is deleted from the $A$ and $B$ lines, leaving no exact test for either $A$ or $B$ unless the assumption is made that $\sigma_{ab}^2$ equals zero, which would be questionable without knowledge to substantiate this. It is possible to use $AB$ as a denominator realizing the fact that it is, on the average, too large and results in the $F$ tests being too small.

Frequently, the $AB$ interaction effects are ignored and the expectations for this situation are simply written

| Source | df | E(MS) |
|--------|----|----|
| $A$ | 7 | $\sigma_e^2 + 2\sigma_a^2$ |
| $B$ | 1 | $\sigma_e^2 + 8\sigma_b^2$ |
| $E$ | 7 | $\sigma_e^2$ |

When $A$ is a random set of effects and $B$ is a fixed set of effects a test is made for the $B$ effects by $\text{MS}_B/\text{MS}_E$, but no test should be made for $A$ owing to the reasons above. This is frequently the case with the randomized complete block, a common statistical design.

**Two-Way Classification: Multiple Observations per Subclass**  Suppose we have four levels of $A$, two levels of $B$, and seven observations per subclass.

| Source | df | E(MS) |
|--------|----|----|
| $A$ | 3 | $\sigma_w^2 + 7\sigma_{ab}^2 + 14\sigma_a^2$ |
| $B$ | 1 | $\sigma_w^2 + 7\sigma_{ab}^2 + 28\sigma_b^2$ |
| $AB$ | 3 | $\sigma_w^2 + 7\sigma_{ab}^2$ |
| $W$ | 48 | $\sigma_w^2$ |

By rule A(1), we have $\sigma_a^2$, $\sigma_b^2$, $\sigma_{ab}^2$, and $\sigma_w^2$ ($n_a = 4$, $n_b = 2$, $n_w = 7$, and $n_a n_b n_w = N = 56$).

By rule A(2), $\sigma_w^2$ appears in every line.

By rule A(3), $\sigma_{ab}^2$ appears in lines for $AB$, $B$, and $A$; $\sigma_b^2$ appears in the line for $B$; and $\sigma_a^2$ appears in the line for $A$.

By rule A(4), $\sigma_{ab}^2$ has a coefficient of $N/n_a n_b = 7$, $\sigma_b^2$ has a coefficient of $N/n_b = 28$, and $\sigma_a^2$ has a coefficient of $N/n_a = 14$.

By rule B, no components are deleted if both $A$ and $B$ are random classifications and the appropriate $F$ ratios are $\text{MS}_{AB}/\text{MS}_W$, $\text{MS}_B/\text{MS}_{AB}$, and $\text{MS}_A/\text{MS}_{AB}$. If $B$ is considered to be a fixed set of effects, the component $7\,\sigma_{ab}^2$ is deleted from the $A$ line and $A$ is now tested by $MS_A/MS_W$. If both $A$ and $B$ are fixed classifications, the component 7 $\sigma_{ab}^2$ is deleted from both the $A$ and $B$ lines and $A$, $B$, and $AB$ are all tested by the $W$ mean square.

**Three-Way Classification: Multiple Observations per Subclass**  Suppose we have three levels of $A$, four levels of $B$, five levels of $C$, and four observations per subclass.

| Source | df | $E(MS)$ |
|--------|-----|---------|
| $A$ | 2 | $\sigma_w^2 + 4\sigma_{abc}^2 + 16\sigma_{ac}^2 + 20\sigma_{ab}^2 + 80\sigma_a^2$ |
| $B$ | 3 | $\sigma_w^2 + 4\sigma_{abc}^2 + 12\sigma_{bc}^2 + 20\sigma_{ab}^2 + 60\sigma_b^2$ |
| $C$ | 4 | $\sigma_w^2 + 4\sigma_{abc}^2 + 12\sigma_{bc}^2 + 16\sigma_{ac}^2 + 48\sigma^2$ |
| $AB$ | 6 | $\sigma_w^2 + 4\sigma_{abc}^2 + 20\sigma_{ab}^2$ |
| $AC$ | 8 | $\sigma_w^2 + 4\sigma_{abc}^2 + 16\sigma_{ac}^2$ |
| $BC$ | 12 | $\sigma_w^2 + 4\sigma_{abc}^2 + 12\sigma_{bc}^2$ |
| $ABC$ | 24 | $\sigma_w^2 + 4\sigma_{abc}^2$ |
| $W$ | 180 | $\sigma_w^2$ |

By rule A(1), we have $\sigma_a^2$, $\sigma_b^2$, $\sigma_c^2$, $\sigma_{ab}^2$, $\sigma_{ac}^2$, $\sigma_{bc}^2$, $\sigma_{abc}^2$, and $\sigma_w^2$ ($n_a = 3$, $n_b = 4$, $n_c = 5$, $n_w = 4$, and $n_a n_b n_c n_w = N = 240$).

By rule A(2), $\sigma_w^2$ appears in each line.

By rule A(3), $\sigma_{abc}^2$ appears in all lines except $W$; $\sigma_{bc}^2$ appears in the lines for $BC$, $C$, and $B$; $\sigma_{ac}^2$ appears in the lines for $AC$, $C$, and $A$; $\sigma_{ab}^2$ appears in the lines for $AB$, $A$, and $B$; $\sigma_c^2$ appears in the line for $C$; $\sigma_b^2$ appears in the line for $B$; and $\sigma_a^2$ appears in the line for $A$.

By rule A(4), $\sigma_{abc}^2$ has a coefficient of $N/n_a n_b n_c = 4$; $\sigma_{bc}^2$ has a coefficient of $N/n_b n_c = 12$; $\sigma_{ac}^2$ has a coefficient of $N/n_a n_c = 16$; $\sigma_{ab}^2$ has a coefficient of $N/n_a n_b = 20$; $\sigma_c^2$ has a coefficient of $N/n_c = 48$; $\sigma_b^2$ has a coefficient of $N/n_b = 60$; and $\sigma_a^2$ has a coefficient of $N/n_a = 80$.

By rule B, no components are deleted if this is a random model, in which case $ABC$ is tested by $W$; $AB$, $AC$, and $BC$ are tested by $ABC$; and there are no exact tests for $A$, $B$, and $C$. Should $C$ be a fixed set of effects, the components $\sigma_{abc}^2$ would be deleted from the $AB$, $A$, and $B$ lines; $\sigma_{bc}^2$ would be deleted from the $B$ line; and $\sigma_{ac}^2$ would be deleted from the $A$ line. $AB$ would now be tested by $W$, $B$ would be tested by $AB$, and $A$ would be tested by $AB$. Were $A$, $B$, and $C$ all sets of fixed effects, all lines in the table above $W$ would be tested by $W$.

**Nested Classifications: Balanced Design**  Suppose we have three levels of $A$, two levels of $B$ within each level of $A$, and four observations per $AB$ subclass.

| Source | df | $E(MS)$ |
|--------|-----|---------|
| $A$ | 2 | $\sigma_w^2 + 4\sigma_{b:a}^2 + 8\sigma_a^2$ |
| $B:A$ | 3 | $\sigma_w^2 + 4\sigma_{b:a}^2$ |
| $W$ | 18 | $\sigma_w^2$ |

By rule A(1), we have $\sigma_a^2$, $\sigma_{b:a}^2$, and $\sigma_w^2$ ($n_a = 3$, $n_b = 2$, $n_w = 4$, and $n_a n_b n_w = N = 24$).

By rule A(2), $\sigma_w^2$ appears in each line.

By rule A(3), $\sigma_{b:a}^2$ appears in the line for $B:A$ and the line for $A$ and $\sigma_a^2$ appears in the line for $A$.

By rule A(4), $\sigma_{b:a}^2$ has a coefficient of $N/n_a n_b = 4$ and $\sigma_a^2$ has a coefficient of $N/n_a = 8$.

By rule B, all components remain in the expectation if this is a random model. Frequently, in a design of this type $B$ is a random classification and $A$ is a fixed classification. In this case, no components are deleted and $B:A$ is tested by $W$ and $A$ is tested by $B:A$.

**Main Classifications and Nested Classification**  Suppose we have three levels of $A$, two levels of $B$, four levels of $C$ within each level of $B$, and five observations in each $ABC$ subclass.

| Source | df | $E(MS)$ |
|---|---|---|
| $A$ | 2 | $\sigma_w^2 + 5\sigma_{ac:b}^2 + 20\sigma_{ab}^2 + 40\sigma_a^2$ |
| $B$ | 1 | $\sigma_w^2 + 5\sigma_{ac:b}^2 + 15\sigma_{c:b}^2 + 20\sigma_{ab}^2 + 60\sigma_b^2$ |
| $AB$ | 2 | $\sigma_w^2 + 5\sigma_{ac:b}^2 + 20\sigma_{ab}^2$ |
| $C:B$ | 6 | $\sigma_w^2 + 5\sigma_{ac:b}^2 + 15\sigma_{c:b}^2$ |
| $AC:B$ | 12 | $\sigma_w^2 + 5\sigma_{ac:b}^2$ |
| $W$ | 96 | $\sigma_w^2$ |

By rule A(1), we have $\sigma_a^2$, $\sigma_b^2$, $\sigma_{ab}^2$, $\sigma_{c:b}^2$, $\sigma_{ac:b}^2$, and $\sigma_w^2$ ($n_a = 3$, $n_b = 2$, $n_c = 4$, $n_w = 5$, and $n_a n_b n_c n_w = N = 120$).

By rule A(2), $\sigma_w^2$ appears in each line.

By rule A(3), $\sigma_{ac:b}^2$ appears in the lines for $AC:B$, $C:B$, $AB$, $B$, and $A$; $\sigma_{c:b}^2$ appears in the lines for $C:B$ and $B$; $\sigma_{ab}^2$ appears in the lines for $AB$, $B$, and $A$; $\sigma_b^2$ appears in the line for $B$; and $\sigma_a^2$ appears in the line for $A$.

By rule A(4), $\sigma_{ac:b}^2$ has a coefficient of $N/n_a n_c n_b = 5$; $\sigma_{c:b}^2$ has a coefficient of $N/n_c n_b = 15$; $\sigma_{ab}^2$ has a coefficient of $N/n_a n_b = 20$; $\sigma_b^2$ has a coefficient of $N/n_b = 60$; and $\sigma_a^2$ has a coefficient of $N/n_a = 40$.

By rule B, all components would remain in the expectations in the random model. Let us assume $A$ and $B$ to be fixed classifications and $C$ a random classification. In this situation $\sigma_{ac:b}^2$ would be deleted from the $C:B$ and $B$ lines and $\sigma_{ab}^2$ would be deleted from the $B$ and $A$ lines. The appropriate $F$ ratios would then be $MS_A/MS_{AC:B}$, $MS_B/MS_{C:B}$, $MS_A/MS_{AC:B}$, $MS_{C:B}/MS_W$, and $MS_{AC:B}/MS_W$.

**Nested Classification: Unbalanced Design**

| Levels of $A$ | | $A_1$ | | $A_2$ | | $A_3$ | |
|---|---|---|---|---|---|---|---|
| Levels of $B:A$ | $B_1$ | $B_2$ | $B_1$ | $B_2$ | $B_1$ | $B_2$ | $B_3$ |
| Number of observations | 2 | 3 | 4 | 6 | 1 | 2 | 4 |

| Source | df | $E(MS)$ |
|---|---|---|
| $A$ | $n_a - 1 = 3 - 1 = 2$ | $\sigma_w^2 + k_2\sigma_{b:a}^2 + k_3\sigma_a^2$ |
| $B:A$ | $n_{ab} - n_a = 7 - 3 = 4$ | $\sigma_w^2 + k_1\sigma_{b:a}^2$ |
| $W$ | $N - n_{ab} = 22 - 7 = 15$ | $\sigma_w^2$ |

While we are no longer able to use all the rules which were used for the balanced design, $\sigma_w^2$ appears in each line with a coefficient of 1; $\sigma_{b:a}^2$ appears in the $B:A$ and the $A$ lines; and $\sigma_a^2$ appears in the $A$ line. With the unequal or disproportionate subclass numbers, the coefficient for $\sigma_{b:a}^2$ will differ in the two lines in which it appears. This has the result of yielding only an approximate $F$ test for $A$ since after omitting $k_3\sigma_a^2$ from the $A$ line, the lines for $A$ and $B:A$ have the same variance components *but* a different coefficient for $\sigma_{b:a}^2$. The $k$ values are then an estimate of the number of observations per class or subclass which has information on a component. The computations for the three $k$'s are shown below.

$$k_1 = \frac{n.. - \Sigma_i \dfrac{\Sigma_j n_{ij}^2}{n_{i.}}}{n_{ab} - n_a} = \frac{22 - \left(\dfrac{4+9}{5} + \dfrac{16+36}{10} + \dfrac{1+4+16}{7}\right)}{7 - 3}$$

$$= \frac{22 - 10.8}{4} = \frac{11.2}{4} = 2.800 \tag{3.30}$$

where $n..$ is the total number of observations, $n_i$ is the number of observations in the $i$th level of $A$, and $n_{ij}$ is the number of observations in the $AB_{ij}$ subclass.

$$k_2 = \frac{\Sigma_i \dfrac{\Sigma_j n_{ij}^2}{n_{i.}} - \dfrac{\Sigma_i \Sigma_j n_{ij}^2}{n..}}{n_a - 1} = \frac{10.8 - \dfrac{4+9+16+36+1+4+16}{22}}{3 - 1}$$

$$= \frac{10.8 - 3.9091}{2} = 3.445 \tag{3.31}$$

$$k_3 = \frac{n.. - \dfrac{\Sigma_i n_{i.}^2}{n..}}{n_a - 1} = \frac{22 - \dfrac{25 + 100 + 49}{22}}{2}$$

$$= \frac{22 - 174/22}{2} = \frac{22 - 7.9091}{2} = 7.045 \tag{3.32}$$

## 3.12 TWO-WAY CLASSIFICATION USING DUMMY VARIABLES

It is a rather common situation when the dependent variable does not yield a measurement on a continuous scale such as inches, pounds, or centimeters. Instead, the observations on the experimental units fall into two or more categories, which can be coded as 0 and 1 and constitute a binomial population, or 0, 1, 2, and so on which would constitute a multinomial population. The binomial population arises frequently and occurs in such situations as when subjects either respond or do not respond, whether subjects are alive or dead, male or female.

Many textbooks present the analysis of discrete data from a binomial population in terms of a chi-square test. However, Li (1957, Chap. 21) gives an extensive discussion of the similarity of the chi-square test and the analysis of variance of coded or dummy variables which is summarized in the following paragraphs.

The codes of 0 and 1 represent the occurrence or nonoccurrence of an event and are termed dummy variables. Li points out that either method may be used and that the conclusions reached by the two methods are usually the same. Furthermore, both methods yield approximate tests in the sense that the validity of the chi-square test requires large samples and the validity of the $F$ test requires normal distributions. However, neither of these requirements is entirely fulfilled in sampling from binomial populations. Harvey (1982a) deals with the analysis of discrete data by means of least-squares procedures, pointing out that there is greater flexibility in the models which may be used under these procedures than by other methods.

It is common to think of each of the categories in a binomial population as being a success or a failure. An observation $Y$ can then be defined as the *number of successes for each observation.* In dealing, for example, with seeds which are viable or nonviable, the observation $Y$ may be defined as the number of viable seeds *for each seed observed.* If a seed is viable $Y$ is 1, and if the seed is nonviable $Y$ is 0. We are then dealing with a binomial population made up of 1's and 0's.

The mean of a binomial population, $\mu$, is equal to the relative frequency of the successes, and the variance $\sigma^2$ is equal to the product of the relative frequencies of successes and failures, $\mu(1 - \mu)$. Sample means drawn from such a population are expected to follow the normal distribution approximately, with mean equal to $\mu$ and variance equal to $\sigma^2/n$, where $n$ is the sample size and $\sigma^2$ is equal to $\mu(1 - \mu)$. However, the sample means do not follow the normal distribution closely unless the samples are large. To decide whether the sample size is adequate, a commonly used working rule is that $n\mu$ and $n(1 - \mu)$ should be equal to or greater than 5.

A sample mean $\bar{y}_i$ is really a proportion and tests between or among proportions are frequently conducted by the chi-square test. Two of the more common methods of carrying out the chi-square test are shown in Table 3.19, using a set of hypothetical data. In the table, the letter $o$ represents the number observed in a category and the letter $c$ represents the number expected or calculated in a category. The number expected in a success category is the total number of successes divided by the total number of observations, $\bar{p}$, multiplied by the number per sample. The number expected in a failure category is the total number of failures divided by the total number of observations, $\bar{q} = 1 - \bar{p}$, multiplied by the number per sample. In the lower portion of Table 3.19, $n_i$ is the size of the sample, $f_i$ is the number of successes per sample, $N$ is the total number of observations, and $F$ is the total number of successes. With $k$ samples, there are $k - 1$ degrees of freedom.

The number of successes divided by the number of observations in a sample is, of course, the sample mean. The sample means $\bar{y}_i$ may be considered to be a sample of $k$ observations drawn from a normal population having a mean of $\mu$ and a variance of $\sigma^2/n$, where $n$ is the sample size. Then the statistic

$$\chi^2 = \frac{\Sigma_k(\bar{y}_i - \bar{y}.)^2}{\sigma^2/n} = \frac{n\,\Sigma_k(\bar{y}_i - \bar{y}.)^2}{\sigma^2} = \frac{\text{Among-sample SS}}{\mu(1 - \mu)} \qquad (3.33)$$

where $\bar{y}.$ is the overall mean, follows the $\chi^2$ distribution with $k - 1$ degrees of freedom. Since the variance of the binomial distribution is equal to $\mu(1 - \mu)$, the quantity $\bar{y}.(1 - \bar{y}.)$ may be used as a pooled estimate of $\sigma^2$, leading to the statistic

**TABLE 3.19** $\chi^2$ TEST AMONG PROPORTIONS

| Sample | Observation | $o$ | $c$ | $o - c$ | $\dfrac{(o - c)^2}{c}$ |
|--------|-------------|------|------|---------|------------------------|
| 1 | Success | 35 | 31.5 | 3.5 | 0.39 |
|   | Failure | 15 | 18.5 | −3.5 | 0.66 |
| 2 | Success | 40 | 31.5 | 8.5 | 2.29 |
|   | Failure | 10 | 18.5 | −8.5 | 3.91 |
| 3 | Success | 25 | 31.5 | −6.5 | 1.34 |
|   | Failure | 25 | 18.5 | 6.5 | 2.28 |
| 4 | Success | 26 | 31.5 | −5.5 | 0.96 |
|   | Failure | 24 | 18.5 | 5.5 | 1.64 |
| Total |  | 200 | 200.0 |  | $\chi^2 = 13.47$ |

$$\chi^2_{0.05} = 7.81, \ 3 \ df$$

| $n_i$ | $f_i$ | $p_i = \dfrac{f_i}{n_i}$ | |
|-------|-------|--------------------------|---|
| 50 | 35 | 0.70 | $\chi^2 = \dfrac{\Sigma_i \ f_i p_i - F\bar{p}}{\bar{p}\bar{q}}$ |
| 50 | 40 | 0.80 | $= \dfrac{82.52 - 79.38}{(0.63)(0.37)}$ |
| 50 | 25 | 0.50 | $= \dfrac{3.14}{0.2331} = 13.47$ |
| 50 | 26 | 0.52 | |
| $N = 200$ | $F = 126$ | $\bar{p} = 0.63$ | |

$$\chi^2 = \frac{\Sigma_k \ n(\bar{y}_k - \bar{y}.)^2}{\bar{y}.(1 - \bar{y}.)} = \frac{\text{Among-sample SS}}{\bar{y}.(1 - \bar{y}.)} \qquad (3.34)$$

which follows approximately the $\chi^2$ distribution with $k - 1$ degrees of freedom.

Using formula (3.34), we can calculate the among-sample sum of squares in the usual method of the analysis of variance for the numerator, resulting in

$$\chi^2 = \frac{\Sigma_k \ n(\bar{y}_k - \bar{y}.)^2}{\bar{y}.(1 - \bar{y}.)} = \frac{\dfrac{\Sigma_k \ Y^2_{k.}}{n} - \dfrac{Y^2_{..}}{N}}{\bar{y}.(1 - \bar{y}.)}$$

$$= \frac{\dfrac{(35)^2 + (40)^2 + (25)^2 + (26)^2}{50} - \dfrac{(126)^2}{200}}{(0.63)(0.37)}$$

$$= \frac{82.52 - 79.38}{0.2331} = \frac{3.14}{0.2331} = 13.47$$

The statistic used in testing the hypothesis that $k$ population means are equal, when concerned with normal populations, is

$$F = \frac{\dfrac{\text{Among-sample SS}}{k - 1}}{\dfrac{\text{Within-sample SS}}{N - k}} = \frac{\text{Among-sample MS}}{\text{Within-sample MS}} \qquad (3.35)$$

**TABLE 3.22**  ANALYSIS OF VARIANCE OF DATA SHOWN IN TABLE 3.19

| Source | df | SS | MS | E(MS) | F |
|--------|-----|---------|--------|-------|------|
| E | 3 | 0.9667 | 0.3222 | $\sigma_w^2 + 60\sigma_e^2$ | 1.39 |
| T | 2 | 3.3583 | 1.6792 | $\sigma_w^2 + 20\sigma_{et}^2 + 80\sigma_t^2$ | 5.90 |
| ET | 6 | 1.7084 | 0.2847 | $\sigma_w^2 + 20\sigma_{et}^2$ | 1.23 |
| W | 228 | 52.9000 | 0.2320 | $\sigma_w^2$ | |

$$SS_W = \sum_i \sum_j \sum_k Y_{ijk}^2 - \frac{\Sigma_i \, \Sigma_j \, Y_{ij\cdot}^2}{n_w} = \sum_i \sum_j \sum_k Y_{ijk}^2 - et$$

$$= 104 - 51.1000 = 52.9000 \tag{3.41}$$

The analysis of variance for this set of data is shown in Table 3.22, where it can be seen that there are significant differences among the treatment means.

The coding could be reversed, with those chickens responding negatively to the microflora treatments being given a value of 1 and the chickens responding positively given a value of 0. The results of this analysis would be exactly the same as that shown in Table 3.22. One could also change each of the cell ratios to percentages and analyze these values by means of the analysis of variance, although such an analysis would yield no information regarding the size or importance of the interaction. With data such as these a transformation is frequently necessary before analysis, as discussed in Section 3.16.2. When data are classified in more than two ways, such as 0, 1, and 2, the methods of computation for the analysis of variance follow those used when the data are classified as 0 and 1. There are occasions when the independent variable is the dummy variable, particularly in regression analyses.

## 3.13  ASSUMPTIONS UNDERLYING THE ANALYSIS OF VARIANCE

We touched earlier on some of the assumptions of the analysis of variance but a general discussion has been delayed until this point because it is felt that some experience with the various types of analysis would be helpful in considering the assumptions. Fortunately, it has been found that minor departures from the underlying assumptions do not disturb the conclusions or significance tests to any important degree. Major violations of the assumptions, of course, are most likely to invalidate conclusions and must be avoided. In many situations encountered in the analysis of variance, departures from the assumptions can be overcome by what is termed a *transformation* of the data. The transformations most commonly used will be discussed later in this chapter.

One of the assumptions of the analysis of variance, really a series of assumptions, is that *the random errors of any mathematical model are normally distributed with a mean of zero and a variance of $\sigma^2$*. In a one-way classification, or a multiway classification with more than one observation per subclass, the error variance is the variance among observations within the classes or subclasses. In other models, the error variance is a measure of the variation of random errors associated with the individual observations, a residual variation not accounted for by the effects in the model. The error term in a model is considered to be an independent random variable. The assumptions are made that the random errors follow the normal distribution and are independently distributed. The term independently distributed means that the errors are expected to appear in a random sequence, not a long series of errors greater than zero, followed by a long series of values less than zero (or any other such pattern), zero being the mean of the random errors. Lack of independence can result from a faulty recorder or scale, or possibly from a poor allocation of treatments to experimental units. There is no method for adjusting for lack of independence and the experiment would need to be redesigned if the problem were encountered. If lack of independence is suspected, a "runs test" (Sokal and Rohlf, 1969) can be carried out. If a test of significance is carried out with data in which independence of errors does not hold, these tests are approximate at best.

A portion of the statement of the assumptions in the preceding paragraph is that the random errors have a variance of $\sigma^2$. If we think of this in terms of the classes in a one-way analysis of variance or the subclasses in a multiway analysis of variance, the variance calculated within each class or subclass is expected to be an estimate of the same population variance $\sigma^2$. Since this is so, the variances among classes or subclasses are not expected to differ significantly from one another. This property is usually expressed as *homogeneity of variances.* Tests for the homogeneity of variances are presented in Section 3.14. This is an extremely important assumption of the analysis of variance and, fortunately, when this assumption does not hold, the problem often can be remedied by one of the data transformations discussed in later sections.

A further assumption, and one perhaps not emphasized as much as it should be in this and other statistical writings, is that experimental *units must be assigned* to *treatments at random.* As an example, if the effects of a particular exercise program are to be evaluated, consideration must be given as to what population the conclusions are to be applicable. If the concern is for a college-age population of students in general, care must be taken that all segments of the population can be represented in the sample to be selected, and that the group or groups to be evaluated are not made up solely of males or females, athletes or nonathletes, or any other particular segment. In a field experiment where soil differences are suspected, caution must be taken that treatments are not favored or penalized by their locations in the field. Proper sampling, or assigning of experimental units to treatments, is a vital part of experimentation, so much so that many courses are offered and textbooks written dealing solely with this aspect of statistical design.

A final assumption is that of *additivity.* If we consider a two-way classification, our assumption is that any variate is a result of the overall mean plus the additive effects of each of the main classifications as well as the random error component. Note that this means that there is no interaction between the main classifications involved, since a significant interaction violates the assumption of additivity. When a significant interaction exists, $F$ tests of main classifications are not very efficient and frequently not very meaningful. An interaction effect indicates that the levels of treatment $A$ do not give the same response at the different levels of treatment $B$ or vice versa, so caution is necessary in making statements about the overall effects of main classifications. Frequently, the interaction is an expected response to the treatments involved

**TABLE 3.23** EXAMPLES OF ADDITIVE AND MULTIPLICATIVE TREATMENT EFFECTS

| Contribution of B treatments | Additive effects | | | Multiplicative effects | | | Logarithms of multiplicative effects | | |
| | Contribution of A treatments | | | Contribution of A treatments | | | | | |
| | $A_1 = 2$ | $A_2 = 6$ | $A_3 = 10$ | $A_1 = 2$ | $A_2 = 6$ | $A_3 = 10$ | $A_1$ | $A_2$ | $A_3$ |
|---|---|---|---|---|---|---|---|---|---|
| $B_1 = 4$ | 6 | 10 | 14 | 8 | 24 | 40 | 0.90 | 1.38 | 1.60 |
| $B_2 = 8$ | 10 | 14 | 18 | 16 | 48 | 80 | 1.20 | 1.68 | 1.90 |

in an experiment and is treated as such. Usually, this results in the experimenter looking at or testing the effects of $A$ at each level of $B$ separately or the effects of $B$ at different levels of $A$; these effects are referred to as simple effects. On other occasions, one or more atypical values can bring about an interaction. Tests can be made to see if such values are truly what are known as "outliers" (Dixon, 1950, 1969) and if they prove to be so, they can be omitted from the data and an analysis completed on the remaining data. A rather common cause of an interaction is the result of multiplicative, rather than additive, effects of the main classifications. When an interaction exists because of multiplicative treatment effects, the data can be made additive by the logarithmic transformation. Table 3.23 presents a hypothetical situation of additive treatment effects, multiplicative treatment effects, and the transformation from values showing multiplicative effects to logarithms transforming the multiplicative effects to additive effects.

   If the hypothetical population had a mean of zero before treatments were applied, Table 3.23, adapted from Steel and Torrie (1980, p. 168), shows the expected subclass means under the additive and multiplicative models. In the additive situation the contributions of the particular levels are merely added, such as $2 + 4 = 6$ for the $A_1B_1$ subclass, or $6 + 8 = 14$ for the $A_2B_2$ subclass. Since the effects give the same responses at each level of $B$ there is no interaction in this part of the table. For the multiplicative situation the values for the individual levels are multiplied to obtain the expected means, giving $6 \times 4 = 24$ for the $A_2B_1$ subclass mean, and $10 \times 8 = 80$ for the $A_3B_2$ subclass mean. Since the $A$ effects give greatly different responses at the different levels of $A$, there is a noticeable interaction in this segment of the table. However, transformation to logarithms of the multiplicative means shows that this transformation has eliminated the interaction.

## 3.14 HOMOGENEITY OF VARIANCES

It is a frequent occurrence when dealing with research data to note that the variation within groups in a one-way classification seems to differ markedly from group to group. The same is true of variation within subclasses in the multiway classification. Since, as we have noted earlier, one of the assumptions of the analysis of variance is homogeneity of variances between or among the groups, where this assumption appears to be violated it is well to conduct a test to see if the variances involved are indeed homogeneous.

   Before we conduct such a test let us first consider the situation of a one-way classification with only two treatments or levels. As we have seen, we can use either the $t$ test or the $F$ test to make a test of significance between the two treatment means. The validity of both of these tests rests on the assumption that the variances within the two groups are equal. While it has not been pointed out earlier, in testing between the two treatment means the $t$ test that is used is a *two-tailed test* while the $F$ test is a *one-tailed test*. It now becomes necessary to distinguish between these two types of test because, to test for the significance of the difference between two treatment variances, $H_0: \sigma_1^2 = \sigma_2^2$, we shall be using a *two-tailed F test*. In a later section we shall be dealing as well with a *one-tailed t test*.

Tests conducted earlier between the means of two groups were not concerned with whether one mean was greater than another, only whether the two means differed and therefore our null hypothesis was $H_0: \mu_1 = \mu_2$, with $H_1: \mu_1 \neq \mu_2$. Let us consider an example with 10 observations in each of two groups. If we wish to conduct a $t$ test for the significance of the difference between the means of the two groups we would calculate a $t$ value and see if it exceeds the value of $t$ in the 5 percent (0.05) column of the $t$ table for 18 degrees of freedom which is 2.101. If our value is equal to or greater than 2.101 *or equal to or less than* $-2.101$ we declare significance at the 5 percent level. Thus, by using both tails of the curve, we are evaluating our $t$ value at the 5 percent level, even though the positive tabulated value represents only the 2.5 percent level. This situation is displayed in Figure 3.2. The same reasoning applies for testing at the 1 percent level which has a value of 2.878. Even though the tabulated values give only positive numbers, the two-tailed test makes use of both the positive and negative values, thus using the 5 and 1 percent areas under the curve.

If we change the situation and now test whether the larger mean is significantly *greater* than the smaller mean, our hypothesis would be $H_0: \mu_1 = \mu_2$ and $H_1: \mu_1 > \mu_2$, where $\mu_1$ is the larger mean. We shall now be concerned only with the positive side of the $t$ distribution and the tabulated value in the $t$ table because the same degrees of freedom will only yield a test at the 2.5 percent level since we are not interested in that portion of the distribution that deals with $\mu_1$ being smaller than $\mu_2$. To test at the 5 percent level (or include 5 percent of the curve), we must double the probability; therefore, we must now use the column for $t_{0.10}$ (0.10), which in our example is 1.734. If our $t$ value equals or exceeds this value, we declare a significant difference at the 5 percent level. If we wish to test at the 1 percent level, we would use the tabulated 2 percent (0.02) column, and if our value equals or exceeds 2.445 we would declare that $\mu$ is highly significantly greater than $\mu_2$. This test is referred to as a *one-tailed t test*.

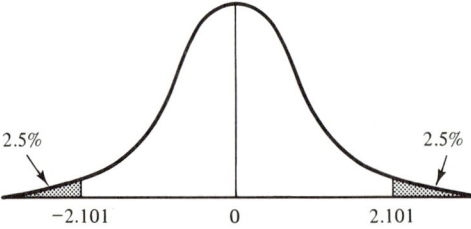

**Figure 3.2**  Approximate distribution of $t$ with 18 degrees of freedom.

If we wish to make the test of significance between the two means (not being concerned as to which mean is greater) by the $F$ test, we will be using a *one-tailed test,* in contrast to the two-tailed $t$ test used for the same purpose. This is because we are testing whether the numerator mean square (that for between groups) is *greater* than the denominator mean square (that for within groups) because of a difference between the two means, *without regard to whether one mean is greater than the other.* Thus, in our example with 18 degrees of freedom for error, our $F$ values would have to equal or exceed 4.41 and 8.28 for significance at the 5 and 1 percent levels. Note that $F$ would equal $t^2$. The $F$ test, of course, is a ratio of two variances, that between groups and that within groups, with the null hypothesis $H_0$: $\sigma_1^2 = \sigma_2^2$ and $H_1$: $\sigma_1^2 > \sigma_2^2$, where $\sigma_1^2$ is the variance between groups and $\sigma_2^2$ is the variance within groups.

Our attention now shifts to the case which prompted this discussion of one- and two-tailed tests, the test to see whether two variances differ from one another, in contrast to the test of whether one variance is greater than the other. Our null hypothesis is now $H_0$: $\sigma_1^2 = \sigma_2^2$ and $H_1$: $\sigma_1^2 \neq \sigma_2^2$. If the two variances are equal, the ratio of sample variances would be expected not to differ greatly from 1. If they do differ from one another, the ratio can vary considerably in either a positive or negative direction. Therefore, we must now take into account both tails of the $F$ distribution, as shown in Figure 3.3, making this a *two-tailed test.* We can deal with only the upper portion of the curve as recorded in the $F$ table by arranging the variance ratio so that the greater mean square or variance is in the numerator. In doing so, we shall use the $F_{0.025}$ and $F_{0.005}$ values for the 5 and 1 percent levels, respectively, since the $F$ tables are tabulated for the situation of the greater mean square being the numerator.

As an example of the test of significance of the difference between the variances of two groups (a test of their homogeneity), we shall use a selection of data from an experiment conducted by Dr. Christine M. Moffitt at the University of Massachusetts concerned with the ovary weight of shad (*Alspa sapidissima*). The weights of right

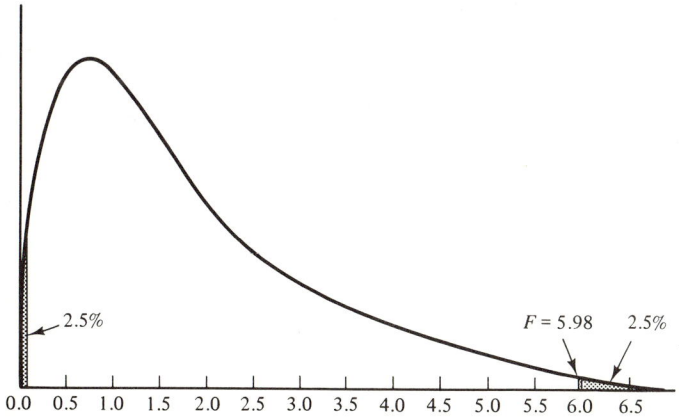

**Figure 3.3** Approximate distribution of $F$ with 1 and 18 degrees of freedom.

and left ovaries of shad from different locations and different years were recorded in grams, and the two groups showing the greatest extremes in variation were selected for this test. The first group had a variance of 834 and the second a variance of 1521, each based on 31 observations. Our test of the significance of the difference between the two variances, $H_0: \sigma_1^2 = \sigma_2^2$, is

$$F = \frac{s_1^2}{s_2^2}$$

$$= \frac{1521}{834} = 1.824 \tag{3.42}$$

where $s_1^2$ is the value for the group with the larger variance. Since this is a two-tailed $F$ test, we look up values for $n_1 - 1$ and $n_2 - 1$ degrees of freedom for $F_{0.025}$ and $F_{0.005}$. While the differences between the two variances is quite large, the $F$ value does not reach the value of 2.07 needed for significance at the 5 percent level.

When more than two groups are involved, the simple variance ratio shown in formula (3.42) can no longer be used. Instead, the test is carried out by an approximate, but usually satisfactory, test proposed by Bartlett (1937). The test takes the form $M/C$ which follows the $\chi^2$ (chi-square) distribution with $(k - 1)$ degrees of freedom, $k$ being the number of groups in the test. Table A.4 contains the cumulative distribution of chi-square, a distribution of great importance in statistical analyses. With equal numbers in the groups

$$M = 2.3026(n - 1)\left(k \log \bar{s}^2 - \sum \log s_i^2\right) \tag{3.43}$$

where log is to the base 10, 2.3026 is a constant ($\log_e 10$) not needed if natural logarithms (base $e$) are used, $(n - 1)$ are the degrees of freedom in each group, $\bar{s} = \sum_i s_i^2/k$, $s_i^2$ is the variance or mean square within each group, and

$$C = 1 + \frac{k + 1}{3k(n - 1)} \tag{3.44}$$

$C$ is a correction factor and is always a little greater than 1. It is usually used when $M$, which in itself is considered to be a crude $\chi^2$, is close to a critical value.

As an example of the test involving more than two groups we have selected four of the groups of shad from Dr. Moffitt's study, consisting of the right and left ovary weights in grams at two fishway facilities. The test of the null hypothesis that the variances among the four groups do not differ is shown in Table 3.24. There were 31 observations in each group.

When the groups have unequal numbers, the formula is changed slightly:

$$M = (2.3026)\left(\sum_i (n_i - 1) \log \bar{s}^2 - \sum_i (n_i - 1) \log s_i^2\right) \tag{3.45}$$

$$C = \frac{1}{3(k - 1)}\left(\sum_i \frac{1}{n_i - 1} - \frac{1}{\sum_i (n_i - 1)}\right) \tag{3.46}$$

As a second example, this one with unequal numbers in the groups, we will use a selection of data from an experiment directed by Dr. Ernest M. Buck of the De-

**TABLE 3.24**  BARTLETT'S TEST OF HOMOGENEITY
OF VARIANCE AMONG FOUR
GROUPS OF EQUAL SIZE

| Group | $s_i^2$ | $\log s_i^2$ |
|:-----:|:-------:|:------------:|
| 1 | 866 | 2.93752 |
| 2 | 834 | 2.92117 |
| 3 | 1521 | 3.18213 |
| 4 | 1264 | 3.10175 |
| Totals | 4485 | 12.14257 |

$$\bar{s}^2 = 1121.25 \qquad \log \bar{s}^2 = 3.04970$$

$$\chi^2 = M = (2.3026)(30)[(4)(3.04970) - 12.14257]$$

$$= (69.0780)(0.05623) = 3.884 \quad (df = 3)$$

$$C = 1 + \frac{5}{(3)(4)(30)} = 1 + 0.014 = 1.014$$

$$\text{Corrected } \chi^2 = \frac{3.884}{1.014} = 3.830 \quad (df = 3), P > 0.25$$

partment of Food and Nutrition at the University of Massachusetts dealing with different methods of tenderizing roasts of beef. Table 3.25 presents the test of homogeneity of variance for these data.

Should the test for the homogeneity of variances prove to be significant, this indicates that some type of transformation should be applied to the data so that homogeneity of variance prevails. The different types of transformation used will be discussed in Section 3.16.

## 3.15  TUKEY'S TEST FOR NONADDITIVITY

As pointed out in Section 3.14, one of the assumptions of the analysis of variance is that the effects in the model are additive. In the discussion of this assumption it was noted that one of the ways in which nonadditivity can arise is by the effects in the model being multiplicative, which results in an interaction between the effects involved. If this situation is suspected in a set of data, it is possible to test whether the interaction, or a sizable portion of it, is due to multiplicative effects. The test is one proposed by Tukey (1949). The assumption of additivity implies that, in a two-way classification, if we obtain a certain linear response to the levels of $B$ in the first level of $A$ we will obtain the same response at other levels of $A$. The same would be true of the linear response of $A$ at different levels of $B$. Tukey's test is essentially a test of the interaction of the linear effects of the treatments, that is, whether the linear response of one set of effects changes as one goes from one level to another of the second effect. The test partitions the interaction sum of squares into a portion due to nonadditivity and a residual.

We shall consider first the two-way classification with a single observation per subclass, usually a randomized complete block design. As an example we shall consider

**TABLE 3.25**  BARTLETT'S TEST OF HOMOGENEITY OF VARIANCES
AMONG FIVE GROUPS OF UNEQUAL SIZES

| Treatment | $n_i - 1$ | $\Sigma y^2 = (n_i - 1)s_i^2$ | $s_i^2$ | $\log s_i^2$ | $(n_i - 1) \log s_i^2$ | $\dfrac{1}{(n_i - 1)}$ |
|---|---|---|---|---|---|---|
| 1 | 9 | 61,245 | 6,805 | 3.8328 | 34.4952 | 0.1111 |
| 2 | 12 | 326,923 | 27,244 | 4.4353 | 53.2236 | 0.0833 |
| 3 | 7 | 85,000 | 12,143 | 4.0843 | 28.5901 | 0.1429 |
| 4 | 14 | 143,333 | 10,238 | 4.0102 | 56.1428 | 0.0714 |
| 5 | 16 | 322,353 | 20,147 | 4.3042 | 68.8672 | 0.0625 |
| Totals | 58 | 938,859 | | | 241.3189 | 0.4712 |

$$\bar{s}^2 = \frac{\Sigma(n_i - 1)s_i^2}{\Sigma(n_i - 1)} = \frac{938,859}{58} = 16,187.22$$

$$\Sigma(n_i - 1) \log \bar{s}^2 = (58)(4.2092) = 244.1336$$

$$\chi^2 = M = (2.3026)\left( \sum_i (n_i - 1) \log \bar{s}^2 - \sum_i (n_i - 1) \log s_i^2 \right)$$

$$= (2.3026)(244.1336 - 241.3189) = 6.481 \quad (\text{df} = 4), \ P > 0.10$$

$$C = 1 + \frac{1}{3(k-1)}\left( \sum_i \frac{1}{n_i - 1} - \frac{1}{\Sigma_i(n_i - 1)} \right)$$

$$= 1 + \frac{1}{(3)(4)}(0.4712 - 0.0172) = 1.0378$$

$$\text{Corrected } \chi^2 = \frac{6.481}{1.0378} = 6.245 \quad (\text{df} = 4), \ P > 0.10$$

the data shown in Table 3.1, using the model previously shown, $Y_{ij} = \mu + r_i + t_j + e_{ij}$, where the $r_i$ are replication effects and the $t_j$ are treatment effects. The necessary calculations are shown in Table 3.26.

The results of these calculations can be put in an analysis of variance as shown in Table 3.27. The sum of squares for nonadditivity is assigned one degree of freedom and is tested by a residual, the difference between the interaction sum of squares and the sum of squares due to additivity. The very small value of $F$ indicates that there is no hint of multiplicative effects of the sources of variation.

As a second example, let us use the data shown in Table 3.7. Our only difference from the method of calculation shown in Table 3.23 is that we now use subclass totals rather than individual observations. The calculations necessary to complete the test are shown in Table 3.28.

It can be seen that although the $F$ value is quite large it does not approach significance since we have only one degree of freedom for both numerator and denominator. Significance would have been surprising in view of the fact that an $F$ value of 161 would have been necessary. However, since it is a very common occurrence in experimentation that very few degrees of freedom are available for interaction, it does not seem reasonable to avoid Tukey's procedure for this reason. The calculations can be very revealing, as in this example where they do show that a very large portion of the interaction sum of squares is due to the multiplicative effects of the two main

**TABLE 3.26** CALCULATIONS FOR TUKEY'S TEST FOR NONADDITIVITY IN THE TWO-WAY CLASSIFICATION WITH A SINGLE OBSERVATION PER SUBCLASS

| Replication ($i = 1, \ldots, 8$) | Treatment ($j = 1, \ldots, 4$) | | | | $\bar{y}_i.$ | $\bar{y}_i. - \bar{y}..$ |
|---|---|---|---|---|---|---|
| | 1 | 2 | 3 | 4 | | |
| 1 | 47 | 58 | 29 | 10 | 36.00 | 3.50 |
| 2 | 51 | 45 | 32 | 6 | 33.50 | 1.00 |
| 3 | 37 | 55 | 11 | 11 | 28.50 | −4.00 |
| 4 | 41 | 51 | 29 | 2 | 30.75 | −1.75 |
| 5 | 49 | 45 | 28 | 19 | 35.25 | 2.75 |
| 6 | 52 | 37 | 31 | 17 | 34.25 | 1.75 |
| 7 | 40 | 55 | 28 | 8 | 32.75 | 0.25 |
| 8 | 38 | 42 | 26 | 10 | 29.00 | −3.50 |
| $\bar{y}._j$ | 44.375 | 48.500 | 26.750 | 10.375 | 32.50 | |
| $\bar{y}._j - \bar{y}..$ | 11.875 | 16.000 | −5.750 | −22.125 | | 0.00 |
| $Q_j = \Sigma_i(\bar{y}_i. - \bar{y}..)Y_{ij}$ | 98.500 | −6.000 | 86.000 | 42.500 | | |

The sum of squares for nonadditivity is calculated as

$$SS = \frac{Q^2}{D}$$

where

$$Q = \sum_j(\bar{y}._j - \bar{y}..)\sum_i(\bar{y}_i. - \bar{y}..)Y_{ij} = \sum_j(\bar{y}._j - \bar{y}..)Q_j$$

$$= (11.875)(98.500) + (16.000)(-6.000) + (-5.750)(86.000)$$

$$+ (-22.125)(42.500) = -361.125 \tag{3.47}$$

and

$$D = n_w \sum_i(\bar{y}_i. - \bar{y}..)^2 \sum_j(\bar{y}._j - \bar{y}..)^2$$

$$= (1)(55.25)(919.59375) = 50,807.55 \tag{3.48}$$

$$SS = \frac{Q^2}{D} = \frac{(-361.125)^2}{50,807.55} = 2.57$$

classifications. In fact, if we were to use the within mean square to test each portion of the interaction sum of squares, we would find that the portion due to nonadditivity is nearly highly significant and the residual portion does not approach significance. In view of these results, it would be reasonable to transform the original data to logarithms before completing an analysis of variance.

**TABLE 3.27** ANALYSIS OF VARIANCE AND TEST FOR NONADDITIVITY

| Source | df | SS | MS | F |
|---|---|---|---|---|
| R | 7 | 221.00 | 31.57 | |
| T | 3 | 7356.75 | 2452.25 | 55.01 |
| RT | 21 | 936.25 | 44.58 | |
| Nonadditivity | 1 | 2.57 | 2.57 | 0.06 |
| Residual | 20 | 933.68 | 46.68 | |

**TABLE 3.28**　CALCULATIONS FOR TUKEY'S TEST FOR NONADDITIVITY IN THE
TWO-WAY CLASSIFICATION WITH THREE OBSERVATIONS
PER SUBCLASS

| Source ($i = 1, 2$) | Level ($j = 1, \ldots, 3$) | | | $\bar{y}_i.$ | $\bar{y}_i. - \bar{y}..$ |
| | 1 | 2 | 3 | | |
| --- | --- | --- | --- | --- | --- |
| 1 | 45.0 | 100.7 | 134.5 | 31.1333 | −8.8111 |
| 2 | 55.1 | 146.1 | 237.6 | 48.7556 | 8.8112 |
| $\bar{y}._j$ | 16.6833 | 41.1333 | 62.0167 | 39.9444 | |
| $\bar{y}._j - \bar{y}..$ | −23.2611 | 1.1889 | 22.0723 | | 0.0001 |
| $Q_j = \Sigma_i(\bar{y}_i. - \bar{y}..)Y_{ij}.$ | 88.9976 | 400.0386 | 908.4482 | | |

$$Q = \sum_j(\bar{y}._j - \bar{y}..)\sum_i(\bar{y}_i. - \bar{y}..)Y_{ij}. = \sum_j(\bar{y}._j - \bar{y}..)Q_j$$

$$= (-23.2611)(88.9976) + (1.1889)(400.0386) + (22.0723)(908.4482) = 18{,}457$$

$$D = n_w \sum_i(\bar{y}_i. - \bar{y}..)^2\sum_i(\bar{y}._j - \bar{y}..)^2$$

$$= (3)(155.2727)(1029.6787) = 479{,}643$$

$$\text{SS} = \frac{Q^2}{D} = \frac{(18{,}457)^2}{479{,}643} = 710.2383$$

| Test of nonadditivity | | | | |
| Source | df | SS | MS | F |
| --- | --- | --- | --- | --- |
| NL | 2 | 734.6878 | 367.3439 | |
| Nonadditivity | 1 | 710.2383 | 710.2383 | 29.05 |
| Residual | 1 | 24.4495 | 24.4495 | |
| $W$ | 12 | 927.9600 | 77.3300 | |

## 3.16 TRANSFORMATIONS

It is a reasonably common situation with research data that the assumptions of the analysis of variance are violated, primarily the assumptions of additivity and homogeneity of variances. We shall consider three common methods of transforming data for the purpose of satisfying these assumptions: the first method is concerned usually with the lack of additivity in a set of data due to multiplicative effects being involved, and the last two methods are concerned with equalizing unequal variances. It should be pointed out that it is not always possible to equalize the variances, a situation occurring often when controls are included in an experiment. Very frequently the variation within the control group is so much greater or so much less than the variation within the treated groups that transformation to equalize variances is impossible. Usually, in such situations, it is best to omit such greatly divergent groups from the analysis, although it may be possible to subdivide the error sum of squares in such a way as to develop an appropriate error mean square for use in making comparisons among the means, as discussed by Cochran and Cox (1957).

## 3.16.1 Logarithmic Transformation

The most commonly used transformation is that of the transformation of individual observations to logarithms, usually common logarithms, although natural logarithms can be used. As pointed out earlier, when an interaction is due primarily to the multiplicative effects of the classifications, the logarithmic transformation usually will equalize the variances and remove the effects of nonadditivity. When there is a multiplicative action involved, the standard deviations in the groups are proportional to the means, that is, the coefficients of variation are similar from group to group. When this occurs, for whatever reason, the logarithmic transformation will usually be effective and produce variances that are independent of the mean. In making the transformation, if some zeros occur in the data this problem can be overcome by using the $\log(Y + 1)$ for all values.

Since we have already found in Table 3.28 that the data in Table 3.7 contain a multiplicative effect, let us transform these data to logarithms and carry out an analysis of variance. It should be mentioned that a Bartlett's test of homogeneity conducted on these data showed a significant $\chi^2$ value, indicating that the variances in subclasses were not homogeneous. The logarithms of the numbers from Table 3.7 are shown in Table 3.29.

The analysis of variance is shown in Table 3.30 where it can be seen that with the transformed data there is no significant interaction, so additivity now prevails. In addition, a test of homogeneity now shows that the variances within subclasses are homogeneous. If further tests are to be made with the data, such as the mean separation procedures shown in Chapter 4, the transformed data must be used. The means to be presented from such an analysis are the antilogarithms of the logarithmic means found

**TABLE 3.29** LOGARITHMS OF DATA FROM TABLE 3.7

| Levels | Source of nitrogen | | Totals |
| --- | --- | --- | --- |
| | Ammonium sulfate | Nitrate | |
| 1 | 1.155 | 1.246 | |
| | 1.201 | 1.380 | |
| | 1.170 | 1.130 | |
| Subtotals | 3.526 | 3.756 | 7.282 |
| 2 | 1.574 | 1.636 | |
| | 1.468 | 1.728 | |
| | 1.529 | 1.693 | |
| Subtotals | 4.571 | 5.057 | 9.628 |
| 3 | 1.617 | 1.777 | |
| | 1.698 | 1.993 | |
| | 1.635 | 1.900 | |
| Subtotals | 4.950 | 5.670 | 10.620 |
| Totals | 13.047 | 14.483 | 27.530 |

**TABLE 3.30**   ANALYSIS OF VARIANCE OF LOGARITHMS
                 OF TABLE 3.29

| Source | df | SS | MS | F |
|--------|-----|--------|--------|---------|
| $N$ | 1 | 0.1146 | 0.1146 | 19.76** |
| $L$ | 2 | 0.9795 | 0.4898 | 84.45** |
| $NL$ | 2 | 0.0199 | 0.0100 | 1.72 |
| $W$ | 12 | 0.0695 | 0.0058 | |

in the analysis. If confidence intervals about the means are to be presented they must first be calculated in terms of logarithms and then retransformed into antilogarithms. Thus, the mean of the first source of nitrogen would be

$$\bar{y}'_1. = \text{antilog } \bar{y}_1. = \text{antilog}(1.450) = 28.19$$

and the confidence interval about this mean would be

$$\text{CI} = \text{antilog}\left(\bar{y}_1. \pm t_{0.05}\sqrt{\frac{s^2}{9}}\right) = \text{antilog}\left(1.450 \pm (2.179)\sqrt{\frac{0.0058}{9}}\right)$$

$$= \text{antilog}(1.450 \pm 0.0553)$$

$$= 24.82, 32.01$$

Note that the interval is not symmetric about the mean as it has been with previous calculations.

   Another and perhaps more common use of the logarithmic transformation is when a group of positive numbers covers a very wide range. As an example of this situation let us consider some data resulting from an experiment conducted by Professor Olga Weinack of the Department of Veterinary and Animal Sciences at the University of Massachusetts, shown in Table 3.31. The experiment was concerned with the effect of a challenge dose of a naladixic-acid-resistant strain of *Salmonella typhimurium* on untreated and treated chickens and turkeys. A three-way classification of the data is involved. The details of this type of analysis will be discussed later in this section.

   The analysis of the logarithms of the counts is shown in Table 3.32. Since all the effects in the model are considered to be fixed classifications the bottom line (*GTD*) of the ANOVA table must be used as the denominator in all $F$ ratios. In doing so we are assuming that the *GTD* interaction does not exist. If this assumption is not true the $F$ values will all be underestimated.

### 3.16.2 Arcsin Transformation

This is a commonly used transformation, finding its utility when dealing with fractions or percentages. Percentages are expected to follow the binomial distribution, which is reasonably symmetric when $p$, the probability of an event occurring, is not greatly different from $q$, where $q = 1 - p$, but becomes quite skewed when $p$ does differ from $q$. This transformation gives a value designated as $\Theta$ which is equal to arcsin $\sqrt{p}$, where $p$ is a proportion or a percentage. This means the angle whose sine is the square root

**TABLE 3.31** COUNT OF COLONY-FORMING UNITS PER GRAM OF FECES

| Dose | Treatment | Chickens | | Turkeys | | Log totals |
|---|---|---|---|---|---|---|
| | | Count | Log | Count | Log | |
| 1 | 1 | 9,340 | 3.97 | 2,090 | 3.32 | 7.29 |
| 2 | | 2,760,000 | 6.44 | 75,500 | 4.88 | 11.32 |
| 3 | | 2,460,000 | 6.39 | 275,500 | 5.44 | 11.83 |
| 4 | | 32,400,000 | 7.51 | 501,200 | 5.70 | 13.21 |
| 5 | | 4,680,000 | 6.67 | 2,188,000 | 6.34 | 13.01 |
| 6 | | 3,810,000 | 6.58 | 691,900 | 5.84 | 12.42 |
| 7 | | 2,355,000 | 6.37 | 5,012,000 | 6.70 | 13.07 |
| Subtotals | | | 43.93 | | 38.22 | 82.15 |
| 1 | 2 | 80 | 1.90 | 52 | 1.72 | 3.62 |
| 2 | | 94 | 1.97 | 220 | 2.34 | 4.31 |
| 3 | | 159 | 2.20 | 871 | 2.94 | 5.14 |
| 4 | | 19,060 | 4.28 | 18,200 | 4.26 | 8.54 |
| 5 | | 4,266 | 3.63 | 87,100 | 4.94 | 8.57 |
| 6 | | 10,480 | 4.02 | 3,163 | 3.50 | 7.52 |
| 7 | | 51,290 | 4.71 | 48,980 | 4.69 | 9.40 |
| Subtotals | | | 22.71 | | 24.39 | 47.10 |
| 1 | 3 | 17 | 1.23 | 21 | 1.32 | 2.55 |
| 2 | | 15 | 1.18 | 97 | 1.99 | 3.17 |
| 3 | | 417 | 2.62 | 78 | 1.89 | 4.51 |
| 4 | | 4,710 | 3.67 | 70,800 | 4.85 | 8.52 |
| 5 | | 45,710 | 4.66 | 174 | 2.24 | 6.90 |
| 6 | | 9,121 | 3.96 | 8,129 | 3.91 | 7.87 |
| 7 | | 398,200 | 5.60 | 158,500 | 5.20 | 10.80 |
| Subtotals | | | 22.92 | | 21.40 | 44.32 |
| Totals | | | 89.56 | | 84.01 | 173.57 |

of the percentage to be transformed. This transformation leads to a nearly normal distribution, spreading out both tails of the original distribution and compacting the middle. The transformation is not expected to be useful if the percentages range between 30 and 70 but is recommended when the percentages are clustered at one end or the

**TABLE 3.32** ANALYSIS OF VARIANCE OF LOGARITHMS OF DATA SHOWN IN TABLE 3.31

| Source | df | SS | MS | F |
|---|---|---|---|---|
| Groups ($G$) | 1 | 0.7334 | 0.7334 | 1.33 |
| Treatments ($T$) | 2 | 63.5081 | 31.7540 | 57.48** |
| Doses ($D$) | 6 | 49.9803 | 8.3300 | 15.08** |
| $GT$ | 2 | 1.9621 | 0.9810 | 1.78 |
| $GD$ | 6 | 0.2326 | 0.0388 | 0.07 |
| $TD$ | 12 | 6.7813 | 0.5651 | 1.02 |
| $GTD$ | 12 | 6.6282 | 0.5524 | |

**TABLE 3.33**  PERCENTAGES AND ARCSINES OF TURKEYS POSITIVE TO TREATMENT

| Treatment | Experiment 1 % | 1 Arcsin | 2 % | 2 Arcsin | 3 % | 3 Arcsin | 4 % | 4 Arcsin | Total of arcsin |
|-----------|-----|--------|-----|--------|-----|--------|-----|--------|--------|
| 1 | 60.0 | 50.77 | 10.0 | 18.44 | 33.3 | 35.24 | 40.0 | 39.23 | 143.68 |
| 2 | 26.7 | 31.11 | 6.7 | 15.00 | 26.6 | 31.05 | 50.0 | 45.00 | 122.16 |
| 3 | 17.7 | 24.88 | 3.3 | 10.47 | 46.6 | 43.05 | 13.3 | 21.39 | 99.79 |
| Totals | | 106.76 | | 43.91 | | 109.34 | | 105.62 | 365.63 |

other of the range, that is, from 0 to 30 percent or 70 to 100 percent. A wider range is frequently included, particularly when several values are near the extremes.

When tests beyond the analysis of variance are conducted the transformed data should be used, but when the means and/or standard errors are reported they should be transformed back to percentages. The arcsins can be found in Table A.9 and the means can be converted back to percentages by the same table. The analyses of data assume equal numbers in each of the percentages. Small departures from equality are not likely to invalidate an analysis but if the numbers are considerably discrepant, a weighted analysis such as suggested by Harvey (1982b) should be used.

As an example of this transformation we shall use some data adapted from an experiment conducted by Professor Olga Weinack of the Department of Veterinary and Animal Sciences of the University of Massachusetts concerned with the cumulative culture results in turkeys subjected to three microflora treatments. The results are recorded as the percentage of birds reacting positively to the treatments. There were 30 turkeys in each group. Since many of the percentages were below 30 it was appropriate to transform the data to arcsins. The data are shown in Table 3.33, together with the arcsin transformations. The analysis of this set of data is shown in Table 3.34 where it can be seen that the treatment effects were not significant. Had it not been necessary to transform the data, analyses either of the original percentages or of 0's and 1's as discussed in Section 3.12 could have been conducted. Both of these analyses would have given the same results for the test of significance of the treatment effects. However, since the 0 and 1 analysis would allow a test of the significance of the interaction, it would be preferable.

### 3.16.3  Square Root Transformation

The square root transformation is useful in dealing with enumeration data, such as numbers of insects on a plant or number of cells in a culture, which frequently follow

**TABLE 3.34**  ANALYSIS OF ARCSIN TRANSFORMATION OF DATA IN TABLE 3.30

| Source | df | SS | MS | F |
|--------|----|------|------|-----|
| Experiments (E) | 3 | 1005.10 | 335.03 | |
| Treatments (T) | 2 | 240.82 | 120.41 | 1.35 |
| ET | 6 | 533.52 | 88.92 | |

the Poisson distribution. In this distribution the mean and variance are equal and therefore not independent of one another as they should be to fulfill the assumptions of the analysis of variance. Transformation of the counts to square roots is usually effective in making the variances within groups independent of the means of the groups. When several zero counts are involved the transformation to $\sqrt{Y+1}$ or $\sqrt{Y+\frac{1}{2}}$ is recommended. The range of counts may vary from zero to a few hundred but does not show the extreme range of values shown with the logarithmic transformation.

As an example of the square root transformation we can use the data shown in Table 3.35 which resulted from an experiment conducted by Joann McLaughlin in the Department of Plant and Soil Sciences at the University of Massachusetts. The study was concerned with the effect of different treatments on the number of dead seeds in the apples. One group of nine trees was sprayed with the chemical Daminozide

**TABLE 3.35**  NUMBER OF DEAD SEEDS IN 30 APPLE SAMPLES

| | No Daminozide | | | | | | | |
|---|---|---|---|---|---|---|---|---|
| | Treatment | | | | | | | |
| | 1 | | 2 | | 3 | | 4 | |
| Tree | Y | $\sqrt{Y}$ | Y | $\sqrt{Y}$ | Y | $\sqrt{Y}$ | Y | $\sqrt{Y}$ |
| 1 | 7 | 2.65 | 29 | 5.39 | 49 | 7.00 | 183 | 13.53 |
| 2 | 8 | 2.83 | 25 | 5.00 | 50 | 7.07 | 109 | 10.44 |
| 3 | 7 | 2.65 | 42 | 6.48 | 62 | 7.87 | 87 | 9.33 |
| 4 | 12 | 3.46 | 10 | 3.16 | 86 | 9.27 | 155 | 12.45 |
| 5 | 13 | 3.61 | 24 | 4.90 | 36 | 6.00 | 156 | 12.49 |
| 6 | 14 | 3.74 | 1 | 1.00 | 41 | 6.40 | 106 | 10.30 |
| 7 | 8 | 2.83 | 26 | 5.10 | 39 | 6.24 | 69 | 8.31 |
| 8 | 19 | 4.36 | 27 | 5.20 | 96 | 9.80 | 77 | 8.77 |
| 9 | 8 | 2.83 | 31 | 5.57 | 69 | 8.31 | 129 | 11.36 |

| | Daminozide | | | | | | | |
|---|---|---|---|---|---|---|---|---|
| | Treatment | | | | | | | |
| | 1 | | 2 | | 3 | | 4 | |
| Tree | Y | $\sqrt{Y}$ | Y | $\sqrt{Y}$ | Y | $\sqrt{Y}$ | Y | $\sqrt{Y}$ |
| 1 | 18 | 4.24 | 23 | 4.80 | 20 | 4.47 | 129 | 11.36 |
| 2 | 8 | 2.83 | 26 | 5.10 | 39 | 6.24 | 69 | 8.31 |
| 3 | 12 | 3.46 | 22 | 4.69 | 106 | 10.30 | 84 | 9.17 |
| 4 | 11 | 3.32 | 22 | 4.69 | 50 | 7.07 | 166 | 12.88 |
| 5 | 2 | 1.41 | 14 | 3.74 | 40 | 6.32 | 52 | 7.21 |
| 6 | 7 | 2.65 | 11 | 3.32 | 32 | 5.66 | 66 | 8.12 |
| 7 | 7 | 2.65 | 15 | 3.87 | 31 | 5.57 | 153 | 12.37 |
| 8 | 6 | 2.45 | 18 | 4.24 | 26 | 5.10 | 182 | 13.49 |
| 9 | 7 | 2.65 | 30 | 5.48 | 46 | 6.78 | 98 | 9.90 |
| Mean | 9.67 | 3.03 | 22.00 | 4.54 | 51.00 | 6.97 | 115.00 | 10.54 |
| Range | 17 | 2.95 | 41 | 5.48 | 86 | 5.83 | 131 | 6.32 |

TABLE 3.36  ANALYSIS OF VARIANCE OF SQUARE ROOTS IN TABLE 3.35

| Source | df | SS | MS | $E(MS)$ | F |
|--------|----|----|----|---------|---|
| $B$ | 1 | 5.4395 | 5.4395 | $\sigma_w^2 + \sigma_{s:b}^2 + 36\sigma_b^2$ | 2.56 |
| $S:B$ | 16 | 34.0083 | 2.1255 | $\sigma_w^2 + \sigma_{s:b}^2$ | |
| $T$ | 3 | 579.9002 | 193.3001 | $\sigma_w^2 + \sigma_{st:b}^2 + 18\sigma_t^2$ | 89.29** |
| $BT$ | 3 | 2.3927 | 0.7976 | $\sigma_w^2 + \sigma_{st:b}^2 + 9\sigma_{bt}^2$ | 0.37 |
| $ST:B$ | 48 | 103.9118 | 2.1648 | $\sigma_w^2 + \sigma_{st:b}^2$ | |

and a second group of nine trees was left unsprayed. Within each tree four limbs were selected, one limb used as a control, a second limb treated with Benzyladine, a third limb treated with Gibberellin, and a fourth limb treated with a combination of Benzyladine and Gibberellin. A sample of 30 apples from each limb was selected and the number of dead seeds recorded. The table shows both the original counts and the square roots of the counts. The bottom of the table shows the means and the ranges of the counts, and it can be seen that the ranges vary widely before transformation but are quite similar after transformation. Thus, the means and variances appear to be independent of one another after transformation, indicating a homogeneity of the variances.

The analysis of variance of these data is shown in Table 3.36, where $B$ represents the two groups of nine trees, $S:B$ the trees within the two groups, and $T$ the four treatments. The $B$ and $T$ effects are considered to be fixed and the $S$ effects considered to be random. It can be seen that there are highly significant differences among treatments. Any mean separation procedures would be carried out using the transformed data. When the means are presented they would be the squares of the average square root for each block or treatment.

## 3.17  THREE-WAY CLASSIFICATION: SINGLE OBSERVATION PER SUBCLASS

The analysis of the three-way classification with a single observation per subclass is a simple extension of the preceding methods, with additional interaction effects included. As an example, let us consider an experiment conducted by Laura Groves, a graduate student in the Department of Food Science and Nutrition at the University of Massachusetts. The study was concerned with an examination of the repeatability of the evaluation of food samples made by 11 panelists on 8 different sample formulations with 2 repetitions of the experiment. The analysis considered is only a portion of the entire study. The mathematical model was

$$Y_{ijk} = \mu + p_i + s_j + r_k + ps_{ij} + pr_{ik} + sr_{jk} + e_{ijk}$$

where $p_i$ represents panelists, $s_j$ represents samples, and $r_k$ represents replicates. The individual values are not given here but the necessary calculations would be made as follows:

$$\text{Total SS} = \sum_i \sum_j \sum_k Y_{ijk}^2 - \frac{Y_{...}^2}{n_p n_s n_r = N} = psr - (1) \tag{3.49}$$

$$\text{Panelist SS} = \text{SS}_P = \frac{\Sigma_i \, Y_{i..}^2}{n_s n_r} - \frac{Y_{...}^2}{N} = p - (1) \tag{3.50}$$

$$\text{Sample SS} = \text{SS}_S = \frac{\Sigma_j \, Y_{.j.}^2}{n_p n_r} - \frac{Y_{...}^2}{N} = s - (1) \tag{3.51}$$

$$\text{Replicates SS} = \text{SS}_R = \frac{\Sigma_k \, Y_{..k}^2}{n_p n_s} - \frac{Y_{...}^2}{N} = r - (1) \tag{3.52}$$

$$\text{SS}_{PS} = \frac{\Sigma_i \, \Sigma_j \, Y_{ij.}^2}{n_r} - \frac{\Sigma_i \, Y_{i..}^2}{n_s n_r} - \frac{\Sigma_j \, Y_{.j.}^2}{n_p n_r} + \frac{Y_{...}^2}{N}$$

$$= ps - p - s + (1) \tag{3.53}$$

$$\text{SS}_{PR} = \frac{\Sigma_i \, \Sigma_k \, Y_{i.k}^2}{n_s} - \frac{\Sigma_i \, Y_{i..}^2}{n_s n_r} - \frac{\Sigma_k \, Y_{..k}^2}{n_p n_s} + \frac{Y_{...}^2}{N}$$

$$= pr - p - r + (1) \tag{3.54}$$

$$\text{SS}_{SR} = \frac{\Sigma_j \, \Sigma_k \, Y_{.jk}^2}{n_p} - \frac{\Sigma_j \, Y_{.j.}^2}{n_p n_r} - \frac{\Sigma_k \, Y_{..k}^2}{n_p n_s} + \frac{Y_{...}^2}{N}$$

$$= sr - s - r + (1) \tag{3.55}$$

$$\text{SS}_{PSR} = \text{Total SS} - \text{SS}_P - \text{SS}_S - \text{SS}_R - \text{SS}_{PS} - \text{SS}_{PR} - \text{SS}_{SR} \tag{3.56}$$

The analysis of variance for this example is shown in Table 3.37 together with the expectations of mean squares. It can be seen from the expectations of mean squares that there are exact tests of significance for $P$, $R$, $PS$, and $SR$ but not for $S$ and $PR$, considering $P$ and $R$ to be random effects and $S$ to be a fixed effect. Looking first at $S$, we can develop an approximate test for this (Satterthwaite, 1946; Cochran, 1951). We shall again be looking for a numerator which differs from the denominator only by containing the variance component to be tested. In our example, the numerator would be $\text{MS}_S$ and the appropriate denominator would be $\text{MS}_{PS} + \text{MS}_{SR} - \text{MS}_{PSR}$. To eliminate the negative values in the denominator the mean square that has been subtracted may be transposed to the numerator, giving a ratio of expectations:

$$F' = \frac{\text{MS}_S + \text{MS}_{PSR}}{\text{MS}_{PS} + \text{MS}_{SR}} = \frac{2\sigma_w^2 + 2\sigma_{psr}^2 + 11\sigma_{sr}^2 + 2\sigma_{ps}^2 + 22\sigma_s^2}{2\sigma_w^2 + 2\sigma_{psr}^2 + 11\sigma_{sr}^2 + 2\sigma_{ps}^2} \tag{3.57}$$

**TABLE 3.37** ANALYSIS OF VARIANCE AND EXPECTATIONS OF MEAN SQUARES

| Source | df | SS | MS | E(MS) |
|--------|----|------|------|-------|
| $P$ | 10 | 124.4090 | 12.4409 | $\sigma_w^2 + 8\sigma_{pr}^2 + 16\sigma_p^2$ |
| $S$ | 7 | 486.6134 | 69.5162 | $\sigma_w^2 + \sigma_{psr}^2 + 11\sigma_{sr}^2 + 2\sigma_{ps}^2 + 22\sigma_s^2$ |
| $R$ | 1 | 1.1136 | 1.1136 | $\sigma_w^2 + 8\sigma_{pr}^2 + 88\sigma_r^2$ |
| $PS$ | 70 | 190.1340 | 2.7162 | $\sigma_w^2 + \sigma_{psr}^2 + 2\sigma_{ps}^2$ |
| $PR$ | 10 | 32.1360 | 3.2136 | $\sigma_w^2 + 8\sigma_{pr}^2$ |
| $SR$ | 7 | 11.3407 | 1.6201 | $\sigma_w^2 + \sigma_{psr}^2 + 11\sigma_{sr}^2$ |
| $PSR$ | 70 | 166.4110 | 2.3773 | $\sigma_w^2 + \sigma_{psr}^2$ |

The approximate degrees of freedom are

$$n_1 = \frac{(MS_S + MS_{PSR})^2}{MS_S^2/f_S + MS_{PSR}^2/f_{PSR}} \tag{3.58}$$

and

$$n_2 = \frac{(MS_{PS} + MS_{SR})^2}{MS_{PS}^2/f_{PS} + MS_{SR}^2/f_{SR}} \tag{3.59}$$

Substituting our values in formulas (3.53), (3.54), and (3.55) we have

$$F' = \frac{69.5162 + 2.3773}{2.7162 + 1.6201} = \frac{71.8935}{4.3363} = 16.58$$

$$n_1 = \frac{(69.5162 + 2.3773)^2}{(69.5162)^2/7 + (2.3773)^2/70} = \frac{5168.6753}{690.4381} = 7.49$$

$$n_2 = \frac{(2.7162 + 1.6201)^2}{(2.7162)^2/70 + (1.6201)^2/7} = \frac{18.8035}{0.4804} = 39.14$$

When degrees of freedom of 7 and 40 are used, the $F$ table shows a highly significant value of $F'$.

Unless we are willing to assume $\sigma_{psr}^2 = 0$ there is no exact test of significance for $PR$. However, we can use $PSR$ to test $PR$ knowing that the $F$ value is expected to be too small.

## 3.18 THREE-WAY CLASSIFICATION: MULTIPLE OBSERVATIONS PER SUBCLASS

The inclusion of more than one observation per subclass changes the mathematical model because we now have a second-order interaction included. Should this three-way or second-order interaction show significance, then the results of other tests of significance in the analysis of variance must be considered with caution. The interpretation of a second-order interaction is a little more difficult than that of a first-order interaction but still requires a study of two-way tables. With a significant interaction we would look at the two-way tables of $B$ and $C$ at each level of $A$, the two-way tables of $A$ and $B$ at each level of $C$, or the two-way tables of $A$ and $C$ at each level of $B$ to gain an understanding of where the interaction is occurring. At times it is reasonable to treat a three-way classification with a significant second-order interaction as a one-way classification (each three-way subclass being one level), although this is not a generally acceptable procedure because it tends to ignore the purposes of the original treatments. In other cases, some sort of nesting is required to make a satisfactory analysis of the data.

As an example of the calculations involved in a three-way classification, we shall use the data resulting from an experiment with pepper plants conducted by William M. Clapham of the Department of Plant and Soil Sciences at the University of Massachusetts. The data, which are shown in Table 3.38, are the accumulative weights of peppers after the eighth harvest. Individual observations are not shown, only subclass

**TABLE 3.38** ACCUMULATIVE WEIGHTS OF PEPPERS (IN GRAMS) AFTER THE EIGHT HARVESTS

|  |  | $C_1$ | $C_2$ | $C_3$ | Totals |
|---|---|---|---|---|---|
| $B_1$ | $S_1$ | 1,507 | 2,852 | 1,406 | 5,765 |
|  | $S_2$ | 1,679 | 1,464 | 1,278 | 4,421 |
|  | $S_3$ | 1,157 | 444 | 837 | 2,438 |
|  |  | 4,343 | 4,760 | 3,521 | 12,624 |
| $B_2$ | $S_1$ | 1,848 | 2,919 | 410 | 5,177 |
|  | $S_2$ | 301 | 1,269 | 566 | 2,136 |
|  | $S_3$ | 1,805 | 819 | 717 | 3,341 |
|  |  | 3,954 | 5,007 | 1,693 | 10,654 |
| $B_3$ | $S_1$ | 1,723 | 788 | 1,552 | 4,063 |
|  | $S_2$ | 1,258 | 2,816 | 2,254 | 6,328 |
|  | $S_3$ | 200 | 934 | 1,171 | 2,305 |
|  |  | 3,181 | 4,538 | 4,977 | 12,696 |
| $B_4$ | $S_1$ | 780 | 233 | 1,287 | 2,300 |
|  | $S_2$ | 842 | 2,403 | 1,156 | 4,401 |
|  | $S_3$ | 384 | 1,346 | 1,864 | 3,594 |
|  |  | 2,006 | 3,982 | 4,307 | 10,295 |
| $B_5$ | $S_1$ | 1,044 | 1,591 | 441 | 3,076 |
|  | $S_2$ | 1.792 | 1,286 | 1,685 | 4,763 |
|  | $S_3$ | 110 | 920 | 1,478 | 2,508 |
|  |  | 2,946 | 3,797 | 3,604 | 10,347 |
| $B_6$ | $S_1$ | 1,790 | 1,755 | 339 | 3,884 |
|  | $S_2$ | 1,258 | 839 | 756 | 2,853 |
|  | $S_3$ | 1,142 | 1,264 | 867 | 3,273 |
|  |  | 4,190 | 3,858 | 1,962 | 10,010 |
| $B_7$ | $S_1$ | 1,057 | 1,981 | 1,864 | 4,902 |
|  | $S_2$ | 1,260 | 1,172 | 1,724 | 4,156 |
|  | $S_3$ | 1,401 | 2,563 | 764 | 4,728 |
|  |  | 3,718 | 5,716 | 4,352 | 13,786 |

Totals

$C_1 = 24,338 \qquad C_2 = 31,658 \qquad C_3 = 24,416$

$S_1 = 29,167 \qquad S_2 = 29,058 \qquad S_3 = 22,187$

totals. The experiment was conducted in a randomized block design with three seeding dates ($S$) and three container sizes ($C$). Each value in the table is a sum of three observations. The total uncorrected sum of squares of the individual observations is shown in the calculations. The mathematical model is

$$Y_{ijkl} = \mu + b_i + s_j + c_k + bs_{ij} + bc_{ik} + sc_{jk} + bsc_{ijk} + e_{ijkl} \qquad (3.60)$$

where the $b_i$ are the block effects, $s_j$ are the seeding date effects, and $c_k$ are the container size effects. The calculations necessary for the analysis of variance follow:

$$\text{Total SS} = \sum_i \sum_j \sum_k \sum_l Y_{ijkl}^2 - \frac{Y_{\cdots}^2}{N} = \sum_i \sum_j \sum_k \sum_l Y_{ijkl}^2 - (1)$$

$$= 53{,}714{,}914 - \frac{(80{,}412)^2}{(7)(3)(3)(3)}$$

$$= 53{,}714{,}914 - 34{,}212{,}115 = 19{,}502{,}799 \tag{3.61}$$

$$SS_B = \frac{\sum_i Y_{i\cdots}^2}{n_s n_c n_w} - \frac{Y_{\cdots}^2}{N} = b - (1)$$

$$= \frac{(12{,}624)^2 + (10{,}654)^2 + \cdots + (13{,}786)^2}{27} - (1)$$

$$= 34{,}717{,}142 - (1) = 505{,}027 \tag{3.62}$$

$$SS_S = \frac{\sum_j Y_{\cdot j\cdot\cdot}^2}{n_b n_c n_w} - \frac{Y_{\cdots}^2}{N} = s - (1)$$

$$= \frac{(29{,}167)^2 + (29{,}058)^2 + (22{,}187)^2}{63} - (1)$$

$$= 34{,}719{,}750 - (1) = 507{,}635 \tag{3.63}$$

$$SS_C = \frac{\sum_k Y_{\cdot\cdot k\cdot}^2}{n_b n_s n_w} - \frac{Y_{\cdots}^2}{N} = c - (1)$$

$$= \frac{(24{,}338)^2 + (31{,}658)^2 + (24{,}416)^2}{63} - (1)$$

$$= 34{,}773{,}147 - (1) = 561{,}032 \tag{3.64}$$

$$SS_{BS} = \frac{\sum_i \sum_j Y_{ij\cdot\cdot}^2}{n_c n_w} - \frac{\sum_i Y_{i\cdots}^2}{n_s n_c n_w} - \frac{\sum_j Y_{\cdot j\cdot\cdot}^2}{n_b n_c n_w} + \frac{Y_{\cdots}^2}{N} = bs - b - s + (1)$$

$$= \frac{(5765)^2 + (4421)^2 + \cdots + (4728)^2}{9} - b - s + (1)$$

$$= 37{,}413{,}573 - b - s + (1)$$

$$= 2{,}188{,}796 \tag{3.65}$$

$$SS_{BC} = \frac{\sum_i \sum_k Y_{i\cdot k\cdot}^2}{n_s n_w} - \frac{\sum_i Y_{i\cdots}^2}{n_s n_c n_w} - \frac{\sum_k Y_{\cdot\cdot k\cdot}^2}{n_b n_s n_w} + \frac{Y_{\cdots}^2}{N} = bc - b - c + (1)$$

$$= \frac{(4343)^2 + (4760)^2 + \cdots + (4352)^2}{9} - b - c + (1)$$

$$= 36{,}579{,}020 - b - c + (1)$$

$$= 1{,}300{,}846 \tag{3.66}$$

$$SS_{SC} = \frac{\Sigma_j \Sigma_k Y^2_{\cdot jk\cdot}}{n_b n_w} - \frac{\Sigma_j Y^2_{\cdot j\cdot\cdot}}{n_b n_c n_w} - \frac{\Sigma_k Y^2_{\cdot\cdot k\cdot}}{n_b n_s n_w} + \frac{Y^2_{\cdot\cdot\cdot\cdot}}{N} = sc - s - c + (1)$$

$$= \frac{(9749)^2 + (12{,}119)^2 + \cdots + (7698)^2}{21} - s - c + (1)$$

$$= 35{,}583{,}292 - s - c + (1)$$

$$= 302{,}510 \tag{3.67}$$

$$SS_{BSC} = \frac{\Sigma_i \Sigma_j \Sigma_k Y^2_{ijk\cdot}}{n_w} - \frac{\Sigma_i \Sigma_j Y^2_{ij\cdot\cdot}}{n_c n_w} - \frac{\Sigma_i \Sigma_k Y^2_{i\cdot k\cdot}}{n_s n_w} - \frac{\Sigma_j \Sigma_k Y^2_{\cdot jk\cdot}}{n_b n_w}$$

$$+ \frac{\Sigma_i Y^2_{i\cdot\cdot\cdot}}{n_s n_c n_w} + \frac{\Sigma_j Y^2_{\cdot j\cdot\cdot}}{n_b n_c n_w} + \frac{\Sigma_k Y^2_{\cdot\cdot k\cdot}}{n_b n_s n_w} + \frac{Y^2_{\cdot\cdot\cdot\cdot}}{N}$$

$$= bsc - bs - bc - sc + b + s + c - (1)$$

$$= \frac{(1507)^2 + (2852)^2 + \cdots + (764)^2}{3}$$

$$- bs - bc - sc + b + s + c - (1)$$

$$= 43{,}089{,}194 - bs - bc - sc + b + s + c - (1) = 3{,}511{,}233 \tag{3.68}$$

$$SS_W = \sum_i \sum_j \sum_k \sum_l Y^2_{ijk} - \frac{\Sigma_i \Sigma_j \Sigma_k Y^2_{ijk\cdot}}{n_w} = \sum_i \sum_j \sum_k \sum_l Y^2_{ijk} - bsc$$

$$= 53{,}714{,}914 - bsc = 10{,}625{,}720 \tag{3.69}$$

There is one more subscript in the numerator in this example than in the previous example because we now have more than one observation per subclass. Most of the calculations are similar to those previously carried out with the exception of the second-order interaction. In symbolic notation this can be expressed as

$$(b - 1)(s - 1)(c - 1) = bsc - bs - bc - sc + b + s + c - 1$$

Or, since the sum of squares among $BSC$ subclasses contains the sums of squares for $B$, $S$, $C$, $BS$, $BC$, $SC$, and $BSC$ the sum of squares for $BSC$ is equal to the sum of squares among the $BSC$ subclasses minus the sums of squares for $B$, $S$, $C$, $BS$, $BC$, and $SC$. The sum of squares among the $BSC$ subclasses is, of course, the uncorrected sum of squares for the $BSC$ subclasses minus the correction term. The analysis of variance for this set of data is shown in Table 3.39. The block effects are considered to be random and the seeding date and container size are considered fixed.

## 3.19 ESTIMATION AND USE OF VARIANCE COMPONENTS

As was pointed out in Section 3.1, an investigator is usually not interested in the specific effects for random sets of effects. Instead, the investigator is interested in some

**TABLE 3.39**   ANALYSIS OF VARIANCE OF DATA IN TABLE 3.34

| Source | df | SS | MS | E(MS) | F |
|--------|----|----|----|----|----|
| B | 6 | 505,027 | 84,171 | $\sigma_w^2 + 27\sigma_b^2$ | <1 |
| S | 2 | 507,635 | 253,818 | $\sigma_w^2 + 9\sigma_{bs}^2 + 63\sigma_s^2$ | 1.39 |
| C | 2 | 561,032 | 280,516 | $\sigma_w^2 + 9\sigma_{bc}^2 + 63\sigma_c^2$ | 2.59 |
| BS | 12 | 2,188,796 | 182,400 | $\sigma_w^2 + 9\sigma_{bs}^2$ | 2.16 |
| BC | 12 | 1,300,846 | 108,404 | $\sigma_w^2 + 9\sigma_{bc}^2$ | 1.29 |
| SC | 4 | 302,510 | 75,628 | $\sigma_w^2 + 3\sigma_{bsc}^2 + 21\sigma_{sc}^2$ | <1 |
| BSC | 24 | 3,511,233 | 146,301 | $\sigma_w^2 + 3\sigma_{bsc}^2$ | 1.73* |
| W | 126 | 10,625,720 | 84,331 | $\sigma_w^2$ | |

measure of the amount of variability among such effects. A common statistic used to measure this variation is the *variance component*. Often, one is more interested in the estimation of a variance component than in testing the significance of the differences observed among random effects.

When data are analyzed under a fixed model there is only one variance component to be estimated, that is, $\sigma_e^2$. In this case, the mean square for error, the lowest line in the ANOVA, provides an estimate of $\sigma_e^2$, as seen from the $E(MS)$ values for the examples given in Section 3.11. Estimates of $\sigma_e^2$ are used to compute tests of significance, standard errors, and confidence intervals. For example, the mean square for error in the ANOVA of the data given in Table 3.7 is 77.33 (Table 3.10). Thus,

$$\hat{\sigma}_e^2 = s^2 = 77.33$$

Since both nitrogen sources ($N$) and levels of nitrogen ($L$) are fixed sources of variation, the mean square for error (77.33) is an appropriate denominator mean square for testing the significance of differences due to $N, L,$ or $NL$. The class and subclass means for these data are given in Table 3.40. The standard error of each of the $NL$ subclass means is $\sqrt{\hat{\sigma}_e^2/3} = \sqrt{77.33/3} = 5.08$, since each of these means is an average of three observations. The standard error of each of the means for the level of nitrogen is $\sqrt{\hat{\sigma}_e^2/6} = \sqrt{77.33/6} = 3.59,$ and the standard error of the two means for a nitrogen source is $\sqrt{\hat{\sigma}_e^2/9} = \sqrt{77.33/9} = 2.93$. The reader should note that the ANOVA (Table 3.10) shows that the interaction effects are highly significant statistically. Therefore, the investigator would be most interested in the differences exhibited by the $NL$ subclass means. Differences due to the level of nitrogen are considerably greater for the nitrate

**TABLE 3.40**   TREATMENT MEANS OF NITROGEN
ACCUMULATION (IN GRAMS)
IN SPINACH PLANTS

| | Source of nitrogen | | |
|--------|--------|--------|--------|
| Levels | Ammonium sulfate | Nitrate | Means |
| 1 | 15.00 | 18.37 | 16.68 |
| 2 | 33.57 | 48.70 | 41.13 |
| 3 | 44.83 | 79.20 | 62.02 |
| Means | 31.13 | 48.76 | 39.94 |

source of nitrogen than for the ammonium S source of nitrogen. A closer examination of the interaction effects will be considered in Sections 3.20.2 and 3.20.3.

When data are analyzed under a random model or a mixed model there will be at least two variance components that may be estimated. Variance component estimates are used (1) to determine the relative importance of different random sources of variation, (2) to obtain estimates of intraclass correlations, (3) to compute standard errors of fixed class and subclass means, and (4) to plan future experiments. The estimation and use of variance components will now be illustrated from analyses of data under different experimental designs.

### 3.19.1 Examples of Analyses Under Random Models

**One-Way Classification: Equal Numbers**  In a study reported by Harvey et al. (1953) 25 Jersey heifers were classified subjectively by three judges on overall type characteristics into 1 of 18 classes from "low poor" (1) to "high excellent" (18). For illustration we will ignore judge effects in this section. The one-way ANOVA is presented in Table 3.41.

From this analysis

$$\hat{\sigma}_e^2 = 0.920$$

$$\hat{\sigma}_h^2 = \frac{7.375 - 0.920}{3}$$

$$= 2.152$$

and the intraclass correlation is

$$r_I = \frac{\hat{\sigma}_h^2}{\hat{\sigma}_h^2 + \hat{\sigma}_e^2}$$

$$= \frac{2.152}{3.072}$$

$$= 0.70 \tag{3.70}$$

which estimates the average correlation between scores given to the same heifer by the different judges. The model underlying this analysis is

$$Y_{ij} = \mu + h_i + e_{ij} \qquad i = 1, \ldots, 25; \qquad j = 1, 2, 3$$

$$E(h_i^2) = \sigma_h^2 \qquad\qquad E(h_i) = 0 \tag{3.71}$$

$$E(e_{ij}^2) = \sigma_e^2 \qquad\qquad E(e_{ij}) = 0$$

**TABLE 3.41**  ONE-WAY ANOVA, EQUAL NUMBERS

| Source | df | Mean square | E(MS) |
|---|---|---|---|
| Heifers | 24 | 7.375 | $\sigma_e^2 + 3\sigma_h^2$ |
| Within heifers | 50 | 0.920 | $\sigma_e^2$ |

Therefore,

$$V(Y_{ij}) = E(\mu + h_i + e_{ij})^2 - [E(\mu + h_i + e_{ij})]^2$$

$$= \sigma_h^2 + \sigma_e^2 \tag{3.72}$$

and

$$V(\bar{y}) = \frac{\sigma_e^2}{75} + \frac{\sigma_h^2}{25}$$

$$= \frac{1}{75}(\sigma_e^2 + 3\sigma_h^2) \tag{3.73}$$

where $V$ stands for variance. It will be noted that $\sigma_e^2 + 3\sigma_h^2$ is the expected composition of the mean square for heifers. Hence, the estimate of $V(\bar{y})$ from these data is $7.375/75 = 0.0983$ and the standard error is $\sqrt{0.0983} = 0.31$. The mean score given to these 25 heifers was 9.65. One can therefore conclude that the probability is about 0.68 that the mean score for the population from which these heifers were drawn is $9.65 \pm 0.31$.

In some applications the investigator may want to consider judge effects as fixed and adjust for mean differences between judges. In this case, the appropriate analysis is under a two-way classification mixed model and this analysis will be considered later.

**One-Way Classification: Unequal Numbers**  Ochoa et al. (1981) obtained the weaning weights (in kilograms) for 730 heifer calves out of 410 Hereford cows during the years 1969–1976. The ANOVA is presented in Table 3.42.

The coefficient of $\sigma_c^2$ ($k_0$) was computed as follows:

$$k_0 = \frac{1}{\text{df (cows)}}\left(N - \frac{\Sigma_i n_i^2}{N}\right)$$

$$= \frac{1}{409}\left(730 - \frac{16{,}374}{730}\right)$$

$$= \frac{1}{409}(730 - 22.4301)$$

$$= 1.73 \tag{3.74}$$

where $\Sigma_i n_i^2$ is the sum of squares of the number of calves out of each cow. From this analysis

**TABLE 3.42  ONE-WAY ANOVA, UNEQUAL NUMBERS**

| Source | df | Mean square | E(MS) |
|--------|-----|-------------|-------|
| Cows ($C$) | 409 | 620.1 | $\sigma_e^2 + 1.73\sigma_c^2$ |
| Within cows | 320 | 230.4 | $\sigma_e^2$ |

$$\hat{\sigma}_e^2 = 230.4$$

$$\hat{\sigma}_c^2 = \frac{620.1 - 230.4}{1.73}$$

$$= 225.3$$

and the intraclass correlation is

$$r_I = \frac{\hat{\sigma}_c^2}{\hat{\sigma}_c^2 + \hat{\sigma}_e^2}$$

$$= \frac{225.3}{455.7}$$

$$= 0.49 \tag{3.75}$$

This correlation is an estimate of the average correlation among weaning weights of heifer calves produced by the same cow in the population represented by these 410 cows. Of course, there are important biological reasons why weaning weights of calves out of the same cow are expected to be correlated. However, these will not be discussed here.

With unequal numbers the variance of the overall mean is

$$V(\bar{y}.) = \frac{1}{N} (\sigma_e^2 + k\sigma_c^2) \tag{3.76}$$

where

$$k = \frac{\Sigma_i n_i^2}{N} \tag{3.77}$$

In these data, $k = 22.43$. Therefore, the estimate of the variance of the overall mean weaning weight is

$$V(\bar{y}.) = \frac{1}{730} [230.4 + (22.43)(225.3)]$$

$$= 7.238$$

and the standard error is $\sqrt{7.238} = 2.7$ kg.

**Three-Stage Nested Design: Random Model and Equal Numbers**  A preliminary study was conducted to determine the number of locks of wool and fibers per lock that should be sampled from a given fleece to measure accurately the average fiber length of a fleece (Harvey et al., 1968). Twenty-five locks were randomly selected from each of 38 randomly selected ram fleeces and two fibers were randomly selected from the midpoint area of each lock. The ANOVA for the unstretched fiber length measurements (in centimeters) is presented in Table 3.43.

The variance component estimates are

$$\hat{\sigma}_e^2 = 1.7490$$

$$\hat{\sigma}_{l:r}^2 = \frac{2.8088 - 1.7490}{2}$$

$$= 0.5299$$

**TABLE 3.43** THREE-STAGE NESTED DESIGN ANOVA, EQUAL NUMBERS

| Source | df | Mean square | E(MS) |
|---|---|---|---|
| Rams (R) | 37 | 20.1988 | $\sigma_e^2 + 2\sigma_{l:r}^2 + 50\sigma_r^2$ |
| Locks:R | 912 | 2.8088 | $\sigma_e^2 + 2\sigma_{l:r}^2$ |
| Fibers | 950 | 1.7490 | $\sigma_e^2$ |

and

$$\hat{\sigma}_r^2 = \frac{20.1988 - 2.8088}{50}$$

$$= 0.3478$$

Two intraclass correlations that may be computed from estimates of these variance components are

$$r_1 = \frac{\hat{\sigma}_r^2 + \hat{\sigma}_{l:r}^2}{\hat{\sigma}_r^2 + \hat{\sigma}_{l:r}^2 + \hat{\sigma}_e^2}$$

$$= \frac{0.8777}{2.6267}$$

$$= 0.33 \tag{3.78}$$

and

$$r_2 = \frac{\hat{\sigma}_r^2}{\hat{\sigma}_r^2 + \hat{\sigma}_{l:r}^2 + \hat{\sigma}_e^2}$$

$$= \frac{0.3478}{2.6267}$$

$$= 0.13 \tag{3.79}$$

The first intraclass correlation, 0.33, is the correlation between the fiber length measurements on the two fibers from the same lock. The second intraclass correlation, 0.13, is the correlation between fiber length measurements on individual fibers from two randomly chosen locks taken from the same fleece.

The variance of a fleece mean is

$$V(\bar{y}_i) = \frac{\sigma_e^2}{50} + \frac{\sigma_{l:r}^2}{25}$$

$$= \frac{1}{50}(\sigma_e^2 + 2\sigma_{l:r}^2)$$

and this is estimated as $2.8088/50 = 0.0562$. The standard error of a fleece mean is $\sqrt{0.0562} = 0.24$ cm, which in these data was about 3.5 percent of the overall mean of 6.82. The variance of the overall mean, 6.82, is

$$V(\bar{y}..) = \frac{\sigma_e^2}{1900} + \frac{\sigma_{l:r}^2}{950} + \frac{\sigma_r^2}{38}$$

$$= \frac{1}{1900}(\sigma_e^2 + 2\sigma_{l:r}^2 + 50\sigma_r^2) \tag{3.80}$$

Therefore, the estimate of the variance of the overall mean is $20.1988/1900 = 0.01063$ and the standard error is $\sqrt{0.01063} = 0.10$ or about 1.5 percent of the overall mean of 6.82.

The estimates of $\sigma_e^2$ and $\sigma_{l:r}^2$ may be used to obtain an estimate of the number of locks to sample from each fleece to obtain a given amount of accuracy. If $n_f$ fibers are to be measured from each lock then the number of locks and fibers to sample per fleece to achieve a given degree of accuracy is computed from the formula

$$n = \frac{\hat{\sigma}_e^2 + n_f \hat{\sigma}_{l:r}^2}{V(\bar{y}_i)}$$

which is derived from the formula for $V(\bar{y}_i)$ given above. For example, suppose the standard error desired is 2 percent of the mean, that is, $(0.02)(6.82) = 0.1364$. Now if only one fiber is to be measured per lock then the number of locks and fibers to sample per fleece to achieve this degree of accuracy is

$$n = \frac{1.7490 + 0.5299}{(0.1364)^2}$$

$$= 123$$

If two fibers are to be measured per lock then the total number of fibers to sample per fleece is

$$n = \frac{1.7490 + (2)(0.5299)}{(0.1364)^2} = 152 \tag{3.81}$$

from 76 locks.

**Three-Stage Nested Design: Random Model and Unequal Numbers**  Weaning weights (in pounds) were obtained for 172 pigs sired by 15 sires and out of 29 dams. Since the number of progeny in each litter varied, unequal class and subclass frequencies existed. The ANOVA is presented in Table 3.44.

The coefficients of the variance components were computed from the usual formulas as given in Section 3.11.2, Nested Classification: Unbalanced Design. Estimates of the three variance components are

$$\hat{\sigma}_e^2 = 29.47$$

$$\hat{\sigma}_{d:s}^2 = \frac{82.91 - 29.47}{5.478}$$

$$= 9.75$$

**TABLE 3.44**  THREE-STAGE NESTED DESIGN ANOVA, UNEQUAL NUMBERS

| Source | df | Mean square | E(MS) |
|--------|-----|-------------|-------|
| Sires (S) | 14 | 164.63 | $\sigma_e^2 + 6.321\sigma_{d:s}^2 + 11.394\sigma_s^2$ |
| Dams:S | 14 | 82.91 | $\sigma_e^2 + 5.478\sigma_{d:s}^2$ |
| Full sibs | 143 | 29.47 | $\sigma_e^2$ |

and
$$\hat{\sigma}_s^2 = \frac{164.63 - 29.47 - (6.321)(9.75)}{11.394}$$

$$= 6.45$$

Three intraclass correlations may be computed from these variance component estimates as follows:

$$\text{Correlation among full sibs} = \frac{\hat{\sigma}_s^2 + \hat{\sigma}_{d:s}^2}{\hat{\sigma}_s^2 + \hat{\sigma}_{d:s}^2 + \hat{\sigma}_e^2}$$

$$= \frac{16.20}{45.67}$$

$$= 0.35 \tag{3.82}$$

$$\text{Correlation among paternal half sibs} = \frac{\hat{\sigma}_s^2}{\hat{\sigma}_s^2 + \hat{\sigma}_{d:s}^2 + \hat{\sigma}_e^2}$$

$$= \frac{6.45}{45.67}$$

$$= 0.14 \tag{3.83}$$

$$\text{Correlation among maternal half sibs} = \frac{\hat{\sigma}_{d:s}^2}{\hat{\sigma}_s^2 + \hat{\sigma}_{d:s}^2 + \hat{\sigma}_e^2}$$

$$= \frac{9.75}{45.67}$$

$$= 0.21 \tag{3.84}$$

Researchers in the area of genetics or animal breeding make use of these intraclass correlations to estimate genetic parameters such as heritability, and from the differences among the intraclass correlations certain conclusions can be drawn relative to the importance of maternal effects and dominance effects. Of course, an investigator would need a considerably larger number of degrees of freedom for both sires and dams within sires than available in this example to obtain accurate estimates of the intraclass correlations.

Oftentimes, when one uses the three-stage nested design and unequal frequencies exist, interest is centered on the estimation of the overall mean $\bar{y}$. Hence, the investigator needs to know how to compute the standard error of $\bar{y}$. With equal frequencies we saw that $V(\bar{y}..)$ was simply computed by dividing the mean square for the main classification by the total number of observations. With unequal frequencies use is made of the variance component estimates and two coefficients as follows:

$$V(\bar{y}..) = \frac{1}{N}(\sigma_e^2 + k_1\sigma_s^2 + k_2\sigma_{d:s}^2) \tag{3.85}$$

where
$$k_1 = \frac{\Sigma_i\, n_{i.}^2}{N}$$

$$k_2 = \frac{\Sigma_i \Sigma_j\, n_{ij}^2}{N}$$

The coefficients of the variance components in the $E(\text{MS})$ were computed as follows:

$$5.478 = \frac{1}{14}\left(172 - \sum_i \frac{\sum_j n_{ij}^2}{n_{i.}}\right) \quad \text{the coefficient for } \sigma_{d:s}^2 \text{ in } E(\text{MS}) \text{ for dams:sire}$$

$$6.321 = \frac{1}{14}\left(\sum_i \frac{\sum_j n_{ij}^2}{n_{i.}} - \frac{\sum_i \sum_j n_{ij}^2}{172}\right) \quad \text{the coefficient for } \sigma_{d:s}^2 \text{ in } E(\text{MS}) \text{ for sires}$$

$$11.394 = \frac{1}{14}\left(172 - \frac{\sum_i n_{i.}^2}{172}\right) \quad \text{the coefficient for } \sigma_s^2 \text{ in } E(\text{MS}) \text{ for sires}$$

Therefore, $k_1$ and $k_2$ may be computed indirectly as follows:

$$k_1 = 172 - (14)(11.394)$$

$$= 12.48$$

and

$$\sum_i \frac{\sum_j n_{ij}^2}{n_{i.}} = 172 - (14)(5.478)$$

$$= 95.31$$

Hence,

$$k_2 = 95.31 - (14)(6.321)$$

$$= 6.81$$

The overall mean for weaning weights in these data was 33.45 lb. Therefore, the standard error of this mean is

$$s_{\bar{y}} = \sqrt{\frac{1}{172}[29.47 + (12.48)(6.45) + (6.81)(9.75)]}$$

$$= \sqrt{\frac{176.36}{172}}$$

$$= 1.01$$

which is 1.7 percent of the mean.

## 3.19.2 Analyses Under Mixed Models: Balanced Designs

**Two-Way Classification: Single Observation per Subclass** When classifier (judge) effects are considered for the analysis of the data described for the first example under Section 3.19.1, the model is

$$Y_{ij} = \mu + h_i + c_j + e_{ij} \qquad i = 1, \ldots, 25 \qquad j = 1, 2, 3 \qquad (3.86)$$

where the $h_i$ are the heifer effects (random), $c_j$ refers to the classifier effects (fixed), and $e_{ij} = hc_{ij} + w_{ij}$ represents the heifer by classifier interaction effects and the within-subclass effects. The analysis of variance of the overall type classification scores under this mixed model is presented in Table 3.45.

**TABLE 3.45**   TWO-WAY CLASSIFICATION: ONE OBSERVATION
PER SUBCLASS ANOVA

| Source | df | Mean square | E(MS) |
|---|---|---|---|
| Heifers ($H$) | 24 | 7.375 | $\sigma_w^2 + \sigma_{hc}^2 + 3\sigma_h^2$ |
| Classifiers ($C$) | 2 | 3.455 | $\sigma_w^2 + \sigma_{hc}^2 + 5\sigma_c^2$ |
| $HC$ | 48 | 0.814 | $\sigma_w^2 + \sigma_{hc}^2$ |

Note that the variance component for the interaction, $\sigma_{hc}^2$, appears in the $E(MS)$ for heifers, contrary to rule 3.11.1B, even though classifier effects are considered as fixed. The reason this is necessary, when our interest is centered on estimation and use of variance components, will become clear when we consider examples with unequal subclass numbers and the use to be made of variance components.

The average correlation among scores given the same heifer by the different classifiers, adjusting for the mean differences between classifiers, is given by the intraclass correlation

$$r_1 = \frac{\hat{\sigma}_h^2}{\hat{\sigma}_h^2 + \hat{\sigma}_{hc}^2 + \hat{\sigma}_w^2} \tag{3.87}$$

whereas the average correlation between repeated scores given to the same heifer by the same classifier would be the intraclass correlation

$$r_2 = \frac{\hat{\sigma}_h^2 + \hat{\sigma}_{hc}^2}{\hat{\sigma}_h^2 + \hat{\sigma}_{hc}^2 + \hat{\sigma}_w^2} \tag{3.88}$$

Of course, with only one observation per subclass, $\sigma_w^2$ and $\sigma_{hc}^2$ cannot be estimated separately. Therefore, from these data we can estimate only $r_1$ and this is done as follows:

$$\hat{\sigma}_w^2 + \hat{\sigma}_{hs}^2 = 0.814$$

$$\hat{\sigma}_h^2 = \frac{7.375 - 0.814}{3}$$

$$= 2.187$$

$$r_1 = \frac{2.187}{2.187 + 0.814}$$

$$= \frac{2.187}{3.001}$$

$$= 0.73$$

Adjusting for the significant differences among the means for classifiers (10.08, 9.40, and 9.48) increases the correlation among scores on the same heifer by different classifiers from 0.70 to 0.73.

With fixed classifier effects,

$$V(Y_{ij}) = \sigma_h^2 + \sigma_e^2$$

$$= \sigma_h^2 + \sigma_{hc}^2 + \sigma_w^2$$

and
$$V(\bar{y}..) = \frac{\sigma_w^2 + \sigma_{hc}^2}{75} + \frac{\sigma_h^2}{25}$$

$$= \frac{1}{75}(\sigma_w^2 + \sigma_{hc}^2 + 3\sigma_h^2) \tag{3.89}$$

Since the expected mean square for heifers is $\sigma_h^2 + \sigma_{hc}^2 + 3\sigma_h^2$, $V(\bar{y}..)$ is estimated as $7.375/75 = 0.0983$, the same value obtained when classifier effects were ignored.

**Two-Way Classification: Multiple Observations per Subclass**   The type classification scores on the 25 Jersey heifers for the analysis given above were obtained when the heifers were 18–23 months of age. Each of the three classifiers also classified these same 25 heifers at 6–11 months of age and at 12–17 months of age. The effects of age will be considered first so that the model for the analysis of the $75 \times 3 = 225$ observations will be the two-way classification model with repeated observations per subclass, that is,

$$Y_{ijk} = \mu + h_i + c_j + hc_{ij} + e_{ijk} \tag{3.90}$$

where the $h_i$ are the heifer effects (random), the $c_j$ the classifier effects (fixed), and the $e_{ijk}$ the within-subclass effects. The analysis of variance of the overall type classification scores under this model, when adjustment was made for the age of heifer effects, is presented in Table 3.46.

The mean square given here for "Within" is the mean square obtained for the three-factor interaction when age of heifer is considered (see next example, Three-Way Classification: Single Observation per Subclass). Estimates of the variance components are computed as follows:

$$\sigma_w^2 = 0.771$$

$$\sigma_{hc}^2 = \frac{1.129 - 0.771}{3} = 0.119$$

$$\sigma_h^2 = \frac{13.177 - 1.129}{9}$$

$$= 1.339$$

The average correlation among scores given the same heifer by the different classifiers is estimated using Equation (3.87),

$$r_1 = \frac{1.339}{1.339 + 0.119 + 0.771}$$

$$= 0.60$$

**TABLE 3.46**   TWO-WAY CLASSIFICATION: MULTIPLE
OBSERVATIONS PER SUBCLASS ANOVA

| Source | df | Mean square | E(MS) |
|---|---|---|---|
| Heifers ($H$) | 24 | 13.177 | $\sigma_w^2 + 3\sigma_{hc}^2 + 9\sigma_h^2$ |
| Classifiers ($C$) | 2 | 2.350 | $\sigma_w^2 + 3\sigma_{hc}^2 + 75\sigma_c^2$ |
| $HC$ | 48 | 1.129 | $\sigma_w^2 + 3\sigma_{hc}^2$ |
| Within | 150 | 0.771 | $\sigma_w^2$ |

and the average correlation between scores given to the same heifer at different ages by the same classifier would be estimated by Equation (3.88),

$$r_2 = \frac{1.339 + 0.119}{1.339 + 0.119 + 0.771}$$

$$= 0.65$$

The variance of the overall mean, which was 9.38 in these data, is

$$V(\bar{y}..) = \frac{\hat{\sigma}_w^2}{225} + \frac{\hat{\sigma}_{hc}^2}{75} + \frac{\hat{\sigma}_h^2}{25}$$

$$= \frac{1}{225}(\hat{\sigma}_w^2 + 3\hat{\sigma}_{hc}^2 + 9\hat{\sigma}_h^2)$$

$$= \frac{1}{225}(\text{mean square for heifers})$$

$$= \frac{1}{225}(13.177)$$

$$= 0.0586 \qquad\qquad (3.91)$$

and the standard error is $\sqrt{0.0586} = 0.24$.

The variance of the mean score for each classifier is

$$V(\bar{y}._j) = \frac{\hat{\sigma}_w^2}{75} + \frac{\hat{\sigma}_{hc}^2}{25}$$

$$= \frac{1}{75}(\hat{\sigma}_w^2 + 3\hat{\sigma}_{hc}^2)$$

$$= \frac{1}{75}(\text{mean square for } HC)$$

$$= \frac{1.129}{75}$$

$$= 0.0151$$

and the standard error for each mean is $\sqrt{0.0151} = 0.12$.

**Three-Way Classification: Single Observation per Subclass**   The appropriate model for the analysis of the heifer scores when age effects are considered is

$$Y_{ijk} = \mu + h_i + c_j + a_k + hc_{ij} + ha_{ik} + ca_{jk} + e_{ijk} \qquad (3.92)$$

where the $h_i$ are the random heifer effects, the $c_j$ the fixed classifier effects, and the $a_k$ the fixed age-of-heifer effects. Note that the $e_{ijk} = w_{ijk} + hca_{ijk}$. The analysis of variance in overall type classification scores is given in Table 3.47. It will again be noted that rule 3.11.1B for excluding certain $\sigma^2$ values from $E(MS)$ when some sets of effects are fixed is violated in order to estimate the variance components. Under rule 3.11.1B the $E(MS)$ values for the random sources of variation are as follows:

**TABLE 3.47**  ANALYSIS OF VARIANCE IN CLASSIFICATION SCORES OF HEIFERS

| Source | df | Mean square | E(MS) |
|--------|-----|-------------|-------|
| Heifers ($H$) | 24 | 13.177 | $\sigma_w^2 + \sigma_{hca}^2 + 3\sigma_{ha}^2 + 3\sigma_{hc}^2 + 9\sigma_h^2$ |
| Classifiers ($C$) | 2 | 2.350 | $\sigma_w^2 + 3\sigma_{hc'}^2 + 75\sigma_c^2$ |
| Ages ($A$) | 2 | 4.165 | $\sigma_w^2 + 3\sigma_{ha'}^2 + 75\sigma_a^2$ |
| $HC$ | 48 | 1.129 | $\sigma_w^2 + \sigma_{hca}^2 + 3\sigma_{hc}^2$ |
| $HA$ | 48 | 2.970 | $\sigma_w^2 + \sigma_{hca}^2 + 3\sigma_{ha}^2$ |
| $CA$ | 4 | 0.765 | $\sigma_w^2 + \sigma_{hca}^2 + 25\sigma_{ca}^2$ |
| $HCA$ | 96 | 0.771 | $\sigma_w^2 + \sigma_{hca}^2$ |

$$\text{Heifers } (H) \quad \sigma_w^2 + 9\sigma_{h'}^2$$

$$HC \quad \sigma_w^2 + 3\sigma_{hc'}^2$$

$$HA \quad \sigma_w^2 + 3\sigma_{ha'}^2$$

The prime on the subscripts of the random variance components is used to show that the effects to be tested for significance are different from the variance components to be estimated. From the two sets of $E(MS)$ values for the random sources of variation, it can be seen that

$$\hat{\sigma}_{h'}^2 = \tfrac{1}{9}\hat{\sigma}_{hca}^2 + \tfrac{1}{3}\hat{\sigma}_{ha}^2 + \tfrac{1}{3}\hat{\sigma}_{ha}^2 + \hat{\sigma}_h^2 \tag{3.93}$$

$$\hat{\sigma}_{hc'}^2 = \tfrac{1}{3}\hat{\sigma}_{hca}^2 + \hat{\sigma}_{hc}^2 \tag{3.94}$$

$$\hat{\sigma}_{ha'}^2 = \tfrac{1}{3}\hat{\sigma}_{hca}^2 + \hat{\sigma}_{ha}^2 \tag{3.95}$$

Therefore, the heifer effects that are to be tested for significance are those effects of heifers obtained by averaging across the fixed classes of classifier and the fixed classes of ages. With other classifiers and/or other ages the effects of heifers would be expected to be different when the $HC$ and $HA$ interactions exist. The estimates of the variance components from the analysis of variance are

$$\hat{\sigma}_w^2 + \hat{\sigma}_{hca}^2 = 0.771$$

$$\hat{\sigma}_{ha}^2 = \frac{2.970 - 0.771}{3}$$

$$= 0.7330$$

$$\hat{\sigma}_{hc}^2 = \frac{1.129 - 0.771}{3}$$

$$= 0.1193$$

$$\hat{\sigma}_h^2 = \frac{13.177 - 2.970 - (3)(0.1193)}{9}$$

$$= 1.0943$$

Four intraclass correlations may now be estimated. The first three are estimated as follows:

$$r_1 = \frac{\hat{\sigma}_h^2}{\hat{\sigma}_h^2 + \hat{\sigma}_{ha}^2 + \hat{\sigma}_{hc}^2 + \hat{\sigma}_{hca}^2 + \hat{\sigma}_w^2 = \hat{\sigma}^2}$$

$$= \frac{1.0943}{1.0943 + 0.7330 + 0.1193 + 0.7710}$$

$$= 0.40$$

$$r_2 = \frac{\hat{\sigma}_h^2 + \hat{\sigma}_{ha}^2}{\hat{\sigma}^2}$$

$$= \frac{1.0943 + 0.7330}{2.7176}$$

$$= 0.67$$

$$r_3 = \frac{\hat{\sigma}_h^2 + \hat{\sigma}_{hc}^2}{\hat{\sigma}^2}$$

$$= \frac{1.0943 + 0.1193}{2.7176}$$

$$= 0.45$$

The first correlation $r_1$ estimates the correlation between scores on the same heifer when both classifier and age are different, $r_2$ estimates the correlation between heifer scores when the age is the same but the classifier is different, and $r_3$ estimates the correlation between heifer scores when the classifier is the same but the age is different. The estimate of the correlation between scores on the same heifer by the same classifier at the same age would be estimated from

$$r_4 = \frac{\hat{\sigma}_h^2 + \hat{\sigma}_{ha}^2 + \hat{\sigma}_{hc}^2 + \hat{\sigma}_{hca}^2}{\hat{\sigma}^2}$$

If one assumes that $\hat{\sigma}_{hca}^2 = 0$, then an estimate of $r_4$ from these data is

$$r_4 = \frac{1.0943 + 0.7330 + 0.1193}{2.7176}$$

$$= 0.72$$

Tests of significance can be made for all fixed sets of effects even though the three factor interaction exists. The mean square for $HCA$ is the appropriate error term for testing the significance of the fixed $CA$ effects, the mean square for $HC$ is the error term for the $C$ effects, and the mean square for $HA$ is the error term for the $A$ effects. The standard errors for the classifier and age means are

$$\hat{\sigma}_{\bar{y}\cdot j\cdot} = \sqrt{\frac{\hat{\sigma}_w^2}{75} + \frac{\hat{\sigma}_{hc'}^2}{25}}$$

$$= \sqrt{\frac{1}{75}(\hat{\sigma}_w^2 + 3\hat{\sigma}_{hc'}^2)}$$

$$= \sqrt{\frac{1.129}{75}}$$

$$= 0.12$$

$$\hat{\sigma}_{\bar{y}..k} = \sqrt{\frac{1}{75}(\hat{\sigma}_w^2 + 3\hat{\sigma}_{ha'}^2)}$$

$$= \sqrt{\frac{2.970}{75}}$$

$$= 0.20$$

The means for each of the classifiers were 9.59, 9.28, and 9.28 and these do not differ significantly with $F = 2.350/1.129 = 2.08$ with 2 and 48 degrees of freedom ($P > 0.10$). The means for the three ages were 9.23, 9.27, and 9.65 and these also do not differ significantly with $F = 4.165/2.970 = 1.40$ with 2 and 48 degrees of freedom ($P > 0.25$).

**Combination of Main and Nested Classifications**  In the study involving fiber lengths in fleeces referred to in Section 3.19.1, Three-Stage Nested Design: Random Model and Equal Numbers, the entire set of data consisted of measurements from 15,200 fibers taken by 2 technicians on 2 fibers from each of 4 locations on each of 25 locks sampled from 38 ram fleeces. The model underlying the analysis of variance was

$$Y_{ijklm} = \mu + r_i + t_j + rt_{ij} + l_{ijk} + p_l + rp_{il} + tp_{jl} + rtp_{ijl} + lp_{ijkl} + e_{ijklm} \qquad (3.96)$$

where the $r_i$ are ram effects (random), the $t_j$ are technician effects (fixed), the $l_{ijk}$ represent lock effects (random), and the $p_l$ represent place or location effects (fixed). The analysis of variance in unstretched fiber length measurements with the expected mean squares is given in Table 3.48. For making tests of significance the $E$(MS) for the random sources of variation are as given in Table 3.49. From these two tables it can be seen that the estimates of parameters (linear functions of variance components) which are tested for significance under the null hypothesis are as follows:

**TABLE 3.48**  ANALYSIS OF VARIANCE IN FIBER LENGTH MEASUREMENTS

| Source | df | Mean square | $E$(MS) |
|---|---|---|---|
| $R$ | 37 | 157.132 | $\sigma_w^2 + 2\sigma_{lp:rt}^2 + 8\sigma_{l:rt}^2 + 200\sigma_{rt}^2 + 100\sigma_{rp}^2 + 25\sigma_{rtp}^2 + 400\sigma_r^2$ |
| $T$ | 1 | 482.770 | $\sigma_w^2 + 8\sigma_{l:rt'}^2 + 200\sigma_{rt'}^2 + 7600\sigma_t^2$ |
| $RT$ | 37 | 17.998 | $\sigma_w^2 + 2\sigma_{lp:rt}^2 + 8\sigma_{l:rt}^2 + 25\sigma_{rtp}^2 + 200\sigma_{rt}^2$ |
| $L{:}RT$ | 1824 | 5.988 | $\sigma_w^2 + 2\sigma_{lp:rt}^2 + 8\sigma_{l:rt}^2$ |
| $P$ | 3 | 37.637 | $\sigma_w^2 + 2\sigma_{lp:rt}^2 + 100\sigma_{rp}^2 + 3800\sigma_p^2$ |
| $RP$ | 111 | 2.385 | $\sigma_w^2 + 2\sigma_{lp:rt}^2 + 25\sigma_{rtp}^2 + 100\sigma_{rp}^2$ |
| $TP$ | 3 | 12.417 | $\sigma_w^2 + 2\sigma_{lp:rt}^2 + 25\sigma_{rtp}^2 + 1900\sigma_{tp}^2$ |
| $RTP$ | 111 | 2.288 | $\sigma_w^2 + 2\sigma_{lp:rt}^2 + 25\sigma_{rtp}^2$ |
| $LP{:}RT$ | 5472 | 1.942 | $\sigma_w^2 + 2\sigma_{lp:rt}^2$ |
| Error | 7600 | 1.749 | $\sigma_w^2$ |

**TABLE 3.49** EXPECTED COMPOSITION OF MEAN
SQUARES FOR RANDOM SOURCES
OF VARIATION FOR MAKING
TESTS OF SIGNIFICANCE

| Source | df | $E(MS)$ |
|--------|-----|---------|
| $R$ | 37 | $\sigma_w^2 + 8\sigma_{l:rt'}^2 + 400\sigma_{r'}^2$ |
| $RT$ | 37 | $\sigma_w^2 + 8\sigma_{l:rt'}^2 + 200\sigma_{rt'}^2$ |
| $L:RT$ | 1824 | $\sigma_w^2 + 8\sigma_{l:rt'}^2$ |
| $RP$ | 111 | $\sigma_w^2 + 2\sigma_{lp:rt}^2 + 100\sigma_{rp'}^2$ |
| $RTP$ | 111 | $\sigma_w^2 + 2\sigma_{lp:rt}^2 + 25\sigma_{rtp}^2$ |
| $LP:RT$ | 5472 | $\sigma_w^2 + 2\sigma_{lp:rt}^2$ |
| Error | 7600 | $\sigma_w^2$ |

$$\hat{\sigma}_{r'}^2 = \tfrac{1}{2}\hat{\sigma}_{rt}^2 + \tfrac{1}{4}\hat{\sigma}_{rp}^2 + \tfrac{1}{16}\hat{\sigma}_{rtp}^2 + \hat{\sigma}_r^2$$

$$\hat{\sigma}_{rt'}^2 = \tfrac{1}{8}\hat{\sigma}_{rtp}^2 + \hat{\sigma}_{rt}^2$$

$$\hat{\sigma}_{l:rt'}^2 = \tfrac{1}{4}\hat{\sigma}_{lp:rt}^2 + \hat{\sigma}_{l:rt}^2$$

$$\hat{\sigma}_{rp'}^2 = \tfrac{1}{4}\hat{\sigma}_{rtp}^2 + \hat{\sigma}_{rp}^2$$

$$\hat{\sigma}_{rtp}^2 \quad \text{and} \quad \hat{\sigma}_{lp:rt}^2$$

The estimates of the variance components are computed as follows from Table 3.48:

$$\hat{\sigma}_w^2 = 1.749$$

$$\hat{\sigma}_{lp:rt}^2 = \frac{1.942 - 1.749}{2} = 0.0965$$

$$\hat{\sigma}_{rtp}^2 = \frac{2.288 - 1.942}{25} = 0.0138$$

$$\hat{\sigma}_{rp}^2 = \frac{2.385 - 2.288}{100} = 0.00097$$

$$\hat{\sigma}_{l:rt}^2 = \frac{5.988 - 1.942}{8} = 0.50575$$

$$\hat{\sigma}_{rt}^2 = \frac{17.998 - 5.988 - (25)(0.0138)}{200} = 0.05832$$

$$\hat{\sigma}_r^2 = \frac{157.132 - 17.998 - (100)(0.00097)}{400} = 0.34759$$

To indicate the relative importance of the random sources of variation, investigators
often give the variance component estimates in a table along with the percentage of
the total random variability accounted for by each component of variance. These
results are given for the analysis considered here in Table 3.50. One may also compute
a number of estimates of intraclass correlations from these variance component es-

**TABLE 3.50** VARIANCE COMPONENT ESTIMATES AND PERCENTAGES OF TOTAL VARIANCE

| Variance component | Estimate | Percentage of total variance |
|---|---|---|
| Rams ($\sigma_r^2$) | 0.34759 | 12.54 |
| Rams $\times$ technician ($\sigma_{rt}^2$) | 0.05832 | 2.10 |
| Locks:$RT$ ($\sigma_{l:rt}^2$) | 0.50575 | 18.25 |
| Rams $\times$ place ($\sigma_{rp}^2$) | 0.00097 | 0.03 |
| Rams $\times$ technician $\times$ place ($\sigma_{rtp}^2$) | 0.01380 | 0.50 |
| Locks $\times$ place:$RT$ ($\sigma_{lp:rt}^2$) | 0.09650 | 3.48 |
| Sampling error ($\sigma_w^2$) | 1.74900 | 63.10 |
| Totals ($\sigma^2$) | 2.77193 | 100.00 |

timates, if desired. For example, the estimated correlation between fiber lengths measured by the same technician from fibers randomly selected from the same location on a lock that has been randomly chosen from a ram's fleece is

$$r_I = \frac{2.77193 - 1.74900}{2.77193} = 0.37$$

The estimated correlation between lengths of fibers randomly selected from the same location on two locks that are randomly chosen from the same fleece and measured by the same technician is

$$r_I = \frac{\hat{\sigma}_r^2 + \hat{\sigma}_{rt}^2 + \hat{\sigma}_{rp}^2 + \hat{\sigma}_{rtp}^2}{\hat{\sigma}^2}$$

$$= \frac{0.42068}{2.77193} = 0.15$$

The variance of an individual fleece mean is

$$V(\bar{y}_{i}...) = \frac{\hat{\sigma}_w^2}{400} + \frac{\hat{\sigma}_{lp:rt}^2}{200} + \frac{\hat{\sigma}_{l:rt}^2}{50}$$

$$= \frac{1}{400}(\hat{\sigma}_w^2 + 2\hat{\sigma}_{lp:rt}^2 + 8\hat{\sigma}_{l:rt}^2)$$

$$= \frac{1}{400}(\hat{\sigma}_w^2 + 8\hat{\sigma}_{l:rt'}^2)$$

$$= \frac{1}{400}(\text{MS for } L{:}RT)$$

$$= \frac{1}{400}(5.988)$$

$$= 0.01497$$

and the standard error is $\sqrt{0.01497} = 0.12$, which is 1.8 percent of the overall mean of 6.82 cm. If only 10 locks were sampled per fleece and 2 fibers are measured from each of the 4 locations on each lock, the variance of an individual fleece mean is estimated as

$$V(\bar{y}_i\ldots) = \frac{\hat{\sigma}_w^2}{80} + \frac{\hat{\sigma}_{lp:rt}^2}{40} + \frac{\hat{\sigma}_{l:rt}^2}{10}$$

$$= \frac{1}{80}(\hat{\sigma}_w^2 + 2\hat{\sigma}_{lp:rt}^2 + 8\hat{\sigma}_{l:rt}^2)$$

$$= \frac{1}{80}(5.988)$$

$$= 0.07485$$

and the estimated standard error is $\sqrt{0.07485} = 0.27$, which is 4.0 percent of the overall mean. Calculations such as these can be made to determine the number of locks and fibers to measure per fleece to achieve a given degree of accuracy.

## 3.20 FIXED MODEL ANALYSIS: DISPROPORTIONATE SUBCLASS FREQUENCIES

The general linear model for analysis of data where only fixed effects (other than random errors) are involved is often written in matrix notation as

$$\mathbf{Y} = \mathbf{X}\beta + \mathbf{e} \qquad (3.97)$$

where $\mathbf{Y}$ is an $n \times 1$ column vector of observations for the dependent variable, $\mathbf{X}$ is an $n \times d$ design matrix of 0's and 1's and is often referred to as the "incidence" matrix, $\beta$ is a $d \times 1$ column vector of the fixed effects (constants), and $\mathbf{e}$ is an $n \times 1$ column vector of the random errors. In most analyses the errors are assumed to be uncorrelated (i.e., independent) and normally distributed with a mean of zero [NID(0, $\sigma_e^2$)].

The least-squares normal equations are

$$\mathbf{X'X}\hat{\beta} = \mathbf{X'Y} \qquad (3.98)$$

where $\mathbf{X'}$ is the transpose of $\mathbf{X}$ and $\hat{\beta}$ is a column vector of the fixed effects. However, it is well known that no unique solution exists for this set of equations if $\beta$ includes effects for one or more discrete independent variables. The fact that no unique solution exists, in this case, means that only linear functions of the fixed effects (constants) are estimable, as was pointed out in Section 2.10.2. The restrictions applied to obtain a solution to the least-squares equations in the example to be considered in this section will be the "summation" restrictions [see Equations (2.42) and (2.43)]. Computationally, it has been found that such restrictions can most easily be imposed in the $\mathbf{X}$ matrix rather than in the $\mathbf{X'X}$ matrix. Therefore, the reparametrized or "working" model is

$$\mathbf{Y} = \mathbf{XLM}\beta + \mathbf{e} \qquad (3.99)$$

where $\mathbf{M}$ is the transformation matrix of size $d' \times d$ which defines $d'$ linear functions of $\beta$ to be estimated and $\mathbf{L}$ is a $d \times d'$ matrix which is chosen so that $\mathbf{ML} = \mathbf{I}$, the identity matrix. To simplify the notation we let $\mathbf{X}_r = \mathbf{XL}$ and $\mathbf{B} = \mathbf{M}\beta$. Thus, the "reduced" least-squares equations are

$$X'_rX_r\hat{B} = X'_rY \tag{3.100}$$

and since there is now a unique solution

$$\hat{B} = (X'_rX_r)^{-1}X'_rY \tag{3.101}$$

The total reduction in sum of squares due to fitting all constants is

$$R(B) = \hat{B}'X'_rY \tag{3.102}$$

and the residual or error sum of squares with $n - d'$ degrees of freedom is

$$Y'Y - R(B) \tag{3.103}$$

The estimated variances of estimates of the linear functions of the parameters (effects) are the diagonal elements of $(X'_rX_r)^{-1}\hat{\sigma}_e^2$, where

$$\hat{\sigma}_e^2 = \frac{1}{n - d'} [Y'Y - R(B)] \tag{3.104}$$

and the covariances among these estimates are the off-diagonal elements of the $(X'_rX_r)^{-1}\sigma_e^2$ matrix. For example, $V(\hat{B}_1) = c^{11}\hat{\sigma}_e^2$, where $C = X'_rX_r$ and $C^{-1} = (X'_rX_r)^{-1}, V(\hat{B}_2) = c^{22}\hat{\sigma}_e^2$, and $Cov(B_1, B_2) = c^{12}\hat{\sigma}_e^2$.

The estimated variance of any linear function of the $\hat{B}_i$, say $L = \Sigma_i \lambda_i\hat{B}_i$ is

$$V(L) = (\lambda_1^2c^{11} + \lambda_2^2c^{22} + \cdots + \lambda_{d'}^2c^{d'd'} + 2\lambda_1\lambda_2c^{12}$$

$$+ 2\lambda_1\lambda_3c^{13} + \cdots + 2\lambda_{d'-1}\lambda_{d'}c^{d'-1,d'})\hat{\sigma}_e^2 \tag{3.105}$$

For example, if one wanted to test the significance of the difference between $(\hat{B}_2 + \hat{B}_3)/2$ and $\hat{B}_4$

$$L = \tfrac{1}{2}\hat{B}_2 + \tfrac{1}{2}\hat{B}_3 - \hat{B}_4$$

$$V(L) = (\tfrac{1}{4}c^{22} + \tfrac{1}{4}c^{33} + c^{44} + \tfrac{1}{2}c^{23} - c^{24} - c^{34})\hat{\sigma}_e^2$$

The standard error of this contrast is $\sqrt{V(L)}$ and

$$t = \frac{\tfrac{1}{2}\hat{B}_2 + \tfrac{1}{2}\hat{B}_3 - \hat{B}_4}{\sqrt{V(L)}} = \frac{L}{\sqrt{V(L)}} \tag{3.106}$$

with $n - d'$ degrees of freedom.

To show how one may obtain the least-squares sum of squares for each set of effects included in the model, we let

$$B' = (B_\mu B_A B_B \cdots B_Z) \tag{3.107}$$

where $B_\mu$ refers to the overall mean, $B_A$ refers to the linear functions to be estimated for the $A$ set of effects, and so on. Now,

$$SS(B_\mu) = R(B_\mu|B_A B_B \cdots B_Z)$$

$$= R(B) - R(B_A B_B \cdots B_Z)$$

$$= \hat{B}_\mu(c^{\mu\mu})^{-1}\hat{B}_\mu$$

$$= \frac{\hat{B}_\mu^2}{c^{\mu\mu}}$$

**TABLE 3.51**  NUMBER OF RATS AND GAINS IN WEIGHT (IN GRAMS) CLASSIFIED
BY SEX AND GENERATION

|  | Male | | | | Female | | | | |
|---|---|---|---|---|---|---|---|---|---|
| Generation | 1 | 2 | 3 | 4 | 1 | 2 | 3 | 4 | Total |
| Number of rats | 21 | 15 | 12 | 7 | 27 | 25 | 23 | 19 | 149 |
| Total gain | 3,716 | 2,422 | 1,868 | 1,197 | 2,957 | 2,852 | 2,496 | 2,029 | 19,537 |
| Mean gain | 177 | 161 | 156 | 171 | 110 | 114 | 109 | 107 | 131 |

where $c^{\mu\mu}$ is the inverse diagonal element for $\mu$. Likewise,

$$\text{SS}(\mathbf{B}_A) = R(\mathbf{B}_A | B_\mu \mathbf{B}_B \cdots \mathbf{B}_Z) \tag{3.108}$$

$$= R(\mathbf{B}) - R(B_\mu \mathbf{B}_B \cdots \mathbf{B}_Z) \quad \text{``indirect''} \tag{3.109}$$

$$= \hat{\mathbf{B}}'_A \mathbf{Z}_A^{-1} \hat{\mathbf{B}}_A \quad \text{``direct''} \tag{3.110}$$

where $\mathbf{Z}_A$ is the square segment of the $(\mathbf{X}'_r\mathbf{X}_r)^{-1}$ inverse matrix, which corresponds by row and column to the $A$ set of effects. In general, the sum of squares for $\mathbf{B}_j$ set of effects may be computed by the "direct" procedure from

$$\text{SS}(\mathbf{B}_j) = \hat{\mathbf{B}}'_j \mathbf{Z}_j^{-1} \hat{\mathbf{B}}_j \tag{3.111}$$

The direct method of computing the sum of squares will usually involve less computer time with fixed models than if one uses the "indirect" procedure (i.e., differences in reductions in sum of squares under different models). This is the method that will be used to compute adjusted or "least-squares" sums of squares for examples to be considered.

## 3.20.1 Two-Way Classification: No Interaction

The data given in Table 3.51 are from Snedecor and Cox (1935). Gains in weight (grams) of 149 Wistar rats, 55 males and 94 females, were obtained over a six week period for four successive generations. The number of rats measured in each generation for each sex is given in the table along with the total and mean gains. The total uncorrected sum of squares for the 149 gains in weight was 2,738,543.

The model for the least-squares analysis of these data is

$$Y_{ijk} = \mu + g_i + s_j + e_{ijk}$$

$$i = 1, 2, 3, 4$$

$$j = 1, 2$$

$$k = 1, 2, \ldots, n_{ij} \tag{3.112}$$

where $g_i$ are the generation effects, $s_j$ are the sex effects, and $e_{ijk}$ are the random errors under the assumption of no interaction. The rows of the $\mathbf{X}$ matrix are the same for all observations in the same generation by sex subclass. Therefore, there are only eight different rows in $\mathbf{X}$. These are as follows:

| Subclass | $\mu$ | $g_1$ | $g_2$ | $g_3$ | $g_4$ | $s_1$ | $s_2$ |
|----------|-------|-------|-------|-------|-------|-------|-------|
| 11 | 1 | 1 | 0 | 0 | 0 | 1 | 0 |
| 12 | 1 | 1 | 0 | 0 | 0 | 0 | 1 |
| 21 | 1 | 0 | 1 | 0 | 0 | 1 | 0 |
| 22 | 1 | 0 | 1 | 0 | 0 | 0 | 1 |
| 31 | 1 | 0 | 0 | 1 | 0 | 1 | 0 |
| 32 | 1 | 0 | 0 | 1 | 0 | 0 | 1 |
| 41 | 1 | 0 | 0 | 0 | 1 | 1 | 0 |
| 42 | 1 | 0 | 0 | 0 | 1 | 0 | 1 |

To impose the summation restrictions, we subtract the elements in the $g_4$ column from each of the elements in the other $g_i$ columns, and the elements in the $s_2$ column from the elements in the $s_1$ column. These restrictions will force the estimates of the $g_i$ effects ($\hat{g}_i$) and the $s_j$ effects ($\hat{s}_j$) to sum to zero. This gives the following eight row vectors in the $\mathbf{X}_r$ matrix:

| Subclass | $\mu$ | $g_1$ | $g_2$ | $g_3$ | $s_1$ | $n_{ij}$ | $Y_{ij.}$ |
|----------|-------|-------|-------|-------|-------|----------|-----------|
| 11 | 1 | 1 | 0 | 0 | 1 | 21 | 3716 |
| 12 | 1 | 1 | 0 | 0 | $-1$ | 27 | 2957 |
| 21 | 1 | 0 | 1 | 0 | 1 | 15 | 2422 |
| 22 | 1 | 0 | 1 | 0 | $-1$ | 25 | 2852 |
| 31 | 1 | 0 | 0 | 1 | 1 | 12 | 1868 |
| 32 | 1 | 0 | 0 | 1 | $-1$ | 23 | 2496 |
| 41 | 1 | $-1$ | $-1$ | $-1$ | 1 | 7 | 1197 |
| 42 | 1 | $-1$ | $-1$ | $-1$ | $-1$ | 19 | 2029 |

The corresponding subclass numbers and totals of gain are given for each subclass for convenience in setting up the least-squares equations. The reader will recall that the least-squares equations to be set up are $\mathbf{X}_r'\mathbf{X}_r\hat{\mathbf{B}} = \mathbf{X}_r'\mathbf{Y}$. If the $8 \times 5$ matrix of 0's, 1's, and $-1$'s above is $\mathbf{K}$ and if the column of $n_{ij}$ is arranged on the diagonals of an $8 \times 8$ matrix and identified as the $\mathbf{N}$ matrix then

$$\mathbf{X}_r'\mathbf{X}_r = \mathbf{K}'\mathbf{N}\mathbf{K}$$

$$= \begin{bmatrix} 1 & 1 & 1 & 1 & 1 & 1 & 1 & 1 \\ 1 & 1 & 0 & 0 & 0 & 0 & -1 & -1 \\ 0 & 0 & 1 & 1 & 0 & 0 & -1 & -1 \\ 0 & 0 & 0 & 0 & 1 & 1 & -1 & -1 \\ 1 & -1 & 1 & -1 & 1 & -1 & 1 & -1 \end{bmatrix} \begin{bmatrix} 21 & 0 & 0 & 0 & 0 & 0 & 0 & 0 \\ 0 & 27 & 0 & 0 & 0 & 0 & 0 & 0 \\ 0 & 0 & 15 & 0 & 0 & 0 & 0 & 0 \\ 0 & 0 & 0 & 25 & 0 & 0 & 0 & 0 \\ 0 & 0 & 0 & 0 & 12 & 0 & 0 & 0 \\ 0 & 0 & 0 & 0 & 0 & 23 & 0 & 0 \\ 0 & 0 & 0 & 0 & 0 & 0 & 7 & 0 \\ 0 & 0 & 0 & 0 & 0 & 0 & 0 & 19 \end{bmatrix}$$

$$\times \begin{bmatrix} 1 & 1 & 0 & 0 & 1 \\ 1 & 1 & 0 & 0 & -1 \\ 1 & 0 & 1 & 0 & 1 \\ 1 & 0 & 1 & 0 & -1 \\ 1 & 0 & 0 & 1 & 1 \\ 1 & 0 & 0 & 1 & -1 \\ 1 & -1 & -1 & -1 & 1 \\ 1 & -1 & -1 & -1 & -1 \end{bmatrix}$$

$$= \begin{bmatrix} 21 & 27 & 15 & 25 & 12 & 23 & 7 & 19 \\ 21 & 27 & 0 & 0 & 0 & 0 & -7 & -19 \\ 0 & 0 & 15 & 25 & 0 & 0 & -7 & -19 \\ 0 & 0 & 0 & 0 & 12 & 23 & -7 & -19 \\ 21 & -27 & 15 & -25 & 12 & -23 & 7 & -19 \end{bmatrix} \begin{bmatrix} 1 & 1 & 0 & 0 & 1 \\ 1 & 1 & 0 & 0 & -1 \\ 1 & 0 & 1 & 0 & 1 \\ 1 & 0 & 1 & 0 & -1 \\ 1 & 0 & 0 & 1 & 1 \\ 1 & 0 & 0 & 1 & -1 \\ 1 & -1 & -1 & -1 & 1 \\ 1 & -1 & -1 & -1 & -1 \end{bmatrix}$$

$$= \begin{bmatrix} 149 & 22 & 14 & 9 & -39 \\ 22 & 74 & 26 & 26 & 6 \\ 14 & 26 & 66 & 26 & 2 \\ 9 & 26 & 26 & 61 & 1 \\ -39 & 6 & 2 & 1 & 149 \end{bmatrix}$$

Now if we let the column vector of the $Y_{ij.}$ be denoted by $\mathbf{Y^*}$, then

$$\mathbf{X'_rY = K'Y^*}$$

$$= \begin{bmatrix} 1 & 1 & 1 & 1 & 1 & 1 & 1 & 1 \\ 1 & 1 & 0 & 0 & 0 & 0 & -1 & -1 \\ 0 & 0 & 1 & 1 & 0 & 0 & -1 & -1 \\ 0 & 0 & 0 & 0 & 1 & 1 & -1 & -1 \\ 1 & -1 & 1 & -1 & 1 & -1 & 1 & -1 \end{bmatrix} \begin{bmatrix} 3716 \\ 2957 \\ 2422 \\ 2852 \\ 1868 \\ 2496 \\ 1197 \\ 2029 \end{bmatrix} = \begin{bmatrix} 19{,}537 \\ 3{,}447 \\ 2{,}048 \\ 1{,}138 \\ -1{,}131 \end{bmatrix}$$

and the least-squares equations are

$$\underbrace{\begin{bmatrix} 149 & 22 & 14 & 9 & -39 \\ 22 & 74 & 26 & 26 & 6 \\ 14 & 26 & 66 & 26 & 2 \\ 9 & 26 & 26 & 61 & 1 \\ -39 & 6 & 2 & 1 & 149 \end{bmatrix}}_{\mathbf{X'_rX_r}} \underbrace{\begin{bmatrix} \hat{\mu} \\ \hat{g}_1 \\ \hat{g}_2 \\ \hat{g}_3 \\ \hat{s}_1 \end{bmatrix}}_{\hat{\mathbf{B}}} = \underbrace{\begin{bmatrix} 19{,}537 \\ 3{,}447 \\ 2{,}048 \\ 1{,}138 \\ -1{,}131 \end{bmatrix}}_{\mathbf{X'_rY}}$$

For simplicity of notation, the $*$ is omitted from the estimates of $\mu$, $g_i$, and $s_j$. The inverse of the coefficient matrix $\mathbf{X'_rX_r}$, which may be computed by numerous computational methods, is

$$\mathbf{(X'_rX_r)^{-1}} = \begin{bmatrix} 0.0076594 & -0.0021881 & -0.0008835 & 0.0001446 & 0.0021038 \\ -0.0021881 & 0.0176644 & -0.0043593 & -0.0053287 & -0.0011898 \\ -0.0008835 & -0.0043593 & 0.0195645 & -0.0063460 & -0.0002757 \\ 0.0001446 & -0.0053287 & -0.0063460 & 0.0213450 & 0.0001944 \\ 0.0021038 & -0.0011898 & -0.0002757 & 0.0001944 & 0.0073124 \end{bmatrix}$$

Estimates of the linear functions of the effects are computed from

$$\hat{\mathbf{B}} = \mathbf{(X'_rX_r)^{-1}X'Y}$$

and, for our example,

$$
\begin{bmatrix} \hat{\mu} \\ \hat{g}_1 \\ \hat{g}_2 \\ \hat{g}_3 \\ \hat{s}_1 \end{bmatrix} = \begin{bmatrix} 138.075572 \\ 4.493677 \\ 0.871259 \\ -4.468128 \\ 28.387321 \end{bmatrix} = (\mathbf{X}'_r\mathbf{X}_r)^{-1} \begin{bmatrix} 19,537 \\ 3,447 \\ 2,048 \\ 1,138 \\ -1,131 \end{bmatrix}
$$

The total reduction in sum of squares due to fitting all constants is, in general,

$$R(\mathbf{B}) = \hat{\mathbf{B}}'\mathbf{X}'_r\mathbf{Y}$$

and, for our example,

$$R(\mu, g, s) = 2,677,665.69$$

The sum of squares for the remainder is

$$\mathbf{Y}'\mathbf{Y} - R(\mathbf{B})$$

and, for our example,

$$
\begin{aligned}
\text{Remainder SS} &= \mathbf{Y}'\mathbf{Y} - R(\mu, g, s) \\
&= 2,738,543 - 2,677,665.69 \\
&= 60,877.31
\end{aligned}
$$

The sum of squares for the mean is

$$
\begin{aligned}
R(\mu \mid g, s) &= R(\mu, g, s) - R(g, s) \\
&= \hat{\mathbf{B}}'_\mu \mathbf{Z}_\mu^{-1} \hat{\mathbf{B}}_\mu \\
&= \frac{\hat{\mu}^2}{c^{\mu\mu}} \\
&= \frac{(138.0756)^2}{0.007659424} = 2,489,072.65
\end{aligned}
$$

and it will be noted that this is not equal to the overall correction term, that is, $(19,537)^2/149 = 2,561,707.17$. The sums of squares for generations and sex, adjusted for unequal subclass frequencies, are computed by the direct procedure as follows:

$$\text{SS Generations} = R(g \mid \mu, s) = \hat{\mathbf{B}}'_G\mathbf{Z}_G^{-1}\hat{\mathbf{B}}_G = 1672.42$$

where

$$
\hat{\mathbf{B}}_G = \begin{bmatrix} 4.493677 \\ 0.871259 \\ -4.468130 \end{bmatrix}
$$

$$
\mathbf{Z}_G = \begin{bmatrix} 0.0176644 & -0.0043593 & -0.0053287 \\ -0.0043593 & 0.0195645 & -0.0063460 \\ -0.0053287 & -0.0063460 & 0.0213450 \end{bmatrix}
$$

$$
\begin{aligned}
\text{SS Sex} = R(s \mid \mu, g) &= \hat{\mathbf{B}}'_S\mathbf{Z}_S^{-1}\hat{\mathbf{B}}_S \\
&= \frac{(28.387321)^2}{0.0073124} \\
&= 110,202.16
\end{aligned}
$$

The analysis of variance can now be set up as shown in Table 3.52.

**TABLE 3.52**  ANALYSIS OF VARIANCE OF WEIGHT GAINS
OF RATS NO. 1

| Source | df | SS | MS | $F$ |
|--------|-----|----------|------------|-----------|
| Generations | 3 | 1,672.42 | 557.47 | 1.32 |
| Sex | 1 | 110,202.16 | 110,202.16 | 260.65** |
| Remainder | 144 | 60,877.31 | 422.76 | |

We have assumed that there is no interaction of generations and sex. In the two-way classification the least-squares sum of squares for the interaction can easily be computed. Therefore, one should first obtain this sum of squares and the sum of squares within the two-way subclasses and test for the significance of the interaction rather than simply "assume" that it is nonexistent. Following the general indirect procedure of obtaining an adjusted sum of squares [Equation (3.109)],

$$\text{SS Generations by sex} = R(\mu, g, s, gs) - R(\mu, g, s)$$

$$= \sum_i \sum_j \frac{Y_{ij\cdot}^2}{n_{ij}} - R(\mu, g, s)$$

$$= 2{,}680{,}848 - 2{,}677{,}665.69$$

$$= 3182$$

with three degrees of freedom. The within-subclass sum of squares is

$$\text{Error SS} = 2{,}738{,}543 - 2{,}680{,}848$$

$$= 57{,}695$$

with 141 degrees of freedom. Therefore,

$$F_{G\times S} = \frac{3182/3}{57{,}695/141}$$

$$= \frac{1060.7}{409.2}$$

$$= 2.59$$

and this exceeds the tabulated $F$ value at the 0.10 level of probability, suggesting that an interaction of generations and sex does exist. Thus, the most appropriate least-squares analysis of these data must be completed under the two-way classification model where interaction is included. It is suggested from the work of Bozivich et al. (1956) that the latter analysis be completed if significance is detected for the interaction at the 0.2 level of probability. This analysis will be completed for the present example in Section 3.20.2.

If the interaction of the two sets of effects is found to be unimportant the investigator is interested in the adjusted or "least-squares" means. In our example,

$$\hat{\mu} = 138.08$$

$$\hat{g}_1 = 4.49$$

$$\hat{g}_2 = 0.87$$

$$\hat{g}_3 = -4.47$$

$$\hat{g}_4 = -(4.49 + 0.87 - 4.47) = -0.89$$

$$\hat{s}_1 = 28.39$$

$$\hat{s}_2 = -\hat{s}_1 = -28.39$$

Therefore, the least-squares means, that is, the means that would have been expected if we had equal subclass frequencies, are given in Table 3.53. Standard errors of the least-squares means are computed from the variance of a linear function [Equation (3.105)] as follows:

$$V(\hat{\mu} + \hat{g}_1) = [0.0076594 + 0.0176644 - (2)(0.0021881)](422.8)$$

$$= (0.0209476)(422.8)$$

$$= 8.8566$$

$$SE(\hat{\mu} + \hat{g}_1) = \sqrt{8.8566} = 2.98$$

$$V(\hat{\mu} + \hat{g}_2) = [0.0076594 + 0.0195645 - (2)(0.0008835)](422.8)$$

$$= (0.0254569)(422.8)$$

$$= 10.7632$$

$$SE(\hat{\mu} + \hat{g}_2) = \sqrt{10.7632} = 3.28$$

$$V(\hat{\mu} + \hat{g}_3) = [0.0076594 + 0.0213450 + (2)(0.0001446)](422.8)$$

$$= (0.0292936)(422.8)$$

$$= 12.3853$$

$$SE(\hat{\mu} + \hat{g}_3) = \sqrt{12.3853} = 3.52$$

$$V(\hat{\mu} + \hat{g}_4) = V(\hat{\mu} - \hat{g}_1 - \hat{g}_2 - \hat{g}_3)$$

**TABLE 3.53**  LEAST-SQUARES MEANS FOR GENERATIONS AND SEXES

| Generations | Linear function | LS mean (g) |
|---|---|---|
| 1 | $\hat{\mu} + \hat{g}_1$ | 142.57 |
| 2 | $\hat{\mu} + \hat{g}_2$ | 138.95 |
| 3 | $\hat{\mu} + \hat{g}_3$ | 133.61 |
| 4 | $\hat{\mu} + \hat{g}_4$ | 137.19 |
| Sexes | Linear function | LS mean (g) |
| Males | $\hat{\mu} + \hat{s}_1$ | 166.47 |
| Females | $\hat{\mu} + \hat{s}_2$ | 109.69 |

$$= [0.0076594 + 0.0176644 + 0.0195645$$
$$+ 0.0213450 + (2)(0.0021881)$$
$$+ (2)(0.0008835) - (2)(0.0001446) - (2)(0.0043593)$$
$$- (2)(0.0053287) - (2)(0.0063460)](422.8)$$
$$= (0.0400193)(422.8)$$
$$= 16.9202$$
$$\text{SE}(\hat{\mu} + \hat{g}_4) = \sqrt{16.9202} = 4.11$$
$$V(\hat{\mu} + \hat{s}_1) = [0.0076594 + 0.0073124 + (2)(0.0021038)](422.8)$$
$$= (0.0191794)(422.8)$$
$$= 8.1091$$
$$\text{SE}(\hat{\mu} + \hat{s}_1) = \sqrt{8.1091} = 2.85$$
$$V(\hat{\mu} + \hat{s}_2) = V(\hat{\mu} - \hat{s}_1)$$
$$= [0.0076594 + 0.0073124 - (2)(0.0021038)](422.8)$$
$$= (0.0107642)(422.8)$$
$$= 4.5511$$
$$\text{SE}(\hat{\mu} + \hat{s}_2) = \sqrt{4.5511} = 2.13$$

To illustrate the use of Equation (3.106) to test the significance of the difference between two least-squares means, we compute the standard error of the difference of the two sex means, that is, $166.47 - 109.69 = 56.78$:

$$V(\hat{\mu} + \hat{s}_1 - \hat{\mu} - \hat{s}_2) = V(\hat{s}_1 - \hat{s}_2)$$
$$= V[\hat{s}_1 - (-\hat{s}_1)]$$
$$= V(2\hat{s}_1)$$
$$= 4V(\hat{s}_1)$$
$$= (4)(0.0073124)(422.8)$$
$$= 12.366731$$
$$\text{SE}(\hat{\mu} + \hat{s}_1 - \hat{\mu} - \hat{s}_2) = \sqrt{12.366731} = 3.5166$$

We can now use the $t$ distribution to test the significance of the difference among sexes.

$$t = \frac{56.78}{3.5166}$$
$$= 16.15$$

It should be noted that $t^2 = 260.70$, which is the value found for $F$ in the analysis of variance for sexes, within rounding errors.

### 3.20.2 Two-Way Classification: With Interaction

We shall again consider the data in Table 3.51. The model we now want to use for the analysis of these data is

$$Y_{ijk} = \mu + g_i + s_j + gs_{ij} + e_{ijk}$$

$$i = 1, 2, 3, 4$$

$$j = 1, 2$$

$$k = 1, 2, \ldots, n_{ij} \qquad (3.113)$$

where the $g_i$ are generation effects, the $s_j$ are sex effects, the $gs_{ij}$ are interaction effects, and the $e_{ijk}$ are random errors. Although there are still only eight rows in the **X** matrix we now must have columns for the incidence of the $gs_{ij}$ interaction effects. These eight rows are as follows:

| Subclass | $\mu$ | $g_1$ | $g_2$ | $g_3$ | $g_4$ | $s_1$ | $s_2$ | $gs_{11}$ | $gs_{12}$ | $gs_{21}$ | $gs_{22}$ | $gs_{31}$ | $gs_{32}$ | $gs_{41}$ | $gs_{42}$ |
|---|---|---|---|---|---|---|---|---|---|---|---|---|---|---|---|
| 11 | 1 | 1 | 0 | 0 | 0 | 1 | 0 | 1 | 0 | 0 | 0 | 0 | 0 | 0 | 0 |
| 12 | 1 | 1 | 0 | 0 | 0 | 0 | 1 | 0 | 1 | 0 | 0 | 0 | 0 | 0 | 0 |
| 21 | 1 | 0 | 1 | 0 | 0 | 1 | 0 | 0 | 0 | 1 | 0 | 0 | 0 | 0 | 0 |
| 22 | 1 | 0 | 1 | 0 | 0 | 0 | 1 | 0 | 0 | 0 | 1 | 0 | 0 | 0 | 0 |
| 31 | 1 | 0 | 0 | 1 | 0 | 1 | 0 | 0 | 0 | 0 | 0 | 1 | 0 | 0 | 0 |
| 32 | 1 | 0 | 0 | 1 | 0 | 0 | 1 | 0 | 0 | 0 | 0 | 0 | 1 | 0 | 0 |
| 41 | 1 | 0 | 0 | 0 | 1 | 1 | 0 | 0 | 0 | 0 | 0 | 0 | 0 | 1 | 0 |
| 42 | 1 | 0 | 0 | 0 | 1 | 0 | 1 | 0 | 0 | 0 | 0 | 0 | 0 | 0 | 1 |

It will be noted that there are eight interaction terms but the interaction has only three degrees of freedom. Hence, only three columns for the $gs_{ij}$ effects can remain in the reduced **X** matrix. This is accomplished when using the summation restrictions by requiring that the $\hat{gs}_{ij}$, the estimated linear functions of the interaction effects, sum to zero by row and column, that is, $\Sigma_i \hat{gs}_{ij} = \Sigma_j \hat{gs}_{ij} = 0$. The eight rows of the reduced **X** matrix, $\mathbf{X}_r$, are given below along with columns for $n_{ij}$ and $Y_{ij}$. . Note that the coefficients for the $gs_{ij}$ are obtained by computing products of corresponding coefficients in the $g$ and $s$ columns.

| Subclass | $\mu$ | $g_1$ | $g_2$ | $g_3$ | $s_1$ | $gs_{11}$ | $gs_{21}$ | $gs_{31}$ | $n_{ij}$ | $Y_{ij}$. |
|---|---|---|---|---|---|---|---|---|---|---|
| 11 | 1 | 1 | 0 | 0 | 1 | 1 | 0 | 0 | 21 | 3716 |
| 12 | 1 | 1 | 0 | 0 | -1 | -1 | 0 | 0 | 27 | 2957 |
| 21 | 1 | 0 | 1 | 0 | 1 | 0 | 1 | 0 | 15 | 2422 |
| 22 | 1 | 0 | 1 | 0 | -1 | 0 | -1 | 0 | 25 | 2852 |
| 31 | 1 | 0 | 0 | 1 | 1 | 0 | 0 | 1 | 12 | 1868 |
| 32 | 1 | 0 | 0 | 1 | -1 | 0 | 0 | -1 | 23 | 2496 |
| 41 | 1 | -1 | -1 | -1 | 1 | -1 | -1 | -1 | 7 | 1197 |
| 42 | 1 | -1 | -1 | -1 | -1 | 1 | 1 | 1 | 19 | 2029 |

The coefficients of the least-squares equations, $X_r'X_r$, and the right-hand members for these equations, $X_r'Y$, are calculated in the same manner as described above for the two-way classification without interaction. The equations are then found to be as follows:

$$
\underbrace{\begin{bmatrix}
149 & 22 & 14 & 9 & -39 & 6 & 2 & 1 \\
22 & 74 & 26 & 26 & 6 & -18 & -12 & -12 \\
14 & 26 & 66 & 26 & 2 & -12 & -22 & -12 \\
9 & 26 & 26 & 61 & 1 & -12 & -12 & -23 \\
-39 & 6 & 2 & 1 & 149 & 22 & 14 & 9 \\
6 & -18 & -12 & -12 & 22 & 74 & 26 & 26 \\
2 & -12 & -22 & -12 & 14 & 26 & 66 & 26 \\
1 & -12 & -12 & -23 & 9 & 26 & 26 & 61
\end{bmatrix}}_{X_r'X_r}
\underbrace{\begin{bmatrix}
\hat{\mu} \\ \hat{g}_1 \\ \hat{g}_2 \\ \hat{g}_3 \\ \hat{s}_1 \\ \hat{gs}_{11} \\ \hat{gs}_{21} \\ \hat{gs}_{31}
\end{bmatrix}}_{\hat{B}}
=
\underbrace{\begin{bmatrix}
19{,}537 \\ 3{,}447 \\ 2{,}048 \\ 1{,}138 \\ -1{,}131 \\ 1{,}591 \\ 402 \\ 204
\end{bmatrix}}_{X_r'Y}
$$

The inverse of the coefficient matrix $X_r'X_r$ is given in Table 3.54. The solution of the set of least-squares equations yields the following estimates:

$$\hat{\mu} = 137.9994 \qquad\qquad \hat{gs}_{11} = 5.444933$$

$$\hat{g}_1 = 5.236019 \qquad\qquad \hat{gs}_{12} = -5.444933$$

$$\hat{g}_2 = -0.226097 \qquad\qquad \hat{gs}_{21} = -4.57866$$

$$\hat{g}_3 = -5.90523 \qquad\qquad \hat{gs}_{22} = 4.57866$$

$$\hat{g}_4 = -(\hat{g}_1 + \hat{g}_2 + \hat{g}_3) \qquad\qquad \hat{gs}_{31} = -4.69953$$

$$= 0.895308 \qquad\qquad \hat{gs}_{32} = 4.69953$$

$$\hat{s}_1 = 28.272 \qquad\qquad \hat{gs}_{41} = -(\hat{gs}_{11} + \hat{gs}_{21} + \hat{gs}_{31})$$

$$\hat{s}_2 = -\hat{s}_1 = -28.272 \qquad\qquad = 3.833257$$

$$\hat{gs}_{42} = -(\hat{gs}_{12} + \hat{gs}_{22} + \hat{gs}_{32})$$

$$= -3.833257$$

**TABLE 3.54**  INVERSE OF THE COEFFICIENT MATRIX FOR THE TWO-WAY CLASSIFICATION WITH INTERACTION DATA SET

|  | $\hat{\mu}$ | $\hat{g}_1$ | $\hat{g}_2$ | $\hat{g}_3$ | $\hat{s}_1$ | $\hat{gs}_{11}$ | $\hat{gs}_{21}$ | $\hat{gs}_{31}$ |
|---|---|---|---|---|---|---|---|---|
| $\mu$: | 0.0080254 | −0.0027344 | −0.0013587 | −0.0000996 | 0.0026145 | −0.0019531 | −0.0009479 | −0.0001236 |
| $g_1$: | | 0.0186074 | −0.0039323 | −0.0051914 | −0.0019531 | 0.0039373 | 0.0002865 | −0.0005378 |
| $g_2$: | | | 0.0213587 | −0.0065670 | −0.0009479 | 0.0002865 | 0.0059479 | −0.0015431 |
| $g_3$: | | | | 0.0238768 | −0.0012358 | −0.0005378 | −0.0015431 | 0.0075964 |
| $s_1$: | | | | | 0.0080254 | −0.0012358 | −0.0027344 | −0.0000996 |
| $gs_{11}$: | | | | | | 0.0186074 | −0.0039323 | −0.0051914 |
| $gs_{21}$: | | *Symmetric* | | | | | 0.0213587 | −0.0065670 |
| $gs_{31}$: | | | | | | | | 0.0238768 |

The total reduction in sum of squares due to fitting all constants is

$$R(\mu, g, s, gs) = 2{,}680{,}848$$

$$= \sum_i \sum_j \frac{Y_{ij\cdot}^2}{n_{ij}}$$

the uncorrected sum of squares for the generation by sex subclasses. When a set of estimable linear functions (constants) are fitted for all degrees of freedom for subclasses, the total reduction in the sum of squares, of course, must equal the subclass sum of squares. The error, or within-subclass, sum of squares is $\mathbf{Y'Y} - R(\mathbf{B})$, and in this example

$$\text{Error SS} = \mathbf{Y'Y} - R(\mu, g, s, gs)$$

$$= 2{,}738{,}543 - 2{,}680{,}848$$

$$= 57{,}695$$

The sum of squares due to the mean, which seldom has any practical usefulness, is

$$R(\mu \mid g, s, gs) = \hat{\mathbf{B}}_\mu' \mathbf{Z}_\mu^{-1} \hat{\mathbf{B}}_\mu$$

$$= \frac{(137.9994)^2}{0.00802536}$$

$$= 2{,}372{,}958$$

The sums of squares for generations, sex, and the generation by sex interaction are computed by the direct procedure and, under this analysis with interaction included in the model, these sums of squares are obtained as follows:

$$\text{SS Generations} = R(g \mid \mu, s, gs)$$

$$= \hat{\mathbf{B}}_G' \mathbf{Z}_G^{-1} \hat{\mathbf{B}}_G$$

$$= 2376.84$$

$$\text{SS Sex} = R(s \mid \mu, g, gs)$$

$$= \hat{\mathbf{B}}_S' \mathbf{Z}_S^{-1} \hat{\mathbf{B}}_S$$

$$= 99{,}597.50$$

$$\text{SS } G \times S = R(gs \mid \mu, g, s)$$

$$= \hat{\mathbf{B}}_{GS}' \mathbf{Z}_{GS}^{-1} \hat{\mathbf{B}}_{GS}$$

$$= 3182.48$$

The analysis of variance is given in Table 3.55. The test of significance for interaction is the same as obtained in Section 3.20.1. However, the $F$ values for testing the significance of the main effects have both changed; one has increased and one has decreased. The estimates obtained for the generation and sex effects from this analysis have now been adjusted for unequal frequencies with respect to the interaction effects.

**TABLE 3.55**  ANALYSIS OF VARIANCE OF WEIGHT GAINS
IN RATS NO. 2

| Source | df | SS | MS | F |
|---|---|---|---|---|
| Generation (G) | 3 | 2,376.84 | 792.3 | 1.94 |
| Sex (S) | 1 | 99,597.50 | 99,597.5 | 243.39** |
| GS | 3 | 3,182.48 | 1,060.8 | 2.59 |
| Error | 141 | 57,695.00 | 409.2 | |

Let us examine the adjusted or least-squares means which are actually being compared in making these tests of significance. These are given in Table 3.56 along with the standard errors that have been calculated using Equation (3.105).

The least-squares means for the generation by sex subclasses may be computed from the following equation

$$LSM(GS_{ij}) = \hat{\mu} + \hat{g}_i + \hat{s}_j + \hat{gs}_{ij}$$

or from the simple subclass means, since constants for all degrees of freedom for the GS subclasses are being fitted and no other effects are included in the model. For example,

$$LSM(GS_{11}) = 138.0 + 5.24 + 28.27 + 5.44$$

$$= 177.0$$

$$= \frac{3716}{21}$$

**TABLE 3.56**  LEAST-SQUARES MEANS AND STANDARD
ERRORS FOR THE TWO-WAY
CLASSIFICATION ANALYSIS
WITH INTERACTION

| Classification | | Number of observations | Mean gains (g) | SE |
|---|---|---|---|---|
| Overall | | 149 | 138.0 | 1.8 |
| | | Generations | | |
| 1 | | 48 | 143.2 | 2.9 |
| 2 | | 40 | 137.8 | 3.3 |
| 3 | | 35 | 132.1 | 3.6 |
| 4 | | 26 | 138.9 | 4.5 |
| | | Sex | | |
| Males | | 55 | 166.3 | 3.0 |
| Females | | 94 | 109.7 | 2.1 |
| | | G × S subclass | | |
| 1 | M | 21 | 177.0 | 4.4 |
| | F | 27 | 109.5 | 3.9 |
| 2 | M | 15 | 161.5 | 5.2 |
| | F | 25 | 114.1 | 4.0 |
| 3 | M | 12 | 155.7 | 5.8 |
| | F | 23 | 108.5 | 4.2 |
| 4 | M | 7 | 171.0 | 7.6 |
| | F | 19 | 106.8 | 4.6 |

Also, the standard errors for the generation by sex subclass least-squares means may be obtained from Equation (3.105) or more simply from the subclass numbers and the mean square for error. For example,

$$V[\text{LSM}(GS_{11})] = [0.0080254 + 0.0186074 + 0.0080254 + 0.0186074$$

$$- (2)(0.0027344) + (2)(0.0026145)$$

$$- (2)(0.0019531) - (2)(0.0019531) + (2)(0.0039373)$$

$$- (2)(0.0027344)](409.2)$$

$$= (0.047619)(409.2)$$

$$= 19.4857$$

$$= (\tfrac{1}{21})(409.2)$$

$$\text{SE}[\text{LSM}(GS_{11})] = \sqrt{19.4857} = 4.4$$

The differences in weight gains between sexes for generations 1–4 are 67.5, 47.4, 49.2, and 67.5 g. These differences among sexes from generation to generation are significantly different at the 0.10 level of probability, as shown by the $F$ value of 2.59 for interaction in the analysis of variance with 3 and 141 degrees of freedom. For some reason the difference between the sexes was greater in the first and fourth generations than in the second and third generations. The investigator would probably have an explanation for these differences.

Another feature of the least-squares means that should be noted in Table 3.56 is that the means for generations and sexes are simple averages, that is, unweighted means, of the appropriate subclass means. For example,

$$\text{LSM}(G_1) = \hat{\mu} + \hat{g}_1$$

$$= 138.00 + 5.24$$

$$= 143.2$$

$$= \tfrac{1}{2}(177.0 + 109.5)$$

$$\text{LSM}(S_1) = \hat{\mu} + \hat{s}_1$$

$$= 138.00 + 28.27$$

$$= 166.3$$

$$= \tfrac{1}{4}(177.0 + 161.5 + 155.7 + 171.0)$$

This tells us that, when interaction exists, the best estimate of the marginal means is obtained by computing the unweighted means of the subclass means. Of course, when interaction exists the investigator should examine the nature of the interaction by comparing the subclass means. However, in many cases the investigator is also interested in the main effects even though interaction does exist. For example, if a cattle feeder has only one feed lot and he must feed both heifers and steers together, the feeder is interested in locating the ration that will maximize gains on the average across sexes. The fact that one ration may be best for one sex and another ration best for the other sex is of little concern to such a cattle feeder.

It should be pointed out that the results obtained in this section using methods of least squares are identical with the results obtained when one uses the "weighted squares of means" procedure described by Yates (1934) and given in many textbooks on statistical methods. The general least-squares procedure allows extension to analyses under more complex models, whereas the weighted squares of means procedure is primarily useful for an analysis under the two-way classification model with interaction. With the availability of general least-squares computer programs such as LSML76 by Harvey (1977b and 1982b) and GLM and HARVEY in the Statistical Analysis System (1982) there is little need to use the weighted squares of means procedure. Therefore, this procedure will not be described in this text.

### 3.20.3 Multiway Classification

The data in Table 3.57 are from Bancroft (1968) and were obtained by Professor Edmund Hoffman at the University of Georgia in 1949. The table contains the subclass totals and frequencies of weights of turkeys (in pounds) classified by sex, breed, and hatch. The total uncorrected sum of squares for the 326 observations was 62,501.

The complete model for the analysis of these data is

$$Y_{ijkl} = \mu + s_i + b_j + h_k + sb_{ij} + sh_{ik} + bh_{jk} + sbh_{ijk} + e_{ijkl}$$

$$i = 1, 2 \qquad k = 1, 2, 3$$

$$j = 1, 2, 3, 4 \qquad l = 1, 2, \ldots, n_{ijk} \tag{3.114}$$

where $s_i$ are the sex effects, $b_j$ are the breed effects, $h_k$ are the hatch effects, $sb_{ij}$ are the sex by breed interaction effects, $sh_{ik}$ are the sex by hatch interaction effects, $bh_{jk}$ are the breed by hatch interaction effects, $sbh_{ijk}$ are the sex by breed by hatch interaction effects, and $e_{ijkl}$ are the random errors. The rows of the **X** matrix are the same for all observations in the same sex by breed by hatch subclass. Therefore, there are only 24 different rows in the **X** matrix. However, there are $1 + 2 + 4 + 8 + 6 + 12 + 24 = 60$ columns in the **X** matrix. After imposing the usual restrictions there are only $1 + 1 + 3 + 2 + 3 + 2 + 6 + 6 = 24$ columns in the reduced **X** matrix. The 24 rows and columns of the reduced **X** matrix are given in Table 3.58.

**TABLE 3.57**  NUMBER OF TURKEYS AND TOTAL WEIGHT (IN POUNDS) CLASSIFIED BY SEX, BREED, AND HATCH

| Breed | | 1 | | 2 | | 3 | | 4 | |
|---|---|---|---|---|---|---|---|---|---|
| Hatch \ Sex | | Male | Female | Male | Female | Male | Female | Male | Female |
| 1 | Number of birds | 18 | 23 | 22 | 20 | 17 | 21 | 26 | 25 |
| | Total weight | 237.0 | 190.0 | 361.0 | 214.0 | 319.4 | 225.8 | 413.7 | 255.0 |
| 2 | Number of birds | 11 | 7 | 9 | 8 | 8 | 4 | 7 | 17 |
| | Total weight | 164.2 | 62.2 | 150.8 | 85.5 | 170.2 | 47.8 | 126.8 | 179.0 |
| 3 | Number of birds | 11 | 10 | 6 | 7 | 9 | 5 | 18 | 17 |
| | Total weight | 156.8 | 89.8 | 88.3 | 77.8 | 170.5 | 61.8 | 308.3 | 176.8 |

**TABLE 3.58** REDUCED **X** MATRIX FOR THREE-WAY CLASSIFICATION

| Sub-class | $\mu$ | $s_1$ | $b_1$ | $b_2$ | $b_3$ | $h_1$ | $h_2$ | $sb_{11}$ | $sb_{12}$ | $sb_{13}$ | $sh_{11}$ | $sh_{12}$ | $bh_{11}$ | $bh_{12}$ | $bh_{21}$ | $bh_{22}$ | $bh_{31}$ | $bh_{32}$ | $sbh_{111}$ | $sbh_{112}$ | $sbh_{121}$ | $sbh_{122}$ | $sbh_{131}$ | $sbh_{132}$ |
|---|---|---|---|---|---|---|---|---|---|---|---|---|---|---|---|---|---|---|---|---|---|---|---|---|
| 111 | 1 | 1 | 1 | 0 | 0 | 1 | 0 | 1 | 0 | 0 | 1 | 0 | 1 | 0 | 0 | 0 | 0 | 0 | 1 | 0 | 0 | 0 | 0 | 0 |
| 112 | 1 | 1 | 1 | 0 | 0 | 0 | 1 | 1 | 0 | 0 | 0 | 1 | 0 | 1 | 0 | 0 | 0 | 0 | 0 | 1 | 0 | 0 | 0 | 0 |
| 113 | 1 | 1 | 1 | 0 | 0 | -1 | -1 | 1 | 0 | 0 | -1 | -1 | -1 | -1 | 0 | 0 | 0 | 0 | -1 | -1 | 0 | 0 | 0 | 0 |
| 121 | 1 | 1 | 0 | 1 | 0 | 1 | 0 | 0 | 1 | 0 | 1 | 0 | 0 | 0 | 1 | 0 | 0 | 0 | 0 | 0 | 1 | 0 | 0 | 0 |
| 122 | 1 | 1 | 0 | 1 | 0 | 0 | 1 | 0 | 1 | 0 | 0 | 1 | 0 | 0 | 0 | 1 | 0 | 0 | 0 | 0 | 0 | 1 | 0 | 0 |
| 123 | 1 | 1 | 0 | 1 | 0 | -1 | -1 | 0 | 1 | 0 | -1 | -1 | 0 | 0 | -1 | -1 | 0 | 0 | 0 | 0 | -1 | -1 | 0 | 0 |
| 131 | 1 | 1 | 0 | 0 | 1 | 1 | 0 | 0 | 0 | 1 | 1 | 0 | 0 | 0 | 0 | 0 | 1 | 0 | 0 | 0 | 0 | 0 | 1 | 0 |
| 132 | 1 | 1 | 0 | 0 | 1 | 0 | 1 | 0 | 0 | 1 | 0 | 1 | 0 | 0 | 0 | 0 | 0 | 1 | 0 | 0 | 0 | 0 | 0 | 1 |
| 133 | 1 | 1 | 0 | 0 | 1 | -1 | -1 | 0 | 0 | 1 | -1 | -1 | 0 | 0 | 0 | 0 | -1 | -1 | 0 | 0 | 0 | 0 | -1 | -1 |
| 141 | 1 | 1 | -1 | -1 | -1 | 1 | 0 | -1 | -1 | -1 | 1 | 0 | -1 | 0 | -1 | 0 | -1 | 0 | -1 | 0 | -1 | 0 | -1 | 0 |
| 142 | 1 | 1 | -1 | -1 | -1 | 0 | 1 | -1 | -1 | -1 | 0 | 1 | 0 | -1 | 0 | -1 | 0 | -1 | 0 | -1 | 0 | -1 | 0 | -1 |
| 143 | 1 | 1 | -1 | -1 | -1 | -1 | -1 | -1 | -1 | -1 | -1 | -1 | 1 | 1 | 1 | 1 | 1 | 1 | 1 | 1 | 1 | 1 | 1 | 1 |
| 211 | 1 | -1 | 1 | 0 | 0 | 1 | 0 | -1 | 0 | 0 | -1 | 0 | 1 | 0 | 0 | 0 | 0 | 0 | -1 | 0 | 0 | 0 | 0 | 0 |
| 212 | 1 | -1 | 1 | 0 | 0 | 0 | 1 | -1 | 0 | 0 | 0 | -1 | 0 | 1 | 0 | 0 | 0 | 0 | 0 | -1 | 0 | 0 | 0 | 0 |
| 213 | 1 | -1 | 1 | 0 | 0 | -1 | -1 | -1 | 0 | 0 | 1 | 1 | -1 | -1 | 0 | 0 | 0 | 0 | 1 | 1 | 0 | 0 | 0 | 0 |
| 221 | 1 | -1 | 0 | 1 | 0 | 1 | 0 | 0 | -1 | 0 | -1 | 0 | 0 | 0 | 1 | 0 | 0 | 0 | 0 | 0 | -1 | 0 | 0 | 0 |
| 222 | 1 | -1 | 0 | 1 | 0 | 0 | 1 | 0 | -1 | 0 | 0 | -1 | 0 | 0 | 0 | 1 | 0 | 0 | 0 | 0 | 0 | -1 | 0 | 0 |
| 223 | 1 | -1 | 0 | 1 | 0 | -1 | -1 | 0 | -1 | 0 | 1 | 1 | 0 | 0 | -1 | -1 | 0 | 0 | 0 | 0 | 1 | 1 | 0 | 0 |
| 231 | 1 | -1 | 0 | 0 | 1 | 1 | 0 | 0 | 0 | -1 | -1 | 0 | 0 | 0 | 0 | 0 | 1 | 0 | 0 | 0 | 0 | 0 | -1 | 0 |
| 232 | 1 | -1 | 0 | 0 | 1 | 0 | 1 | 0 | 0 | -1 | 0 | -1 | 0 | 0 | 0 | 0 | 0 | 1 | 0 | 0 | 0 | 0 | 0 | -1 |
| 233 | 1 | -1 | 0 | 0 | 1 | -1 | -1 | 0 | 0 | -1 | 1 | 1 | 0 | 0 | 0 | 0 | -1 | -1 | 0 | 0 | 0 | 0 | 1 | 1 |
| 241 | 1 | -1 | -1 | -1 | -1 | 1 | 0 | 1 | 1 | 1 | -1 | 0 | -1 | 0 | -1 | 0 | -1 | 0 | 1 | 0 | 1 | 0 | 1 | 0 |
| 242 | 1 | -1 | -1 | -1 | -1 | 0 | 1 | 1 | 1 | 1 | 0 | -1 | 0 | -1 | 0 | -1 | 0 | -1 | 0 | 1 | 0 | 1 | 0 | 1 |
| 243 | 1 | -1 | -1 | -1 | -1 | -1 | -1 | 1 | 1 | 1 | 1 | 1 | 1 | 1 | 1 | 1 | 1 | 1 | -1 | -1 | -1 | -1 | -1 | -1 |

**TABLE 3.59**  ANALYSIS OF VARIANCE FOR WEIGHTS GIVEN IN TABLE 3.57

| Source of variation | df | Sum of square | Mean square | F |
|---|---|---|---|---|
| Sexes (S) | 1 | 2428.26 | 2428.26 | 864.15 |
| Breeds (B) | 3 | 516.22 | 172.07 | 61.24 |
| Hatches (H) | 2 | 57.47 | 28.74 | 10.23 |
| SB | 3 | 68.08 | 22.69 | 8.08 |
| SH | 2 | 24.88 | 12.44 | 4.43 |
| BH | 6 | 26.71 | 4.45 | 1.58 |
| SBH | 6 | 22.69 | 3.78 | 1.35 |
| Within | 302 | 848.08 | 2.81 | |

The coefficients for the reduced set of least-squares equations $(X'_r X_r)$ are obtained by computing weighted sums of squares and cross-products for the 24 columns of 0's, 1's, and −1's given in Table 3.58 as was shown for the example problem in Section 3.20.1. Each square or product is weighted by the number of observations in that subclass, that is, $n_{ijk}$. The right-hand members of the reduced set of equations $(X'_r Y)$ are obtained by computing products of the columns of **X** given in Table 3.58 and the subclass total weights given in Table 3.57. When this is done and the analysis is completed by procedures that have previously been described, the analysis of variance given in Table 3.59 is obtained. It is clear from the ANOVA obtained under the full model that the three-factor interaction effects do not exist $(P > 0.24)$. Therefore, a step-down procedure should be followed to estimate more accurately those effects that really do affect these turkey weights. Elimination of the three-factor interaction from the model and solution of the least-squares equations with the last six equations deleted yields the ANOVA given in Table 3.60.

The $F$ value for the breed by hatch interaction effects is 1.62 and the probability of obtaining an $F$ value this large by chance alone is less than 0.2. Therefore, it is suggested that one should not proceed with any further step-down. The least-squares means and standard errors obtained from LSML76 by Harvey (1977b) for all classes and subclasses under this reduced model are given in Table 3.61.

Examination of these means will reveal the nature of the interaction effects. As indicated from the tests of significance the breed by hatch interaction effects are not as large as the sex by breed or the sex by hatch interaction effects. Nevertheless, it can be seen that the average weights of birds from hatches 2 and 3 differed less for breed 1 than for the other three breeds. Also, the difference between the means for hatches

**TABLE 3.60**  ANALYSIS OF VARIANCE FOR WEIGHTS EXCLUDING
              *SBH* INTERACTION

| Source of variation | df | Sum of squares | Mean square | F |
|---|---|---|---|---|
| Sexes (S) | 1 | 2613.91 | 2613.91 | 924.57** |
| Breeds (B) | 3 | 530.95 | 176.98 | 62.60** |
| Hatches (H) | 2 | 59.36 | 29.68 | 10.50** |
| SB | 3 | 80.41 | 26.80 | 9.48** |
| SH | 2 | 19.56 | 9.78 | 3.46** |
| BH | 6 | 27.43 | 4.57 | 1.62 |
| Remainder | 308 | 870.77 | 2.83 | |

**TABLE 3.61**   LEAST-SQUARES MEANS AND STANDARD ERRORS
FOR TURKEY WEIGHTS

| Classification | Number | LS means | SE | Classification | | Number | LS means | SE |
|---|---|---|---|---|---|---|---|---|
| Overall | 326 | 13.54 | 0.12 | $S \times H$ subclasses | | | * | |
| Sexes (S) | | ** | | Males 1 | | 83 | 16.05 | 0.22 |
| Males | 162 | 16.74 | 0.17 | Males 2 | | 35 | 17.75 | 0.33 |
| Females | 164 | 10.35 | 0.18 | Males 3 | | 44 | 16.41 | 0.31 |
| | | | | Females 1 | | 89 | 10.00 | 0.21 |
| Breeds (B) | | ** | | Females 2 | | 36 | 10.53 | 0.35 |
| 1 | 80 | 11.41 | 0.23 | Females 3 | | 39 | 10.52 | 0.33 |
| 2 | 72 | 13.42 | 0.26 | $B \times H$ subclasses | | | $P < 0.2$ | |
| 3 | 64 | 15.65 | 0.28 | 1 | 1 | 41 | 10.72 | 0.31 |
| 4 | 110 | 13.70 | 0.20 | 1 | 2 | 18 | 11.89 | 0.46 |
| Hatches (H) | | ** | | 1 | 3 | 21 | 11.63 | 0.42 |
| 1 | 172 | 13.02 | 0.15 | 2 | 1 | 42 | 13.57 | 0.30 |
| 2 | 71 | 14.14 | 0.24 | 2 | 2 | 17 | 13.71 | 0.47 |
| 3 | 83 | 13.47 | 0.23 | 2 | 3 | 13 | 12.97 | 0.54 |
| $S \times B$ subclasses | | ** | | 3 | 1 | 38 | 14.76 | 0.32 |
| Males 1 | 40 | 14.10 | 0.31 | 3 | 2 | 12 | 16.67 | 0.57 |
| Males 2 | 37 | 16.17 | 0.35 | 3 | 3 | 14 | 15.50 | 0.53 |
| Males 3 | 34 | 19.71 | 0.35 | 4 | 1 | 51 | 13.05 | 0.27 |
| Males 4 | 51 | 16.99 | 0.30 | 4 | 2 | 24 | 14.28 | 0.42 |
| Females 1 | 40 | 8.72 | 0.33 | 4 | 3 | 35 | 13.77 | 0.33 |
| Females 2 | 35 | 10.67 | 0.35 | | | | | |
| Females 3 | 30 | 11.58 | 0.41 | | | | | |
| Females 4 | 59 | 10.42 | 0.26 | | | | | |

\* $P \leq 0.05$.

\*\* $P \leq 0.01$.

1 and 2 was less for breed 2 than for the other breeds. For some reason, the difference between sex means was greatest for birds from hatch 2. Males from hatch 2 outweighed males from the other two hatches by 1.52 lb; whereas, females from hatch 2 only outweighed females from the other two hatches by 0.27 lb. Differences between sexes for the four breeds are 5.38, 5.50, 8.13, and 6.57 lb. The test of significance for the sex by breed interaction effects is a test of the significance of the differences among these differences.

## 3.21  MIXED MODEL ANALYSES: DISPROPORTIONATE SUBCLASS FREQUENCIES

When a mixed model is appropriate and there are disproportionate subclass frequencies the investigator is usually interested in obtaining the same statistics for the fixed effects as may be computed under the fixed linear model. In addition, the investigator will usually want to obtain estimates of the variance components for the random sources of variation. Actually, there are currently several alternative methods of analysis for obtaining unbiased estimates of fixed effects and variance components under the mixed model. Several of these methods use differences among random classes or subclasses that are partially confounded with fixed effects to estimate more accurately the linear function desired for fixed effects. However, to do this one must have reasonably accurate a priori estimates of the variance components for all sets of random effects or these must be estimated iteratively with the estimation of the fixed effects.

The estimation procedure to be described here for mixed models might be referred to as the "least-squares" procedure. The estimates of the "fixed effects" are those obtained from Method 3 analysis of Henderson (1953) as defined by Harvey (1977b, 1981, 1982b). In these analyses the sums of squares for each line in the ANOVA are computed by adjusting for unequal subclass frequencies with respect to only the other effects in the model sufficient to achieve the same expected mean squares as would be the case if equal subclass frequencies existed, except for the coefficients of the variance components.

In mixed models where interactions do not exist among sets of random effects or among random and fixed effects all tests of significance are completed in the ANOVA with the same denominator mean square. However, random sets of effects often have large numbers of degrees of freedom and, of course, this is necessary if accurate estimates of the variance components are to be obtained. This means that to set up the full set of least-squares equations and work with these in the manner described above for fixed linear models will often be prohibitive, even with large computers. Fortunately, shortcut computational procedures have been developed that make it possible for investigators to handle large sets of random effects with reduced computing costs. Examples of the use of these procedures will be described in this section.

If a model includes nested random effects or if interactions exist among random sets of effects it will be seen that only approximate tests of significance exist for some of the sources of variation in the ANOVA. This is not particularly important for random sets of effects since the investigator is primarily interested in obtaining estimates of variance components for these sources of variation.

### 3.21.1 Two-Way Classification: No Interaction

In practice one should have a large number of degrees of freedom for the random set of effects if variance components are to be estimated (say, 50 or more). However, to illustrate the computations required to complete the analysis only a small example set of data will be used. The data given in Table 3.62 are the average daily gains of 65 steers from 9 sires and 3 ages of dam. These are a sample of the 616 steers that were actually fed out in this experiment (Shelby et al., 1963). The model to be considered for the analysis of these data in this section is as follows:

$$Y_{ijk} = \mu + s_i + a_j + e_{ijk}$$

$$i = 1, 2, \ldots, 9$$

$$j = 1, 2, 3$$

$$k = 1, 2, \ldots, n_{ij} \tag{3.115}$$

where $s_i$ are the sire effects (random) and $a_j$ are the age-of-dam effects (fixed).

**TABLE 3.62** AVERAGE DAILY GAIN (IN POUNDS) DURING FEED LOT TEST FOR 65 STEERS FED AT MILES CITY, MONTANA

| Sire | Age dam (years) | | | Sire | Age dam (years) | | |
|------|------|------|------|------|------|------|------|
| | 3 | 4 | 5-up | | 3 | 4 | 5-up |
| 1 | 2.24 | 2.41 | 2.58 | 6 | | 3.00 | 2.25 |
| | 2.65 | 2.25 | 2.67 | | | 2.49 | 2.49 |
| | | | 2.71 | | | | 2.02 |
| | | | 2.47 | | | | 2.31 |
| 2 | | 2.29 | 1.97 | 7 | 2.57 | 2.64 | 2.37 |
| | | 2.26 | 2.14 | | 2.37 | | 2.22 |
| | | | 2.44 | | | | 1.90 |
| | | | 2.52 | | | | 2.61 |
| | | | 1.72 | | | | 2.13 |
| | | | 2.75 | 8 | 2.16 | 2.45 | 1.44 |
| 3 | 2.38 | 2.46 | 2.29 | | 2.33 | | 1.72 |
| | | | 2.30 | | 2.52 | | 2.17 |
| | | | 2.94 | 9 | 2.68 | 2.43 | 2.66 |
| 4 | 2.50 | 2.44 | 2.54 | | | 2.36 | 2.46 |
| | 2.44 | 2.15 | 2.74 | | | 2.44 | 2.52 |
| | | | 2.50 | | | | 2.42 |
| | | | 2.54 | | | | |
| 5 | 2.65 | 2.52 | 2.79 | | | | |
| | | 2.67 | 2.33 | | | | |
| | | | 2.67 | | | | |
| | | | 2.69 | | | | |

**Direct Analysis**   The reduced set of least-squares equations for the data given in Table 3.62 is as follows when the summation restrictions are imposed:

$$
\begin{bmatrix}
65 & 0 & 0 & -3 & 0 & -1 & -2 & 0 & -1 & -25 & -21 \\
 & 16 & 8 & 8 & 8 & 8 & 8 & 8 & 8 & 1 & -1 \\
 & & 16 & 8 & 8 & 8 & 8 & 8 & 8 & -3 & -3 \\
 & & & 13 & 8 & 8 & 8 & 8 & 8 & 1 & -1 \\
 & & & & 16 & 8 & 8 & 8 & 8 & 1 & -1 \\
 & & & & & 15 & 8 & 8 & 8 & 0 & -1 \\
 & \text{Symmetric} & & & & & 14 & 8 & 8 & -1 & -1 \\
 & & & & & & & 16 & 8 & 0 & -3 \\
 & & & & & & & & 15 & 3 & -1 \\
 & & & & & & & & & 49 & 37 \\
 & & & & & & & & & & 53
\end{bmatrix}
\begin{bmatrix}
\hat{\mu} \\ \hat{s}_1 \\ \hat{s}_2 \\ \hat{s}_3 \\ \hat{s}_4 \\ \hat{s}_5 \\ \hat{s}_6 \\ \hat{s}_7 \\ \hat{s}_8 \\ \hat{a}_1 \\ \hat{a}_2
\end{bmatrix}
=
\begin{bmatrix}
156.74 \\ 0.01 \\ -1.88 \\ -7.60 \\ -0.12 \\ -1.65 \\ -5.41 \\ -1.16 \\ -5.18 \\ -58.50 \\ -48.73
\end{bmatrix}
$$

The inverse of the coefficient matrix is given in Table 3.63. Estimates of $\mu$, $s_i$, and $a_j$ are as follows:

$$\hat{\mu} = 2.43501 \qquad \hat{s}_6 = 0.03078$$

$$\hat{s}_1 = 0.07629 \qquad \hat{s}_7 = -0.06392$$

$$\hat{s}_2 = -0.13061 \qquad \hat{s}_8 = -0.32414$$

**TABLE 3.63** INVERSE OF THE REDUCED COEFFICIENT MATRIX[a]

| Equation | $\hat{\mu}$ | $\hat{s}_1$ | $\hat{s}_2$ | $\hat{s}_3$ | $\hat{s}_4$ | $\hat{s}_5$ | $\hat{s}_6$ | $\hat{s}_7$ | $\hat{s}_8$ | $\hat{a}_1$ | $\hat{a}_2$ |
|---|---|---|---|---|---|---|---|---|---|---|---|
| $\mu$: | 19,932 | −3,319 | 1,851 | 6,649 | −3,319 | 452 | 5,480 | −1,818 | −3,742 | 9,785 | 1,109 |
| $\hat{s}_1$: | | 113,558 | −13,516 | −20,299 | −11,442 | −14,238 | −17,965 | −11,713 | −12,177 | −4,809 | 1,320 |
| $\hat{s}_2$: | | | 117,037 | −20,515 | −13,516 | −13,280 | −12,965 | −12,957 | −18,988 | 13,319 | −4,424 |
| $\hat{s}_3$: | | | | 171,511 | −20,299 | −22,509 | −25,456 | −19,832 | −21,728 | −1,822 | 2,778 |
| $\hat{s}_4$: | | | | | 113,558 | −14,238 | −17,965 | −11,713 | −12,177 | −4,809 | 1,320 |
| $\hat{s}_5$: | | | | | | 127,138 | −17,694 | −14,561 | −17,183 | 3,417 | −3,003 |
| $\hat{s}_6$: | | | | | | | 149,335 | −18,359 | −23,859 | 14,384 | −8,768 |
| $\hat{s}_7$: | | | *Symmetric* | | | | | 114,445 | −11,816 | −6,407 | 7,837 |
| $\hat{s}_8$: | | | | | | | | | 133,871 | −19,127 | 11,009 |
| $\hat{a}_1$: | | | | | | | | | | 52,602 | −32,694 |
| $\hat{a}_2$: | | | | | | | | | | | 42,413 |

[a] Each element has been multiplied by $10^6$.

$$\hat{s}_3 = 0.06108 \qquad \hat{s}_9 = -\sum_{i=1}^{8} \hat{s}_i = 0.08368$$

$$\hat{s}_4 = 0.06004 \qquad \hat{a}_1 = 0.06216$$

$$\hat{s}_5 = 0.20679 \qquad \hat{a}_2 = -0.00693$$

$$\hat{a}_3 = -(0.06216 - 0.00693) = -0.05523$$

The sums of squares for sires and ages of dam are computed by the direct procedure, that is, from $\hat{\mathbf{B}}'\mathbf{Z}^{-1}\hat{\mathbf{B}}$. The inverse of the segment of the inverse in Table 3.63 corresponding by row and column to the sire effects is given in Table 3.64. When this inverse is pre- and postmultiplied by the estimates of the linear functions for sire effects, that is, $\hat{\mathbf{B}}'_S\mathbf{Z}_S^{-1}\hat{\mathbf{B}}_S$, the sum of squares for sires adjusted for unequal frequencies with respect to age of dam is found to be 1.330171. The sum of squares for ages of dam is computed as follows:

$$\text{SS}(A) = (0.06216 \quad -0.00693)\begin{bmatrix} 0.052602 & -0.032694 \\ -0.032694 & 0.042413 \end{bmatrix}^{-1}\begin{bmatrix} 0.06216 \\ -0.00693 \end{bmatrix}$$

$$= (0.06216 \quad -0.00693)\begin{bmatrix} 36.4976 & 28.1345 \\ 28.1345 & 45.2655 \end{bmatrix}\begin{bmatrix} 0.06216 \\ -0.00693 \end{bmatrix}$$

$$= 0.118942$$

**TABLE 3.64** INVERSE OF $\mathbf{Z}_S$

| Row | $\hat{s}_1$ | $\hat{s}_2$ | $\hat{s}_3$ | $\hat{s}_4$ | $\hat{s}_5$ | $\hat{s}_6$ | $\hat{s}_7$ | $\hat{s}_8$ |
|---|---|---|---|---|---|---|---|---|
| $\hat{s}_1$ | 15.8542 | 8.0208 | 7.8750 | 7.8542 | 7.9375 | 8.0208 | 7.7917 | 7.7083 |
| $\hat{s}_2$ | | 15.7461 | 7.9291 | 8.0208 | 7.9375 | 7.8542 | 7.9043 | 8.0957 |
| $\hat{s}_3$ | | | 12.7230 | 7.8750 | 7.8750 | 7.8750 | 7.7770 | 7.7230 |
| $\hat{s}_4$ | | | | 15.8542 | 7.9375 | 8.0208 | 7.7917 | 7.7083 |
| $\hat{s}_5$ | | *Symmetric* | | | 14.9375 | 7.9375 | 7.8750 | 7.8750 |
| $\hat{s}_6$ | | | | | | 13.8542 | 7.9583 | 8.0417 |
| $\hat{s}_7$ | | | | | | | 15.6396 | 7.6104 |
| $\hat{s}_8$ | | | | | | | | 14.3896 |

The error or remainder sum of squares is

$$\mathbf{Y'Y} - R(\mu, s, a) = 382.899800 - 379.385803$$

$$= 3.513997$$

The reduction in sum of squares due to fitting all constants, $R(\mu, s, a)$, is obtained from

$$\hat{\mathbf{B}}'\mathbf{X}'_r\mathbf{Y} = (2.43501 \quad 0.07629 \quad \cdots \quad -0.00693) \begin{bmatrix} 156.74 \\ 0.01 \\ \vdots \\ -48.73 \end{bmatrix}$$

$$= 379.385803$$

in the usual manner. The ANOVA is set up in Table 3.65.

Harvey (1970) has shown that when one uses the summation restrictions

$$k_1 = \frac{1}{p-1}\left(\sum_i z_s^{ii} - \frac{1}{p}\sum_i \sum_{i'} z_s^{ii'}\right) \tag{3.116}$$

where the $z_s^{ii'}$ are elements of $\mathbf{Z}_s^{-1}$. Therefore, for this example,

$$k_1 = \tfrac{1}{8}[15.8542 + 15.7461 + \cdots + 14.3896$$

$$- \tfrac{1}{9}(15.8542 + 8.0208 + 7.8750 + \cdots + 14.3896)]$$

$$= \tfrac{1}{8}(56.7024)$$

$$= 7.0878$$

The estimate of the variance component for sires is

$$\hat{\sigma}_s^2 = \frac{0.16627 - 0.06507}{7.0878}$$

$$= 0.0143$$

Usually, in an analysis such as this, one would delete the effects of ages of dam and rerun the analysis since these effects are not significant at the 0.2 level of probability. However, since the sample of data was chosen simply to illustrate computing procedures for the two-way classification mixed model without interaction we shall not follow this procedure here.

**Indirect Analysis** When the random set of effects has a large number of degrees of freedom (say, several hundred) one must use indirect procedures to obtain the adjusted sum of squares for the random set of effects and the coefficient of the variance com-

**TABLE 3.65** ANALYSIS OF VARIANCE IN AVERAGE DAILY GAINS OF STEERS

| Source | df | SS | MS | E(MS) |
|---|---|---|---|---|
| Sires | 8 | 1.330171 | 0.16627* | $\sigma_e^2 + k_1\sigma_s^2$ |
| Age of dam | 2 | 0.118942 | 0.05947 | $\sigma_e^2 + k\sigma_a^2$ |
| Remainder | 54 | 3.513997 | 0.06507 | $\sigma_e^2$ |

ponent in the expected mean squares. The key computational device used with this procedure is the sweeping out, by an "absorption" technique, of the equations for random classes. To describe this procedure let the reparametrized or "working" model [Equation (3.99)] be partitioned as follows:

$$\mathbf{Y} = \mathbf{W}\mathbf{B}_1 + \mathbf{X}_r\mathbf{B}_2 + \mathbf{e} \qquad (3.117)$$

where $\mathbf{B}_1$ contains the $(\hat{\mu} + s_i)$ effects and $\mathbf{B}_2$ contains the linear functions to be estimated for ages of dam. With this partitioning it can be seen that $\mathbf{W'W}$ is a diagonal matrix of the $n_{i.}$, the numbers of progeny for the different sires. If $\mathbf{D} = \mathbf{W'W}$, $\mathbf{N} = \mathbf{W'X}_r$, $\mathbf{S} = \mathbf{X}_r'\mathbf{X}_r$, $\mathbf{Y}_1 = \mathbf{W'Y}$, and $\mathbf{Y}_2 = \mathbf{X}_r'\mathbf{Y}$, then the full set of least-squares equations may be written as follows:

$$\begin{bmatrix} \mathbf{D} & \mathbf{N} \\ \mathbf{N'} & \mathbf{S} \end{bmatrix}\begin{bmatrix} \hat{\mathbf{B}}_1 \\ \hat{\mathbf{B}}_2 \end{bmatrix} = \begin{bmatrix} \mathbf{Y}_1 \\ \mathbf{Y}_2 \end{bmatrix} \qquad (3.118)$$

It is well known that

$$\hat{\mathbf{B}}_2 = (\mathbf{S} - \mathbf{N'D}^{-1}\mathbf{N})^{-1}(\mathbf{Y}_2 - \mathbf{N'D}^{-1}\mathbf{Y}_1) \qquad (3.119)$$

and the computation of $\mathbf{S} - \mathbf{N'D}^{-1}\mathbf{N}$ and $\mathbf{Y}_2 - \mathbf{N'D}^{-1}\mathbf{Y}_1$ is referred to as the "absorption process." Since $\mathbf{D}$ is diagonal the computation of $\mathbf{N'D}^{-1}\mathbf{N}$ and $\mathbf{N'D}^{-1}\mathbf{Y}_1$ reduces to the computation of "between" uncorrected sums of squares and cross-products of the 0, 1, and $-1$ independent variables (sometimes called "dummy" variables) and the cross-products of the independent variables with the dependent variable, respectively. If $n_{ij}$ is an element in the $\mathbf{N}$ matrix then

$$\text{Diagonals of } \mathbf{N'D}^{-1}\mathbf{N} = \sum_i \frac{n_{ij}^2}{n_{i.}}$$

$$\text{Off-diagonals of } \mathbf{N'D}^{-1}\mathbf{N} = \sum_i \frac{n_{ij}n_{ij'}}{n_{i.}} \qquad j \neq j'$$

$$\text{Elements of } \mathbf{N'D}^{-1}\mathbf{Y}_1 = \sum_i \frac{n_{ij}Y_{i..}}{n_{i.}}$$

Now let

$$\mathbf{R}^{-1} = \begin{bmatrix} \mathbf{D} & \mathbf{N} \\ \mathbf{N'} & \mathbf{S} \end{bmatrix}^{-1} = \begin{bmatrix} \mathbf{A} & \mathbf{G} \\ \mathbf{G'} & \mathbf{C} \end{bmatrix} \qquad (3.120)$$

then $\mathbf{C} = (\mathbf{S} - \mathbf{N'D}^{-1}\mathbf{N})^{-1}$. The reader should note that in Section 2.10.2 $\mathbf{C}$ was defined as the entire coefficient matrix $(\mathbf{X}_r'\mathbf{X}_r)$ for the fixed model. Since

$$R(\mathbf{B}_1, \mathbf{B}_2) - R(\mathbf{B}_1) = \hat{\mathbf{B}}_2'\mathbf{C}^{-1}\hat{\mathbf{B}}_2$$

one may compute the total reduction in sum of squares as

$$\begin{aligned} R(\mathbf{B}_1, \mathbf{B}_2) &= R(\mathbf{B}_1) + \hat{\mathbf{B}}_2'\mathbf{C}^{-1}\hat{\mathbf{B}}_2 \\ &= \mathbf{Y}_1'\mathbf{D}^{-1}\mathbf{Y}_1 + \hat{\mathbf{B}}_2'(\mathbf{S} - \mathbf{N'D}^{-1}\mathbf{N})\hat{\mathbf{B}}_2 \\ &= \mathbf{Y}_1'\mathbf{D}^{-1}\mathbf{Y}_1 + \hat{\mathbf{B}}_2'(\mathbf{Y}_2 - \mathbf{N'D}^{-1}\mathbf{Y}_1) \\ &= \mathbf{Y}_1'\mathbf{D}^{-1}\mathbf{Y}_1 + R(\mathbf{B}_2|\mathbf{B}_1) \qquad (3.121) \end{aligned}$$

and for the example problem this reduces to

$$\sum_i \frac{Y_{i..}^2}{n_{i.}} + R(a \mid \mu, s)$$

where $R(a \mid \mu, s) = R(\mu, s, a) - R(\mu, s)$, the sum of squares for ages of dam adjusted for $\mu$ and sire effects.

The sum of squares for the random set of effects (sires in our example) adjusted for the fixed effects may now be obtained as follows:

$$R(s \mid \mu, a) = R(\mu, s, a) - R(\mu, a)$$

$$= \mathbf{Y}_S' \mathbf{D}_S^{-1} \mathbf{Y}_S + R(a \mid \mu, s) - \mathbf{Y}_A' \mathbf{D}_A^{-1} \mathbf{Y}_A$$

$$= \sum_i \frac{Y_{i..}^2}{n_{i.}} + R(a \mid \mu, s) - \sum_j \frac{Y_{.j.}^2}{n_{.j}} \qquad (3.122)$$

where $\mathbf{Y}_S = \mathbf{W}'\mathbf{Y}$, $\mathbf{D}_S = \mathbf{W}'\mathbf{W}$, $\mathbf{Y}_A = \mathbf{X}'\mathbf{Y}$, and $\mathbf{D}_A = \mathbf{X}'\mathbf{X}$. The expectation of the mean square for sires is

$$\frac{1}{p-1} [\text{tr } \mathbf{D} - \text{tr } \mathbf{N}_s' \mathbf{D}_B^{-1} \mathbf{N}_s] \sigma_s^2 + \sigma_e^2 \qquad (3.123)$$

where tr refers to trace and $\mathbf{N}_s = \mathbf{W}'\mathbf{X}$, the matrix which gives the distribution of the random effects with the fixed sets of effects. In the two-way classification the coefficient of $\sigma_s^2$, $k_1$, reduces to

$$\frac{1}{p-1} \left[ n_{..} - \sum_j \left( \frac{\sum_i n_{ij}^2}{n_{.j}} \right) \right] \qquad (3.124)$$

The error sum of squares is

$$\text{Error SS} = \mathbf{Y}'\mathbf{Y} - \sum_i \frac{Y_{i..}^2}{n_{i.}} - R(a \mid \mu, s) \qquad (3.125)$$

**Example Problem**  The partitioned set of least-squares equations for the data given in Table 3.62 is as follows:

$$
\begin{bmatrix}
8 & 0 & 0 & 0 & 0 & 0 & 0 & 0 & 0 & -2 & -2 \\
0 & 8 & 0 & 0 & 0 & 0 & 0 & 0 & 0 & -6 & -4 \\
0 & 0 & 5 & 0 & 0 & 0 & 0 & 0 & 0 & -2 & -2 \\
0 & 0 & 0 & 8 & 0 & 0 & 0 & 0 & 0 & -2 & -2 \\
0 & 0 & 0 & 0 & 7 & 0 & 0 & 0 & 0 & -3 & -2 \\
0 & 0 & 0 & 0 & 0 & 6 & 0 & 0 & 0 & -4 & -2 \\
0 & 0 & 0 & 0 & 0 & 0 & 8 & 0 & 0 & -3 & -4 \\
0 & 0 & 0 & 0 & 0 & 0 & 0 & 7 & 0 & 0 & -2 \\
0 & 0 & 0 & 0 & 0 & 0 & 0 & 0 & 8 & -3 & -1 \\
\hline
-2 & -6 & -2 & -2 & -3 & -4 & -3 & 0 & -3 & 49 & 37 \\
-2 & -4 & -2 & -2 & -2 & -2 & -4 & -2 & -1 & 37 & 53
\end{bmatrix}
\begin{bmatrix}
\hat{\mu} + \hat{s}_1 \\
\hat{\mu} + \hat{s}_2 \\
\hat{\mu} + \hat{s}_3 \\
\hat{\mu} + \hat{s}_4 \\
\hat{\mu} + \hat{s}_5 \\
\hat{\mu} + \hat{s}_6 \\
\hat{\mu} + \hat{s}_7 \\
\hat{\mu} + \hat{s}_8 \\
\hat{\mu} + \hat{s}_9 \\
\hline
\hat{a}_1 \\
\hat{a}_2
\end{bmatrix}
=
\begin{bmatrix}
19.98 \\
18.09 \\
12.37 \\
19.85 \\
18.32 \\
14.56 \\
18.81 \\
14.79 \\
19.97 \\
\hline
-58.50 \\
-48.73
\end{bmatrix}
$$

Note that

$$\mathbf{S} = \begin{bmatrix} 49 & 37 \\ 37 & 53 \end{bmatrix}$$

and

$$\mathbf{Y}_2 = \begin{bmatrix} -58.50 \\ -48.73 \end{bmatrix}$$

The elements in these two matrices are adjusted by the absorption process as follows:

$$49 - \frac{(-2)^2}{8} - \frac{(-6)^2}{8} - \cdots - \frac{(-3)^2}{8} = 36.4976$$

$$37 - \frac{(-2)(-2)}{8} - \frac{(-6)(-4)}{8} - \cdots - \frac{(-3)(-1)}{8} = 28.1345$$

$$53 - \frac{(-2)^2}{8} - \frac{(-4)^2}{8} - \cdots - \frac{(-1)^2}{8} = 45.2655$$

$$-58.50 - \frac{(-2)(19.98)}{8} - \frac{(-6)(18.09)}{8} - \cdots - \frac{(-3)(19.97)}{8} = 2.0736$$

$$-48.73 - \frac{(-2)(19.98)}{8} - \frac{(-4)(18.09)}{8} - \cdots - \frac{(-1)(19.97)}{8} = 1.4351$$

Therefore, the equations for the age of dam effects after absorption of the $\mu + s_i$ equations are

$$(\mathbf{S} - \mathbf{N'D}^{-1}\mathbf{N})\hat{a}_j = (\mathbf{Y}_2 - \mathbf{N'D}^{-1}\mathbf{Y}_1)$$

and for these data the equations are as follows:

$$\begin{bmatrix} 36.4976 & 28.1345 \\ 28.1345 & 45.2655 \end{bmatrix}\begin{bmatrix} \hat{a}_1 \\ \hat{a}_2 \end{bmatrix} = \begin{bmatrix} 2.0736 \\ 1.4351 \end{bmatrix}$$

The within sire sum of squares is

$$\mathbf{Y'Y} - \sum_i \frac{Y_{i..}^2}{n_{i.}} = 382.8998 - 379.2669 = 3.6329$$

Estimates of the $a_j$ are found from the solution of the equations

$$\begin{bmatrix} \hat{a}_1 \\ \hat{a}_2 \end{bmatrix} = \begin{bmatrix} 36.4976 & 28.1345 \\ 28.1345 & 45.2655 \end{bmatrix}^{-1}\begin{bmatrix} 2.0736 \\ 1.4351 \end{bmatrix}$$

$$= \begin{bmatrix} 0.052602 & -0.032694 \\ -0.032694 & 0.042413 \end{bmatrix}\begin{bmatrix} 2.0736 \\ 1.4351 \end{bmatrix}$$

$$= \begin{bmatrix} 0.062156 \\ -0.006929 \end{bmatrix}$$

and $\hat{a}_3 = -\hat{a}_1 - \hat{a}_3 = 0.055227$.

The reduction in sum of squares due to fitting age-of-dam effects on a within-sire basis is

$$R(a \mid \mu, s) = (0.062156 \quad -0.006929)\begin{bmatrix} 2.0736 \\ 1.4351 \end{bmatrix}$$

$$= 0.118943$$

and the sum of squares for error or remainder is

$$\mathbf{Y'Y} - \sum_i \frac{Y_{i..}^2}{n_{i.}} - R(a \mid \mu, s) = 3.6329 - 0.11894$$

$$= 3.5140$$

At this point the reader should note that the estimates obtained for the $a_j$, the sum of squares for ages of dam and the remainder sum of squares, are identical with the results obtained using the direct procedure.

All that remains to complete the analysis by the indirect procedure is to compute the sum of squares for sires and the coefficient of $\sigma_s^2(k_1)$ in $E(MS)$ for sires. From Equation (3.122)

$$\text{SS Sires} = R(s \mid \mu, a) = \sum_i \frac{Y_{i..}^2}{n_{i.}} + R(a \mid \mu, s) - \sum_j \frac{Y_{.j.}^2}{n_{.j}}$$

$$= 379.2669 + 0.1189 - \frac{(29.49)^2}{12} - \frac{(39.26)^2}{16} - \frac{(87.99)^2}{37}$$

$$= 379.2669 + 0.1189 - 378.0556$$

$$= 1.3302$$

From Equation (3.124)

$$k_1 = \frac{1}{p-1}\left[n_{..} - \sum_j\left(\frac{\Sigma_i\, n_{ij}^2}{n_{.j}}\right)\right]$$

$$= \frac{1}{8}\left(65 - \frac{2^2 + 1^2 + \cdots + 1^2}{12} - \frac{2^2 + 2^2 + \cdots + 3^2}{16} - \frac{4^2 + 6^2 + \cdots + 4^2)}{37}\right)$$

$$= \tfrac{1}{8}(65 - 2.000 - 2.000 - 4.2973)$$

$$= \tfrac{1}{8}(56.7027)$$

$$= 7.0878$$

The ANOVA is set up in Table 3.66, and, of course, the same results are obtained as given for the direct procedure, that is, $\sigma_s^2 = 0.0143$ and the estimates of the fixed effects are the same, as are all tests of significance.

TABLE 3.66   ANOVA IN AVERAGE DAILY GAINS OF STEERS
FROM INDIRECT ANALYSIS

| Squares | df | SS | MS | E(MS) |
|---|---|---|---|---|
| Sires | 8 | 1.3302 | 0.1663* | $\sigma_e^2 + 7.0878\sigma_s^2$ |
| A dam | 2 | 0.1189 | 0.0595 | $\sigma_e^2 + k\sigma_a^2$ |
| Remainder | 54 | 3.5140 | 0.0651 | $\sigma_e^2$ |

* $P \le 0.05$.

**Least-Squares Means for Fixed Discrete Effects** In a mixed model one can estimate the overall mean as was done under the direct procedure above. However, several authors have indicated that if the variance component for the random set of effects is small relative to the error variance component then it is more accurate not to adjust for unequal frequencies with respect to the random classes. In this latter case, as in our example problem, a more accurate estimate of $\mu$ is obtained as follows:

$$\tilde{\mu} = \tfrac{1}{65}[156.74 - (12)(0.062156) - (16)(-0.006929) - (37)(-0.055227)]$$

$$= \tfrac{1}{65}(156.74 + 1.4084)$$

$$= 2.4331$$

Least-squares or adjusted means for age-of-dam classes are now computed as follows:

$$\tilde{\mu} + \hat{a}_1 = 2.4331 + 0.0622 = 2.495$$

$$\tilde{\mu} + \hat{a}_2 = 2.4331 - 0.0069 = 2.426$$

$$\tilde{\mu} + \hat{a}_3 = 2.4331 - 0.0552 = 2.378$$

The calculation of the standard errors of these least-squares means must take into account the variance of $\tilde{\mu}$, the variance of the $\hat{a}_j$, and the covariance between $\hat{\mu}$ and $\hat{a}_j$, that is,

$$V(\tilde{\mu} + \hat{a}_j) = V(\tilde{\mu}) + V(\hat{a}_j) + 2\,\text{Cov}(\tilde{\mu}, \hat{a}_j)$$

The $V(\hat{a}_j) = \mathbf{C}\hat{\sigma}_e^2$, where $\mathbf{C} = (\mathbf{S} - \mathbf{N'D^{-1}N})^{-1}$ and for our example

$$\mathbf{C} = \begin{bmatrix} 0.052602 & -0.032694 \\ -0.032694 & 0.042413 \end{bmatrix}$$

and $\hat{\sigma}_e^2 = 0.0651$. However, $V(\tilde{\mu})$ and $\text{Cov}(\tilde{\mu}, \hat{a}_j)$ are not available from the analyses described. If one fits the reduced model

$$Y_{jk} = \mu + a_j + e_{jk} \tag{3.126}$$

by least squares, approximations of $V(\tilde{\mu})$ and $\text{Cov}(\tilde{\mu}, \hat{a}_j)$ can be obtained. In this case, the equations are

$$\begin{bmatrix} 65 & -25 & -21 \\ -25 & 49 & 37 \\ -21 & 37 & 53 \end{bmatrix} \begin{bmatrix} \hat{\mu} \\ \hat{a}_1 \\ \hat{a}_2 \end{bmatrix} = \begin{bmatrix} 156.74 \\ -58.50 \\ -48.73 \end{bmatrix}$$

and

$$\begin{bmatrix} \hat{\mu} \\ \hat{a}_1 \\ \hat{a}_2 \end{bmatrix} = \begin{bmatrix} 65 & -25 & -21 \\ -25 & 49 & 37 \\ -21 & 37 & 53 \end{bmatrix}^{-1} \begin{bmatrix} 156.74 \\ -58.50 \\ -48.73 \end{bmatrix}$$

$$= \begin{bmatrix} 0.019207 & 0.008571 & 0.001627 \\ 0.008571 & 0.046984 & -0.029404 \\ 0.001627 & -0.029404 & 0.040040 \end{bmatrix} \begin{bmatrix} 156.74 \\ -58.50 \\ -48.73 \end{bmatrix}$$

$$= \begin{bmatrix} 2.4298 \\ 0.0277 \\ 0.0240 \end{bmatrix}$$

$V(\tilde{\mu})$ is now estimated as $R^{\mu\mu}(\hat{\sigma}_e^2 + k\hat{\sigma}_s^2)$, where $R^{\mu\mu} = 0.019207$, $\hat{\sigma}_e^2 = 0.0651$, $\hat{\sigma}_s^2 = 0.0143$, and

$$k = \frac{\Sigma_i\, n_{i.}^2}{n_{..}} = \frac{8^2 + 8^2 + \cdots + 8^2}{65} = 7.3692$$

Hence, our estimate of $V(\tilde{\mu})$ is

$$(0.019207)[0.0651 + 7.3692)(0.0143)] = (0.019207)(0.170480)$$

$$= 0.003274$$

The covariances of $\tilde{\mu}$ and $\hat{a}_j$ are estimated from the formula

$$\text{Cov}(\tilde{\mu}, \hat{a}_j) \simeq R^{\mu a_j}\hat{\sigma}_e^2$$

and, for our example,

$$\text{Cov}(\tilde{\mu}, \hat{a}_1) \simeq (0.008571)(0.0651)$$

$$\simeq 0.000558$$

$$\text{Cov}(\tilde{\mu}, \hat{a}_2) \simeq (0.001627)(0.0651)$$

$$\simeq 0.000106$$

$$\text{Cov}(\tilde{\mu}, \hat{a}_3) \simeq (-0.010198)(0.0651)$$

$$\simeq -0.000664$$

Therefore, the approximate variance of a least-squares mean for an age-of-dam class is

$$V(\tilde{\mu} + \hat{a}_j) \simeq R^{\mu\mu}(\hat{\sigma}_e^2 + k\hat{\sigma}_s^2) + C_{a_j a_j}\hat{\sigma}_e^2 + 2R^{\mu a_j}\hat{\sigma}_e^2$$

$$\simeq (R^{\mu\mu} + C_{a_j a_j} + 2R^{\mu a_j})\hat{\sigma}_e^2 + R^{\mu\mu}k\hat{\sigma}_s^2 \qquad (3.127)$$

and  $V(\tilde{\mu} + \hat{a}_1) \simeq [0.019207 + 0.052602 + (2)(0.008571)](0.0651)$

$$+ (0.019207)(7.3692)(0.0143)$$

$$\simeq (0.088951)(0.0651) + (0.141540)(0.0143)$$

$$\simeq 0.005791 + 0.002024$$

$$\simeq 0.007815$$

$$\text{SE}(\tilde{\mu} + \hat{a}_1) \simeq \sqrt{0.007815} = 0.088$$

$$V(\tilde{\mu} + \hat{a}_2) \simeq [0.019207 + 0.042413 + (2)(0.001627)](0.0651)$$

$$+ (0.019207)(7.3692)(0.0143)$$

$$\simeq (0.064960)(0.0651) + 0.002024$$

$$\simeq 0.006253$$

$$\text{SE}(\tilde{\mu} + \hat{a}_2) \simeq \sqrt{0.006253} = 0.079$$

$$V(\tilde{\mu} + \hat{a}_3) \simeq [0.019207 + 0.029627 + (2)(-0.010198)](0.0651)$$

$$+ 0.002024$$

$$\simeq (0.028438)(0.0651) + 0.002024$$

$$\simeq 0.003875$$

$$\mathrm{SE}(\tilde{\mu} + \hat{a}_3) \simeq \sqrt{0.003875} = 0.062$$

In summary, the least-squares means for age-of-dam classes and standard errors are presented in Table 3.67. It should be pointed out that in mixed models the standard errors of adjusted means for fixed classes or subclasses must take into consideration the extent to which these means are expected to vary due to the sampling of all random variables. On the other hand, since the fixed effects are estimated on an intrarandom class basis, the accuracy of comparisons among these estimates is determined only by the random errors. Hence, in the case of mixed model analyses, one should not use standard errors of least-squares means to compare differences among the means as can be done when all sets of effects are fixed. For example, one might compare the mean for 3-year-old dams with the mean for dams that were 5 years and older $(2.495 - 2.378)$ and incorrectly conclude from the standard errors that this difference would be significant if it exceeds $0.088 + 0.062 = 0.150$. Actually, the difference required for significance at the 0.05 level of probability is the square root of

$$t_{0.05}^2(C_{a_1a_1} + C_{a_3a_3} - 2C_{a_1a_3})\hat{\sigma}_e^2 = (2.005)^2[0.052602 + 0.029627$$

$$- (2)(-0.019908)](0.0651)$$

$$= (4.020)(0.122045)(0.0651)$$

$$= 0.031940$$

which is 0.179, a larger value in this case, because of the large negative covariance between $\hat{a}_1$ and $\hat{a}_3$ $(-0.019908)$.

## 3.21.2 Two-Way Classification: With Interaction

When interaction exists in the two-way classification mixed model there will usually be a large number of degrees of freedom for the random set of effects and for the interaction. Therefore, only the indirect method of analysis is feasible in most cases. Hence, this is the analytical procedure that will now be illustrated with the data given in Table 3.62.

The model for this analysis is

$$Y_{ijk} = \mu + s_i + a_j + sa_{ij} + e_{ijk}$$

$$i = 1, 2, \ldots, 9$$

TABLE 3.67  LEAST-SQUARES MEANS
FOR AGE-OF-DAM CLASSES

| Age of dam | LS mean | SE |
|------------|---------|-------|
| 3 | 2.495 | 0.088 |
| 4 | 2.426 | 0.079 |
| 5-up | 2.378 | 0.062 |

$$j = 1, 2, 3$$

$$k = 1, 2, \ldots, n_{ij} \tag{3.128}$$

where $s_i$ are the sire effects (random) and $a_j$ are the fixed age-of-dam effects. The ANOVA desired under this model is presented in Table 3.68.

To complete this ANOVA by the indirect procedure and obtain least-squares means for the fixed classes, analyses must be completed under three models. In the first analysis constants are fitted only for the fixed effects to obtain an appropriate estimate for $\mu$, the remainder sum of squares $\mathbf{Y'Y} - R(\mu, a)$, and the number of $\sigma_s^2$ and $\sigma_{sa}^2$ components in $R(\mu, a)$. The coefficient matrix for $\mu$ and the $a_j$ is given below along with the matrix ($\mathbf{N}_s$) which associates the sire effects with the fixed effects.

|  | $\hat{\mu}$ | $\hat{a}_1$ | $\hat{a}_2$ | $\hat{s}_1$ | $\hat{s}_2$ | $\hat{s}_3$ | $\hat{s}_4$ | $\hat{s}_5$ | $\hat{s}_6$ | $\hat{s}_7$ | $\hat{s}_8$ | $\hat{s}_9$ |
|---|---|---|---|---|---|---|---|---|---|---|---|---|
| $\mu$: | 65 | −25 | −21 | 8 | 8 | 5 | 8 | 7 | 6 | 8 | 7 | 8 |
| $a_1$: | −25 | 49 | 37 | −2 | −6 | −2 | −2 | −3 | −4 | −3 | 0 | −3 |
| $a_2$: | −21 | 37 | 53 | −2 | −4 | −2 | −2 | −2 | −2 | −4 | −2 | −1 |
|  |  | **R** |  |  |  |  | **N**$_s$ |  |  |  |  |  |

The right-hand members for the reduced set of equations for $\mu$ and the $a_j$ are as follows:

$$\mu: 156.74$$

$$a_1: -58.50$$

$$a_2: -48.73$$

The inverse of the coefficient matrix is

$$\mathbf{R}^{-1} = \begin{bmatrix} 0.019207 & 0.008571 & 0.001627 \\ 0.008571 & 0.046984 & -0.029404 \\ 0.001627 & -0.029404 & 0.040040 \end{bmatrix}$$

and the solution of the least-squares equations is

$$\begin{bmatrix} \hat{\mu} \\ \hat{a}_1 \\ \hat{a}_2 \end{bmatrix} = \begin{bmatrix} 2.429786 \\ 0.027714 \\ 0.023964 \end{bmatrix}$$

The remainder sum of squares is

$$E_1 = \mathbf{Y'Y} - R(\mu, a)$$

$$= 382.8998 - 378.0556$$

$$= 4.8442$$

**TABLE 3.68**  ANOVA WHEN INTERACTION IS CONSIDERED

| Source | df | SS | E(MS) |
|---|---|---|---|
| Sires (S) | 8 | $R(\mu, s, a) - R(\mu, a)$ | $\sigma_w^2 + k_2\sigma_{sa}^2 + k_3\sigma_s^2$ |
| A dam (A) | 2 | $R(\mu, s, a) - R(\mu, s)$ | $\sigma_w^2 + k\sigma_{sa}^2 + k\sigma_a^2$ |
| SA | 14 | $R(\mu, s, a, sa) - R(\mu, s, a)$ | $\sigma_w^2 + k_1\sigma_{sa}^2$ |
| Within | 40 | $\mathbf{Y'Y} - R(\mu, s, a, sa)$ | $\sigma_w^2$ |

The number of $\sigma_s^2$ and $\sigma_{sa}^2$ components in $R(\mu, a)$, in general, is computed from

$$\text{tr}(\mathbf{N}_s'\mathbf{R}^{-1}\mathbf{N}_s) = \text{tr}(\mathbf{R}^{-1}\mathbf{N}_s\mathbf{N}_s')$$

$$= \sum_i \sum_j R^{ij}N_{ij} \qquad (3.129)$$

where $N_{ij}$ represents elements in the $\mathbf{N}_s\mathbf{N}_s'$ matrix. Now

$$\mathbf{N}_s\mathbf{N}_s' = \begin{bmatrix} 8 & 8 & 5 & 8 & 7 & 6 & 8 & 7 & 8 \\ -2 & -6 & -2 & -2 & -3 & -4 & -3 & 0 & -3 \\ -2 & -4 & -2 & -2 & -2 & -2 & -4 & -2 & -1 \end{bmatrix} \begin{bmatrix} 8 & -2 & -2 \\ 8 & -6 & -4 \\ 5 & -2 & -2 \\ 8 & -2 & -2 \\ 7 & -3 & -2 \\ 6 & -4 & -2 \\ 8 & -3 & -4 \\ 7 & 0 & -2 \\ 8 & -3 & -1 \end{bmatrix}$$

$$= \begin{bmatrix} 479 & -183 & -154 \\ -183 & 91 & 65 \\ -154 & 65 & 57 \end{bmatrix}$$

Therefore,

$$\text{tr}(\mathbf{R}^{-1}\mathbf{N}_s\mathbf{N}_s') = 8.2973$$

and, as pointed out in Section 3.21.1, this equals $\sum_j [(\sum_i n_{ij}^2)/n_{.j}]$ for the two-way classification. However, if other fixed effects exist in the model one must use the general procedure illustrated here to obtain the number of $\sigma_s^2$ and $\sigma_{sa}^2$ in the total reduction in the sum of squares due to the fixed effects.

The second analysis is completed under the reduced model

$$Y_{ijk} = \mu + s_i + a_j + e_{ijk} \qquad (3.130)$$

using the indirect procedure, that is, the equations for $\mu + s_i$ are absorbed into the equations for $a_j$ as was done in Section 3.21.1. From that analysis we found that the remainder sum of squares, which will be referred to as $E_2$, was

$$E_2 = \mathbf{Y}'\mathbf{Y} - R(\mu, s, a)$$

$$= 3.5140$$

In addition, from this analysis we need to compute the number of $\sigma_{sa}^2$ components in $R(\mu, s, a)$. To do this, using the procedure described above, we would need the inverse of the coefficient matrix ($\mathbf{R}$) for the $\mu$, $s_i$, and $a_j$ effects and the matrix ($\mathbf{N}_s$) which associates these effects with the sire $\times$ age-of-dam subclass effects. However, with the $s_i$ being random, the $\mathbf{R}$ matrix will often be very large. Hence, a "shortcut" computational procedure was developed by Harvey (1970) for use in his LSML76 computer program to handle this problem. Let us partition the coefficient matrix into four parts, as was done in Section 3.21.1, to show the absorption procedure, and let us partition the $\mathbf{N}_s$ matrix into two parts as follows:

|          | $\mu + s_i$ | $a_j$ | $sa_{ij}$ |
|----------|-------------|-------|-----------|
| $\mu + s_i$ | **D**    | **N** | **P**     |
| $a_j$    | **N'**      | **S** | **Q**     |
| $sa_{ij}$ | **P'**     | **Q'** | —        |

$$\text{(3.131)}$$

It was shown in Harvey's 1970 paper that

$$\text{tr}(\mathbf{R}^{-1}\mathbf{N}_s\mathbf{N}_s') = \text{tr}(\mathbf{D}^{-1}\mathbf{PP}') + \text{tr}(\mathbf{CF}) \tag{3.132}$$

where

$$\mathbf{F} = \mathbf{QQ}' + \mathbf{N}'\mathbf{D}^{-1}(\mathbf{PP}'\mathbf{D}^{-1}\mathbf{N} - 2\mathbf{PQ}')$$

$$\mathbf{C} = (\mathbf{S} - \mathbf{N}'\mathbf{D}^{-1}\mathbf{N})^{-1} \tag{3.133}$$

It is easily shown that

$$\text{tr}(\mathbf{D}^{-1}\mathbf{PP}') = \sum_i \frac{\sum_j n_{ij}^2}{n_{i.}} \tag{3.134}$$

and, for our present example,

$$\sum_i \frac{\sum_j n_{ij}^2}{n_{i.}} = \frac{2^2 + 2^2 + 4^2}{8} + \frac{2^2 + 6^2}{8} + \cdots + \frac{1^2 + 3^2 + 4^2}{8}$$

$$= 29.2476$$

From the indirect analysis completed in Section 3.21.1

$$\mathbf{C} = (\mathbf{S} - \mathbf{N}'\mathbf{D}^{-1}\mathbf{N})^{-1} = \begin{bmatrix} 0.05260182 & -0.03269439 \\ -0.03269439 & 0.04241292 \end{bmatrix}$$

and the matrix $\mathbf{F}$, as computed by LSML76 (Harvey, 1977b), is as follows:

$$\mathbf{F} = \begin{bmatrix} 97.290556 & 85.457718 \\ 85.457718 & 123.575210 \end{bmatrix}$$

and $\text{tr}(\mathbf{CF}) = 4.7709$. Therefore, the number of $\sigma_{sa}^2$ components in $R(\mu, s, a)$ is

$$29.2476 + 4.7709 = 34.0185$$

The sum of squares obtained for ages of dam in Section 3.21.1 where the interaction was ignored is the appropriate sum of squares since it equals $R(\mu, s, a) - R(\mu, s)$ and this was found to be 0.1189.

The third analysis needed is under the full model and the equations for $\mu + s_i + a_j + sa_{ij}$ (the sire by age-of-dam subclass effects) are absorbed. The remainder from this analysis is

$$E_3 = \mathbf{Y}'\mathbf{Y} - R(\mu, s, a, sa)$$

$$= \mathbf{Y}'\mathbf{Y} - \sum_i \sum_j \frac{Y_{ij.}^2}{n_{ij}}$$

$$= 382.8998 - 380.6325$$

$$= 2.2673$$

The coefficients of the variance components in the expected mean squares are now computed as follows:

$$k_1 = \tfrac{1}{14}(65 - 34.0185)$$

$$= \tfrac{1}{14}(30.9815)$$

$$= 2.2130$$

$$k_2 = \tfrac{1}{8}(34.0185 - 8.2973)$$

$$= \tfrac{1}{8}(25.7212)$$

$$= 3.2152$$

$$k_3 = \tfrac{1}{8}(65 - 8.2973)$$

$$= \tfrac{1}{8}(56.7027)$$

$$= 7.0878$$

The sums of squares for the random sources of variation are obtained as follows:

$$\text{SS Sires} = E_1 - E_2$$

$$= 4.8442 - 3.5140$$

$$= 1.3302 \tag{3.135}$$

$$\text{Sires by age-of-dam SS} = E_2 - E_3$$

$$= 3.5140 - 2.2673$$

$$= 1.2467 \tag{3.136}$$

The ANOVA is now set up in Table 3.69.

The variance component estimates are computed as follows:

$$\hat{\sigma}_w^2 = 0.0567$$

$$\hat{\sigma}_{sa}^2 = \frac{0.0891 - 0.0567}{2.213}$$

$$= 0.0146$$

$$\hat{\sigma}_s^2 = \frac{0.1663 - (3.215)(0.0146) - 0.0567}{7.088}$$

$$= 0.0088$$

**TABLE 3.69**   ANALYSIS OF VARIANCE IN AVERAGE DAILY GAIN OF STEERS

| Source | df | SS | MS | E(MS) |
|--------|----|------|------|-------|
| Sires ($S$) | 8 | 1.3302 | 0.1663 | $\sigma_w^2 + 3.215\sigma_{sa}^2 + 7.088\sigma_s^2$ |
| Age of dam ($A$) | 2 | 0.1189 | 0.0595 | $\sigma_w^2 + k\sigma_{sa}^2 + k\sigma_a^2$ |
| $SA$ | 14 | 1.2467 | 0.0891 | $\sigma_w^2 + 2.213\sigma_{sa}^2$ |
| Within | 40 | 2.2673 | 0.0567 | $\sigma_w^2$ |

Least-squares means for age-of-dam classes are computed by using the estimate of $\mu$ when the overall mean is adjusted for age-of-dam effects by using the estimates of the age-of-dam effects $(a_j)$ from the second analysis, where adjustment was made for unequal frequencies with respect to sire effects. The least-squares means are therefore the same as given in Section 3.21.1, where interaction was ignored. However, consideration must now be given to the variance due to the interaction in calculating the standard errors. For example,

$$V(\tilde{\mu}) \simeq R^{\mu\mu}(\hat{\sigma}_w^2 + k_1\hat{\sigma}_s^2 + k_2\hat{\sigma}_{sa}^2) \tag{3.137}$$

where $R^{\mu\mu}$ is the inverse diagonal element for $\mu$ from the fixed model analysis and

$$k_1 = \frac{\sum_i n_{i.}^2}{n_{..}}$$

$$k_2 = \frac{\sum_i \sum_j n_{ij}^2}{n_{..}}$$

Also,
$$V(\tilde{\mu} + \hat{a}_j) \simeq (R^{\mu\mu} + C_{a_j a_j} + 2R^{\mu a_j})(\hat{\sigma}_w^2 + k_j\hat{\sigma}_{sa}^2) + R^{\mu\mu}k_1\hat{\sigma}_s^2 \tag{3.138}$$

Here **C** is the inverse of the coefficient matrix after absorbing the $\mu + a_i$ equations in the second analysis and

$$k_j = \frac{\sum_i n_{ij}^2}{n_{.j.}} \tag{3.139}$$

For the example problem the standard errors of the least-squares means are therefore computed as follows:

$$V(\tilde{\mu} + \hat{a}_1) \simeq [0.019207 + 0.052602 + (2)(0.008571)]$$

$$\times [0.0567 + (2.0000)(0.0146)] + (0.019207)(7.3692)(0.0088)$$

$$\simeq (0.088951)(0.085900) + 0.001246$$

$$\simeq 0.007641 + 0.001246$$

$$\simeq 0.008887$$

$$\mathrm{SE}(\tilde{\mu} + \hat{a}_1) \simeq 0.094$$

$$V(\tilde{\mu} + \hat{a}_2) \simeq [0.019207 + 0.042413 + (2)(0.001627)]$$

$$\times [0.0567 + (2.0000)(0.0146)] + (0.019207)(7.3692)(0.0088)$$

$$\simeq 0.005573 + 0.001246$$

$$\simeq 0.006819$$

$$\mathrm{SE}(\tilde{\mu} + \hat{a}_2) \simeq 0.083$$

$$V(\tilde{\mu} + \hat{a}_3) \simeq [0.019207 + 0.029626 + (2)(-0.010198)]$$

$$\times [0.0567 + (4.2973)(0.0146)] + 0.001246$$

$$\simeq 0.003397 + 0.001246$$

$$\simeq 0.004643$$

$$\mathrm{SE}(\tilde{\mu} + \hat{a}_3) \simeq 0.068$$

**TABLE 3.70**  LEAST-SQUARES MEANS AND STANDARD
ERRORS FOR AGE-OF-DAM CLASSES

| Age of dam | LS mean | Standard errors | |
|---|---|---|---|
| | | Interaction ignored | Interaction considered |
| 3 | 2.495 | 0.088 | 0.094 |
| 4 | 2.426 | 0.079 | 0.083 |
| 5-up | 2.378 | 0.062 | 0.068 |

For comparison, the least-squares means for age-of-dam classes are given in Table 3.70 with the approximate standard errors when interaction is ignored and when it is considered as a source of variation.

It should be pointed out that in practice one should use step-down procedures when unequal subclass frequencies exist to estimate more accurately those effects and variance components which are really important. In this sample of data it will be noted that age-of-dam effects do not seem to exist ($P > 0.5$). However, the test of significance for the interaction of sires with age of dam suggests the possibility that $\sigma_{sa}^2 > 0$ ($P > 0.13$). Of course, if the interaction is to be left in the model the main effects must also be left in the model even though such effects, on the average, are not significantly different from zero.

Another point that should be noted from the analysis in this section is the fact that there are missing subclasses. This presents no difficulties in the computational procedures when these indirect procedures are followed and the interaction effects are random (see Harvey, 1977a). It is important to recognize that to obtain unbiased estimates of the variance components one must assume that the loss of observations, including the loss of entire subclasses, occurred at random. If the frequencies are associated with the subclass means the methods described here will not yield unbiased estimates. Computational procedures described in this section and Section 3.21.1 are easily extended to analyses under more complex mixed models. The interested reader is referred to publications by Harvey (1960, 1970, 1977a, 1977b, 1981, and 1982b) for analyses under more complex mixed models.

## EXERCISES

**3.1.** In a study of the effect of the chemical Daminozide on Golden Delicious apples, Joann McLaughlin of the Department of Plant and Soil Sciences at the University of Massachusetts established 9 two-tree blocks in which one tree served as a control and the other was treated with Daminozide. An average weight per apple was calculated from 30 apples harvested from each tree. Carry out an analysis of variance for this two-way classification.

| Treatment | Block | | | | | | | | |
|---|---|---|---|---|---|---|---|---|---|
| | 1 | 2 | 3 | 4 | 5 | 6 | 7 | 8 | 9 |
| Control | 0.115 | 0.104 | 0.093 | 0.087 | 0.082 | 0.107 | 0.091 | 0.104 | 0.089 |
| Daminozide | 0.093 | 0.099 | 0.083 | 0.078 | 0.092 | 0.095 | 0.110 | 0.102 | 0.092 |

**3.2.** Since the preceding experiment was made up of blocks of two trees in each block, making up nine pairs, the data can be analyzed by means of a paired $t$ test. Carry out this test, showing that $t^2$ equals $F$ for the test of significance of the treatments.

**3.3.** A portion of the data resulting from an experiment by Mills et al. (1976) is shown below, where the observations are the fresh weights of radish plants in grams. Ammonium sulfate and potassium nitrate were two different sources of nitrogen. N+ indicates the application of nitrapyrin and N− indicates no nitrapyrin. Complete the appropriate analysis of variance for this set of data, using the $W$ mean square to test for the significance of all sources of variation. Calculate the estimates of the interaction constants and show that the sum of squares for the interaction can be completed by use of the formula $n_w \Sigma_i \Sigma_j ab_{ij}^2$, where $ab$ represents the interaction between source of nitrogen and nitrapyrin treatment, $W$ is the within-subclass sum of squares, and $n_w$ is the number of observations in a subclass.

|     | Ammonium sulfate | | | | Potassium nitrate | | | |
|-----|------|------|------|------|------|------|------|------|
| N+  | 14.3 | 15.9 | 14.8 | 20.8 | 17.6 | 24.0 | 13.5 | 17.9 |
| N−  | 9.9  | 7.7  | 13.2 | 10.5 | 12.2 | 11.4 | 10.8 | 9.6  |

**3.4.** R. A. Coler and B. Tease of the Department of Environmental Sciences at the University of Massachusetts conducted an experiment concerned with the amount of oxygen generated in river water. Samples of the water were collected in 3 tubes in the river at each of 2 stations on 10 different days. The data are milligrams of oxygen generated per liter. The tubes in the river are considered to be nested within the stations (upstream and downstream). Complete an analysis of variance of this set of data.

|       | Upstream | | | Downstream | | |
|-------|--------|--------|--------|--------|--------|--------|
| Date  | Tube 1 | Tube 2 | Tube 3 | Tube 1 | Tube 2 | Tube 3 |
| 8/26  | 1.3 | 1.5 | 2.2 | 2.3 | 1.6 | 1.5 |
| 8/28  | 0.7 | 0.3 | 1.1 | 1.0 | 1.0 | 1.1 |
| 8/29  | 0.4 | 0.3 | 0.3 | 1.3 | 0.5 | 0.7 |
| 9/5   | 0.3 | 0.3 | 0.2 | 1.4 | 0.8 | 0.9 |
| 9/6   | 0.4 | 0.8 | 0.6 | 1.6 | 1.2 | 1.5 |
| 9/7   | 1.1 | 1.3 | 1.1 | 3.0 | 2.1 | 2.2 |
| 9/9   | 1.0 | 1.0 | 1.1 | 1.9 | 1.8 | 1.1 |
| 9/11  | 1.3 | 1.9 | 1.8 | 2.8 | 2.2 | 2.4 |
| 9/12  | 1.3 | 1.1 | 1.1 | 3.1 | 3.3 | 1.9 |
| 9/16  | 0.9 | 0.8 | 0.9 | 2.5 | 1.9 | 1.2 |

**3.5.** Write out the analysis of variance tables for the following examples, including in the tables all sources of variation, the degrees of freedom, the formulas *in symbolic notation* for the sums of squares (except for the within), and the expectations of mean squares. Indicate the appropriate $F$ ratios for tests of significance where they exist.

  **(a)** Given the classifications of $A$, $B$, and $C:AB$, with four levels of $A$, three levels of $B$, five levels of $C$, and two observations in the smallest subclass. $A$ and $B$ effects are fixed and $C$ is random.

  **(b)** Given the classifications of $A$, $B:A$, and $C$ with five levels of $A$, four levels of $B$, three levels of $C$, and seven observations in the smallest subclass. $A$ and $C$ effects are fixed and $B$ is random.

(c) Given the classifications of $A$, $B$, $C:AB$, and $D:ABC$, with three levels of $A$, five levels of $B$, four levels of $C$, two levels of $D$, and one observation in the smallest subclass. $A$ and $B$ effects are fixed; $C$ and $D$ effects are random.

(d) Given the classifications of $A$, $B:A$, $C$, and $D:C$, with four levels of $A$, three levels of $B$, five levels of $C$, two levels of $D$, and six observations in the smallest subclass. $A$ and $C$ effects are fixed; $B$ and $D$ effects are random.

(e) Given the classifications of $A$, $B:A$, $C:A$, and $D$ with five levels of $A$, four levels of $B$, three levels of $C$, six levels of $D$, and four observations in the smallest subclass. $A$ and $D$ effects are fixed; $B$ and $C$ effects are random.

**3.6.** Suppose that in a test for the effectiveness of two different chemical sprays in the prevention of a fungus disease in fruit trees, 30 trees are subjected to the first spray and 30 to the second spray. Consider that 22 of the first 30 trees and 6 of the second 30 trees become affected. Coding the live trees as 0's and the infected trees as 1's (dummy variables), complete an analysis of variance to test for the significance of the difference between the two chemicals.

**3.7.** The yields in bushels of nine apple trees in 1975 and 1982 are shown below. Test for the significance of the difference between the variance among the nine trees in 1975 and the variance in 1982 by means of the two-tailed $F$ test.

|       |      |      |       | Tree  |       |      |       |      |      |
|-------|------|------|-------|-------|-------|------|-------|------|------|
| Year  | 1    | 2    | 3     | 4     | 5     | 6    | 7     | 8    | 9    |
| 1975  | 1.00 | 1.00 | 0.75  | 1.00  | 0.75  | 0.25 | 1.00  | 1.00 | 0.50 |
| 1982  | 9.25 | 7.25 | 10.00 | 12.50 | 11.00 | 8.50 | 11.00 | 9.00 | 7.25 |

**3.8.** The yields in bushels of the same eight apple trees in four different years are shown below. Test for the homogeneity of variance among the four groups by means of Bartlett's test of homogeneity.

| 1976 | 1978 | 1980 | 1982 |
|------|------|------|------|
| 2.00 | 4.50 | 4.00 | 4.50 |
| 0.50 | 1.75 | 3.50 | 5.75 |
| 1.25 | 3.00 | 4.00 | 8.00 |
| 1.25 | 4.25 | 5.75 | 8.00 |
| 1.50 | 4.25 | 5.75 | 9.25 |
| 2.00 | 4.50 | 6.25 | 7.50 |
| 2.00 | 4.50 | 6.25 | 8.25 |
| 1.50 | 4.25 | 3.25 | 7.25 |

**3.9.** A portion of the data resulting from an experiment conducted by Mills et al. (1976) is shown below where the observations are the fresh weights of radish plants in grams. Ammonium sulfate and potassium nitrate were two different sources of nitrogen, each applied at 0, 400, and 800 mg/kg of soil. Carry out a complete analysis of variance for this set of data. Now test for the assumption of additivity in these data by means of Tukey's test.

| Level of nitrogen | Ammonium sulfate | Potassium nitrate |
|:---:|:---:|:---:|
| 0 | 14.3 | 17.6 |
| | 15.9 | 24.0 |
| | 14.8 | 13.5 |
| | 20.8 | 17.9 |
| 400 | 35.8 | 47.3 |
| | 26.2 | 65.6 |
| | 43.2 | 69.8 |
| | 40.1 | 57.5 |
| 800 | 65.7 | 95.2 |
| | 61.9 | 92.7 |
| | 66.1 | 82.9 |
| | 66.8 | 91.6 |

**3.10.** Transform the individual observations shown in Exercise 3.9 to logarithms and complete an analysis of variance on the transformed data, noting any differences between the two analyses.

**3.11.** Consider an experiment comparing three different methods of teaching, conducted in four different schools. In each school classes of 30 students were used and the results recorded in terms of percent of class receiving a passing grade in the experiment. The data are shown below. Complete an analysis of variance for this set of data considering the schools to be a random effect and the teaching methods a fixed effect. Now transform the data to arcsines and complete a second analysis of variance, noting the drastic change in the size of the $F$ value for methods.

| School | Teaching method | | |
|:---:|:---:|:---:|:---:|
| | 1 | 2 | 3 |
| 1 | 100 | 82 | 97 |
| 2 | 92 | 71 | 92 |
| 3 | 88 | 69 | 85 |
| 4 | 93 | 74 | 93 |

**3.12.** Given the following numbers of gypsy moth caterpillars found on raspberry plants during a seven-day period after treatment with four different chemical sprays, carry out an analysis of variance for this set of data and then transform each count to its square root and complete a second analysis of variance, this one on the transformed data, comparing the results.

| Block | Treatment | | | |
|:---:|:---:|:---:|:---:|:---:|
| | 1 | 2 | 3 | 4 |
| 1 | 6 | 74 | 29 | 72 |
| 2 | 8 | 68 | 34 | 100 |
| 3 | 14 | 51 | 21 | 124 |
| 4 | 22 | 80 | 48 | 98 |

**3.13.** The expectations of mean squares for the analysis of variance of the data displayed in Exercise 3.4 reveal that no exact test of significance is available for the stations effect, considering the stations as fixed effects and the sites and dates as random effects. An approximate $F$ test can be completed for the stations by the methods suggested by Satterthwaite and Cochran (Section 3.17). Show that the numerator mean square would be a pooling of the station and the day $\times$ tube within-station interaction mean squares and that the denominator mean square is a pooling of the station $\times$ day interaction and the tube within-station mean squares. Show further that the degrees of freedom for the numerator mean square would be one and for the denominator mean square eight.

**3.14.** The expectations of mean squares for Exercise 3.5(e) show that there are no exact $F$ tests for $A$, $D$, and $AD$. Using the method referred to in Exercise 3.13, show what combinations of mean squares would be used in the numerator and denominators for testing these effects.

**3.15.** The weaning weights (in pounds) of 37 male calves from 6 sires in a Hereford herd are given in the table below. Assuming that the one-way classification random model is appropriate, test the significance of differences among sires with ANOVA and estimate $\sigma_s^2$ and $\sigma_e^2$. What is the estimate of the correlation among paternal half sibs? What is the standard error of the overall mean?

| | | Sire number | | | |
|---|---|---|---|---|---|
| 1 | 2 | 3 | 4 | 5 | 6 |
| 485 | 413 | 409 | 498 | 413 | 492 |
| 436 | 469 | 396 | 431 | 446 | 462 |
| 422 | 520 | 392 | 500 | 444 | 483 |
| 461 | 432 | 426 | 443 | | 499 |
| | 426 | 399 | 456 | | 467 |
| | 478 | 444 | 468 | | 429 |
| | 405 | | 429 | | 464 |
| | 482 | | 433 | | |
| | 504 | | | | |

**3.16.** Red maple trees from half-sib families were grown in a plantation at the Nursery Crops Research Lab, USDA, Delaware, Ohio, under the direction of Dr. A. M. Townsend. Trees from each family were grown in six randomly chosen plots. A small sample of the data collected on tree heights (in centimeters) at 9 years of age is given in the table below. Complete the analysis of variance for these data and obtain estimates of $\sigma_f^2$, $\sigma_{p:f}^2$, and $\sigma_e^2$. From these estimates compute the estimate of $r_1$, the correlation among the heights of trees in the same family and plot, and $r_2$, the correlation among the heights of trees in the same family from one plot to another. Calculate the overall mean $\bar{y}$ and its standard error. Interpret the results.

| | | Plots | | | | | |
|---|---|---|---|---|---|---|---|
| Family | Tree | 1 | 2 | 3 | 4 | 5 | 6 |
| 1 | 1 | 318 | 390 | 160 | 352 | 272 | 190 |
| | 2 | 346 | 370 | 245 | 244 | 293 | |
| | 3 | 212 | 231 | | 130 | 376 | |
| | 4 | | 180 | | 287 | | |
| 2 | 1 | 157 | 204 | 260 | 330 | 107 | 164 |
| | 2 | 203 | 163 | 241 | 177 | 150 | 300 |
| | 3 | 241 | 210 | | 220 | 158 | 311 |
| | 4 | | | | 280 | 260 | 345 |
| | 5 | | | | 100 | | |
| 3 | 1 | 184 | 257 | 235 | 168 | 300 | 245 |
| | 2 | 277 | 244 | 273 | 186 | 130 | 182 |
| | 3 | 263 | 186 | 203 | 220 | | 203 |
| | 4 | 234 | 349 | 320 | 287 | | 103 |
| | 5 | 257 | | 210 | | | 200 |
| 4 | 1 | 210 | 197 | 217 | 180 | 237 | 317 |
| | 2 | 170 | 110 | 173 | 170 | 344 | 177 |
| | 3 | 325 | 150 | 252 | | 296 | 214 |
| | 4 | 247 | 185 | 184 | | | 249 |
| | 5 | | | | | | 300 |

**3.17.** Suppose the following observations were missing in Table 3.7:

$$Y_{112} \ (15.9) \qquad Y_{223} \ (49.3)$$

$$Y_{221} \ (43.3) \qquad Y_{321} \ (59.8)$$

Complete a least-squares analysis of the remaining data as described in Section 3.20.

# Chapter 4

# Mean Separation Procedures

## 4.1 INTRODUCTION

Up to now we have considered many types of data and the appropriate tests of significance, usually the $F$ test. However, the $F$ test, as we have used it, frequently does not give us the amount of information we want. Usually, there is a great deal more information to be gained beyond the result of the preliminary $F$ test which tells us that there are or are not significant differences, on the average, among the treatment means. There are several different methods of extracting information beyond the original $F$ test, some of which will be discussed in this section. It is to be emphasized that the particular method or methods to be used are dependent on the type of treatments applied, or expressed differently, on the questions posed by the experimenter when selecting the various treatments.

In some cases, we can glean a great deal more information by making tests between or among particular treatment means or combinations of treatment means. To assist us in exploring this method of obtaining the maximum amount of information from our data, we shall make use of some data resulting from a study by Dr. Robyn Rufner at the University of Massachusetts on the effects of nitrogen fertilizer and nitrapyrin on the ultrastructure of mesophyll chloroplasts of radish. Six different fertilizer treatments were included in the experiment, each treatment occurring without the addition of nitrapyrin and each occurring with the addition of 10 ppm of nitrapyrin, giving a total of 12 treatment combinations. Three inorganic fertilizers were used, potassium nitrate, urea, and ammonium sulfate. Of these, potassium nitrate is a nitrate fertilizer and urea and ammonium sulfate are ammoniacal fertilizers. The other three fertilizers, bovung, milorganite, and mergreen are organic fertilizers. Bovung is a dehydrated cow manure and milorganite and mergreen are sludges from two different sources. A summary of the data to be used is presented in Table 4.1.

**TABLE 4.1** TOTAL AREA DATA FOR RADISH MESOPHYLL CHLOROPLASTS (18 OBSERVATIONS PER TREATMENT)

| Treatment | Fertilizer | Nitrapyrin (ppm) | Total area ($\mu m^2$) | Mean |
|-----------|------------|------------------|------------------------|------|
| A | KNO₃ | 0 | 217.32 | 12.0733 |
| B | KNO₃ | 10 | 191.07 | 10.6150 |
| C | Urea | 0 | 167.75 | 9.3194 |
| D | Urea | 10 | 226.49 | 12.5828 |
| E | (NH₄)₂SO₄ | 0 | 218.49 | 12.1383 |
| F | (NH₄)₂SO₄ | 10 | 169.85 | 9.4361 |
| G | Bovung | 0 | 179.21 | 9.9561 |
| H | Bovung | 10 | 166.45 | 9.2472 |
| I | Milorganite | 0 | 205.80 | 11.4333 |
| J | Milorganite | 10 | 109.48 | 6.0822 |
| K | Mergreen | 0 | 197.03 | 10.9461 |
| L | Mergreen | 10 | 192.75 | 10.7083 |

The analysis of variance, treating the data as a one-way classification, is shown in Table 4.2. While the $F$ test tells us that there are highly significant differences among the treatment means, this information does not help us a great deal in interpreting the results. Methods discussed in this chapter and in Chapter 5 will allow us to make far more informative statements regarding the results of this experiment. We shall start by developing procedures for making meaningful comparisons among means or groups of means.

# 4.2 SINGLE DEGREE OF FREEDOM COMPONENTS OR COMPARISONS: EQUAL NUMBERS

One of the most informative procedures for extracting information from a set of data beyond the original analysis of variance is the use of single degree of freedom comparisons between two treatment means or between two sets of treatment means. For example, if we have five treatments, we might wish to test for the significance of the difference between the first two treatment means, $\bar{y}_1 - \bar{y}_2$, or between the average of the first two treatment means and the average of the last three treatment means, $(\bar{y}_1 + \bar{y}_2)/2 - (\bar{y}_3 + \bar{y}_4 + \bar{y}_5)/3$. The second difference can be expressed as $\frac{1}{2}\bar{y}_1 + \frac{1}{2}\bar{y}_2 - \frac{1}{3}\bar{y}_3 - \frac{1}{3}\bar{y}_4 - \frac{1}{3}\bar{y}_5$ or as $3\bar{y}_1 + 3\bar{y}_2 - 2\bar{y}_3 - 2\bar{y}_4 - 2\bar{y}_5$. It is usually more convenient to make use of totals rather than means when carrying out the computations necessary for testing for the significance of such differences.

**TABLE 4.2** ANALYSIS OF VARIANCE FOR CHLOROPLAST DATA

| Source | df | SS | MS | F |
|--------|-----|----------|---------|---------|
| Among treatments | 11 | 618.3682 | 56.2153 | 18.32** |
| Within treatments | 204 | 626.0816 | 3.0690 | |

## 4.3 LINEAR FUNCTIONS AND COMPARISONS

In developing single degree of freedom comparisons such as shown in Section 4.2, we are making use of *linear functions* of the treatment means. Using values of $c_{ij}$ to represent the coefficients and $T_j$ to represent treatment totals, we can generalize a linear function as

$$Q_i = c_{i1}T_1 + c_{i2}T_2 + \cdots + c_{ip}T_p = \sum_j c_{ij}T_j \tag{4.1}$$

where $j = 1, \ldots, p$ and $p$ is the number of treatments. The linear function shown in Equation (4.1) is also called a *comparison* provided that the sum of the coefficients is zero, that is, $\sum_j c_{ij} = 0$. Linear functions that are not comparisons have frequent utility in statistical analyses.

For example, using the data in Table 4.1,

$$(1)T_A - (1)T_B = (1)(217.32) - (1)(191.07) = 26.25$$

is a linear function of the $T$'s and is also a comparison, and

$$T_A - \frac{T_C + T_E}{2} = 217.32 - \frac{167.75 + 218.49}{2} = 24.20$$

is a linear function of the $T$'s and is also a comparison; however,

$$T_B + T_D + T_F = 191.07 + 226.49 + 169.85 = 587.41$$

is a linear function but not a comparison.

In order to develop the sum of squares attributable to a single degree of freedom comparison, needed for a test of significance, we make use of the square of the $Q$ value found in Equation (4.1) together with

$$D_i = r(c_{i1}^2 + c_{i2}^2 + \cdots + c_{ip}^2) = r \sum_i c_{ij}^2 \tag{4.2}$$

where $r$ is the number of observations per treatment. Putting these values together in the form $Q_i^2/D_i$, we arrive at a sum of squares that is a portion, or component, of the treatment sum of squares.

EXAMPLE 1 ────────────────────────────────────────────────────

$$Q_i = (1)T_G - (1)T_H = 179.21 - 166.45 = 12.76$$

$$D_i = r(c_{i1}^2 + c_{i2}^2) = (18)(1 + 1) = 36$$

$$SS_{Q_i} = \frac{Q_i^2}{D_i} = \frac{(12.76)^2}{36} = 4.5227 \tag{4.3}$$

EXAMPLE 2 ────────────────────────────────────────────────────

$$Q_i = T_H - \frac{T_J + T_L}{2} = (1)(166.45) - (\tfrac{1}{2})(109.48) - (\tfrac{1}{2})(192.75) = 15.335$$

$$D_i = r(c_{i1}^2 + c_{i2}^2 + c_{i3}^2) = (18)(1 + \tfrac{1}{4} + \tfrac{1}{4}) = (18)(1.5) = 27$$

$$SS_{Q_i} = \frac{Q_i^2}{D_i} = \frac{(15.335)^2}{27} = 8.7097$$

The last comparison can also be calculated by using whole numbers for coefficients rather than fractions as follows:

$$Q_i = (2)T_H - (1)T_J - (1)T_L = (2)(166.45) - (1)(109.48) - (1)(192.75) = 30.67$$

$$D_i = r(c_{i1}^2 + c_{i2}^2 + c_{i3}^2) = (18)(4 + 1 + 1) = 108$$

$$SS_{Q_i} = \frac{Q_i^2}{D_i} = \frac{(30.67)^2}{108} = 8.7097$$

## 4.4 ORTHOGONALITY OF COMPARISONS

We normally make comparisons among treatment means that are independent of one another, which means that there is no confounding or entanglement of treatment effects. To be independent of one another, comparisons must be what is termed *orthogonal*. For comparisons to be orthogonal, the sum of the products of corresponding coefficients in each comparison must be zero, that is,

$$c_{11}c_{21} + c_{12}c_{22} + \cdots + c_{1p}c_{2p} = \sum_j c_{1j}c_{2j} = 0 \tag{4.4}$$

When two comparisons are orthogonal, the sums of squares are said to be additive, that is, each is a portion or a component of the treatment sum of squares.

**EXAMPLE 3**

| Treatment | $T_A$ | $T_C$ | $T_E$ |
|-----------|-------|-------|-------|
| Total | 217.32 | 167.75 | 218.49 |
| $c_{1j}$ | 2 | −1 | −1 |
| $c_{2j}$ | 0 | 1 | −1 |
| $\sum_j c_{1j}c_{2j} = (2)(0) + (-1)(+1) + (-1)(-1) = 0$ | | | |

Normally, when it is desired to use more than one single degree of freedom comparison, the second one is selected so that it is orthogonal to the first. If more than two comparisons (up to the $p - 1$ degrees of freedom available) are used, then as each new one is selected it must be checked for orthogonality with *each* of the previous comparisons in turn if orthogonality is to be maintained. If we take $p - 1$ single degree of freedom comparisons, each orthogonal to every other comparison, we have

$$\text{Treatment SS} = \frac{Q_1^2}{D_1} + \frac{Q_2^2}{D_2} + \cdots + \frac{Q_{p-1}^2}{D_{p-1}} \tag{4.5}$$

Since the sums of squares for the $p - 1$ comparisons add to the treatment sum of squares they are said to be additive.

A complete set of orthogonal, single degree of freedom comparisons for the chloroplast data is shown in Table 4.3.

**TABLE 4.3** ORTHOGONAL, SINGLE DEGREE OF FREEDOM COMPARISONS OR CONTRASTS

| Treat-ment total | A 217.32 | B 191.07 | C 167.75 | D 226.49 | E 218.49 | F 169.85 | G 179.21 | H 166.45 | I 205.80 | J 109.48 | K 197.03 | L 198.75 |
|---|---|---|---|---|---|---|---|---|---|---|---|---|
| $c_{1j}$ | 1 | −1 | 1 | −1 | 1 | −1 | 1 | −1 | 1 | −1 | 1 | −1 |
| $c_{2j}$ | 1 | 1 | 1 | 1 | 1 | 1 | −1 | −1 | −1 | −1 | −1 | −1 |
| $c_{3j}$ | 2 | 2 | −1 | −1 | −1 | −1 | 0 | 0 | 0 | 0 | 0 | 0 |
| $c_{4j}$ | 0 | 0 | 1 | 1 | −1 | −1 | 0 | 0 | 0 | 0 | 0 | 0 |
| $c_{5j}$ | 0 | 0 | 0 | 0 | 0 | 0 | 2 | 2 | −1 | −1 | −1 | −1 |
| $c_{6j}$ | 0 | 0 | 0 | 0 | 0 | 0 | 0 | 0 | 1 | 1 | −1 | −1 |
| $c_{7j}$ | 1 | −1 | 1 | −1 | 1 | −1 | −1 | 1 | −1 | 1 | −1 | 1 |
| $c_{8j}$ | 2 | −2 | −1 | 1 | −1 | 1 | 0 | 0 | 0 | 0 | 0 | 0 |
| $c_{9j}$ | 0 | 0 | 1 | −1 | −1 | 1 | 0 | 0 | 0 | 0 | 0 | 0 |
| $c_{10j}$ | 0 | 0 | 0 | 0 | 0 | 0 | 2 | −2 | −1 | 1 | −1 | 1 |
| $c_{11j}$ | 0 | 0 | 0 | 0 | 0 | 0 | 0 | 0 | 1 | −1 | −1 | 1 |

$$\frac{Q_1^2}{D_1} = \frac{(129.50)^2}{(18)(12)} = 77.6520^{**} \qquad \frac{Q_2^2}{D_2} = \frac{(140.25)^2}{(18)(12)} = 91.0651^{**}$$

$$\frac{Q_3^2}{D_3} = \frac{(34.20)^2}{(18)(12)} = 5.4150 \qquad \frac{Q_4^2}{D_4} = \frac{(5.90)^2}{(18)(4)} = 0.4835$$

$$\frac{Q_5^2}{D_5} = \frac{(-13.74)^2}{(18)(12)} = 0.8740 \qquad \frac{Q_6^2}{D_6} = \frac{(-74.50)^2}{(18)(4)} = 77.0868^{**}$$

$$\frac{Q_7^2}{D_7} = \frac{(-97.21)^2}{(18)(12)} = 43.7490^{**} \qquad \frac{Q_8^2}{D_8} = \frac{(62.60)^2}{(18)(12)} = 18.1424^{**}$$

$$\frac{Q_9^2}{D_9} = \frac{(107.38)^2}{(18)(4)} = 160.1453^{**} \qquad \frac{Q_{10}^2}{D_{10}} = \frac{(-75.08)^2}{(18)(12)} = 26.0973^{**}$$

$$\frac{Q_{11}^2}{D_{11}} = \frac{(92.04)^2}{(18)(4)} = 117.6578^{**} \qquad \sum_{p-1} \frac{Q_i^2}{D_i} = 618.3682$$

Reference to Table 4.1 will help in differentiating the different contrasts. The first comparison $Q_1$ can be seen to be between 0 and 10 ppm of nitrapyrin. $Q_2$ is a comparison between the inorganic fertilizers and the organic ones. $Q_3$ is a comparison between the nitrate fertilizer and the average of the two ammoniacal fertilizers. $Q_4$ compares the two ammoniacal fertilizers. $Q_5$ compares the bovung fertilizer with the average of the two sludges and $Q_6$ tests between the two sludges. The remaining $Q$'s are interactions between level of nitrapyrin and the succeeding comparisons, $Q_2$ through $Q_6$. The $c_{ij}$ values for $Q_7$ through $Q_{11}$ are the products of $c_{1j}$ and $c_{ij}$ for $i$ going from 2 through 6. Since there is one degree of freedom for each comparison, the mean squares are equal to the sums of squares. Division of the mean square for each contrast or comparison by the error mean square from Table 4.2 results in the asterisks or lack thereof in Table 4.3.

The sizes of the $F$ values for $Q_7$ through $Q_{11}$ indicate that there is a large difference in response to the levels of nitrapyrin in each of the other comparisons. This means that the comparisons do not give the same response at the two levels of nitrapyrin.

Since there were six different fertilizer treatments, each occurring without the addition of nitrapyrin and each occurring with the addition of nitrapyrin, this example can also be viewed as a $6 \times 2$ factorial experiment. We would then have the sources of variation of fertilizers with 5 degrees of freedom, nitrapyrin with 1 degree of freedom, interaction of fertilizers and nitrapyrin with 5 degrees of freedom, and within with 204 degrees of freedom. The sum of squares for $Q_1$ would be the sum of squares for the nitrapyrin source, the sums of squares for $Q_2$ through $Q_6$ would equal the sum of squares for fertilizers, and the sum of squares for $Q_7$ through $Q_{11}$ would equal the sum of squares for the interaction.

In this section we have considered planned, orthogonal comparisons. However, there are occasions when an experimenter is concerned in advance with certain non-orthogonal comparisons. This situation calls for a different method of analysis than the one presented here, as does the testing of contrasts or comparisons not planned in advance but suggested by the data resulting from an experiment. For a discussion of the methods recommended for these situations see Gill (1978, vol. 1, pp. 176–178).

## 4.5 MULTIPLE DEGREE OF FREEDOM COMPONENTS

Cochran and Cox (1957) have pointed out that it is not necessary to use only single degree of freedom comparisons in partitioning the treatment sum of squares and have suggested the following method of partitioning, making use of multiple degree of freedom components. This involves the grouping of means into different subsets. If we have a subset of $p$ treatments, a portion of the treatment sum of squares could be

$$ P = \frac{T_1^2}{n_1} + \frac{T_2^2}{n_2} + \cdots + \frac{T_p^2}{n_p} - \frac{(T_1 + T_2 + \cdots + T_p)^2}{n_1 + n_2 + \cdots + n_p} \tag{4.6} $$

where $P$ is a portion of the treatment sum of squares with $p - 1$ degrees of freedom and $n_i$ is the number of observations in the $i$th treatment, $i = 1, \ldots, p$.

EXAMPLE 4

The sum of squares among treatments $I$, $J$, $K$, and $L$ shown in Table 4.1 would be (equal numbers per treatment)

$$ P = \frac{T_I^2 + T_J^2 + T_K^2 + T_L^2}{n} - \frac{(T_I + T_J + T_K + T_L)^2}{4n} $$

$$ = \frac{(205.80)^2 + (109.48)^2 + (197.03)^2 + (192.75)^2}{18} - \frac{(705.06)^2}{72} $$

$$ = 7239.6052 - 6904.3000 = 335.3052 \tag{4.7} $$

This sum of squares is associated with three degrees of freedom giving a mean square of 111.7684 which could be tested for significance with the error mean square. The remaining eight degrees of freedom could be used for single degree of freedom comparisons. Referring back to Table 4.3, suppose that instead of the single degree of freedom comparisons shown there we had used a different set:

1. Comparison of the nonsludge and the sludge, with coefficients of 1, 1, 1, 1, 1, 1, 1, 1, −2, −2, −2, −2, yielding a sum of squares of 37.0481. The remaining comparisons will deal only with the first eight treatments, the nonsludge treatments.

2. Comparison of the 0 level and 10 ppm of nitrapyrin with coefficients of 1, −1, 1, −1, 1, −1, 1, −1, yielding a sum of squares of 5.8041.

3. Comparison of inorganic and organic fertilizers with coefficients 1, 1, 1, 1, 1, 1, −3, −3, yielding a sum of squares of 54.8910.

4. Comparison of the nitrate fertilizer with the ammoniacal fertilizers with coefficients of 2, 2, −1, −1, −1, −1, 0, 0, yielding a sum of squares of 5.4150.

5. Comparison of the two ammoniacal fertilizers with coefficients of 0, 0, 1, 1, −1, −1, 0, 0, yielding a sum of squares of 0.4835.

6. Three more single degree of freedom comparisons representing the interactions of the level of nitrapyrin (comparison 2) with the comparisons shown in 3, 4, and 5 above. The coefficients would be 1, −1, 1, −1, 1, −1, 3, −3 for the first of these comparisons, 2, −2, 1, −1, 1, −1 for the second, and 0, 0, 1, −1, −1, 1, 0, 0 for the third. The sums of squares would be 1.1337, 18.1424, and 160.1453 for the three interaction components, respectively.

The total of the sums of squares for the eight single degree of freedom comparisons is 283.0631. Adding this to the sum of squares of 335.3052 found earlier among the last four treatments, we find a sum of squares of 618.3683, showing that we have partitioned the 11 degrees of freedom and the corresponding sum of squares into one 3 degrees of freedom component and eight 1 degree of freedom components.

As pointed out by Cochran and Cox (1957), there are situations where the treatments (or part of the treatments) can be divided into a number of groups. The number of treatments does not have to be the same in each group. Let $T_{ij}$ be the total for the $j$th treatment in the $i$th group and let

$$S_i = T_{i1} + T_{i2} + \cdots + T_{ip_i} \tag{4.8}$$

and
$$N_i = n_{i1} + n_{i2} + \cdots + n_{ip_i} \tag{4.9}$$

where $i = 1, \ldots, g$ and $g$ is the number of groups or subsets and $j = 1, \ldots, p_i$ and $p_i$ is the number of treatments in the $i$th group. Then let

$$P_i = \frac{T_{i1}^2}{n_{i1}} + \frac{T_{i2}^2}{n_{i2}} + \cdots + \frac{T_{ip_i}^2}{n_{ip_i}} - \frac{S_i^2}{N_i} \tag{4.10}$$

This amounts to the sum of squares among the treatments in a particular group or subset of treatments. This can be repeated for the $g$ groups. Then

$$P_{g+1} = \frac{S_1^2}{N_1} + \frac{S_2^2}{N_2} + \cdots + \frac{S_g^2}{N_g} - \frac{(S_1 + S_2 + \cdots + S_g)^2}{N_1 + N_2 + \cdots + N_g} \tag{4.11}$$

The $P_{g+1}$ portion will be the sum of squares *among the groups or subsets* of treatments.

**EXAMPLE 5** ────────────────────────────────────────────────────────

Let us partition the treatments shown in Table 4.1 into three groups: the first group is made up of treatments *A, B, C,* and *D,* the second group is made up of treatments

*E, F, G,* and *H,* and the third group is made up of treatments *I, J, K,* and *L.* We can then partition the treatment sum of squares into the sums of squares among the treatments in each of the three groups or subsets plus the sum of squares among the three groups.

$$P_1 = \frac{T_A^2 + T_B^2 + T_C^2 + T_D^2}{n} - \frac{(T_A + T_B + T_C + T_D)^2}{4n = N_1}$$

$$= \frac{(217.32)^2 + (191.07)^2 + (167.75)^2 + (226.49)^2}{18} - \frac{(802.63)^2}{72}$$

$$= 9065.1950 - 8947.4294 = 117.7656$$

$$P_2 = \frac{T_E^2 + T_F^2 + T_G^2 + T_H^2}{n} - \frac{(T_E + T_F + T_G + T_H)^2}{4n = N_2}$$

$$= \frac{(218.49)^2 + (169.85)^2 + (179.21)^2 + (166.45)^2}{18} - \frac{(734.00)^2}{72}$$

$$= 7578.2627 - 7482.7222 = 95.5405$$

$$P_3 = \frac{T_I^2 + T_J^2 + T_K^2 + T_L^2}{n} - \frac{(T_I + T_J + T_K + T_L)^2}{4n = N_3}$$

$$= \frac{(205.80)^2 + (109.48)^2 + (197.03)^2 + (192.75)^2}{18} - \frac{(705.06)^2}{72}$$

$$= 7239.6052 - 6904.3000 = 335.3052$$

$$P_4 = \frac{S_1^2}{N_1} + \frac{S_2^2}{N_2} + \frac{S_3^2}{N_3} - \frac{(S_1 + S_2 + S_3)^2}{N_1 + N_2 + N_3}$$

$$= \frac{(802.63)^2}{72} + \frac{(734.00)^2}{72} + \frac{(705.06)^2}{72} - \frac{(2241.69)^2}{216}$$

$$= 23{,}334.4517 - 23{,}264.6947 = 69.7570$$

where $n$ is the number of observations per treatment. The four portions then sum to the treatment sum of squares:

$$117.7656 + 95.5405 + 335.3052 + 69.7570 = 618.3683$$

The results of this partitioning of treatment sum of squares are presented in Table 4.4. Other mean separation procedures discussed in this and other sections may be applied with the several groups into which the treatments and their sums of squares have been partitioned.

## 4.6 STANDARD ERRORS FOR LINEAR FUNCTIONS OF TREATMENT MEANS

We have been discussing various linear functions, or comparisons, among treatment means and methods of obtaining sums of squares attributable to such effects in order

**TABLE 4.4  PARTITIONING OF TREATMENT SUM OF SQUARES**

| Source | df | SS | MS | F |
|---|---|---|---|---|
| Among treatments | 11 | 618.3682 | 56.2153 | 18.32 |
| Among $A, B, C, D$ | 3 | 117.7656 | 39.2552 | 12.79 |
| Among $E, F, G, H$ | 3 | 95.5405 | 31.8468 | 10.38 |
| Among $I, J, K, L$ | 3 | 335.3052 | 111.7684 | 36.42 |
| Among $(A, B, C, D), (E, F, G, H), (I, J, K, L)$ | 2 | 69.7570 | 34.8785 | 11.36 |
| Within treatments | 204 | 626.0816 | 3.0690 | |

to conduct the appropriate $F$ tests of significance. It is frequently easier to calculate the standard errors for the single degree of freedom comparisons and use the $t$ test. We shall be looking at the comparisons as linear functions of the treatment means.

The standard error of any linear function (as shown by Cochran and Cox (1957))

$$Q = c_1 Y_1 + c_2 Y_2 + \cdots + c_n Y_n \tag{4.12}$$

of *individual observations* is

$$\sigma_Q = \sigma \sqrt{c_1^2 + c_2^2 + \cdots + c_n^2} \tag{4.13}$$

The *estimated* standard error is determined by substituting $s$ for $\sigma$, $s$ being the square root of the error mean square in the analysis of variance.

**EXAMPLE 6**

Using the four observations, $Y_1 = 2$, $Y_2 = 3$, $Y_3 = 4$, and $Y_4 = 6$, a linear function of these observations could be

$$Q = \frac{1}{n} Y_1 + \frac{1}{n} Y_2 + \frac{1}{n} Y_3 + \frac{1}{n} Y_4 = \frac{1}{n} \sum Y = \bar{y}$$

$$= \tfrac{1}{4}(2) + \tfrac{1}{4}(3) + \tfrac{1}{4}(4) + \tfrac{1}{4}(6) = \tfrac{1}{4}(15) = 3.75$$

The estimated standard error of this linear function, which is seen to be the mean, would be

$$s_Q = s\sqrt{(\tfrac{1}{4})^2 + (\tfrac{1}{4})^2 + (\tfrac{1}{4})^2 + (\tfrac{1}{4})^2} = s\sqrt{\tfrac{4}{16}} = s\sqrt{\tfrac{1}{4}} = \frac{s}{\sqrt{4}} = \frac{s}{\sqrt{n}}$$

or the usual formula that is used for the standard error of a mean.

It follows, then, that the estimated standard error of a linear function

$$Q = c_1 \bar{y}_1 + c_2 \bar{y}_2 + c_2 \bar{y}_2 + \cdots + c_k \bar{y}_k \tag{4.14}$$

of *treatment means* is

$$s_Q = \frac{s}{\sqrt{n}} \sqrt{c_1^2 + c_2^2 + \cdots + c_k^2} \tag{4.15}$$

**EXAMPLE 7**

Using the data from Table 4.1 again, suppose that we wish to compare the nitrate fertilizer $KNO_3$ with the ammoniacal fertilizers urea and $(NH_4)_2SO_4$. Our linear function, or comparison, would be

$$Q = (2)\bar{y}_A + (2)\bar{y}_B - (1)\bar{y}_C - (1)\bar{y}_D - (1)\bar{y}_E - (1)\bar{y}_F$$

$$= (2)(12.0733) + (2)(10.6150) - (1)(9.3194) - (1)(12.5828) - (1)(12.1383)$$

$$- (1)(9.4361)$$

$$= 1.9000$$

$$s_Q = \sqrt{\frac{3.0690}{18}} \sqrt{(2)^2 + (2)^2 + (-1)^2 + (-1)^2 + (-1)^2 + (-1)^2}$$

$$= \sqrt{\frac{3.0690}{18}} \sqrt{12} = 1.4304$$

The $s$ value is the square root of the error mean square that is used to test for the significance of differences among the treatment means in the original analysis of variance. Thus,

$$t = \frac{Q}{s_Q} = \frac{1.9000}{1.4304} = 1.328$$

The linear function we have just obtained is the same comparison as $Q_3$ shown in Table 4.3. In that case, we found a sum of squares of 5.4150 which, with one degree of freedom, equals the mean square, yielding

$$F = \frac{5.4150}{3.0690} = 1.764$$

Since a single degree of freedom is in the numerator, $t^2$ should equal $F$; thus,

$$t^2 = (1.328)^2 = 1.764 = F$$

**EXAMPLE 8**

Again, using data from the same source, let us compare the two ammoniacal fertilizers. Our linear function would be

$$Q = \bar{y}_C + \bar{y}_D - \bar{y}_E - \bar{y}_F$$

$$= 9.3194 + 12.5828 - 12.1383 - 9.4361 = 0.3278$$

$$s_Q = \sqrt{\frac{3.0690}{18}} \sqrt{(1)^2 + (1)^2 + (-1)^2 + (-1)^2} = \sqrt{\frac{3.0690}{18}} \sqrt{4} = 0.8258$$

Then $\quad t = \frac{Q}{s_Q} = \frac{0.3278}{0.8258} = 0.397$

This second linear function is the same as comparison $Q_4$ in Table 4.3, which yields

$$F = \frac{0.4835}{3.0690} = 0.157 = t^2$$

While we could have used coefficients of $\frac{1}{2}$, $\frac{1}{2}$, $-\frac{1}{2}$, and $-\frac{1}{2}$ in the second linear function to give the precise difference between the average of the first two means and the average of the last two means, the test of significance is exactly the same using coefficients of 1, 1, −1, and −1.

When, in a one-way classification, the treatments have unequal numbers of observations, the standard error of $Q$ is

$$s_Q = s \sqrt{\frac{c_1^2}{n_1} + \frac{c_2^2}{n_2} + \cdots + \frac{c_k^2}{n_k}} \qquad i = 1, \ldots, k \qquad (4.16)$$

**EXAMPLE 9** ———————————————————————————————————————————

Suppose we are given the following information:

| Treatments | A | B | C | D |
|---|---|---|---|---|
| Totals | 35 | 70 | 84 | 160 |
| $n_i$ | 5 | 7 | 8 | 10 |
| Means | 7.00 | 10.00 | 10.50 | 16.00 |

One linear function could be

$$Q_1 = \frac{\bar{y}_A + \bar{y}_B}{2} - \frac{\bar{y}_C + \bar{y}_D}{2} = -4.75$$

$$s_{Q_1} = s \sqrt{\frac{1/4}{5} + \frac{1/4}{7} + \frac{1/4}{8} + \frac{1/4}{10}} = s \sqrt{\frac{159}{1120}} = 0.37678s$$

$$t = \frac{4.75}{0.37678s} = \frac{12.607}{s}$$

Since with one degree of freedom, $t^2 = F = SS/s^2$, then $SS = t^2 s^2$ and in our function

$$SS_{Q_1} = \left(\frac{12.607}{s}\right)^2 s^2 = 158.94$$

A second linear function could be

$$Q_2 = \bar{y}_A - \bar{y}_B = 7.00 - 10.00 = -3.00$$

$$s_{Q_2} = s \sqrt{\frac{(1)^2}{5} + \frac{(-1)^2}{7}} = s \sqrt{\frac{12}{35}} = 0.58554s$$

$$t = \frac{3.00}{0.58554s} = \frac{5.1235}{s} \qquad \text{and} \quad SS_{Q_2} = t^2 s^2 = 26.25$$

A third linear function could be

$$Q_3 = \bar{y}_C - \bar{y}_D = 10.50 - 16.00 = -5.50$$

$$s_{Q_3} = s \sqrt{\frac{(1)^2}{8} + \frac{(-1)^2}{10}} = s \sqrt{\frac{18}{80}} = 0.47434s$$

$$t = \frac{5.50}{0.47434s} = \frac{11.595}{s} \quad \text{and} \quad SS_{Q_3} = t^2 s^2 = 134.44$$

In the above calculations $s^2$ is the within-treatment, or error, mean square. The sum of squares among the treatments in this set of data is found to be 326.97 so it can be seen that the sums of squares of $158.94 + 26.25 + 134.44 + 319.63$ shown above do not sum up to the treatment sum of squares due to the unequal numbers involved.

An alternative method of completing the test of significance of a comparison, which consists of constructing a linear function of *linear functions,* has been shown by Cochran and Cox (1957). If $Q_1$ and $Q_2$ are orthogonal to one another, the standard error of a linear function of these two linear functions

$$Q = c_1 Q_1 + c_2 Q_2 \tag{4.17}$$

is
$$s_Q = \sqrt{c_1^2 s_1^2 + c_2^2 s_2^2} \tag{4.18}$$

where $s_i$ is the standard error of the $i$th linear function.

**EXAMPLE 10** ─────────────────────────────────────────────────

Let us once more use the data from Table 4.1 which gives us the following information:

| Treatment | A | B | C | D | E | F | G | H |
|---|---|---|---|---|---|---|---|---|
| Mean | 12.0733 | 10.6150 | 9.3194 | 12.5828 | 12.1383 | 9.4361 | 9.9561 | 9.2472 |

As a first linear function let us take

$$Q_1 = -3\bar{y}_A - 1\bar{y}_B + 1\bar{y}_C + 3\bar{y}_D = 0.2329$$

and for a second function let us take

$$Q_2 = -3\bar{y}_E - 1\bar{y}_F + 1\bar{y}_G + 3\bar{y}_H = -8.1533$$

Now as a linear function of these two functions we could have

$$Q = (1)Q_1 - (1)Q_2 = (0.2329) - (-8.1533) = 8.3862$$

$$s_{Q_1} = \sqrt{\frac{3.0690}{18}} \sqrt{(-3)^2 + (-1)^2 + (1)^2 + (3)^2} = \sqrt{\frac{3.0690}{18}} \sqrt{20}$$

$$= \sqrt{\frac{61.3800}{18}} = 1.8466$$

$$s_{Q_2} = 1.8466$$

giving

$$s_Q = \sqrt{(1)^2 \left(\frac{61.3800}{18}\right) + (1)^2 \left(\frac{61.3800}{18}\right)} = \sqrt{\frac{122.7600}{18}} = 2.6115$$

Since our linear function is also a comparison, we can test this comparison for the significance of the difference between the two original functions by

$$t = \frac{Q}{s_Q} = \frac{8.3862}{2.6115} = 3.211 \quad \text{and} \quad SS_Q = t^2 s^2 = (3.211)^2 (3.0690) = 31.6430$$

The appropriate degrees of freedom for the $t$ tests in the preceding examples are those corresponding to the error mean square (the mean square used to test the treatment mean square).

This same comparison $Q$ could also be tested using the $F$ test with a single degree of freedom comparison using the coefficients of $-3, -1, 1, 3, 3, 1, -1, -3$ with the totals for the treatments $A$ through $H$. The sum of squares for this comparison would be $Q^2/D$, where

$$Q = \sum_i c_i T_i \quad \text{and} \quad D = n_w \sum_i c_i^2$$

yielding

$$SS_Q = \frac{(150.95)^2}{(18)(40)} = 31.6471 = t^2 s^2$$

which leads to

$$F = \frac{31.6471}{3.0690} = 10.312 = t^2$$

## 4.7 PAIRWISE COMPARISONS OF TREATMENT MEANS

Sections 4.1–4.6 dealt with partitioning the treatment sum of squares, with particular emphasis on orthogonal, single degree of freedom comparisons among the means. In some cases, comparisons were made between two means; in other cases, the number of means differed in the two groups to be compared. We shall now turn our attention to the situation where we are concerned solely with making pairwise comparisons among the means.

### 4.7.1 Least Significant Difference

The *least significant difference* (LSD) is a test criterion that has been used widely for many years. It is simple to calculate and to apply but has been subject to considerable abuse, being used in situations where it is not applicable, primarily in conducting tests suggested by the data. The LSD is used in an a priori test only for comparisons between pairs of means that have been planned *before* the experiment is conducted. Therefore, it is not to be used to test between any two means whose difference is noticeable from the results of the experiment. The planned comparisons do not have to be orthogonal. This test is in contrast to a posteriori tests, those which have not been planned before the experiment is conducted.

The least significant difference is a value which the difference between two means must equal or exceed to be declared significant. We have seen that with the $t$ test for the difference between two means

$$t = \frac{\bar{d}}{s_{\bar{d}}}$$

which means that for significance,

$$\bar{d} \geq t_\alpha s_{\bar{d}} \tag{4.19}$$

The $\alpha$ value is the level of $t$ to be used from the $t$ table, usually the 5 or 1 percent level. Thus, the LSD is

$$LSD = t_\alpha s_{\bar{d}} = t_\alpha \sqrt{\frac{2s^2}{n}} \tag{4.20}$$

where $s^2$ is the mean square used as the denominator in the $F$ test for treatments in the analysis of variance, and $n$ is the number of observations in any treatment.

As an example of the LSD, we can use the data in Table 4.1. If we wish to test for the significance of the difference between treatments $A$ and $B$ by means of the LSD, we would use the experimental error mean square from Table 4.2 and find

$$LSD_{0.05} = t_{0.05} \sqrt{\frac{2s^2}{n}} \quad \text{with 204 degrees of freedom}$$

$$= (1.960) \sqrt{\frac{(2)(3.0690)}{18}} = (1.960)(0.5840) = 1.1446$$

Since the difference between the two means is $12.0733 - 10.6150 = 1.4583$, we would find the difference to be statistically significant at the 5 percent level.

With unequal numbers in the one-way classification the LSD would be calculated as

$$LSD = t_\alpha s_{\bar{d}} = t_\alpha s \sqrt{\frac{1}{n_1} + \frac{1}{n_2}} = t_\alpha s \sqrt{\frac{n_1 + n_2}{n_1 n_2}} \tag{4.21}$$

where $n_1$ and $n_2$ are the numbers in each of the two treatments.

## 4.7.2 Duncan's New Multiple-Range Test

Probably the most frequently used test for pairwise comparisons among the treatment means, Duncan's new multiple-range test (1955, 1957), uses the concept of a significant difference, but instead of one value it has a series of values of increasing size. In Duncan's test, the means are ranked and the significant differences are developed such that with two adjacent means the test would be the same as an LSD but as the means become farther apart in the ranking, the differences necessary to declare significance increase. Duncan has termed these differences *shortest significant ranges.*

As with the LSD, Duncan's test is based on finding how large a difference must exist between two means to declare a significant difference between them. If we have just two treatments in our experiment, we can test for the significance of the difference between them by use of the $t$ test, where

$$t = \frac{\bar{d}}{s_{\bar{d}}} \quad \text{or} \quad \frac{\bar{d}}{s_{\bar{d}}} \geq t_\alpha \quad \text{or} \quad \bar{d} \geq t_\alpha s_{\bar{d}} \tag{4.22}$$

showing that the difference must equal or exceed the value of $t$ at a selected probability level times the standard error of the mean difference in order to be declared significant. In the general case, this would be

$$\bar{d} \geq t_\alpha s \sqrt{\frac{1}{n_1} + \frac{1}{n_2}} \quad \text{and} \quad s \sqrt{\frac{1}{n_1} + \frac{1}{n_2}} = s_{\bar{d}} \tag{4.23}$$

**TABLE 4.5**   ANALYSIS OF VARIANCE

| Source | df | MS | F |
|---|---|---|---|
| Blocks ($B$) | 5 | 141.95 | |
| Treatments ($T$) | 6 | 366.97 | 4.61** |
| Error ($BT$) | 30 | 79.64 | |

| | | | Varietal means ranked in order | | | |
|---|---|---|---|---|---|---|
| $A$ | $F$ | $G$ | $D$ | $C$ | $B$ | $E$ |
| 49.6 | 58.1 | 61.0 | 61.5 | 67.6 | 71.2 | 71.3 |

where $n_1$ and $n_2$ are the number of observations in treatments 1 and 2, respectively. With equal numbers this becomes

$$\bar{d} \geq t_\alpha s \sqrt{\frac{2}{n}} \geq t_\alpha \sqrt{2}\,\frac{s}{\sqrt{n}} \geq t_\alpha \sqrt{2} s_{\bar{y}} \tag{4.24}$$

The value $t_\alpha\sqrt{2}$ has been designated by Duncan as $r_p$ and a table of these values, *adjusted for the number of means included in the set to be tested,* has been prepared for the 5 and 1 percent levels of probability (Table A.5). This table is entered under the degrees of freedom for error and the number of means included in the set. The product of the value in the table, $r_p$, times the $s_{\bar{y}}$, yields $R_p$ which is called the *shortest significant range.* The subscript $p$ equals the size of the range, the number of means included in the set.

Note that the mean square and degrees of freedom for error $s^2$ are those associated with the denominator used in the $F$ test among the treatment means. It is recommended that the range test be applied only when a significant $F$ value has been found.

To illustrate Duncan's test, we will use the example presented by Duncan when this test was proposed in the *Journal of Biometrics* (1955). The data resulted from a randomized block design in which seven varieties of barley were tested, the results being measured in bushels of yield. Table 4.5 shows the resulting analysis, together with the ranked variety means. The lines beneath the means will be explained shortly.

To proceed with the pairwise comparison of all possible pairs of means, we first calculate the standard error of a varietal mean,

$$s_{\bar{y}} = \frac{s}{\sqrt{n}} = \sqrt{\frac{79.64}{6}} = 3.643$$

Now using the "significant Studentized range" table (Table A.5) for the 5 percent level, we find values for 30 degrees of freedom as shown below in the row $r_p$. These values have been multiplied by the standard error of the mean, 3.643, to obtain the values in the row $R_p$, the shortest significant ranges.

| $p$: | 2 | 3 | 4 | 5 | 6 | 7 |
|---|---|---|---|---|---|---|
| $r_p$: | 2.89 | 3.04 | 3.12 | 3.20 | 3.25 | 3.29 |
| $R_p$: | 10.53 | 11.07 | 11.37 | 11.66 | 11.84 | 11.99 |

The variety mean differences are compared with the shortest significant ranges appropriate to the $p$ under test in the following order: the largest minus the smallest, up to the largest minus the second largest, or until a nonsignificant difference is found; then the second largest minus the smallest, the second largest minus the second smallest, and so on, as shown in Table 4.6. Following this procedure, we would first find the difference between treatment $E$ and treatment $A$ which is 21.7. Since there are 7 means in this set or range, we would compare 21.7 with 11.99, and since 21.7 exceeds 11.99 a significant difference is declared between the two means. We would next find the difference between treatment $E$ and treatment $F$ and compare that difference, 13.2, with the $R_p$ for a set of 6 which is 11.84 and finding it to be larger, we declare this difference significant. Following this procedure, we arrive at the results shown in Table 4.6, where (ns) indicates nonsignificant. The same procedure would be followed if it were desired to test at the 1 percent level rather than the 5 percent level.

While Table 4.6 does present the results of the range test, it is only one of the ways to present the results. One of the most common methods is the use of lines as shown beneath the means presented in Table 4.5. These lines are accompanied by a footnote explaining that any two means not underscored by the same line are significantly different and any two means underscored by the same line are not significantly different. The lines, of course, are developed from the results shown in Table 4.6.

The results can also be presented by the use of parentheses placed beneath the means, including the appropriate treatments. In this example, we would have $(E, B, C, D, G)$, $(C, D, G, F)$, and $(F, A)$. Another convention commonly used, particularly when the ranked means are presented in a column rather than a row, is that of following each mean with a lowercase letter such that those means followed by the same lowercase letter do not differ significantly and those means not followed by the same letter differ significantly. In our present example, $E$, $B$, $C$, $D$, and $G$ would be followed by a lowercase $a$; $C$, $D$, $G$, and $F$ would be followed by a lowercase $b$; and $F$ and $A$ would be followed by a lowercase $c$. It should always be made clear, when presenting results, what level of probability was used.

One point to be noted in using Duncan's test is that no difference can be declared significant if the two means concerned are both contained in a set of means which has a nonsignificant range. The term set may include the complete group of means

**TABLE 4.6**  RESULT OF PAIRWISE COMPARISONS AMONG MEANS

| Comparison | Range = $p$ | Difference | Significance and $R_p$ |
|------------|:-----------:|------------|------------------------|
| $E$–$A$ | 7 | $71.3 - 49.6 = 21.7$ | $>11.99$* |
| $E$–$F$ | 6 | $71.3 - 58.1 = 13.2$ | $>11.84$* |
| $E$–$G$ | 5 | $71.3 - 61.0 = 10.3$ | $<11.66$ (ns) |
| $B$–$A$ | 6 | $71.2 - 49.6 = 21.6$ | $>11.84$* |
| $B$–$F$ | 5 | $71.2 - 58.1 = 13.1$ | $>11.66$* |
| $B$–$G$ | 4 | $71.2 - 61.0 = 10.2$ | $<11.37$ (ns) |
| $C$–$A$ | 5 | $67.6 - 49.6 = 18.0$ | $>11.66$* |
| $C$–$F$ | 4 | $67.6 - 58.1 =\ \ 9.5$ | $<11.37$ (ns) |
| $D$–$A$ | 4 | $61.5 - 49.6 = 11.9$ | $>11.37$* |
| $D$–$F$ | 3 | $61.5 - 58.1 =\ \ 3.4$ | $<11.07$ (ns) |
| $G$–$A$ | 3 | $61.0 - 49.5 = 11.4$ | $<11.07$* |
| $G$–$F$ | 2 | $61.0 - 58.1 =\ \ 2.9$ | $<10.53$ (ns) |
| $F$–$A$ | 2 | $58.1 - 49.6 =\ \ 8.5$ | $<10.53$ (ns) |

where necessary. As an example of this rule, suppose that the mean $G$ were 59.8 instead of 61.0. Now $E - G = 71.3 - 59.8 = 11.5$ with a range of 5. Since this difference does not equal or exceed the shortest significant range of 11.66, we declare that the two means do not differ; that is, it is a nonsignificant range. However, if we take the difference $B - G = 71.2 - 59.8 = 11.4$ with a range of 4, we find that this difference exceeds the shortest significant range of 11.37. Now we follow the first statement in this paragraph; since this comparison is contained within a set of means $E$ through $G$ which has a nonsignificant range, the difference $B - G$ is not declared to be significant.

Although the example we have discussed was a two-way classification, the same procedure can be followed for any multiway analysis with equal numbers. Caution must be taken to use the appropriate mean square, which is always the same mean square used as the denominator in the $F$ test for the significance of the treatments for which the range test is being conducted.

Duncan's test can also be used for unbalanced data. We shall concern ourselves now with its use in the one-way classification with unequal numbers. The test takes a little different form in this situation because means with different numbers of observations have different standard errors, so we must deal with standard errors of mean differences.

Starting from the expression used previously, we can write

$$\bar{d} \geq t_\alpha s_{\bar{d}} = t_\alpha s \sqrt{\frac{1}{n_1} + \frac{1}{n_2}} = t_\alpha s \frac{\sqrt{2}}{\sqrt{2}} \sqrt{\frac{n_1 + n_2}{n_1 n_2}} = t_\alpha \sqrt{2} s \sqrt{\frac{n_1 + n_2}{2 n_1 n_2}}$$

remembering that $t_\alpha \sqrt{2} = r_p$. This can be written in the form

$$\frac{\bar{d}}{\sqrt{(n_1 + n_2)/2 n_1 n_2}} \geq t_\alpha \sqrt{2} s$$

or

$$\bar{d} \sqrt{\frac{2 n_1 n_2}{n_1 + n_2}} \geq t_\alpha \sqrt{2} s \geq r_p s \qquad (4.25)$$

where $r_p s$ is designated as $R'_p$, the *significant range factor*. As an example of the test with unequal numbers in the one-way classification, we shall use the example presented by Clyde Kramer (1956) when he proposed this extension to the original test of Duncan's which dealt with the two-way classification with equal numbers. Table 4.7 shows the analysis of variance as well as the ranked means used by Kramer.

The analysis of variance yields an $s$ value of $\sqrt{2397.00} = 48.96$ with 16 degrees of freedom which is used in conjunction with the Studentized $t$ table (Table A.5) to develop $R'_p$ values, the significant range factors, as shown below.

| $p$: | 2 | 3 | 4 | 5 | 6 |
|------|------|------|------|------|------|
| $r_p$: | 3.00 | 3.15 | 3.23 | 3.30 | 3.34 |
| $R'_p$: | 146.88 | 154.22 | 158.14 | 161.57 | 163.53 |

Following the previously outlined procedures for testing the largest versus the smallest, and so on and using formula (4.25), we can proceed to make the appropriate calculations, as shown in Table 4.8.

**TABLE 4.7** ANALYSIS OF VARIANCE AND RANKED TREATMENT MEANS

| Source | df | MS | F |
|---|---|---|---|
| Treatments | 5 | 9306.17 | 3.88* |
| Error | 16 | 2397.00 | |

Ranked treatment means and number of replications

| | F | D | A | B | C | E |
|---|---|---|---|---|---|---|
| $\bar{y}$ | 458 | 498 | 521 | 528 | 564 | 630 |
| $n$ | 3 | 5 | 4 | 3 | 5 | 2 |

The results displayed in Table 4.8 can then be presented by the method of underlining as seen in Table 4.7 or by means of the other procedures discussed previously. Again, a statement explaining the method used should be included in any presentation. Contrary to the exception rule cited earlier, owing to the unequal number of replications, significant differences may be declared within a set for which a nonsignificant range has been found. When this occurs, broken lines may be used to present the results.

## 4.7.3 Student–Newman–Keuls (SNK) Test

As the name of this test implies, the men by whom the test is known contributed to its development, although it is frequently referred to as the Newman–Keuls test (Newman, 1939; Keuls, 1952). The procedure is similar to Duncan's new multiple-range test in that it uses different ranges for testing the sets of means for homogeneity. As in Duncan's test, the criteria used in testing the sets of means depend on the number of means included in the set. This test makes use of what is known as the *studentized range,* the distribution of $q_\alpha = (\bar{y}_{\max} - \bar{y}_{\min})/s_{\bar{y}}$. Table A.10 gives this distribution for $\alpha$ levels of 0.05 and 0.01.

**TABLE 4.8** COMPARISONS BETWEEN PAIRS OF MEANS

| $\bar{d}\sqrt{2n_1n_2/(n_1 + n_2)}$ | Range | Significant range factor | Significance |
|---|---|---|---|
| $(\bar{y}_E - \bar{y}_F)\sqrt{(2)(2)(3)/(2 + 3)} = 266.43$ | 6 | 163.53 | * |
| $(\bar{y}_E - \bar{y}_D)\sqrt{(2)(2)(5)/(2 + 5)} = 223.08$ | 5 | 161.57 | * |
| $(\bar{y}_E - \bar{y}_A)\sqrt{(2)(2)(4)/(2 + 4)} = 178.00$ | 4 | 158.14 | * |
| $(\bar{y}_E - \bar{y}_B)\sqrt{(2)(2)(3)/(2 + 3)} = 158.00$ | 3 | 154.22 | * |
| $(\bar{y}_E - \bar{y}_C)\sqrt{(2)(2)(5)/(2 + 5)} = 111.54$ | 2 | 146.88 | ns |
| $(\bar{y}_C - \bar{y}_F)\sqrt{(2)(5)(3)/(5 + 3)} = 205.22$ | 5 | 161.57 | * |
| $(\bar{y}_C - \bar{y}_D)\sqrt{(2)(5)(5)/(5 + 5)} = 147.58$ | 4 | 158.14 | ns |
| $(\bar{y}_B - \bar{y}_F)\sqrt{(2)(3)(3)/(3 + 3)} = 121.24$ | 4 | 158.14 | ns |

Making use of Table A.10, we can calculate values of

$$W_p = q_\alpha(p, f)s_{\bar{y}} \tag{4.26}$$

where $q$ is the tabulated value for $p$, the number of treatments in the set under test, and $f$ represents the degrees of freedom for the experimental error. The value of $p$ will range from two to the total number of treatments (or levels of treatment) involved. $W_p$ is often referred to as a least significant range (LSR).

If we apply this test to the example used by Duncan for the one-way classification (Table 4.5), we would have the following:

| $p$: | 2 | 3 | 4 | 5 | 6 | 7 |
|------|------|------|------|------|------|------|
| $q_{0.05}$: | 2.89 | 3.49 | 3.84 | 4.10 | 4.30 | 4.46 |
| $W_p$: | 10.53 | 12.71 | 13.99 | 14.94 | 15.66 | 16.25 |

The $s_{\bar{y}} = 3.643$ as with the previous example. The testing procedure would follow that shown for Duncan's new multiple-range test. It can be seen that while the test criterion is the same when the range of means in the set is only two, the remaining criteria (least significant ranges) are noticeably more conservative than those of Duncan's test. With only two means in the range both tests yield the same value as the least significant difference (LSD) of Section 4.7.1. The SNK test is seen to be similar to Duncan's new multiple-range test except that it uses a different table of critical values.

### 4.7.4 Tukey's *w* Procedure

This test proposed by Tukey (1953) differs from Duncan's new multiple-range test and the Student–Newman–Keuls test in that the test criterion does not change as the number of means in the set varies. One value is determined, that value being the same as the value for the largest range in the SNK test. The test criterion is

$$w = q_\alpha(p, f)s_{\bar{y}} \tag{4.27}$$

where $q_\alpha$ is derived from Table A.10, $p$ is the number of treatments (or levels of treatments) to be tested, and $f$ is the degrees of freedom for the experimental error.

Again, using the example from Duncan's test (Section 4.7.2), we find

$$w = (4.46)(3.643) = 16.25$$

Testing would proceed as explained in Section 4.7.2, only with this test the same criterion $w$ is used regardless of the number of means in the set under test. Obviously, this is a more conservative test than either Duncan's test or the SNK test.

### 4.7.5 Scheffé's Test

Another method for making pairwise comparisons among treatment means is that due to Scheffé (1953). This method is similar to that of Tukey's *w* procedure in that one test criterion is established against which all ranges are tested. In this method the value

$$S = \sqrt{f_t F_\alpha(f_t, f_e)} \tag{4.28}$$

is computed, where $f_t$ and $f_e$ are the degrees of freedom for treatments and error, respectively, and $F_\alpha$ is the tabulated $F$ for the $\alpha$ level. The critical value is calculated by using this $S$ value, the critical value being

$$\text{Scheffé's value} = Ss_{\bar{d}} \tag{4.29}$$

the product of $S$ times the standard error of the mean difference. Again using the example in Section 4.7.2, we find

$$S = \sqrt{(6)(2.42)} = 3.81$$

for 6 and 30 degrees of freedom. Now

$$s_{\bar{d}} = \sqrt{\frac{2s^2}{n}} = \sqrt{\frac{(2)(79.64)}{6}} = 5.15$$

giving

$$Ss_{\bar{d}} = (3.81)(5.15) = 19.62$$

Testing would proceed as shown in Section 4.7.2. However, this test is so much more conservative than the others mentioned that it is not often used.

## 4.7.6 Dunnett's Procedure: Testing Treatment Means Against a Control

It is a frequent occurrence that an experiment is conducted with the purpose of testing whether a series of treatments give different responses than a control. In such an experiment, differences among the treated groups are of secondary importance. In some cases, the experimenter may want to know whether the treatments give a greater response, and in some cases, whether the treatments give a lesser response than the control. In other situations, the only concern is whether there is a difference between each treated group and the control, regardless of which gives the greater response. While the LSD test could be used for these planned comparisons, Dunnett (1955) proposed a test which has found a wider acceptance.

If the experimenter wishes to test whether a treated group is significantly greater than the control, a one-tailed test is used. The same is true if the test is whether the treated group is smaller than the control. However, if the concern is only whether there is a significant difference between the treated group and the control, regardless of sign, a two-tailed test is used. For the one-tailed test, values from Table A.11A are used and for the two-tailed test values from Table A.11B are used.

As an example of this test, we shall consider some unpublished data from an experiment conducted by J. E. Simon, D. R. Decoteau, and L. E. Craker of the Plant and Soil Sciences Department at the University of Massachusetts. The original analysis of the data has been altered slightly for the purpose of this example. The experiment was concerned with the detection of phytotoxic compounds by using stress-ethylene and ethane production by plants. In this study, the herbicide paraquat was applied to wheat seedlings growing on an agar medium in test tubes. Four methods of application and four concentrations were included in the experiment, the zero concentration being a control. The dependent variable of amount of gas produced in each test tube resulted in a very wide range of values (0–1420) so the data were transformed to logarithms

**TABLE 4.9**   LOGARITHMS OF ETHANE GAS MEASUREMENTS

| Concentrations | Methods | | | | Means |
|---|---|---|---|---|---|
| | 1 | 2 | 3 | 4 | |
| 1 | 1.825 | 0.114 | 1.492 | 0.934 | 1.087 |
| | 0.544 | 1.745 | 0.000 | 2.043 | |
| 2 | 2.222 | 2.497 | 1.672 | 2.514 | 2.254 |
| | 2.490 | 2.381 | 2.204 | 2.056 | |
| 3 | 2.514 | 2.339 | 2.795 | 2.334 | 2.440 |
| | 2.241 | 2.602 | 2.056 | 2.643 | |
| 4 | 2.597 | 2.680 | 2.375 | 2.614 | 2.592 |
| | 2.643 | 2.849 | 2.526 | 2.448 | |

before analysis. The logarithms for the values found are shown in Table 4.9, each value being based on four observations.

The analysis of variance for this set of data is shown in Table 4.10 where it can be seen that there are highly significant differences among the concentrations. The analysis is for a fixed model.

Dunnett's procedure will be used to determine whether the effect of each concentration of paraquat is significantly greater than that of the control. This then will be a one-sided or one-tailed test. To conduct this test, we enter Table A.11A under $p$, the number of treatments excluding the control, and the degrees of freedom for the experimental error, and the probability level at which we are testing. For $p = 3$ and degrees of freedom of 16 at the 0.05 level, we find

$$d = t(\text{Dunnett})s_{\bar{d}} = t(\text{Dunnett}) \sqrt{\frac{2s^2}{n}}$$

$$= 2.23 \sqrt{\frac{(2)(0.2876)}{8}} = (2.23)(0.268) = 0.598 \qquad (4.30)$$

Looking at the means and control at the side of Table 4.9, we can see that the response in each of the treated groups is significantly greater than that of the control. If we were to test at the 0.01 level, we would multiply 0.268 by 3.05, giving a value of 0.817. This value shows that the differences are significant at the 0.01 level.

## 4.8  TREND ANALYSIS

There is one additional important method of partitioning treatment sums of squares which will be dealt with in Chapter 5 and which deals with regression analysis. A very

**TABLE 4.10**   ANALYSIS OF VARIANCE OF DATA IN TABLE 4.9

| Source | df | SS | MS | F |
|---|---|---|---|---|
| Methods ($M$) | 3 | 0.4589 | 0.1530 | <1 |
| Concentrations ($C$) | 3 | 11.2570 | 3.7523 | 13.05** |
| $MC$ | 9 | 0.5789 | 0.0643 | <1 |
| $W$ | 16 | 4.6008 | 0.2876 | |

common procedure in experimentation is to apply treatments in increasing levels or dosages. We are then interested in a trend response, that is, whether we obtain a straight line response or some type of nonlinear response. With factorial designs, we are frequently interested in what are called response surfaces. Methods of regression analysis using orthogonal coefficients will be presented as means of partitioning the sums of squares. Since a knowledge of regression is required in understanding this procedure it is postponed until Chapter 5.

## 4.9 SELECTION OF PROCEDURE TO BE USED

We have presented or mentioned several methods or procedures for extracting information in addition to the original analysis. We wish to emphasize that it is important that the researcher be careful to use the method appropriate to the data involved. The greatest error that occurs is in the indiscriminate use of the range test or pairwise comparisons. These tests are to be used when the experiment is concerned with variety trials or pesticide trials or similar experiments where one is attempting to select the better performers and making groups among the treatments. Such treatments as methods of application of a chemical or methods of training athletes would also be amenable to this type of analysis. However, if the treatments have been applied with increasing levels or dosages, then some form of trend analysis using regression methods is in order. If orthogonal linear functions or comparisons have been planned, then the methods in the first part of this section should be used. It is entirely possible that a combination of methods may be used with the same set of treatments.

## 4.10 COMPARISON OF LEAST-SQUARES MEANS: DISPROPORTIONATE SUBCLASS FREQUENCIES

The errors associated with treatment means that have been adjusted by least-squares procedures for unequal frequencies, with respect to other sets of effects or for covariates, are correlated. Therefore, to obtain tests of significance for comparisons among such means we must use the inverse matrix $C^{-1}$, as was explained in Section 3.20. When the summation restrictions have been imposed, the least-squares mean in a set of cross-classified effects, say treatments, is $\hat{\mu} + \hat{t}_i$. The variance–covariance matrix for $\hat{\mu}$ and the $\hat{t}_i$ is

$$
\begin{bmatrix}
c^{\mu\mu} & c^{\mu t_1} & c^{\mu t_2} & \cdots & c^{\mu t_p} \\
 & c^{t_1 t_1} & c^{t_1 t_2} & \cdots & c^{t_1 t_p} \\
 & & c^{t_2 t_2} & \cdots & c^{t_2 t_p} \\
 & \text{\textit{Symmetric}} & & & \vdots \\
 & & & & c^{t_p t_p}
\end{bmatrix} \hat{\sigma}_e^2
\tag{4.31}
$$

where the $c^{ij}$ are elements from the inverse of the least-squares coefficient matrix and $\hat{\sigma}_e^2$ is the estimate of the error variance. Since $\hat{\mu}$ is common to all means, a comparison among least-squares means is the same as a comparison among the corresponding $\hat{t}_i$.

**EXAMPLE 11** ————————————————————————————————————————————

$$Q = \hat{\mu} + \hat{t}_1 - \hat{\mu} - \hat{t}_2$$

$$= \hat{t}_1 - \hat{t}_2$$

$$V(Q) = V(\hat{t}_1) + V(\hat{t}_2) - 2\,\mathrm{Cov}(\hat{t}_1, \hat{t}_2)$$

$$= c^{t_1 t_1}\hat{\sigma}_e^2 + c^{t_2 t_2}\hat{\sigma}_e^2 - 2c^{t_1 t_2}\hat{\sigma}_e^2$$

$$= (c^{t_1 t_1} + c^{t_2 t_2} - 2c^{t_1 t_2})\hat{\sigma}_e^2$$

In general, if

$$Q = \lambda_1 \hat{t}_1 + \lambda_2 \hat{t}_2 + \cdots + \lambda_p \hat{t}_p = \sum_i \lambda_i \hat{t}_i \qquad (4.32)$$

then

$$V(Q) = (\lambda_1^2 c^{t_1 t_1} + \lambda_2^2 c^{t_2 t_2} + \cdots + \lambda_p^2 c^{t_p t_p} + 2\lambda_1\lambda_2 c^{t_1 t_2}$$

$$+ 2\lambda_1\lambda_3 c^{t_1 t_3} + \cdots + 2\lambda_{p-1}\lambda_p c^{t_{p-1} t_p})\hat{\sigma}_e^2 \qquad (4.33)$$

As previously pointed out, if the sum of the coefficients (the $\lambda$'s) in $Q$ equals zero, then $Q$ is a comparison or contrast among treatment effects. Otherwise, $Q$ is a linear function but not a comparison of the treatment effects.

### 4.10.1 Linear Comparisons: *t* Test

As pointed out in Section 4.6

$$t = \frac{Q}{s_Q}$$

where

$$s_Q = \sqrt{V(Q)}$$

Also, since

$$t^2 = \frac{Q^2}{s_Q^2} = F = \frac{\mathrm{SS}(Q)}{\hat{\sigma}_e^2}$$

then

$$\mathrm{SS}(Q) = \frac{\hat{\sigma}_e^2 Q^2}{V(Q)} \qquad (4.34)$$

Frequently, it is desired to give tests of significance for selected linear contrasts or comparisons in the analysis of variance table. In this case, one needs the $\mathrm{SS}(Q)$ and it is easily computed from Equation (4.34).

To illustrate the computation of $t$ and the sum of squares for linear contrasts, let us use the analysis of the average daily gains of 65 steers that were from nine sires and out of three ages of dam given in Section 3.21.1. Suppose the investigator is interested in the following set of comparisons among the three ages of dam:

|       | $\hat{a}_1$ | $\hat{a}_2$ | $\hat{a}_3$ |
|-------|------|------|------|
| $Q_1$ | 2    | −1   | −1   |
| $Q_2$ | 0    | 1    | −1   |

Now the variance–covariance matrix for the $\hat{a}_1$ and $\hat{a}_2$ effects when the restriction was imposed that $\Sigma_j\,\hat{a}_j = 0$ was as follows (Table 3.63):

$$\begin{bmatrix} 0.052602 & -0.032694 \\ -0.032694 & 0.042413 \end{bmatrix}\hat{\sigma}_e^2$$

Using formula (4.33), we find

$$V(Q_1) = [(2)^2 c^{a_1 a_1} + (-1)^2 c^{a_2 a_2} + (-1)^2 c^{a_3 a_3} + (2)(2)(-1)c^{a_1 a_2}$$

$$+ (2)(2)(-1)c^{a_1 a_3} + (2)(-1)(-1)c^{a_2 a_3}]\hat{\sigma}_e^2$$

$$V(Q_2) = [(1)^2 c^{a_2 a_2} + (-1)^2 c^{a_3 a_3} + (2)(1)(-1)c^{a_2 a_3}]\hat{\sigma}_e^2$$

The inverse elements $c^{a_1 a_3}$, $c^{a_2 a_3}$, and $c^{a_3 a_3}$ may be computed as follows:

$$c^{a_1 a_3} = -(0.052602 - 0.032694) = -0.019908$$

$$c^{a_2 a_3} = -(-0.032694 + 0.042413) = -0.009719$$

$$c^{a_3 a_3} = -(-0.019908 - 0.009719) = 0.029627$$

The variances, standard errors, and sums of squares for the two linear contrasts may now be computed.

$$V(Q_1) = [(4)(0.052602) + 0.042413 + 0.029627 + (4)(0.032694)$$

$$+ (4)(0.019908) - (2)(0.009719)](0.065074)$$

$$= 0.030807$$

$$V(Q_2) = [0.042413 + 0.029627 + (2)(0.009719)](0.065074)$$

$$= 0.005953$$

$$s_{Q_1} = \sqrt{V(Q_1)} = 0.175519$$

$$s_{Q_2} = \sqrt{V(Q_2)} = 0.077156$$

$$SS(Q_1) = \frac{(0.065074)[(2)(0.06216) + 0.00693 + 0.05523]^2}{0.030807}$$

$$= \frac{(0.065074)(0.18648)^2}{0.030807}$$

$$= 0.07346$$

$$SS(Q_2) = \frac{(0.065074)(-0.00693 + 0.05523)^2}{0.005953}$$

$$= \frac{(0.065074)(0.04830)^2}{0.005953}$$

$$= 0.02550$$

It should be noted that with disproportionate subclass frequencies the sum of the sums of squares for $Q_1$ and $Q_2$ is not equal to the SS for ages of dam (0.11894).

The $t$ tests and $F$ tests are completed as follows:

$$Q_1: \quad t = \frac{0.18648}{0.175519} = 1.062 \quad df = 54$$

$$Q_2: \quad t = \frac{0.04830}{0.077156} = 0.626 \quad df = 54$$

$$Q_1: \quad F = \frac{0.07346}{0.065074} = 1.129 \quad df = 1, 54$$

$$Q_2: \quad F = \frac{0.02550}{0.065074} = 0.392 \quad df = 1, 54$$

Of course, $t^2 = F$ and therefore the results are the same whether the $t$ or $F$ test is used, in this case.

A simpler method of computing $V(Q_1)$ and $V(Q_2)$ will now be illustrated. Since $\hat{a}_3 = -\hat{a}_1 - \hat{a}_2$, the linear contrasts may be expressed only in terms of $\hat{a}_1$ and $\hat{a}_2$; that is,

$$2\hat{a}_1 - \hat{a}_2 - \hat{a}_3 = 2\hat{a}_1 - \hat{a}_2 - (-\hat{a}_1 - \hat{a}_2) = 3\hat{a}_1$$

and

$$\hat{a}_2 - \hat{a}_3 = \hat{a}_2 - (-\hat{a}_1 - \hat{a}_2) = \hat{a}_1 + 2\hat{a}_2$$

Let the matrix of $\lambda$'s for the two comparisons [Equation (4.32)] be

$$\lambda = \begin{bmatrix} 2 & -1 & -1 \\ 0 & 1 & -1 \end{bmatrix}$$

The coefficients for only $\hat{a}_1$ and $\hat{a}_2$ are obtained by subtracting the last column in $\lambda$ from the other two columns to give

$$\lambda_r = \begin{bmatrix} 3 & 0 \\ 1 & 2 \end{bmatrix}$$

The variances of the two linear comparisons may now be computed from the following matrix multiplication:

$$V(Q) = \lambda_r \mathbf{Z}_a \lambda_r' \hat{\sigma}_e^2$$

$$= \begin{bmatrix} 3 & 0 \\ 1 & 2 \end{bmatrix} \begin{bmatrix} 0.052602 & -0.032694 \\ -0.032694 & 0.042413 \end{bmatrix} \begin{bmatrix} 3 & 1 \\ 0 & 2 \end{bmatrix} \hat{\sigma}_e^2$$

$$= \begin{bmatrix} 0.157806 & -0.098082 \\ -0.012786 & 0.052132 \end{bmatrix} \begin{bmatrix} 3 & 1 \\ 0 & 2 \end{bmatrix} [0.065074]$$

$$= \begin{bmatrix} 0.473418 \\ 0.091478 \end{bmatrix} [0.065074]$$

$$= \begin{bmatrix} 0.030807 \\ 0.005953 \end{bmatrix} = \begin{bmatrix} V(Q_1) \\ V(Q_2) \end{bmatrix} \tag{4.35}$$

If the fixed set of effects for which linear comparisons are desired interact with a random set of effects (Section 3.21.2), then the variance due to the interaction as well as the error variance must be considered in computing the variance of a linear comparison among the estimates of the fixed effects (Harvey, 1970). When disproportionate frequencies exist, use is made of the elements of the variance–covariance matrix and the variance component estimates to compute the variance of these linear comparisons. In this case, the variance of $\hat{t}_i$ is

$$V(\hat{t}_i) = c^{t_i t_i}(\hat{\sigma}_e^2 + k_i \hat{\sigma}_{ta}^2)$$

where $\hat{\sigma}_{ta}^2$ is the estimate of the interaction variance component for treatments and the random source of variation and

$$k_i = \frac{\sum_j n_{ij}^2}{n_{i.}}$$

with $n_{ij}$ being the number of observations in the $j$th random class for the $i$th treatment. The covariance of $\hat{t}_i$ and $\hat{t}_{i'}$ is

$$\text{Cov}(\hat{t}_i, \hat{t}_{i'}) = c^{t_i t_{i'}}[\hat{\sigma}_e^2 + \tfrac{1}{2}(k_i + k_{i'})\hat{\sigma}_{ta}^2] \qquad i \neq i'$$

Therefore, the variance of any linear function of the treatment effects ($Q$) is

$$V(Q) = \left( \sum_i \lambda_i^2 c^{t_i t_i} - 2 \sum_{i \neq i'} \lambda_i \lambda_{i'} c^{t_i t_{i'}} \right) \hat{\sigma}_e^2$$

$$+ \left[ \sum_i \lambda_i^2 c^{t_i t_i} k_i - \sum_{i \neq i'} \lambda_i \lambda_{i'} c^{t_i t_{i'}}(k_i + k_{i'}) \right] \hat{\sigma}_{ta}^2 \qquad (4.36)$$

For illustration, let us consider the three age-of-dam effects estimated for average daily gain of the 65 steers in the data analyzed in Section 3.21.2. The interaction of age-of-dam with sires (random set of effects) approached significance at the 0.10 level of probability. The approximate $F$ test for age-of-dam effects indicated no significance. However, if logical comparisons were planned among age-of-dam classes prior to conducting the experiment the investigator would be interested in whether the individual comparisons show significance or not. In this case there would be no particular reason for even computing the $F$ test for differences among all three ages of dam. Suppose the same two comparisons among the three ages of dam are desired as were considered above, that is,

$$Q_1 = 2\hat{a}_1 - \hat{a}_2 - \hat{a}_3$$

$$Q_2 = \hat{a}_2 - \hat{a}_3$$

In this case,

$$V(Q_1) = 4V(\hat{a}_1) + V(\hat{a}_2) + V(\hat{a}_3) - 4\,\text{Cov}(\hat{a}_1, \hat{a}_2) - 4\,\text{Cov}(\hat{a}_1, \hat{a}_3) + 2\,\text{Cov}(\hat{a}_2, \hat{a}_3)$$

$$= 4c^{a_1 a_1}(\hat{\sigma}_e^2 + k_1 \hat{\sigma}_{sa}^2) + c^{a_2 a_2}(\hat{\sigma}_e^2 + k_2 \hat{\sigma}_{sa}^2) + c^{a_3 a_3}(\hat{\sigma}_e^2 + k_3 \hat{\sigma}_{sa}^2)$$

$$- 4c^{a_1 a_2}[\hat{\sigma}_e^2 + \tfrac{1}{2}(k_1 + k_2)\hat{\sigma}_{sa}^2] - 4c^{a_1 a_3}[\hat{\sigma}_e^2 + \tfrac{1}{2}(k_1 + k_3)\hat{\sigma}_{sa}^2]$$

$$+ 2c^{a_2 a_3}[\hat{\sigma}_e^2 + \tfrac{1}{2}(k_2 + k_3)\hat{\sigma}_{sa}^2]$$

$$= (4c^{a_1 a_1} + c^{a_2 a_2} + c^{a_3 a_3} - 4c^{a_1 a_2} - 4c^{a_1 a_3} + 2c^{a_2 a_3})\hat{\sigma}_e^2$$

$$+ [4c^{a_1 a_1} k_1 + c^{a_2 a_2} k_2 + c^{a_3 a_3} k_3 - 2c^{a_1 a_2}(k_1 + k_2)$$

$$- 2c^{a_1 a_3}(k_1 + k_3) + c^{a_2 a_3}(k_2 + k_3)]\hat{\sigma}_{sa}^2$$

and
$$V(Q_2) = V(\hat{a}_2) + V(\hat{a}_3) - 2 \operatorname{Cov}(\hat{a}_2, \hat{a}_3)$$

$$= c^{a_2a_2}(\hat{\sigma}_e^2 + k_2\hat{\sigma}_{sa}^2) + c^{a_3a_3}(\hat{\sigma}_e^2 + k_3\hat{\sigma}_{sa}^2)$$

$$- 2c^{a_2a_3}[\hat{\sigma}_e^2 + \tfrac{1}{2}(k_2 + k_3)\hat{\sigma}_{sa}^2]$$

$$= (c^{a_2a_2} + c^{a_3a_3} - 2c^{a_2a_3})\hat{\sigma}_e^2$$

$$+ [c^{a_2a_2}k_2 + c^{a_3a_3}k_3 - c^{a_2a_3}(k_2 + k_3)]\hat{\sigma}_{sa}^2$$

The information needed for the calculation of these two variances is obtained from Section 3.21.2 and is summarized below:

$$\mathbf{Z}_a = \begin{bmatrix} 0.052602 & -0.032694 & -0.019908 \\ -0.032694 & 0.042413 & -0.009719 \\ -0.019908 & -0.009719 & 0.029627 \end{bmatrix}$$

$$\hat{\sigma}_e^2 = 0.056682 \qquad \hat{\sigma}_{sa}^2 = 0.014627$$

$$k_1 = 2.000000 \qquad k_2 = 2.000000 \qquad k_3 = 4.29730$$

$$Q_1 = (2)(0.062156) - (-0.006929) - (-0.055227)$$

$$= 0.186468$$

$$Q_2 = -0.006929 - (-0.055227)$$

$$= 0.048298$$

The variances and standard errors of the two linear comparisons are computed as follows:

$$V(Q_1) = [(4)(0.052602) + 0.042413 + 0.029627 - (4)(-0.032694)$$

$$- (4)(-0.019908) + (2)(-0.009719)](0.056682)$$

$$+ [(4)(0.052602)(2) + (0.042413)(2) + (0.029627)(4.29730)$$

$$- (2)(-0.032694)(2 + 2) - (2)(-0.019908)(2 + 4.29730)$$

$$+ (-0.009719)(2 + 4.29730)](0.014627)$$

$$= (0.473418)(0.056682) + (1.084040)(0.014627)$$

$$= 0.026834 + 0.015856$$

$$= 0.042690$$

$$s_{Q_1} = 0.206616$$

$$V(Q_2) = [0.042413 + 0.029627 - (2)(-0.009719)](0.056682)$$

$$+ [(0.042413)(2) + (0.029627)(4.29730)$$

$$- (-0.009719)(2 + 4.29730)](0.014627)$$

$$= (0.091478)(0.056682) + (0.273346)(0.014627)$$

$$= 0.005185 + 0.003998$$

$$= 0.009183$$

$$s_{Q_1} = 0.095829$$

The $t$ values may now be computed as follows:

$$Q_1: \quad t = \frac{0.186468}{0.206616} = 0.902$$

$$Q_2: \quad t = \frac{0.048298}{0.095829} = 0.504$$

The number of degrees of freedom associated with these $t$ values is unknown but it lies somewhere between 14 (df for the $SA$ interaction) and 40 (df for error). Of course, in this example, it is clear that neither comparison differs significantly even if there were 40 degrees of freedom. Perhaps a simple average of the two degrees of freedom (27) would be accurate enough for most applications.

A simpler method of computing approximate tests of significance for the two comparisons is to use the mean square for the interaction and the same linear function of the inverse elements as used when both sets of effects are fixed. The computations required in this case are as follows:

$$V(Q) = \begin{bmatrix} 3 & 0 \\ 1 & 2 \end{bmatrix} \begin{bmatrix} 0.052602 & -0.032694 \\ -0.032694 & 0.042413 \end{bmatrix} \begin{bmatrix} 3 & 1 \\ 0 & 2 \end{bmatrix} MS(SA)$$

$$= \begin{bmatrix} 0.473418 \\ 0.091478 \end{bmatrix} [0.089051]$$

$$= \begin{bmatrix} 0.042158 \\ 0.008146 \end{bmatrix}$$

$$s_Q = \begin{bmatrix} 0.205324 \\ 0.090256 \end{bmatrix}$$

$$Q_1: \quad t = \frac{0.186468}{0.205324} = 0.908 \quad df = 14$$

$$Q_2: \quad t = \frac{0.048298}{0.090256} = 0.535 \quad df = 14$$

This approximation assumes that $k_1 = k_2 = k_3$ and that this is the coefficient of $\sigma_{sa}^2$ in the expected mean square for $SA$. The more variable the $k_i$, the less accurate will be these tests of significance.

## 4.10.2  Approximate Mean Separation Procedures

When disproportionate subclass frequencies exist there is no common least significant difference (LSD) that can be used to determine the significance of all pairwise differences among treatment means. The difference needed for significance at the $\alpha$ level of probability is $t_\alpha s_d$, where $s_d$ is the standard error of the difference between the two treatment means being compared, that is,

$$s_d^2 = V(\hat{\mu} + \hat{t}_i) + V(\hat{\mu} + \hat{t}_j) - 2\,\text{Cov}[(\hat{\mu} + \hat{t}_i), (\hat{\mu} + \hat{t}_j)]$$

$$= V(\hat{t}_i) + V(\hat{t}_j) - 2\,\text{Cov}(\hat{t}_i, \hat{t}_j)$$

$$= (c^{t_it_i} + c^{t_jt_j} - 2c^{t_it_j})\hat{\sigma}_e^2$$

for the fixed linear model or when the treatment effects do not interact with any set of random effects. It is advisable to use the $t$ test as a mean separation procedure only when the $F$ test in the ANOVA is significant (see Carmer and Swanson, 1973).

Duncan's multiple-range test, as modified by Kramer (1957), may be used to test all pairwise comparisons. If the value

$$(\hat{t}_i - \hat{t}_j)\sqrt{\frac{2}{c^{ii} + c^{jj} - 2c^{ij}}}$$

is greater than $\hat{\sigma}_e z_{p,n_2}$, the difference is significant; where $z_{p,n_2}$ is the significant Studentized range value in Duncan's tables, $p$ is the number of means in the range chosen, and $n_2$ is the number of degrees of freedom for error.

To illustrate these two procedures of mean separation let us consider the means for breeds obtained in the analysis of turkey weights in Section 3.20.3. The least-squares means for the four breeds were found to be as follows:

| | |
|---|---|
| Breed 1: 11.412 | Breed 3: 15.645 |
| Breed 2: 13.417 | Breed 4: 13.703 |

These means were found to be highly significantly different in the ANOVA (Table 3.60). The segment of the inverse matrix needed to complete the computations is as follows:

$$\mathbf{Z}_b = \begin{bmatrix} 0.011071 & -0.004060 & -0.004716 & -0.002295 \\ -0.004060 & 0.012822 & -0.005710 & -0.003052 \\ -0.004716 & -0.005710 & 0.014378 & -0.003952 \\ -0.002295 & -0.003052 & -0.003952 & 0.009299 \end{bmatrix}$$

For computational convenience the computations can be carried out as shown in Table 4.11, first for the $t$ tests and then for Duncan's multiple-range test.

For this example, both the $t$ tests and Duncan's multiple-range test indicate that all pairwise comparisons except breeds 2 and 4 differ significantly in turkey weights at the 0.01 level of probability.

In mixed models with disproportionate subclass frequencies, where the fixed set of effects interacts with a random set of effects, the same procedures can be used as described above except $\hat{\sigma}_e$ would be the square root of the mean square for the interaction. In this case, the pairwise tests of significance would be approximate. The same is true if the approximate error term for the fixed set of effects is a random set of effects that is nested within the fixed effects.

**TABLE 4.11**   MEAN SEPARATION BY TWO METHODS

$t$ Test

| Comparison | $\hat{b}_i - \hat{b}_j$ | $\sqrt{\dfrac{1}{c^{ii} + c^{jj} - 2c^{ij}}}$ | Product | $t_{0.05}\hat{\sigma}_e$ | $t_{0.01}\hat{\sigma}_e$ |
|---|---|---|---|---|---|
| $\hat{b}_1 - \hat{b}_2$ | $-2.005$ | 5.5890 | $-11.21$** | 3.31 | 4.37 |
| $\hat{b}_1 - \hat{b}_3$ | $-4.233$ | 5.3543 | $-22.66$** | 3.31 | 4.37 |
| $\hat{b}_1 - \hat{b}_4$ | $-2.291$ | 6.3296 | $-14.50$** | 3.31 | 4.37 |
| $\hat{b}_2 - \hat{b}_3$ | $-2.228$ | 5.0885 | $-11.34$** | 3.31 | 4.37 |
| $\hat{b}_2 - \hat{b}_4$ | $-0.286$ | 5.9523 | 1.70 | 3.31 | 4.37 |
| $\hat{b}_3 - \hat{b}_4$ | 1.942 | 5.6271 | $10.93$** | 3.31 | 4.37 |

Duncan's multiple-range test

| Comparison | $\hat{b}_i - \hat{b}_j$ | $\sqrt{\dfrac{2}{c^{ii} + c^{jj} - 2c^{ij}}}$ | Product | $\hat{\sigma}_e z_{p,n_2}$ $(\alpha = 0.01)$ |
|---|---|---|---|---|
| $\hat{b}_1 - \hat{b}_2$ | $-2.005$ | 7.9041 | $-15.85$** | 6.18 |
| $\hat{b}_1 - \hat{b}_3$ | $-4.233$ | 7.5722 | $-32.05$** | 6.67 |
| $\hat{b}_1 - \hat{b}_4$ | $-2.291$ | 8.9514 | $-20.51$** | 6.49 |
| $\hat{b}_2 - \hat{b}_3$ | $-2.228$ | 7.1963 | $-16.03$** | 6.49 |
| $\hat{b}_2 - \hat{b}_4$ | $-0.286$ | 8.4178 | $-2.41$ | 6.18 |
| $\hat{b}_3 - \hat{b}_4$ | 1.942 | 7.9580 | $15.45$** | 6.18 |

## EXERCISES

**4.1.** The data presented below have been adapted from an experiment conducted by David Weeks in the Department of Forestry and Wildlife Management at the University of Massachusetts involving the measurement of phosphorus in the foliage of trees subjected to four different treatments. A control group was included in the experiment and the observations are percentage of phosphorus by weight. Complete an analysis of variance for this set of data. Calculate the sums of squares for the following single degree of freedom comparisons and test each for significance.

**(a)** Control versus treated groups.
**(b)** $T_1 + T_2$ versus $T_3 + T_4$ (root cutting versus no root cutting).
**(c)** $T_1 + T_3$ versus $T_2 + T_4$ (soil removal versus stem girdling).
**(d)** $T_1 + T_4$ versus $T_2 + T_3$ [interaction between (b) and (c)].

Treatments

| Control | Soil removal | Stem girdling | Root cutting and soil removal | Root cutting and stem girdling |
|---|---|---|---|---|
| 0.123 | 0.137 | 0.121 | 0.115 | 0.110 |
| 0.121 | 0.137 | 0.145 | 0.115 | 0.118 |
| 0.123 | 0.163 | 0.118 | 0.118 | 0.121 |
| 0.121 | 0.113 | 0.143 | 0.123 | 0.118 |
| 0.121 | 0.151 | 0.121 | 0.134 | 0.129 |
| 0.110 | 0.140 | 0.131 | 0.126 | 0.129 |

**4.2.** The selected, unpublished data presented below are drawn from an experiment conducted by Dr. William J. Lord of the Department of Plant and Soil Sciences at the University of Massachusetts. The experiment included eight blocks and four strains of McIntosh apple trees and four strains of Delicious apple trees. The observations are trunk circumference in centimeters. Complete an analysis of variance for this two-way classification and then partition the sum of squares among the eight strains into the following:

**(a)** The sum of squares between the first four strains and the last four strains.
**(b)** The sum of squares among the first four strains.
**(c)** The sum of squares among the second four strains.
Test for the significance of each of these components.

| | | | | Block | | | | |
|---|---|---|---|---|---|---|---|---|
| Strain | 1 | 2 | 3 | 4 | 5 | 6 | 7 | 8 |
| 1 | 13.5 | 15.2 | 11.2 | 15.2 | 12.3 | 14.8 | 10.0 | 15.0 |
| 2 | 11.1 | 12.8 | 15.1 | 12.4 | 8.3 | 9.9 | 13.5 | 12.3 |
| 3 | 12.5 | 12.9 | 17.8 | 12.8 | 12.5 | 18.8 | 16.0 | 10.6 |
| 4 | 10.8 | 10.0 | 8.4 | 8.6 | 8.7 | 8.0 | 11.4 | 9.4 |
| 5 | 10.2 | 10.0 | 11.0 | 11.0 | 9.8 | 11.0 | 10.1 | 14.1 |
| 6 | 10.7 | 11.6 | 10.8 | 10.9 | 10.3 | 14.2 | 8.2 | 10.2 |
| 7 | 11.8 | 10.9 | 11.2 | 8.6 | 10.0 | 9.3 | 10.1 | 7.0 |
| 8 | 12.9 | 11.2 | 11.5 | 11.5 | 12.6 | 12.5 | 11.5 | 11.7 |

**4.3.** Referring back to Exercise 4.1, write out the linear functions of the means involved in comparisons (b) and (c), calculate the standard errors of these linear functions, and test for the significance of these linear functions by means of the $t$ test. Check that $t^2 = F$ for each test.

**4.4.** The selected shear tests on beef presented below are from an experiment conducted under the direction of Dr. Ernest M. Buck of the Department of Food Sciences and Nutrition at the University of Massachusetts in which five treatments were involved. The measurements are grams of force required to shear the samples of beef. Complete an analysis of variance for this set of data. Write out the following linear functions, calculate the standard errors of the functions, and test for their significance by means of the $t$ test.

**(a)** Treatment 1 versus all other treatments.
**(b)** Treatments 2 and 3 versus treatments 4 and 5.
**(c)** Treatment 2 versus treatment 3.
**(d)** Treatment 4 versus treatment 5.

| | | Treatment | | |
|---|---|---|---|---|
| 1 | 2 | 3 | 4 | 5 |
| 300 | 700 | 500 | 650 | 550 |
| 400 | 650 | 550 | 600 | 200 |
| 450 | 550 | 600 | 650 | 350 |
| 400 | 650 | 600 | 700 | 600 |
| 350 | 500 | 450 | | 500 |
| 350 | | 550 | | |
| 500 | | | | |

**4.5.** Using the data shown in Table 3.7, calculate a linear function using the mean of the first level of nitrogen and the mean of the third level of nitrogen for the source of ammonium sulfate, with the coefficients $-1$ and 1. Similarly, calculate a linear function for the first and third levels of nitrogen for the nitrate source using the same coefficients. Calculate the standard error of each function. Now calculate a linear function of the difference between those two functions, $Q = Q_1 - Q_2$, and test for its significance. This is a test for the significance of the difference between the linear regression lines for the levels of nitrogen for the two sources of nitrogen.

**4.6.** Dr. Anne Katz, formerly of the Department of Veterinary and Animal Sciences at the University of Massachusetts conducted an experiment in animal physiology involving five treatments with five rabbits on each treatment. A few selected observations are shown below. The observations represent measures of fluorescence in tissue sections, in average counts per a specified grid area. Complete an analysis of variance for this set of data. Then make pairwise comparisons among all treatment means, using the following methods, at the 5 percent level:
**(a)** Least significant difference (LSD).
**(b)** Duncan's new multiple-range test.
**(c)** Student–Newman–Keuls test.
**(d)** Tukey's $w$ procedure.
**(e)** Scheffé's test.
Note the differences in conclusions using the various tests.

| Treatment | | | | |
|---|---|---|---|---|
| 1 | 2 | 3 | 4 | 5 |
| 33.0 | 52.4 | 73.4 | 21.4 | 44.2 |
| 29.0 | 31.0 | 50.8 | 21.2 | 50.2 |
| 29.6 | 31.0 | 51.6 | 33.8 | 54.0 |
| 32.4 | 48.6 | 56.8 | 23.0 | 55.4 |
| 27.6 | 28.0 | 60.0 | 25.0 | 54.0 |

**4.7.** A portion of an experiment conducted by Dr. William J. Lord of the Department of Plant and Soil Sciences at the University of Massachusetts in which five strains of McIntosh apples were included, four spur strains and one nonspur strain, is shown below. The data are the lateral branching per meter of limb length on the trees. Complete an analysis of variance for this set of data and then, making use of Dunnett's procedure, test for the significance of the difference between the nonspur and each of the spur tree species in turn. Carry out this two-tailed test at the 5 percent level.

| Strain | 1 | 2 | 3 | 4 | 5 | 6 | 7 | 8 |
|---|---|---|---|---|---|---|---|---|
| Nonspur | 0.74 | 2.90 | 2.65 | 4.61 | 2.56 | 4.00 | 1.75 | 3.38 |
| Spur 1 | 3.55 | 1.92 | 3.64 | 0.63 | 1.16 | 2.01 | 1.50 | 2.46 |
| Spur 2 | 1.35 | 0.00 | 0.42 | 0.00 | 0.04 | 0.06 | 2.13 | 0.10 |
| Spur 3 | 3.58 | 2.09 | 2.47 | 2.44 | 0.55 | 0.81 | 2.22 | 1.39 |
| Spur 4 | 0.74 | 1.41 | 2.10 | 3.20 | 0.52 | 4.46 | 2.04 | 0.37 |

**4.8.** Grease fleece weights were obtained from 670 ewes of the same breed over a four year period. The distribution of numbers and fleece weight totals (in grams) by age of ewe and year are given in the table below. The total uncorrected sum of squares for fleece weight was 1,371,239,778. Complete the least-squares analysis for these data under the two-way classification fixed model without interaction (Section 3.20.1). Use the $t$ test to test for the significance of all differences among the three ages of ewes (Section 4.10.2). Prepare a table giving the least-squares means for years and ages of ewes with standard errors. Interpret your results.

| Year | | 2 | 3 | 4-up | Totals |
|------|------|------|------|------|------|
| | | | **Age** | | |
| 1969 | Number | 61 | 59 | 42 | 162 |
| | Total | 90,892 | 84,787 | 56,564 | 232,243 |
| | Mean | 1,490 | 1,437 | 1,347 | 1,434 |
| 1970 | Number | 49 | 40 | 23 | 112 |
| | Total | 65,635 | 58,878 | 33,251 | 157,764 |
| | Mean | 1,339 | 1,472 | 1,446 | 1,409 |
| 1971 | Number | 86 | 58 | 69 | 213 |
| | Total | 117,486 | 82,506 | 98,613 | 298,605 |
| | Mean | 1,336 | 1,423 | 1,429 | 1,402 |
| 1972 | Number | 6 | 65 | 112 | 183 |
| | Total | 7,847 | 92,440 | 163,265 | 263,552 |
| | Mean | 1,308 | 1,422 | 1,458 | 1,440 |
| Totals | Number | 202 | 222 | 246 | 670 |
| | Total | 281,860 | 318,611 | 351,693 | 952,164 |
| | Mean | 1,395 | 1,435 | 1,430 | 1,421 |

# Regression and Correlation Analyses

## 5.1 INTRODUCTION

In Chapters 1–4 we considered only a single measurement on each individual or experimental unit. We are now going to concern ourselves with inferences based on two or more observations on each experimental unit. Such measurements could be age and weight of a plant, an animal, or a human; aptitude scores and performance grades of a group of students; or initial weight and amount of growth of a group of animals. In dealing with data of these types, we use the method of analysis known as regression. When conducting a regression analysis we are concerned with the dependence of one variable $Y$ on another variable $X$. Thus, $Y$ is considered to be a *dependent variable* and $X$ an *independent variable.* In the examples above we would expect the weight of a plant to be dependent to a large extent on the age, the performance grades to be dependent in part on aptitude, and the amount of growth to be dependent in part on initial weight.

While the uses of regression are many, we shall be concerned primarily with three. First, regression analysis can help inform us as to what has happened in a given situation and allow us to display the results graphically. Second, regression allows us to predict what we can expect to occur under particular conditions. Third, regression can be used to yield a degree of statistical control in an experiment by reducing the unexplained experimental error.

## 5.2 LINEAR REGRESSION

Theoretically, we are concerned with one of two types of model in regression analysis. However, since the solutions to both models are similar in almost every respect, little note is made of the type of model in either discussing an analysis or presenting the results.

Model I of regression is concerned with the situation where the $X$ values are specified by the experimental design and are considered as fixed $X$'s. The $Y$ values are selected from several populations (or subpopulations), each of which is determined by the selected $X$. These subpopulations of $Y$ are assumed to be normally distributed and have a common variance, and the $Y$ values are selected at random. The $X$ values are considered to be measured without error.

In Model II the $X$'s as well as the $Y$'s are random and we are dealing with random samples of pairs drawn from a bivariate distribution. Constant (homoscedastic) variance is assumed for $X$ at each value of $Y$.

The linear regression analysis is concerned with selecting the straight line that fits the data "best," usually expressed as the *best-fitting straight line.* Our criterion for the best-fitting straight line is that the sum of squares of the deviations of the observed values of $Y$ from the regression line is a minimum. This property is frequently referred to as a *least-squares solution.*

As a first example of regression let us consider the data shown in Table 5.1, supplied by Edward Pickering of the Department of Plant and Soil Sciences at the University of Massachusetts. The data are a selection from a much larger set of observations.

Our interest in this set of data is to measure the change in strength of the soil (a measurement of force) for a given change in the weight of the water in the soil. We would like to know, first, whether the strength of the soil changes as the weight of the water changes and if so to what degree. The change can be described as a moving average, the average change in $Y$ for a given change in $X$, and is known as the *coefficient of regression* or *regression coefficient,* usually designated as $\beta$ or $b$. To arrive at the value of the regression coefficient, we use the model

$$Y_i = \mu + \beta(X_i - \bar{x}) + e_i \tag{5.1}$$

**TABLE 5.1**   WEIGHT OF WATER IN SOIL
              ($X$, IN GRAMS) AND
              STRENGTH OF SOIL
              ($Y$, IN KILOGRAMS/CENTIMETER$^2$)

| X | Y | X | Y |
|------|-------|------|-------|
| 10.6 | 0.752 | 10.2 | 0.713 |
| 10.3 | 0.909 | 11.6 | 0.424 |
| 8.6 | 0.939 | 11.1 | 0.513 |
| 11.8 | 0.256 | 9.3 | 0.987 |
| 9.8 | 0.383 | 7.7 | 0.987 |
| 9.1 | 0.833 | 9.5 | 0.518 |
| 10.7 | 0.669 | 9.9 | 0.934 |
| 10.4 | 0.924 | 8.5 | 0.968 |
| 8.8 | 0.932 | 9.0 | 0.936 |
| 8.7 | 0.849 | 10.8 | 0.799 |

$$\Sigma X = 196.4 \qquad \Sigma Y = 15.225$$
$$\Sigma X^2 = 1952.02 \qquad \Sigma Y^2 = 12.562235$$
$$\Sigma XY = 146.2327$$

where $Y_i$ is any value of the dependent variable, $\mu$ is the population mean of the $Y$'s, $\beta$ is the coefficient of regression, $X_i$ is the value of the independent variable corresponding to $Y_i$, $\bar{x}$ is the mean of the independent variable, and $e_i$ is the random error associated with the $i$th value of $Y$. The random errors are assumed to be drawn from a normal population with a mean of 0 and a variance of $\sigma^2$.

The data presented in Table 5.1 can be displayed in a scatter diagram as shown in Figure 5.1. The straight line passing through the points making up the scatter diagram in this figure is the linear regression line which is developed by substituting our sample statistics $\bar{y}$, $b$, and $\bar{x}$ in formula (5.1), giving

$$Y_i = \bar{y} + b(X_i - \bar{x}) + \hat{e}_i \tag{5.2}$$

We can use this relationship to estimate values of $Y$ as

$$\hat{Y}_i = \bar{y} + b(X_i - \bar{x}) = \bar{y} - b\bar{x} + bX_i$$

and letting $a = \bar{y} - b\bar{x}$ we have the prediction equation

$$\hat{Y}_i = a + bX_i \tag{5.3}$$

where $\hat{Y}_i$ is a predicted value, $a$ is the *intercept* or the point where the regression line passes through the $Y$ axis when $X = 0$, $b$ is the regression coefficient, and $X_i$ is any selected value of the independent variable. This equation is frequently written without the subscripts for convenience. The value of $b$ is given by

$$b = \frac{\Sigma(X - \bar{x})(Y - \bar{y})}{\Sigma(X - \bar{x})^2} = \frac{\Sigma\,xy}{\Sigma\,x^2} = \frac{\Sigma\,XY - (\Sigma\,X)(\Sigma\,Y)/n}{\Sigma\,X^2 - (\Sigma\,X)^2/n} \tag{5.4}$$

where $n$ is the number of pairs of $X$'s and $Y$'s.

The calculations necessary to yield the prediction equation for the data displayed in Table 5.1 are shown below.

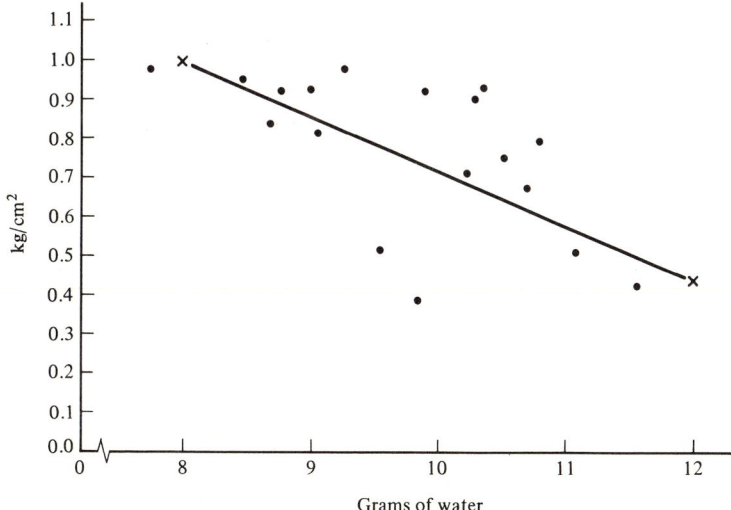

**Figure 5.1** Scatter diagram and linear regression line for data presented in Table 5.1.

$$\bar{x} = \frac{\Sigma X}{n} = \frac{196.4}{20} = 9.82 \qquad \bar{y} = \frac{\Sigma Y}{n} = \frac{15.225}{20} = 0.76125$$

$$\Sigma xy = \Sigma XY - \frac{(\Sigma X)(\Sigma Y)}{n} = 146.2327 - \frac{(196.4)(15.225)}{20} = -3.2768$$

$$\Sigma x^2 = \Sigma X^2 - \frac{(\Sigma X)^2}{n} = 1952.02 - \frac{(196.4)^2}{20} = 23.372$$

$$b = \frac{\Sigma xy}{\Sigma x^2} = \frac{-3.2768}{23.372} = -0.140202$$

$$a = \bar{y} - b\bar{x} = 0.76125 - (-0.140202)(9.82) = 2.138034$$

$$\hat{Y}_i = 2.138034 - 0.140202X_i$$

To draw a straight line regression through the scatter points in Figure 5.1 all that we need to do is predict two points, one near each extreme of the range of values of $X$. Values of $X$ are shown on the horizontal axis, known as the *abscissa,* and values of $Y$ are shown on the perpendicular axis, known as the *ordinate.* Using the values of $X = 8$ and $X = 12$ we find

| $X_i$ | $\hat{Y}_i = 2.138034 - 0.140202X_i$ |
|---|---|
| 8 | $2.138034 - (0.140202)(8) \ \ = 1.02$ |
| 12 | $2.138034 - (0.140202)(12) = 0.46$ |

Plotting these two points, represented by small ×'s, in Figure 5.1 and connecting them by a straight line, we have the linear regression line shown in that figure. The $a$ and $b$ constants have been carried out to more decimal places than necessary to reduce rounding errors in later calculations.

### 5.2.1  Test of Significance of Linear Regression

We are usually interested in knowing whether the regression coefficient that we have calculated differs from a given value. Most commonly, we test for the significance of the difference between our calculated regression coefficient and a coefficient of zero. Thus, we are testing to see whether the slope of the line differs significantly from a slope of zero, which would mean that there is no change in $Y$ for a given change in $X$. To perform this test we can make use of the $t$ test in the form

$$t = \frac{b - 0}{s_b} \quad \text{or} \quad t = \frac{b}{s_b} \tag{5.5}$$

where $s_b$ is the *standard error of the regression coefficient,* given by

$$s_b = \sqrt{\frac{s_{y \cdot x}^2}{\Sigma x^2}} = \frac{s_{y \cdot x}}{\sqrt{\Sigma x^2}} \tag{5.6}$$

where $s_{y \cdot x}^2$ is the *variance from regression*. The square root of $s_{y \cdot x}^2$ is the *standard deviation from regression* which is commonly referred to as the *standard error of estimate*. The variance from regression is calculated from the formula

$$s_{y \cdot x}^2 = \frac{\Sigma d_{y \cdot x}^2}{n - 2} \tag{5.7}$$

where $\Sigma d_{y \cdot x}^2$ is the *sum of the squares of the deviations from regression* and $n$ is the number of pairs of $X$'s and $Y$'s. The sum of the squares of the deviations from regression can be found from

$$\Sigma d_{y \cdot x}^2 = \Sigma (Y - \hat{Y})^2 = \Sigma y^2 - \frac{(\Sigma xy)^2}{\Sigma x^2} \tag{5.8}$$

where $\Sigma y^2$ is the total sum of squares and $(\Sigma xy)^2 / \Sigma x^2$ is the *sum of squares due to regression*. We are thus partitioning the total sum of squares among the $Y$'s into a portion due to deviations *from* regression and a portion *due to* regression. In formula (5.7) the degrees of freedom are $n - 2$ since one degree of freedom is used to estimate the mean and one degree of freedom is used to estimate the regression coefficient.

The total sum of squares in our example is

$$\Sigma y^2 = \Sigma (Y - \bar{y})^2 = \Sigma Y^2 - \frac{(\Sigma Y)^2}{n} = 12.562235 - \frac{(15.225)^2}{20}$$

$$= 12.562235 - 11.590031 = 0.972204$$

and the sum of squares due to regression is

$$\frac{(\Sigma xy)^2}{\Sigma x^2} = \frac{(-3.2768)^2}{23.372} = 0.459414$$

Using formulas (5.8), (5.7), and (5.6), respectively, we obtain

$$\Sigma d_{y \cdot x}^2 = \Sigma (Y - \hat{Y})^2 = \Sigma y^2 - \frac{(\Sigma xy)^2}{\Sigma x^2} = 0.972204 - 0.459414 = 0.512790$$

$$s_{y \cdot x}^2 = \frac{\Sigma d_{y \cdot x}^2}{n - 2} = \frac{0.512790}{18} = 0.028488$$

$$s_b = \sqrt{\frac{s_{y \cdot x}^2}{\Sigma x^2}} = \sqrt{\frac{0.028488}{23.372}} = 0.034913$$

From these calculations we arrive at

$$t = \frac{b}{s_b} = \frac{0.1402}{0.0349} = 4.02**$$

Using the $t$ table with 18 degrees of freedom we find that the regression coefficient is highly significantly different from zero; so there is a highly significant negative slope to the line.

Formula (5.8) defines $d_{y \cdot x}$ as $Y - \hat{Y}$, indicating that the sum of squares of the deviations from regression could be calculated by finding each of the deviations, squar-

**TABLE 5.2** CALCULATIONS OF THE SUM OF SQUARES OF THE INDIVIDUAL
DEVIATIONS FROM REGRESSION AND THE SUM OF SQUARES
OF THE INDIVIDUAL DEVIATIONS DUE TO REGRESSION

| | X | Y | $\hat{Y}$ | $Y - \hat{Y} = d_{y \cdot x}$ | $(Y - \hat{Y})^2$ | $\hat{Y} - \bar{y}$ | $(\hat{Y} - \bar{y})^2$ |
|---|---|---|---|---|---|---|---|
| | 10.6 | 0.752 | 0.6519 | 0.1001 | 0.010020 | −0.1094 | 0.011968 |
| | 10.3 | 0.909 | 0.6940 | 0.2150 | 0.046225 | −0.0672 | 0.004516 |
| | 8.6 | 0.939 | 0.9323 | 0.0067 | 0.000045 | 0.1710 | 0.029241 |
| | 11.8 | 0.256 | 0.4837 | −0.2277 | 0.051847 | −0.2776 | 0.077062 |
| | 9.8 | 0.383 | 0.7641 | −0.3811 | 0.145237 | 0.0028 | 0.000008 |
| | 9.1 | 0.833 | 0.8622 | −0.0292 | 0.000853 | 0.1010 | 0.010201 |
| | . | . | . | . | . | . | . |
| | . | . | . | . | . | . | . |
| | . | . | . | . | . | . | . |
| | 10.8 | 0.799 | 0.6239 | 0.1751 | 0.030660 | −0.1374 | 0.018879 |
| Totals | 196.4 | 15.225 | 15.2253 | −0.0003 | 0.512822 | 0.0000 | 0.459380 |

ing the deviations, and summing as shown in Table 5.2. This table also shows that
the sum of squares due to regression can be calculated from individual deviations
as $\Sigma(\hat{Y} - \bar{Y})^2$. There is, of course, some slight rounding error in the totals. While it
would be a very tedious task to calculate $\Sigma d_{y \cdot x}^2$ as $\Sigma(Y - \hat{Y})^2$ and $(\Sigma xy)^2 / \Sigma x^2$ as
$\Sigma(\hat{Y} - \bar{y})^2$, these formulas show more clearly what is involved in the process of obtaining
these sums of squares.

The relationships dealt with in the preceding presentation may perhaps be clarified
by inspection of Figure 5.2. This graphic display is intended to show clearly the division
of the deviation $y = Y - \bar{y}$ into its two portions, the deviation $\hat{Y} - \bar{y}$ which is that
portion accounted for by the regression line and the deviation $Y - \hat{Y}$ which is that
portion not accounted for by regression. Thus,

$$Y - \bar{y} = (\hat{Y} - \bar{y}) + (Y - \hat{Y}) \tag{5.9}$$

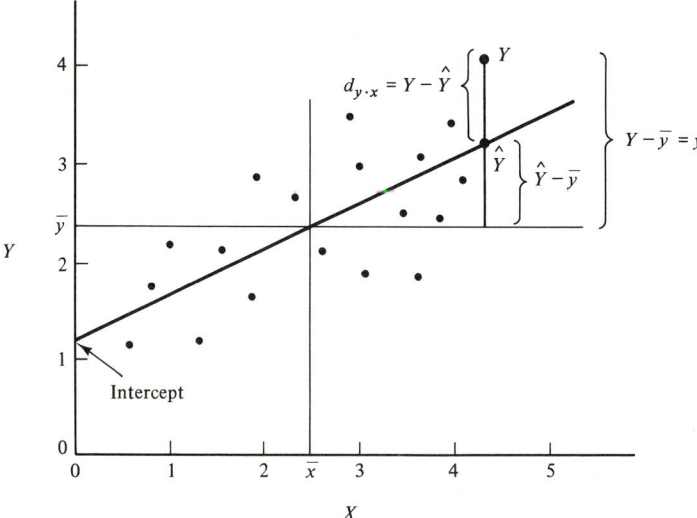

**Figure 5.2** Relationship of deviations involved in regression.

**TABLE 5.3**   ANALYSIS OF VARIANCE FOR REGRESSION

| Source | df | SS | MS | F |
|---|---|---|---|---|
| Total | 19 | 0.9722 | | |
| Linear regression | 1 | 0.4594 | 0.4594 | 16.12** |
| Deviations from regression | 18 | 0.5128 | 0.0285 | |

The point of interception of the $\bar{x}$ and $\bar{y}$ lines in such a graph is known as the *origin of deviations* and the (0, 0) point is known as the *origin of coordinates.* Any point along the regression line could represent a $\hat{Y}$, only one of which is shown represented by a large, solid circle.

The test for the significance of the regression coefficient can also be carried out by means of the $F$ test in an analysis of variance as shown in Table 5.3. It can be seen that

$$F = \frac{0.4594}{0.0285} = 16.12 = t^2 \simeq (4.02)^2 = 16.16$$

As is always the case when the same test is made by both the $t$ test and the $F$ test, $t^2 = F$, when there is one degree of freedom in the numerator.

## 5.2.2 Confidence Interval for the Regression Coefficient

In addition to the test of significance of the regression coefficient, we may wish to calculate a confidence interval about the estimated regression coefficient to show the range which we are confident, to a given degree, includes the population regression coefficient. The degree to which we are confident depends on the confidence limits which are set up, usually 95 percent. We have noted previously that the standard error of a regression coefficient is estimated by

$$s_b = \frac{s_{y \cdot x}}{\sqrt{\Sigma x^2}} \tag{5.10}$$

The 95 percent confidence interval, which is made up of the statistic plus or minus the 95 percent confidence limits, is calculated as

$$\text{CI}_{(b)} = b \pm t_{0.05} s_b = b \pm t_{0.05} \frac{s_{y \cdot x}}{\sqrt{\Sigma x^2}} \tag{5.11}$$

Using the data from Table 5.1 we have

$$\text{CI}_{(b)} = -0.1402 \pm (2.101)(0.0349) = -0.1402 \pm 0.0733$$

$$= (-0.2135, -0.0669)$$

We conclude that we are 95 percent confident that the interval from $-0.2135$ to $-0.0669$ includes the true population regression coefficient. The $t$ value is that for 18 degrees of freedom, or $n - 2$.

### 5.2.3  Standard Error and Confidence Interval for a Subpopulation Mean and Confidence Band About the Regression Line

As noted earlier, one of the assumptions of regression analysis is that for a given $X$ there is a mean value of $Y$, frequently designated as $\mu_{y \cdot x}$ which is given by $\mu_{y \cdot x} = \alpha + \beta X_i$ when $\mu_{y \cdot x}$ is the population mean of $Y$ for $X_i$, $\alpha$ is the population regression line intercept, and $\beta$ is the population regression coefficient. Also, it should be noted that the $\mu_{y \cdot x}$ values fall on a straight line. It can be seen that the equation used earlier

$$\hat{Y} = \bar{y} + b(X - \bar{x}) = a + bX \tag{5.12}$$

yields an estimate of a subpopulation mean $\mu_{y \cdot x}$ for any given $X$. By a subpopulation we mean a population of $Y$ values for a given value of $X$.

Since a subpopulation mean is estimated by Equation (5.12), its variance must take into account both the variance of $\bar{y}$ and the variance of $b(X - \bar{X})$. The covariance between $\bar{y}$ and $b$ can be shown to be zero, so the variance of the subpopulation mean $Y$ is the sum of the variances of $\bar{y}$ and $b(X - \bar{x})$, giving a standard error of

$$s_{\hat{Y}} = \sqrt{\frac{s_{y \cdot x}^2}{n} + s_{y \cdot x}^2 \frac{(X - \bar{x})^2}{\Sigma(X - \bar{x})^2}} = s_{y \cdot x} \sqrt{\frac{1}{n} + \frac{x^2}{\Sigma x^2}} \tag{5.13}$$

If we wish to establish a confidence interval for a given subpopulation mean, the standard error calculated by Equation (5.13) is used. For the 95 percent confidence limits we would have

$$\text{CI}_{(\hat{Y})} = \bar{y} + b(X - \bar{x}) \pm t_{0.05} s_{\hat{Y}} = \hat{Y} \pm t_{0.05} s_{\hat{Y}} \tag{5.14}$$

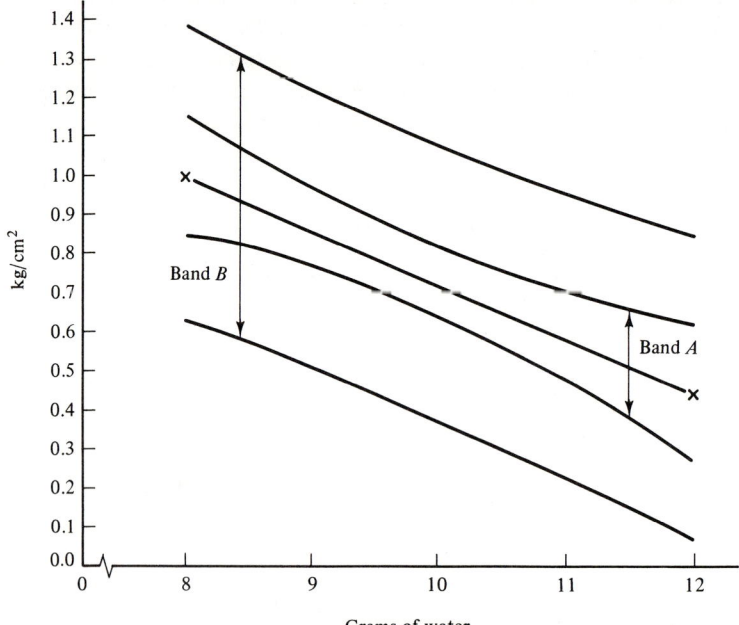

**Figure 5.3**  Regression line for data presented in Table 5.1 with confidence bands for the regression line (*A*) and predicted values (*B*).

where the $t$ value is the tabulated value for $n - 2$ degrees of freedom. Using the data from Table 5.1, we calculate the confidence interval of the subpopulation mean where $X = 10.2$ as

$$CI_{(\hat{Y} \text{ for } X = 10.2)} = 0.7612 - (0.1402)(10.2 - 9.82) \pm (2.101)(0.1688)\sqrt{\frac{1}{20} + \frac{(0.38)^2}{23.372}}$$

$$= 0.7612 - 0.0533 \pm (0.3546)\sqrt{0.05 + 0.006178}$$

$$= 0.7079 \pm (0.3546)(0.2370) = 0.7079 \pm 0.0840$$

$$= (0.6239, 0.7919)$$

We are then 95 percent confident that the interval from 0.6239 to 0.7919 includes the subpopulation true mean value.

If a series of such confidence intervals is calculated, a confidence belt or band can be constructed about the entire regression line. Because of the term $(X - \bar{x})^2/\Sigma(X - \bar{x})^2$ in the standard error of $\hat{Y}$, the confidence limits about the $\hat{Y}$'s increase as the distance between $X$ and $\bar{x}$ increases, resulting in a concave band. The 95 percent confidence belt for the data in Table 5.1 is shown in Figure 5.3.

## 5.2.4 Standard Error and Confidence Interval for a Predicted Value and Confidence Band for Predicted Values

In addition to its utility in describing the response of $Y$ for a series of $X$ values, regression analysis also allows for the prediction of some future event. While we still use formula (5.3) to predict such an event, there is a different concept involved because we are now attempting to predict such an event or to make a statement as to the reliability of this single predicted value rather than that of a mean $Y$, which we considered previously. It is now a matter of developing a variance, or standard error, appropriate to an individual rather than a mean. The standard error of this predicted value is estimated as

$$s_{\hat{Y}_P} = \sqrt{s_{y \cdot x}^2 + \frac{s_{y \cdot x}^2}{n} + s_{y \cdot x}^2 \frac{(X - \bar{x})^2}{\Sigma(X - \bar{x})^2}} = s_{y \cdot x}\sqrt{1 + \frac{1}{n} + \frac{x^2}{\Sigma x^2}} \quad (5.15)$$

where the subscript $P$ is used to indicate that we are now concerned with a predicted value rather than a subpopulation mean.

Again using the data from Table 5.1, if we wish to predict the strength of the soil when there are 10 g of water in the soil, we have

$$\hat{Y} = 2.1380 - (0.1402)(10) = 0.7360$$

The standard error of this value is

$$s_{\hat{Y}_P} = (0.1688)\sqrt{1 + \frac{1}{20} + \frac{(0.18)^2}{23.372}}$$

$$= (0.1688)\sqrt{1.0514} = (0.1688)(1.0254) = 0.1731$$

If we wish to calculate the 95 percent confidence interval for this predicted value, the formula is

$$\text{CI}_{\hat{Y}_P} = \bar{y} + b(X - \bar{x}) \pm t_{0.05}s_{\hat{Y}_P} = \hat{Y} \pm t_{0.05}s_{\hat{Y}_P} \qquad (5.16)$$

Substituting our values in formula (5.16), we have for $X = 10$

$$\text{CI}_{\hat{Y}_P} = 0.7612 - (0.1402)(0.18) \pm (2.101)(0.1731)$$

$$= 0.7360 \pm 0.3637$$

$$= (0.3723, 1.0997)$$

A confidence belt can be constructed for the predicted values (which would, of course, fall on the regression line) by calculating a series of confidence intervals for several predicted values. Since a greater amount of variation is involved with the predicted values than with the subpopulation means, this belt will lie outside the confidence belt for the regression, as shown in Figure 5.3. It can be seen that the component $x^2/\sum x^2$ assumes a smaller portion for the standard error of the predicted values than for the standard error of the subpopulation means. This results in a confidence belt whose lines more closely approach straight lines, although they are still slightly concave. When the published results of a regression analysis include a graph, it is normally the confidence belt for the subpopulation means, discussed in Section 5.2.3, which is presented.

## 5.2.5 Test of Significance Between Two Linear Regression Lines or Coefficients

It is a very common occurrence to test whether there is a significant difference between the slopes of two linear regression lines (which is the same as testing between the two regression coefficients giving rise to these lines). As an example of this test let us look at the data in Table 5.4 which represent the initial weights and amounts of gain of

**TABLE 5.4**   INITIAL WEIGHTS (X, IN GRAMS)
AND GAINS IN WEIGHT OVER A
10-WEEK PERIOD (Y, IN GRAMS)
OF TWO DIFFERENT STRAINS IN RATS

|        | Strain 1 | | Strain 2 | |
|--------|------|------|------|------|
|        | X | Y | X | Y |
|        | 63 | 157 | 72 | 143 |
|        | 74 | 156 | 51 | 105 |
|        | 62 | 114 | 81 | 140 |
|        | 69 | 123 | 70 | 115 |
|        | 55 | 128 | 55 | 102 |
|        | 58 | 108 | 74 | 124 |
|        | 48 | 108 | 61 | 117 |
|        | 54 | 107 | 67 | 104 |
|        |    |     | 54 | 111 |
|        |    |     | 62 | 121 |
| Totals | 483 | 1001 | 647 | 1182 |

two different strains of rats raised on the same diet. If we wish to test whether the initial weight causes a different linear response in weight gains we can calculate the regression coefficient in each group and, by means of the $t$ test shown in formula (5.17), complete the desired test.

$$t = \frac{b_1 - b_2}{s_{b_1 - b_2}} = \frac{b_1 - b_2}{\sqrt{s_p^2(1/\Sigma \, x_1^2 + 1/\Sigma \, x_2^2)}} \tag{5.17}$$

where

$$s_p^2 = \frac{(\Sigma \, y_1^2 - (\Sigma \, x_1 y_1)^2/\Sigma \, x_1^2) + (\Sigma \, y_2^2 - (\Sigma \, x_2 y_2)^2/\Sigma \, x_2^2)}{(n_1 - 2) + (n_2 - 2)} \tag{5.18}$$

The $s_p^2$ has the subscript $p$ to indicate that the variance being estimated is calculated from the pooled sums of squares of the deviations about the two regression lines. The four degrees of freedom subtracted from the total number of observations are the result of calculating two means and two regression coefficients.

From the data shown in Table 5.4 we can calculate the following values:

$$\Sigma \, x_1^2 = 497.875 \qquad \Sigma \, y_1^2 = 3020.875 \qquad \Sigma \, x_1 y_1 = 820.625$$

$$\Sigma \, x_2^2 = 856.100 \qquad \Sigma \, y_2^2 = 1833.600 \qquad \Sigma \, x_2 y_2 = 952.600$$

$$b_1 = \frac{\Sigma \, x_1 y_1}{\Sigma \, x_1^2} = 1.6483 \qquad b_2 = \frac{\Sigma \, x_2 y_2}{\Sigma \, x_2^2} = 1.1127$$

We can now calculate the variance from regression using the pooled sum of squares of deviations from regression and the pooled degrees of freedom.

$$s_p^2 = \frac{1668.2757 + 773.6225}{8 - 2 + 10 - 2} = 174.4213$$

The standard error of the mean difference is then

$$s_{b_1 - b_2} = \sqrt{(174.4213)\left(\frac{1}{497.875} + \frac{1}{856.100}\right)} = \sqrt{(174.4213)(0.00317662)}$$

$$= \sqrt{0.554070} = 0.7444$$

Now,

$$t = \frac{b_1 - b_2}{s_{b_1 - b_2}} = \frac{0.5356}{0.7444} = 0.72$$

The appropriate degrees of freedom for entering the $t$ table are $n_1 - 2 + n_2 - 2 = 14$ and the result shows that there is no significant difference between the two slopes or regression coefficients. This same test can be completed by an $F$ test, as demonstrated in Section 5.2.7, which must be used when testing for the significance of differences among three or more coefficients.

## 5.2.6 Test of Significance Between Two Intercepts or Adjusted Means

While the test of significance between two slopes tells us they do not differ from one another, this does not tell us whether the regression lines are essentially the same since there may be a difference between the elevations of the two lines. Since the two slopes did not differ significantly, we assume that the slopes are the same and develop a single regression coefficient for the two groups, pooling the information as

$$b = \frac{\Sigma\, x_1 y_1 + \Sigma\, x_2 y_2}{\Sigma\, x_1^2 + \Sigma\, x_2^2} = \frac{1773.225}{1353.975} = 1.3096 \qquad (5.19)$$

The difference in elevations of the two lines can be measured by the difference in the intercepts of the two lines, *or* by the difference between the two weight-gain means adjusted with the common regression to the average initial weight. The intercepts are calculated as

$$a_1 = \bar{y}_1 - b\bar{x}_1 = 125.1250 - (1.3096)(60.3750) = 46.0579$$

$$a_2 = \bar{y}_2 - b\bar{x}_2 = 118.2000 - (1.3096)(64.7000) = 33.4689$$

which, of course, are the values where the regression lines pass through the $Y$ axis at $X = 0$. The difference between these two is 12.5890.

If we wish to adjust the means to what they would have been if each group had had the same average initial weights we do this as

$$\bar{y}_{1a} = \bar{y}_1 - b(\bar{x}_1 - \bar{x}) = 125.1250 - (1.3096)(-2.4028) = 128.2717$$

$$\bar{y}_{2a} = \bar{y}_2 - b(\bar{x}_2 - \bar{x}) = 118.2000 - (1.3096)(1.9222) = 115.6827$$

where $\bar{y}_{ia}$ represents an adjusted mean and $\bar{x}$ represents the overall mean of the $X$'s. It can be seen that the difference between the two means adjusted to the same average initial weight is 12.5890, the same difference as that between the two intercepts. The two regression lines must be parallel since both have the same slope so the difference between them is the same whether $X = 0$ or $X = \bar{x}$. Thus, it can be seen that testing between intercepts is the same as testing between adjusted means when there is a common regression.

Since we have the difference between the adjusted means or the intercepts we now need the standard error of this difference to complete a $t$ test. This standard error can be found by the formula

$$s_{\bar{y}_{1a}-\bar{y}_{2a}} = \sqrt{s_{y\cdot x}^2 \left( \frac{1}{n_1} + \frac{1}{n_2} + \frac{(\bar{x}_1 - \bar{x}_2)^2}{\Sigma\, x_1^2 + \Sigma\, x_2^2} \right)} \qquad (5.20)$$

where $n_1$ and $n_2$ are the number of pairs in each group and $s_{y\cdot x}^2$ is calculated by

$$s_{y\cdot x}^2 = \frac{\Sigma\, y_1^2 + \Sigma\, y_2^2 - (\Sigma\, x_1 y_1 + \Sigma\, x_2 y_2)^2 / (\Sigma\, x_1^2 + \Sigma\, x_2^2)}{n_1 + n_2 - 3} \qquad (5.21)$$

Substituting in formulas (5.21) and (5.20) the values found earlier, we have

$$s_{y\cdot x}^2 = \frac{4854.475 - (1773.225)^2 / 1353.975}{8 + 10 - 3} = 168.8121$$

and

$$s_{\bar{y}_{1a}-\bar{y}_{2a}} = \sqrt{168.8121 \left( \frac{1}{8} + \frac{1}{10} + \frac{(-4.325)^2}{1353.975} \right)} = 6.3492$$

The $t$ test for the difference between the two intercepts or the two adjusted means is

$$t = \frac{\bar{y}_{1a} - \bar{y}_{2a}}{s_{\bar{y}_{1a} - \bar{y}_{2a}}} = \frac{12.5890}{6.3492} = 1.98$$

with 15 degrees of freedom, a nonsignificant result. Three degrees of freedom are subtracted from the total due to calculation of the two means and one regression coefficient. The same results can be obtained by using the $F$ test, as demonstrated in Section 5.2.8.

## 5.2.7  Test of Significance Among Several Regression Lines

The situation frequently arises where it is desired to know whether there are significant differences among the slopes of several linear regression lines. In this situation a $t$ test no longer suffices and instead we make use of the $F$ test. As an example of such a problem let us look at Table 5.5 which presents the Kjeldahl nitrogen values and dry weights of plants from different treatment groups. The data were supplied by Kevin Duffy of the Department of Plant and Soil Sciences at the University of Massachusetts. To test for the significance of the differences among the linear regressions of dry weight on the Kjeldahl nitrogen in the four groups we can make the necessary calculations and set up Table 5.6. The first four lines of the table are developed from within each group separately, the sum of squares of deviations from regression being calculated by the formula

$$\sum d^2_{y \cdot x} = \sum y^2 - \frac{(\sum xy)^2}{\sum x^2} \tag{5.22}$$

The sums of squares from each of the four regression lines are totaled, giving a value of 1233.6195 with 24 degrees of freedom from which an average variance from regression of 51.4008 is calculated. Now all the values calculated for sums of squares and

**TABLE 5.5**  KJELDAHL NITROGEN VALUES ($X$, IN PERCENT) AND DRY WEIGHTS ($Y$, IN GRAMS) OF PLANTS FROM FOUR DIFFERENT TREATMENT GROUPS

|  | Group 1 | | Group 2 | | Group 3 | | Group 4 | |
|---|---|---|---|---|---|---|---|---|
|  | KjN | DW | KjN | DW | KjN | DW | KjN | DW |
|  | 2.54 | 6.64 | 2.52 | 7.03 | 1.86 | 9.54 | 2.85 | 12.15 |
|  | 2.69 | 10.29 | 2.05 | 4.46 | 2.32 | 5.41 | 2.28 | 9.62 |
|  | 3.14 | 18.16 | 1.95 | 6.86 | 2.08 | 7.52 | 2.30 | 10.06 |
|  | 2.86 | 19.17 | 2.66 | 3.61 | 2.50 | 10.09 | 2.67 | 18.93 |
|  | 2.80 | 5.21 | 2.38 | 9.88 | 2.93 | 21.84 | 2.67 | 13.00 |
|  | 2.19 | 9.24 | 2.37 | 9.46 | 2.53 | 21.23 | 2.32 | 5.89 |
|  | 2.98 | 6.99 | 2.60 | 12.78 | 2.70 | 27.78 | 2.73 | 13.51 |
|  | 2.45 | 20.60 | 2.70 | 19.38 | 2.54 | 28.28 | 2.54 | 34.56 |
| $\sum X, \sum Y$ | 21.65 | 96.30 | 19.23 | 73.46 | 19.46 | 131.69 | 20.36 | 117.72 |
| $\sum X^2, \sum Y^2$ | 59.2459 | 1432.9900 | 46.7623 | 855.4230 | 48.1458 | 2777.8235 | 52.1556 | 2280.3212 |
| $\sum XY$ | 262.5181 | | 181.3268 | | 335.7025 | | 303.2817 | |

**TABLE 5.6**   COMPUTATIONS FOR TESTS OF SIGNIFICANCE AMONG
REGRESSION COEFFICIENTS AND ADJUSTED MEANS

| Line | Group | df | $\Sigma x^2$ | $\Sigma xy$ | $\Sigma y^2$ | $b$ | df | SS | MS |
|------|-------|-----|--------|---------|-----------|---------|-----|-----------|---------|
| | | | Sums of squares and cross-products | | | | Deviations from regression | | |
| 1 | 1 | 7 | 0.6556 | 1.9062 | 273.7788 | 2.9076 | 6 | 268.2364 | |
| 2 | 2 | 7 | 0.5382 | 4.7473 | 180.8766 | 8.8207 | 6 | 139.0021 | |
| 3 | 3 | 7 | 0.8094 | 15.3666 | 610.0415 | 18.9852 | 6 | 318.3039 | |
| 4 | 4 | 7 | 0.3394 | 3.6843 | 548.0714 | 10.8553 | 6 | 508.0771 | |
| 5 | Pooled deviations | | | | | | 24 | 1233.6195 | 51.4008 |
| 6 | Common | 28 | 2.3426 | 25.7044 | 1612.7683 | 10.9726 | 27 | 1330.7243 | 49.2861 |
| 7 | Regression coefficients (6-5) | | | | | | 3 | 97.1048 | 32.3683 |
| 8 | Total | 31 | 2.7943 | 25.7348 | 1855.8237 | | 30 | 1618.8127 | |
| 9 | Adjusted means (8-6) | | | | | | 3 | 288.0884 | 96.0295 |

cross-products for the individual regression lines, as well as the degrees of freedom, are summed and a common regression line is calculated. This common (or pooled) regression line is a weighted average of the four individual regression lines, the weights being the $\Sigma x^2$ values. Thus, the common or pooled regression is

$$\bar{b} = \frac{\Sigma w_i b_i}{\Sigma w_i} = \frac{25.7044}{2.3426} = 10.9726 \tag{5.23}$$

where $w_i = \Sigma x_i^2$. The sum of squares of deviations about the common regression line is now calculated in the usual fashion in line 6 of Table 5.6. The sum of squares for testing for the significance of the differences among the individual regression coefficients or slopes is now found by the difference between the sum of squares of the deviations about the individual regression lines and the sum of squares of deviations about the common regression line $1330.7243 - 1233.6195 = 97.1048$ with three degrees of freedom. This difference can also be found as

$$\Sigma w_i(b_i - \bar{b})^2 = 97.1048 \tag{5.24}$$

While it is evident that a better fit to a scatter of points is made by developing several regression lines rather than just one, yielding a smaller sum of the squares of the deviations, our test of significance is a question of whether the difference between these two sums of squares is significant. The test of significance among the regression lines uses the sum of squares of deviations about the individual regression lines as an error term, 1233.6195. A comparison of the corresponding mean squares yields a test statistic,

$$F = \frac{32.3683}{51.4008} = 0.63$$

with 3 and 24 degrees of freedom. The $F$ value tells us that there are no significant differences among the four regression coefficients so our interest now focuses on whether there are significant differences among the adjusted means or intercepts. Had there been significant differences among the regression coefficients, we would proceed to test for the significance of the differences among means that have been adjusted to a common $X$ by using the separate regression coefficients.

## 5.2.8  Tests of Significance Among Several Intercepts
### or Adjusted Means

The test of significance among the means adjusted with the common or pooled regression makes use of the last four lines of Table 5.6. Line 8 is made up of the total degrees of freedom and total sums of squares and cross-products together with the sum of squares of deviations about a regression line calculated from these total values. The difference, between the sum of squares of deviations of lines 8 and 6, of 288.0884 is obtained and represents the sum of squares among the adjusted means with three degrees of freedom. The mean square of 96.0295 calculated from these values is tested by the mean square from line 6, giving

$$F = \frac{96.0295}{49.2861} = 1.95$$

with 3 and 27 degrees of freedom.

The $F$ value forces us to conclude that there are no significant differences among the intercepts or the adjusted means. The test among these adjusted means is known as the analysis of covariance and will be discussed in Section 8.2.

## 5.3  CORRELATION COEFFICIENT

A parameter closely related to the regression coefficient is the *coefficient of correlation* or *correlation coefficient*. This value yields a measure of what is usually referred to as concomitant variation, that is, the degree to which two variables vary together. Whereas in regression analysis one variable is said to depend on another, often implying a cause and effect relationship, in correlation analysis no distinction is made between independent and dependent variable. Examples of the use of correlation could be to measure the degree of relationship between the heights of brother and sister, intelligence quotients of mother and number of children in family, and heights of quarterbacks and percentages of passes completed. The correlation coefficient $r$ is an estimate of the population parameter $\rho$ and is calculated as

$$r = \frac{\sum x_i x_2}{\sqrt{\sum x_1^2 x_2^2}} = \frac{\sum X_1 X_2 - (\sum X_1)(\sum X_2)/n}{\sqrt{[\sum X_1^2 - (\sum X_1)^2/n][\sum X_2^2 - (\sum X_2)^2/n]}} \qquad (5.25)$$

This is referred to as the Pearson product-moment formula.

As an example of correlation let us look at the data presented in Table 5.7 which represent the grades students received in two different examinations. If we wish to examine the relationship between the grades of the students in the two examinations we can calculate the correlation, using formula (5.25), as

$$r = \frac{\sum x_1 x_2}{\sqrt{\sum x_1^2 \sum x_2^2}} = \frac{68{,}751 - (811)(841)/10}{\sqrt{[66{,}587 - (811)^2/10][71{,}253 - (841)^2/10]}}$$

$$= \frac{545.9}{654.0191} = 0.8347$$

**TABLE 5.7**   GRADES RECEIVED BY 10 STUDENTS IN
EXAMINATIONS IN CALCULUS ($X_1$) AND
ANCIENT GREEK ($X_2$)

| Student number | $X_1$ | $X_2$ |
|---|---|---|
| 1 | 74 | 81 |
| 2 | 85 | 86 |
| 3 | 94 | 92 |
| 4 | 62 | 74 |
| 5 | 87 | 91 |
| 6 | 78 | 75 |
| 7 | 82 | 86 |
| 8 | 74 | 81 |
| 9 | 83 | 78 |
| 10 | 92 | 97 |

$$\Sigma X_1 = 811 \qquad \Sigma X_2 = 841$$
$$\Sigma X_1^2 = 66{,}587 \qquad \Sigma X_2^2 = 71{,}253$$
$$\Sigma X_1 X_2 = 68{,}751$$

It is obvious from the calculations used for the correlation coefficient that there is a relationship between the correlation and regression coefficients. If we were to consider $X_1$ as being dependent on $X_2$ first and then $X_2$ dependent on $X_1$, we could calculate the corresponding regression coefficients

$$b_{12} = \frac{\Sigma x_1 x_2}{\Sigma x_2^2} = \frac{545.9}{524.9} = 1.0400$$

and

$$b_{21} = \frac{\Sigma x_1 x_2}{\Sigma x_1^2} = \frac{545.9}{814.9} = 0.6699$$

Taking the geometric mean of these two values we have

$$\sqrt{b_{12} b_{21}} = \sqrt{\frac{\Sigma x_1 x_2}{\Sigma x_2^2} \frac{\Sigma x_1 x_2}{\Sigma x_1^2}} = \frac{\Sigma x_1 x_2}{\sqrt{\Sigma x_1^2 \Sigma x_2^2}} = r \qquad (5.26)$$

and using the data from our example we have

$$r = \sqrt{b_{12} b_{21}} = \sqrt{(1.0400)(0.6699)} = 0.8347$$

showing that the correlation coefficient is the geometric mean of the two regression coefficients.

One of the properties of the correlation coefficient is that it has a range from $-1$ to $+1$. Using the $X$ and $Y$ notation as we did previously in regression analysis, we can write

$$\left(\Sigma xy\right)^2 = \frac{(\Sigma xy)(\Sigma xy)}{\Sigma x^2 \Sigma y^2} \Sigma x^2 \Sigma y^2 = r^2 \Sigma x^2 \Sigma y^2$$

and substituting this value in the equation for the sum of squares of deviations from regression $\Sigma y^2 - (\Sigma xy)^2 / \Sigma x^2$, we have

$$\Sigma d_{y \cdot x}^2 = \Sigma y^2 - \frac{r^2 \Sigma x^2 \Sigma y^2}{\Sigma x^2} = \Sigma y^2 - r^2 \Sigma y^2 = (1 - r^2) \Sigma y^2 \qquad (5.27)$$

It can be seen from Equation (5.27) that if $r$ exceeds $-1$ or $+1$ the $\Sigma\, d_{y\cdot x}^2$ would be negative, which, of course, is not possible; thus, the range of the correlation coefficient is limited.

Knowing this range we can see that our correlation of 0.8347 represents a reasonably high relationship between the examinations in the two subjects. A positive correlation indicates that as the examination scores in one subject increase, there is a corresponding increase in scores in the other subject. A negative correlation indicates that as the values in one independent variable increase, the corresponding values in the other variable decrease. A logical next step would be to test for significance of the correlation coefficient.

### 5.3.1 Test of Significance of the Correlation Coefficient

The correlation coefficient can be tested for significance, a test of the null hypothesis that $\rho = 0$, by means of a $t$ test as

$$t = \frac{r}{s_r} \tag{5.28}$$

where $s_r$ is the standard error of the correlation coefficient, which is given by

$$s_r = \sqrt{\frac{1 - r^2}{n - 2}} \tag{5.29}$$

It should be noted that this standard error can only be used when $\rho = 0$ and cannot be used for setting confidence limits or testing for a hypothesis that $\rho$ equals some value other than zero. This is a result of the fact that, due to the limits of $-1$ and $+1$ of the correlation coefficient, the distribution of $r$ is not symmetric when $\rho$ has some value other than zero.

In our example, the standard error of the correlation coefficient is

$$s_r = \sqrt{\frac{1 - (0.8347)^2}{10 - 2}} = \sqrt{0.0379} = 0.1947$$

leading to

$$t = \frac{0.8347}{0.1947} = 4.29$$

with $n - 2 = 8$ degrees of freedom, a highly significant value.

The test of significance of the correlation between $X_1$ and $X_2$ is exactly the same as the test for the significance of the regression of $X_1$ on $X_2$ or the regression of $X_2$ on $X_1$. It can be seen from Equations (5.30), (5.31), and (5.32) that the $t$ tests for all three statistics yield identical values:

$$t = \frac{b_{12}}{s_{b12}} = \frac{(\Sigma\, x_1 x_2)/\Sigma\, x_2^2}{\sqrt{s_{x1\cdot x2}^2/\Sigma\, x_2^2}} = \frac{(\Sigma\, x_1 x_2)/\Sigma\, x_2^2}{\sqrt{[\Sigma\, x_1^2 - (\Sigma\, x_1 x_2)^2/\Sigma\, x_2^2]/(n - 2)\Sigma\, x_2^2}}$$

$$= \frac{\Sigma\, x_1 x_2}{\Sigma\, x_2^2}\sqrt{\frac{(n - 2)(\Sigma\, x_2^2)^2}{\Sigma\, x_1^2 \,\Sigma\, x_2^2 - (\Sigma\, x_1 x_2)^2}} = \sqrt{\frac{(\Sigma\, x_1 x_2)^2(n - 2)}{\Sigma\, x_1^2 \,\Sigma\, x_2^2 - (\Sigma\, x_1 x_2)^2}} \tag{5.30}$$

$$t = \frac{b_{21}}{s_{b_{21}}} = \frac{\Sigma\, x_1 x_2 / \Sigma\, x_1^2}{\sqrt{s_{x_2 \cdot x_1}^2 / \Sigma\, x_1^2}} = \frac{(\Sigma\, x_1 x_2)/\Sigma\, x_1^2}{\sqrt{[\Sigma\, x_2^2 - (\Sigma\, x_1 x_2)^2 / \Sigma\, x_1^2]/(n-2)\Sigma\, x_1^2}}$$

$$= \frac{\Sigma\, x_1 x_2}{\Sigma\, x_1^2} \sqrt{\frac{(n-2)(\Sigma\, x_1^2)^2}{\Sigma\, x_1^2\, \Sigma\, x_2^2 - (\Sigma\, x_1 x_2)^2}} = \sqrt{\frac{(\Sigma\, x_1 x_2)^2 (n-2)}{\Sigma\, x_1^2\, \Sigma\, x_2^2 - (\Sigma\, x_1 x_2)^2}} \qquad (5.31)$$

$$t = \frac{r}{s_r} = \frac{\Sigma\, x_1 x_2 / \sqrt{\Sigma\, x_1^2\, \Sigma\, x_2^2}}{\sqrt{(1-r^2)/(n-2)}} = \frac{\Sigma\, x_1 x_2}{\sqrt{\Sigma\, x_1^2\, \Sigma\, x_2^2}} \sqrt{\frac{n-2}{1 - (\Sigma\, x_1 x_2)^2 / \Sigma\, x_1^2\, \Sigma\, x_2^2}}$$

$$= \frac{\Sigma\, x_1 x_2}{\sqrt{\Sigma\, x_1^2\, \Sigma\, x_2^2}} \sqrt{\frac{n-2}{[\Sigma\, x_1^2\, \Sigma\, x_2^2 - (\Sigma\, x_1 x_2)^2]/\Sigma\, x_1^2\, \Sigma\, x_2^2}}$$

$$= \sqrt{\frac{(\Sigma\, x_1 x_2)^2 (n-2)}{\Sigma\, x_1^2\, \Sigma\, x_2^2 - (\Sigma\, x_1 x_2)^2}} \qquad (5.32)$$

Using formula (5.10), we find $s_{b_{12}} = 0.2426$ and $s_{b_{21}} = 0.1563$:

$$\text{for } b_{12} \quad t = \frac{1.0400}{0.2426} = 4.29$$

$$\text{for } b_{21} \quad t = \frac{0.6699}{0.1563} = 4.29$$

Since the $t$ test for the significance of the correlation coefficient is seen to be [from Equations (5.28) and (5.29)] a function of the size of the coefficient and the degrees of freedom, we can determine from this relationship the size of the correlation coefficient necessary to declare significance for a given number of degrees of freedom. In our example, the $t$ value must equal or exceed 2.306 since we have eight degrees of freedom. Therefore, to declare significance at the 5 percent level,

$$\frac{r}{\sqrt{(1-r^2)/(n-2)}} \geq 2.306$$

and squaring both sides we have

$$\frac{r^2}{(1-r^2)/(8)} \geq 5.3176 \qquad r^2 \geq 0.3993 \quad \text{and} \quad r \geq 0.632$$

Table A.6 has been developed from this relationship and is a convenient table for testing for the significance of correlation coefficients.

The problem of being able to use the standard error of the correlation coefficient, as shown in formula (5.29), only when testing the null hypothesis that $\rho$ equals zero was solved by R. A. Fisher. This solution involved the transformation of $r$ to a value $z$ defined as

$$z = \tfrac{1}{2}[\log_e(1 + r) - \log_e(1 - r)] \qquad (5.33)$$

The quantity $z$ follows a distribution which is almost normal and has a standard deviation of

$$\sigma_z = \frac{1}{\sqrt{n-3}} \qquad (5.34)$$

The parametric standard deviation of $z$ is the standard error of $z$. This standard deviation is independent of the size of the correlation in the population from which the sample is drawn. Fortunately, we do not have to use formula (5.33) each time we wish to transform from $r$ to $z$ or $z$ or $r$. Instead, we can use Table A.7 when we wish to go from $r$ to $z$ and Table A.8 when we wish to go from $z$ to $r$. Inspection of these tables shows that for large values of $r$, large differences between $r$ and $z$ occur. An $r$ value of 0.250 yields a $z$ value of 0.255, a very small difference, while an $r$ value of 0.980 yields a $z$ value of 2.443, a much greater difference.

As one example of the use of the $z$ transformation, suppose that we wish to test whether the correlation coefficient of 0.835 that we calculated earlier differs from a hypothesized value of 0.50. The standard error used for testing for the significance of the difference between a sample $r$ and a population $\rho$ of some value other than zero is the standard error of the transformation or $z$ value of the sample $r$. This standard error is used in a $t$ test as before. For our example, $z_r = 1.204$ for $r = 0.835$ and $z_\rho = 0.549$ for $\rho = 0.50$ and $n = 10$. Therefore,

$$t = \frac{z_r - z_\rho}{s_{z_r}} = \frac{1.204 - 0.549}{\sqrt{1/7}} = \frac{0.655}{0.378} = 1.73$$

Since $z$ follows a normal distribution and we are using the standard error of a population value, we use degrees of freedom of infinity for the $t$ test. Our conclusion is that the correlation coefficient in our sample does not differ significantly from 0.50.

The fact that the significance of a correlation coefficient is affected by both the sample size and the size of the coefficient is of concern when interpreting the importance of correlation coefficients with large samples. Very frequently, a correlation coefficient is declared to be highly significant only because of the large number of degrees of freedom, although its size may be so small as to be meaningless. Interpretation of the importance of the correlation coefficient can be aided by calculating the percentage of the total sum of squares accounted for when either variable is considered as independent. If this percentage is small, the correlation coefficient may have no meaning or use, in spite of its being highly significant.

### 5.3.2 Confidence Interval for the Correlation Coefficient

In setting confidence limits on the correlation coefficient, we again make use of the $z$ transformation. Let us set 95 percent confidence limits on the coefficient of 0.835 which we have calculated. The $z$ value for this $r$ is 1.204 and $\sigma_z = \sqrt{1/(n - 3)} = 0.378$. As we have noted, $z$ follows the normal distribution closely regardless of the size of the sample, so we therefore use the $t$ value for an infinite number of degrees of freedom, 1.96. We then have a confidence interval of

$$\text{CI}_z = z \pm t_{0.05}\sigma_z$$

$$= 1.204 \pm (1.96)(0.378) = 1.204 \pm 0.741$$

$$= (0.463, 1.945) \tag{5.35}$$

Converting these $z$ values to $r$ values by use of Table A.8, we have for the 95 percent confidence interval about the correlation coefficient $r = 0.835$

$$\text{CI}_r = (0.432, 0.960)$$

### 5.3.3 Test of Significance Between Two Correlation Coefficients

The data displayed in Table 5.8 are part of a larger set of similar data collected by Dr. Walter P. Kroll of the Department of Exercise Science at the University of Massachusetts. The correlation between two independent variables for the power athletes was 0.649 and for the endurance athletes it was 0.429. Let us conduct a test of significance of the difference between these two correlation coefficients. Transforming the $r$ values to $z$ values, we find $z_1 = 0.773$ and $z_2 = 0.459$. The standard error for $\sigma_{z_1-z_2}$ takes the form

$$\sigma_{z_1-z_2} = \sqrt{\sigma_{z_1}^2 + \sigma_{z_2}^2} = \sqrt{\frac{1}{n_1 - 3} + \frac{1}{n_2 - 3}} \tag{5.36}$$

which leads to

$$t = \frac{z_1 - z_2}{s_{z_1-z_2}} = \frac{z_1 - z_2}{\sqrt{1/(n_1 - 3) + 1/(n_2 - 3)}} \tag{5.37}$$

For our example, we have

$$t = \frac{0.773 - 0.459}{\sqrt{1/5 + 1/5}} = \frac{0.314}{0.632} = 0.50 \quad \text{with} \quad df = \infty$$

leading to the conclusion that there is no significant difference between the two correlation coefficients.

In this test, as well as the following test for the homogeneity of several correlation coefficients, we are testing between or among correlations of the same two variables in different groups or populations. For a discussion of the problem of comparing dependent correlations, such as $r_{12}$ and $r_{13}$, where one variable is common to both correlations, see Dunn and Clark (1971).

**TABLE 5.8** ISOMETRIC KNEE EXTENSION STRENGTH IN POUNDS ($X_1$) AND PERCENT VASTUS LATERALIS (SLOW TWITCH) MUSCLE ($X_2$) IN POWER ATHLETES AND ENDURANCE ATHLETES

| Power athletes | | Endurance athletes | |
|---|---|---|---|
| $X_1$ | $X_2$ | $X_1$ | $X_2$ |
| 196.0 | 56.0 | 161.0 | 98.4 |
| 183.0 | 28.8 | 142.0 | 70.8 |
| 295.0 | 57.2 | 122.5 | 35.4 |
| 203.0 | 46.0 | 123.0 | 74.5 |
| 195.0 | 35.5 | 176.5 | 79.5 |
| 289.0 | 58.5 | 156.0 | 62.1 |
| 198.0 | 41.4 | 126.0 | 74.3 |
| 206.9 | 21.6 | 95.0 | 67.7 |

$\Sigma X_1 = 1{,}765.9 \quad \Sigma X_2 = 345.0 \quad \Sigma X_1 = 1{,}102.0 \quad \Sigma X_2 = 562.7$
$\Sigma X_1^2 = 403{,}696.61 \quad \Sigma X_2^2 = 16{,}216.30 \quad \Sigma X_1^2 = 156{,}609.50 \quad \Sigma X_2^2 = 41{,}779.05$
$\Sigma X_1 X_2 = 78{,}953.64 \quad\quad \Sigma X_1 X_2 = 78{,}908.65$

## 5.3.4 Test of Significance for Homogeneity of Correlation Coefficients

In addition to correlations from data collected by Dr. Kroll as shown in Section 5.3.3, correlations were also made between plantar flexion and percent gastrocnemius muscle in both power athletes and endurance athletes. The correlation for power athletes was $-0.938$ and for endurance athletes it was $-0.044$. If we wish to test whether the four correlation coefficients in the study were homogeneous or drawn from the same population, we can no longer use the $t$ test but instead must turn to the $\chi^2$ distribution. This test uses the fact that the quantity

$$\sum w_i(z_i - \bar{z}_w)^2 = \sum w_i z_i^2 - \frac{(\sum w_i z_i)^2}{\sum w_i} \tag{5.38}$$

follows the $\chi^2$ distribution with $k - 1$ degrees of freedom, where $k$ is the number of $z$'s, the symbol $w_i$ represents a weight and is defined as the reciprocal of the variance, $1/[1/(n_i - 3)] = n_i - 3$, and $\bar{z}_w$ represents the sum of the weighted $z$'s divided by the sum of the weights, $\sum w_i z_i/\sum w_i = \Sigma(n - 3)z_i/\sum w_i$. The test is more easily carried out in the form of Table 5.9.

The average $z$ is found as

$$\bar{z}_w = \frac{\Sigma(n_i - 3)z_i}{\sum w_i} = \frac{0.821}{17} = 0.048 \tag{5.39}$$

and the $\chi^2$ value is found as

$$\chi^2 = \sum w_i z_i^2 - \frac{(\sum w_i z_i)^2}{\sum w_i} = \Sigma(n_i - 3)z_i^2 - \frac{[\Sigma(n_i - 3)z_i]^2}{\Sigma(n_i - 3)}$$

$$= 12.935 - \frac{(0.821)^2}{17} = 12.935 - 0.040 = 12.895 \tag{5.40}$$

With $k - 1 = 4 - 1$ degrees of freedom we have a highly significant $\chi^2$ (Table A.4) and conclude that the probability is extremely small ($<0.01$) that the four $r$'s were drawn from the same population. Actually, there were eight pairs in each sample originally but two were dropped from sample 3 and one from sample 4 to emphasize the fact that equal numbers in the samples are not necessary.

**TABLE 5.9**  TEST OF SIGNIFICANCE AMONG CORRELATION COEFFICIENTS

| Sample | $n$ | $w = n - 3$ | $r$ | $z$ | $(n-3)z$ | $(n-3)z^2$ |
|--------|-----|-------------|--------|--------|----------|------------|
| 1 | 8 | 5 | 0.649 | 0.773 | 3.865 | 2.988 |
| 2 | 8 | 5 | 0.429 | 0.459 | 2.295 | 1.053 |
| 3 | 6 | 3 | $-0.938$ | $-1.721$ | $-5.163$ | 8.886 |
| 4 | 7 | 4 | $-0.044$ | $-0.044$ | $-0.176$ | 0.008 |
| Sums | | 17 | | | 0.821 | 12.935 |

$$[\Sigma(n-3)z_i]^2/\Sigma(n-3) = 0.040 \qquad \text{Average } z = 0.048$$

$$\chi^2 = 12.895$$

Had the $\chi^2$ value not been significant we could have made an estimate of a common population correlation coefficient merely by transforming the average $z$ to its corresponding $r$ value.

## 5.4 COEFFICIENT OF DETERMINATION

While the correlation coefficient has great utility in providing a measure of the degree of relationship between two independent variables, a form of the correlation coefficient finds frequent usage in interpreting the importance of a regression coefficient. With a large number of pairs of observations it is possible to find a highly significant regression coefficient which really is of no great importance. As an aid in assessing the importance of the regression coefficient we can make use of the *coefficient of determination,* which is merely the square of the correlation coefficient. The coefficient of determination yields a measure of the proportion of the total variation in $Y$ accounted for by the regression of $Y$ on $X$.

Referring back to Table 5.1 we can calculate the coefficient of correlation between $X$ and $Y$ as

$$r = \frac{\Sigma\,xy}{\sqrt{\Sigma\,x^2\,\Sigma\,y^2}} = \frac{-3.2768}{\sqrt{(23.372)(0.9722)}} = \frac{-3.2768}{4.7668} = -0.6874$$

and the coefficient of determination is

$$r^2 = (-0.6874)^2 = 0.4725$$

This tells us that 47.25 percent of the variation in $Y$ is accounted for by the regression of $Y$ on $X$. Since the coefficient of determination is the proportion of the total variation in $Y$ accounted for by regression it can also be calculated as

$$r^2 = \frac{(\Sigma\,xy)^2/\Sigma\,x^2}{\Sigma\,y^2} = \frac{SS_b}{\Sigma\,y^2} = \frac{0.4594}{0.9722} = 0.4725$$

The sum of squares due to regression is frequently written $r^2\,\Sigma\,y^2$ since

$$r^2\,\Sigma\,y^2 = \frac{(\Sigma\,xy)^2}{\Sigma\,x^2\,\Sigma\,y^2}\,\Sigma\,y^2 = \frac{(\Sigma\,xy)^2}{\Sigma\,x^2} \tag{5.41}$$

and the sum of squares of deviations from regression is frequently written $(1 - r^2)\Sigma\,y^2$ since

$$\Sigma\,y^2 - r^2\,\Sigma\,y^2 = \Sigma\,y^2 - \frac{(\Sigma\,xy)^2}{\Sigma\,x^2\,\Sigma\,y^2}\,\Sigma\,y^2 = \Sigma\,y^2 - \frac{(\Sigma\,xy)^2}{\Sigma\,x^2} \tag{5.42}$$

The term $(1 - r^2)$ is known as the *coefficient of alienation,* or the *coefficient of non-determination.*

## 5.5 QUADRATIC REGRESSION

We can see from the coefficient of determination calculated in Section 5.4 that the linear regression of $Y$ on $X$ for the data in Table 5.1 accounted for only 47.25 percent of the total variation of $Y$. A next logical step would be to fit a curvilinear line to the data and see whether we improve the fit of the curve significantly. Each additional degree of the fitted curve will improve the fit but the improvement may be so slight as to be meaningless. Linear regression is referred to as a *first degree polynomial*. The second degree polynomial is known as the *quadratic regression,* which we shall now consider. An additional degree of polynomial can be fitted for each degree of freedom available.

## 5.5.1 Method of Calculation

To fit the quadratic regression line, we use the square of $X$ as a second independent variable. This is described by the model

$$Y_i = \mu + \beta_l(X_{1i} - \bar{x}_1) + \beta_q(X_{2i} - \bar{x}_2) + e_i \tag{5.43}$$

where $X_2$ is the square of $X_1$, $\beta_l$ is the linear regression coefficient, $\beta_q$ is the quadratic regression coefficient, and $e_i$ is the random error. This model leads to the prediction equation

$$\hat{Y}_i = \bar{y} + b_l(X_{1i} - \bar{x}_1) + b_q(X_{2i} - \bar{x}_2) = a + b_l X_{1i} + b_q X_{2i} \tag{5.44}$$

where $$a = \bar{y} - b_l \bar{x}_1 - b_q \bar{x}_2$$

Values for $b_l$ and $b_q$ are calculated by the following pair of simultaneous equations:

$$\sum x_1^2 b_l + \sum x_1 x_2 b_q = \sum x_1 y$$
$$\sum x_1 x_2 b_l + \sum x_2^2 b_q = \sum x_2 y \tag{5.45}$$

The elements necessary in this pair of equations can be obtained by setting up a table such as Table 5.10. We can solve for the unknown column vector **B** of the constants $b_l$ and $b_q$ by inverting the matrix $\mathbf{X'X}$ and performing the matrix multiplication $(\mathbf{X'X})^{-1}\mathbf{Y}$. For the methods of matrix inversion and matrix multiplication see Chapter 9.

$$(\mathbf{X'X}) = \begin{bmatrix} 23.372 & 459.7756 \\ 459.7756 & 9076.4842 \end{bmatrix} \text{ and } (\mathbf{X'X})^{-1} = \begin{bmatrix} 12.23268435 & -0.61965511 \\ -0.61965511 & 0.03149923 \end{bmatrix}$$

$$(\mathbf{X'X})^{-1}Y = \begin{bmatrix} 12.23268435 & -0.61965511 \\ -0.61965511 & 0.03149923 \end{bmatrix} \begin{bmatrix} -3.2768 \\ -65.652475 \end{bmatrix}$$

$$= \begin{bmatrix} 0.59783154 = b_l \\ -0.03751655 = b_q \end{bmatrix} = \mathbf{B}$$

Having found the values for $b_l$ and $b_q$, the linear and quadratic regression coefficients, we can now solve for $a$, where

$$a = \bar{y} - b_l \bar{x}_1 - b_q \bar{x}_2$$

$$= 0.76125 - (0.597832)(9.82) - (-0.037517)(97.601)$$

$$= 0.76125 - 5.870710 + 3.661697 = -1.447763$$

leading to the prediction equation

**TABLE 5.10** COMPUTATION OF VALUES NECESSARY FOR SOLVING EQUATIONS (5.45)

| $X_1$ | $X_2 = X_1^2$ | $Y$ | $X_1^2$ | $X_2^2$ | $X_1 X_2$ | $X_1 Y$ | $X_2 Y$ |
|---|---|---|---|---|---|---|---|
| 10.6 | 112.36 | 0.752 | 112.36 | 12,624.7696 | 1191.016 | 7.9712 | 84.49472 |
| 10.3 | 106.09 | 0.909 | 106.09 | 11,255.0881 | 1092.727 | 9.3627 | 96.43581 |
| 8.6 | 73.96 | 0.939 | 73.96 | 5,470.0816 | 636.056 | 8.0754 | 69.44844 |
| 11.8 | 139.24 | 0.256 | 139.24 | 19,387.7776 | 1643.032 | 3.0208 | 35.64544 |
| 9.8 | 96.04 | 0.383 | 96.04 | 9,223.6816 | 941.192 | 3.7534 | 36.78332 |
| 9.1 | 82.81 | 0.833 | 82.81 | 6,857.4961 | 753.571 | 7.5803 | 68.98073 |
| 10.7 | 114.49 | 0.669 | 114.49 | 13,107.9601 | 1225.043 | 7.1583 | 76.59381 |
| 10.4 | 108.16 | 0.924 | 108.16 | 11,698.5856 | 1124.864 | 9.6096 | 99.93984 |
| 8.8 | 77.44 | 0.932 | 77.44 | 5,996.9536 | 681.472 | 8.2016 | 72.17408 |
| 8.7 | 75.69 | 0.849 | 75.69 | 5,728.9761 | 658.503 | 7.3863 | 64.26081 |
| 10.2 | 104.04 | 0.713 | 104.04 | 10,824.3216 | 1061.208 | 7.2726 | 74.18052 |
| 11.6 | 134.56 | 0.424 | 134.56 | 18,106.3936 | 1560.896 | 4.9184 | 57.05344 |
| 11.1 | 123.21 | 0.513 | 123.21 | 15,180.7041 | 1367.631 | 5.6943 | 63.20673 |
| 9.3 | 86.49 | 0.987 | 86.49 | 7,480.5201 | 804.357 | 9.1791 | 85.36563 |
| 7.7 | 59.29 | 0.987 | 59.29 | 3,515.3041 | 456.533 | 7.5999 | 58.51923 |
| 9.5 | 90.25 | 0.518 | 90.25 | 8,145.0625 | 857.375 | 4.9210 | 46.74950 |
| 9.9 | 98.01 | 0.934 | 98.01 | 9,605.9601 | 970.299 | 9.2466 | 91.54134 |
| 8.5 | 72.25 | 0.968 | 72.25 | 5,220.0625 | 614.125 | 8.2280 | 66.93800 |
| 9.0 | 81.00 | 0.936 | 81.00 | 6,561.0000 | 729.000 | 8.4240 | 75.81600 |
| 10.8 | 116.64 | 0.799 | 116.64 | 13,604.8896 | 1259.712 | 8.6292 | 93.19536 |
| Totals 196.4 | 1952.02 | 15.225 | 1952.02 | 199,595.5882 | 19,628.612 | 146.2327 | 1420.32275 |

$$\Sigma x_1^2 = \Sigma X_1^2 - \frac{(\Sigma X_1)^2}{n} = 1952.02 - \frac{(196.4)^2}{20} = 1952.02 - 1928.648 = 23.372$$

$$\Sigma x_2^2 = \Sigma X_2^2 - \frac{(\Sigma X_2)^2}{n} = 199,595.5882 - \frac{(1952.02)^2}{20}$$

$$= 199,595.5882 - 190,519.1040 = 9076.4842$$

$$\Sigma x_1 x_2 = \Sigma X_1 X_2 - \frac{(\Sigma X_1)(\Sigma X_2)}{n} = 19,628.612 - \frac{(196.4)(1952.02)}{20}$$

$$= 19,628.612 - 19,168.8364 = 459.7756$$

$$\Sigma x_1 y = \Sigma X_1 Y - \frac{(\Sigma X_1)(\Sigma Y)}{n} = 146.2327 - \frac{(196.4)(15.225)}{20}$$

$$= 146.2327 - 149.5095 = -3.2768$$

$$\Sigma x_2 y = \Sigma X_2 Y - \frac{(\Sigma X_2)(\Sigma Y)}{n} = 1420.32275 - \frac{(1952.02)(15.225)}{20}$$

$$= 1420.32275 - 1485.975225 = -65.652475$$

$$\Sigma y^2 = \Sigma Y^2 - \frac{(\Sigma Y)^2}{n} = 12.562235 - \frac{(15.225)^2}{20} = 12.562235 - 11.590031 = 0.972204$$

We now let

$$\mathbf{X'X} = \begin{bmatrix} \Sigma x_1^2 & \Sigma x_1 x_2 \\ \Sigma x_1 x_2 & \Sigma x_2^2 \end{bmatrix} \quad \mathbf{B} = \begin{bmatrix} b_l \\ b_q \end{bmatrix} \quad \text{and} \quad \mathbf{Y} = \begin{bmatrix} \Sigma x_1 y \\ \Sigma x_2 y \end{bmatrix}$$

$$\hat{Y}_i = -1.447763 + 0.597832 X_{1i} - 0.037517 X_{2i} \tag{5.46}$$

Using Equation (5.46), we can now develop a quadratic regression line. To do this, we need to select enough $X$ values to depict the regression line accurately. The

**TABLE 5.11** PREDICTED VALUES FOR
QUADRATIC REGRESSION
FOR DATA SHOWN IN TABLE 5.1

| $X_1$ | $X_2$ | $\hat{Y}$ |
|------|--------|-------|
| 8.0 | 64.00 | 0.934 |
| 8.5 | 72.25 | 0.923 |
| 9.0 | 81.00 | 0.894 |
| 9.5 | 90.25 | 0.846 |
| 10.0 | 100.00 | 0.779 |
| 10.5 | 110.25 | 0.693 |
| 11.0 | 121.00 | 0.589 |
| 11.5 | 132.25 | 0.466 |
| 12.0 | 144.00 | 0.324 |

selected $X$ values do not necessarily have to be those appearing in the data. We can select several points covering the range of $X$ values in the data. In this example, we have chosen the values shown in Table 5.11, which presents a series of predicted values.

The predicted values shown in Table 5.11 are plotted in Figure 5.4 to form the quadratic curve. This figure also shows a cubic curve which will be discussed further.

### 5.5.2 Test for the Significance of Quadratic Regression

One could guess from looking at Figure 5.4 that the quadratic line fits the data better than the linear regression line of Figure 5.1. However, we would like to know whether this is a *significant* improvement and we can test for this by testing for the additional sum of squares accounted for by fitting the quadratic regression line.

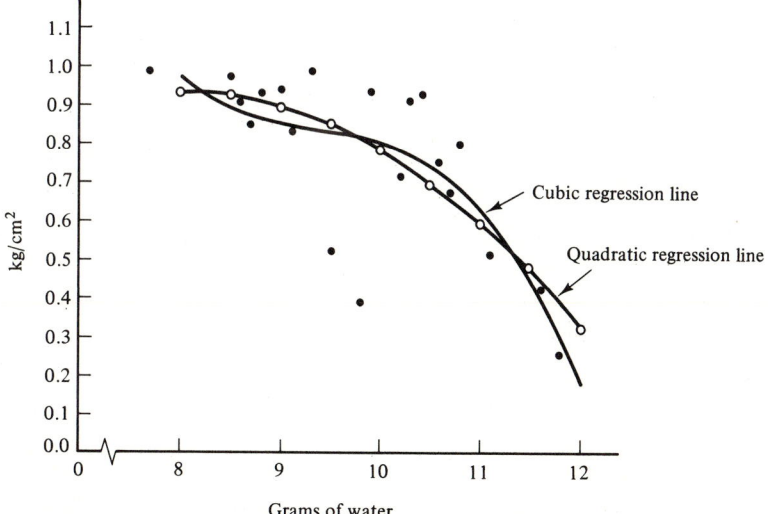

**Figure 5.4** Scatter diagram and quadratic and cubic regression lines for data presented in Table 5.1.

The sum of squares due to the quadratic portion can be found by calculating the difference between the sum of squares accounted for by the combined linear and quadratic regression and the sum of squares due to the linear regression alone. We previously found that the sum of squares due to linear regression is

$$SS_l = \frac{(\sum xy)^2}{\sum x^2} = b \sum xy = \frac{(-3.2768)^2}{23.372} = 0.4594$$

The sum of squares due to linear and quadratic regression combined can be found as

$$SS_{(l,q)} = b_l \sum x_1 y + b_q \sum x_2 y$$

$$= (0.597832)(-3.2768) + (-0.037517)(-65.652475)$$

$$= 0.5041 \tag{5.47}$$

The sum of squares due to the quadratic regression can now be found as

$$SS_q = SS_{(l,q)} - SS_l = 0.5041 - 0.4594 = 0.0447$$

To test for the significance of the quadratic portion of the sum of squares we can set up Table 5.12 in a manner similar to Table 5.3. The $F$ of 1.625 with 1 and 17 degrees of freedom is not significant. Note that one degree of freedom is lost from the error term for each degree of polynomial fitted.

It is of interest that the sum of squares for the highest-order regression in the model can be found by squaring the regression coefficient and dividing by the corresponding inverse element. By this is meant the diagonal for the last row and column of the inverse matrix. In our example,

$$SS_q = \frac{b_q^2}{c^{22}} = \frac{(-0.037517)^2}{0.031499} = 0.0447 \tag{5.48}$$

The test of significance for the quadratic regression shown in Table 5.12 can also be carried out by means of a $t$ test in the usual form of

$$t = \frac{b_q}{s_{b_q}}$$

where $s_{b_q} = \sqrt{c^{22}s^2}$ and $s^2$ is the error mean square as shown in Table 5.12. For our example, we have

$$t = \frac{b_q}{s_{b_q}} = \frac{0.037517}{\sqrt{(0.031499)(0.0275)}} = \frac{0.037517}{0.029432} = 1.2747$$

**TABLE 5.12**   ANALYSIS OF VARIANCE FOR TESTING
FOR THE SIGNIFICANCE OF
QUADRATIC REGRESSION

| Source | df | SS | MS | F |
|--------|-----|--------|--------|-------|
| Total | 19 | 0.9722 | | |
| Linear | 1 | 0.4594 | 0.4594 | |
| Quadratic | 1 | 0.0447 | 0.0447 | 1.625 |
| Error | 17 | 0.4681 | 0.0275 | |

with $n - 3 = 17$ degrees of freedom. One degree of freedom is accounted for by the calculation of the mean and one degree of freedom for each regression coefficient. The value of $t^2 = (1.274)^2 = 1.625 = F$. The $F$ test and the $t$ test both lead to the same conclusion, that we have not improved the fit of the line significantly.

## 5.6 CUBIC REGRESSION

### 5.6.1 Method of Calculation

Since there is still a sizable sum of squares in the residual or error line as seen in Table 5.12, it is possible that we may improve the fit significantly by fitting a third-degree polynomial, the cubic regression where $X_3 = X_1^3$. To do this, we must set up a new set of equations, adding the cubic regression to the equations previously shown in (5.45). The model for the cubic regression would add the component $\beta_c(X_{3i} - \bar{x}_3)$ to Equation (5.43). These equations would be

$$\sum x_1^2 b_l + \sum x_1 x_2 b_q + \sum x_1 x_3 b_c = \sum x_1 y$$

$$\sum x_1 x_2 b_l + \sum x_2^2 b_q + \sum x_2 x_3 b_c = \sum x_2 y \qquad (5.49)$$

$$\sum x_1 x_3 b_l + \sum x_2 x_3 b_q + \sum x_3^2 b_c = \sum x_3 y$$

Putting these equations in matrix form we have

$$\mathbf{X'X} = \begin{bmatrix} \sum x_1^2 & \sum x_1 x_2 & \sum x_1 x_3 \\ \sum x_1 x_2 & \sum x_2^2 & \sum x_2 x_3 \\ \sum x_1 x_3 & \sum x_2 x_3 & \sum x_3^2 \end{bmatrix} \quad \mathbf{B} = \begin{bmatrix} b_l \\ b_q \\ b_c \end{bmatrix} \quad \mathbf{Y} = \begin{bmatrix} \sum x_1 y \\ \sum x_2 y \\ \sum x_3 y \end{bmatrix}$$

By using and expanding Table 5.10 (the expansion is not shown here), we can calculate (note that $X_1 X_2 = X_3$)

$$\sum x_3^2 = \sum X_3^2 - \frac{(\sum X_3)^2}{n} = 21,294,906.82 - \frac{(19,628.612)^2}{20}$$

$$= 21,294,906.82 - 19,264,120.45 = 2,030,786.37$$

$$\sum x_1 x_3 = \sum X_1 X_3 - \frac{(\sum X_1)(\sum X_3)}{n} = 199,595.5882 - \frac{(196.4)(19,628.612)}{20}$$

$$= 199,595.5882 - 192,752.9699 = 6,842.6183$$

$$\sum x_2 x_3 = \sum X_2 X_3 - \frac{(\sum X_2)(\sum X_3)}{n} = 2,051,313.662 - \frac{(1,952.02)(19,628.612)}{20}$$

$$= 2,051,313.662 - 1,915,772.160 = 135,541.502$$

$$\sum x_3 y = \sum X_3 Y - \frac{(\sum X_3)(\sum Y)}{n} = 13,946.39662 - \frac{(19,628.612)(15.225)}{20}$$

$$= 13,946.39662 - 14,942.28088 = -995.88426$$

We now have

$$\mathbf{X'X} = \begin{bmatrix} 23.372 & 459.7756 & 6{,}842.6183 \\ 459.7756 & 9{,}076.4842 & 135{,}541.502 \\ 6{,}842.6183 & 135{,}541.502 & 2{,}030{,}786.37 \end{bmatrix}$$

and

$$\mathbf{Y} = \begin{bmatrix} -3.2768 \\ -65.652475 \\ -995.88426 \end{bmatrix}$$

Inverting the $\mathbf{X'X}$ matrix we have

$$(\mathbf{X'X})^{-1} = \begin{bmatrix} 1594.3431 & -163.54670 & 5.5436085 \\ -163.54670 & 16.809863 & -0.57088542 \\ 5.5436085 & -0.57088542 & 0.01942443 \end{bmatrix}$$

and making the multiplication $(\mathbf{X'X})^{-1}\mathbf{Y}$ we obtain

$$\mathbf{B'} = [-7.88995986 = b_l, \quad 0.83656302 = b_q, \quad -0.02974065 = b_c]$$

The inversion and matrix multiplication were done by computer.

We can now calculate the sum of squares accounted for by the combined linear, quadratic, and cubic regression as

$$\mathrm{SS}_{(l,q,c)} = b_l \sum x_1 y + b_q \sum x_2 y + b_c \sum x_3 y$$

$$= (-7.889960)(-3.2768) + (0.836563)(-65.652475)$$

$$\quad + (-0.029741)(-995.88426)$$

$$= 0.5500 \tag{5.50}$$

The sum of squares due to cubic regression can now be found by

$$\mathrm{SS}_c = \mathrm{SS}_{(l,q,c)} - \mathrm{SS}_{l,q}$$

$$= 0.5500 - 0.5041 = 0.0459 \tag{5.51}$$

While the $\mathbf{X'X}$ matrix including the cubic regression can be inverted using a desk calculator, it is an "ill-conditioned" matrix due to the disparity in size of the elements in the matrix, resulting in a great deal of rounding error in the inversion process. This problem can be surmounted by using $(X - \bar{x})$, $(X - \bar{x})^2$, and $(X - \bar{x})^3$ in place of $X_{1_3}$, $X_2$, and $X_3$ for the independent variables, where $X_2 = X^2$ and $X_3 = X^3$. This results in the following $\mathbf{X'X}$ and $\mathbf{Y}$ matrices:

$$\mathbf{X'X} = \begin{bmatrix} 23.372 & 0.74952 & 59.0833424 \\ 0.74952 & 31.7708232 & -0.32525115 \\ 59.0833424 & -0.32525115 & 200.9967763 \end{bmatrix}$$

and

$$\mathbf{Y} = \begin{bmatrix} -3.2768 \\ -1.296123 \\ -9.73140744 \end{bmatrix}$$

Inverting this $\mathbf{X'X}$ matrix and solving for $\mathbf{B}$, the column vector of constants, we have

$$(\mathbf{X'X})^{-1} = \begin{bmatrix} 0.16717180 & -0.00444697 & -0.04914763 \\ -0.00444697 & 0.03159424 & 0.00135832 \\ -0.0494763 & 0.00135832 & 0.01942443 \end{bmatrix}$$

and
$$\mathbf{B} = \begin{bmatrix} -0.06374912 \\ -0.03959656 \\ -0.02974064 \end{bmatrix}$$

The sum of squares due to fitting the combined linear, quadratic, and cubic regressions, found by $\mathbf{B'Y}$, yields a value of 0.5496 which differs from the value of 0.5500 found previously only by rounding error. Although the linear and quadratic regression coefficients and corresponding inverse elements differ from those found earlier, the cubic coefficient of $-0.02984064$ is the same as that found previously as is the inverse element $c^{33} = 0.01942443$.

For the prediction equation the values $(X - \bar{x})$, $(X - \bar{x})^2$, and $(X - \bar{x})^3$ would be used in place of $X_1$, $X_1^2$, and $X_1^3$. The intercept would be calculated using these values together with the new regression coefficients. The prediction equation would be

$$\hat{Y}_i = 0.80863710 - (0.06374912)(X_{1i} - \bar{x}_1)$$
$$- (0.03959656)(X_{1i} - \bar{x}_1)^2 - (0.02974064)(X_{1i} - \bar{x}_1)^3$$

The prediction equation for the analysis not using the deviations is shown in Equation (5.52). Substituting, as an example, $X = 10$ in each of the prediction equations yields a value of 0.7957.

The same methods would apply if dealing with a quadratic regression or a regression of higher order than the cubic, as well as with multiple regression.

## 5.6.2 Test for the Significance of Cubic Regression

We can test for the significance of the sum of squares for the cubic regression as shown in Table 5.13. The $F$ test tells us that we have not improved the fit of the line to the data significantly. If the $F$ value had proved to be significant, we would first solve for $a$,

$$a = \bar{y} - b_l\bar{x}_1 - b_q\bar{x}_2 - b_c\bar{x}_3$$

$$= 0.76125 - (-7.889960)(9.82) - (0.836563)(97.601) - (-0.029741)(981.4306)$$

$$= 0.76125 + 77.479407 - 81.649385 + 29.188727$$

$$= 25.779999$$

and then write out the prediction equation as

$$\hat{Y}_i = 25.779999 - 7.889960X_{1i} + 0.836563X_{2i} - 0.029741X_{3i} \qquad (5.52)$$

Although the cubic component is not significant, the cubic regression line is plotted for illustration using Equation (5.52) and is shown in Figure 5.4.

**TABLE 5.13**   ANALYSIS OF VARIANCE FOR TESTING
FOR THE SIGNIFICANCE OF
THE CUBIC REGRESSION

| Source | df | SS | MS | F |
|--------|-----|--------|--------|------|
| Total | 19 | 0.9722 | | |
| Linear | 1 | 0.4594 | | |
| Quadratic | 1 | 0.0447 | | |
| Cubic | 1 | 0.0459 | 0.0459 | 1.79 |
| Error | 16 | 0.4222 | 0.0264 | |

As with the quadratic regression, we can also test for the significance of the cubic regression as shown in Section 5.5.1, giving for the cubic test in our example

$$t = \frac{b_c}{s_{b_c}} = \frac{b_c}{\sqrt{c^{33}s^2}} = \frac{0.029741}{\sqrt{(0.01942443)(0.0264)}} = \frac{0.029741}{0.022645} = 1.313$$

with $n - 4 = 20 - 14 = 16$ degrees of freedom, a nonsignificant value. Again, $t^2 = (1.313)^2 = 1.72 = F$.

Similar equations can be set up if more degrees of curvature are desired, $X_i$ taking on an additional power for each additional degree of polynomial.

Referring back to the linear regression analysis, the linear regression coefficient was 0.140202. When the quadratic regression analysis was made the linear regression coefficient changed to 0.597832, and we had a quadratic regression coefficient of $-0.037517$. Completing the cubic regression analysis, we found a linear coefficient of $-7.889960$, another change, and a quadratic coefficient of 0.836563, a change from $-0.037517$. Thus, as additional degrees are added to the analysis, all previous coefficients undergo a change. In addition to the change in the coefficients, it can be seen that the intercept also changes with a change in degree of polynomial fitted. Of course, to develop the least-squares regression the linear term must remain in a quadratic regression equation, the linear and quadratic terms must remain in a cubic regression, and so on.

## 5.7  LINEAR REGRESSION THROUGH TREATMENT MEANS

### 5.7.1  Method of Calculation

The calculations made in the preceding example followed those that would be made for a Model II regression or a Model I regression where either the intervals between the $X$'s or the number of $Y$ values for each $X$ were unequal, or both. The $X$'s will frequently represent different levels of a set of treatments. The calculations can be shortened when dealing with an equal number of observations at each $X$. In this case, we can fit a regression line to the treatment means (the means of $Y$ for each level of $X$), weighting the sums of squares appropriately. As an example, let us make use of the data shown in Table 5.14 furnished by Dr. William J. Lord of the Plant and Soil Sciences Department at the University of Massachusetts. Table 5.15 presents an analysis

**TABLE 5.14**  EFFECTS OF DIFFERING LEVELS OF SIMAZINE TREATMENTS
AT DIFFERENT LOCATIONS (PERCENT NITROGEN IN ROOTED
APPLE STOCKS)

| Soils | Hinckley | | | | Woodbridge | | | |
|---|---|---|---|---|---|---|---|---|
| Simazine | 0.0 | 0.2 | 0.4 | 0.6 | 0.0 | 0.2 | 0.4 | 0.6 |
| | 1.90 | 2.10 | 2.05 | 1.92 | 1.87 | 2.02 | 2.37 | 1.84 |
| | 1.88 | 1.99 | 1.98 | 1.90 | 1.91 | 1.97 | 1.96 | 1.95 |
| | 1.85 | 2.00 | 1.84 | 1.89 | 1.93 | 1.89 | 2.05 | 1.90 |
| | 1.68 | 1.85 | 1.90 | 1.85 | 1.79 | 1.95 | 2.04 | 1.86 |
| | 1.76 | 1.92 | 1.97 | 1.91 | 1.80 | 2.05 | 2.18 | 2.02 |
| Totals | 9.07 | 9.86 | 9.74 | 9.47 | 9.30 | 9.88 | 10.60 | 9.57 |

of these data showing highly significant differences among the simazine treatment
means. Since the simazine treatments have been applied in increasing levels it is of
interest to study the trend response in terms of regression. We can set up a table as
we did previously and make the calculations necessary for fitting a linear regression
line to the treatment means. We shall temporarily ignore the fact that there are 10
observations per mean and take this into account later in the calculations. The computations necessary for estimating a linear regression line are shown in Table 5.16.

## 5.7.2  Test for the Significance of Linear Regression Through Treatment Means

Before we proceed with calculating the prediction equation, let us test for the significance of the slope (i.e., the regression coefficient). To do this, we first calculate the
sum of squares due to regression as

$$SS_b = \frac{n(\Sigma\ xy)^2}{\Sigma\ x^2} = \frac{(10)(0.0006812)}{0.20} = \frac{0.006812}{0.20} = 0.0341$$

where $n$ is the number of observations per treatment. Multiplication of $(\Sigma\ xy)^2/\Sigma\ x^2$
by $n$ now takes into account the fact that there were 10 observations per treatment.
It must be emphasized that the only time that $(\Sigma\ xy)^2/\Sigma\ x^2$ is multiplied by $n$ is when
we have grouped more than one $Y$ per $X$. Were $n$ used in the earlier calculation of
the sum of squares, it would, of course, equal 1. The test of significance can be cast
into the form of an analysis of variance as shown in Table 5.17. It can be seen from

**TABLE 5.15**  ANALYSIS OF DATA PRESENTED
IN TABLE 5.9

| Source | df | SS | MS | F |
|---|---|---|---|---|
| Soils (*S*) | 1 | 0.0366 | 0.0366 | 4.63* |
| Simazine (*T*) | 3 | 0.2187 | 0.0729 | 9.23** |
| *ST* | 3 | 0.0437 | 0.0146 | 1.85 |
| *W* | 32 | 0.2534 | 0.0079 | |

**TABLE 5.16** CALCULATIONS OF LINEAR REGRESSION THROUGH TREATMENT MEANS

| Level of simazine ($X$) | Percent nitrogen mean ($Y$) | $X^2$ | $Y^2$ | $XY$ |
|---|---|---|---|---|
| 0.0 | 1.837 | 0.00 | 3.374569 | 0.0000 |
| 0.2 | 1.974 | 0.04 | 3.896676 | 0.3948 |
| 0.4 | 2.034 | 0.16 | 4.137156 | 0.8136 |
| 0.6 | 1.904 | 0.36 | 3.625216 | 1.1424 |
| $\Sigma X = 1.2$ | $\Sigma Y = 7.749$ | $\Sigma X^2 = 0.56$ | $\Sigma Y^2 = 15.033617$ | $\Sigma XY = 2.3508$ |

$$b = \frac{\Sigma xy}{\Sigma x^2} = \frac{\Sigma XY - (\Sigma X)(\Sigma Y)/k}{\Sigma X^2 - (\Sigma X)^2/k} = \frac{2.3508 - (1.2)(7.749)/4}{0.56 - (1.2)^2/4}$$

$$= \frac{2.3508 - 2.3247}{0.56 - 0.36} = \frac{0.0261}{0.20} = 0.1305$$

where $k$ equals the number of treatment means, in this case 4

this table that the single degree of freedom for linear regression represents a partitioning of the degrees of freedom among the simazine treatments, accomplished by the partitioning of the corresponding sum of squares.

As we have seen earlier the same test of significance can be carried out by means of a $t$ test, in this case,

$$t = \frac{b}{s_b} \sqrt{n} \tag{5.53}$$

where $s_b = s^2/\Sigma x^2$ and $s^2$ is the error mean square from Table 5.17, the mean square that was used in the $F$ test. The multiplication by $\sqrt{n}$ is necessitated by the fact that we are dealing with means based on $n = 10$ observations. In our example, we have

$$t = \frac{b}{s_b} \sqrt{n} = \frac{b\sqrt{n}}{\sqrt{s^2/\Sigma x^2}} = \frac{(0.1305)(3.1623)}{\sqrt{(0.0079)/(0.20)}} = \frac{0.4127}{0.1987} = 2.077*$$

and $t^2 = 4.31 = F$ (with a slight rounding error). The degrees of freedom for the $t$ test are 32, the degrees of freedom corresponding to the error mean square.

Having found that the linear regression is significant, we can develop a prediction equation and plot the linear regression line. We have already calculated $b = 0.1305$ and can find $a = \bar{y} - b\bar{x} = 1.93725 - (0.1305)(0.3000) = 1.8981$, yielding the prediction equation

$$\hat{Y}_i = a + bX_i = 1.8981 + 0.1305X_i \tag{5.54}$$

**TABLE 5.17** TEST OF SIGNIFICANCE OF THE LINEAR REGRESSION THROUGH TREATMENT MEANS

| Source | df | SS | MS | $F$ |
|---|---|---|---|---|
| Soils ($S$) | 1 | 0.0366 | 0.0366 | 4.63* |
| Simazine ($T$) | 3 | 0.2187 | 0.0729 | 9.23** |
| Linear regression | 1 | 0.0341 | 0.0341 | 4.32* |
| $ST$ | 3 | 0.0437 | 0.0146 | |
| $W$ | 32 | 0.2534 | 0.0079 | |

**TABLE 5.18**   ACTUAL AND PREDICTED SIMAZINE LEVEL MEANS

| Simazine level = $X$ | Actual mean = $\bar{y}$ | Predicted mean = $\hat{\bar{y}}$ |
|---|---|---|
| 0.0 | 1.837 | 1.8981 |
| 0.2 | 1.974 | 1.9242 |
| 0.4 | 2.034 | 1.9503 |
| 0.6 | 1.904 | 1.9764 |

Substituting the levels of simazine for $X$, we find the predicted values shown in Table 5.18.

The linear regression line is plotted in Figure 5.5 where the circles represent the treatment means and the small $\times$'s represent the predicted means. The curve connecting the dots (a quadratic curve) will be discussed later.


## 5.8  CURVILINEAR REGRESSION THROUGH TREATMENT MEANS

### 5.8.1  Method of Calculation

Even though we found that there is a significant linear regression through the treatment means, it can be seen clearly from Figure 5.5 that the linear does not fit the data at all well. We will therefore fit a quadratic line through the treatment means in a manner similar to that shown in Section 5.5. We again use the square of $X$ as a second independent variable and have the model

$$Y_i = \mu + \beta_l(X_{1i} - \bar{x}_1) + \beta_q(X_{2i} - \bar{x}_2) + e_i \tag{5.55}$$

where $X_1 = X$ and $X_2 = X_1^2$ and the prediction equation

$$\hat{Y}_i = \bar{y} + b_l(X_{1i} - \bar{x}_1) + b_q(X_{2i} - \bar{x}_2) = a + b_l X_{1i} + b_q X_{2i}$$

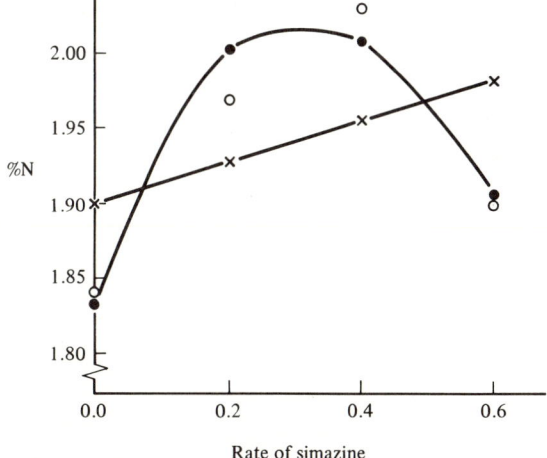

**Figure 5.5**   Linear and quadratic regression lines through simazine treatment means.

with $a = \bar{y} - b_l\bar{x}_1 - b_q\bar{x}_2$. We again need to calculate values for $b_l$ and $b_q$ so we use the relationships

$$\sum x_1^2 b_l + \sum x_1 x_2 b_q = \sum x_1 y$$
$$\sum x_1 x_2 b_l + \sum x_2^2 b_q = \sum x_2 y$$

(5.56)

We can now set up a table to help us in these calculations and this is done in Table 5.19.

Again let

$$\mathbf{X'X} = \begin{bmatrix} \sum x_1^2 & \sum x_1 x_2 \\ \sum x_1 x_2 & \sum x_2^2 \end{bmatrix} \quad \mathbf{B} = \begin{bmatrix} b_l \\ b_q \end{bmatrix} \quad \mathbf{Y} = \begin{bmatrix} \sum x_1 y \\ \sum x_2 y \end{bmatrix}$$

We solve for $b_l$ and $b_q$ as before:

$$\mathbf{X'X} = \begin{bmatrix} 0.20 & 0.120 \\ 0.120 & 0.0784 \end{bmatrix} \quad \text{and} \quad (\mathbf{X'X})^{-1} = \begin{bmatrix} 61.25 & -93.75 \\ -93.75 & 156.25 \end{bmatrix}$$

$$(\mathbf{X'X})^{-1}\mathbf{Y} = \begin{bmatrix} 61.25 & -93.75 \\ -93.75 & 156.25 \end{bmatrix}\begin{bmatrix} 0.0261 \\ 0.00498 \end{bmatrix} = \begin{bmatrix} 1.13175 = b_l \\ -1.66875 = b_q \end{bmatrix} = \mathbf{B}$$

We now calculate $a$ as

$$a = \bar{y} - b_l\bar{x}_1 - b_q\bar{x}_2$$

$$= 1.93725 - (1.13175)(0.30000) - (-1.66875)(0.1400)$$

$$= 1.93725 - 0.33952 + 0.23362 = 1.83135$$

which gives us the prediction equation

**TABLE 5.19**  COMPUTATION OF VALUES NECESSARY FOR SOLVING EQUATIONS (5.56)

| $X_1$ | $X_2 = X_1^2$ | $Y$ | $X_1^2$ | $X_2^2$ | $X_1 X_2$ | $X_1 Y$ | $X_2 Y$ |
|---|---|---|---|---|---|---|---|
| 0.0 | 0.00 | 1.837 | 0.00 | 0.0000 | 0.000 | 0.0000 | 0.00000 |
| 0.2 | 0.04 | 1.974 | 0.04 | 0.0016 | 0.008 | 0.3948 | 0.07896 |
| 0.4 | 0.16 | 2.034 | 0.16 | 0.0256 | 0.064 | 0.8136 | 0.32544 |
| 0.6 | 0.36 | 1.904 | 0.36 | 0.1296 | 0.216 | 1.1424 | 0.68544 |
| Totals  1.2 | 0.56 | 7.749 | 0.56 | 0.1568 | 0.288 | 2.3508 | 1.08984 |

$$\sum x_1^2 = \sum X_1^2 - \frac{(\sum X_1)^2}{k} = 0.56 - \frac{(1.2)^2}{4} = 0.56 - 0.36 = 0.20$$

$$\sum x_2^2 = \sum X_2^2 - \frac{(\sum X_2)^2}{k} = 0.1568 - \frac{(0.56)^2}{4} = 0.1568 - 0.0784 = 0.0784$$

$$\sum x_1 x_2 = \sum X_1 X_2 - \frac{(\sum X_1)(\sum X_2)}{k} = 0.288 - \frac{(1.2)(0.56)}{4} = 0.288 - 0.168 = 0.120$$

$$\sum x_1 y = \sum X_1 Y - \frac{(\sum X_1)(\sum Y)}{k} = 2.3508 - \frac{(1.2)(7.749)}{4} = 2.3508 - 2.3247 = 0.0261$$

$$\sum x_2 y = \sum X_2 Y - \frac{(\sum X_2)(\sum Y)}{k} = 1.08984 - \frac{(0.56)(7.749)}{4} = 1.08984 - 1.08486 = 0.00498$$

**TABLE 5.20**   PREDICTED VALUES FOR QUADRATIC REGRESSION

| $X$ | $Y$ | $\hat{Y}$ | $Y - \hat{Y} = d_{y \cdot x}$ | $(Y - \hat{Y})^2 = d_{y \cdot x}^2$ |
|------|-------|---------|------------------------|-------------------------|
| 0.0 | 1.837 | 1.83135 | 0.00565 | 0.000032 |
| 0.2 | 1.974 | 1.99095 | −0.01695 | 0.000287 |
| 0.4 | 2.034 | 2.01705 | 0.01695 | 0.000287 |
| 0.6 | 1.904 | 1.90965 | −0.00565 | 0.000032 |
| Totals | 7.749 | 7.74900 | 0.00000 | 0.000638(10) = 0.0064 |

$$\hat{Y}_i = 1.83135 + 1.13175X_{1i} - 1.66875X_{2i} \tag{5.57}$$

Inserting the simazine levels for the $X$ values in this equation, we can now predict values that will allow us to fit a quadratic regression line to the treatment means. Table 5.20 presents the values predicted from this equation, together with the deviations from $Y$ which will be discussed later.

The predicted points have been plotted as dots in Figure 5.5 and connected by a curve which depicts the quadratic regression of percent $N$ on simazine level. It is obvious that this curved line comes much closer to fitting the data points than a straight line, so by fitting a quadratic regression line to the data we have improved the fitting of the line to the actual means. We would like to know whether this is a statistically significant improvement and we can test for this by testing for the significance of the additional portion of the treatment sum of squares accounted for by fitting the quadratic line. To do this, we need a sum of squares due to quadratic regression which will be developed in the next section.

## 5.8.2  Test for the Significance of Curvilinear Regression Through Treatment Means

The sum of squares due to the quadratic portion of the fitted line can be found by calculating the difference between the sum of squares accounted for by the combined linear and quadratic regression and the sum of squares due to the linear regression alone. We have previously found that the sum of squares due to linear regression is

$$SS_l = \frac{(\sum xy)^2}{\sum x^2}(n) = b \sum xy(n) = \frac{(0.0261)^2}{0.20}(10) = 0.0341$$

where $n$ is the number of observations in each treatment. The sum of squares due to the linear and quadratic regression combined can be found as

$$SS_{(l,q)} = (b_l \sum x_1 y + b_q \sum x_2 y)n$$

$$= [(1.13175)(0.0261) - (1.66875)(0.00498)](10) = 0.2123$$

Again it is emphasized that $n$ is used only because we are dealing with treatment means and must take into account the fact that each $X$ value is accompanied by 10 $Y$ values. We can now calculate the sum of squares due to quadratic regression as

$$SS_q = SS_{(l,q)} - SS_l = 0.2123 - 0.0341 = 0.1782$$

Since there is one degree of freedom for the quadratic regression, the mean square is the same as the sum of squares. We can now make an $F$ test of the quadratic mean square, using as a denominator the same mean square as was used to test the mean square for differences among the simazine levels. When partitioning treatment sums of squares into the different degrees of polynomial regression, the same mean square is used to test each of them, which differs from the situation discussed earlier in Sections 5.2.1, 5.5.1, and 5.6.1 when a different error term was used for each degree of polynomial. Our test of significance is then

$$F = \frac{0.1782}{0.0079} = 22.56**$$

with 1 and 32 degrees of freedom. As anticipated, we find a highly significant effect of the quadratic term in the regression equation.

The sum of squares for the quadratic regression can also be calculated as

$$SS_q = \frac{b_q^2}{c^{22}}(n) \tag{5.58}$$

where $b_q$ is the quadratic regression coefficient, $c^{22}$ is the inverse element relating to the quadratic coefficient in the $(\mathbf{X'X})^{-1}$ matrix, and $n$ is the number of observations per treatment. In our example,

$$SS_q = \frac{b_q^2}{c^{22}}(n) = \frac{(1.66875)^2}{156.25}(10) = 0.1782$$

The test of significance of the quadratic regression can also be made by means of a $t$ test with

$$t = \frac{b_q}{s_{b_q}}\sqrt{n} \quad \text{where} \quad s_{b_q} = \sqrt{c^{22}s^2}$$

$$= \frac{1.66875}{\sqrt{(156.25)(0.0079)}}\sqrt{10} = \frac{5.27705}{1.11102} = 4.7497 \tag{5.59}$$

and $t^2 = 22.56 = F$. The $s^2$ in the standard error of $b$ is the error mean square used to test treatments.

We could test further for the third degree of curvature, the cubic regression. *In this example,* the sum of squares due to cubic regression can be obtained merely by finding the difference between the sum of squares among treatments and the sum of squares due to linear and quadratic regressions. This is so because fitting the cubic regression would require all three degrees of freedom among the four treatments, thus accounting for all the treatment sum of squares. The sum of squares due to cubic regression would then be

$$SS_c = SS_{\text{among treatments}} - SS_{(l,q)} = 0.2187 - 0.2123 = 0.0064$$

Referring back to Table 5.20, we see that the figure 0.0064 is the sum of the squared deviations of treatment means from the predicted means when a quadratic regression line is fitted. This value of 0.0064 is the sum of squares not previously accounted for and now ascribed to cubic regression.

This sum of squares and its corresponding mean square can be tested for significance by the $F$ test as

$$F = \frac{0.0064}{0.0079} = 0.81$$

which leads to the conclusion that there is no significant effect due to cubic regression. We can also state that the cubic regression does not result in a significantly better fit than the quadratic regression.

Were there five or more means, the sum of squares for cubic regression could not be found by a difference. Instead, a new set of equations would have to be developed as shown in (5.47), using $X^3$ as the third independent variable. In matrix form, the values necessary would be

$$\mathbf{X'X} = \begin{bmatrix} \sum x_1^2 & \sum x_1 x_2 & \sum x_1 x_3 \\ \sum x_1 x_2 & \sum x_2^2 & \sum x_2 x_3 \\ \sum x_1 x_3 & \sum x_2 x_3 & \sum x_3^2 \end{bmatrix} \quad \mathbf{B} = \begin{bmatrix} b_l \\ b_q \\ b_c \end{bmatrix} \quad \mathbf{Y} = \begin{bmatrix} \sum x_1 y \\ \sum x_2 y \\ \sum x_3 y \end{bmatrix} \quad (5.60)$$

Using Table 5.19 (and noting that $X_1 X_2 = X_3$), we can calculate

$$\sum x_3^2 = \sum X_3^2 - \frac{(\sum X_3)^2}{k} = 0.050816 - 0.020736 = 0.030080$$

$$\sum x_1 x_3 = \sum X_1 X_3 - \frac{(\sum X_1)(\sum X_3)}{k} = 0.1568 - 0.0864 = 0.0704$$

$$\sum x_2 x_3 = \sum X_2 X_3 - \frac{(\sum X_2)(\sum X_3)}{k} = 0.08832 - 0.04032 = 0.0480$$

$$\sum x_3 y = \sum X_3 Y - \frac{(\sum X_3)(\sum Y)}{k} = 0.557232 - 0.557928 = -0.000696$$

leading to

$$\mathbf{X'X} = \begin{bmatrix} 0.20 & 0.120 & 0.0704 \\ 0.120 & 0.0784 & 0.0480 \\ 0.0704 & 0.0480 & 0.03008 \end{bmatrix} \quad \text{and} \quad \mathbf{Y} = \begin{bmatrix} 0.0261 \\ 0.00498 \\ -0.000696 \end{bmatrix} \quad (5.61)$$

Inverting the matrix $\mathbf{X'X}$ yields

$$(\mathbf{X'X})^{-1} = \begin{bmatrix} 368.055556 & -1562.500000 & 1631.944444 \\ -1562.500000 & 7187.500000 & -7812.500000 \\ 1631.944444 & -7812.500000 & 8680.555556 \end{bmatrix} \quad (5.62)$$

Now making the matrix multiplication $(\mathbf{X'X})^{-1}\mathbf{Y}$, we obtain

$$\mathbf{B'} = [0.689167 = b_l, \; 0.450000 = b_q, \; -2.354167 = b_c]$$

We can now calculate the sum of squares accounted for by the combined linear, quadratic, and cubic regressions as

$$\mathrm{SS}_{(l,q,c)} = [b_l \sum x_1 y + b_q \sum x_2 y + b_c \sum x_3 y]n$$

$$= [(0.689167)(0.0261) + (0.450000)(0.00498)$$

$$+ (-2.354167)(-0.000696)](10)$$

$$= (0.021867)(10) = 0.2187 \quad (5.63)$$

which is, as it should be, the sum of squares among treatments since we have accounted for all three degrees of freedom.

The sum of squares due to cubic regression can now be found as

$$SS_c = SS_{(l,q,c)} - SS_{(l,q)} = 0.2187 - 0.2123 = 0.0064$$

As shown previously, the sum of squares for the highest-order regression coefficient in a polynomial regression analysis can be found by squaring the regression coefficient and dividing by the corresponding inverse element. Again, only because we are dealing with treatment means, we include $n$ in the formula, giving

$$SS_c = \frac{b_c^3}{c^{33}}(n) = \frac{(-2.354167)^2}{8680.555556}(10) = (0.00064)(10) = 0.0064$$

The test of significance of the cubic regression component can also be carried out by the $t$ test as

$$t = \frac{b_c}{s_{b_c}}\sqrt{n} \quad \text{where} \quad s_{b_c} = \sqrt{c^{33}s^2} \tag{5.64}$$

The $s^2$ in the standard error of the regression coefficient is again the mean square used as the denominator in the $F$ test for testing differences among the treatments in the analysis of variance. In our example,

$$t = \frac{b_c}{s_{b_c}}\sqrt{n} = \frac{2.354167}{\sqrt{(8680.555556)(0.0079)}}\sqrt{10} = \frac{7.44453}{8.28109} = 0.8990$$

and $t^2 = 0.81 = F$.

Similar equations can be set up if more degrees of curvature are desired, $X_i$ taking on an additional power for each degree.

If we had wished to use the cubic regression equation for prediction we would have had the equation

$$Y_i = a + b_l X_{1i} + b_q X_{2i} + b_c X_{3i}$$

$$= 1.8370 + 0.689176X_{1i} + 0.450000X_{2i} - 2.354167X_{3i} \tag{5.65}$$

where $a = \bar{y} - b_l\bar{x}_1 - b_q\bar{x}_2 - b_c\bar{x}_3$.

## 5.9 ORTHOGONAL POLYNOMIAL COEFFICIENTS

While it is possible to use the methods described in Section 5.8.1 to partition the sum of squares among treatments into single degree of freedom components due to linear, quadratic, and higher-order regressions, a special method is available for use when there are equal numbers and equally spaced intervals between treatment levels. This method uses *orthogonal polynomial coefficients* whose values are found in many statistics texts or books of statistical tables. An abbreviated set of orthogonal polynomial coefficients is shown in Table 5.21. The coefficients for levels from 8 to 15 are shown in Table A.12. More complete sets of coefficients are available in statistical tables such as those of Anderson and Houseman (1942) or Beyer (1979).

Should values be desired for levels not included in available tables, they can be developed for the first three degrees of polynomial regression from the following for-

TABLE 5.21  ORTHOGONAL POLYNOMIAL COEFFICIENTS AND DIVISORS FOR EQUALLY SPACED INTERVALS

| Number of levels | Regression component | Treatment totals | | | | | | $\lambda$ | Divisors $\Sigma c_{ij}^2$ |
|---|---|---|---|---|---|---|---|---|---|
| | | $T_1$ | $T_2$ | $T_3$ | $T_4$ | $T_5$ | $T_6$ | | |
| 2 | Linear | $-1$ | $+1$ | | | | | | 2 |
| 3 | Linear | $-1$ | $0$ | $+1$ | | | | 1 | 2 |
| | Quadratic | $+1$ | $-2$ | $+1$ | | | | 3 | 6 |
| 4 | Linear | $-3$ | $-1$ | $+1$ | $+3$ | | | 2 | 20 |
| | Quadratic | $+1$ | $-1$ | $-1$ | $+1$ | | | 1 | 4 |
| | Cubic | $-1$ | $+3$ | $-3$ | $+1$ | | | $\frac{10}{3}$ | 20 |
| 5 | Linear | $-2$ | $-1$ | $0$ | $+1$ | $+2$ | | 1 | 10 |
| | Quadratic | $+2$ | $-1$ | $-2$ | $-1$ | $+2$ | | 1 | 14 |
| | Cubic | $-1$ | $+2$ | $0$ | $-2$ | $+1$ | | $\frac{5}{6}$ | 10 |
| | Quartic | $+1$ | $-4$ | $+6$ | $-4$ | $+1$ | | $\frac{35}{12}$ | 70 |
| 6 | Linear | $-5$ | $-3$ | $-1$ | $+1$ | $+3$ | $+5$ | 2 | 70 |
| | Quadratic | $+5$ | $-1$ | $-4$ | $-4$ | $-1$ | $+5$ | $\frac{3}{2}$ | 84 |
| | Cubic | $-5$ | $+7$ | $+4$ | $-4$ | $-7$ | $+5$ | $\frac{5}{3}$ | 180 |
| | Quartic | $+1$ | $-3$ | $+2$ | $+2$ | $-3$ | $+1$ | $\frac{7}{12}$ | 28 |
| | Quintic | $-1$ | $+5$ | $-10$ | $+10$ | $-5$ | $+1$ | $\frac{21}{10}$ | 252 |
| 7 | Linear | $-3$ | $-2$ | $-1$ | $0$ | $+1$ | $+2$ | $+3$ | 1 | 28 |
| | Quadratic | $+5$ | $0$ | $-3$ | $-4$ | $-3$ | $0$ | $+5$ | 1 | 84 |
| | Cubic | $-1$ | $+1$ | $+1$ | $0$ | $-1$ | $-1$ | $-1$ | $\frac{1}{6}$ | 6 |
| | Quartic | $+3$ | $-7$ | $+1$ | $+6$ | $+1$ | $-7$ | $+3$ | $\frac{7}{12}$ | 154 |
| | Quintic | $-1$ | $+4$ | $-5$ | $0$ | $+5$ | $-4$ | $+1$ | $\frac{7}{20}$ | 84 |

mulas, where $X_j$ is the value for the level of treatment, $n$ is the number of levels, $d$ is the spacing between any two levels, $c_{ij}$ is the $j$th coefficient for the $i$th degree polynomial, and $\lambda$ is the coefficient necessary as a multiplier to obtain the smallest set of whole numbers for the coefficients. See Grandage (1958) for a discussion of this method, which can be carried out as follows:

$$\text{Linear } c_{1j} = (X_j + a_1)\frac{\lambda_1}{d} \quad \text{where} \quad a_1 = -\frac{\Sigma X_j}{n} \tag{5.66}$$

$$\text{Quadratic } c_{2j} = (X_j^2 + b_2 X_j + a_2)\frac{\lambda_2}{d^2} \tag{5.67}$$

where
$$b_2 = -\frac{\Sigma c_{1j} X_j^2}{\Sigma c_{1j} X_j} \quad \text{and} \quad a_2 = -\frac{\Sigma X_j^2 + b_2 \Sigma X_j}{n}$$

$$\text{Cubic } c_{3j} = (X_j^3 + c_3 X_j^2 + b_3 X_j + a_3)\frac{\lambda_3}{d^3} \tag{5.68}$$

where
$$c_3 = -\frac{\Sigma c_{2j} X_j^3}{\Sigma c_{2j} X_j^2} \qquad b_3 = -\frac{\Sigma c_{1j} X_j^3 + \Sigma c_{1j} X_j^2 c_3}{\Sigma c_{1j} X_j}$$

and
$$a_3 = -\frac{\Sigma X_j^3 + \Sigma X_j^2 c_3 + \Sigma X_j b_3}{n}$$

As an example, suppose that we had an experiment in which plants were subjected to 8, 12, 16, 20, and 24 hour photoperiods. If we wish to calculate the orthogonal polynomials for the first three degrees we can set up Table 5.22, which we can start by writing down the first three columns.

We can now develop the coefficients for linear regression by first calculating

$$a_1 = -\frac{\Sigma X_j}{n} = \frac{-80}{5} = -16$$

Substituting this value in the numerator of Equation (5.66) with each $X_j$ in turn and dividing by $d = 4$, we arrive at values for $c_{1j}$ of $-2, -1, 0, 1,$ and $2$ where $\lambda$ is equal to 1.

To develop coefficients for the quadratic regression we start by calculating columns (5) and (6) of Table 5.22 for $c_{ij}X_j$ and $c_{1j}X_j^2$. We can now find that

$$b_2 = -\frac{\Sigma c_{1j}X_j^2}{\Sigma c_{1j}X_j} = -\frac{1280}{40} = -32$$

and

$$a = -\frac{\Sigma X_j^2 + b_2 \Sigma X_j}{n} = -\frac{1440 - (32)(80)}{5} = \frac{1120}{5} = 224$$

Substituting each of these values in the numerator of Equation (5.67) with each $X_j$ in turn and dividing by $d^2 = 16$ we arrive at values of $2, -1, -2, -1,$ and $2$ which can be seen to be the same values shown in Table 5.21. $\lambda_1$ is equal to 1. We can now enter these values in column (7) of Table 5.22.

To find the coefficients for the cubic regression we now produce columns (8), (9), and (10) by the indicated multiplication and addition. We are now able to calculate

$$c_3 = -\frac{\Sigma c_{2j}X_j^3}{\Sigma c_{2j}X_j^2} = -\frac{10,752}{224} = -48$$

$$b_3 = -\frac{\Sigma c_{1j}X_j^3 + \Sigma c_{1j}X_j^2 c_3}{\Sigma c_{1j}X_j} = -\frac{32,896 + (1280)(-48)}{40} = 713.6$$

$$a_3 = -\frac{\Sigma X_j^3 + \Sigma X_j^2 c_3 + \Sigma X_j b_3}{n}$$

$$= -\frac{28,160 + (1440)(-48) + (80)(713.6)}{5} = -3225.6$$

**TABLE 5.22** VALUES USED IN DEVELOPING ORTHOGONAL POLYNOMIAL COEFFICIENTS

| (1) | (2) | (3) | (4) | (5) | (6) | (7) | (8) | (9) | (10) |
|---|---|---|---|---|---|---|---|---|---|
| $X_j$ | $X_j^2$ | $X_j^3$ | $c_{1j}$ | $c_{1j}X_j$ | $c_{1j}X_j^2$ | $c_{2j}$ | $c_{2j}X_j^3$ | $c_{2j}X_j^2$ | $c_{1j}X_j^3$ |
| 8 | 64 | 512 | $-2$ | $-16$ | $-128$ | 2 | 1,024 | 128 | $-1,024$ |
| 12 | 144 | 1,728 | $-1$ | $-12$ | $-144$ | $-1$ | $-1,728$ | $-144$ | $-1,728$ |
| 16 | 256 | 4,096 | 0 | 0 | 0 | $-2$ | $-8,192$ | $-512$ | 0 |
| 20 | 400 | 8,000 | 1 | 20 | 400 | $-1$ | $-8,000$ | $-400$ | 8,000 |
| 24 | 576 | 13,824 | 2 | 48 | 1152 | 2 | 27,648 | 1152 | 27,648 |
| 80 | 1440 | 28,160 | 0 | 40 | 1280 | 0 | 10,752 | 224 | 32,896 |

**TABLE 5.23** CALCULATION OF SUMS OF SQUARES FOR LINEAR, QUADRATIC, AND CUBIC REGRESSIONS BY MEANS OF ORTHOGONAL POLYNOMIAL COEFFICIENTS

| Simazine levels | 0.0 | 0.2 | 0.4 | 0.6 | $Q_i$ | $D_i$ | SS | |
|---|---|---|---|---|---|---|---|---|
| Total | 18.37 | 19.74 | 20.34 | 19.04 | $\Sigma c_{ij} T_j$ | $r \Sigma c_{ij}^2$ | $Q_i^2 / D_i$ | $F$ |
| $Q_l$ | $-3$ | $-1$ | $+1$ | $+3$ | 2.61 | 200 | 0.0341 | 4.32* |
| $Q_q$ | $+1$ | $-1$ | $-1$ | $+1$ | $-2.67$ | 40 | 0.1782 | 22.56** |
| $Q_c$ | $-1$ | $+3$ | $-3$ | $+1$ | $-1.13$ | 200 | 0.0064 | 0.81 |

Substituting these values in Equation (5.68) with each $X_j$ successively and dividing by $d^3 = 64$ we arrive at values of $-1.20, 2.40, 0.00, -2.40,$ and $1.20$. Multiplying these values by a $\lambda$ of $1/1.2$ or $\frac{5}{6}$, we can arrive at the smallest set of whole numbers of $-1$, 2, 0, $-2$, and 1, which are those shown in Table 5.21. The equations for additional degrees of polynomial have not been developed here because they are quite lengthy and awkward and it is a rare event when the higher polynomials yield additional information in an analysis. The computations can be made simpler by subtracting the smallest level from all others, dividing each level by the interval $d$, and omitting $d$ from formulas (5.66), (5.67), and (5.68).

The coefficients shown in Table 5.21 are directly comparable to those discussed in Chapter 4, when we dealt with linear functions and single degree of freedom comparisons, and conform to the same definitions of linear functions and orthogonal comparisons. We use these coefficients in our calculations exactly as before. Suppose we wish to partition the sums of squares for the simazine treatments shown in Table 5.15 into linear, quadratic, and cubic regression components; we can set up calculations as shown in Table 5.23. Since there are four levels of simazine we would select the coefficients from Table 5.21 as shown in Table 5.23. It can be seen that the sums of squares for the linear, quadratic, and cubic effects are the same as those calculated previously and are found in a fraction of the time previously used with the nonorthogonal polynomial regression methods. The $F$ values shown in Table 5.23 result from using the error mean square shown in Table 5.15, the analysis of variance table. Since each degree of polynomial is assigned one degree of freedom, the numerator mean squares in the $F$ ratios are the same as the sums of squares. The $F$ table is entered, of course, with 1 and 32 degrees of freedom. The results of this analysis would be displayed in analysis of variance form as shown in Table 5.17 with an additional two lines for the quadratic and cubic regressions.

## 5.9.1 Prediction Equations Using Orthogonal Polynomial Coefficients

The prediction equations using these coefficients take on a different and a little more simple form than those used previously. For linear regression we would have

$$\hat{\bar{y}}_j = \bar{y} + b_l c_{1j} \tag{5.69}$$

where $\bar{y}$ is the overall mean of the $Y$ values, $c_{1j}$ is the coefficient for the linear regression corresponding to the $j$th mean, and $b_l$ is the linear regression coefficient calculated from

$$b_l = \frac{\Sigma c_{1j}T_j}{r \Sigma c_{1j}^2} = \frac{2.61}{200} = 0.01305 \tag{5.70}$$

Using the overall mean of $\bar{y}.. = Y.../N = 77.49/40 = 1.93725$ and the appropriate orthogonal polynomial coefficients, our predicted values for linear regression would be

$$\hat{\bar{y}}_1 = 1.93725 + (0.01305)(-3) = 1.8981$$

$$\hat{\bar{y}}_2 = 1.93725 + (0.01305)(-1) = 1.9242$$

$$\hat{\bar{y}}_3 = 1.93725 + (0.01305)(1) = 1.9503$$

$$\hat{\bar{y}}_4 = 1.93725 + (0.01305)(3) = 1.9764$$

which, of course, agree with those presented in Table 5.18 but were calculated by a different formula.

For quadratic regression we would have

$$\hat{\bar{y}}_j = \bar{y} + b_l c_{1j} + b_q c_{2j} \tag{5.71}$$

where
$$b_q = \frac{\Sigma c_{2j}T_j}{r \Sigma c_{2j}^2} = \frac{-2.67}{40} = -0.06675 \tag{5.72}$$

The predicted values for the quadratic regression would be

$$\hat{\bar{y}}_1 = 1.93725 + (0.01305)(-3) - (0.06675)(1) = 1.83135$$

$$\hat{\bar{y}}_2 = 1.93725 + (0.01305)(-1) - (0.06675)(-1) = 1.99095$$

$$\hat{\bar{y}}_3 = 1.93725 + (0.01305)(1) - (0.06675)(-1) = 2.01705$$

$$\hat{\bar{y}}_4 = 1.93725 + (0.01305)(3) - (0.06675)(1) = 1.90965$$

It can be seen that the predicted values for quadratic regression are the same as those shown in Table 5.20 but found by a different procedure.

Prediction for cubic regression would follow the same methods as for linear and quadratic regressions, merely adding one more coefficient multiplied by the cubic regression coefficient of $-1.13/200 = -0.00565$. Note that when prediction is carried out by means of orthogonal polynomial coefficients the same coefficient is used for the linear portion of the equation regardless of the degree of polynomial, contrary to the previous method where the coefficient is changed. The same is true of higher-order coefficients.

### 5.9.2  Conversion from Orthogonal Polynomial Prediction Equation to Nonorthogonal Prediction Equation

While the prediction equations using the nonorthogonal formulas and those using the orthogonal polynomial coefficients appear different, it is possible to change the prediction equation obtained by using orthogonal coefficients to the equation that would have been obtained had the previous formula been used as shown by Anderson and Houseman (1942). To convert these equations we can make use of the relationships shown in Equations (5.66), (5.67), and (5.68), where the $\lambda$ values come from Table 5.21, $d$ is the difference between any two treatment levels, and $n$ is the number of levels. Transforming the linear prediction equation (5.69) with the $X_j$ (simazine levels) and $c_{ij}$ (orthogonal polynomial coefficients) values from Table 5.24 we have

$$\hat{Y}_j = \bar{y} + b_l(X_j + a_1)\frac{\lambda_1}{d} \tag{5.73}$$

where $\lambda_1 = 2$, $d = 0.2$, $n = 4$, $a_1 = -\Sigma X_j/n = -1.2/4 = -0.3$, and $\bar{y} = 1.93725$, giving

$$\hat{Y}_j = 1.93725 + (0.01305)(X_j - 0.3)(10) = 1.8981 + 0.1305\, X_j$$

and yielding the same equation shown in Equation (5.53).

Transforming the quadratic regression prediction equations leads to

$$\hat{Y}_j = \bar{y} + b_l(X_j + a_1)\frac{\lambda_1}{d} + b_q(X_j^2 + b_2 X_j + a_2)\frac{\lambda_2}{d^2} \tag{5.74}$$

where
$$\lambda_2 = 1 \qquad b_2 = -\frac{\Sigma c_{1j}X_j^2}{\Sigma c_{1j}X_j} = -\frac{1.20}{2.00} = -0.60$$

and
$$a_2 = -\frac{\Sigma X_j^2 + b_2 \Sigma X_j}{n} = -\frac{0.56 - 0.72}{4} = 0.04$$

$$\hat{Y}_j = 1.93725 + (0.01305)(X_j - 0.3)(10) - (0.06675)(X_j^2 - 0.60X_j + 0.04)(25)$$

$$= 1.83135 + 1.13175X_j - 1.66875X_j^2$$

which is the same as the equation shown in (5.57) with minor notation differences.

For the cubic regression transformation we have

$$\hat{Y}_j = \bar{y} + b_l(X_j + a_1)\frac{\lambda_1}{d} + b_q(X_j^2 + b_2 X_j + a_2)\frac{\lambda_2}{d^2} + b_c(X_j^3 + c_3 X_j^2 + b_3 X_i + a_3)\frac{\lambda_3}{d^3}$$

where
$$\lambda_3 = \frac{10}{3} \qquad c_3 = -\frac{\Sigma c_{2j}X_j^3}{\Sigma c_{2j}X_j^2} = -\frac{0.144}{0.16} = -0.9 \tag{5.75}$$

**TABLE 5.24** VALUES USED IN TRANSFORMING FROM ORTHOGONAL POLYNOMIAL PREDICTION EQUATION TO NONORTHOGONAL PREDICTION EQUATION

| (1) | (2) | (3) | (4) | (5) | (6) | (7) | (8) | (9) | (10) |
|---|---|---|---|---|---|---|---|---|---|
| $X_j$ | $X_j^2$ | $X_j^3$ | $c_{1j}$ | $c_{1j}X_j$ | $c_{1j}X_j^2$ | $c_{2j}$ | $c_{2j}X_j^3$ | $c_{2j}X_j^2$ | $c_{1j}X_j^3$ |
| 0.0 | 0.00 | 0.000 | $-3$ | 0.0 | 0.00 | 1 | 0.000 | 0.00 | 0.000 |
| 0.2 | 0.04 | 0.008 | $-1$ | 0.2 | $-0.04$ | $-1$ | $-0.008$ | $-0.04$ | $-0.008$ |
| 0.4 | 0.16 | 0.064 | 1 | 0.4 | 0.16 | $-1$ | $-0.064$ | $-0.16$ | 0.064 |
| 0.6 | 0.36 | 0.216 | 3 | 1.8 | 1.08 | 1 | 0.216 | 0.36 | 0.648 |
| 1.2 | 0.56 | 0.388 | 0 | 2.0 | 1.20 | 0 | 0.144 | 0.16 | 0.704 |

$$b_3 = -\frac{\Sigma\, c_{1j}X_j^3 + \Sigma\, c_{1j}X_j^2 c_3}{\Sigma\, c_{1j}X_j} = -\frac{0.704 + (1.20)(-0.9)}{2.0} = 0.188 \qquad (5.76)$$

$$a_3 = -\frac{\Sigma\, X_j^3 + \Sigma\, x_j^2 c_3 + \Sigma\, X_j b_3}{n}$$

$$= -\frac{0.288 + (0.56)(-0.9) + (1.2)(0.188)}{4} = 0.0024 \qquad (5.77)$$

$$\hat{Y}_j = 1.83135 + 1.13175X_j - 1.66875X_j^2$$

$$- (0.00565)(X_j^3 - 0.9X_j^2 + 0.188X_j - 0.0024)\left(\frac{10}{0.024}\right)$$

$$= 1.83125 + 1.13715X_j - 1.66875X_j^2 - 2.354167X_j^3$$

$$+ 2.11875X_j^2 - 0.442583X_j + 0.00565$$

$$= 1.8370 + 0.689167X_j + 0.450000X_j^2 - 2.354167X_j^3 \qquad (5.78)$$

which is the same equation as shown in (5.65), again with a slightly different notation.

### 5.9.3 Orthogonal Polynomial Coefficients with Unequal Intervals

When, for some reason, treatment levels do not have equal intervals between them the orthogonal polynomial coefficients shown in Table 5.21 cannot be used. However, it is possible to develop such coefficients using the relationships shown previously as shown by Grandage (1958). Again, we shall concern ourselves only with the first three degree polynomials. The linear, quadratic, and cubic coefficients can all be developed using Equations (5.66), (5.67), and (5.68) by omitting the $d$ values in these equations, since when unequal intervals exist $d$ values do not.

As an example, suppose that we had levels of 8, 12, 16, and 24 hour photoperiods. We can prepare Table 5.25 as we did Table 5.22. The value of $n$ in this example is 4. Solving for the linear regression coefficients we find

$$a_1 = \frac{-\Sigma\, X_j}{n} = \frac{-60}{4} = -15 \qquad (5.79)$$

**TABLE 5.25**　VALUES USED IN DEVELOPING ORTHOGONAL POLYNOMIAL COEFFICIENTS WITH UNEQUAL INTERVALS

| (1) | (2) | (3) | (4) | (5) | (6) | (7) | (8) | (9) | (10) |
|---|---|---|---|---|---|---|---|---|---|
| $X_j$ | $X_j^2$ | $X_j^3$ | $c_{1j}$ | $c_{1j}X_j$ | $c_{1j}X_j^2$ | $c_{2j}$ | $c_{2j}X_j^3$ | $c_{2j}X_j^2$ | $c_{1j}X_j^3$ |
| 8 | 64 | 512 | −7 | −56 | −448 | 7 | 3,584 | 448 | −3,584 |
| 12 | 144 | 1,728 | −3 | −36 | −432 | −4 | −6,912 | −576 | −5,184 |
| 16 | 256 | 4,096 | 1 | 16 | 256 | −8 | −32,768 | −2,048 | 4,096 |
| 24 | 576 | 13,824 | 9 | 216 | 5,184 | 5 | 69,120 | 2,880 | 124,416 |
| 60 | 1,040 | 20,160 | 0 | 140 | 4,560 | 0 | 33,024 | 704 | 119,744 |

Using this $a$ value in Equation (5.66) (and omitting the $d$) with each $X_j$ in turn, we find values of $-7, -3, 1,$ and $9$. Since this is the smallest set of whole numbers possible, $\lambda_1 = 1$.

After calculating columns (5) and (6) for $\Sigma\, c_{1j}X_j$ and $c_{1j}X_j^2$, we can find

$$b_2 = -\frac{\Sigma\, c_{1j}X_j^2}{\Sigma\, c_{1j}X_j} = -\frac{4560}{140} = -\frac{228}{7} \tag{5.80}$$

and $\quad a_2 = -\dfrac{\Sigma\, X_j^2 + b_2 \Sigma\, X_j}{n} = -\dfrac{1040 + (-228/7)(60)}{4} = \dfrac{6400}{28} = \dfrac{1600}{7}$ $\tag{5.81}$

When we substitute each of these values in Equation (5.67) (again omitting the $d$ value) with each $X_j$ value in turn, we find values of $224/7, -128/7, -256/7,$ and $160/7$. Using a $\lambda_2$ of $7/32$ we arrive at coefficients of $7, -4, -8,$ and $5$. These values sum to zero as they should and $\Sigma\, c_{ij}c_{2j} = 0$, ensuring that these coefficients for the quadratic regression are orthogonal to the linear coefficients.

To calculate the coefficients for the cubic regression, we can first develop columns (8), (9), and (10) as shown. We can now calculate

$$c_3 = -\frac{\Sigma\, c_{2j}X_j^3}{\Sigma\, c_{2j}X_j^2} = -\frac{33,024}{704} = -\frac{516}{11} \tag{5.82}$$

$$b_3 = -\frac{\Sigma\, c_{1j}X_j^3 + \Sigma\, c_{1j}X_j^2 c_3}{\Sigma\, c_{1j}X_j}$$

$$= -\frac{119,744 + (4560)(-516/11)}{140} = \frac{36,992}{55} \tag{5.83}$$

$$a_3 = -\frac{\Sigma\, X_j^3 + \Sigma\, X_j^2 c_3 + \Sigma\, X_j b_3}{n}$$

$$= -\frac{20,160 + (1040)(-516/11) + (60)(36,992/55)}{4} = -\frac{32,256}{11} \tag{5.84}$$

Substituting $a_3, b_3,$ and $c_3$ in Equation (5.68) with each $X_j$, we arrive at values of $-2304/55, 6144/55, -4608/55,$ and $768/55$. We can then find that a $\lambda_3$ of $55/768$ reduces these figures to $-3, 8, -6,$ and $1$, the smallest set of whole numbers. We can then check that $\Sigma\, c_{1j}c_{3j}$ and $\Sigma\, c_{2j}c_{3j}$ are zero and see that the cubic coefficients are orthogonal to the coefficients previously calculated. We can now proceed to use these coefficients to partition the treatment sum of squares into linear, quadratic, and cubic components. The coefficients would also be used in prediction equations in the same way as were those for equal intervals.

## 5.9.4 Orthogonal Polynomial Coefficients with Unequal Numbers in the One-Way Classification

Frequently, the situation arises in a one-way classification where the treatments have unequal numbers of observations and we wish to fit a regression line to the treatment means. Let us consider the following hypothetical set of data representing the average

**TABLE 5.26** AVERAGE DAILY GAINS (IN POUNDS) OF PIGS ON FOUR DIFFERENT LEVELS OF PROTEIN

| | Level | | |
|---|---|---|---|
| 1 | 3 | 4 | 6 |
| 0.75 | 0.91 | 1.45 | 1.10 |
| 0.81 | 0.87 | 1.21 | 1.15 |
| 0.69 | 0.93 | 1.36 | 1.08 |
| 0.74 | 0.81 | 1.42 | 1.01 |
| 0.68 | 0.86 | 1.20 | 1.22 |
| | 0.83 | 1.18 | 1.13 |
| | 0.92 | 1.31 | 1.02 |
| | 0.98 | | 1.11 |
| | | | 1.03 |
| | | | 1.05 |
| 3.67 | 7.11 | 9.13 | 10.90 |

daily gains of four different groups of pigs fed four different levels of protein, which we will consider as levels 1, 3, 4, and 6. The data are shown in Table 5.26.

The analysis of variance for this set of data is shown in Table 5.27.

If we wished to fit a regression line to the means of this experiment to evaluate the change in daily gain over the four different protein levels by means of orthogonal polynomial coefficients, we would have not only the problem of unequal numbers in the groups but also the problem of unequal intervals between treatment levels. Each of these situations renders a table of coefficients such as Table 5.21 unusable. We could handle the unequal interval difficulty as shown in Section 5.9.3 but would have to adapt the formulas used to accommodate the unequal numbers. The method that we shall use for the unequal number situation is exactly the same whether or not we have equal intervals, and is presented by Carmer and Seif (1963).

We can set up Table 5.28 as we did in previous computations, writing out columns (1), (2), (3), (4), and (5), where $k_j$ is the number of observations in the $j$th treatment. Defining the linear regression coefficients as

$$c_{1j} = X_j + a_1 \tag{5.85}$$

we can calculate

$$a_1 = -\frac{\Sigma k_j X_j}{\Sigma k_j} = -\frac{117}{30} = -\frac{39}{10} \tag{5.86}$$

**TABLE 5.27** ANOVA FOR DATA SHOWN IN TABLE 5.24

| Source | df | SS | MS | F |
|---|---|---|---|---|
| Treatments | 3 | 1.1600 | 0.3867 | 69.05** |
| Error | 26 | 0.1454 | 0.0056 | |

**TABLE 5.28** COMPUTATION OF VALUES USED FOR REGRESSION THROUGH TREATMENT MEANS BASED ON UNEQUAL NUMBERS AND HAVING UNEQUAL INTERVALS

| (1) | (2) | (3) | (4) | (5) | (6) | (7) | (8) | (9) | (10) | (11) | (12) | (13) | (14) |
|---|---|---|---|---|---|---|---|---|---|---|---|---|---|
| $X_j$ | $X_j^2$ | $X_j^3$ | $k_j$ | $k_jX_j$ | $c_{1j}$ | $k_jc_{1j}X_j^2$ | $k_jc_{1j}X_j$ | $k_jX_j^2$ | $c_{2j}$ | $k_jc_{2j}X_j^3$ | $k_jc_{2j}X_j^2$ | $k_jc_{1j}X_j^3$ | $k_jX_j^3$ |
| 1 | 1 | 1 | 5 | 5 | −29 | −145 | −145 | 5 | 652 | 3,260 | 3,260 | −145 | 5 |
| 3 | 9 | 27 | 8 | 24 | −9 | −648 | −216 | 72 | −405 | −87,480 | −29,160 | −1,944 | 216 |
| 4 | 16 | 64 | 7 | 28 | 1 | 112 | 28 | 112 | −470 | −210,560 | −52,640 | 448 | 448 |
| 6 | 36 | 216 | 10 | 60 | 21 | 7560 | 1260 | 360 | 327 | 706,320 | 117,720 | 45,360 | 2160 |
| 14 | | | 30 | 117 | | 6879 | 927 | 549 | | 411,540 | 39,180 | 43,719 | 2829 |

and subtracting $a_1$ from each $X_j$ in turn, we arrive at values of $-29/10$, $-9/10$, $1/10$, and $21/10$. Multiplying the values of $\lambda_1 = 10$, we have for the coefficients of linear regression $-29$, $-9$, $1$, and $21$ which we can enter in column (6).

For the quadratic regression we can develop columns (7), (8), and (9) and defining the quadratic regression coefficients as

$$c_{2j} = X_j^2 + X_jb_2 + a_2 \tag{5.87}$$

we calculate

$$b_2 = -\frac{\Sigma\, k_jc_{1j}X_j^2}{\Sigma\, k_jc_{1j}X_j} = -\frac{6879}{927} = -\frac{2293}{309} \tag{5.88}$$

$$a_2 = -\frac{\Sigma\, k_jX_j^2 + b_2\, \Sigma\, k_jX_j}{\Sigma\, k_j} = -\frac{549 - (2293/309)(117)}{30} = \frac{3288}{309} \tag{5.89}$$

Substituting these values of $b_2$ and $a_2$ in Equation (5.86) with each $X_j$, we find values of $1304/309$, $-810/309$, $-940/309$, and $654/309$. Multiplying these values by $\lambda_2 = 309/2$, we have the quadratic coefficients of $652$, $-405$, $-470$, and $327$. Ascertaining that the coefficients are orthogonal to the linear coefficients, we find that $\Sigma\, k_jc_{1j}c_{2j} = 0$. We can now enter the quadratic coefficients in column (10) of Table 5.28.

Proceeding with the development of the cubic coefficients, we can develop columns (10), (11), (12), (13), and (14). Using these and other values in Table 5.28 and defining the cubic regression coefficients as

$$c_{3j} = X_j^3 + X_j^2c_3 + X_jb_3 + a_3 \tag{5.90}$$

we calculate

$$c_3 = -\frac{\Sigma\, k_jc_{2j}X_j^3}{\Sigma\, k_jc_{2j}X_j^2} = -\frac{411,540}{39,180} = -\frac{6859}{653} \tag{5.91}$$

$$b_3 = -\frac{\Sigma\, k_jc_{1j}X_j^3 + \Sigma\, k_jc_{1j}X_j^2c_3}{\Sigma\, k_jc_{1j}X_j}$$

$$= -\frac{43,719 + 6879\,(-6859/653)}{927} = \frac{20,102}{653} \tag{5.92}$$

$$a_3 = -\frac{\Sigma \, k_j X_j^3 + \Sigma \, k_j X_j^2 c_3 + \Sigma \, k_j X_j b_3}{\Sigma \, k_j}$$

$$= -\frac{2829 + (549)(-6859/653) + (117)(20{,}102/653)}{30}$$

$$= -\frac{433{,}680}{(30)(653)} = -\frac{14{,}456}{653} \tag{5.93}$$

We can now substitute the values that we have found for $c_3$, $b_3$, and $a_3$ in Equation (5.90) and calculate values of $-560/653$, $1750/653$, $-2000/653$, and $280/653$. Multiplication of these values by a $\lambda_3$ value of $653/10$ reduces them to the smallest set of whole numbers, $-56$, $175$, $-200$, and $28$ which are the $c_{3j}$ or coefficients for the cubic regression. We can make certain that these coefficients for the linear and quadratic regression coefficients are orthogonal to the cubic coefficients by finding that $\Sigma \, k_j c_{1j} c_{3j} = \Sigma \, k_j c_{2j} c_{3j} = 0$.

Having calculated the orthogonal polynomial coefficients for the three degrees of regression, we proceed to partition the treatment sum of squares into the corresponding sums of squares. We can set up Table 5.29 and complete the computations shown in the table. Each of the sums of squares has one degree of freedom so the mean squares are equal to the sums of squares. Each mean square is tested for significance by the same mean square (as is always the case) that was used for testing for the significance of treatment effects, 0.0056. The $F$ values have 1 and 26 degrees of freedom and show a highly significant effect for each degree of polynomial.

If we wished to predict values by means of the orthogonal polynomial coefficients we would want to use the cubic regression since this was highly significant. We can calculate the regression coefficients as

$$b_i = \frac{\Sigma \, c_{ij} T_j}{\Sigma \, k_j c_{ij}^2} \tag{5.94}$$

giving for the linear, quadratic, and cubic coefficients 0.007293, $-0.000200$, and $-0.000879$. Our prediction equation for cubic regression would be

$$\hat{Y}_j = \bar{y} + c_{1j} b_l + c_{2j} b_q + c_{3j} b_c \tag{5.95}$$

and using this equation to predict we would find, of course, that our predicted values would be the treatment means since we have used all the degrees of freedom.

**TABLE 5.29**  CALCULATION OF SUMS OF SQUARES FOR LINEAR, QUADRATIC, AND CUBIC REGRESSIONS

| Treatment number | 1 | 2 | 3 | 4 | | | |
|---|---|---|---|---|---|---|---|
| Totals | 3.67 | 7.11 | 9.13 | 10.90 | | | |
| Number of observations $= k_j$ | 5 | 8 | 7 | 10 | $Q_j = \Sigma \, c_{ij} T_j$ | $D_j = \Sigma \, k_j c_{ij}^2$ | $SS_{Qj} = Q_j^2/D_j$ |
| Linear | $-29$ | $-9$ | 1 | 21 | 67.61 | 9,270 | 0.4931** |
| Quadratic | 652 | $-405$ | $-470$ | 327 | $-1213.51$ | 6,053,310 | 0.2433** |
| Cubic | $-56$ | 175 | $-200$ | 28 | $-482.07$ | 548,520 | 0.4237** |

If the nonorthogonal regression formulas were preferred, the regression through the treatment means could be carried out by that method with the same results. In that case, there would be 5 values of $X = 1$, 8 values of $X = 3$, 7 values of $X = 4$, and 10 values of $X = 6$, all accompanied by a corresponding $Y$.

## 5.10 MULTIPLE REGRESSION

One of the most powerful tools in the evaluation of research data is multiple regression analysis. This type of analysis is a logical extension of the methods discussed earlier in this section and is comparable to the methods of curvilinear regression. However, with multiple regression each $X$ is a separate independent variable rather than a function of one variable. The interest lies in determining which of a series of independent variables have a significant influence in determining the value of a given dependent variable. For example, one might be interested in examining the influence of age, weight, and height on the performance of a particular task or the influence of initial age and weight on rate of gain of mice.

As an example of multiple regression let us consider the data displayed in Table 5.30, supplied by Roland Vosburgh, graduate student in the Department of Forestry and Wildlife Management at the University of Massachusetts.

The model describing this set of data is

$$Y_i = \mu + \beta_1(X_{1i} - \bar{x}_1) + \beta_2(X_{2i} - \bar{x}_2) + \beta_3(X_{3i} - \bar{x}_3) + e_i \tag{5.96}$$

This model is sometimes written

$$Y_i = \mu + \beta_{Y1.23}(X_{1i} - \bar{x}_1) + \beta_{Y2.13}(X_{2i} - \bar{x}_2) + \beta_{Y3.12}(X_{3i} - \bar{x}_3) + e_i \tag{5.97}$$

The $\beta$'s are known as partial regression coefficients. $\beta_{Y1.23}$, for example, is read as the regression of $Y$ on $X_1$ independent of $X_2$ and $X_3$, or holding $X_2$ and $X_3$ constant.

The model expressed in (5.97) then leads to the prediction equation

$$\hat{Y}_i = a + b_{Y1.23}X_{1i} + b_{Y2.13}X_{2i} + b_{Y3.12}X_{3i} \tag{5.98}$$

where $$a = \bar{y} - b_{Y1.23}\bar{x}_1 - b_{Y2.13}\bar{x}_2 - b_{Y3.12}\bar{x}_3 \tag{5.99}$$

Calculation of the partial regression coefficients involves the solution of a set of equations quite similar to those previously developed for the curvilinear regression equation. The appropriate set of simultaneous (least-squares) normal equations is

$$b_{Y1.23} \sum x_1^2 + b_{Y2.13} \sum x_1x_2 + b_{Y3.12} \sum x_1x_3 = \sum x_1y$$

$$b_{Y1.23} \sum x_1x_2 + b_{Y2.13} \sum x_2^2 + b_{Y3.12} \sum x_2x_3 = \sum x_2y \tag{5.100}$$

$$b_{Y1.23} \sum x_1x_3 + b_{Y2.13} \sum x_2x_3 + b_{Y3.12} \sum x_3^2 = \sum x_3y$$

Substituting in these equations the values found at the foot of Table 5.29 and putting the equations in matrix form, we have

**TABLE 5.30**   DIAMETER ($X_1$, IN MILLIMETERS), BASAL AREA
($X_2$, IN METER$^2$/HECTARE), DEPTH OF WATER AROUND BASE
OF TREE IN APRIL ($X_3$, IN METERS), AND WIDTH OF ANNUAL
GROWTH OF SILVER MAPLE TREES GROWING AROUND LAKE
CHAMPLAIN WETLANDS Y: A FEW SELECTED OBSERVATIONS

| $X_1$ | $X_2$ | $X_3$ | $Y$ | $\hat{Y}$ |
|---|---|---|---|---|
| 213 | 23 | 0.76 | 1.510 | 1.379 |
| 231 | 37 | 0.46 | 2.765 | 2.318 |
| 183 | 23 | 0.66 | 2.195 | 1.936 |
| 267 | 21 | 0.66 | 2.543 | 2.551 |
| 368 | 14 | 0.76 | 4.426 | 3.003 |
| 165 | 18 | 0.76 | 1.243 | 1.592 |
| 196 | 14 | 0.66 | 0.651 | 2.830 |
| 129 | 28 | 0.76 | 0.753 | 0.488 |
| 307 | 21 | 0.66 | 1.009 | 2.757 |
| 323 | 21 | 0.43 | 4.969 | 4.474 |
| 297 | 14 | 0.53 | 6.863 | 4.273 |
| 183 | 21 | 0.53 | 2.252 | 3.044 |
| 312 | 16 | 0.53 | 3.933 | 4.166 |
| 248 | 16 | 0.63 | 3.075 | 3.126 |
| 305 | 28 | 0.63 | 2.622 | 2.317 |
| 203 | 28 | 0.53 | 3.263 | 2.503 |
| 338 | 7 | 0.43 | 6.563 | 5.837 |
| 241 | 14 | 0.33 | 5.653 | 5.406 |
| 422 | 21 | 0.36 | 3.909 | 5.480 |
| 420 | 14 | 0.46 | 4.683 | 5.402 |

$\Sigma X_1 = \ \ 5,351$   $\Sigma X_2 = \ 399$   $\Sigma X_3 = 11.53$   $\Sigma Y = \ 64.880$   $\Sigma \hat{Y} = \ 64.882$
$\Sigma X_1^2 = 1,561,381$   $\Sigma X_2^2 = 8849$   $\Sigma X_3^2 = \ \ 7.0017$   $\Sigma Y^2 = 275.903164$   $\Sigma \hat{Y}^2 = 253.740508$
$\Sigma x_1^2 = \ \ 129,720.95$   $\Sigma x_2^2 = 888.95$   $\Sigma x_3^2 = \ \ 0.354655$   $\Sigma y^2 = \ 65.432444$   $\Sigma \hat{y}^2 = \ 43.256812$

$\Sigma X_1 X_2 = 102,637$   $\Sigma X_1 X_3 = 2984.86$   $\Sigma X_2 X_3 = 232.49$
$\Sigma x_1 x_2 = -4,115.45$   $\Sigma x_1 x_3 = \ -99.9915$   $\Sigma x_2 x_3 = \ \ 2.4665$
$\Sigma X_1 Y = \ 19,113.704$   $\Sigma X_2 Y = 1174.013$   $\Sigma X_3 Y = \ 34.14251$
$\Sigma x_1 y = \ \ 1,755.060$   $\Sigma x_2 y = -120.343$   $\Sigma x_3 y = \ -3.26081$

$$
\underset{\textbf{X'X}}{\begin{bmatrix} 129,720.95 & -4,115.45 & -99.9915 \\ -4,115.45 & 888.95 & 2.4665 \\ -99.9915 & 2.4665 & 0.354655 \end{bmatrix}} \underset{\textbf{B}}{\begin{bmatrix} b_{Y1 \cdot 23} \\ b_{Y2 \cdot 13} \\ b_{Y3 \cdot 12} \end{bmatrix}} = \underset{\textbf{Y}}{\begin{bmatrix} 1755.06 \\ -120.343 \\ -3.26081 \end{bmatrix}} \quad (5.101)
$$

Inverting the matrix $\textbf{X'X}$ by the method shown in Chapter 9, we find the inverse

$$
(\textbf{X'X})^{-1} = \begin{bmatrix} 0.00001135 = c^{11} & 0.00004452 = c^{12} & 0.00289014 = c^{13} \\ 0.00004452 = c^{21} & 0.00132174 = c^{22} & 0.00336059 = c^{23} \\ 0.00289014 = c^{31} & 0.00336059 = c^{32} & 3.61111544 = c^{33} \end{bmatrix} \quad (5.102)
$$

where the $c^{ij}$, as before, are the inverse elements. The appropriate matrix multiplication
of $(\textbf{X'X})^{-1}\textbf{Y}$ yields the column vector of partial regression coefficients as follows:

$$b_{Y1 \cdot 23} = (0.00001135)(1755.06) + (0.00004452)(-120.343)$$

$$+ (0.00289014)(-3.26081) = 0.005138$$

$$b_{Y2 \cdot 13} = (0.00004452)(1755.06) + (0.00132174)(-120.343)$$

$$+ (0.00336059)(-3.26081) = -0.091885$$

$$b_{Y3 \cdot 12} = (0.00289014)(1755.06) + (0.00336059)(-120.343)$$

$$+ (3.61111544)(-3.26081) = -7.107216$$

Substituting in Equation (5.99), we can calculate the intercept $a$ as

$$a = \frac{64.880}{20} - (0.005138)\left(\frac{5351}{20}\right) - (-0.091885)\left(\frac{399}{20}\right)$$

$$- (-7.107216)\left(\frac{11.53}{20}\right) = 7.799744$$

using the data from Table 5.29. We can now write the prediction equation as

$$\hat{Y}_i = 7.799744 + 0.005138X_{1i} - 0.091885X_{2i} - 7.107216X_{3i} \qquad (5.103)$$

Having completed the above computations, our interest now lies in the interpretation of the results. As a first step, let us test for the significance of the three regressions considered jointly or the multiple regression. This test requires that we obtain the appropriate sum of squares accounted for by these three regressions which is found by

$$\text{Reduction SS} = b_{Y1 \cdot 23} \sum x_1 y + b_{Y2 \cdot 13} \sum x_2 y + b_{Y3 \cdot 12} \sum x_3 y = \sum_i b_i \sum x_i y$$

$$= (0.005138)(1755.06) + (-0.091885)(-120.343)$$

$$+ (-7.107216)(-3.26081)$$

$$= 43.250496 \qquad (5.104)$$

This reduction is frequently presented in the form

$$\text{Reduction SS} = R^2_{Y \cdot 123} \sum y^2 \qquad (5.105)$$

where $R^2$ is defined as the *coefficient of multiple determination*, and $\sum y^2$ is the total sum of squares. $R^2$ is directly comparable to $r^2$, the coefficient of determination, which is used when only one independent variable is involved. From Equation (5.104)

$$\text{Reduction SS} = 43.250496 = R^2_{Y \cdot 123}(65.432444)$$

and

$$R^2_{Y \cdot 123} = \frac{43.250496}{65.432444} = 0.6610$$

$R^2$ is then a measure of the proportion of the total variation among the $Y$'s accounted for by the three regressions combined. $R_{Y \cdot 123}$ is known as the *multiple correlation coefficient*, a measure of the correlation between $Y$ and $\hat{Y}$. Table 5.30 shows that $\Sigma(\hat{Y} - \bar{\hat{y}})^2 = \Sigma \hat{y}^2 = 43.256812$, which is (within rounding error) the reduction in the sum of squares found by Equation (5.104), the sum of squares accounted for by the regression of $Y$ on the independent variables.

### 5.10.1 Tests of Significance in Multiple Regression

We can now place the appropriate values in an analysis of variance table and test for the significance in the variation accounted for by the three regressions as shown in Table 5.31. We can see from this table that a highly significant amount of variation among the $Y$ values is accounted for by the three independent variables. One method of evaluating their relative importance is by comparison of *standard partial regression coefficients.*

The standard partial regression coefficients are the partial regression coefficients when all the variables are expressed in standard measure. In standard measure the deviation of any variate from its mean is expressed in units of the standard deviation of that variable. Thus,

$$Y'_i = \frac{Y_i - \bar{y}_i}{s_y} \tag{5.106}$$

is a standard deviate, $s$ being the standard deviation. Standard partial regression coefficients are also distinguished by a prime. As an example, the first coefficient is designated as $b'_{Y1 \cdot 23} = b'_1$. The coefficients can be calculated as

$$b'_i = b_i \frac{s_i}{s_y} = b_i \sqrt{\frac{\Sigma x_i^2}{\Sigma y^2}}, \tag{5.107}$$

where $s_i$ is the standard deviation of the $i$th independent variable. The second term of the equality is the form usually used. Thus,

$$b'_{Y1 \cdot 23} = b_{Y1 \cdot 23} \sqrt{\frac{\Sigma x_1^2}{\Sigma y^2}} \qquad b'_{Y2 \cdot 13} = b_{Y2 \cdot 13} \sqrt{\frac{\Sigma x_2^2}{\Sigma y^2}}$$

$$b'_{Y3 \cdot 12} = b_{Y3 \cdot 12} \sqrt{\frac{\Sigma x_3^2}{\Sigma y^2}}$$

The standard partial coefficients for the example under study are

**TABLE 5.31**   TEST OF SIGNIFICANCE OF PARTIAL
REGRESSION COEFFICIENTS
CONSIDERED JOINTLY
(MULTIPLE REGRESSION)

| Source | df | SS | MS | F |
|--------|-----|---------|---------|---------|
| Total | 19 | 65.4324 | | |
| Regressions | 3 | 43.2505 | 14.4168 | 10.40** |
| Error | 16 | 22.1819 | 1.3864 | |

$$b'_{Y1 \cdot 23} = 0.005138 \sqrt{\frac{129,720.25}{65.432444}} = 0.2288$$

$$b'_{Y2 \cdot 13} = -0.091885 \sqrt{\frac{888.95}{65.432444}} = -0.3387$$

$$b'_{Y3 \cdot 12} = -7.107216 \sqrt{\frac{0.354655}{65.432444}} = -0.5232$$

Multiplication of the $b_i$ by $s_i/s_y$ makes the comparisons among the coefficients possible without regard to the original units of measurement. The $b'_i$ values estimate the change in the dependent variable $Y$, as a fraction of $s_y$, for a change of one standard deviation of $X_i$. A comparison of the absolute values of these coefficients yields a ranking of their relative importance. Thus, the depth of water around the base of the tree in April is seen to be the most important of the three independent variables in determining the width of the annual growth, indicating the detrimental effect of too much water around the base of the tree.

It is of interest to see that these standard partial regression coefficients can be calculated by a different method. We can start by using the calculations at the base of Table 5.30 and developing what is called a *correlation matrix*. A correlation matrix is one made up of 1's on the diagonals and correlation coefficients on the off-diagonals. Furthermore, instead of considering $Y$ as a dependent variable, it is treated as a fourth variable to be correlated with the three $X$ variables. The correlation matrix is

$$\mathbf{C} = \begin{matrix} & X_1 & X_2 & X_3 & Y \\ X_1 \\ X_2 \\ X_3 \\ Y \end{matrix} \begin{bmatrix} 1 & r_{12} & r_{13} & r_{1Y} \\ r_{12} & 1 & r_{23} & r_{2Y} \\ r_{13} & r_{23} & 1 & r_{3Y} \\ r_{1Y} & r_{2Y} & r_{3Y} & 1 \end{bmatrix}$$

$$= \begin{bmatrix} 1.00000000 & -0.38324245 & -0.46618144 & 0.60240746 \\ -0.38324245 & 1.00000000 & 0.13891189 & -0.49898282 \\ -0.46618144 & 0.13891189 & 1.00000000 & -0.67690213 \\ 0.60240746 & -0.49898282 & -0.67690213 & 1.00000000 \end{bmatrix}$$

The inverse of this matrix is

$$\mathbf{C}^{-1} = \begin{bmatrix} 1.62646896 = c^{11} & 0.24968880 = c^{12} & 0.26698558 = c^{13} & -0.67448351 = c^{14} \\ 0.24968880 = c^{21} & 1.51318188 = c^{22} & 0.58228055 = c^{23} & 0.99878430 = c^{24} \\ 0.26698558 = c^{31} & 0.58228055 = c^{32} & 2.08817701 = c^{33} & 1.54320535 = c^{34} \\ -0.67448351 = c^{41} & 0.99878430 = c^{42} & 1.54320534 = c^{43} & 2.94928910 = c^{44} \end{bmatrix}$$

We are now able to calculate the standard partial regression coefficients, defined in terms of elements of the correlation matrix inverse, as

$$b'_i = \frac{-c^{ij}}{c^{ii}} \tag{5.108}$$

where the $i$ superscript refers to the column for the variable $Y$ and the $j$ superscript refers to the column for any of the other variables. Thus, the standard partial coefficients would be

$$b_1' = b_{Y1 \cdot 23}' = \frac{-c^{41}}{c^{44}} = -\frac{-0.67448351}{2.94928910} = 0.2287$$

$$b_2' = b_{Y2 \cdot 13}' = \frac{-c^{42}}{c^{44}} = -\frac{0.99878430}{2.94928910} = -0.3387$$

$$b_3' = b_{Y3 \cdot 12}' = \frac{-c^{43}}{c^{44}} = -\frac{1.54320535}{2.94928910} = -0.5232$$

which are the same as those calculated previously.

The standard partial regression coefficients can also be found by setting up the equations

$$\overset{\mathbf{C}}{\begin{bmatrix} 1 & r_{12} & r_{13} \\ r_{12} & 1 & r_{23} \\ r_{13} & r_{23} & 1 \end{bmatrix}} \overset{\mathbf{B}}{\begin{bmatrix} b_1' \\ b_2' \\ b_3' \end{bmatrix}} = \overset{\mathbf{Y}}{\begin{bmatrix} r_{1y} \\ r_{2y} \\ r_{3y} \end{bmatrix}}$$

and solving for $\mathbf{B}$ by $\mathbf{C}^{-1}\mathbf{Y}$. For our example, we would have

$$\overset{\mathbf{C}}{\begin{bmatrix} 1 & -0.38324245 & -0.46618144 \\ -0.38324245 & 1 & 0.13891189 \\ -0.46618144 & 0.13891189 & 1 \end{bmatrix}} \overset{\mathbf{Y}}{\begin{bmatrix} 0.60240746 \\ -0.49898282 \\ -0.67690213 \end{bmatrix}}$$

Inverting the $\mathbf{C}$ matrix yields

$$\mathbf{C}^{-1} = \begin{bmatrix} 1.47221891 & 0.47810437 & 0.61990675 \\ 0.47810437 & 1.17494101 & 0.05967011 \\ 0.61990675 & 0.05960711 & 1.28070013 \end{bmatrix}$$

and performing the matrix multiplication $\mathbf{C}^{-1}\mathbf{Y}$, we find the standard partial regression coefficients

$$b_{Y1 \cdot 23}' = 0.2287 \qquad b_{Y2 \cdot 13}' = -0.3387 \qquad b_{Y3 \cdot 12}' = -0.5232$$

## 5.10.2 Tests of Significance for Partial Regression Coefficients

Since we found the effect of the three regressions considered jointly to be highly significant, we can now proceed to see which of the individual partial regressions are significant. This can be accomplished by the use of an $F$ test or a $t$ test since only one degree of freedom is involved for each regression. To conduct the $t$ test we need an estimate of the common standard deviation from regression, or the *standard error of estimate,* which can be defined as

$$S_{y \cdot 1 \ldots k} = \sqrt{\frac{\Sigma(Y - \hat{Y})^2}{n - k - 1}} \qquad (5.109)$$

where $n$ is the total number of $Y$'s and $k$ is the number of independent variables. The $\Sigma(Y - \hat{Y})^2$ can be found as

$$\Sigma(Y - \hat{Y})^2 = \Sigma y^2 - \sum_i b_i\left(\Sigma x_i y\right) \qquad (5.110)$$

The term $\Sigma_i b_i(\Sigma x_i y)$, referred to as the reduction SS, was found to be 43.2505 for the example problem, the sum of squares accounted for by regression. If we subtract this from the total sum of squares $\Sigma y^2$, we have a residual or an error sum of squares, that is, the sum of squares not accounted for by regression. Solving Equation (5.109) for the example, we have

$$\Sigma(Y - \hat{Y})^2 = 65.4324 - 43.2505 = 22.1819$$

as has already been shown in Table 5.30. We can then calculate

$$S_{y \cdot 123} = \sqrt{\frac{22.1819}{20 - 3 - 1}} = \sqrt{1.3864} = 1.1775$$

The standard error of a $b_i$ value in general is given by

$$S_{b_i} = \sqrt{c^{ii} s_{y \cdot 123}^2} \qquad (5.111)$$

where $c^{ii}$ is the diagonal inverse element corresponding to the $b_i$. For the preceding partial regression coefficients,

$$S_{b_1} = \sqrt{c^{11} s_{y \cdot 123}^2} = \sqrt{(0.00001135)(1.3864)} = 0.003967$$

$$S_{b_2} = \sqrt{c^{22} s_{y \cdot 123}^2} = \sqrt{(0.00132174)(1.3864)} = 0.042807$$

$$S_{b_3} = \sqrt{c^{33} s_{y \cdot 123}^2} = \sqrt{(3.61111544)(1.3864)} = 2.237510$$

The inverse elements come from the matrix shown in (5.102). Our $t$ test would then test for the significance of the difference between the given partial regression coefficient and a value of zero. Our tests are

$$t_1 = \frac{b_1}{S_{b_1}} = \frac{0.005138}{0.003968} = 1.295$$

$$t_2 = \frac{b_2}{S_{b_2}} = \frac{0.091885}{0.042807} = 2.146*$$

$$t_3 = \frac{b_3}{S_{b_3}} = \frac{7.107216}{2.237510} = 3.176**$$

All the $t$ values have the 16 degrees of freedom for error shown in Table 5.30. It can be seen that the sizes of the $t$ values reflect the same relative importance as shown previously by the standard partial regression coefficients.

The $F$ tests require the calculation of the sum of squares due to each coefficient, found by

$$SS_{b_i} = \frac{b_i^2}{c^{ii}} \quad \text{or} \quad t_i^2 s_{y \cdot 123}^2 = SS_{b_i} \tag{5.112}$$

and they use the same error mean square used in the $t$ tests. Thus,

$$SS_{b_1} = \frac{(0.005138)^2}{0.00001135} = 2.325903 \qquad F = \frac{2.325903}{1.3864} = 1.678$$

$$SS_{b_2} = \frac{(-0.091885)^2}{0.00132174} = 6.387681 \qquad F = \frac{6.387681}{1.3864} = 4.607*$$

$$SS_{b_3} = \frac{(-7.107216)^2}{3.61111544} = 13.988065 \qquad F = \frac{13.988065}{1.3864} = 10.089**$$

The $F$ table would be entered with 1 and 16 degrees of freedom.

While the ranking of the importance of the independent variables would be the same whether one considered the size of the $t$ value or the standard partial regression coefficients for the example problem this is not always the case. Evaluation of relative importance appears to be more reliable by consideration of the standard partial regression coefficients.

### 5.10.3  Tests of Significance Among Partial Regression Coefficients

It may be of interest to test for the significance of differences among the partial regression coefficients. The standard error for the difference between any two of the partial regression coefficients may be found as

$$s_{b_i - b_j} = \sqrt{s^2} \sqrt{c^{ii} + c^{jj} - 2c^{ij}} \tag{5.113}$$

where $s^2$ is the error mean square in the analysis of variance. If we wished to test the difference between $b_{Y1 \cdot 23}$ and $b_{Y2 \cdot 13}$ we would have

$$b_1 - b_2 = 0.005138 + 0.091885 = 0.097023$$

with 
$$s_{b_1 - b_2} = \sqrt{1.3864} \sqrt{0.00001135 + 0.00132174 - 2(0.00004452)}$$

$$= \sqrt{1.3864} \sqrt{0.00124405} = 0.041530$$

We then have

$$t = \frac{b_1 - b_2}{s_{b_1 - b_2}} = \frac{0.097023}{0.041530} = 2.336*$$

With 16 degrees of freedom we find that we have a significant difference between the two partial regression coefficients.

## 5.11  ELIMINATION OF ONE OR MORE INDEPENDENT VARIABLES

Very commonly, in multiple regression analyses a large number of independent variables are included in the original analysis with the idea of eliminating those variables which seem to have little or no effect on the variability of the dependent variable. Decisions on which variables to omit are most commonly based on the size of the $F$ test or $t$ test of the individual regression coefficients. Once a decision is made to eliminate variables, it is not necessary to start anew with the calculations because variables may be eliminated by some matrix operations which will yield the partial regression coefficients and the inverse matrix that would have been obtained had the variable to be eliminated not been included in the original analysis.

The matrix operations necessary to eliminate one or more of the variables are

$$\mathbf{C} = \mathbf{S} - \mathbf{N}'\mathbf{D}^{-1}\mathbf{N} \quad \text{and} \quad \mathbf{B} = \mathbf{B}_2 - \mathbf{N}'\mathbf{D}^{-1}\mathbf{B}_1 \qquad (5.114)$$

where $\mathbf{C}$ is the new matrix inverse, $\mathbf{S}$ is the original matrix inverse omitting the rows and columns for the variables to be omitted, $\mathbf{N}'$ is a column vector of the inverse elements for the variables, $\mathbf{D}$ is the segment of the matrix inverse corresponding to the variables to be deleted, $\mathbf{B}$ is the new set of partial regression coefficients, $\mathbf{B}_2$ is a column vector of the coefficients for the variables to be retained, and $\mathbf{B}_1$ is a column vector of the coefficients of the variables to be eliminated.

Suppose that in our example we wish to delete the $X_1$ variable. We would have [referring back to Equation (5.102) and the corresponding partial regression coefficients]

$$\mathbf{S} = \begin{bmatrix} 0.00132174 & 0.00336059 \\ 0.00336059 & 3.61111544 \end{bmatrix} \qquad \mathbf{N}' = \begin{bmatrix} 0.00004452 \\ 0.00289014 \end{bmatrix}$$

$$\mathbf{D}^{-1} = \begin{bmatrix} \dfrac{1}{0.00001135} = 88105.72687 \end{bmatrix} \qquad \mathbf{B}_2 = \begin{bmatrix} -0.09188514 \\ -7.10721571 \end{bmatrix}$$

$$\mathbf{B}_1 = [0.00513806]$$

Following the matrix operations outlined, we arrive at

$$\mathbf{N}'\mathbf{D}^{-1}\mathbf{N} = \begin{bmatrix} 0.00017463 & 0.01133648 \\ 0.01133648 & 0.73593914 \end{bmatrix} \qquad \mathbf{N}'\mathbf{D}^{-1}\mathbf{B}_1 = \begin{bmatrix} 0.02015387 \\ 1.30834473 \end{bmatrix}$$

$$\mathbf{C} = \mathbf{S} - \mathbf{N}'\mathbf{D}^{-1}\mathbf{N} = \begin{bmatrix} 0.00114711 & -0.00797589 \\ -0.00797589 & 2.87517630 \end{bmatrix}$$

$$\mathbf{B} = \mathbf{B}_2 - \mathbf{N}'\mathbf{D}^{-1}\mathbf{B}_1 = \begin{bmatrix} -0.11203901 \\ -8.41556044 \end{bmatrix}$$

The matrices $\mathbf{C}$ and $\mathbf{B}$ then yield the inverse matrix and the partial regression coefficients that would have been obtained had only $X_2$ and $X_3$ been included in the model.

## 5.12 INTERACTION OF INDEPENDENT VARIABLES IN MULTIPLE REGRESSION

Just as we have noted the effect of an interaction between two or more classifications in the analysis of variance, we may also be concerned with interaction in a multiple regression analysis. To investigate the effect of an interaction in a multiple regression analysis we can create a variable which is the product of the two independent variables with which we are concerned. Let us consider the data shown in Table 5.32 which presents a portion of the results of a site study in a red pine (Pinus resinosa) plantation conducted by Dr. Donald L. Mader of the Forestry and Wildlife Department at the University of Massachusetts. The study was conducted to investigate the influence of organic matter and nitrogen content in the $A$ horizon of the soil over a five-year period. It was felt that there might be an interaction between the organic matter and the nitrogen, so a third independent variable was created, $X_3$, as a product of $X_1$ and $X_2$ and included in the analysis of this portion of the data. It may be noted in passing that the curvilinear effects of the independent variables may be included in a multiple regression analysis resulting in what would be an analysis of multiple curvilinear regression.

Using the values shown at the bottom of Table 5.32, calculated from the data, we can set up the following equations as shown in Equations (5.100) and (5.101):

$$\mathbf{X'X} = \begin{bmatrix} 60.22933333 & 1.89595333 & 25.30041670 \\ 1.89595333 & 0.06946973 & 0.84445877 \\ 25.30041670 & 0.84445877 & 11.11624717 \end{bmatrix}$$

$$\mathbf{B} = \begin{bmatrix} b_{Y1 \cdot 23} \\ b_{Y2 \cdot 13} \\ b_{Y3 \cdot 12} \end{bmatrix} \quad \mathbf{Y} = \begin{bmatrix} 460.600 \\ 6.272 \\ -35.326 \end{bmatrix}$$

**TABLE 5.32**  CUBIC FEET PER ACRE INCREASE IN RED PINE OVER A FIVE-YEAR PERIOD ($Y$) ORGANIC MATTER PERCENT BY WEIGHT IN THE $A$ HORIZON ($X_1$), NITROGEN PERCENT BY WEIGHT IN THE $A$ HORIZON ($X_2$), AND INTERACTION BETWEEN $X_1$ AND $X_2$ ($X_3$)

| $X_1$ | $X_2$ | $X_3 = X_1X_2$ | $Y$ |
|-------|-------|----------------|-----|
| 5.9 | 0.200 | 1.1800 | 1,800 |
| 4.6 | 0.128 | 0.5888 | 1,230 |
| 8.0 | 0.260 | 2.0800 | 1,150 |
| 4.7 | 0.137 | 0.6439 | 1,160 |
| 10.0 | 0.342 | 3.4200 | 1,150 |
| 5.7 | 0.240 | 1.3680 | 1,130 |
| 3.4 | 0.075 | 0.2550 | 1,070 |
| 4.9 | 0.162 | 0.7938 | 1,340 |
| 7.1 | 0.214 | 1.5194 | 1,480 |
| 8.4 | 0.255 | 2.1420 | 1,350 |
| 4.1 | 0.157 | 0.6437 | 1,100 |
| 5.8 | 0.196 | 1.1368 | 1,460 |
| 4.8 | 0.147 | 0.7056 | 1,200 |
| 9.7 | 0.260 | 2.5220 | 1,320 |
| 4.3 | 0.115 | 0.4945 | 1,500 |

$\Sigma X_1 = 91.4$  $\Sigma X_2 = 2.888$  $\Sigma X_3 = 19.4935$  $\Sigma Y = 19,440$
$\Sigma X_1^2 = 617.16$  $\Sigma X_2^2 = 0.625506$  $\Sigma X_3^2 = 36.44934999$  $\Sigma Y^2 = 25,747,800$
$\Sigma x_1^2 = 60.22933333$  $\Sigma x_2^2 = 0.06946973$  $\Sigma x_3^2 = 11.11624717$  $\Sigma y^2 = 553,560$

$\Sigma X_1X_2 = 19.4935$  $\Sigma X_1X_3 = 144.08081$  $\Sigma X_2X_3 = 4.5976073$
$\Sigma x_1x_2 = 1.89595333$  $\Sigma x_1x_3 = 25.30041670$  $\Sigma x_2x_3 = 0.84445877$
$\Sigma X_1Y = 118,915.0$  $\Sigma X_2Y = 3749.120$  $\Sigma X_3Y = 25,228.25$
$\Sigma x_1y = 460.6$  $\Sigma x_2y = 6.272$  $\Sigma x_3y = -35.326$

Inverting the $\mathbf{X'X}$ matrix, we have

$$(\mathbf{X'X})^{-1} = \begin{bmatrix} 0.39703818 & 1.94231905 & -1.05120366 \\ 1.94231905 & 197.4923540 & -19.42342846 \\ -1.05120366 & -19.42342845 & 3.95800619 \end{bmatrix}$$

Performing the matrix multiplication $(\mathbf{X'X})^{-1}\mathbf{Y}$, we find

$$\mathbf{B} = \begin{bmatrix} 232.192831 = b_{Y1 \cdot 23} \\ 2819.456232 = b_{Y2 \cdot 13} \\ -745.828676 = b_{Y3 \cdot 12} \end{bmatrix}$$

To obtain the reduction in sum of squares accounted for by the three regressions, we can calculate

$$\mathbf{B'Y} = \sum_i b_i \sum x_i y = 150{,}978.7913$$

Using Equation (5.112), we can calculate the sums of squares for each of the independent variables as

$$SS_{b_1} = \frac{b_1^2}{c^{11}} = \frac{(232.192831)^2}{0.39703818} = 135{,}789.2351$$

$$SS_{b_2} = \frac{b_2^2}{c^{22}} = \frac{(2819.456232)^2}{197.4923540} = 40{,}251.3479$$

$$SS_{b_3} = \frac{b_3^2}{c^{33}} = \frac{(-745.828676)^2}{3.95800619} = 140{,}540.5619$$

To test for the significance of the combined regression and the individual partial regressions, we need to find the appropriate error term. We can develop the error term and the mean square by writing out an analysis of variance table. We have the total sum of squares, $\sum y^2 = 553{,}560$, from Table 5.32 and can find the error sum of squares by removing the reduction due to the combined regression, 150,978.7913, from the total sum of squares. We can then set up Table 5.33 which presents the results in analysis of variance form.

While it is more interesting to discuss statistical effects when they prove to be significant, this set of data, unfortunately, does not afford us that opportunity since none of the effects is significant. However, this does serve as an example of how the interaction between two independent variables can be incorporated into a regression

**TABLE 5.33** REGRESSION ANALYSIS OF DATA SHOWN IN TABLE 5.32

| Source | df | SS | MS | F |
|---|---|---|---|---|
| Total | 14 | 553,560.0000 | | |
| Regression $X_1, X_2, X_3$ | 3 | 150,978.7913 | 50,326.2638 | 1.38 |
| $X_1$ | 1 | 135,789.2351 | 135,789.2351 | 3.71 |
| $X_2$ | 1 | 40,251.3479 | 40,251.3479 | 1.10 |
| $X_3 = (X_1 X_2)$ | 1 | 140,540.5619 | 140,540.5619 | 3.84 |
| Error | 11 | 402,581.2087 | 36,598.2917 | |

analysis. This is of more than passing interest since there are occasions when an interaction is of vital concern to the experimenter.

## 5.13 PARTIAL AND MULTIPLE CORRELATION

Just as we have investigated the relationship between a dependent variable and each of two or more independent variables by means of partial regression coefficients, we can study the relationship between any two of a set of three or more variables, when we have no reason to consider the variables as dependent or independent, by means of *partial correlation coefficients*. The partial correlation coefficient measures the correlation between two of a set of three or more variables holding the third, or additional variables, constant. By holding these variables constant we mean that we are estimating what the correlation between $X_1$ and $X_2$ would be if each $X_3$ had the same value, if each $X_4$ had the same value, and so on. This is sometimes referred to as the correlation between two variables adjusted for a third or additional variables. The purpose of this statistic is to measure that part of the correlation between $X_1$ and $X_2$ that is free of their relationships with the remaining variables.

The correlation of $X_1$ and $X_2$ holding $X_3$ constant is designated as $r_{12 \cdot 3}$ and is measured as

$$r_{12 \cdot 3} = \frac{r_{12} - r_{13}r_{23}}{\sqrt{(1 - r_{13}^2)(1 - r_{23}^2)}} \tag{5.115}$$

with $r_{13 \cdot 2}$ and $r_{23 \cdot 1}$ defined similarly with a rearrangement of the correlations.

The correlations of $X_1$ and $X_2$ holding $X_3$ and $X_4$ constant is designated as $r_{12 \cdot 34}$ and is measured as

$$r_{12 \cdot 34} = \frac{r_{12 \cdot 3} - r_{14 \cdot 3}r_{24 \cdot 3}}{\sqrt{(1 - r_{14 \cdot 3}^2)(1 - r_{24 \cdot 3}^2)}} = \frac{r_{12 \cdot 4} - r_{13 \cdot 4}r_{23 \cdot 4}}{\sqrt{(1 - r_{13 \cdot 4}^2)(1 - r_{23 \cdot 4}^2)}} \tag{5.116}$$

Formulas for partial correlations between other variables are formed similarly.

The significance of a partial correlation involving three variables can be tested by using Table A.6 with $n - 3$ degrees of freedom, that of a correlation involving four variables with $n - 4$ degrees of freedom, and so on. The tests can also be carried out by means of a $t$ test as

$$t = \frac{r}{s_r}$$

where
$$s_{r_{12 \cdot 3}} = \sqrt{\frac{1 - r_{12 \cdot 3}^2}{n - 3}} \quad \text{and} \quad s_{r_{12 \cdot 34}} = \sqrt{\frac{1 - r_{12 \cdot 34}^2}{n - 4}} \tag{5.117}$$

Confidence intervals for partial correlation coefficients can be calculated by making use of the standard errors shown in Equation (5.117) as

$$r_{12 \cdot 3} \pm t_{0.05}s_{r_{12 \cdot 3}} \quad \text{with } n - 3 \text{ degrees of freedom}$$

and
$$r_{12 \cdot 34} \pm t_{0.05}s_{r_{12 \cdot 34}} \quad \text{with } n - 4 \text{ degrees of freedom}$$

for a 95 percent confidence interval.

As an example of a partial correlation analysis let us look at part of the results of a study conducted by Dr. Walter P. Kroll of the Department of Exercise Science at the University of Massachusetts, which was concerned with four different traits measured on power athletes at the University. The traits were body weight in pounds, percent vastus lateralis muscle, plantar flexion, and percent gastrocnemius muscle, which we shall designate as $X_1$, $X_2$, $X_3$, and $X_4$, respectively. A correlation matrix among these traits is shown in Table 5.34. The values have been carried out to more decimal places than would normally be the case to cut down on rounding error in further computations.

To gain a little further insight into the relationship between $X_1$ and $X_2$, for example, we can calculate $r_{12 \cdot 34}$. As Equation (5.116) shows, we will need $r_{12 \cdot 3}$, $r_{14 \cdot 3}$, and $r_{24 \cdot 3}$ to calculate the coefficient by the first form shown and $r_{12 \cdot 4}$, $r_{13 \cdot 4}$, and $r_{23 \cdot 4}$ to calculate it by the second form. Using Equation (5.115), we can calculate

$$r_{12 \cdot 3} = 0.603140 \qquad r_{14 \cdot 3} = 0.654406 \qquad r_{24 \cdot 3} = 0.504653$$

$$r_{12 \cdot 4} = 0.645622 \qquad r_{13 \cdot 4} = 0.825311 \qquad r_{23 \cdot 4} = 0.541891$$

Using these values we can now compute, using the two forms of Equation (5.116),

$$r_{12 \cdot 34} = \frac{0.603140 - (0.654406)(0.504653)}{\sqrt{[1 - (0.654406)^2][1 - (0.504653)^2]}} = 0.4180$$

$$= \frac{0.645622 - (0.825311)(0.541891)}{\sqrt{[1 - (0.825311)^2][1 - (0.541891)^2]}} = 0.4180$$

If one wishes to obtain the partial correlation coefficients when four or more variables are concerned, it can be seen that the formulas presented would require a formidable amount of work. Let us first look at an alternate method of calculating the partial coefficients when only three variables are concerned. Considering the correlation matrix containing only the first three variables, $X_1$, $X_2$, and $X_3$, as seen in Table 5.34, we can invert this matrix and find the inverse to be

$$\begin{bmatrix} 4.59919123 = c^{11} & -1.66729402 = c^{12} & -3.34394320 = c^{13} \\ -1.66729402 = c^{21} & 1.66152497 = c^{22} & 0.96656081 = c^{23} \\ -3.34394319 = c^{31} & 0.96656081 = c^{32} & 3.48838690 = c^{33} \end{bmatrix}$$

Now defining the partial correlation coefficient as

$$r_{ij \cdot k} = \frac{-c^{ij}}{\sqrt{c^{ii} c^{jj}}} \qquad\qquad (5.118)$$

**TABLE 5.34**  SIMPLE CORRELATIONS AMONG FOUR TRAITS MEASURED ON EIGHT POWER ATHLETES

|       | $X_1$ | $X_2$ | $X_3$ | $X_4$ |
|-------|-------|-------|-------|-------|
| $X_1$ | 1.00000000 | 0.53149882 | 0.81132556 | -0.62225688 |
| $X_2$ | 0.53149882 | 1.00000000 | 0.23241148 | -0.04268021 |
| $X_3$ | 0.81132556 | 0.23241148 | 1.00000000 | -0.93463830 |
| $X_4$ | -0.62225688 | -0.04268021 | -0.93463830 | 1.00000000 |

we can calculate

$$r_{12 \cdot 3} = \frac{1.66729402}{\sqrt{(4.59919123)(1.66152497)}} = 0.603140$$

$$r_{13 \cdot 2} = \frac{3.34394320}{\sqrt{(4.59919123)(3.48838690)}} = 0.834845$$

$$r_{23 \cdot 1} = \frac{0.96656081}{\sqrt{(1.66152497)(3.48838690)}} = 0.401480$$

Thus, if only three variables are involved, the partial correlation coefficients could be calculated either by the use of Equation (5.114) or by the matrix inversion procedure with little difference in the amount of work involved. However, if four or more variables are involved, a great deal of work can be avoided by using the matrix inversion procedure. Again using the data from the power athletes, we can invert the correlation matrix shown in Table 5.34, obtaining

$$\begin{bmatrix} 6.20152947 = c^{11} & -1.36476062 = c^{12} & -9.18928949 = c^{13} & -4.78796583 = c^{14} \\ -1.36476061 = c^{21} & 1.71864561 = c^{22} & -0.13708457 = c^{23} & -0.90400402 = c^{24} \\ -9.18928949 = c^{31} & -0.13708457 = c^{32} & 24.81227346 = c^{33} & 17.46655178 = c^{34} \\ -4.78796583 = c^{41} & -0.90400402 = c^{42} & 17.46655178 = c^{43} & 14.30698059 = c^{44} \end{bmatrix}$$

Using the formula

$$r_{ij \cdot kl} = \frac{-c^{ij}}{\sqrt{c^{ii} c^{jj}}} \tag{5.119}$$

we can calculate

$$r_{12 \cdot 34} = \frac{1.36476062}{\sqrt{(6.20152947)(1.71864561)}} = 0.418036$$

The remaining coefficients would be

$$r_{13 \cdot 24} = 0.740797 \qquad r_{14 \cdot 23} = 0.508309 \qquad r_{23 \cdot 14} = 0.020992$$

$$r_{24 \cdot 13} = 0.182307 \qquad r_{34 \cdot 12} = -0.927043$$

A comparison of the partial correlation coefficients with the simple correlations does show some interesting changes, particularly for the correlation between the body weight and the gastrocnemius muscle which has made a large change, including that of sign.

As noted in Section 5.10, the multiple correlation coefficient measures the correlation between the observed $Y$ values in a multiple regression analysis and their corresponding $\hat{Y}$ values predicted from the equation implied in the mathematical model used. As an example, let us consider the data presented in Table 5.30, the last column of which contains the values predicted from Equation (5.95). We can calculate the correlation between the actual values and the predicted values as

$$R_{Y \cdot 123} = R_{Y\hat{Y}} = \frac{\Sigma(Y - \bar{y})(\hat{Y} - \hat{\bar{y}})}{\sqrt{[(\Sigma(Y - \bar{y})^2/n)(\Sigma(\hat{Y} - \hat{\bar{y}})^2/n)]}} = \frac{\Sigma y\hat{y}}{\sqrt{\Sigma y^2 \, \Sigma \hat{y}^2}}$$

$$= \frac{\Sigma Y\hat{Y} - (\Sigma Y)(\Sigma \hat{Y})/n}{\sqrt{[\Sigma Y^2 - (\Sigma Y)^2/n][\Sigma \hat{Y}^2 - (\Sigma \hat{Y})^2/n]}}$$

$$= \frac{253.729033 - (64.880)(64.882)/20}{\sqrt{[275.903164 - (64.880)^2/20][253.740508 - (64.882)^2/20]}}$$

$$= \frac{43.251825}{53.201494} = 0.812981$$

the correlation between $Y$ and the $\hat{Y}$ which is predicted from $X_1$, $X_2$, and $X_3$. While the multiple correlation coefficient has not found a great usage, the square of this value, known as the *coefficient of multiple determination* is widely used. The coefficient of multiple determination, as noted in Section 5.10, yields the proportion of the total variation (sum of squares) of the dependent variable accounted for by the regression of $Y$ on the independent variables. We have previously found that the reduction in the sum of squares of $Y$ due to the multiple regression was 43.250496 and the total (corrected) sum of squares of $Y$ was 65.432444. Now,

$$R^2_{Y \cdot 123} = R^2_{Y\hat{Y}} = (0.812981)^2 = 0.6609$$

$$= \frac{\text{Reduction SS}}{\text{Total SS}} = \frac{43.250496}{65.432444} = 0.6610$$

The slight rounding error is due to the fact that $\Sigma \hat{Y} = 64.882$ rather than 64.880. We see that the regression of $Y$ on $X_1$, $X_2$, and $X_3$ accounts for 66.10 percent the variation in $Y$.

## 5.14 RELATIONSHIP OF REGRESSION ANALYSIS TO THE ANALYSIS OF VARIANCE

Both regression analyses and analyses of variance are defined by mathematical models. These mathematical models may be transformed to prediction equations merely by dropping the error terms and replacing the parameters with the estimates of the parameters. We can then use the results of an analysis of variance of a factorial design to predict values for different treatments or treatment combinations. It is also possible to cast the same observations into a regression model and develop a prediction equation from that model. Let us look at the analysis of variance model first, using the data from the example in Section 3.4 which deal with nitrate accumulation in spinach plants.

### 5.14.1 Prediction Using the Analysis of Variance

The sample model used for the data shown in Table 3.7 was

$$Y_{ijk} = \bar{y} + \hat{n}_i + \hat{l}_j + \hat{nl}_{ij} + \hat{e}_{ijk} \tag{5.120}$$

where $n_i$ represents two sources of nitrogen and $l_j$ represents three levels of application. Table 5.35 shows the analysis of variance for this set of data.

To predict from model (5.120) we can use the formula

$$\hat{y}_{ij} = \bar{y} + b_1\hat{n}_i + b_2\hat{l}_j + b_3\hat{nl}_{ij} \tag{5.121}$$

**TABLE 5.35** ANALYSIS OF DATA SHOWN IN TABLE 3.7

| Source | df | SS | MS | F |
|---|---|---|---|---|
| Nitrogen ($N$) | 1 | 1397.4422 | 1397.4422 | 18.07** |
| Levels ($L$) | 2 | 6178.0544 | 3089.0272 | 39.95** |
| $NL$ | 2 | 734.6878 | 367.3439 | 4.75* |
| Error | 12 | 927.9600 | 77.3300 | |

where the coefficients (the $b$'s) are made up of 0's or 1's depending on whether the corresponding effect (or constant) is involved in the prediction. The estimated constants resulting from the data in Table 3.7 are

$$\bar{y} = 39.9444 \qquad \hat{n}_1 = -8.8111 \qquad \hat{n}_2 = 8.8111$$

$$\hat{l}_1 = -23.2611 \qquad \hat{l}_2 = 1.1889 \qquad \hat{l}_3 = 22.0722$$

$$\hat{nl}_{11} = 7.1278 \qquad \hat{nl}_{12} = 1.2445 \qquad \hat{nl}_{13} = -8.3722$$

$$\hat{nl}_{21} = -7.1278 \qquad \hat{nl}_{22} = -1.2445 \qquad \hat{nl}_{23} = 8.3723$$

The estimated constants are calculated as shown in Sections 3.2 and 3.5.

If we wish to predict the grams of nitrate in the subclass containing the first nitrogen source and the first level of application, we would have

$$\hat{y}_{11} = \bar{y} + (1)\hat{n}_1 + (1)\hat{l}_1 + (1)\hat{nl}_{11}$$

$$= 39.9444 + (1)(-8.8111) + (1)(-23.2611) + (1)(7.1278)$$

$$= 15.0000 \tag{5.122}$$

which is, of course, $\bar{y}_{11}$, the subclass mean. It can be seen then that the subclass mean is a linear function of the overall mean, the effect of the level of nitrogen, the level of application, and the interaction between the two. If we wish to predict the grams of nitrate in a main class, rather than a subclass, we need only be concerned with the mean and the main class effect. For example, we estimate the grams of nitrate in the third level of application as

$$\hat{y}_{.3} = \bar{y} + (1)\hat{l}_3$$

$$= 39.9444 + (1)(22.0722) = 62.0167 \tag{5.123}$$

Since a significant interaction exists between source of nitrogen and level of nitrogen, caution should be taken in discussing the overall treatment effects because the interaction indicates different responses of the levels of nitrogen in the two sources of nitrogen.

## 5.14.2 Prediction Using Nonorthogonal Polynomial Coefficients as Dummy Variables

Putting the results of this experiment into the form necessary for a regression analysis requires the use of *dummy variables*. A dummy variable is a type of coding in which a number is assigned to represent a particular level of a treatment. For example, we could use the dummy variables of 0 and 1, or 1 and 2, for the two sources of nitrogen, the first number representing the first source and the second number the second source. For the levels of application we could assign the dummy variables of 1, 2, and 3 for the different levels. Since we are planning a regression analysis for a comparison with the analysis of variance, we should start by fitting the complete model which would include the linear effect of the source of nitrogen, the linear *and* quadratic effects of the levels of application, the interaction of the linear effect of source with the linear effect of levels, and the interaction of the linear effect of source with the quadratic effect of levels. Since there is a single degree of freedom for each of these, this would account for the five degrees of freedom among the six subclasses.

Using the dummy variables discussed above, the quadratic effects of levels would have the codes (or dummy variables) of 1, 4, and 9, the squares of the linear codes. The interaction of the linear effect of sources with the linear and quadratic effects of levels would call for dummy variable values that would be the products of the linear × linear and linear × quadratic values. Table 5.36 shows the appropriate values for the dummy variables together with their corresponding $Y$ values. After coding the data with this set of dummy variables, we can develop the necessary normal equations and complete the regression analysis which is shown in Table 5.37. The values that have been used for the dummy variables are, in fact, nonorthogonal polynomial coefficients and because of the nonorthogonality involved only the mean squares for the linear × quadratic effect and the error term are meaningful. The sums of squares in

**TABLE 5.36** DUMMY VARIABLES FOR REGRESSION ANALYSIS OF DATA SHOWN IN TABLE 3.7: $X_1$ IS THE LINEAR EFFECT OF SOURCE, $X_2$ IS THE LINEAR EFFECT OF LEVELS, $X_3$ IS THE QUADRATIC EFFECT OF LEVELS, $X_4 = X_1X_2$, AND $X_5 = X_1X_3$

| Level of nitrogen = $i$ | Level of application = $j$ | $X_1$ | $X_2$ | $X_3$ | $X_4$ | $X_5$ | $Y$ |
|---|---|---|---|---|---|---|---|
| 1 | 1 | 1 | 1 | 1 | 1 | 1 | 14.3 |
|   |   | 1 | 1 | 1 | 1 | 1 | 15.9 |
|   |   | 1 | 1 | 1 | 1 | 1 | 14.8 |
|   | 2 | 1 | 2 | 4 | 2 | 4 | 37.5 |
|   |   | 1 | 2 | 4 | 2 | 4 | 29.4 |
|   |   | 1 | 2 | 4 | 2 | 4 | 33.8 |
|   | 3 | 1 | 3 | 9 | 3 | 9 | 41.4 |
|   |   | 1 | 3 | 9 | 3 | 9 | 49.9 |
|   |   | 1 | 3 | 9 | 3 | 9 | 43.2 |
| 2 | 1 | 2 | 1 | 1 | 2 | 2 | 17.6 |
|   |   | 2 | 1 | 1 | 2 | 2 | 24.0 |
|   |   | 2 | 1 | 1 | 2 | 2 | 13.5 |
|   | 2 | 2 | 2 | 4 | 4 | 8 | 43.3 |
|   |   | 2 | 2 | 4 | 4 | 8 | 53.5 |
|   |   | 2 | 2 | 4 | 4 | 8 | 49.3 |
|   | 3 | 2 | 3 | 9 | 6 | 18 | 59.8 |
|   |   | 2 | 3 | 9 | 6 | 18 | 98.4 |
|   |   | 2 | 3 | 9 | 6 | 18 | 79.4 |

**TABLE 5.37**   REGRESSION ANALYSIS USING NONORTHOGONAL
                 POLYNOMIAL COEFFICIENTS

| Source | df | SS | MS | F |
|---|---|---|---|---|
| Linear regression ($N$) | 1 | 0.0688 | 0.0688 | |
| Linear regression ($L$) | 1 | 20.5250 | 20.5250 | |
| Quadratic regression ($L$) | 1 | 21.8054 | 21.8054 | |
| Linear ($N$) × linear ($L$) | 1 | 0.0197 | 0.0197 | |
| Linear ($N$) × quadratic ($L$) | 1 | 13.9378 | 13.9378 | 0.18 |
| Error | 12 | 927.9600 | 77.3300 | |

the analysis, of course, are not additive but it can be found from the completed analysis that the sum of squares accounted for by the five regressions is 8310.1844 which is the sum of squares among the six subclasses. This is, of course, the sum of squares accounted for by the $N$, $L$, and $NL$ in the analysis of variance shown in Table 5.34. To obtain the sums of squares for the remaining effects, new models must be fitted omitting the higher-order effects in turn. The model omitting the linear × quadratic effect would yield a sum of squares of 720.75 for the linear × linear effect and a sum of squares of 12.7211 for the quadratic effect, the correct sums of squares. Now a model including only the two linear effects will give the sum of squares of 1397.4422 for the linear effect of sources and 6165.3333 for the linear effect of levels. The analysis with appropriate sums of squares is shown in Table 5.38.

The analysis including all five $X$'s yielded coefficients of $b_1 = -0.933333$, $b_2 = 28.95$, $b_3 = 7.383333$, $b_4 = 0.566667$, and $b_5 = 3.733333$. If we wish to predict using this complete model, we would use these coefficients together with the appropriate nonorthogonal polynomial coefficients shown in Table 5.36. In addition, we will need to compute the intercept $a$, which is done as shown previously,

$$a = \bar{y} - b_1\bar{x}_1 - b_2\bar{x}_2 - b_3\bar{x}_3 - b_4\bar{x}_4 - b_5\bar{x}_5$$

$$= -9.933333 \tag{5.124}$$

where the $\bar{y}$ and $\bar{x}$ values can be obtained from Table 5.36. Our prediction equation would be

$$\hat{y}_{ij} = a + b_1X_{1ij} + b_2X_{2ij} + b_3X_{3ij} + b_4X_{4ij} + b_5X_{5ij} \tag{5.125}$$

where the $b$'s are the regression coefficients, the $X_{ij}$'s are the nonorthogonal polynomial

**TABLE 5.38**   SUMMARY OF REGRESSION ANALYSES FOR OBTAINING
                 SUMS OF SQUARES

| Source | df | SS | MS | F |
|---|---|---|---|---|
| Linear regression ($N$) | 1 | 1397.4422 | 1397.4422 | 18.07** |
| Linear regression ($L$) | 1 | 6165.3333 | 6165.3333 | 79.73** |
| Quadratic regression ($L$) | 1 | 12.7211 | 12.7211 | <1 |
| Linear ($N$) × linear ($L$) | 1 | 720.7500 | 720.7500 | 9.32* |
| Linear ($N$) × quadratic ($L$) | 1 | 13.9378 | 13.9378 | <1 |
| Error | 12 | 927.9600 | 77.3300 | |

coefficients, the $i$ defines the level of nitrogen, and the $j$ defines the level of application. If we wish to predict the first subclass mean we would have

$$\hat{y}_{11} = -9.933333 + (-0.933333)(1) + (28.95)(1)$$

$$+ (-7.383333)(1) + (0.566667)(1) + (3.733333)(1)$$

$$= 15.0000$$

which is the same value we arrived at when using the results of an analysis of variance for prediction.

A comparison of Table 5.35 with Table 5.38 has several points of interest. It can be seen that the sum of squares for sources of nitrogen is exactly the sum of squares for the linear effect of sources, which, of course, it should be since there is only one degree of freedom involved. The sum of squares for the linear effect of levels of application plus the sum of squares for the quadratic effect of levels add up to the sum of squares among the levels of application in the analysis of variance, so we have merely partitioned this sum of squares into linear and quadratic components. The sum of squares of linear × linear and linear × quadratic add up to the sum of squares for the interaction in the analysis of variance, a second partitioning.

While the regression analysis with the complete model (using all five degrees of freedom) accounts for exactly the same total sum of squares as does the analysis of variance previously completed, it does shed a little more light on the underlying causes of variation. While we saw earlier that there were highly significant differences among levels of nitrogen application, we can now see that the response to the increasing levels is overwhelmingly linear. We can further see from the significant linear × linear interaction that the linear response to the three levels of application is different in the two sources of nitrogen.

Since we can see from Table 5.38 that the quadratic and linear × quadratic effects account for a very small portion of the total sum of squares, it would be reasonable to omit these effects and fit a model including only the linear and linear × linear effects. The results of such an analysis are shown in Table 5.39.

It is emphasized that only the sums of squares for the linear × linear effect and the error term are correct. An analysis including only the two linear effects alone would yield the sums of squares for these effects. However, had there been only two levels of $N$ and two levels of $L$ originally, an analysis including linear and linear × linear effects would have given the correct sums of squares for all three effects. The regression coefficients from this analysis are $b_1 = -13.377778$,

**TABLE 5.39** NONORTHOGONAL POLYNOMIAL REGRESSION ANALYSIS WITH THREE VARIABLES

| Source | df | SS | MS | F |
|---|---|---|---|---|
| Linear ($N$) | 1 | 115.0489 | 115.0489 | |
| Linear ($L$) | 1 | 0.4083 | 0.4083 | |
| Linear ($N$) × linear ($L$) | 1 | 720.7500 | 720.7500 | 10.57** |
| Error | 14 | 954.6188 | 68.1871 | |

**TABLE 5.40** MODELS NEEDED TO OBTAIN SUMS OF SQUARES

| Effects included in model | Sums of squares obtained |
|---|---|
| $l_1, q_1, l_2, q_2, l_1l_2, l_1q_2, q_1l_2, q_1q_2$ | $q_1q_2$ |
| $l_1, q_1, l_2, q_2, l_1l_2, l_1q_2, q_1l_2$ | $l_1q_2, q_1l_2$ |
| $l_1, q_1, l_2, q_2, l_1l_2$ | $l_1l_2, q_1, q_2$ |
| $l_1, l_2$ | $l_1, l_2$ |

$b_2 = -0.583333$, and $b_3 = 15.5$. If we wish to predict using these three variables we can first calculate

$$a = \bar{y} - b_1\bar{x}_1 - b_2\bar{x}_2 - b_3\bar{x}_4$$

$$= 39.944444 + 20.066667 + 1.166666 - 46.5$$

$$= 14.677777$$

Our prediction equation would now be

$$\hat{y}_{ij} = 14.677777 - (13.377778)X_{1ij} - (0.583333)X_{2ij} + (15.5)X_{4ij}$$

where the $X$'s refer to those in Table 5.36, with $i$ and $j$ again defining the level of nitrogen and level of application, respectively. If we wish to predict the value for the first subclass, we would have

$$\hat{y}_{11} = 14.677777 - (13.377778)(1) - (0.583333)(1) + (15.5)(1) = 16.2167$$

If our original analysis of variance had included three sources of nitrogen of increasing quality (instead of two) as well as three levels of application, the same procedures could be followed. However, it would not be necessary to fit eight different models to obtain all the correct sums of squares. Table 5.40 shows the models that could be fitted to obtain all the sums of squares.

### 5.14.3 Prediction Using Orthogonal Polynomial Coefficients as Dummy Variables

A much easier method for calculation, interpretation, and utilization involves the substitution of orthogonal polynomial coefficients in place of the nonorthogonal coefficients used in the previous section. Thus, since there are two levels for source of nitrogen, the coefficients would be $-1$ and $1$. The coefficients for the linear effect of levels of application would be $-1$, $0$, and $1$. The interactions of the linear effect of sources with the linear and quadratic effects of levels would be measured by the products of the appropriate coefficients. Using these coefficients we can set up Table 5.41 and the sample model would be

$$Y_{ijk} = \bar{y} + b_1X_{1ij} + b_2X_{2ij} + b_3X_{3ij} + b_4X_{4ij} + b_5X_{5ij} + \hat{e}_{ijk} \qquad (5.126)$$

Note that means are not subtracted from the $X$'s since the means of the orthogonal polynomials are zero which is not true with nonorthogonal polynomials. Thus, when

**TABLE 5.41** ORTHOGONAL POLYNOMIAL COEFFICIENTS USED AS DUMMY VARIABLES IN A REGRESSION ANALYSIS OF DATA SHOWN IN TABLE 3.7: $X_1$ IS THE LINEAR EFFECT OF SOURCE, $X_2$ IS THE LINEAR EFFECT OF LEVELS, $X_3$ IS THE QUADRATIC EFFECT OF LEVELS, $X_4 = X_1X_2$, AND $X_5 = X_1X_3$

| Level of nitrogen = $i$ | Level of application = $j$ | $X_1$ | $X_2$ | $X_3$ | $X_4$ | $X_5$ | $Y$ |
|---|---|---|---|---|---|---|---|
| 1 | 1 | −1 | −1 | 1 | 1 | −1 | 14.3 |
|   |   | −1 | −1 | 1 | 1 | −1 | 15.9 |
|   |   | −1 | −1 | 1 | 1 | −1 | 14.8 |
|   | 2 | −1 | 0 | −2 | 0 | 2 | 37.5 |
|   |   | −1 | 0 | −2 | 0 | 2 | 29.4 |
|   |   | −1 | 0 | −2 | 0 | 2 | 33.8 |
|   | 3 | −1 | 1 | 1 | −1 | −1 | 41.4 |
|   |   | −1 | 1 | 1 | −1 | −1 | 49.9 |
|   |   | −1 | 1 | 1 | −1 | −1 | 43.2 |
| 2 | 1 | 1 | −1 | 1 | −1 | 1 | 17.6 |
|   |   | 1 | −1 | 1 | −1 | 1 | 24.0 |
|   |   | 1 | −1 | 1 | −1 | 1 | 13.5 |
|   | 2 | 1 | 0 | −2 | 0 | −2 | 43.3 |
|   |   | 1 | 0 | −2 | 0 | −2 | 53.5 |
|   |   | 1 | 0 | −2 | 0 | −2 | 49.3 |
|   | 3 | 1 | 1 | 1 | 1 | 1 | 59.8 |
|   |   | 1 | 1 | 1 | 1 | 1 | 98.4 |
|   |   | 1 | 1 | 1 | 1 | 1 | 79.4 |

a prediction equation is developed from Equation (5.126), $a = \bar{y}$. The $b$ values in Equation (5.126) represent regression coefficients for the effects discussed previously and the $X$'s are the orthogonal polynomial coefficients, or products of the coefficients.

Since we are dealing with orthogonal polynomial coefficients, we need not develop the usual set of normal equations involved in a regression analysis. We can solve for the regression coefficients simply as

$$b_1 = \frac{\Sigma_i \, \Sigma_j \, \Sigma_k \, X_{1ijk}Y_{ijk}}{\Sigma_i \, \Sigma_j \, \Sigma_k \, X_{1ijk}^2} = \frac{158.6}{18} = 8.811$$

$$b_2 = \frac{\Sigma_i \, \Sigma_j \, \Sigma_k \, X_{2ijk}Y_{ijk}}{\Sigma_i \, \Sigma_j \, \Sigma_k \, X_{2ijk}^2} = \frac{272.0}{12} = 22.6667$$

$$b_3 = \frac{\Sigma_i \, \Sigma_j \, \Sigma_k \, X_{3ijk}Y_{ijk}}{\Sigma_i \, \Sigma_j \, \Sigma_k \, X_{3ijk}^2} = \frac{-21.4}{36} = -0.5944$$

$$b_4 = \frac{\Sigma_i \, \Sigma_j \, \Sigma_k \, X_{4ijk}Y_{ijk}}{\Sigma_i \, \Sigma_j \, \Sigma_k \, X_{4ijk}^2} = \frac{93.0}{12} = 7.7500$$

$$b_5 = \frac{\Sigma_i \, \Sigma_j \, \Sigma_k \, X_{5ijk}Y_{ijk}}{\Sigma_i \, \Sigma_j \, \Sigma_k \, X_{5ijk}^2} = \frac{22.4}{36} = 0.6222$$

We are also able to calculate readily the sums of squares for these effects by merely squaring the numerator in each expression, yielding

$$SS_{b_1} = \frac{(158.6)^2}{18} = 1397.4222 \qquad SS_{b_2} = \frac{(272.0)^2}{12} = 6165.3333$$

$$SS_{b_3} = \frac{(-21.4)^2}{36} = 12.7211 \qquad SS_{b_4} = \frac{(93.0)^2}{12} = 720.7500$$

$$SS_{b_5} = \frac{(22.4)^2}{36} = 13.9378$$

The sums of squares obtained are the same as calculated before, using a series of analyses with nonorthogonal polynomial coefficients, and are presented in Table 5.38. We have then obtained the sums of squares by a much quicker and simpler method. The coefficients have also been calculated much more readily, but they do differ from the previous coefficients since we have used orthogonal polynomials. If we wish to predict the value for grams of nitrate in the first subclass we would have the prediction equation

$$\hat{y}_{11} = \bar{y} + b_1 X_{111} + b_2 X_{211} + b_3 X_{311}$$

$$+ b_4 X_{411} + b_5 X_{511}$$

$$= 39.9444 + (8.8111)(-1) + (22.6667)(-1) + (-0.5944)(1)$$

$$+ (7.7500)(1) + (0.6222)(-1)$$

$$= 15.0000 \tag{5.127}$$

which is the same value as found by previous methods. The $X_{ij}$ values in Equation (5.127) are the coefficients shown in Table 5.41 and would change according to the level of source of nitrogen and level of application involved.

If we wish to predict the grams of nitrate in the third level of application, we need only be concerned with the mean and the level of application coefficients (both linear and quadratic) and we would have the prediction equation

$$\hat{y}_j = \bar{y} + b_2 X_{2j} + b_3 X_{3j}$$

giving

$$\hat{y}_3 = \bar{y} + b_2 X_{23} + b_3 X_{33}$$

$$= 39.9444 + (22.6667)(1) + (-0.5944)(1)$$

$$= 62.0167$$

as found before in the analysis of variance prediction. If we wished to predict the value for a particular source of nitrogen, all we would be concerned with would be the mean and the source of nitrogen.

As noted in Section 5.14.2, a reasonable model would be one that omits the quadratic components of the original,

$$Y_{ijk} = \bar{y} + b_1 X_{1ij} + b_2 X_{2ij} + b_4 X_{4ij} + \hat{e}_{ijk} \tag{5.128}$$

where $b_1$ represents the linear effect of the source of nitrogen, $b_2$ represents the linear effects of application, and $b_4$ represents the interaction of the two linear effects. This

**TABLE 5.42**   MULTIPLE REGRESSION ANALYSIS WITH THREE VARIABLES

| Source | df | SS | MS | F |
|---|---|---|---|---|
| Linear ($N$) | 1 | 1397.4422 | 1397.4422 | 20.49** |
| Linear ($L$) | 1 | 6165.3333 | 6165.3333 | 90.42** |
| Linear ($N$) $\times$ linear ($L$) | 1 | 720.7500 | 720.7500 | 10.57** |
| Error | 14 | 954.6189 | 68.1871 | |

would correspond with $X_1$, $X_2$, and $X_4$ in Table 5.41. We can write this analysis readily from the results obtained in this section as shown in Table 5.42.

The presence of the highly significant interaction of the two linear effects indicates that caution should be taken in making statements regarding the main effects or linear effects alone and consideration should be given to the fact that there is a different linear response to the levels of application in the two sources of nitrogen.

Using formula (5.128) to predict a value for the first subclass, we would have

$$\hat{\bar{y}}_{11} = 39.9444 + (8.8111)(-1) + (22.6667)(-1) + (7.7500)(1) = 16.2166$$

which is the same value that we obtained using nonorthogonal polynomial coefficients.

## 5.15  DECISION ON FINAL MULTIPLE REGRESSION MODEL

There are no simple rules for making a decision as to which independent variables should be included in a regression model. If several independent variables are to be considered in a model one could fit all possible combinations to see which combination of variables is the most appropriate in terms of accounting for the variation among the $Y$'s. However, with many variables involved, this could be quite an adventure in computing and is not commonly done. Generally, one of two alternative methods are used.

In the first method, all potential independent variables are included in the analysis. Then, the independent variable yielding the smallest sum of squares is examined and a decision made as to whether to include or omit this variable. One rule that is frequently used is that if the $F$ value is less than 1 the variable is omitted. If the variable is omitted a new model is fitted and the same procedure followed until a satisfactory model is developed.

With the second method, the correlations of $Y$ with each of the possible $X$'s are calculated and the variable showing the highest correlation (without regard to sign) is selected to be retained in the model as long as the $F$ value meets some preselected level. Next, the partial correlation coefficients of each of the remaining variables with the dependent variable are calculated, holding the selected independent variable constant. That variable showing the highest partial correlation is selected as a second independent variable to be included in the model. This procedure continues with the calculation of partial correlation coefficients of the remaining variables, holding constant the first two variables selected. Again, that variable showing the highest partial correlation is selected to be added to the model. This procedure continues with the calculation of a new series of partial regression coefficients until it is decided that a new variable does not account for a sufficient sum of squares to be included, a decision based on an $F$ test of the partial correlation coefficient. The process described in this second method is known as *stepwise regression* and many computer programs are available for this type of analysis.

Let us apply the stepwise regression to the data shown in Table 5.30. Making use of the computations at the foot of the table, we find the values of $r_{Y1} = 0.6024$, $r_{Y2} = 0.4990$, and $r_{Y3} = 0.6769$. Since $X_3$ has the highest correlation and has an $F$ value of 15.22 this is the first variable included in the model. The partial correlations $r_{Y1.3} = 0.4405$ and $r_{Y2.3} = 0.5555$ are calculated next. Since the partial correlation of $Y$ with $X_2$ is the larger with an $F$ value of 7.59 it is the next variable entered. With more variables this procedure would be continued until the $F$ value for a partial correlation coefficient that was considered is deemed to be too small. The $F$ values can be calculated by squaring the $t$ values found by using formulas (5.28) and (5.117) with $n - 3$ degrees of freedom for the first partial coefficients, although, of course, $t$ values could be used as readily as $F$ values.

## 5.16 DISCRIMINANT FUNCTION

A method of analysis quite similar to multiple regression is the discriminant function analysis. While this is a topic generally considered in the area of multivariate analysis, a brief introduction is given here. The necessity for this analysis arises when one wishes to select one or more variables which will help in distinguishing one group or population from another. For example, one might wish to study the traits which distinguish high-yielding blueberry plants from low-yielding plants. Several traits, such as size of leaves, number of leaves, annual growth of shoots, height of plant, spread of plant, and so on, could be measured and recorded. The problem would then be to make use of these measurements in some way to distinguish between the higher- and lower-yielding plants. This problem was solved by R. A. Fisher with the development of discriminant function analysis. While it is possible to discriminate among several groups we will limit our discussion and example to two groups.

The problem, in general, is to develop a function

$$Z = \lambda_1 X_1 + \lambda_2 X_2 + \cdots + \lambda_k X_k \tag{5.129}$$

in which the $X$'s are variables measured in the two groups and the $\lambda$'s are weights or coefficients that will allow for a maximum separation between the two groups. The method of analysis in developing this discriminant function is such that it results in the smallest probability of misclassification. In the example mentioned with the blueberry plants, the $\lambda$'s would be selected or calculated so that together with the measurements made on a plant they would allow us to predict as accurately as possible whether a plant would be a high-yielding one or a low-yielding one.

Given that we decided to use three variables for discriminating purposes, the function could be calculated, as shown by Fisher (1936), by solving the simultaneous equations

$$\sigma_{11}\lambda_1 + \sigma_{12}\lambda_2 + \sigma_{13}\lambda_3 = \delta_1$$

$$\sigma_{12}\lambda_1 + \sigma_{22}\lambda_2 + \sigma_{23}\lambda_3 = \delta_2 \qquad (5.130)$$

$$\sigma_{13}\lambda_1 + \sigma_{23}\lambda_2 + \sigma_{22}\lambda_3 = \delta_2$$

where $\sigma_{ii}$ is the variance of $X_i$, $\sigma_{ij}$ is the covariance of $X_i$ and $X_j$, $\lambda_i$ is the discriminant function coefficient, and $\delta_i$ is the difference between the means of the two groups of $X_i$, $\mu_{1i} - \mu_{2i}$.

As an example of this analysis we shall use the data displayed in Table 5.43 furnished by Gregory Ruark of the Department of Forestry and Wildlife at the University of Massachusetts. The data were collected during a study of the Sugar Maple

**TABLE 5.43** SUGAR MAPLE DECLINE ON COMPACTED SITES

| | Compacted/decline | | | Compacted/nondecline | | |
|---|---|---|---|---|---|---|
| | Sodium (mEq/100 g) $X_1$ | Carbon/nitrogen $X_2$ | Sand (%) $X_3$ | Sodium (mEq/100 g) $X_1$ | Carbon/nitrogen $X_2$ | Sand (%) $X_3$ |
| | 0.048 | 25.2 | 73.69 | 0.204 | 20.7 | 67.09 |
| | 0.074 | 21.5 | 71.92 | 0.039 | 22.0 | 70.53 |
| | 0.070 | 23.8 | 69.23 | 0.048 | 20.8 | 70.05 |
| | 0.061 | 22.0 | 69.37 | 0.035 | 21.7 | 71.93 |
| | 0.161 | 17.0 | 58.53 | 0.030 | 26.4 | 75.67 |
| | 0.161 | 20.1 | 62.81 | 0.035 | 26.4 | 67.66 |
| | 0.391 | 23.6 | 42.69 | 0.030 | 23.6 | 66.99 |
| | 0.135 | 19.7 | 41.80 | 0.039 | 29.7 | 66.50 |
| | 0.113 | 24.1 | 56.89 | 0.461 | 29.5 | 54.22 |
| | 0.135 | 24.1 | 60.50 | 0.170 | 30.9 | 62.49 |
| | 0.065 | 20.5 | 44.69 | 0.148 | 19.3 | 59.86 |
| | 0.139 | 19.7 | 45.03 | 0.122 | 22.6 | 61.61 |
| | 0.117 | 27.3 | 45.39 | 0.035 | 34.2 | 63.36 |
| | 0.200 | 23.8 | 44.25 | 0.043 | 36.8 | 69.01 |
| | 0.096 | 20.8 | 43.73 | 0.074 | 21.1 | 50.36 |
| | 0.109 | 21.5 | 53.28 | 0.122 | 19.4 | 52.06 |
| | 0.304 | 24.9 | 55.13 | 0.091 | 25.0 | 51.21 |
| | 0.335 | 22.2 | 55.27 | 0.096 | 16.3 | 53.52 |
| | 0.270 | 27.9 | 49.26 | 0.113 | 32.7 | 59.36 |
| | 0.313 | 32.8 | 49.57 | 0.122 | 33.6 | 59.74 |
| | 0.100 | 21.2 | 43.91 | 0.100 | 26.3 | 50.05 |
| | 0.183 | 23.9 | 36.21 | 0.148 | 37.4 | 48.65 |
| | 0.113 | 20.5 | 55.04 | 0.152 | 21.3 | 70.19 |
| | 0.174 | 20.0 | 43.73 | 0.144 | 22.0 | 69.29 |
| | 0.078 | 22.7 | 75.88 | 0.134 | 35.0 | 70.66 |
| | 0.052 | 31.8 | 77.41 | 0.165 | 29.4 | 70.76 |
| $\Sigma X$ | 3.997 | 602.6 | 1,425.21 | 2.900 | 684.1 | 1,632.82 |
| $\Sigma X^2$ | 0.839507 | 14,296.9 | 81,748.1019 | 0.518930 | 18,909.19 | 104,195.9842 |
| $\Sigma X_1X_2$ | | 94.5312 | | | 77.3406 | |
| $\Sigma X_1X_3$ | | | 206.61423 | | 176.60546 | |
| $\Sigma X_2X_3$ | | | 33,228.579 | | 42,965.083 | |

decline problem in urban areas subjected to surface soil compaction. Only three of the many soil variables included in the study are used for this example. These variables were thought to be the most highly involved by the researchers. The ratio of the weights for percent total carbon over percent total soil nitrogen accounted for the largest contribution to the discriminant function. The percent sand, by weight, was slightly less important, while the amount of sodium, expressed in milliequivalents per 100 g of sieved soil, provided the third variable for the function.

To solve for the discriminant function coefficients we develop a pooled covariance matrix

$$\mathbf{S}_w = \frac{(n_a - 1)\mathbf{S}_a + (n_b - 1)\mathbf{S}_b}{n_a + n_b - 2} \tag{5.131}$$

where 
$$\mathbf{S}_a = \frac{1}{n_a - 1} \begin{bmatrix} \sum x_{a1}^2 & \sum x_{a1}x_{a2} & \sum x_{a1}x_{a3} \\ \sum x_{a1}x_{a2} & \sum x_{a2}^2 & \sum x_{a2}x_{a3} \\ \sum x_{a1}x_{a3} & \sum x_{a2}x_{a3} & \sum x_{a3}^2 \end{bmatrix} \tag{5.132}$$

and likewise for $\mathbf{S}_b$, where $a$ refers to the first group and $b$ the second. We can now calculate the coefficients $\mathbf{B}_x$ as

$$\mathbf{B}_x = \mathbf{S}_w^{-1}\mathbf{Y} \tag{5.133}$$

where $\mathbf{Y}$ is a column vector of group mean differences and $d_i = \bar{x}_{ai} - \bar{x}_{bi}$. It should be noted that the $(n_a - 1)$ and $(n_b - 1)$ values in the numerator of Equation (5.131) are canceled by the denominators in the scalars of $\mathbf{S}_a$ and $\mathbf{S}_b$, so when the $\mathbf{S}_a$ and $\mathbf{S}_b$ matrices are developed it is not necessary to multiply the elements of these matrices by the scalars.

Using the values at the foot of Table 5.42, we can calculate the following:

$$\sum x_{a1}^2 = 0.22504512 \quad \sum x_{a2}^2 = 330.4861500 \quad \sum x_{a3}^2 = 3624.119440$$

$$\sum x_{a1}x_{a2} = 1.89303846 \sum x_{a1}x_{a3} = -12.4843996 \sum x_{a2}x_{a3} = 196.5964600$$

$$\sum x_{b1}^2 = 0.19546846 \quad \sum x_{b2}^2 = 909.4665400 \quad \sum x_{b3}^2 = 1653.632200$$

$$\sum x_{b1}x_{b2} = 1.03713846 \sum x_{b1}x_{b3} = -5.51677080 \sum x_{b2}x_{b3} = 3.07677000$$

$$d_1 = 0.04219231 \qquad d_2 = -3.13461538 \qquad d_3 = -7.98500000$$

We now can calculate $\mathbf{S}_w$ using formula (5.131) by adding the corresponding sums of squares and products and dividing each by $n_a + n_b - 2 = 50$:

$$\mathbf{S}_w = \begin{bmatrix} 0.00841027 & 0.05860354 & -0.36002341 \\ 0.05860354 & 24.79905380 & 3.99346460 \\ -0.36002341 & 3.99346460 & 105.55503280 \end{bmatrix}$$

and $\mathbf{Y} = \begin{bmatrix} 0.04219231 = d_1 \\ -3.13461538 = d_2 \\ -7.98500000 = d_3 \end{bmatrix}$

Using the relationship shown in Equation (5.133), we find

$$S_w^{-1} = \begin{bmatrix} 143.4580137 & -0.42036530 & 0.50520546 \\ -0.42036551 & 0.04180307 & -0.00301530 \\ 0.50520526 & -0.00301530 & 0.01131095 \end{bmatrix}$$

and $\quad B_x = \begin{bmatrix} 3.336443 = \lambda_1 \\ -0.124696 = \lambda_2 \\ -0.059550 = \lambda_3 \end{bmatrix}$

With these $\lambda$ values we have the discriminant function

$$Z = 3.336443X_1 - 0.124696X_2 - 0.059550X_3 \qquad (5.134)$$

Substituting the mean values of the three variables from the first group in Equation (5.134), we find

$$\bar{Z}_a = \lambda_1(0.153731) + \lambda_2(23.176923) + \lambda_3(54.815769) = -5.641434$$

and similarly for the second group

$$\bar{Z}_b = \lambda_1(0.111538) + \lambda_2(26.311538) + \lambda_3(62.800769) = -6.648589$$

A test of significance of the linear function can be made by calculating the sum of squares for within groups as

$$D = \sum \lambda_i d_i = \bar{Z}_a - \bar{Z}_b = 1.007155 \qquad (5.135)$$

and the sum of squares between groups as

$$\frac{n_1 n_2 (\sum \lambda_i d_i)^2}{(n_1 + n_2)(n_1 + n_2 - 2)} = \frac{n_1 n_2 D^2}{(n_1 + n_2)(n_1 + n_2 - 2)} = \frac{685.70817}{2600}$$

$$= 0.263734 \qquad (5.136)$$

It should be noted that if a pooled sums of cross-products matrix is used rather than a pooled covariance matrix, the term $(n_1 + n_2 - 2)$ in the denominator of Equation (5.134) would not be included.

These values with $n_1 + n_2 - k - 1$ and $k$ degrees of freedom, respectively, are entered into Table 5.44 where an $F$ value of 4.19 is found, a value that is just short of being highly significant, indicating that the function is effective in discriminating between the two groups.

The test of significance shown in Table 5.43 may be put in the form

**TABLE 5.44**   ANALYSIS OF VARIANCE OF DISCRIMINANT FUNCTION

| Source | df | SS | MS | F |
|---|---|---|---|---|
| Between groups | 3 | 0.263734 | 0.087911 | 4.19* |
| Within groups | 48 | 1.007155 | 0.020982 | |

$$F = \frac{n_1 n_2 (n_1 + n_2 - k - 1)}{(n_1 + n_2)(n_1 + n_2 - 2)k} D \tag{5.137}$$

with $k$ and $n_1 + n_2 - k - 1$ degrees of freedom.

It is possible to obtain an idea of the accuracy of the function by classifying the original set of cases to see how many are classified correctly, using the three variables included in the study. To do this, we can develop a constant

$$c = \frac{\bar{Z}_a + \bar{Z}_b}{2} = \frac{-5.641434 - 6.648589}{2} = -6.145012 \tag{5.138}$$

such that if a value predicted from Equation (5.134) is $\geq c$ it is classified as belonging to population $A$ and if the value is $< c$ it is classified into population or group $B$.

Classification is made a little easier by the expediency of subtracting $c$ from Equation (5.134) before using it for prediction. Then, any predicted value $\geq 0$ falls into group $A$ and any predicted value $< 0$ falls into group $B$. Our prediction equation would be

$$\hat{\bar{Z}}' = \lambda_1 X_1 + \lambda_2 X_2 + \lambda_3 X_3 - c \tag{5.139}$$

Using this equation for the first set of $X$'s in group $A$, we would have

$$\hat{\bar{Z}}' = (3.336443)(0.048) - (0.124696)(25.2) - (0.059550)(73.69) + 6.145012$$

$$= -1.225418$$

The negative value would incorrectly classify this unit into group $B$. When all cases are predicted it is found that 19 out of 26 in the first group are predicted correctly and 17 out of 26 in the second group are predicted correctly.

Since the $X$'s are measured on different scales, it is not possible to state which of them are the most important in the discriminant function. It is possible to standardize the $\lambda$'s by multiplying each of them by the standard deviation of each $X$, calculated from the pooled sums of squares. Thus,

$$s_1 = \sqrt{\frac{\sum x_{a_1}^2 + \sum x_{b_1}^2}{n_1 + n_2 - 2}} = \sqrt{\frac{0.42051358}{50}} = 0.09170753$$

$$s_2 = \sqrt{\frac{\sum x_{a_2}^2 + \sum x_{b_2}^2}{n_1 + n_2 - 2}} = \sqrt{\frac{1239.95269}{50}} = 4.97986484$$

$$s_3 = \sqrt{\frac{\sum x_{a_3}^2 + \sum x_{b_3}^2}{n_1 + n_2 - 2}} = \sqrt{\frac{5277.75164}{50}} = 10.27399790$$

Designating the standardized $\lambda$'s with a prime, we have

$$\lambda_1' = 0.30597695 \qquad \lambda_2' = -0.62096923 \qquad \lambda_3' = -0.61181657$$

We can use these as we have the standard partial regression coefficients to evaluate the relative importance of the $\lambda$'s. We see that while $\lambda_2'$ and $\lambda_3'$ are very similar, both are more important in the function than $\lambda_1'$. The signs, of course, are disregarded.

While our example had equal numbers in the two groups, all the procedures followed allow an unequal number of cases in the two groups. For further elaboration of the subject of discriminant function analysis several textbooks on bivariate and multivariate analysis cover the subject in far greater depth, as do some of the computer program manuals. Some of the sources that are recommended are Cooley and Lohnes (1971), Lachenbruck (1975), Lindeman et al. (1975), and Nie et al. (1975).

## 5.17  CURVE FITTING TO LEAST-SQUARES MEANS: DISPROPORTIONATE SUBCLASS FREQUENCIES AND UNEQUAL INTERVALS

To minimize the sum of squares of deviations from a curve obtained by fitting polynomials, weighted least-squares procedures are required when the means have been adjusted for disproportionate subclass numbers and/or unequal intervals exist. One could apply an orthogonalizing transformation to the variance–covariance matrix of the least-squares means and from this obtain a set of orthogonal polynomial coefficients. However, these coefficients are useful only in computing sums of squares and for plotting the selected curve. The procedure to be presented here uses differences in reductions in sums of squares under different models and nonorthogonal partial regression coefficients to obtain the same results.

Let the general linear model be partitioned as follows:

$$\mathbf{Y} = \mathbf{X}_1\mathbf{A} + \mathbf{X}_2\mathbf{B} + \mathbf{e} \tag{5.140}$$

where $\mathbf{Y}$ is the $n \times 1$ column vector of observations; $\mathbf{A}$ is a $t \times 1$ column vector of constants for one set of effects, such as treatments, years, and ages, for which one wants to fit polynomials; $\mathbf{X}_1$ is the incidence or design matrix of order $n \times t$ with zero and one elements for the $\mathbf{A}$ set of effects; $\mathbf{B}$ is a $q \times 1$ column vector of the constants for all other effects included in the model; $\mathbf{X}_2$ is the $n \times q$ design matrix for the $\mathbf{B}$ effects; and $\mathbf{e}$ is the $n \times 1$ column vector of random errors with $E(\mathbf{ee}') = \mathbf{I}\sigma_e^2$ and the errors are assumed to be normally distributed.

As pointed out in Section 3.20, the specific constants are not estimable for discrete sets of effects. In practice, the experimenter chooses a set of appropriate linear functions of these constants which are estimable. Hence, the working model is

$$\mathbf{Y} = \mathbf{X}_1\mathbf{L}_1\mathbf{M}_1\mathbf{A} + \mathbf{X}_2\mathbf{L}_2\mathbf{M}_2\mathbf{B} + \mathbf{e} \tag{5.141}$$

where $\mathbf{M}_1\mathbf{A} = \mathbf{a}$, the $t - 1$ linear functions of the $A_i$ to be estimated, and $\mathbf{M}_2\mathbf{B} = \mathbf{b}$, which includes the mean and all other linear functions to be estimated for other effects. The matrices $\mathbf{L}_1$ and $\mathbf{L}_2$ are chosen so that both $\mathbf{M}_1\mathbf{L}_1$ and $\mathbf{M}_2\mathbf{L}_2$ will each equal the identity matrix. Now if $\mathbf{X}_1\mathbf{L}_1 = \mathbf{W}_1$ and $\mathbf{X}_2\mathbf{L}_2 = \mathbf{W}_2$ the reduced set of normal equations for an ordinary least-squares analysis are as follows:

$$\begin{bmatrix} \mathbf{W}_1'\mathbf{W}_1 & \mathbf{W}_1'\mathbf{W}_2 \\ \mathbf{W}_2'\mathbf{W}_1 & \mathbf{W}_2'\mathbf{W}_2 \end{bmatrix}\begin{bmatrix} \hat{\mathbf{a}} \\ \hat{\mathbf{b}} \end{bmatrix} = \begin{bmatrix} \mathbf{W}_1'\mathbf{Y} \\ \mathbf{W}_2'\mathbf{Y} \end{bmatrix} \tag{5.142}$$

And if

$$\begin{bmatrix} \mathbf{W}_1'\mathbf{W}_1 & \mathbf{W}_1'\mathbf{W}_2 \\ \mathbf{W}_2'\mathbf{W}_1 & \mathbf{W}_2'\mathbf{W}_2 \end{bmatrix}^{-1} = \begin{bmatrix} \mathbf{Z}_{11} & \mathbf{Z}_{12} \\ \mathbf{Z}_{12}' & \mathbf{Z}_{22} \end{bmatrix} \tag{5.143}$$

one may compute $\hat{\mathbf{a}}$ and the adjusted sum of squares for $\mathbf{a}$ as follows:

$$\hat{\mathbf{a}} = \mathbf{Z}_{11}\mathbf{W}_1'\mathbf{Y} + \mathbf{Z}_{12}\mathbf{W}_2'\mathbf{Y} \tag{5.144}$$

$$\text{SS}(\mathbf{a}) = \hat{\mathbf{a}}'\mathbf{Z}_{11}^{-1}\hat{\mathbf{a}} \tag{5.145}$$

We now want to partition the sum of squares for $\mathbf{a}$ into that which can be accounted for sequentially by each degree in polynomials, provide a suitable test of significance for each degree, and obtain an appropriate prediction equation that may be used to plot the selected curve.

If $\mathbf{M}_1$ is chosen so that

$$E(\hat{a}_i) = a_i - \frac{\Sigma_i\, a_i}{t} \tag{5.146}$$

as is often the case in practice, then $\Sigma_i\, \hat{a}_i = 0$ and if $\mathbf{M}_2$ is chosen in a similar manner for other sets of effects, then the adjusted treatment means will be $\hat{\mu} + \hat{a}_i$. However, with nonorthogonal (or unbalanced) data the variance–covariance matrix of the $\hat{\mu} + \hat{a}_i$ is not simply $\mathbf{I}\sigma_e^2$ or even $\mathbf{D}\sigma_e^2$, where $\mathbf{D}$ is a diagonal matrix, even if equal intervals exist between classes. Therefore, to fit nonorthogonal polynomials to the adjusted means using least-squares procedures, one must use general weighted least-squares techniques. A polynomial curve fitted to the $\hat{a}_i$, of course, will be the same curve as would be obtained if fitted to the $\hat{\mu} + \hat{a}_i$. Also, since there are only $t - 1$ estimable linear functions among the $a_i$, one can fit the weighted nonorthogonal polynomials by using only the $t - 1$ $\hat{a}_i$ which are included in $\hat{\mathbf{a}}$ and the variance–covariance matrix of $\hat{\mathbf{a}}$, $\mathbf{Z}_{11}\sigma_e^2$.

## 5.17.1 Fitting the Linear Regression

Let $\mathbf{P}$ be a $t \times 1$ column vector of the $t$ levels for the $a_i$ set of effects and let $\mathbf{X} = \mathbf{M}_1\mathbf{P}$ be a $(t - 1) \times 1$ column vector. From weighted least-squares theory it is clear that the weighted least-squares equation for the linear regression $\beta_1$ is

$$\mathbf{X}'\mathbf{Z}_{11}^{-1}\mathbf{X}\hat{\beta}_1 = \mathbf{X}'\mathbf{Z}_{11}^{-1}\hat{\mathbf{a}} \tag{5.147}$$

and
$$\hat{\beta}_1 = \frac{\mathbf{X}'\mathbf{Z}_{11}^{-1}\hat{\mathbf{a}}}{\mathbf{X}'\mathbf{Z}_{11}^{-1}\mathbf{X}} \tag{5.148}$$

The sum of squares accounted for by the linear regression is

$$\text{SS}(\beta_1) = \frac{(\mathbf{X}'\mathbf{Z}_{11}^{-1}\hat{\mathbf{a}})^2}{\mathbf{X}'\mathbf{Z}_{11}^{-1}\mathbf{X}} \tag{5.149}$$

and the residual sum of squares for the $A$ classes is

$$\text{Residual SS}(\mathbf{a}) = \text{SS}(\mathbf{a}) - \text{SS}(\beta_1) \tag{5.150}$$

The standard error of $\hat{\beta}_1$ may be computed as

$$s_{\hat{\beta}_1} = \left(\frac{\text{MS(Error)}}{\mathbf{X}'\mathbf{Z}_{11}^{-1}\mathbf{X}}\right)^{1/2} \tag{5.151}$$

and the $F$ values for testing the significance of the linear regression coefficient $\hat{\beta}_1$ and the deviations from linearity—the residual $A$ effects—may be computed as follows:

$$F \text{ for } \hat{\beta}_1 = \frac{SS(\beta_1)}{MS(\text{Error})}$$

$$F \text{ for Residual (\textbf{a})} = \frac{MS(\text{Residual } \mathbf{a})}{MS(\text{Error})}$$

Note that the mean square for error from the least-squares analysis is used instead of the residual mean square for $\mathbf{a}$ to obtain $s_{\hat{\beta}_1}$ and to test the significance of $\hat{\beta}_1$. This procedure is preferred because (1) the error mean square will usually have more degrees of freedom and (2) the residual mean square for $\mathbf{a}$ is likely to contain quadratic and higher-degree polynomial effects.

## 5.17.2 General Polynomial Fitting

The fitting of additional degrees in polynomials may be done stepwise, adding one additional equation in each step. When this procedure is followed, the sum of squares for each successive polynomial and hence the test of significance will be the same as would be obtained if orthogonal polynomial coefficients were first computed with some orthogonalizing transformation (see Harvey and Swiger, 1978).

If $\mathbf{P}_1$ is the column vector of levels for the $A$ classes, $\mathbf{P}_2$ is the column vector of the levels squared, and so on, then

$$\mathbf{X}_i = \mathbf{M}_1 \mathbf{P}_i \tag{5.152}$$

is the $(t - 1) \times 1$ column vector of the transformed $\mathbf{P}_i$ vector and the set of equations for $q$ polynomial regression coefficients may be set up as follows:

$$
\begin{aligned}
\mathbf{X}'\mathbf{Z}_{11}^{-1}\mathbf{X}_1\hat{\beta}_1 + \mathbf{X}_1'\mathbf{Z}_{11}^{-1}\mathbf{X}_2\hat{\beta}_2 + \cdots &= \mathbf{X}_1'\mathbf{Z}_{11}^{-1}\hat{\mathbf{a}} \\
\mathbf{X}_2'\mathbf{Z}_{11}^{-1}\mathbf{X}_1\hat{\beta}_1 + \mathbf{X}_2'\mathbf{Z}_{11}^{-1}\mathbf{X}_2\hat{\beta}_2 + \cdots &= \mathbf{X}_2'\mathbf{Z}_{11}^{-1}\hat{\mathbf{a}} \\
\vdots \qquad\qquad \vdots \qquad\qquad &\quad\ \vdots \\
\mathbf{X}_q'\mathbf{Z}_{11}^{-1}\mathbf{X}_1\hat{\beta}_1 + \mathbf{X}_q'\mathbf{Z}_{11}^{-1}\mathbf{X}_2\hat{\beta}_2 + \cdots &= \mathbf{X}_q'\mathbf{Z}_{11}^{-1}\hat{\mathbf{a}}
\end{aligned}
\tag{5.153}
$$

Now if we let $\mathbf{C}^{-1}$ be the inverse of the coefficient matrix and $\mathbf{R}$ the column vector of right-hand members, then

$$\hat{\beta}_i = \sum_j c^{ij} r_j \tag{5.154}$$

and the total reduction due to fitting the $q$th degree polynomial is $\hat{\beta}'\mathbf{R}$, where $\hat{\beta}'$ is a row vector of the $\hat{\beta}_i$. The residual sum of squares for $\mathbf{a}$ is

$$\hat{\mathbf{a}}'\mathbf{Z}_{11}^{-1}\hat{\mathbf{a}} - \hat{\beta}'\mathbf{R} \tag{5.155}$$

with $t - 1 - q$ degrees of freedom. The sum of squares for the $q$th degree polynomial is

$$SS(\beta_q) = \frac{\hat{\beta}_q^2}{c^{qq}} \tag{5.156}$$

and, of course, this is equal to the difference between the reduction in the sum of squares when the $q$th degree polynomial is fitted and the reduction in the sum of squares when the $(q - 1)$th degree polynomial is fitted. The mean square for error from the least-squares analysis should again be used as the denominator for computing $F$ to test for the significance of each polynomial and the residual.

When all $t - 1$ degrees of freedom among the classes or levels in the $a_i$ set of effects are fitted with polynomials, the sum of squares obtained for each degree in this stepwise procedure will sum to the sum of squares for $\mathbf{a}$, that is, to $\hat{\mathbf{a}}'\mathbf{Z}_{11}^{-1}\hat{\mathbf{a}}$. This will be true regardless of the confounding of the $\mathbf{a}$ and $\mathbf{b}$ effects and regardless of whether or not equal intervals exist.

Rounding errors in computations may be reduced materially by first expressing the levels of the classes as deviations from the mean of all levels. When this is done, the prediction equation for plotting the curve is

$$\hat{Y}_j = \hat{\beta}_0 + \hat{\beta}_1(p_j - \bar{p}) + \hat{\beta}_2(p_j - \bar{p})^2 + \cdots \qquad (5.157)$$

where

$$\hat{\beta}_0 = \hat{\mu} - \frac{1}{t} \sum_{i=1}^{q} \hat{\beta}_i \sum_{j=1}^{t} (p_j - \bar{p})^i \qquad (5.158)$$

$p_j$ is the $j$th level, and $\bar{p}$ is the unweighted mean of all class levels.

If one prefers to have the prediction equation in the more conventional form

$$\hat{Y}_j = \hat{\beta}_0^* + \hat{\beta}_1^* p_j + \hat{\beta}_2^* p_j^2 + \cdots \qquad (5.159)$$

then it is easily shown that

$$\hat{\beta}_j^* = \sum_{i=j}^{q} (-1)^{i-j} \binom{i}{i-j} \hat{\beta}_i (\bar{p})^{i-j} \qquad (5.160)$$

where

$$\binom{i}{i-j} = \frac{i!}{(i-j)!j!}$$

is the standard combinatorial formula.

### 5.17.3 Illustration of Computing Procedures

Suppose the following data were available from a two-way classification experiment where there were two levels of $A$ and three levels of $B$:

|  | $b_1$ | $b_2$ | $b_3$ |
|---|---|---|---|
|  | 5 | 2 | 3 |
|  | 6 | 3 |  |
| $a_1$ |  | 5 |  |
|  |  | 6 |  |
|  |  | 7 |  |
| Subtotals | 11 | 23 | 3 |

|        | $b_1$ | $b_2$ | $b_3$ |
|--------|-------|-------|-------|
|        | 2     | 8     | 4     |
|        | 3     | 8     | 4     |
| $a_2$  |       | 9     | 6     |
|        |       |       | 6     |
|        |       |       | 7     |
| Subtotals | 5  | 25    | 27    |
| Totals | 16    | 48    | 30    |

Now if these data are analyzed by least squares under the model

$$Y_{ijk} = \mu + a_i + b_j + ab_{ij} + e_{ijk} \tag{5.161}$$

as described in Section 3.20.2 with the usual or summation restrictions imposed, the inverse segment of the inverse of the least-squares equations involving the $\hat{b}_j(\mathbf{Z}_{11})$ is found to be

$$\mathbf{Z}_{11} = \begin{bmatrix} 0.159259 & -0.051852 \\ -0.051852 & 0.120370 \end{bmatrix}$$

Estimates of $b_1$ and $b_2$ are

$$\hat{b}_1 = -0.888889 \quad \text{and} \quad \hat{b}_2 = 1.577778$$

and the sum of squares for the $b_j$ set of effects that is to be partitioned into linear and quadratic effects is

$$\hat{\mathbf{b}}'\mathbf{Z}_{11}^{-1}\hat{\mathbf{b}} = (-0.888889 \quad 1.577778)\begin{bmatrix} 7.3034 & 3.1461 \\ 3.1461 & 9.6630 \end{bmatrix}\begin{bmatrix} -0.888889 \\ 1.577778 \end{bmatrix}$$

$$= 21.001$$

Let us now suppose that the three levels of $B$ are 1, 3, and 8, that is, unequal intervals. Therefore,

$$\mathbf{X}_1 = \mathbf{M}_1\mathbf{P}_1 = \frac{1}{3}\begin{bmatrix} 2 & -1 & -1 \\ -1 & 2 & -1 \end{bmatrix}\begin{bmatrix} -3 \\ -1 \\ 4 \end{bmatrix}$$

$$= \begin{bmatrix} -3 \\ -1 \end{bmatrix}$$

when the $p_i$ are expressed as deviations from the unweighted mean of the $p_i$ and

$$\mathbf{X}_2 = \mathbf{M}_1\mathbf{P}_2 = \frac{1}{3}\begin{bmatrix} 2 & -1 & -1 \\ -1 & 2 & -1 \end{bmatrix}\begin{bmatrix} 9 \\ 1 \\ 16 \end{bmatrix}$$

$$= \begin{bmatrix} \frac{1}{3} \\ -\frac{23}{3} \end{bmatrix}$$

The equation for the linear regression is

$$\mathbf{X}_1'\mathbf{Z}_{11}^{-1}\mathbf{X}_1\hat{\beta}_1 = \mathbf{X}_1'\mathbf{Z}_{11}^{-1}\hat{\mathbf{b}}$$

$$\mathbf{X}_1'\mathbf{Z}_{11}^{-1}\mathbf{X}_1 = (-3 \ \ -1)\mathbf{Z}_{11}^{-1}\begin{bmatrix} -3 \\ -1 \end{bmatrix}$$

$$= 94.2702$$

$$\mathbf{X}_1'\mathbf{Z}_{11}^{-1}\hat{\mathbf{b}} = (-3 \ \ -1)\mathbf{Z}_{11}^{-1}\begin{bmatrix} -0.888889 \\ 1.577778 \end{bmatrix}$$

$$= -7.8653$$

Therefore,

$$\hat{\beta}_1 = \frac{-7.8653}{94.2702} = -0.083434$$

$$\mathrm{SS}(\beta_1) = \frac{(-7.8653)^2}{94.2702} = 0.6562$$

$$s_{\hat{\beta}_1} = \sqrt{\frac{2.1721}{94.2702}} = 0.1518$$

where 2.1721 is the mean square for error from ANOVA.

The following calculations are necessary to set up the equations for both the linear and quadratic regressions:

$$\mathbf{X}_1'\mathbf{Z}_{11}^{-1}\mathbf{X}_2 = (-3 \ \ -1)\mathbf{Z}_{11}^{-1}\begin{bmatrix} \frac{1}{3} \\ -\frac{23}{3} \end{bmatrix}$$

$$= 138.0912$$

$$\mathbf{X}_2'\mathbf{Z}_{11}^{-1}\mathbf{X}_2 = (\tfrac{1}{3} \ \ -\tfrac{23}{3})\mathbf{Z}_{11}^{-1} - \begin{bmatrix} \frac{1}{3} \\ \frac{23}{3} \end{bmatrix}$$

$$= 552.7012$$

$$\mathbf{X}_2'\mathbf{Z}_{11}^{-1}\hat{\mathbf{b}} = (\tfrac{1}{3} \ \ -\tfrac{23}{3})\mathbf{Z}_{11}^{-1}\begin{bmatrix} -0.888889 \\ 1.577778 \end{bmatrix}$$

$$= -95.9558$$

The two equations for the linear and quadratic regressions are

$$\begin{bmatrix} 94.2702 & 138.0912 \\ 138.0912 & 552.7012 \end{bmatrix}\begin{bmatrix} \hat{\beta}_1 \\ \hat{\beta}_2 \end{bmatrix} = \begin{bmatrix} -7.8653 \\ -95.9558 \end{bmatrix}$$

and

$$\begin{bmatrix} 94.2702 & 138.0912 \\ 138.0912 & 552.7012 \end{bmatrix}^{-1} = \begin{bmatrix} 0.016731 & -0.004180 \\ -0.004180 & 0.002854 \end{bmatrix}$$

The solution of the equations yields the estimates of the nonorthogonal regression coefficients,

$$\begin{bmatrix} \hat{\beta}_1 \\ \hat{\beta}_2 \end{bmatrix} = \begin{bmatrix} 0.016731 & -0.004180 \\ -0.004180 & 0.002854 \end{bmatrix} \begin{bmatrix} -7.8653 \\ -95.9558 \end{bmatrix}$$

$$= \begin{bmatrix} 0.2695 \\ -0.2410 \end{bmatrix}$$

The sum of squares for the quadratic regression is

$$SS(\beta_2 \mid \beta_1) = \frac{(-0.2410)^2}{0.002854} = 20.3508$$

At this point it should be noted that

$$\mathbf{b}'\mathbf{Z}_{11}^{-1}\mathbf{b} = SS(\beta_1) + SS(\beta_2 \mid \beta_1)$$

$$= 0.6562 + 20.3508$$

$$= 21.007$$

which agrees with the SS previously obtained for the $b_j$ within rounding errors. The standard error of $\hat{\beta}_2$ is

$$s_{\hat{\beta}_2} = \sqrt{(2.1721)(0.002854)}$$

$$= 0.0787$$

The ANOVA for the $B$ effects may now be set up as follows:

| Source | df | MS | F |
|--------|-----|--------|--------|
| $B$ | 2 | 10.503 | 4.83* |
| $B_L$ | 1 | 0.656 | <1 |
| $B_Q$ | 1 | 20.351 | 9.37** |
| Error | 12 | 2.172 | |

The prediction equation for the quadratic curve is

$$\hat{Y}_j = 6.9776 + 0.2695(p_j - 4) - 0.2410(p_j - 4)^2$$

where 

$$6.9776 = 4.888889 + 0.2410\left(\frac{9 + 1 + 16}{3}\right)$$

and 

$$\hat{\mu} = 4.888889$$

The unweighted mean of the $p_j$ is 4. The predicted $Y$ values for the $B$ classes are 4.0, 6.5, and 4.2 and, of course, these are equal to the least-squares means for the $B$ classes, in this case.

## 5.18 GENERAL SURFACE FITTING PROCEDURES

When there are two sets of treatments for which the fitting of orthogonal polynomials is appropriate, the investigator is often interested in partitioning the sum of squares for the interaction into that due to the single degree of freedom interaction components—linear × linear, linear × quadratic, quadratic × linear, quadratic × quadratic, and so on. To do this efficiently when there is arbitrary spacing and/or correlated subclass means, the investigator must use a weighted least-squares procedure.

To describe the computational procedures involved in this case, let the general linear model now be partitioned as follows:

$$\mathbf{Y} = \mathbf{X}_1\mathbf{S} + \mathbf{X}_2\mathbf{C} + \mathbf{e} \tag{5.162}$$

where $\mathbf{Y}$ is the $n \times 1$ column vector of observations; $\mathbf{S}$ is a $t \times 1$ column vector of constants for the $AB$ subclass effects, such as years × ages, that is, $s_{ij} = a_i + b_j + ab_{ij}$, for which one wants to fit polynomials; $\mathbf{X}_1$ is the incidence or design matrix of order $n \times t$ with 0 and 1 elements for the $\mathbf{S}$ set of effects; $\mathbf{C}$ is a $q \times 1$ column vector of the constants for all other effects included in the model; $\mathbf{X}_2$ is the $n \times q$ design matrix for the $\mathbf{C}$ effects; and $\mathbf{e}$ is the $n \times 1$ column vector of random errors with $E(\mathbf{ee}') = \mathbf{I}\sigma_e^2$ and the errors are assumed to be normally distributed.

If there are $t_1$ classes of $A$ and $t_2$ classes of $B$, then $t = t_1 \times t_2$ when there are no missing cells. With this model $\mathbf{M}_1\mathbf{S} = \mathbf{s}$, the $t - 1$ linear functions of the $s_{ij}$ to be estimated, and $\mathbf{M}_2\mathbf{C} = \mathbf{c}$, which includes $\mu$ and all linear functions of other effects to be estimated. Also, $V(\hat{\mathbf{s}}) = \mathbf{Z}_{11}\sigma_e^2$ and the sum of squares for $s$ is as follows:

$$\text{SS}(\mathbf{s}) = \hat{\mathbf{s}}'\mathbf{Z}_{11}^{-1}\hat{\mathbf{s}} \tag{5.163}$$

Now let the estimates of the subclass effects ($\hat{s}_{ij}$) be arranged in conventional order, that is, $11\ 12 \cdots 1t_2\ 21\ 22 \cdots 2t_2 \cdots t_1 1\ t_1 2 \cdots t_1 t_2$ and let $\mathbf{P}_{11}$ be a column vector of the levels for the $A$ classes, $\mathbf{P}_{12}$ a column vector of the levels for the $A$ classes squared, and so on, $\mathbf{P}_{21}$ a column vector of levels for the $B$ classes, $\mathbf{P}_{22}$ a column vector of the levels of $B$ squared, and so on. Then,

$$\mathbf{X}_{1i} = \mathbf{M}_1(\mathbf{1}_{t_2} \otimes \mathbf{P}_{1i}) \qquad i = 1, 2, \ldots, t_1 \tag{5.164}$$

$$\mathbf{X}_{2i'} = \mathbf{M}_1(\mathbf{P}_{2i'} \otimes \mathbf{1}_{t_1}) \qquad i' = 1, 2, \ldots, t_2 \tag{5.165}$$

$$\mathbf{X}_{3ii'} = \mathbf{M}_1[\text{Diag}(\mathbf{1}_{t_2} \otimes \mathbf{P}_{1i})(\mathbf{P}_{2i'} \otimes \mathbf{1}_{t_1})'] \tag{5.166}$$

where $\mathbf{1}_{t_2}$ is a column vector of $t_2$ ones, $\otimes$ refers to the direct product operation (the direct product operation is described later), $\mathbf{1}_{t_1}$ is a column vector of $t_1$ ones, and Diag means that the diagonal elements of the product of two vectors is to form a column vector. The weighted least-squares set of equations for all $t - 1$ polynomial regression coefficients are as follows:

$$\mathbf{X}'_{11}\mathbf{Z}_{11}^{-1}\mathbf{X}_{11}\hat{\beta}_1 + \mathbf{X}'_{11}\mathbf{Z}_{11}^{-1}\mathbf{X}_{12}\hat{\beta}_2 + \cdots = \mathbf{X}'_{11}\mathbf{Z}_{11}^{-1}\hat{\mathbf{s}}$$

$$\mathbf{X}'_{12}\mathbf{Z}_{11}^{-1}\mathbf{X}_{11}\hat{\beta}_1 + \mathbf{X}'_{12}\mathbf{Z}_{11}^{-1}\mathbf{X}_{12}\hat{\beta}_2 + \cdots = \mathbf{X}'_{12}\mathbf{Z}_{11}^{-1}\hat{\mathbf{s}}$$

$$\vdots \qquad\qquad \vdots \qquad\qquad \vdots \tag{5.167}$$

$$\mathbf{X}'_{3ii'}\mathbf{Z}_{11}^{-1}\mathbf{X}_{11}\hat{\beta}_1 + \mathbf{X}'_{3ii'}\mathbf{Z}_{11}^{-1}\mathbf{X}_{12}\hat{\beta}_2 + \cdots = \mathbf{X}'_{3ii'}\mathbf{Z}_{11}^{-1}\hat{\mathbf{s}}$$

$$\vdots \qquad\qquad \vdots \qquad\qquad \vdots$$

If $\mathbf{C}^{-1} = [c^{ij}]$ is the inverse of the coefficient matrix and $\mathbf{r} = [r_j]$ is the column vector of the right-hand members, then

$$\hat{\beta}_i = \sum_j c^{ij} r_j \tag{5.168}$$

and the total reduction due to fitting all $\beta$'s is $\hat{\boldsymbol{\beta}}'\mathbf{R}$, where $\hat{\boldsymbol{\beta}}'$ is a row vector of the $\hat{\beta}_i$ and this will equal $\hat{\mathbf{s}}'\mathbf{Z}_{11}^{-1}\hat{\mathbf{s}}$, the sum of squares for the $AB$ subclasses.

The sum of squares for the highest-degree polynomial in the interaction is

$$SS(\hat{\beta}_{(t_1-1)(t_2-1)}) = \frac{\hat{\beta}^2(t_1 - 1)(t_2 - 1)}{c^{(t_1-1)(t_2-1)}} \tag{5.169}$$

which, of course, is equal to the difference in the total reduction in sum of squares when the $\beta$'s for all degrees of freedom are fitted and the reduction in sum of squares when all $\beta$'s except the highest-degree polynomial are fitted. The mean square for error from ANOVA may be used to obtain a standard error of the highest-degree polynomial regression coefficient, if desired, in the usual manner.

A sequential step-down procedure that will allow the sums of squares for the polynomials to be additive and that will select a satisfactory prediction equation will now be described. To illustrate the procedure let us suppose there are three levels of $A$ and four levels of $B$. In this case, the $\beta$'s to be fitted are presented in Table 5.45. The order of elimination suggested for the step-down procedure is as follows:

1. $\hat{\beta}_{11}$.
2. Least significant of $\hat{\beta}_8$ and $\hat{\beta}_{10}$.
3. $\hat{\beta}_8$ or $\hat{\beta}_{10}$.
4. Least significant of $\hat{\beta}_5$, $\hat{\beta}_7$, and $\hat{\beta}_9$.
5. Least significant of remaining two $\beta$'s.
6. $\hat{\beta}_5$, $\hat{\beta}_7$, or $\hat{\beta}_9$, whichever is left.
7. Least significant of $\hat{\beta}_2$, $\hat{\beta}_4$, and $\hat{\beta}_6$, and so on.

At each step the sum of squares for the $\beta_i$ to be eliminated is computed from $\hat{\beta}_i^2/c^{ii}$. The prediction equation for plotting the three-dimensional surface is chosen by selecting a priori a probability level for the step-down procedure.

**TABLE 5.45  POLYNOMIALS TO BE FITTED WITH THREE LEVELS OF _A_ AND FOUR LEVELS OF _B_**

| $\beta$ | Description | Degree of polynomial |
|---|---|---|
| $\beta_1$ | Linear $A$ | 1 |
| $\beta_2$ | Quadratic $A$ | 2 |
| $\beta_3$ | Linear $B$ | 1 |
| $\beta_4$ | Quadratic $B$ | 2 |
| $\beta_5$ | Cubic $B$ | 3 |
| $\beta_6$ | Linear $A \times$ linear $B$ | 2 |
| $\beta_7$ | Linear $A \times$ quadratic $B$ | 3 |
| $\beta_8$ | Linear $A \times$ cubic $B$ | 4 |
| $\beta_9$ | Quadratic $A \times$ linear $B$ | 3 |
| $\beta_{10}$ | Quadratic $A \times$ quadratic $B$ | 4 |
| $\beta_{11}$ | Quadratic $A \times$ cubic $B$ | 5 |

As in the case of a single factor, rounding errors may be reduced materially by first expressing the levels as deviations from the mean of all levels for each factor separately. When this is done, the prediction equation is

$$\hat{Y}_{kl} = \hat{\beta}_0 + \sum_{i=1}^{q_1} \hat{\beta}_{i\cdot}(p_k - \bar{p}_A)^i + \sum_{j=1}^{q_2} \hat{\beta}_{\cdot j}(p_l - \bar{p}_B)^j$$

$$+ \sum_{i=1}^{q_1'} \sum_{j=1}^{q_2'} \hat{\beta}_{ij}(p_k - \bar{p}_A)^i (p_l - \bar{p}_B)^j \qquad (5.170)$$

where $q_1$ is the number of degrees in polynomials fitted for $A$, $q_2$ is the number of degrees in polynomials fitted for $B$, $q_1'$ and $q_2'$ are the respective numbers in the interaction polynomials, and

$$\hat{\beta}_0 = \hat{\mu} - \frac{1}{t_1} \sum_{i=1}^{q_1} \hat{\beta}_{i\cdot} \sum_{k=1}^{t_1}(p_k - \bar{p}_A)^i - \frac{1}{t_2} \sum_{j=1}^{t_2} \hat{\beta}_{\cdot j} \sum_{l=1}^{t_2}(p_l - \bar{p}_B)^j$$

$$- \frac{1}{t_1 t_2} \sum_{i=1}^{q_1'} \sum_{j=1}^{q_2'} \hat{\beta}_{ij} \sum_{k=1}^{t_1} \sum_{l=1}^{t_2}(p_k - \bar{p}_A)^i (p_l - \bar{p}_B)^j \qquad (5.171)$$

**Example Problem**  We shall illustrate the computational procedures described in this section with the synthetic data from the $2 \times 3$ factorial experiment given in Section 5.17.3. With this set of data there are five degrees of freedom among the $AB$ subclasses that we want to partition into linear $A$, linear $B$, quadratic $B$, linear $A \times$ linear $B$, and linear $A \times$ quadratic $B$. The $\mathbf{Z}_{11}$ matrix for the $AB$ subclass effects (the $\hat{s}_{ij}$) is

$$\mathbf{Z}_{11} = \begin{bmatrix} 0.40926 & -0.04074 & -0.17407 & -0.09074 & -0.06296 \\ -0.04074 & 0.20926 & -0.12407 & -0.04074 & -0.01296 \\ -0.17407 & -0.12407 & 0.74259 & -0.17407 & -0.14630 \\ -0.09074 & -0.04074 & -0.17407 & 0.40926 & -0.06296 \\ -0.06296 & -0.01296 & -0.14630 & -0.06296 & 0.29815 \end{bmatrix}$$

with

$$\mathbf{M}_1 = \frac{1}{6} \begin{bmatrix} 5 & -1 & -1 & -1 & -1 & -1 \\ -1 & 5 & -1 & -1 & -1 & -1 \\ -1 & -1 & 5 & -1 & -1 & -1 \\ -1 & -1 & -1 & 5 & -1 & -1 \\ -1 & -1 & -1 & -1 & 5 & -1 \end{bmatrix}$$

and the estimates of the subclass effects (the $\hat{s}_{ij}$) are

$$\hat{\mathbf{s}}' = [0.61111 \quad -0.28889 \quad -1.88889 \quad -2.38889 \quad 3.44444]$$

The inverse of $\mathbf{Z}_{11}$ is

$$\mathbf{Z}_{11}^{-1} = \begin{bmatrix} 6.5000 & 5.0000 & 4.3333 & 4.5000 & 4.6667 \\ 5.0000 & 10.0000 & 5.0000 & 5.0000 & 5.0000 \\ 4.3333 & 5.0000 & 5.1111 & 4.3333 & 4.5556 \\ 4.5000 & 5.0000 & 4.3333 & 6.5000 & 4.6667 \\ 4.6667 & 5.0000 & 4.5556 & 4.6667 & 7.7778 \end{bmatrix}$$

and the sum of squares for differences among the $AB$ subclasses is

$$\hat{\mathbf{s}}'\mathbf{Z}_{11}^{-1}\hat{\mathbf{s}} = 51.044$$

When the levels of both $A$ and $B$ (the $p_i$ and $p_j$) are expressed first as deviations from the respective unweighted means (1.5 and 4.0), the $\mathbf{P}'_{ij}$ are computed as follows:

$$\mathbf{P}'_{11} = [-0.5 \quad 0.5]$$

$$\mathbf{P}'_{21} = [-3 \quad -1 \quad 4]$$

$$\mathbf{P}'_{22} = [9 \quad 1 \quad 16]$$

The five $\mathbf{X}$ vectors are now computed as follows:

$$\mathbf{X}'_{11} = [\mathbf{M}_1(\mathbf{1}_3 \otimes \mathbf{P}_{11})]'$$

$$= \left\{ \frac{1}{6} \begin{pmatrix} 5 & -1 & -1 & -1 & -1 & -1 \\ -1 & 5 & -1 & -1 & -1 & -1 \\ -1 & -1 & 5 & -1 & -1 & -1 \\ -1 & -1 & -1 & 5 & -1 & -1 \\ -1 & -1 & -1 & -1 & 5 & -1 \end{pmatrix} \left[ \begin{pmatrix} 1 \\ 1 \\ 1 \end{pmatrix} \otimes \begin{pmatrix} -0.5 \\ 0.5 \end{pmatrix} \right] \right\}'$$

$$= \left[ \frac{1}{6} \begin{pmatrix} 5 & -1 & -1 & -1 & -1 & -1 \\ -1 & 5 & -1 & -1 & -1 & -1 \\ -1 & -1 & 5 & -1 & -1 & -1 \\ -1 & -1 & -1 & 5 & -1 & -1 \\ -1 & -1 & -1 & -1 & 5 & -1 \end{pmatrix} \begin{pmatrix} -0.5 \\ -0.5 \\ -0.5 \\ 0.5 \\ 0.5 \\ 0.5 \end{pmatrix} \right]'$$

$$= \tfrac{1}{2}[-1 \quad -1 \quad -1 \quad 1 \quad 1]$$

$$\mathbf{X}'_{21} = [\mathbf{M}_1(\mathbf{P}_{21} \otimes \mathbf{1}_2)]'$$

$$= \left\{ \mathbf{M}_1 \left[ \begin{pmatrix} -3 \\ -1 \\ 4 \end{pmatrix} \otimes \begin{pmatrix} 1 \\ 1 \end{pmatrix} \right] \right\}'$$

$$= \left[ \mathbf{M}_1 \begin{pmatrix} -3 \\ -1 \\ 4 \\ -3 \\ -1 \\ 4 \end{pmatrix} \right]'$$

$$= [-3 \quad -1 \quad 4 \quad -3 \quad -1]$$

$$\mathbf{X}'_{22} = [\mathbf{M}_1(\mathbf{P}_{22} \otimes \mathbf{1}_2)]'$$

$$= \left\{ \mathbf{M}_1 \left[ \begin{pmatrix} 9 \\ 1 \\ 16 \end{pmatrix} \otimes \begin{pmatrix} 1 \\ 1 \end{pmatrix} \right] \right\}'$$

$$= \left[ \mathbf{M}_1 \begin{pmatrix} 9 \\ 1 \\ 16 \\ 9 \\ 1 \\ 16 \end{pmatrix} \right]'$$

$$= \tfrac{1}{3}[1 \;\; -23 \;\; 22 \;\; 1 \;\; -23]$$

$$\mathbf{X}'_{311} = \{ \mathbf{M}_1 [\mathrm{Diag}(\mathbf{1}_3 \otimes \mathbf{P}_{11})(\mathbf{P}_{21} \otimes \mathbf{1}_2)'] \}'$$

$$= \left\{ \mathbf{M}_1 \left[ \mathrm{Diag} \begin{pmatrix} -0.5 \\ -0.5 \\ -0.5 \\ 0.5 \\ 0.5 \\ 0.5 \end{pmatrix} (-3 \;\; -1 \;\; 4 \;\; -3 \;\; -1 \;\; 4) \right] \right\}'$$

$$= \left[ \mathbf{M}_1 \frac{1}{2} \begin{pmatrix} 3 \\ 1 \\ -4 \\ -3 \\ -1 \\ 4 \end{pmatrix} \right]'$$

$$= \tfrac{1}{2}[3 \;\; 1 \;\; -4 \;\; -3 \;\; -1]$$

$$\mathbf{X}_{312} = \{ \mathbf{M}_1 [\mathrm{Diag}(\mathbf{1}_3 \otimes \mathbf{P}_{11})(\mathbf{P}_{22} \otimes \mathbf{1}_2)'] \}'$$

$$= \left[ \mathbf{M}_1 \frac{1}{2} \begin{pmatrix} -9 \\ -1 \\ -16 \\ 9 \\ 1 \\ 16 \end{pmatrix} \right]'$$

$$= \tfrac{1}{2}[-9 \;\; -1 \;\; -16 \;\; 9 \;\; 1]$$

The coefficients of the five polynomial regression coefficients and the right-hand members for the weighted least-squares set of equations are now computed as follows:

$$\mathbf{X}'_{11}\mathbf{Z}^{-1}_{11}\mathbf{X}_{11} = \tfrac{1}{2}[-1 \;\; -1 \;\; -1 \;\; 1 \;\; 1]\mathbf{Z}^{-1}_{11} \begin{bmatrix} -1 \\ -1 \\ -1 \\ 1 \\ 1 \end{bmatrix} \frac{1}{2} = 4.4444$$

$$\mathbf{X}'_{11}\mathbf{Z}_{11}^{-1}\mathbf{X}_{21} = \tfrac{1}{2}[-1 \;\; -1 \;\; -1 \;\; 1 \;\; 1]\mathbf{Z}_{11}^{-1}\begin{bmatrix} -3 \\ -1 \\ 4 \\ -3 \\ -1 \end{bmatrix} = 8.7778$$

$$\vdots$$

$$\mathbf{X}'_{312}\mathbf{Z}_{11}^{-1}\mathbf{X}_{312} = \tfrac{1}{2}[-9 \;\; -1 \;\; -16 \;\; 9 \;\; 1]\mathbf{Z}_{11}^{-1}\begin{bmatrix} -9 \\ -1 \\ -16 \\ 9 \\ 1 \end{bmatrix}\tfrac{1}{2} = 413.6111$$

$$\mathbf{X}'_{11}\mathbf{Z}_{11}^{-1}\hat{\mathbf{s}} = \tfrac{1}{2}[-1 \;\; -1 \;\; -1 \;\; 1 \;\; 1]\mathbf{Z}_{11}^{-1}\begin{bmatrix} 0.61111 \\ -0.28889 \\ -1.88889 \\ -2.38889 \\ 3.44444 \end{bmatrix} = 4.7778$$

$$\vdots$$

$$\mathbf{X}'_{312}\mathbf{Z}_{11}^{-1}\hat{\mathbf{s}} = \tfrac{1}{2}[-9 \;\; -1 \;\; -16 \;\; 9 \;\; 1]\mathbf{Z}_{11}^{-1}\begin{bmatrix} 0.61111 \\ -0.28889 \\ -1.88889 \\ -2.38889 \\ 3.44444 \end{bmatrix} = 4.1111$$

The least-squares equations for the $\beta$'s are

$$\begin{bmatrix} 4.4444 & 8.7778 & 23.2222 & 0.5 & 33.2778 \\ 8.7778 & 139.1111 & 236.8889 & 29.0 & 122.1111 \\ 23.2222 & 236.8889 & 779.1111 & 59.0 & 269.8889 \\ 0.5000 & 29.0000 & 59.0000 & 30.5 & 51.5000 \\ 33.2778 & 122.1111 & 269.8889 & 51.5 & 413.6111 \end{bmatrix}\begin{bmatrix} \hat{\beta}_1 \\ \hat{\beta}_2 \\ \hat{\beta}_3 \\ \hat{\beta}_4 \\ \hat{\beta}_5 \end{bmatrix} = \begin{bmatrix} 4.7778 \\ 3.1111 \\ -59.1111 \\ 9.0000 \\ 4.1111 \end{bmatrix}$$

The solution of the equations is

$$\hat{\beta}' = [5.5429 \;\; 0.26952 \;\; -0.24095 \;\; 1.2905 \;\; -0.51905]$$

and $R(\beta) = \hat{\beta}'\mathbf{R} = 51.044$, the sum of squares for differences among the $AB$ subclasses. The sum of squares for linear $\times$ quadratic is

$$\frac{\hat{\beta}_5^2}{c^{55}} = \frac{(-0.51905)^2}{0.011415} = 23.602$$

where $c^{55}$ is the diagonal element in the inverse of the coefficient matrix $(\mathbf{C}^{-1})$ corresponding to the regression coefficient for the linear $\times$ quadratic polynomial term. Since the error mean square with 12 degrees of freedom is 2.172,

$$F = \frac{23.602}{2.172} = 10.87**$$

and this $F$ value is highly significant indicating that for these data the best-fitting surface involves all polynomial terms. The prediction equation is

$$\hat{Y}_{kl} = \hat{\beta}_0 + \hat{\beta}_1.(p_k - \bar{p}_A) + \hat{\beta}._1(p_l - \bar{p}_B) + \hat{\beta}._2(p_l - \bar{p}_B)^2$$

$$+ \hat{\beta}_{11}(p_k - \bar{p}_A)(p_l - \bar{p}_B) + \hat{\beta}_{12}(p_k - \bar{p}_A)(p_l - \bar{p}_B)^2$$

$$= 6.9771 + 5.5429(p_k - 1.5) + 2.6952(p_l - 4)$$

$$- 0.24095(p_l - 4)^2 + 1.2905(p_k - 1.5)(p_l - 4)$$

$$- 0.51905(p_k - 1.5)(p_l - 4)^2$$

and
$$\hat{\beta}_0 = 6.9771 = 4.8889 + 0.24095\left(\frac{9 + 1 + 16}{3}\right)$$

If the linear $\times$ quadratic interaction was not significant, the next step would be to eliminate the equation for $\beta_5$ above. When this is done

$$\hat{\beta}' = [1.8384 \quad 0.17982 \quad -0.22549 \quad -0.53015]$$

$$R(\beta) = 27.443$$

$$SS(A_L \times B_L) = 6.624$$

$$SS(B_Q) = 17.880 = R(\beta_3 \mid \beta_1, \beta_2, \beta_4)$$

and if neither the $A_L \times B_L$ interaction nor the $B_Q$ were significant, one would delete the equation for the least significant effect and proceed with the step-down procedure. Omitting the equation for $\beta_4$, the $A_L \times B_L$ interaction, we obtain the following results:

$$\hat{\beta}' = [1.6180 \quad 0.27287 \quad -0.20706]$$

$$R(\beta) = 20.819$$

$$SS(B_Q) = 15.354 = R(\beta_3 \mid \beta_1, \beta_2)$$

If $B_Q$ were not significant, one would delete the equation for $\beta_3$ and continue the step-down procedure. In this case,

$$\hat{\beta}' = [1.1776 \quad -0.05194]$$

$$R(\beta) = 5.465$$

$$SS(B_L) = 0.329$$

$$SS(A_L) = 0.070 = R(\beta_1 \mid \beta_2)$$

and when the equation for $\beta_2$ is omitted

$$\hat{\beta}_1 = 1.0750$$

$$R(\beta_1) = 5.136 = SS(A_L)$$

It should be noted that

$$R(\beta) = R(\beta_1) + R(\beta_2 \mid \beta_1) + R(\beta_3 \mid \beta_1, \beta_2)$$

$$+ R(\beta_4 \mid \beta_1, \beta_2, \beta_3) + R(\beta_5 \mid \beta_1, \beta_2, \beta_3, \beta_4)$$

**TABLE 5.46**   ANALYSIS OF VARIANCE
               FOR POLYNOMIALS

| Source | df | MS | F |
|--------|-----|--------|--------|
| *AB* subclasses | 5 | 10.209 | 4.70* |
| $A_L$ | 1 | 5.136 | 2.36 |
| $B_L$ | 1 | 0.329 | <1 |
| $B_Q$ | 1 | 15.354 | 7.07* |
| $A_L \times B_L$ | 1 | 6.624 | 3.05 |
| $A_L \times B_Q$ | 1 | 23.602 | 10.87** |
| Error | 12 | 2.172 | |

$$= 5.136 + 0.329 + 15.354 + 6.624 + 23.602$$

$$= 51.045$$

The ANOVA is presented in Table 5.46.

# EXERCISES

**5.1.** The data presented below have been gathered in experimental work conducted by Betty McGuire in the Department of Zoology at the University of Massachusetts. Three species of voles were used and the variable shown below represents the number of contacts with pups by dams in the three species. Six female voles were included in each species and the observations $Y_i$ are the number of contacts with pups over a two-day period for the six dams in each species. Ten 2-day periods were included, represented by $X$. These data will be used in several exercises. Using only the data from the first species, meadow voles, calculate the following:

(a) Linear regression coefficient.
(b) Prediction equation for the linear regression, plotting the regression line.
(c) Sum of the squares due to regression.
(d) Sum of the squares due to deviations from regression.
(e) Variance from regression.
(f) Standard error of the regression coefficient.
(g) $t$ test for the significance of the regression coefficient.
(h) $F$ test for the significance of the regression coefficient
(i) Ninety-five percent confidence interval to the regression coefficient.

| Species | | | | | |
|---------|---|---|---|---|---|
| Meadow vole | | Prairie vole | | Pine vole | |
| X | Y | X | Y | X | Y |
| 1 | 37 | 1 | 52 | 1 | 43 |
| 2 | 24 | 2 | 45 | 2 | 47 |
| 3 | 28 | 3 | 40 | 3 | 43 |
| 4 | 19 | 4 | 49 | 4 | 35 |
| 5 | 18 | 5 | 26 | 5 | 34 |
| 6 | 15 | 6 | 12 | 6 | 29 |

| Species | | | | | |
| --- | --- | --- | --- | --- | --- |
| Meadow vole | | Prairie vole | | Pine vole | |
| X | Y | X | Y | X | Y |
| 7 | 18 | 7 | 12 | 7 | 20 |
| 8 | 9 | 8 | 14 | 8 | 22 |
| 9 | 10 | 9 | 8 | 9 | 27 |
| 10 | 0 | 10 | 6 | 10 | 19 |

**5.2.** Using the values found in Exercise 5.1, calculate the individual deviations of the actual values from the predicted values, $Y - \hat{Y}$, and show that the sum of the deviations is zero and that the sum of the squares of the deviations, $\Sigma(Y - \hat{Y})^2$, is equal to $\Sigma\, d^2_{y.x}$ found in Exercise 5.1.

**5.3.** Calculate and plot the 95 percent confidence band about the regression line.

**5.4.** Using the data shown in Exercise 5.1, calculate the regression coefficient for the prairie vole group and test for the significance of the difference between the regression for the meadow voles and the regression for the prairie voles.

**5.5.** Again using the data from Exercise 5.1 and succeeding calculations, test for the significance of the difference between two intercepts or adjusted means (which is the same test) of the regression lines for the first two groups, the meadow voles and the prairie voles.

**5.6.** Including the third species of voles, test, by means of an $F$ test, whether there are significant differences among the regressions from the three species of voles.

**5.7.** The data presented below have been selected from the results of a 120-day feeding trial at the University of Florida by Dr. Joseph P. Tritschler. The weight gain $X_1$ (in kilograms/ day) and the yield grade $X_2$ (in percent) were measured on 10 animals on each of three treatments. Using these data, make the following calculations:
(a) Correlation between $X_1$ and $X_2$ for treatment 1.
(b) Standard error of the correlation coefficient and the $t$ test for the significance of the correlation coefficient.
(c) Confidence interval for the correlation coefficient.
(d) Correlation between $X_1$ and $X_2$ for treatment 2.
(e) $t$ test for the significance of the difference between the two correlation coefficients.
(f) Correlation between $X_1$ and $X_2$ for treatment 3.
(g) Test of significance of the homogeneity of the three correlation coefficients.
(h) Average correlation coefficient.

| Treatment 1 | | Treatment 2 | | Treatment 3 | |
| --- | --- | --- | --- | --- | --- |
| $X_1$ | $X_2$ | $X_1$ | $X_2$ | $X_1$ | $X_2$ |
| 1.33 | 49.6 | 0.99 | 51.3 | 1.24 | 50.2 |
| 1.14 | 50.8 | 0.91 | 50.2 | 0.86 | 52.6 |
| 1.28 | 50.8 | 0.53 | 50.8 | 0.99 | 48.4 |
| 1.07 | 51.8 | 1.01 | 51.1 | 1.05 | 51.0 |
| 0.72 | 49.2 | 0.74 | 51.3 | 1.14 | 49.1 |
| 1.28 | 50.2 | 0.88 | 50.9 | 1.12 | 48.5 |
| 0.99 | 51.2 | 0.88 | 50.1 | 1.49 | 49.3 |

| Treatment 1 | | Treatment 2 | | Treatment 3 | |
|---|---|---|---|---|---|
| $X_1$ | $X_2$ | $X_1$ | $X_2$ | $X_1$ | $X_2$ |
| 1.35 | 46.8 | 0.79 | 50.0 | 0.83 | 50.0 |
| 0.92 | 50.2 | 0.92 | 50.3 | 1.10 | 49.1 |
| 1.42 | 47.8 | 0.84 | 49.6 | 1.06 | 50.1 |

5.8. The data in this exercise were collected by Dr. Heinrich Fenner at the University of Massachusetts in a feeding trial with dairy cattle. Several constituents of rumen fluid were measured, two of which are presented here. Each of the $Y_1$ values is a mean of 16 measurements of $n$-valeric acid percent by weight of total volatile fatty acids, and each of the $Y_2$ values is a mean of 16 measurements of propionic acid percent by weight of total volatile fatty acids. The samples from which the measurements were made were taken from the rumen at hourly intervals from 7 a.m. to 3 p.m. (the $X$ values). Using these data, calculate the linear and quadratic regressions of $Y_1$ on $X$ through use of the cross-products formulas and test for the significance of each by means of a $t$ test and an $F$ test. Develop the appropriate prediction equation and plot the quadratic regression line. Next, calculate the linear, quadratic, and cubic regressions of $Y_2$ on $X$ through use of the cross-products formulas and test for the significance of each by means of a $t$ test and an $F$ test. Develop the appropriate equation and plot the cubic regression line.

| $X$: | 1 | 2 | 3 | 4 | 5 | 6 | 7 | 8 | 9 |
|---|---|---|---|---|---|---|---|---|---|
| $Y_1$: | 1.18 | 1.34 | 1.40 | 1.36 | 1.32 | 1.24 | 1.26 | 1.18 | 1.14 |
| $Y_2$: | 18.76 | 19.67 | 19.61 | 19.36 | 18.81 | 18.85 | 18.69 | 18.55 | 18.66 |

5.9. Orthogonal polynomial coefficients are most frequently used when fitting a regression line to treatment means, since there is generally a limited number of means with equal intervals between them. However, they can also be used when each $X$ is accompanied by a single $Y$ as is the case in Exercise 5.8, when the number of observations is not too great and the intervals are equal. Therefore, using orthogonal polynomial coefficients, calculate the sums of squares for the linear and quadratic regression of $Y_1$ on $X$ and for the linear, quadratic, and cubic regression of $Y_2$ on $X$ for the data in Exercise 5.8. Develop the prediction equation for the quadratic regression of $Y_1$ on $X$ and show that the predicted values are the same as those calculated previously. Similarly, develop the prediction equation for the cubic regression of $Y_2$ on $X$ and show that the predicted values are the same as found earlier.

5.10. Referring to the data presented in Exercise 5.8:
   (a) Calculate the linear prediction equation of $Y_1$ on $X$, using orthogonal polynomial coefficients.
   (b) Calculate the linear prediction equation of $Y_1$ on $X$, using the cross-products formula.
   (c) Show the transformation from the prediction equation using orthogonal polynomial coefficients to the equation found by the use of the cross-products formula.

5.11. Suppose that the levels of treatment in a balanced experiment are 1, 3, 6, and 8. Calculate the orthogonal polynomial coefficients for the linear, quadratic, and cubic regression.

5.12. The data presented below have been selected from an experiment conducted by Kroll et al. (1983) in the field of exercise science. $X_1$ represents the body weight (in kilograms) of 12 female subjects, $X_2$ represents the length of the hand (in centimeters), $X_3$ represents the girth of the deltoid (in centimeters), and $Y$ represents the isometric strength of the forearm extension (in kilograms).

(a) Write out the normal equations for this multiple regression problem, invert the coefficient matrix, and solve for the partial regression coefficients.

(b) Calculate the reduction in the sum of squares, which is the sum of squares for the three independent variables considered simultaneously. Subtract this reduction sum of squares from the total uncorrected sum of squares to find the residual sum of squares and test for the significance of the three regressions by means of the analysis of variance.

(c) Calculate the standard error of each partial regression coefficient and test each coefficient for significance by means of the $t$ test, using the error variance from the analysis of variance table. Calculate the sum of squares for each partial regression coefficient by squaring the coefficient and dividing by the appropriate inverse element. Test for the significance of each coefficient in an analysis of variance table. Check that $F = t^2$.

(d) Complete a multiple regression analysis including only $X_1$ and $X_2$ and show that the difference between the reduction in this analysis and the reduction in the previous analysis is equal to the sum of squares for the regression due to $X_3$.

(e) Rank the partial regression coefficients for their relative importance by calculating the standard partial regression coefficients for $b_{Y1.23}$, $b_{Y2.13}$, and $b_{Y3.12}$.

(f) Calculate the coefficient of multiple determination for the analysis including the three independent variables.

| Subject | $X_1$ | $X_2$ | $X_3$ | $Y$ |
|---------|-------|-------|-------|------|
| 1 | 57.6 | 19.0 | 27.0 | 25.5 |
| 2 | 56.5 | 18.8 | 24.2 | 14.0 |
| 3 | 45.8 | 20.5 | 31.2 | 26.0 |
| 4 | 59.4 | 18.0 | 26.2 | 19.5 |
| 5 | 55.3 | 19.2 | 24.2 | 16.8 |
| 6 | 54.7 | 18.5 | 27.8 | 18.5 |
| 7 | 58.0 | 19.0 | 26.0 | 24.8 |
| 8 | 53.5 | 18.0 | 23.5 | 17.0 |
| 9 | 55.8 | 19.0 | 22.0 | 17.5 |
| 10 | 56.8 | 18.0 | 24.8 | 16.0 |
| 11 | 56.7 | 20.0 | 24.0 | 24.0 |
| 12 | 55.4 | 18.5 | 25.2 | 25.8 |

**5.13.** The data presented below are a portion of the observations collected by Dr. Joseph P. Tritschler at the University of Florida, resulting from a 120-day feeding trial with beef cattle steers. The values from 10 steers are included, where $X_1$ is the weight gain (in kilograms/day), $X_2$ is the fat gain (in kilograms/day), $X_3$ is the yield grade (in percent), and $X_4$ is the backfat thickness (in centimeters). Using these data, calculate the following:

(a) $r_{12}$, $r_{13}$, $r_{14}$, $r_{23}$, $r_{24}$, and $r_{34}$.

(b) The $t$ tests of significance for all correlation coefficients.

(c) The partial correlation coefficients $r_{12.3}$, $r_{14.3}$, $r_{24.3}$, and $r_{12.34}$.

(d) The $t$ tests for the significance of the partial correlation coefficients.

Set up the correlation matrix including only the first three variables, invert this matrix, and, using the inverse elements, calculate $r_{12.3}$, $r_{13.2}$, and $r_{23.1}$. Next, set up the correlation matrix including all four variables, invert the matrix, and, using the inverse elements, calculate $r_{12.34}$, $r_{13.24}$, $r_{14.23}$, $r_{23.14}$, $r_{24.13}$, and $r_{34.12}$.

| Animal number | $X_1$ | $X_2$ | $X_3$ | $X_4$ |
|:---:|:---:|:---:|:---:|:---:|
| 43 | 1.33 | 0.76 | 49.6 | 0.90 |
| 44 | 1.14 | 0.61 | 50.8 | 1.00 |
| 45 | 1.28 | 0.74 | 50.8 | 0.75 |
| 46 | 1.07 | 0.54 | 51.8 | 0.90 |
| 47 | 0.72 | 0.48 | 49.2 | 0.90 |
| 48 | 1.28 | 0.88 | 50.2 | 1.00 |
| 49 | 0.99 | 0.58 | 51.2 | 0.75 |
| 50 | 1.35 | 0.97 | 46.8 | 1.75 |
| 51 | 0.92 | 0.53 | 50.2 | 1.15 |
| 52 | 1.42 | 0.73 | 47.8 | 1.65 |

**5.14.** Using the reduction sum of squares and the total sum of squares found in Exercise 5.12, calculate the coefficient of multiple determination and the multiple correlation coefficient.

**5.15.** The table below presents a portion of the data collected by Charles F. Mancino and Professor Joseph Troll of the Plant and Soil Science Department at the University of Massachusetts in a study of the effect of the source of nitrogen on turf grass. Two rates of nitrogen were applied and the soil was sampled at three depths. The observations are percent of total soil nitrogen.

(a) Complete an analysis of variance for this set of data.

(b) Analyze the data a second time, this time using a multiple regression analysis where $X_1$ is the linear effect of rate of application, $X_2$ is the linear effect of depth, $X_3$ is the quadratic effect of depth, $X_4$ is the linear $\times$ linear interaction, and $X_5$ is the linear $\times$ quadratic interaction. Use the appropriate orthogonal polynomial coefficients as dummy variables for the $X$'s as shown in Section 5.14.3. Test for each of the five partial regression coefficients by means of $F$ tests in an analysis of variance table.

| | Depth | | |
|:---:|:---:|:---:|:---:|
| Rate | 0–8.5 cm | 8.5–17 cm | 17–25.5 cm |
| 1 | 0.10 | 0.19 | 0.09 |
|   | 0.10 | 0.21 | 0.10 |
|   | 0.11 | 0.18 | 0.11 |
|   | 0.10 | 0.18 | 0.11 |
| 2 | 0.09 | 0.19 | 0.12 |
|   | 0.11 | 0.17 | 0.10 |
|   | 0.10 | 0.21 | 0.08 |
|   | 0.18 | 0.20 | 0.10 |

**5.16.** For the data given on fleece weights in Exercise 4.8 complete the least-squares analysis under the two-way classification model with interaction. Give the estimates of constants, least-squares means, and standard errors. Using procedures described in Section 5.17, partition the sums of squares for years and ages (using codes of 2, 3, and 4) by fitting orthogonal polynomials. In each case, give the prediction equation for the "best-fitted" curve. Using procedures described in Section 5.18, partition the sum of squares for year $\times$ age subclasses into single degree polynomials for years, ages, and for the year $\times$ age interaction. Give the prediction equation for the "best-fitted" surface. Use the probability level of 0.10 in the step-down procedure. Interpret your results.

# Chapter 6

# Experimental Designs

## 6.1 INTRODUCTION

Most of the material dealt with in Chapters 1–5 has been leading up to the design of experiments. While the methods of analysis which have been discussed are commonly used for dealing with data collected with no design in mind, it is vastly preferable, whenever possible, to plan in advance a specific design. Such designs must take into account the resources available in terms of such things as materials, equipment, and financial support. Within the restrictions of available resources the researcher must plan the experiment so as to obtain the most desirable precision. By precision we are referring to the estimation of treatment effects with enough accuracy or sensitivity to detect significance, if it exists, at the level at which the null hypothesis is to be tested. It should be the intent of the experimenter to plan the experiment so that if differences of a particular magnitude exist they will be declared significant by the application of a test of the null hypothesis.

It should be noted that while effects in an analysis may be shown to be statistically significant or even highly significant they may not be of any practical importance. In many cases, significance may appear but the proportion of the total sum of squares accounted for by the effect is minor compared to other effects in the model. Thus, the careful interpretation of results is a vital aspect of statistical analysis.

The statistician, of course, always urges that appropriate time be spent in planning the experiment to prevent the frequent waste of time and energy that occurs in conducting experiments which have little hope of yielding the desired conclusion, or answering the questions posed by the experimenter. When any experiment, other than the most simple, is planned, it is advisable to write out the appropriate analysis of variance table including the sources of variation, degrees of freedom, and the expectations of mean squares. On many occasions this will show that there is no exact test for the treatments under consideration or such a poor test in terms of degrees of freedom that a reconsideration of the original plan is necessary.

One of the prime considerations in the selection of an appropriate design is to reduce the experimental error as much as possible to improve the power of a given test of significance. There are many definitions and discussions presented in textbooks of *experimental error,* by which is meant the error variance used for testing for the significance of differences due to the different sources of variation involved in the analysis of variance. The most useful explanation of this error variance (or experimental error) is that presented by R. A. Fisher which states that the correct error variance for testing the variance among a set of treatment means is one which contains all the sources of variation inherent in the variation among treatment means except that portion of the variance due specifically to the treatments themselves. We have made use of this definition in Section 3.11 where we have developed the appropriate *F* ratios for null hypotheses based on the expectations of mean squares.

In a book such as this, it is not possible to treat in detail all the vital considerations of experimental design selection or to discuss the very large number of designs available to the experimenter. Texts which deal specifically with experimental designs may be consulted, primarily those of Cochran and Cox (1957) and Federer (1955).

## 6.2  COMPLETELY RANDOMIZED DESIGN

In the completely randomized design, treatments are allotted or assigned to the experimental units in a random fashion; that is, it is a matter of chance as to which treatment a unit is assigned. This results in a one-way classification of data which has been previously discussed. While randomization can be carried out by different methods, the most common one is by use of a random number table (Table A.1). Random number tables normally present series of five-digit numbers arranged in numbered rows and columns. Suppose that we have available 12 experimental units which we wish to assign at random to 3 different treatments. As a first step we assign each unit a number from 1 to 12. We now go to a table of random numbers and select a starting place by any convenient method, one of which is merely sticking a finger on the page and selecting the nearest five-digit figure. We can use the first two digits for the row in which to start and the third and fourth to select a column. Using the selected starting place, one can proceed down the column or up the column and select the first three digits of succeeding numbers. Three-digit groups are selected rather than two to reduce the opportunity for ties. Suppose that we find the 12 numbers found in Table 6.1. Ranks are now assigned to the selected numbers; 992 being the largest number is ranked first, followed by 905, and so on. The first four are taken to be the numbers of the experimental units to which treatment one is applied, the next four numbers to be used on treatment two, and the last four numbers on treatment three. Other

**TABLE 6.1**  TWELVE RANDOMLY SELECTED NUMBERS WITH THEIR RANKS

| Number | 762 | 832 | 171 | 225 | 992 | 376 | 869 | 100 | 749 | 905 | 543 | 126 |
|--------|-----|-----|-----|-----|-----|-----|-----|-----|-----|-----|-----|-----|
| Rank   | 5   | 4   | 10  | 9   | 1   | 8   | 3   | 12  | 6   | 2   | 7   | 11  |

**TABLE 6.2**   DISTANCE (IN CENTIMETERS) IN THE AIR
TRAVELED BY FISH RESULTING FROM
FORCED AIR IN A SORTING PROCESS

|  | Species 1 | Species 2 | Species 3 |
|---|---|---|---|
|  | 142 | 164 | 182 |
|  | 154 | 168 | 191 |
|  | 151 | 156 | 184 |
|  | 138 | 161 | 170 |
|  | 157 | 172 | 176 |
| Totals | 742 | 821 | 903 |

schemes for using the tables of random numbers for randomization procedures can be developed, of course.

While the experiment may be conducted on experimental units selected from one population to be allotted at random to different treatments, another situation also results in a completely randomized design. In this case, experimental units from different strains, different manufacturing designs, different IQ levels, for example, may be assigned to the same experimental condition. In this case, the different strains, designs, or IQ levels become the treatments. The necessary randomization would be the random selection of units from each of the groups to appear in the experiment. Thus, if it were desired to see which breed of beef cattle performed best under a particular management scheme, animals would be selected from several different breeds by a random procedure.

The design is an easy one to plan and conduct, with the analysis of the results completed readily. No problem is encountered with unequal numbers of observations in the different treatment groups, as is the case with most other designs. In addition, more degrees of freedom are available for experimental error than with any other type of design.

It is possible, however, that other sources of variation may be identified in an experimental situation. Other designs are able to remove additional variation and so must be considered if such variation is believed to exist. As an example of the completely randomized design, let us consider the data displayed in Table 6.2, which were adapted from experimental work conducted by Dr. Lester Whitney of the Department of Food and Agricultural Engineering at the University of Massachusetts. The appropriate analysis of variance is shown in Table 6.3. It can be seen that highly significant differences exist among the treatments.

## 6.3  RANDOMIZED BLOCK DESIGN

One of the primary objectives in experimental design is the reduction of the experimental error and this can often be accomplished by a grouping or blocking of experimental units. For example, if a field experiment is being planned it is usually advan-

**TABLE 6.3**   ANALYSIS OF VARIANCE OF DATA
DISPLAYED IN TABLE 6.2

| Source | df | SS | MS | F |
|---|---|---|---|---|
| Among treatments | 2 | 2592 | 1296 | 23.14** |
| Within treatments | 12 | 670 | 56 | |

tageous to select blocks of land which are then divided into plots to which treatments are applied, rather than having a large number of plots scattered about the field to which treatments are applied at random. The number of plots in such a block is usually equal to the number of treatments or some multiple of the number of treatments. The idea of grouping or blocking is to have differences among the units in a block as small as possible with the differences among the blocks expected to be greater than the differences among units within a block. Thus, since each treatment appears in each block, observed differences will largely be due to treatments. The block differences represent an extraneous source of variation and it is hoped that enough of this variation is removed from the experimental error so that the error is reduced over what it would have been without blocking. Since each treatment does appear in each block, variation among the blocks does not affect treatment differences. Of course, treatments are assigned to the units within a block by some random procedure.

In many experiments that use animals as experimental units, the animals are placed in what are called *outcome groups,* or blocks, on the basis of such characteristics as age, weight, or condition. In setting up outcome groups, the heavier, or older animals are assigned to one group and the treatments are allotted to these animals at random. The next lighter or younger animals are assigned to the second group, and so on.

If the blocking is successful in removing a significant amount of variation from the experimental error, the precision of the experiment will be increased. When controls are included in the experiment, it is sometimes advantageous to have more than one control assigned in each block in order to have a better comparison between the treated and untreated groups. With too many treatments in a block, however, a large amount of variation within a block not brought about by the treatments can occur, causing a large experimental error. In such a case it would be wise to consider another type of design.

It is possible, of course, to assign more than one unit in a block to each treatment, or to have more than one determination or observation on each unit which leads to a sampling error, as well as an experimental error where the blocks are considered as random effects. However, if the blocks are considered as fixed effects, as they may well be in some cases, the mean square resulting from differences among similar treatments in the same block or differences among more than one observation on each unit becomes the experimental error. It is possible to have a random model where both blocks and treatments are considered as random effects but this would be an uncommon circumstance. The expectations of mean squares for the various combinations are shown in Table 6.4. Since we have treated data of a similar nature in Sections 3.2 and 3.3, no example will be given here.

**TABLE 6.4**  EXPECTATIONS OF MEAN SQUARES WITH VARIOUS
MODELS OF THE RANDOMIZED COMPLETE BLOCK DESIGN

### A. Single unit per treatment and single observation per unit

|  | Model | | |
|---|---|---|---|
| Source | Random $E(MS)$ | Mixed $E(MS)$ | Fixed $E(MS)$ |
| Blocks ($B$) | $\sigma_w^2 + \sigma_{bt}^2 + n_t\sigma_b^2$ | $\sigma_w^2 + n_t\sigma_b^2$ | $\sigma_w^2 + n_t\sigma_b^2$ |
| Treatments ($T$) | $\sigma_w^2 + \sigma_{bt}^2 + n_b\sigma_t^2$ | $\sigma_w^2 + \sigma_{bt}^2 + n_b\sigma_t^2$ | $\sigma_w^2 + n_b\sigma_t^2$ |
| $BT$ | $\sigma_w^2 + \sigma_{bt}^2$ | $\sigma_w^2 + \sigma_{bt}^2$ | $\sigma_w^2 + \sigma_{bt}^2$ |

### B. Two or more units per treatment or two or more observations per unit

|  | Model | | |
|---|---|---|---|
| Source | Random $E(MS)$ | Mixed $E(MS)$ | Fixed $E(MS)$ |
| Blocks ($B$) | $\sigma_w^2 + n_w\sigma_{bt}^2 + n_w n_t\sigma_b^2$ | $\sigma_w^2 + n_w n_t\sigma_b^2$ | $\sigma_w^2 + n_w n_t\sigma_b^2$ |
| Treatments ($T$) | $\sigma_w^2 + n_w\sigma_{bt}^2 + n_w n_b\sigma_t^2$ | $\sigma_w^2 + n_w\sigma_{bt}^2 + n_w n_b\sigma_t^2$ | $\sigma_w^2 + n_w n_b\sigma_t^2$ |
| $BT$ | $\sigma_w^2 + n_w\sigma_{bt}^2$ | $\sigma_w^2 + n_w\sigma_{bt}^2$ | $\sigma_w^2 + n_w\sigma_{bt}^2$ |
| $W$ | $\sigma_w^2$ | $\sigma_w^2$ | $\sigma_w^2$ |

## 6.4  GAIN IN EFFICIENCY: RANDOMIZED BLOCK VERSUS COMPLETELY RANDOMIZED DESIGN

When designing an experiment we are concerned with using the most efficient design possible, given the available resources. One method of assuring that we have done so is to estimate the gain in efficiency when selecting one design over another. If an experiment has been carried out using the randomized block design, we can compare its efficiency to the same experiment had it been conducted as a completely randomized design by calculating a ratio of the relative amounts of information from the two designs. The amount of information of an experiment is defined as the reciprocal of the error variance (Federer, 1955). We can compare the relative efficiency of the randomized block to the completely randomized design by methods shown by Cochran and Cox (1957) and Steel and Torrie (1980).

We can estimate the error mean square for the completely randomized design by the formula

$$MS_e(CR) = \frac{f_b MS_b + (f_t + f_e)MS_e}{f_b + f_t + f_e = \text{total df}} \tag{6.1}$$

where $MS_b$ and $MS_e$ are the mean squares for the blocks and error and $f_b$, $f_t$, and $f_e$ are the block, treatment, and error degrees of freedom, respectively. We can now estimate the relative efficiency of the randomized block and the completely randomized design as

$$E(\text{RB to CR}) = \frac{1/\text{MS(RB)}}{1/\text{MS(CR)}} = \frac{\text{MS(CR)}}{\text{MS(RB)}} \qquad (6.2)$$

If we multiply this figure by 100, we obtain what is known as the *precision factor.*

However, when we use the randomized block design the number of degrees of freedom for error has been decreased, making it more difficult to obtain significant differences among the treatments. When the degrees of freedom for error for the randomized block are fewer than 20 this is taken into account by multiplying the precision factor by

$$\frac{(f_1 + 1)/(f_1 + 3)}{(f_2 + 1)/(f_2 + 3)} = \frac{(f_1 + 1)(f_2 + 3)}{(f_1 + 3)(f_2 + 1)} \qquad (6.3)$$

where $f_1$ and $f_2$ are the error degrees of freedom for the randomized block and completely randomized design, respectively.

As an example of this type of comparison let us look at the analysis of a randomized block design shown in Table 3.3. Using formula (6.1), we find an estimate of the value that would have been obtained for the mean square for error had we used a completely randomized design,

$$\text{MS}_e(\text{CR}) = \frac{(7)(31.57) + (3 + 21)(44.58)}{7 + 3 + 21} = 41.64$$

Now using formula (6.2) and multiplying the result by 100, we obtain the precision factor,

$$\frac{\text{MS}_e(\text{CR})}{\text{MS}_e(\text{RB})} (100) = \frac{41.64}{44.58} (100) = 93.41 \text{ percent}$$

Although we do have more than 20 degrees of freedom, let us nevertheless apply the correction for the loss of degrees of freedom by the correction term shown in formula (6.3); we find

$$\frac{(21 + 1)(28 + 3)}{(21 + 3)(28 + 1)} (93.41) = 91.53 \text{ percent}$$

The results indicate that the blocking has not been effective in this case because the randomized block is estimated to be only about 92 percent as efficient as the completely randomized design would have been. This is borne out by noting that the sum of squares accounted for by the blocks is relatively small and not great enough to affect the loss encountered by the fewer degrees of freedom available for error.

## 6.5 MISSING VALUES IN THE RANDOMIZED BLOCK DESIGN

When only one or two observations are missing in a randomized block experiment, these values can be estimated and an approximate analysis of variance completed in the usual fashion, taking into account the missing value(s) where appropriate. When this is done, the assumption is made that the loss of data is not due to treatment or block effects. When more than two values are missing, it is usually easier to use a different method of analysis, for example, the method of fitting constants (see Chapter 2). Yates (1933) has presented a method for estimating missing values as follows.

When a single value is missing, an estimate can be obtained by applying the formula

$$Y = \frac{rB + tT - G}{(r - 1)(t - 1)} \tag{6.4}$$

where $r$ and $t$ are the number of blocks and treatments, $B$ and $T$ are the totals of the observations in the block and treatment containing the missing value, and $G$ is the grand total of all the observations.

The analysis of variance can now be completed including the estimated value. However, one degree of freedom must be subtracted from the total degrees of freedom and the error degrees of freedom because of the estimated value. In the resulting analysis the error sum of squares is the appropriate least-squares sum of squares but the treatment and block sums of squares are biased upward. With only one missing value this bias in the treatment sum of squares can be measured by the equation

$$\frac{[B - (t - 1)Y]^2}{t(t - 1)} \tag{6.5}$$

as shown by Steel and Torrie (1980), where $Y$ is calculated by formula (6.4). The upward bias in block sum of squares can be calculated in a similar way. Subtracting the bias from the calculated sum of squares then allows an exact test of significance.

If there is more than one missing observation, values are first approximated for all units except one. The values can be approximated by

$$\frac{\bar{y}_{i.} + \bar{y}_{.j}}{2} \tag{6.6}$$

where $\bar{y}_{i.}$ and $\bar{y}_{.j}$ are the means of the known values for the treatment and block containing any one of the missing values. After we approximate all values except one by using formula (6.6), the remaining value is estimated by Equation (6.4). Using this approximation and the values previously developed by Equation (6.6) to all but one of the remaining values, we again use Equation (6.4) to approximate this one.

A complete cycle is made so that each missing observation is approximated by Equation (6.4) and then a second approximation is made in the same sequence as previously used. This procedure is continued until the new approximations are essentially the same as the previous ones. Usually, two cycles are sufficient.

The analysis of variance is now completed using the estimated values together with the observed values. One degree of freedom is subtracted from both the total and error degrees of freedom for each missing value.

To illustrate the method used with two or more missing observations, let us use the data shown in Table 6.5, supplied by Dr. William J. Lord of the Plant and Soil Science Department of the University of Massachusetts, altered slightly for this example.

We can estimate the two missing observations as follows:

$$a = \frac{\bar{y}_{1.} + \bar{y}_{2.}}{2} = \frac{3.1850 + 2.2133}{2} = 2.6992$$

**TABLE 6.5**  FRUIT SET OF MUTSU APPLES PER CUBIC CENTIMETER OF LIMB CIRCUMFERENCE

| Treatment | Replicates[a] | | | | | Sums |
|---|---|---|---|---|---|---|
| | 1 | 2 | 3 | 4 | 5 | |
| A | 4.79 | a | 3.42 | 2.38 | 2.15 | 12.74 |
| B | 4.42 | 1.87 | 2.39 | 1.83 | 1.26 | 11.77 |
| C | 3.61 | 3.45 | 2.17 | 1.94 | 2.27 | 13.44 |
| D | 0.61 | 1.32 | 0.57 | b | 1.21 | 3.71 |
| Sums | 13.43 | 6.64 | 8.55 | 6.15 | 6.89 | 41.66 |

[a] $a$ and $b$ represent missing values.

$$b(\text{first cycle}) = \frac{rB + tT - G}{(r-1)(t-1)} = \frac{(5)(6.15) + (4)(3.71) - 44.3592}{(5-1)(4-1)}$$

$$= \frac{0.4559}{12} = 0.0380$$

Note that $G = 41.66 + 2.6992 = 44.3592$.

$$a(\text{first cycle}) = \frac{(5)(6.64) + (4)(12.74) - 41.6980}{12} = \frac{42.4620}{12} = 3.5385$$

Note that $G = 41.66 + 0.0380 = 41.6980$.

$$b(\text{second cycle}) = \frac{(5)(6.15) + (4)(3.71) - 45.1985}{12} = \frac{0.3915}{12} = 0.0326$$

Note that $G = 41.66 + 3.5385 = 45.1985$.

$$a(\text{second cycle}) = \frac{(5)(6.64) + (4)(12.74) - 41.6926}{12} = \frac{42.4674}{12} = 3.5390$$

Note that $G = 41.66 + 0.0326 = 41.6926$.

$$b(\text{third cycle}) = \frac{(5)(6.15) + (4)(3.71) - 45.1990}{12} = \frac{0.3910}{12} = 0.0326$$

Note that $G = 41.66 + 3.5390 = 45.1990$.

$$a(\text{third cycle}) = \frac{(5)(6.64) + (4)(12.74) - 41.6926}{12} = \frac{42.4674}{12} = 3.5390$$

Note that $G = 41.66 + 0.0326 = 41.6926$.

Rounding these two values of $a$ and $b$ to the second decimal place similar to the values in Table 6.4, we can enter them into the table and the analysis of variance is completed. Two degrees of freedom must be subtracted from the total and error degrees of freedom which would be obtained using the estimated values.

It must also be noted that while the error mean square is unbiased, the treatment (as well as block) mean squares are biased upward. This bias should not be great with only one or two missing data but could be noticeable with several missing observations.

An exact test of significance for the treatment and block effects is available by using the method of covariance. However, with the large number of computer programs handling data with disproportionate subclass numbers, the most practical solution is to use one of these programs when missing data are involved.

## 6.6  LATIN SQUARE DESIGN

The Latin square design is used frequently in many areas of research and is extremely useful when applied under the appropriate conditions. While the randomized block design is considered as a single grouping, the Latin square allows for double grouping. This double grouping allows for the removal of two extraneous sources of variation, sometimes referred to as nuisance variables.

In the Latin square design, each row and each column are a complete replicate or block, so each treatment occurs once and only once in each row and each column.

The following are Latin squares:

| | Column | | |  | | Column | | | |
|---|---|---|---|---|---|---|---|---|---|
| Row | 1 | 2 | 3 | | Row | 1 | 2 | 3 | 4 |
| 1 | A | B | C | | 1 | A | B | C | D |
| 2 | B | C | A | | 2 | B | C | D | A |
| 3 | C | A | B | | 3 | C | D | A | B |
| | $3 \times 3$ | | | | 4 | D | A | B | C |
| | | | | | | $4 \times 4$ | | | |

where the letters apply to the different treatments.

These squares could be laid out in a field where rows and columns would account for soil variation in two directions, for example, or in an animal experiment the columns could be animals and the rows time periods.

In the Latin square design no valid test of interaction among the three sources of variation—rows, columns, and treatments—is possible and so unless the experimenter is prepared to assume that these interactions do not exist, this design should not be used.

The primary drawback to the use of the Latin square is that the number of rows, columns, and treatments must be equal. The most common squares are $4 \times 4$ through $8 \times 8$. Very rarely are larger squares used.

Small squares yield very few degrees of freedom for error and so must give a substantial decrease in the experimental error to compensate. However, the degrees of freedom for error can be increased by using more than one square in an experiment.

We will use as an example some data supplied by Dr. Heinrich Fenner of the Department of Veterinary and Animal Sciences of the University of Massachusetts.

The data resulted from an experiment designed to measure the influence of different rations on several rumen fluids in dairy cattle. Each of four fistulated cows were placed on four different rations for a one-month period. At the end of the first period the cows were changed to a different ration, four periods of one month each were used. Near the end of each period samples were taken at 10 a.m. on each of five days. In this analysis we concern ourselves with only the means of the five days. These means are shown in Table 6.6 for one of the 14 dependent variables included in the study.

The mathematical model for a Latin square is written differently by various statisticians. One way of writing the model is

$$Y_{ij(t)} = \mu + r_i + c_j + t_{(t)} + e_{ij} \qquad (6.7)$$

where the $r_i$ represent row effects, the $c_j$ represent column effects, and the $t_{(t)}$ represent treatment effects. The use of $(t)$ implies that the ordinary three-way classification is not involved. The $i$ and $j$ subscripts completely identify an observation as to row and column but not as to treatment. Another way of writing the model is

$$Y_{ijh} = \mu + r_i + c_j + t_h + e_{ijh} \qquad (6.8)$$

Again, the use of the subscript $h$ rather than $k$ is to indicate that the usual three-way classification is not involved. This text will use the model as written in (6.7).

The calculations for this set of data would be

$$\text{Total SS} = \sum_i \sum_j Y_{ij}^2 - \frac{Y_{..}^2}{N} = \sum_i \sum_j Y_{ij}^2 - (1) = rc - (1)$$

$$= 1426.8841 - \frac{(150.5802)^2}{16} = 1426.8841 - 1417.1498 = 9.7343 \qquad (6.9)$$

$$SS_R = \frac{\sum_i Y_{i.}^2}{n_k} - \frac{Y_{..}^2}{N} = r - (1)$$

$$= \frac{(37.3980)^2 + (36.8918)^2 + \cdots + (36.8386)^2}{4} - (1)$$

$$= 1418.2856 - (1) = 1.1358 \qquad (6.10)$$

**TABLE 6.6** MEANS OF TOTAL VOLATILE FATTY ACIDS OF FOUR COWS TAKEN AT FOUR MONTHLY INTERVALS FOR RATIONS A, B, C, AND D

| Periods | Cows 1 | 2 | 3 | 4 | Totals |
|---------|--------|---|---|---|--------|
| 1 | A = 10.0854 | B = 10.2028 | C = 9.1842 | D = 7.9256 | 37.3980 |
| 2 | B = 9.3762 | C = 9.1898 | D = 8.8376 | A = 9.4882 | 36.8918 |
| 3 | C = 9.7070 | D = 10.2938 | A = 10.7468 | B = 8.7042 | 39.4518 |
| 4 | D = 8.4668 | A = 9.9844 | B = 10.0724 | C = 8.3150 | 36.8386 |
| Totals | 37.6354 | 39.6708 | 38.8410 | 34.4330 | 150.5802 |

Ration totals A = 40.3048    B = 38.3556    C = 36.3960    D = 35.5238

$$SS_C = \frac{\sum_j Y^2_{\cdot j}}{n_k} - \frac{Y^2_{\cdot \cdot}}{N} = c - (1)$$

$$= \frac{(37.6354)^2 + (39.6708)^2 + \cdots + (34.4330)^2}{4} - (1)$$

$$= 1421.1126 - (1) = 3.9628 \qquad\qquad (6.11)$$

$$SS_T = \frac{\sum_t Y^2_t}{n_k} - \frac{Y^2_{\cdot \cdot}}{N} = t - (1)$$

$$= \frac{(40.3048)^2 + (38.3556)^2 + \cdots + (35.5238)^2}{4} - (1)$$

$$= 1420.5595 - (1) = 3.4097 \qquad\qquad (6.12)$$

Error SS $= rc - r - c - t + (2)$

$$= SS_{total} - SS_R - SS_C - SS_T = 1.2260 \qquad\qquad (6.13)$$

where $Y_t$ is a treatment total, $n_k$ is the size of the square (number of rows, columns, or treatments), and (2) is two times the correction term.

Note that in the formulas for calculating the sums of squares, the denominators take on a different form owing to the peculiarities of this design. The total number of observations $N$ is now found by $n_k^2$. In spite of having three main effects only two subscripts are used in the numerators and one in the denominators for calculating the sums of squares for periods and cows. For the treatment, or ration effects, we merely use $Y_t$ and $n_k$. Table 6.7 shows the analysis of variance together with the expectations of mean squares.

In the Latin square design we do not have a three-way subclass involving the three classifications as we would have had with the previous analyses, since a three-way interaction is assumed not to exist. Instead, we have an experimental error sum of squares that is a residual, that portion of the total sum of squares not accounted for by the effects in the model. This line is therefore labeled error (sometimes residual or remainder) and the corresponding variance component is labeled $\sigma_e^2$, changing our earlier method of writing out expectations of mean squares. $\sigma_e^2$ appears in expectations of every mean square with a coefficient of 1.

Prior to the Latin square design, we were able to calculate the coefficients of the other variance components either as $N$ divided by the product of the $n$'s corresponding to the letters in the subscript of $\sigma^2$, or as the product of all $n$'s excluding those corre-

**TABLE 6.7**   ANALYSIS OF VARIANCE OF DATA DISPLAYED IN TABLE 6.5

| Source | df | SS | MS | F | E(MS) |
|---|---|---|---|---|---|
| Periods | $n_k - 1 = 3$ | 1.1358 | 0.3786 | 1.85 | $\sigma_e^2 + 4\sigma_r^2$ |
| Cows | $n_k - 1 = 3$ | 3.9628 | 1.3209 | 6.47* | $\sigma_e^2 + 4\sigma_c^2$ |
| Rations | $n_k - 1 = 3$ | 3.4097 | 1.1366 | 5.56* | $\sigma_e^2 + 4\sigma_t^2$ |
| Error | $(n_k - 1)(n_k - 2) = 6$ | 1.2260 | 0.2043 | | $\sigma_e^2$ |

sponding to the letters in the subscript of $\sigma^2$. The Latin square does not allow this option because the coefficients must be calculated as $N$ divided by the product of the $n$'s corresponding to the letters in the subscript of $\sigma^2$, or $n_k$. Thus, for the coefficient of $\sigma_r^2$ we would have $N/n_k = 16/4 = 4$.

The analysis of variance in Table 6.7 shows that there were significant differences among the treatments so we would proceed to use some mean separation procedure to obtain further information. The procedure used would depend on the type of treatments applied.

It is possible to have more than one observation or determination per cell in which case we would have an additional line in the analysis of variance table. For example, if two or more measurements of pH were made for each cow $\times$ period $\times$ ration cell, we would have a sampling error and the model would be

$$Y_{ij(t)k} = \mu + r_i + c_j + t_{(t)} + e_{ij} + d_{ijk} \qquad (6.14)$$

where $e_{ij}$ is the experimental error to be used to test periods, cows, and treatments and $d_{ijk}$ is the sampling error to be compared with the experimental error for a consideration of a future experiment. If we had $m$ random determinations in each cell our analysis would be that presented in Table 6.8, where $n_m$ is the number of observations in each cell. The expectations of mean squares are presented in Table 6.9.

It is a very common situation to use more than one Latin square in an experiment, particularly when small squares are involved. The analyses of such experiments are discussed in some detail by Fisher (1949, Sec. 65) and Federer (1955, p. 148). In most recommended analyses the rows and the columns are nested within the squares. Analyses will differ depending primarily on whether the squares are considered to be a random effect or a fixed effect and also, in the former case, whether there is a significant interaction between the squares and the treatments.

Normally, when a Latin square design is repeated several times to build up a greater number of degrees of freedom for the error mean square, the squares are considered to be a random effect. If the interaction between squares and treatments is not significant, the sum of squares for this interaction is pooled with the residual

**TABLE 6.8** ANALYSIS OF VARIANCE FOR LATIN SQUARE WITH A SAMPLING ERROR

| Source | df | SS |
|---|---|---|
| Periods | $n_k - 1 = 3$ | $\dfrac{\Sigma_i Y_{i..}^2}{n_k n_m} - \dfrac{Y_{...}^2}{n_k^2 n_m = N}$ |
| Cows | $n_k - 1 = 3$ | $\dfrac{\Sigma_i Y_{.j.}^2}{n_k n_m} - \dfrac{Y_{...}^2}{N}$ |
| Rations | $n_k - 1 = 3$ | $\dfrac{\Sigma_t Y_t^2}{n_k n_m} - \dfrac{Y_{...}^2}{N}$ |
| Experimental error | $(n_k - 1)(n_k - 2) = 6$ | $\dfrac{\Sigma_i \Sigma_j Y_{ij.}^2}{n_m} - \dfrac{Y_{...}^2}{N} - SS_P - SS_C - SS_T$ |
| Sampling error | $n_k^2(n_m - 1) = 16$ | $\Sigma_i \Sigma_j \Sigma_k Y_{ijk}^2 - \dfrac{\Sigma_i \Sigma_j Y_{ij.}^2}{n_m}$ |

**TABLE 6.9**   EXPECTATIONS OF MEAN SQUARES FOR LATIN
SQUARE WITH SAMPLING ERROR

| Source | General case | Example |
|---|---|---|
| Periods | $\sigma_d^2 + n_m\sigma_e^2 + \dfrac{N}{n_k}\sigma_r^2$ | $\sigma_d^2 + 2\sigma_e^2 + 8\sigma_r^2$ |
| Cows | $\sigma_d^2 + n_m\sigma_e^2 + \dfrac{N}{n_k}\sigma_c^2$ | $\sigma_d^2 + 2\sigma_e^2 + 8\sigma_c^2$ |
| Rations | $\sigma_d^2 + n_m\sigma_e^2 + \dfrac{N}{n_k}\sigma_t^2$ | $\sigma_d^2 + 2\sigma_e^2 + 8\sigma_t^2$ |
| Experimental error | $\sigma_d^2 + n_m\sigma_e^2$ | $\sigma_d^2 + 2\sigma_e^2$ |
| Sampling error | $\sigma_d^2$ | $\sigma_d^2$ |

sum of squares for an error term. One would usually not pool this sum of squares if significance is indicated at the 0.20 probability level. If this interaction is significant at this or lower levels of probability, the interaction mean square is used to test for the significance of the treatment effects.

If the squares are considered to be a fixed effect, as they are in the example shown in Section 6.7, then the square by treatment interaction sum of squares is not pooled with the residual, nor is it used as an error term.

Let us consider four $3 \times 3$ Latin squares included in each of the situations presented in Table 6.10 and note the differences in the details of each analysis. In all three cases, we have the sources of variation of squares, rows, columns, and treatments where $n_s$ is the number of squares and $n_k$ is the size of each square.

It is important to recognize that in each situation a different error term is used to test for the significance of treatment effects. This is critical not only in the first test of treatment effects, but also in any mean separation procedure that may be used beyond the original $F$ test.

## 6.7  ORTHOGONAL LATIN SQUARES

Frequently, experimenters make use of more than one Latin square to build up degrees of freedom for the error term. In other situations the experiment itself calls for more than one square, sometimes requiring a set of orthogonal squares. A set of Latin squares is said to be orthogonal if, when any two squares are superimposed, each letter of one square occurs once with every letter of the other square (Cochran and Cox, 1957). The use of orthogonal squares ensures that each treatment is preceded an equal number of times by each of the other treatments.

An experiment using three pairs of orthogonal $3 \times 3$ Latin squares was conducted by Timothy C. Quick at The Ohio State University. The study was concerned with the use of three different forms of forage by three different types of animal. The forms of the forage were $A$ = pelleted, $B$ = chopped (2–3 cm), and $C$ = long (15–23 cm). The three different types of animal were dairy goats, hair sheep, and wool sheep. Fistulated animals were used in one square of each pair and nonfistulated animals in

**TABLE 6.10**  DIFFERENT FORMS OF THE ANALYSIS OF VARIANCE FOR DESIGNS WITH SEVERAL LATIN SQUARES

| Source | df | $E$(MS) |
|---|---|---|
| | A.  Squares are a random effect: no interaction between squares and treatments | |
| $S$ | $n_s - 1 = 3$ | $\sigma_e^2 + 9\sigma_s^2$ |
| $R{:}S$ | $(n_k - 1)n_s = 8$ | $\sigma_e^2 + 3\sigma_{r{:}s}^2$ |
| $C{:}S$ | $(n_k - 1)n_s = 8$ | $\sigma_e^2 + 3\sigma_{c{:}s}^2$ |
| $T$ | $n_k - 1 = 2$ | $\sigma_e^2 + 12\sigma_t^2$ |
| Error | $n_s(n_k - 1)^2 - (n_k - 1) = 14$ | $\sigma_e^2$ |
| Total | $n_s n_k^2 - 1 = 35$ | |
| | B.  Squares are a random effect: a significant interaction between squares and treatments | |
| $S$ | $n_s - 1 = 3$ | $\sigma_e^2 + 9\sigma_s^2$ |
| $R{:}S$ | $(n_k - 1)n_s = 8$ | $\sigma_e^2 + 3\sigma_{r{:}s}^2$ |
| $C{:}S$ | $(n_k - 1)n_s = 8$ | $\sigma_e^2 + 3\sigma_{c{:}s}^2$ |
| $T$ | $n_k - 1 = 2$ | $\sigma_e^2 + 3\sigma_{st}^2 + 12\sigma_t^2$ |
| $ST$ | $(n_s - 1)(n_k - 1) = 6$ | $\sigma_e^2 + 3\sigma_{st}^2$ |
| Error | $n_s(n_k - 1)(n_k - 2) = 8$ | $\sigma_e^2$ |
| Total | $n_s n_k^2 - 1 = 35$ | |
| | C.  Squares are a fixed effect | |
| $S$ | $n_s - 1 = 3$ | $\sigma_e^2 + 9\sigma_s^2$ |
| $R{:}S$ | $(n_k - 1)n_s = 8$ | $\sigma_e^2 + 3\sigma_{r{:}s}^2$ |
| $C{:}S$ | $(n_k - 1)n_s = 8$ | $\sigma_e^2 + 3\sigma_{c{:}s}^2$ |
| $T$ | $n_k - 1 = 2$ | $\sigma_e^2 + 12\sigma_t^2$ |
| $ST$ | $(n_s - 1)(n_k - 1) = 6$ | $\sigma_e^2 + 3\sigma_{st}^2$ |
| Error | $n_s(n_k - 1)(n_k - 2) = 8$ | $\sigma_e^2$ |
| Total | $n_s n_k^2 - 1 = 35$ | |

the other. The design and results for one dependent variable are shown in Table 6.11. The mathematical model used for the analysis of this set of data is

$$Y_{ijk(t)} = \mu + s_i + a_{ij} + p_k + sp_{ik} + t_{(t)} + st_{i(t)} + e_{ijk} \qquad (6.15)$$

where the $s_i$ represent square effects, the $a_{ij}$ represent animals within squares, the $p_k$ represent periods, and the $t_{(t)}$ represent treatment effects. The sum of squares among the six squares was partitioned into sums of squares for between the fistulated and nonfistulated animals ($F$), among the types of animal ($K$), and the interaction between the two ($FK$). The partitioning makes this a rather large analysis with many sources of variation, so the development of several tables prior to the analysis is helpful in isolating the appropriate totals and subtotals (see Table 6.12).

The necessary calculations follow.

**TABLE 6.11**  PERCENTAGE OF DRY MATTER DIGESTIBILITY

### Goats

| | Fistulated | | | Nonfistulated | | |
|---|---|---|---|---|---|---|
| | Animal 102 | Animal 203 | Animal 305 | Animal 101 | Animal 204 | Animal 306 |
| Period 1 | A = 49.3 | B = 55.0 | C = 54.0 | A = 50.8 | B = 50.9 | C = 54.9 |
| Period 2 | B = 54.3 | C = 57.0 | A = 54.6 | C = 55.0 | A = 52.0 | B = 49.8 |
| Period 3 | C = 59.6 | A = 60.4 | B = 49.9 | B = 49.5 | C = 55.5 | A = 51.5 |
| Totals | 163.2 | 172.4 | 158.5 | 155.3 | 158.4 | 156.2 |
| | A = 164.3 | B = 159.2 | C = 170.6 | A = 154.3 | B = 150.2 | C = 165.4 |

### Wool sheep

| | Fistulated | | | Nonfistulated | | |
|---|---|---|---|---|---|---|
| | Animal 108 | Animal 210 | Animal 311 | Animal 107 | Animal 209 | Animal 312 |
| Period 1 | A = 53.9 | B = 51.1 | C = 56.6 | A = 47.3 | B = 51.1 | C = 55.3 |
| Period 2 | B = 48.3 | C = 58.2 | A = 52.0 | C = 58.8 | A = 48.8 | B = 52.1 |
| Period 3 | C = 55.8 | A = 53.5 | B = 47.2 | B = 45.4 | C = 55.9 | A = 49.2 |
| Totals | 158.0 | 162.8 | 155.8 | 151.5 | 155.8 | 156.6 |
| | A = 159.4 | B = 146.6 | C = 170.6 | A = 145.3 | B = 148.6 | C = 170.0 |

### Hair sheep

| | Fistulated | | | Nonfistulated | | |
|---|---|---|---|---|---|---|
| | Animal 114 | Animal 216 | Animal 317 | Animal 114 | Animal 215 | Animal 318 |
| Period 1 | A = 49.9 | B = 54.2 | C = 56.3 | A = 49.2 | B = 54.7 | C = 56.6 |
| Period 2 | B = 52.9 | C = 60.9 | A = 51.3 | C = 60.7 | A = 49.2 | B = 46.1 |
| Period 3 | C = 57.7 | A = 46.2 | B = 50.1 | B = 44.3 | C = 55.5 | A – 46.8 |
| Totals | 160.5 | 161.3 | 157.7 | 154.2 | 159.4 | 149.5 |
| | A = 147.4 | B = 157.2 | C = 174.9 | A = 145.2 | B = 145.1 | C = 172.8 |

$$\text{Total SS} = \sum_i \sum_j \sum_k Y^2_{ijk} - \frac{Y^2_{\ldots}}{N} = sap - (1)$$

$$= 151{,}021.65 - \frac{(2847.1)^2}{54} = 151{,}021.65 - 150{,}110.7113$$

$$= 910.9387 \quad \text{where } N = n_s n_k^2 \text{ and } n_k = \text{size of square} \qquad (6.16)$$

$$\text{Squares SS} = \frac{\Sigma_i\, Y^2_{i..}}{n_k^2} - \frac{Y^2_{\ldots}}{N} = s - (1)$$

$$= \frac{(494.1)^2 + (476.6)^2 + \cdots + (463.1)^2}{9} - (1)$$

$$= 150{,}185.9389 - (1)$$

$$= 75.2276 \qquad\qquad (6.17)$$

**TABLE 6.12** TOTALS USED IN ANALYZING THE DATA DISPLAYED IN TABLE 6.11

| | Fistulated | | | | Nonfistulated | | | |
| --- | --- | --- | --- | --- | --- | --- | --- | --- |
| | Type 1 | Type 2 | Type 3 | Totals | Type 1 | Type 2 | Type 3 | Totals |
| Period 1 | 158.3 | 161.6 | 160.4 | (480.3) | 156.6 | 153.7 | 160.5 | (470.8) |
| Period 2 | 165.9 | 158.5 | 165.1 | (489.5) | 156.8 | 159.7 | 156.0 | (472.5) |
| Period 3 | 169.9 | 156.5 | 154.0 | (480.4) | 156.5 | 150.5 | 146.6 | (453.6) |
| Totals | 494.1 | 476.6 | 479.5 | (1,450.2) | 469.9 | 463.9 | 463.1 | (1,396.9) |
| Forage $A$ | 164.3 | 159.4 | 147.4 | (471.1) | 154.3 | 145.3 | 145.2 | (444.8) |
| Forage $B$ | 159.2 | 146.6 | 157.2 | (463.0) | 150.2 | 148.6 | 145.1 | (443.9) |
| Forage $C$ | 170.6 | 170.6 | 174.9 | (516.1) | 165.4 | 170.0 | 172.8 | (508.2) |
| Period 1 | 314.9 | 315.3 | 320.9 | (951.1) | $A$ 318.6 | 304.7 | 292.6 | (915.9) |
| Period 2 | 322.7 | 318.2 | 321.1 | (962.0) | $B$ 309.4 | 295.2 | 302.3 | (906.9) |
| Period 3 | 326.4 | 307.0 | 300.6 | (934.0) | $C$ 336.0 | 340.6 | 347.7 | (1,024.3) |
| | Total uncorrected sum of squares = 151,021.65 | | | | | | | 2,847.1 |

The partitioning of the sum of squares for squares follows.

$$\text{Fistulated versus nonfistulated } (F) = f - 1 = \frac{(1450.2)^2 + (1396.9)^2}{27} - (1)$$

$$= 150,163.3204 - (1) = 52.6091$$

$$\text{Type of animal } (K) = k - 1 = \frac{(964.0)^2 + (940.5)^2 + (942.6)^2}{18} - (1)$$

$$= 150,129.5006 - (1) = 18.7893$$

$$FK = fk - f - k + (1)$$

$$= \frac{(494.1)^2 + (476.6)^2 + \cdots + (463.1)^2}{9} - (1)$$

$$= 150,185.9389 - f - k + (1)$$

$$= 3.8292 \qquad (s = fk)$$

$$\text{Animals within squares SS} = \frac{\sum_i \sum_j Y_{ij\cdot}^2}{n_k} - \frac{\sum_i Y_{i\cdot\cdot}^2}{n_k^2} = as - s$$

$$= \frac{(163.2)^2 + (172.4)^2 + \cdots + (149.5)^2}{3} - s$$

$$= 150,253.25 - s = 67.3111 \qquad (6.18)$$

$$\text{Period SS} = \frac{\sum_k Y_{\cdot\cdot k}^2}{n_s n_k} - \frac{Y_{\cdots}^2}{N} = p - (1)$$

$$= \frac{(951.1)^2 + (962.0)^2 + (934.0)^2}{18} - (1)$$

$$= 150,132.8450 - (1) = 22.1337$$

$$SP\ SS = \frac{\Sigma_i\ \Sigma_j\ Y_{i\cdot k}^2}{n_k} - \frac{\Sigma_i\ Y_{i\cdot\cdot}^2}{n_k^2} - \frac{\Sigma_k\ Y_{\cdot\cdot k}^2}{n_s n_k} + \frac{Y_{\cdot\cdot\cdot}^2}{N}$$

$$= sp - s - p + 1$$

$$= \frac{(158.3)^2 + (165.9)^2 + \cdots + (146.6)^2}{3}$$

$$- s - p + (1)$$

$$= 150,282.2767 - s - p + (1)$$

$$= 74.2041 \tag{6.19}$$

The partitioning of the sum of squares for square $\times$ period interaction follows.

$$FP = fp - f - p + 1 = \frac{(480.3)^2 + (489.5)^2 + \cdots + (453.6)^2}{9}$$

$$- f - p + (1)$$

$$= 150,193.8167 - f - p + (1) = 8.3626$$

$$KP = kp - k - p + 1 = \frac{(314.9)^2 + (322.7)^2 + \cdots + (300.6)^2}{6}$$

$$- k - p + (1)$$

$$= 150,198.4950 - k - p + (1) = 46.8607$$

$$FKP = fkp - fk - fp - kp + f + k + p - 1$$

$$= \frac{(158.3)^2 + (165.9)^2 + \cdots + (146.6)^2}{3}$$

$$- fk - fp - kp + f + k + p - (1)$$

$$= 150,282.2767 - fk - fp - kp + f + k + p - (1) = 18.9808$$

$$\text{Treatment SS} = \frac{\Sigma_t\ Y_t^2}{n_s n_k} - \frac{Y_{\cdot\cdot\cdot}^2}{N} = t - 1$$

$$= \frac{(915.9)^2 + (906.9)^2 + (1024.3)^2}{18} - (1)$$

$$= 150,585.0506 - (1) = 474.3393 \tag{6.20}$$

$$ST\ SS = \frac{\Sigma_i\ \Sigma_t\ Y_{i\cdot\cdot(t)}^2}{n_k} - \frac{\Sigma_i\ Y_{i\cdot\cdot}^2}{n_k^2} - \frac{\Sigma_t\ Y_t^2}{n_s n_k} + \frac{Y_{\cdot\cdot\cdot}^2}{N} = st - s - t + (1)$$

$$= \frac{(164.3)^2 + (159.2)^2 + \cdots + (172.8)^2}{3} - s - t + (1)$$

$$= 150,764.3367 - s - t + (1) = 104.0585 \tag{6.21}$$

The partitioning of the sum of squares for square × treatment interaction follows.

$$FT = ft - f - t + (1) = \frac{(471.1)^2 + (463.0)^2 + \cdots + (508.2)^2}{9} - f - t + (1)$$

$$= 150{,}647.2122 - f - t + (1) = 9.5525$$

$$KT = kt - k - t + (1) = \frac{(318.6)^2 + (309.4)^2 + \cdots + (347.7)^2}{6} - k - t + (1)$$

$$= 150{,}669.8583 - k - t + (1) = 66.0184$$

$$FKT = fkt - fk - ft - kt + f + k + t - 1$$

$$= \frac{(164.3)^2 + (159.4)^2 + \cdots + (172.8)^2}{3} - fk - ft - kt + f + k + t - (1)$$

$$= 150{,}764.3367 - fk - ft - kt + f + k + t - (1) = 28.4876$$

$$\text{Error SS} = sap - as - sp - st + 2s$$

$$= \text{Total SS} - SS_S - SS_{A:S} - SS_P - SS_{SP} - SS_T - SS_{ST}$$

$$= 93.6644 \tag{6.22}$$

The analysis for this set of data is shown in Table 6.13.

The results of the analysis show a highly significant difference between fistulated and nonfistulated animals as well as highly significant differences among treatments. Application of one of the suggested mean separation procedures would yield further information with regard to treatment effects.

The method of calculating degrees of freedom and the expectations of mean squares for the model in (6.10) are shown in Table 6.14. The squares, periods, and treatments are considered to be fixed effects and the animals to be random effects.

**TABLE 6.13**   ANALYSIS OF VARIANCE OF DATA SHOWN IN TABLE 6.7

| Source | df | SS | MS | F |
|---|---|---|---|---|
| Squares | 5 | 75.2276 | 15.0455 | 2.68 |
| Fistulation ($F$) | 1 | 52.6091 | 52.6091 | 9.38** |
| Types ($K$) | 2 | 18.7893 | 9.3946 | 1.67 |
| FK | 2 | 3.8292 | 1.9146 | 0.34 |
| Animals within squares ($A{:}S$) | 12 | 67.3111 | 5.6093 | 0.72 |
| Periods ($P$) | 2 | 22.1337 | 11.0668 | 1.42 |
| SP | 10 | 74.2041 | 7.4204 | 0.95 |
| FP | 2 | 8.3626 | 4.1813 | 0.54 |
| KP | 4 | 46.8607 | 11.7152 | 1.50 |
| FKP | 4 | 18.9808 | 4.7452 | 0.61 |
| Treatments ($T$) | 2 | 474.3393 | 237.1696 | 30.39** |
| ST | 10 | 104.0585 | 10.4058 | 1.33 |
| FT | 2 | 9.5525 | 4.7762 | 0.61 |
| KT | 4 | 66.0184 | 16.5046 | 2.11 |
| FKT | 4 | 28.4876 | 7.1219 | 0.91 |
| Error | 12 | 93.6644 | 7.8054 | |

**TABLE 6.14**   DEGREES OF FREEDOM AND
EXPECTATIONS OF MEAN SQUARES

| Source | df | E(MS) |
|--------|-----|-------|
| $S$ | $n_s - 1$ | $\sigma_e^2 + 3\sigma_{a:s}^2 + 9\sigma_s^2$ |
| $A{:}S$ | $(n_a - 1)n_s$ | $\sigma_e^2 + 3\sigma_{a:s}^2$ |
| $P$ | $n_p - 1$ | $\sigma_e^2 + 18\sigma_p^2$ |
| $SP$ | $(n_s - 1)(n_p - 1)$ | $\sigma_e^2 + 3\sigma_{sp}^2$ |
| $T$ | $n_t - 1$ | $\sigma_e^2 + 18\sigma_t^2$ |
| $ST$ | $(n_s - 1)(n_t - 1)$ | $\sigma_e^2 + 3\sigma_{st}^2$ |
| Error | $n_s(n_p - 1)(n_a - 2)$ | $\sigma_e^2$ |

$N = n_s n_a n_p = 54$     $n_s = 6$     $n_a = 3$     $n_p = 3$     $n_t = 3$

## 6.8 RANDOMIZATION IN THE LATIN SQUARE

Since there are many possible arrangements of Latin squares, it is necessary to make a selection of the square (or squares) to be used by some random method. Cochran and Cox (1957) present a series of standard Latin squares which can be used for this purpose, and methods of selection. A standard Latin square is one in which the first row and the first column are arranged in alphabetical sequence.

For a $3 \times 3$ Latin square:

1. Write out the standard Latin square,

$$A \quad B \quad C$$
$$B \quad C \quad A$$
$$C \quad A \quad B$$

2. Assign each treatment a letter.
3. Arrange the columns at random.
4. Arrange the last two rows at random.

For a $4 \times 4$ Latin square:

1. Use the four standard Latin squares,

$$A \; B \; C \; D \qquad A \; B \; C \; D \qquad A \; B \; C \; D \qquad A \; B \; C \; D$$
$$B \; A \; D \; C \qquad B \; C \; D \; A \qquad B \; D \; A \; C \qquad B \; A \; D \; C$$
$$C \; D \; B \; A \qquad C \; D \; A \; B \qquad C \; A \; D \; B \qquad C \; D \; A \; B$$
$$D \; C \; A \; B \qquad D \; A \; B \; C \qquad D \; C \; B \; A \qquad D \; C \; B \; A$$

2. Assign each treatment a letter.
3. Select one of the above squares at random.
4. Arrange at random all columns and the last three rows or all columns and rows.

For a 5 × 5 and higher Latin squares:

**1.** Use the one square shown by Cochran and Cox,

$$
\begin{array}{ccccc}
A & B & C & D & E \\
B & A & E & C & D \\
C & D & A & E & B \\
D & E & B & A & C \\
E & C & D & B & A
\end{array}
$$

**2.** Assign each treatment a letter.
**3.** Arrange all rows and columns at random.

## 6.9 GAIN IN EFFICIENCY: LATIN SQUARE VERSUS RANDOMIZED BLOCK

In Section 6.4, which deals with the randomized block design, we made an estimate of the gain in efficiency of the randomized block relative to the completely randomized design. We can now do the same type of thing, estimating the efficiency of a Latin square relative to the randomized block. Since either the rows or the columns can be considered as blocks we can make two estimates, one for rows as blocks, the other for column as blocks. The method of estimation as shown by Cochran and Cox (1957) and Steel and Torrie (1980) follows.

If rows are the only blocks, we can estimate the error mean square for the randomized block by means of the formula

$$
\text{MS}_e(\text{RB}) = \frac{f_c \text{MS}_c + (f_t + f_e)\text{MS}_e}{f_c + f_t + f_e} \tag{6.23}
$$

where $\text{MS}_c$ and $\text{MS}_e$ are the mean squares for the columns and error found in the Latin square and $f_c$, $f_t$, and $f_e$ are the degrees of freedom for the columns, treatments, and error of the Latin square.

If we consider columns as the blocks, we can substitute $f_r$ and $\text{MS}_r$ for $f_c$ and $\text{MS}_c$ in the formula (6.23), where $f_r$ and $\text{MS}_r$ are the degrees of freedom and mean square for rows.

Use of the Latin square instead of the randomized block results in fewer degrees of freedom for estimating the error mean square, so this must be taken into account when estimating the relative efficiency of the two designs, particularly if fewer than 20 degrees of freedom are involved in the Latin square error, which is usually the case. The fewer degrees of freedom result in a loss of precision in estimating the mean square of the Latin square as compared to the randomized block design. This can be taken into account by multiplying $E(\text{LS to RB})$ by

$$
\frac{(f_1 + 1)(f_2 + 3)}{(f_1 + 3)(f_2 + 1)} \tag{6.24}
$$

where $f_1$ and $f_2$ are the degrees of freedom for the error in the Latin square and randomized block designs, respectively.

Using the analysis shown in Table 6.7, the estimated error mean square using the rows as the only blocking is

$$MS_e(RB) = \frac{(3)(1.3209) + (3 + 6)(0.2043)}{3 + 3 + 6} = \frac{5.8014}{12} = 0.4834$$

and using the columns as the only blocking the estimated mean square is

$$MS_e(RB) = \frac{(3)(0.3786) + (3 + 6)(0.2043)}{3 + 3 + 6} = \frac{2.9745}{12} = 0.2479$$

The adjustment for the differences in degrees of freedom for the two designs is

$$\frac{(6 + 1)(9 + 3)}{(6 + 3)(9 + 1)} = 0.9333$$

Thus, the estimated relative precision using rows as blocks is

$$E(\text{LS to RB}) = \frac{MS_e(RB)(f_1 + 1)(f_2 + 3)}{MS_e(LS)(f_1 + 3)(f_2 + 1)}(100)$$

$$= \frac{(0.4834)}{(0.2043)}(0.9333)(100) = 220.83 \text{ percent}$$

and using the columns as blocks

$$E(\text{LS to RB}) = \frac{MS_e(RB)}{MS_e(LS)} \frac{(f_1 + 1)(f_2 + 3)}{(f_1 + 3)(f_2 + 1)}(100)$$

$$= \frac{0.2479}{0.2043}(0.9333)(100) = 113.25 \text{ percent}$$

The difference in these two percentages is due to the fact that if the rows (periods) were the blocks in a randomized block design they would not have been too effective in reducing the error sum of squares. The columns (cows), on the other hand, would account for a large portion of the sum of squares.

## 6.10  MISSING VALUES IN THE LATIN SQUARE

We have previously seen that we can estimate one or more missing values in a randomized block design and complete the analysis. The same is true in a Latin square by use of the formula

$$Y = \frac{n_k(R + C + T) - 2G}{(n_k - 1)(n_k - 2)} \tag{6.25}$$

where $n_k$ represents the size of the square, $R$, $C$, and $T$ are the totals of the values in the row, column, and treatment containing the missing value, and $G$ is the overall total of observed values, as shown by Yates (1933). This procedure cannot be used if an entire row or column or treatment is missing. However, if more than one observation is missing not making up a row, column, or treatment, repeated application of formula (6.25) can be made as it was for randomized blocks.

The treatment sum of squares, as was the case in the randomized block, is biased upward. The amount of bias can be estimated by the formula

$$\frac{[G - R - C - (n_k - 1)T]^2}{[(n_k - 1)(n_k - 2)]^2} \tag{6.26}$$

as shown by Steel and Torrie (1980). This amount can be removed from the treatment sum of squares. One degree of freedom is subtracted from the error for each value that has been estimated.

## 6.11 GRAECO-LATIN SQUARE

It is possible to have triple grouping by making use of the Latin square design and adding a further restriction. In the Latin square each treatment appears once in each row and in each column. If we add a subscript to each of the letters in the Latin square, representing an additional source of variation, in such a way that each treatment now appears once in each row, once in each column, and once with each subscript, we now have a Graeco-Latin square. A $3 \times 3$ Graeco-Latin square would look like

$$\begin{array}{ccc} A_1 & B_3 & C_2 \\ B_2 & C_1 & A_3 \\ C_3 & A_2 & B_1 \end{array}$$

Designs for the number of treatments from 3 to 12 with the exception of 6 and 10 are presented by Cochran and Cox (1957).

If a single $3 \times 3$ Graeco-Latin square were used, there would be two degrees of freedom each for rows, columns, letters, and subscripts, leaving no degrees of freedom for residual or error. Therefore, if a $3 \times 3$ design is to be used, more than one square is necessary. The appropriate analyses for such designs follow the methods discussed previously when several Latin squares are included in an experiment. However, there are now two fixed effects, letters and subscripts, making the analyses a bit more complex. Interactions of squares with both letters and subscripts must be considered, leading to several possible analyses. Care must be taken that enough squares are used to yield a reasonable number of degrees of freedom for error.

Although it has not been possible to construct a $6 \times 6$ Graeco-Latin square, a design making use of this basic form has been used by Zulma Garcia of the Department of Exercise Science at the University of Massachusetts. In this design several $6 \times 6$ squares were used, each square consisting of six subjects, six days, and six treatments (letters). Three levels of an additional treatment were included in such a way that each level of this treatment appeared twice with each subject, twice with each day, and twice with each treatment. One portion of the overall experiment is shown in Table 6.15 where the treatments (letters) consisted of six fatigue regimens and the additional

**TABLE 6.15** FOREARM FLEXION TIME (IN MILLISECONDS)

| Days | Subjects | | | | | | | | | | |
|---|---|---|---|---|---|---|---|---|---|---|---|
| | **1** | | **2** | | **3** | | **4** | | **5** | | **6** |
| 1 | $A_0$ | 154.0 | $B_2$ | 179.3 | $C_1$ | 214.0 | $D_0$ | 169.0 | $E_2$ | 201.7 | $F_1$ | 165.3 |
| 2 | $B_1$ | 155.0 | $C_0$ | 147.7 | $D_2$ | 235.3 | $E_1$ | 139.3 | $F_0$ | 169.7 | $A_2$ | 200.7 |
| 3 | $C_2$ | 189.8 | $D_1$ | 167.7 | $E_0$ | 190.0 | $F_2$ | 188.3 | $A_1$ | 174.3 | $B_0$ | 132.0 |
| 4 | $D_0$ | 145.7 | $E_2$ | 181.0 | $F_1$ | 202.3 | $A_0$ | 134.7 | $B_2$ | 183.7 | $C_1$ | 170.3 |
| 5 | $E_1$ | 172.3 | $F_0$ | 139.3 | $A_2$ | 250.0 | $B_1$ | 145.7 | $C_0$ | 147.3 | $D_2$ | 224.3 |
| 6 | $F_2$ | 202.3 | $A_1$ | 157.0 | $B_0$ | 149.3 | $C_2$ | 177.0 | $D_1$ | 160.7 | $E_0$ | 142.3 |

$A = 1070.7 \qquad B = 945.0 \qquad C = 1046.1 \qquad D = 1102.7 \qquad E = 1026.6 \qquad F = 1067.2$

$0 = 1821.0 \qquad 1 = 2023.9 \qquad 2 = 2413.4$

treatments (subscripts) were made up of three loads on the arm. The loads were zero, three times the moment of inertia, and seven times the moment of inertia.

The model for this set of data is

$$Y_{ij} = \mu + c_i + d_j + t_{(t)} + s_{(s)} + e_{ij} \tag{6.27}$$

where the $c_i$ represent column or subject effects, the $d_j$ represent day effects, the $t_{(t)}$ represent treatment (regimen) effects, and the $s_{(s)}$ represent load effects (subscripts). The calculations for the analysis, in which $n_k$ equals the size of the square, follow:

$$\text{Total SS} = \sum_i \sum_j Y_{ij}^2 - \frac{Y_{..}^2}{N} = \sum_i \sum_j Y_{ij}^2 - (1)$$

$$= 1{,}117{,}328.89 - \frac{(6258.3)^2}{36} = 1{,}117{,}328.89 - 1{,}087{,}953.30$$

$$= 29{,}375.59 \tag{6.28}$$

$$\text{Subject SS} = \frac{\sum_i Y_{i.}^2}{n_k} - \frac{Y_{..}^2}{N} = c - (1)$$

$$= \frac{(1019.1)^2 + (972.0)^2 + \cdots + (1034.9)^2}{6} - (1)$$

$$= 1{,}096{,}752.40 - (1) = 8799.10 \tag{6.29}$$

$$\text{Day SS} = \frac{\sum_j Y_{.j}^2}{n_k} - \frac{Y_{..}^2}{N} = d - (1)$$

$$= \frac{(1083.3)^2 + (1047.7)^2 + \cdots + (988.6)^2}{6} - (1)$$

$$= 1{,}089{,}042.51 - (1) = 1089.21 \tag{6.30}$$

$$\text{Letter SS} = \frac{\sum_t Y_t^2}{n_k} - \frac{Y_{..}^2}{N} = t - (1)$$

**TABLE 6.16** ANALYSIS OF VARIANCE OF FOREARM FLEXION DATA

| Source | df | SS | MS | F | E(MS) |
|--------|-----|-----------|----------|---------|--------------------------|
| Subjects | 5 | 8,799.10 | 1,759.82 | 16.54** | $\sigma_e^2 + 6\sigma_c^2$ |
| Days | 5 | 1,089.21 | 217.84 | 2.05 | $\sigma_e^2 + 6\sigma_d^2$ |
| Regimens | 5 | 2,466.60 | 493.32 | 4.64** | $\sigma_e^2 + 6\sigma_t^2$ |
| Loads | 2 | 15,106.01 | 7,553.00 | 71.01** | $\sigma_e^2 + 12\sigma_s^2$ |
| Error | 18 | 1,914.67 | 106.37 | | $\sigma_e^2$ |

$$= \frac{(1070.7)^2 + (945.0)^2 + \cdots + (1067.2)^2}{6} - (1)$$

$$= 1,090,419.90 - (1) = 2466.60 \tag{6.31}$$

$$\text{Subscript SS} = \frac{\Sigma_s Y_s^2}{2n_k} - \frac{Y_{..}^2}{N} = s - (1)$$

$$= \frac{(1821.0)^2 + (2023.9)^2 + (2413.4)^2}{12} - (1)$$

$$= 1,103,059.31 - (1) = 15,106.01 \tag{6.32}$$

$$\text{Error SS} = \text{Total SS} - \text{SS}_C - \text{SS}_D - \text{SS}_T - \text{SS}_S = 1914.67 \tag{6.33}$$

The analysis of variance for these data is shown in Table 6.16 where it can be seen that three of the classifications showed highly significant effects. The complete design from which these data were selected included four $6 \times 6$ squares, two made up of females and two made up of males, thus allowing for a test of the difference between the sexes in response to time.

## 6.12 CHANGEOVER DESIGN TO MEASURE RESIDUAL EFFECTS

In experiments where the treatments are applied in sequence to the same person, animal, store, field plot, and so on, the effects of the treatments may continue after the application of the treatments is discontinued. The effect of a given treatment is, or can be, influenced by the carry-over or residual effect of the previous treatment.

One method of eliminating the effect of the previous treatment is to insert a rest period between the treatment periods. In this way the treatment effects are freed of most or all of the residual effects. However, it is not always possible or desirable to use such a practice.

The alternative is to use an experimental design which yields a measurement of both *residual* and *direct* effects of treatments and an adjustment of the direct effects for the residual effects and vice versa. A design that allows for this in two $3 \times 3$ orthogonal Latin squares is shown in Table 6.17.

**TABLE 6.17** TWO ORTHOGONAL LATIN SQUARES

| Period | Sequence | | | | | |
|--------|---|---|---|---|---|---|
|        | I | II | III | IV | V | VI |
| 1 | A | B | C | A | B | C |
| 2 | B | C | A | C | A | B |
| 3 | C | A | B | B | C | A |

The treatments in this design can produce first direct effects, designated by Cochran and Cox (1957) as $t_a$, $t_b$, and $t_c$ during the period of their application. After the change from one treatment to another, it is possible that residual effects, designated as $r_a$, $r_b$, and $r_c$, can also be produced. Therefore, in the first period only direct effects result from the treatments, but in subsequent periods combinations such as $(t_b + r_a)$ and $(t_a + r_c)$ can occur.

The primary intent of this design is to estimate the direct effects of the treatments after removal of, or adjustment for, the residual effects where they exist. In the design suggested each treatment is preceded by each of the other treatments an equal number of times. When the number of treatments is odd, two Latin squares are required. When the number of treatments is even, the selection of a suitable single Latin square will accomplish the same purpose. However, enough squares are needed to build up a suitable number of degrees of freedom for the experimental error. Cochran and Cox (1957) show the appropriate squares.

An example of this design shown by Cochran and Cox (1957) is presented in Table 6.18. The experiment involved feeding dairy cows with treatment $A$ = roughage, treatment $B$ = limited grain, and treatment $C$ = full grain. The experiment and method of analysis were given by Williams (1949). A somewhat different method of analysis is presented here, taking advantage of matrix algebra to shorten the computations.

The model for the analysis of the experiment is

$$Y_{ijk(t)(r)} = \mu + s_i + c_{ij} + p_{ik} + t_{(t)} + r_{(r)} + e_{ijk} \tag{6.34}$$

where the $s_i$ represent square effects, the $c_{ij}$ represent cows within squares, the $p_{ik}$ represent periods within squares, the $t_{(t)}$ represent direct effects of treatments, and the $r_{(r)}$ represent residual effects of treatments.

**TABLE 6.18** MILK YIELD PER PERIOD (IN LB/10–100)

| Period | Sequence (cow) | | | | | | Totals | Sequence (cow) | | | | | | Totals |
|--------|---|---|---|---|---|---|--------|---|---|---|---|---|---|--------|
|        | I | | II | | III | | | IV | | V | | VI | | |
| 1 | A | 38 | B | 109 | C | 124 | 271 | A | 86 | B | 75 | C | 101 | 262 |
| 2 | B | 25 | C | 86 | A | 72 | 183 | C | 76 | A | 35 | B | 63 | 174 |
| 3 | C | 15 | A | 39 | B | 27 | 81 | B | 46 | C | 34 | A | 1 | 81 |
| Totals | | 78 | | 234 | | 223 | 535 | | 208 | | 144 | | 165 | 517 |

To complete the analysis we calculate the following:

$$\text{Total SS} = \sum_i \sum_j \sum_k Y_{ijk}^2 - \frac{Y_{\cdot\cdot\cdot}^2}{N}$$

$$= 81{,}846 - \frac{(1052)^2}{18} = 81{,}846 - 61{,}483.6$$

$$= 20{,}362.4 \tag{6.35}$$

$$\text{Square SS} = \frac{\sum_i Y_{i\cdot\cdot}^2}{n_k^2} - \frac{Y_{\cdot\cdot\cdot}^2}{N} = s - (1)$$

$$= \frac{(535)^2 + (517)^2}{9} - (1) = 61{,}501.6 - (1)$$

$$= 18.0 \tag{6.36}$$

$$\text{Cows within squares SS} = \frac{\sum_i \sum_j Y_{ij\cdot}^2}{n_k} - \frac{\sum_i Y_{i\cdot\cdot}^2}{n_k^2} = cs - s$$

$$= \frac{(78)^2 + (234)^2 + \cdots + (165)^2}{3} - s$$

$$= \frac{201{,}794}{3} - s$$

$$= 67{,}264.7 - s = 5763.1 \tag{6.37}$$

$$\text{Periods within squares SS} = \frac{\sum_i \sum_k Y_{i\cdot k}^2}{n_k} - \frac{\sum_i Y_{i\cdot\cdot}^2}{n_k^2} = ps - s$$

$$= \frac{(271)^2 + (183)^2 + \cdots + (81)^2}{3} - s$$

$$= \frac{218{,}972}{3} - s$$

$$= 72{,}990.7 - s = 11{,}489.1 \tag{6.38}$$

While the sums of squares we have just obtained were calculated using the usual equal number formulas, the same procedures cannot be used for the direct and residual effects of treatments. The residual effects are confounded with both the direct effects and the column (squares and cows within squares) effects so a different method must be followed. The method used here was to write out the normal equations for the model, combining the mean and column effects and omitting the period effects since they are orthogonal to all other effects. The equations for the direct effects and residual effects were adjusted for the column effects by absorbing the $\hat{\mu} + \hat{c}_i$ equations, by the procedure shown elsewhere (Chapter 9), $C = S - N'D^{-1}N$ and $Y = Y_2 - N'D^{-1}Y$. The resulting matrix was reduced and inverted and is shown below as $C^{-1}$, together with the right-hand members $Y$. The matrix multiplication $C^{-1}Y$ then yields the constants for the direct effects and indirect effects. Of course, the entire set of equations could be solved, including all the effects, if convenient.

$$\frac{1}{72}\begin{bmatrix} 10 & -5 & 6 & -3 \\ -5 & 10 & -3 & 6 \\ 6 & -3 & 18 & -9 \\ -3 & 6 & -9 & 18 \end{bmatrix} \begin{bmatrix} T_A - T_C \\ T_B - T_C \\ R_A - R_C - \frac{1}{3}(S_1 + S_5 - S_2 - S_6) \\ R_B - R_C - \frac{1}{3}(S_1 + S_5 - S_3 - S_4) \end{bmatrix} = \begin{bmatrix} \hat{t}_a \\ \hat{t}_b \\ \hat{r}_a \\ \hat{r}_b \end{bmatrix}$$

$$\mathbf{C^{-1}} \qquad\qquad \mathbf{Y} \qquad\qquad \mathbf{B}$$

The letters $T$, $R$, and $S$ represent totals for the direct effects of treatments, the residual effects of treatments, and the sequences, respectively. The $T$'s come from the treatment totals and the $R$'s come from the yields in periods immediately following the application of the treatments. Thus,

$$T_A = 38 + 39 + 72 + 86 + 35 + 1 = 271$$

$$T_B = 25 + 109 + 27 + 46 + 75 + 63 = 345$$

$$T_C = 15 + 86 + 124 + 76 + 34 + 101 = 436$$

$$R_A = 25 + 27 + 76 + 34 = 162$$

$$R_B = 15 + 86 + 35 + 1 = 137$$

$$R_C = 39 + 72 + 46 + 63 = 220$$

Our **Y** vector would be

$$\begin{bmatrix} 271 - 436 = -165 \\ 345 - 436 = -91 \\ 162 - 220 - \frac{1}{3}(78 + 144 - 234 - 165) = 1 \\ 137 - 220 - \frac{1}{3}(78 + 144 - 223 - 208) = -\frac{40}{3} \end{bmatrix}$$

The multiplication of $\mathbf{C^{-1}Y}$ yields

$$\begin{bmatrix} -\frac{1149}{72} = -15.9583 = \hat{t}_a \\ -\frac{168}{72} = -2.3333 = \hat{t}_h \\ -\frac{579}{72} = -8.0417 = \hat{r}_a \\ -\frac{300}{72} = -4.1667 = \hat{r}_b \end{bmatrix} = \mathbf{B}$$

Since the estimates of the constants for the direct effects and the residual effects must sum to zero, we find

$$\hat{t}_c = -(-\tfrac{1149}{72} - \tfrac{168}{72}) = \tfrac{1317}{72} = 18.2917$$

$$\hat{r}_c = -(-\tfrac{579}{72} - \tfrac{300}{72}) = \tfrac{879}{72} = 12.2083$$

We can now calculate the direct effect means as

$$\hat{\mu} + \hat{t}_a = 58.4444 - 15.9583 = 42.4861$$

$$\hat{\mu} + \hat{t}_b = 58.4444 - 2.3333 = 56.1111$$

$$\hat{\mu} + \hat{t}_c = 58.4444 + 18.2917 = 76.7361$$

where $\hat{\mu} = \frac{1052}{18} = 58.4444$.

Next, we can calculate

$$\text{Direct effect of treatments SS} = \frac{24 \, \Sigma \, \hat{t}^2}{5} = \frac{14{,}272.750}{5} = 2854.6 \qquad (6.39)$$

and

$$\text{Residual effect of treatments SS} = \frac{8 \, \Sigma \, \hat{r}^2}{3} = \frac{1848.5833}{3} = 616.2 \qquad (6.40)$$

Since the sums of squares for the direct and residual effects of treatments are not additive, we need to obtain the sum of squares due to the two effects considered simultaneously which can be found by the matrix multiplication

$$\mathbf{B'Y} = 2893.0$$

The error sum of squares is found by subtracting the sums of squares for squares, cows within squares, periods within squares, and the combined treatments from the total sum of squares giving 199.2. The completed analysis is shown in Table 6.19. It should be noted that the sum of squares for cows within squares has not been adjusted for the confounding with residual effects.

The standard error of a direct effect mean is given by

$$s_{\bar{y}_a} = s\sqrt{\tfrac{1}{18} + \tfrac{10}{72}} = \sqrt{49.8}\sqrt{\tfrac{7}{36}} = 3.112 \qquad (6.41)$$

The standard error for the difference between two direct effect means is given by

$$s_{\hat{t}_i - \hat{t}_j} = s\sqrt{\tfrac{10}{72} + \tfrac{10}{72} - (2)(-\tfrac{5}{72})} = \sqrt{49.8}\sqrt{\tfrac{30}{72}} = 4.555 \qquad (6.42)$$

and for the difference between two residual effects by

$$s_{\hat{r}_i - \hat{r}_j} = s\sqrt{\tfrac{18}{72} + \tfrac{18}{72} - (2)(-\tfrac{9}{72})} = \sqrt{49.8}\sqrt{\tfrac{54}{72}} = 6.111 \qquad (6.43)$$

In this design, estimates of direct effects are measured with more precision than estimates of residual effects since more observations are used for the estimation of

**TABLE 6.19**  ANALYSIS OF VARIANCE FOR DIRECT AND RESIDUAL EFFECTS

| Source | df | SS | MS | F |
|--------|----|------|------|------|
| Squares | 1 | 18.0 | 18.0 | |
| Cows within squares | 4 | 5,763.1 | 1,440.8 | |
| Periods within squares | 4 | 11,489.1 | 2,872.3 | |
| Direct effects | 2 | 2,854.6 | 1,427.3 | 28.66** |
| Residual effects | 2 | 616.2 | 308.1 | 6.19 |
| Error | 4 | 199.2 | 49.8 | |

**TABLE 6.20** DESIGNS BALANCED FOR MEASURING DIRECT AND RESIDUAL EFFECTS

| Four treatments | | | | | Six treatments | | | | | | |
|---|---|---|---|---|---|---|---|---|---|---|---|
| | Sequence | | | | | Sequence | | | | | |
| Period | 1 | 2 | 3 | 4 | Period | 1 | 2 | 3 | 4 | 5 | 6 |
| 1 | A | C | B | D | 1 | A | B | C | D | E | F |
| 2 | B | A | D | C | 2 | C | D | E | F | A | B |
| 3 | C | D | A | B | 3 | B | C | D | E | F | A |
| 4 | D | B | C | A | 4 | E | F | A | B | C | D |
| | | | | | 5 | F | A | B | C | D | E |
| | | | | | 6 | D | E | F | A | B | C |

Five treatments

| | Sequence | | | | | | Sequence | | | | |
|---|---|---|---|---|---|---|---|---|---|---|---|
| Period | 1 | 2 | 3 | 4 | 5 | Period | 1 | 2 | 3 | 4 | 5 |
| 1 | A | B | C | D | E | 1 | A | B | C | D | E |
| 2 | B | C | D | E | A | 2 | C | D | E | A | B |
| 3 | D | E | A | B | C | 3 | B | C | D | E | A |
| 4 | E | A | B | C | D | 4 | E | A | B | C | D |
| 5 | C | D | E | A | B | 5 | D | E | A | B | C |

direct effects. A modification of this design, in which the last row of each square is repeated, allows for an increase in the precision of the residual effect estimation.

Other squares for use in this type of design are, presented by Cochran and Cox (1957), shown in Table 6.20. As noted earlier, with an odd number of treatments two squares must be used for balance.

However, since there would only be three degrees of freedom for error with the 4 × 4 square, it would be wise to use two squares with that design.

If more squares are used, or squares of a different size than that of the example, the method proposed by Williams could be used since it is general. However, a very reasonable alternative is to write out the appropriate normal equations for the model and proceed as shown here or in Chapters 2 and 3.

## 6.13 CROSSOVER DESIGN

There are a number of designs similar to the Latin square that are referred to as crossover, changeover, double changeover, reversal, double reversal, switchback, and perhaps other terms. In these designs, treatments are applied in sequence for two or more periods to individual units, in a manner similar to the Latin square. As in the Latin square each unit must receive all treatments and each treatment must appear an equal number of times in each period.

As an example of a crossover design, suppose that we wished to test the effects of two ration supplements on the rate of gain of eight steers. Results of such an experiment appear in Table 6.21, where it can be seen that this is a series of 2 × 2 Latin

**TABLE 6.21**  AVERAGE DAILY GAIN OF STEERS (IN POUNDS)

| | | | | | | | | | | | | | | | | | Totals |
|---|---|---|---|---|---|---|---|---|---|---|---|---|---|---|---|---|---|
| | | | | | | | Steer | | | | | | | | | | |
| $A$ | 2.62 | $B$ | 1.43 | $B$ | 1.74 | $A$ | 1.91 | $B$ | 2.18 | $A$ | 2.14 | $A$ | 1.92 | $B$ | 2.21 | | 16.15 |
| $B$ | 1.91 | $A$ | 2.39 | $A$ | 2.82 | $B$ | 1.96 | $A$ | 2.43 | $B$ | 2.47 | $B$ | 1.56 | $A$ | 2.78 | | 18.32 |
| Totals | 4.53 | | 3.82 | | 4.56 | | 3.87 | | 4.61 | | 4.61 | | 3.48 | | 4.99 | | 34.47 |

squares, although the columns could have been arranged such that Latin squares were not involved. With an even number of treatments the number of columns is an even number, and with an odd number of treatments an odd number. If we had three treatments, of course, a third row or period would be required. In conducting the above experiment we would be assuming that no residual or carryover effect was involved. The mathematical model for these data would be

$$Y_{ij(t)} = \mu + s_i + p_j + t_{(t)} + e_{ij} \tag{6.44}$$

where the $s_i$ represent steer effects, the $p_j$ represent period effects, and the $t_{(t)}$ represent treatment effects. The analysis would be as follows:

$$\text{Total SS} = \sum_i \sum_j Y_{ij}^2 - \frac{Y_{..}^2}{N} = \sum_i \sum_j Y_{ij}^2 - (1)$$

$$= 76.8095 - \frac{(34.47)^2}{16} = 76.8095 - 74.2613 = 2.5482 \tag{6.45}$$

$$\text{Steer SS} = \frac{\sum_i Y_{i.}^2}{n_p} - \frac{Y_{..}^2}{N} = s - 1$$

$$= \frac{(4.53)^2 + (3.82)^2 + \cdots + (4.99)^2}{2} - (1) = 75.1992 - (1)$$

$$= 0.9379 \tag{6.46}$$

$$\text{Period SS} = \frac{\sum_j Y_{.j}^2}{n_s} - \frac{Y_{..}^2}{N} = p - 1$$

$$= \frac{(16.15)^2 + (18.32)^2}{8} - (1) = 74.5556 - (1) = 0.2943 \tag{6.47}$$

$$\text{Treatment SS} = \frac{\sum_t Y_t^2}{n_s} - \frac{Y_{..}^2}{N} = t - 1$$

$$= \frac{(19.01)^2 + (15.46)^2}{8} - (1) = 75.0490 - (1) = 0.7877 \tag{6.48}$$

$$\text{Error SS} = \text{Total SS} - \text{SS}_S - \text{SS}_P - \text{SS}_T = 0.5283 \tag{6.49}$$

The analysis of variance together with the expectations of mean squares are shown in Table 6.22. We find that there is a significant difference between the two treatments.

**TABLE 6.22**   ANALYSIS OF DATA SHOWN IN TABLE 6.19

| Source | df | SS | MS | F | E(MS) |
|---|---|---|---|---|---|
| Steers | 7 | 0.9379 | 0.1340 | 1.52 | $\sigma_e^2 + 2\sigma_s^2$ |
| Periods | 1 | 0.2943 | 0.2943 | 3.34 | $\sigma_e^2 + 8\sigma_p^2$ |
| Treatments | 1 | 0.7877 | 0.7877 | 8.95* | $\sigma_e^2 + 8\sigma_t^2$ |
| Error | 6 | 0.5283 | 0.0880 | | $\sigma_e^2$ |

$$n_s = 8 \qquad n_p = 2 \qquad n_t = 2 \qquad N = n_s n_p = 16$$

When a design has been set up, as this one has, in a series of Latin squares it is possible to remove an additional source of variation from the error term—that variation due to rows within the squares. If we do this, then we will no longer have the source of the rows overall. Should there be a trend of increasing or decreasing differences between the squares this would be accounted for by the variation between periods more completely than with the single degree of freedom for periods or rows. The model for such an analysis would be

$$Y_{ijk(t)} = \mu + q_i + s_{ij} + p_{ik} + t_{(t)} + e_{ijk} \qquad (6.50)$$

where the $q_i$ represent square effects, the $s_{ij}$ represent steer within square effects, the $p_{ij}$ stand for period within square effects, and the $t_{(t)}$ stand for treatment effects. The necessary calculations follow.

$$\text{Total SS} = \sum_i \sum_j \sum_k Y_{ijk}^2 - \frac{Y_{\cdot\cdot\cdot}^2}{N} = \sum_i \sum_j \sum_k Y_{ijk}^2 - (1)$$

$$= 76.8095 - 74.2613 = 2.5482 \qquad (6.51)$$

$$\text{Square SS} = \frac{\sum_i Y_{i\cdot\cdot}^2}{n_s n_p} - \frac{Y_{\cdot\cdot\cdot}^2}{N} = q - (1)$$

$$= \frac{(8.35)^2 + (8.43)^2 + \cdots + (8.47)^2}{4} - (1)$$

$$= 74.3842 - (1) = 0.1229 \qquad (6.52)$$

$$\text{Steer within squares SS} = \frac{\sum_i \sum_j Y_{ij\cdot}^2}{n_p} - \frac{\sum_i Y_{i\cdot\cdot}^2}{n_s n_p} = sq - q$$

$$= \frac{(4.53)^2 + (3.82)^2 + \cdots + (4.99)^2}{2} - q$$

$$= 75.1992 - q = 0.8150 \qquad (6.53)$$

$$\text{Periods within squares SS} = \frac{\sum_i \sum_k Y_{i\cdot k}^2}{n_s} - \frac{\sum_i Y_{i\cdot\cdot}^2}{n_s n_p} = pq - q$$

$$= \frac{(4.05)^2 + (4.30)^2 + \cdots + (4.34)^2}{2} - q$$

$$= 74.8142 - q = 0.4300 \qquad (6.54)$$

$$\text{Treatments SS} = \frac{\Sigma_t \, Y_t^2}{n_q n_t} - \frac{Y^2 \cdots}{N} = t - (1)$$

$$= \frac{(19.01)^2 + (15.46)^2}{8} - (1)$$

$$= 75.0490 - (1) = 0.7877 \tag{6.55}$$

$$\text{Error SS} = \text{Total SS} - SS_Q - SS_{S:Q} - SS_{P:Q} - SS_T$$

$$= 0.3926 \tag{6.56}$$

The analysis under this model is shown in Table 6.23, where it can be seen that the second model did not improve the analysis since we have a smaller $F$ value for treatments and fewer degrees of freedom for error, resulting in a lack of significance between the two treatment means. The sizes of the differences between the rows in the squares obviously has to be quite sizable to improve the analysis.

A second example of a crossover design comes from the area of exercise science where these designs are used frequently, most often in conjunction with a split-plot or repeated measures design which will be discussed in Chapter 7. The data have been supplied by Sherry Zigon of the Exercise Science Department at the University of Massachusetts and deal with the reaction times of 16 subjects, each subject having undergone two different exercise treatments. Each treatment was presented first an equal number of times to prevent any confounding of learning effect with treatment effects. The split-plot effects have been ignored in this analysis and the data are shown in Table 6.24, where the 1's and 2's in the body of the table represent treatments 1 and 2.

The first row of the table is made up of one group of subjects (males) and the second row of the table is made up of responses from females. The mathematical model for this set of data is

$$Y_{ij(t)k} = \mu + g_i + s_{ij} + d_k + t_{(t)} + gt_{i(t)} + e_{ijk} \tag{6.57}$$

where the $g_i$ represent groups, the $s_{ij}$ represent subjects within groups, the $d_k$ stand for days, and the $t_{(t)}$ stand for treatments. The analysis is as follows:

$$\text{Total SS} = \sum_i \sum_j \sum_k Y_{ijk}^2 - \frac{Y^2 \cdots}{N} = \sum_i \sum_j \sum_k Y_{ijk}^2 - (1)$$

**TABLE 6.23**  ANALYSIS OF VARIANCE OF CROSSOVER UNDER SECOND MODEL

| Source | df | SS | MS | F | E(MS) |
|---|---|---|---|---|---|
| Squares | 3 | 0.1229 | 0.0410 | 0.20 | $\sigma_e^2 + 2\sigma_{s:q}^2 + 4\sigma_q^2$ |
| Steers within squares | 4 | 0.8150 | 0.2038 | 1.56 | $\sigma_e^2 + 2\sigma_{s:q}^2$ |
| Periods within squares | 4 | 0.4300 | 0.1075 | 0.82 | $\sigma_e^2 + 2\sigma_{p:q}^2$ |
| Treatments | 1 | 0.7877 | 0.7877 | 6.02 | $\sigma_e^2 + 8\sigma_t^2$ |
| Error | 3 | 0.3926 | 0.1309 | | $\sigma_e^2$ |
| $n_q = 4$ | $n_s = 2$ | $n_p = 2$ | $n_t = 2$ | $N = n_q n_s n_p = 16$ | |

**TABLE 6.24**   UNRESISTED ANKLE SIMPLE REACTION TIME FOR TWO EXERCISE
                 TREATMENTS (IN MILLISECONDS)

|       |         |         |         | Subject |         |         |         |         |
|-------|---------|---------|---------|---------|---------|---------|---------|---------|
| Day   | 1       | 2       | 3       | 4       | 5       | 6       | 7       | 8       |
| 1     | 1—228.0 | 1—220.5 | 2—195.5 | 2—225.0 | 1—233.5 | 2—224.0 | 2—188.0 | 1—225.5 |
| 2     | 2—230.5 | 2—231.5 | 1—214.0 | 1—223.0 | 2—215.0 | 1—221.0 | 1—193.5 | 2—209.5 |
| Totals| 458.5   | 452.0   | 409.5   | 448.0   | 448.5   | 445.0   | 381.5   | 435.0   |

|       |         |         |         | Subject |         |         |         |         |
|-------|---------|---------|---------|---------|---------|---------|---------|---------|
| Day   | 9       | 10      | 11      | 12      | 13      | 14      | 15      | 16      |
| 1     | 1—230.0 | 2—223.0 | 2—195.0 | 1—187.5 | 1—209.0 | 2—189.5 | 2—198.5 | 1—200.0 |
| 2     | 2—248.0 | 1—242.5 | 1—203.5 | 2—187.5 | 2—206.0 | 1—198.5 | 1—196.0 | 2—203.5 |
| Totals| 478.0   | 465.5   | 398.5   | 375.0   | 415.0   | 388.0   | 394.5   | 403.5   |

$$= 1{,}452{,}337 - \frac{(6796)^2}{32}$$

$$= 1{,}452{,}337 - 1{,}443{,}300.50 = 9036.50 \tag{6.58}$$

$$\text{Group SS} = \frac{\sum_i Y_{i..}^2}{n_s n_d} - \frac{Y_{...}^2}{N} = g - 1$$

$$= \frac{(3478)^2 + (3318)^2}{16} - (1)$$

$$= 1{,}444{,}100.50 - (1) = 800.00 \tag{6.59}$$

$$\text{Subjects within groups SS} = \frac{\sum_i \sum_j Y_{ij.}^2}{n_d} - \frac{\sum_i Y_{i..}^2}{n_s n_d} = sg - g$$

$$= \frac{(458.5)^2 + (452.0)^2 + \cdots + (403.5)^2}{2} - g$$

$$= 1{,}451{,}339.00 - g = 7238.50 \tag{6.60}$$

$$\text{Day SS} = \frac{\sum_k Y_{..k}^2}{n_g n_s} - \frac{Y_{...}^2}{N} = d - 1$$

$$= \frac{(3372.5)^2 + (3423.5)^2}{16} - (1)$$

$$= 1{,}443{,}381.78 - (1) = 81.28 \tag{6.61}$$

$$\text{Treatment SS} = \frac{\sum_t Y_t^2}{n_s n_t} - \frac{Y_{...}^2}{N} = t - 1$$

**TABLE 6.25** ANALYSIS OF VARIANCE FOR CROSSOVER DESIGN

| Source | df | SS | MS | F | E(MS) |
|--------|-----|---------|--------|---------|-------|
| $G$ | 1 | 800.00 | 800.00 | 1.55 | $\sigma_e^2 + 2\sigma_{s:g}^2 + 16\sigma_g^2$ |
| $S{:}G$ | 14 | 7238.50 | 517.04 | 8.39** | $\sigma_e^2 + 2\sigma_{s:g}^2$ |
| $D$ | 1 | 81.28 | 81.28 | 1.32 | $\sigma_e^2 + 16\sigma_d^2$ |
| $T$ | 1 | 98.00 | 98.00 | 1.59 | $\sigma_e^2 + 16\sigma_t^2$ |
| $GT$ | 1 | 18.00 | 18.00 | 0.29 | $\sigma_e^2 + 8\sigma_{gt}^2$ |
| Error | 13 | 800.72 | 61.59 | | $\sigma_e^2$ |

$$n_g = 2 \qquad n_s = 8 \qquad n_d = 2 \qquad n_t = 2 \qquad N = n_g n_s n_d = 32$$

$$= \frac{(3426)^2 + (3370)^2}{16} - (1)$$

$$= 1{,}443{,}398.50 - (1) = 98.00 \tag{6.62}$$

$$\text{Group by treatment SS} = \frac{\Sigma_i \, \Sigma_t \, Y_{i..(t)}^2}{n_s} - \frac{\Sigma_i \, Y_{i..}^2}{n_{sd}} - \frac{\Sigma_t \, Y_t^2}{n_s n_t} + \frac{Y_{...}^2}{N}$$

$$= gt - g - t + 1$$

$$= \frac{(1759)^2 + (1719)^2 + \cdots + (1651)^2}{8} - g - t + (1)$$

$$= 1{,}444{,}216.50 - g - t + (1) = 18.00 \tag{6.63}$$

$$\text{Error SS} = \text{Total SS} - \text{SS}_G - \text{SS}_{S:G} - \text{SS}_D - \text{SS}_T - \text{SS}_{GT}$$

$$= 800.72 \tag{6.64}$$

The analysis of variance together with the expectations of mean squares are shown in Table 6.25. The expectations of mean squares show that all effects are not tested by the bottom line in the analysis of variance table as is common with the more simple crossover designs. In this case we have two errors, the first of which is used as a denominator in the $F$ test for groups and the second of which is used to test all the remaining effects. As the analysis shows, there were highly significant differences among the subjects within the groups but nothing else approached significance.

It should be pointed out that while the subjects (columns) were used to test between the two groups in this analysis, it is possible to have more than just one main classification involved in the columns.

## 6.14 OTHER DESIGNS

While the designs we have discussed are those most commonly used, there are a great many others available. Two useful texts in this area are those by Cochran and Cox (1957) and Federer (1955), each of which presents a large variety of designs. Some of these designs are the incomplete block design, lattice design, balanced and partially balanced incomplete block designs, lattice squares, and incomplete lattice squares. Special features of some of these designs are the use of intentional confounding and fractional replication. A major type of design, and one that is used in conjunction with other designs, is the split-plot design which is presented in Chapter 7.

## 6.15 FACTORIAL EXPERIMENTS

Many texts discuss and treat factorial experiments as a separate experimental design. Since a factorial experiment means that two or more sets of main effects or classifications, each at two or more levels, are included in the experiment and since this occurs or can occur with all types of designs discussed, no separate discussion is needed except for some special methods of treating the data which actually can be used with the common designs. Thus, "factorial" is not an experimental design. For more details concerning factorial experiments see Cochran and Cox (1957) or Federer (1955).

## EXERCISES

**6.1.** Dr. Christine M. Moffitt of the Idaho Cooperative Fisheries Research Unit at the University of Idaho conducted an experiment involving the evaluation of the weights of fish ovaries. Four groups of observations have been selected from her data, containing the weights in grams of the left ovaries of *Alosa sapidissima,* and are shown below. Carry out an analysis of variance for this set of data, a completely randomized design.

|  | Groups | | |
|---|---|---|---|
| 1 | 2 | 3 | 4 |
| 124.9 | 163.0 | 82.0 | 92.6 |
| 121.8 | 119.1 | 134.5 | 89.9 |
| 143.0 | 134.1 | 137.0 | 94.0 |
| 158.3 | 112.6 | 42.9 | 144.9 |
| 167.0 | 168.3 | 82.5 | 129.7 |
| 200.0 | 90.6 | 125.2 | 133.4 |
| 167.0 | 143.8 | 154.1 | 116.5 |
| 195.0 | 95.4 | 146.8 | 181.4 |
| 175.8 | 182.0 | 90.9 | 104.4 |
| 149.8 | 156.0 | 89.9 | 143.0 |

**6.2.** R. L. Anderson (1946) presented the yields of wheat straw in a randomized block experiment, using four blocks and four treatments. A portion of the data are presented here. Complete an analysis of variance for this randomized block experiment, including the expectations of mean squares.

| Treatment | Block | | | |
|---|---|---|---|---|
|  | 1 | 2 | 3 | 4 |
| 1 | 332 | 260 | 202 | 210 |
| 2 | 412 | 384 | 362 | 348 |

|  |  | Block |  |  |
|---|---|---|---|---|
| Treatment | 1 | 2 | 3 | 4 |
| 3 | 542 | 472 | 516 | 458 |
| 4 | 730 | 590 | 294 | 560 |

**6.3.** Using the methods described in Section 6.4, estimate the gain in efficiency in using the randomized design for the example used in Exercise 6.2.

**6.4.** In a feeding trial studying the effect of the level of added okara in a ration on total digestible nutrient, Scott Werme of the Department of Veterinary and Animal Sciences at the University of Massachusetts used four sheep in a $4 \times 4$ Latin square design. The TDN levels (in percent) in the total ration are shown below. Carry out an analysis of variance on this set of data, including the expectations of mean squares.

|  | Sheep |  |  |  |
|---|---|---|---|---|
| Period | 1 | 2 | 3 | 4 |
| 1 | $T_1 = 61.60$ | $T_4 = 75.83$ | $T_3 = 68.17$ | $T_2 = 65.61$ |
| 2 | $T_2 = 62.05$ | $T_1 = 58.63$ | $T_4 = 70.33$ | $T_3 = 67.22$ |
| 3 | $T_3 = 66.91$ | $T_2 = 68.10$ | $T_1 = 58.43$ | $T_4 = 71.98$ |
| 4 | $T_4 = 69.87$ | $T_3 = 67.25$ | $T_2 = 63.32$ | $T_1 = 60.30$ |

**6.5.** Part of a study conducted by Timothy C. Quick at The Ohio State University included two orthogonal $3 \times 3$ Latin squares, the first square containing three fistulated goats and the second square three nonfistulated goats. The observations are changes in weight per period (21 days) in grams per day. The columns in the data presented below represent goat numbers and the rows represent periods. Complete an analysis of variance for this set of data, including the expectations of mean squares considering the square, period, and treatment effects to be fixed.

|  | I |  |  |  | II |  |  |
|---|---|---|---|---|---|---|---|
| Period | Animal 102 | Animal 203 | Animal 305 | Period | Animal 101 | Animal 204 | Animal 306 |
| 1 | $A = 1.36$ | $B = -0.91$ | $C = -2.73$ | 1 | $A = -0.91$ | $B = -0.91$ | $C = -2.27$ |
| 2 | $B = -2.27$ | $C = 0.45$ | $A = -1.36$ | 2 | $C = -0.45$ | $A = -0.91$ | $B = 0.91$ |
| 3 | $C = -3.63$ | $A = -9.09$ | $B = -0.45$ | 3 | $B = -0.45$ | $C = 2.27$ | $A = 0.91$ |

**6.6.** A portion of the data collected by Zulma Garcia of the Department of Exercise Science at the University of Massachusetts was made up of the two Graeco-Latin squares shown below. The columns in each square are subjects, the rows are periods, the letters are exercise regimens, and the subscripts are load conditions. The observations are movement time for forearm flexion in postfatigue subjects. Complete an analysis of variance for this set of data, writing out the expectations of mean squares, considering the period, the exercise regimens, and the load conditions to be fixed effects. Any possible interaction of squares with other effects should be included in the error term.

| I | | | II | | |
|---|---|---|---|---|---|
| $A_0 = 158$ | $C_1 = 197$ | $B_2 = 203$ | $A_0 = 135$ | $C_1 = 172$ | $B_2 = 177$ |
| $B_1 = 159$ | $A_2 = 254$ | $C_0 = 147$ | $B_1 = 150$ | $A_2 = 217$ | $C_0 = 150$ |
| $C_2 = 189$ | $B_0 = 153$ | $A_1 = 167$ | $C_2 = 187$ | $B_0 = 189$ | $A_1 = 184$ |

**6.7.** In a study of the digestibility and feeding value of the soybean by-product okara for sheep and lactating dairy cattle, Scott Werme, at the University of Massachusetts, conducted an experiment with a crossover design using two treatments and two feeding periods of five weeks each. While 38 cows were used in nineteen $2 \times 2$ Latin squares, only 8 cows making up four $2 \times 2$ squares are included here. The data are pounds of fat produced in each five-week period. Complete an analysis of variance for this set of data, accounting for the differences among the squares, among the cows within the squares, between the periods, between the treatments, and the remainder or error. Write out the expectations of mean squares, considering the periods and treatments to be fixed effects.

| | I | | II | | III | | IV | |
|---|---|---|---|---|---|---|---|---|
| Period | Animal 400 | Animal 453 | Animal 470 | Animal 528 | Animal 597 | Animal 607 | Animal 190 | Animal 610 |
| 1 | $A = 6.99$ | $B = 8.10$ | $A = 6.58$ | $B = 6.33$ | $A = 9.38$ | $B = 8.08$ | $A = 6.44$ | $B = 7.32$ |
| 2 | $B = 5.15$ | $A = 7.12$ | $B = 4.64$ | $A = 3.65$ | $B = 8.30$ | $A = 8.41$ | $B = 4.91$ | $A = 5.27$ |

**6.8.** Mills et al. (1976) conducted an experiment with radishes from which a portion of the data has been selected for this exercise. There were two sources of nitrogen—ammonium sulfate and potassium nitrate—and three levels of nitrogen; nitrapyrin or no nitrapyrin were the treatments. The four observations are the fresh weights of the plants in grams/pot. Carry out an analysis of variance for this set of data. Write out the expectations of mean squares, considering this to be a fixed model.

| | Source of nitrogen | | | |
|---|---|---|---|---|
| | Ammonium sulfate | | Potassium nitrate | |
| Level | Nitrapyrin | No nitrapyrin | Nitrapyrin | No nitrapyrin |
| 1 | 14.3 | 17.5 | 17.6 | 13.9 |
| | 15.9 | 16.7 | 24.0 | 16.8 |
| | 14.8 | 15.7 | 13.5 | 17.3 |
| | 20.8 | 15.1 | 17.9 | 12.6 |
| 2 | 37.5 | 39.6 | 43.3 | 45.8 |
| | 29.4 | 33.0 | 53.5 | 46.9 |
| | 33.8 | 52.8 | 49.3 | 48.0 |
| | 33.1 | 36.2 | 49.9 | 47.0 |
| 3 | 41.4 | 52.5 | 59.8 | 77.8 |
| | 49.9 | 53.4 | 98.4 | 87.6 |
| | 43.2 | 51.7 | 79.4 | 83.4 |
| | 40.1 | 52.2 | 80.0 | 84.7 |

# Chapter 7

# Split-Plot Designs

## 7.1 INTRODUCTION

One of the most valuable designs in the biological sciences, the split plot, is very widely used but, unfortunately, it is frequently not recognized by experimenters when planning or analyzing their own experiments. The application of this design is not limited to the biological sciences and it has found wide usage in such areas as psychology, sociology, and economics, sometimes being referred to as a *between and within subject design.* The split-plot design is used in conjunction with all the designs previously discussed and greatly enhances their value.

The split-plot design differs from other designs in that *whole plots* or *whole units* are subjected to one or more sets of treatments and then the whole plots are divided into *split plots* or *subunits* to which one or more additional sets of treatments are applied. Since each whole unit contains all levels of the split-plot treatments, it can be thought of as a complete block as far as the split-plot treatments are concerned. However, the whole plot contains only one level or combination of levels of the whole-plot set of treatments, making this a special kind of incomplete block design (Steel and Torrie, 1980).

Split-plot designs were developed in agronomic experimentation where whole units were plots of land that were split into subunits. However, the concept of a whole plot has expanded well beyond this usage since a whole plot can consist of such items as animals and humans, and in the Latin square and crossover designs, the row $\times$ column cells become the whole plots.

Let us consider a randomized block, split-plot design. Letting $A$ be the treatments applied to the whole plots and $B$ the treatments applied to the subplots, the design would be

| $A_1$ | $A_3$ | $A_2$ | $A_4$ |
|-------|-------|-------|-------|
| $B_2$ | $B_2$ | $B_1$ | $B_2$ |
| $B_1$ | $B_1$ | $B_2$ | $B_1$ |

Block I

| $A_4$ | $A_1$ | $A_3$ | $A_2$ |
|-------|-------|-------|-------|
| $B_1$ | $B_2$ | $B_1$ | $B_2$ |
| $B_2$ | $B_1$ | $B_2$ | $B_1$ |

Block II

| $A_3$ | $A_1$ | $A_4$ | $A_2$ |
|-------|-------|-------|-------|
| $B_2$ | $B_1$ | $B_1$ | $B_2$ |
| $B_1$ | $B_2$ | $B_2$ | $B_1$ |

Block III

It can be seen that the randomization of treatments to the whole plots and to the split plots must occur at two separate instances. First, the levels of treatment $A$ must be randomized among the whole plots in each block. Second, the levels of treatment $B$ are randomized over the subunits in each whole plot separately. Thus, the whole plots are the experimental units for the $A$ treatment and the split plots are the experimental units for the $B$ treatment.

If, instead of using the split-plot design for factors (treatments) $A$ and $B$, we had used only the randomized block design, the layout of the experiment would be quite different as shown below:

| $A_1B_2$ | $A_1B_1$ | $A_2B_2$ | $A_3B_2$ |
|----------|----------|----------|----------|
| $A_2B_1$ | $A_4B_1$ | $A_3B_1$ | $A_4B_2$ |

Block I

| $A_1B_2$ | $A_4B_2$ | $A_2B_1$ | $A_1B_1$ |
|----------|----------|----------|----------|
| $A_4B_1$ | $A_2B_2$ | $A_3B_2$ | $A_3B_1$ |

Block II

| $A_4B_2$ | $A_3B_1$ | $A_3B_2$ | $A_2B_2$ |
|----------|----------|----------|----------|
| $A_2B_1$ | $A_1B_2$ | $A_1B_1$ | $A_4B_1$ |

Block III

In this situation each combination of $A$ and $B$ is treated as a single treatment and the eight combinations are randomized in each of the three random blocks. The randomization is then one staged rather than two staged as pointed out for the split-plot design. In the ANOVA table for the split-plot design the whole-plot effects appear first, followed by the split-plot effects.

The analyses of the two designs are quite similar as an inspection of the expectations of means shows. In the randomized block design the interactions $RA$ and $RB$ are used to test the significance of the $A$ and $B$ treatments, respectively. It is common practice to pool both of these interactions with the error line, yielding an error mean square based on a greater number of degrees of freedom. However, this pooling should not be done automatically but only after tests of significance have shown them to be not significantly different from zero at a selected probability level. Both the 20 and 25 percent levels have been suggested for use in this test. However, the 25 percent level seems to be preferred by most statisticians. Then, if these interactions are not significant in this range, they are pooled. In this way, the data guide in the decision to pool rather than just an automatic pooling without justification.

The expectations of mean squares for the split-plot design show a similar situation regarding the ratios for the tests of significance. Note that the first three lines of the table (the whole-plot portion of the analysis) contain a whole-plot error component $\sigma_d^2$ which has not been introduced earlier in the text. As can be seen for the error ($a$) line of the table, $\sigma_d^2$ is wholly confounded with $\sigma_{ra}^2$. For a further discussion of the significance of the variance component $\sigma_d^2$, see Anderson and McLean (1974). Since there are two subunits in each whole plot, the coefficient for the combined $\sigma_{ra}^2 + \sigma_d^2$ is 2. Because $A$ is a fixed effect, $\sigma_{ra}^2$ is deleted from the expected mean square for

**TABLE 7.1** ANALYSES OF VARIANCE OF TWO STATISTICAL DESIGNS INVOLVING THE SAME TREATMENT FACTORS ($R$ = REPLICATIONS = BLOCKS)

| Randomized block | | Split plot | |
|---|---|---|---|
| Source | df | Source | df |
| $R$ | $n_r - 1 = 2$ | $R$ | $n_r - 1 = 2$ |
| $A$ | $n_a - 1 = 3$ | $A$ | $n_a - 1 = 3$ |
| $B$ | $n_b - 1 = 1$ | Error $(a)$ $(RA)$ | $(n_r - 1)(n_a - 1) = 6$ |
| $RA$ | $(n_r - 1)(n_a - 1) = 6$ | $B$ | $n_b - 1 = 1$ |
| $RB$ | $(n_r - 1)(n_b - 1) = 2$ | $RB$ | $(n_r - 1)(n_b - 1) = 2$ |
| $AB$ | $(n_a - 1)(n_b - 1) = 3$ | $AB$ | $(n_a - 1)(n_b - 1) = 3$ |
| Error $(RAB)$ | $(n_r - 1)(n_a - 1)(n_b - 1) = 6$ | Error $(b)$ $(RAB)$ | $(n_r - 1)(n_a - 1)(n_b - 1) = 6$ |

| Source | E(MS) | Source | E(MS) |
|---|---|---|---|
| $R$ | $\sigma_e^2 + 12\sigma_r^2$ | $R$ | $\sigma_e^2 + 2\sigma_d^2 + 12\sigma_r^2$ |
| $A$ | $\sigma_e^2 + 2\sigma_{ra}^2 + 6\sigma_a^2$ | $A$ | $\sigma_e^2 + 2(\sigma_{ra}^2 + \sigma_d^2) + 6\sigma_a^2$ |
| $B$ | $\sigma_e^2 + 4\sigma_{rb}^2 + 12\sigma_b^2$ | Error $(a)$ $(RA)$ | $\sigma_e^2 + 2(\sigma_{ra}^2 + \sigma_d^2)$ |
| $RA$ | $\sigma_e^2 + 2\sigma_{ra}^2$ | $B$ | $\sigma_e^2 + 4\sigma_{rb}^2 + 12\sigma_b^2$ |
| $RB$ | $\sigma_e^2 + 4\sigma_{rb}^2$ | $RB$ | $\sigma_e^2 + 4\sigma_{rb}^2$ |
| $AB$ | $\sigma_e^2 + 3\sigma_{ab}^2$ | $AB$ | $\sigma_e^2 + 3\sigma_{ab}^2$ |
| Error $(RAB)$ | $\sigma_e^2$ | Error $(b)$ $(RAB)$ | $\sigma_e^2$ |

$$n_r = 3 \qquad n_a = 4 \qquad n_b = 2$$

$R$. In the split-plot portion of the analysis, should the $RB$ interaction be found not significant in the 20–25 percent range, it may be pooled with the error term.

The two error terms in the split-plot design, designated error $(a)$ and error $(b)$, are expected to differ since subunits in the same whole unit may be correlated. In an agronomic experiment, subunits lying side by side in one whole plot are likely to be more similar in fertility than subunits in different whole plots; that is, subplot measurements and their experimental errors in the same whole plot are expected to be correlated. Let us consider the model

$$Y_{ijk} = \mu + r_i + a_j + ra_{ij} + d_{ij} + b_k + rb_{ik} + ab_{jk} + e_{ijk} \qquad (7.1)$$

for the split-plot design where the $r_i$ represent block effects, the $a_j$ represent $A$ treatment effects, and the $b_k$ represent $B$ treatment effects. Assuming that there is a correlation $\rho$ between the experimental errors $e_{ijk}$ and $e_{iju}$ for any two subplots in the same whole plot and no correlation between errors $e_{ijk}$ and $e_{stu}$ for any two subplots in different whole plots, the covariance in the two cases, as pointed out by Cochran and Cox (1957), would be

$$E(e_{ijk}e_{iju}) = \rho\sigma_e^2 \qquad E(e_{ijk}e_{stu}) = 0$$

The value $\rho\sigma^2$ is the whole-plot error component $\sigma_d^2$. If there are two subplots in each whole plot, the error variance of a *unit total* is

$$E(e_{ij1} + e_{ij2})^2 = E(e_{ij1})^2 + E(e_{ij2})^2 + 2E(e_{ij1}e_{ij2})$$

$$= \sigma_e^2 + \sigma_e^2 + 2\rho\sigma_e^2 = 2\sigma_e^2(1 + \rho)$$

The appropriate error variance per subunit for the main effects of $A$ is then $\sigma_e^2(1 + \rho)$. With $\beta$ subplots per whole plot the corresponding variance is $\sigma_e^2[1 + (\beta - 1)\rho]$.

Since the main effects of $B$ arise from differences between the two subunits, the variance of the differences is

$$E(e_{ij1} - e_{ij2})^2 = 2\sigma_e^2(1 - \rho)$$

Therefore, the variance per subunit applicable to the split-plot treatment $B$ is $\sigma_e^2(1 - \rho)$. This variance also applies to comparisons among $B$ effects at different levels of $A$. For a more detailed discussion, see Cochran and Cox (1957, p. 294).

It can be seen then that if a correlation does exist among or between subunits in the same whole units, the variance per subunit for the whole-plot effects is expected to be larger than the variance per subunit for the split-plot effects. Compared to the randomized block design the $B$ treatment effects will be measured with greater precision. However, there is a commensurate loss in the measurement of the whole-plot effects which are measured with less precision. One of the consequences of the difference in error terms is that when certain comparisons of treatment effects are made, they require a weighted average of two error variances or mean squares.

Since the split-plot effects are measured with greater precision than whole-plot effects, this design can be used when smaller differences are expected to arise among the levels of one set of treatments. This set, of course, would be applied to the subunits. In many experiments certain of the treatments require larger areas while other treatments can more conveniently be fitted into smaller areas. For example, in an agronomic experiment testing different depths of tillage or different methods of tillage, reasonably large areas must be used to accommodate the machinery. One can then superimpose a second treatment, such as different varieties, on the different tillage methods, the tillage methods serving as the whole plots. Again, an additional source of variation may be included in an experiment to obtain additional information. In the first example discussed in this chapter, a study was made of the effect of crowding on the survival of mosquitoes under different levels of density. Several replicates were included on each level. Additional information was obtained by recording the survival separately in each replicate for males and females, the effect of sex being a split-plot effect.

When a plot of ground is split into two or more subunits this is generally referred to as a *split plot in space*. However, the same whole plot might instead be measured or evaluated at two or more dates in which case we would have a *split plot in time*. There are occasions when split-plot designs are used but cannot really be categorized as to space or time. Both the split plot in space and the split plot in time can readily occur in the same experiment. Thus, if the subunits (split plots) in space are measured at specific time periods this would result in a second split, so we would have what is called a split-split-plot design. Of course, the split plots could be subdivided rather than measured over time, with an additional factor or treatment applied to the subdivisions which would also result in a split-split-plot design. The number of splits is limited only by the practical considerations of the experiment.

When measurements are made on the same unit at two or more times it is common to refer to the design as a split plot in time. However, it is perhaps more accurate to refer to this type of design as a "repeated measures" design. When designs such as this are analyzed in the same fashion as are the split plots, the assumption is made that the correlations between all time periods are the same. However, it may well be that the correlation between time period 1 and time period 2 is greater than that between time period 1 and time period 3, or even greater than between time period 1 and later time periods. Box (1950) gives a method, based on multivariate theory, to test whether the assumption made in a split-plot analysis is correct. If the assumptions do not hold, Danford et al. (1960) and Cole and Grizzle (1966) present multivariate methods of analysis which do not require these assumptions. For further discussion of the problems involved and possible solutions in the repeated measures designs see the preceding references as well as Snedecor and Cochran (1980).

Designs that include split plots are very similar in nature to designs that include nested classifications. If a whole plot is divided to get two or more random measurements of the treatment effect, then the observations would merely be plots nested within the block by treatment $A$ subclasses. However, if the two plots represent effects of a nonrandom nature (e.g., a.m. versus p.m. and different levels of a treatment), then we would have a split-plot design.

It is common in split-plot designs to calculate the whole-plot sum of squares and the split-plot sum of squares. In the design under discussion, the whole-plot sum of squares is the sum of squares among the whole plots within the blocks. The split-plot sum of squares is that for among split plots within whole plots. In split-split-plot designs the sum of squares for the split-split portion of the analysis is that for among the split-split plots within the split plots.

In some situations, particularly field experiments, it is not feasible or possible to randomize the subunit treatments within the whole units. Instead, the subunit treatments are applied in strips across the whole units. Such a design is shown in Table 7.2 together with the method of analysis and the expectations of mean squares. When

**TABLE 7.2**  RANDOMIZED BLOCK DESIGN WITH SUBUNIT TREATMENTS APPLIED IN STRIPS

|  | Block I |  |  |  |  | Block II |  |  |  |  | Block III |  |  |
|---|---|---|---|---|---|---|---|---|---|---|---|---|---|
|  | $A_4$ | $A_2$ | $A_1$ | $A_3$ |  | $A_1$ | $A_4$ | $A_3$ | $A_2$ |  | $A_2$ | $A_3$ | $A_1$ | $A_4$ |
| $B_2$ |  |  |  |  | $B_1$ |  |  |  |  | $B_3$ |  |  |  |  |
| $B_1$ |  |  |  |  | $B_2$ |  |  |  |  | $B_1$ |  |  |  |  |
| $B_3$ |  |  |  |  | $B_3$ |  |  |  |  | $B_2$ |  |  |  |  |

| Source | df | E(MS) |
|---|---|---|
| Block $(R)$ | $n_r - 1 = \phantom{0}2$ | $\sigma_e^2 + 4\sigma_d^2 + 3\sigma_{gr}^2 + 12\sigma_r^2$ |
| $A$ | $n_a - 1 = \phantom{0}3$ | $\sigma_e^2 + 3(\sigma_{ra}^2 + \sigma_g^2) + 9\sigma_a^2$ |
| Error $(a)$ $(RA)$ | $(n_r - 1)(n_a - 1) = \phantom{0}6$ | $\sigma_e^2 + 3(\sigma_{ra}^2 + \sigma_g^2)$ |
| $B$ | $(n_b - 1) = \phantom{0}2$ | $\sigma_e^2 + 4(\sigma_{rb}^2 + \sigma_d^2) + 12\sigma_b^2$ |
| Error $(b)$ $(RB)$ | $(n_r - 1)(n_b - 1) = \phantom{0}4$ | $\sigma_e^2 + 4(\sigma_{rb}^2 + \sigma_d^2)$ |
| $AB$ | $(n_a - 1)(n_b - 1) = \phantom{0}6$ | $\sigma_e^2 + 3\sigma_{ab}^2$ |
| Error $(c)$ $(RAB)$ | $(n_r - 1)(n_a - 1)(n_b - 1) = 12$ | $\sigma_e^2$ |

conducting an experiment with this design both the $A$ treatments and the $B$ treatments should be randomized independently. This design is not generally recommended because the precision of the measurement on the main effects of $A$ and $B$ is decreased with a concomitant increase in precision for the $AB$ interaction. If the main purpose of the experiment is a study of the interaction, then, of course, the design may be a desirable one.

## 7.2 COMPLETELY RANDOMIZED SPLIT-PLOT DESIGN

As an example of this type of design, let us look at the data resulting from an experiment conducted in the Department of Entomology at the University of Massachusetts by Pedro Barbosa, T. Michael Peters, and N. C. Greenough. The experiment was performed to investigate the effects of overcrowding in larval *Aedes aegypti* (L.) and to illustrate the compensatory response to stress in high-density conditions. Density levels of 160, 280, 400, 520, 640, 960, and 1280 larvae per milliliter of solution were included in the experiment. It was assumed that deviations from a 1:1 ratio of males to females in the different density levels was of a random nature. While several parameters were evaluated, we shall concern ourselves with larval survival. The data are shown in Table 7.3.

The mathematical model can be written

$$Y_{ijk} = \mu + d_i + r_{ij} + s_k + ds_{ik} + e_{ijk} \tag{7.2}$$

where the $d_i$ represent densities, the $r_{ij}$ represent replicates within the density, and the $s_k$ represent sexes. The $r_{ij}$ effects are for the whole-plot error and the $e_{ijk}$ are for the split-plot error, where $e_{ijk} = rsd_{ijk} + w_{ijk}$.

The whole plots in this example are the density $\times$ replicate subclasses and the split plots are the density $\times$ replicate $\times$ sex observations. Since each replicate is split into two sexes, sex is the split-plot effect.

The calculations necessary for the analysis follow.

$$\text{Total SS} = \sum_i \sum_j \sum_k Y_{ijk}^2 - \frac{Y_{\cdots}^2}{N} = drs - (1)$$

$$= 74{,}840.7781 - \frac{(1914.033)^2}{56}$$

$$= 74{,}840.7781 - 65{,}420.0415 = 9420.7366 \tag{7.3}$$

$$\text{Whole-plot SS} = \frac{\sum_i \sum_j Y_{ij\cdot}^2}{n_s} - \frac{Y_{\cdots}^2}{N} = dr - (1)$$

$$= \frac{(89.375)^2 + (95.000)^2 + \cdots + (26.954)^2}{2} - (1)$$

$$= 73{,}700.9610 - (1) = 8280.9195 \tag{7.4}$$

**TABLE 7.3**  PERCENT LARVAL SURVIVAL OF *AEDES AEGYPTI (L.)* UNDER DIFFERING DENSITY CONDITIONS

| Initial density | Replicate | Males | Females | Subtotals |
|---|---|---|---|---|
| 160 | 1 | 51.250 | 38.125 | 89.375 |
| | 2 | 57.500 | 37.500 | 95.000 |
| | 3 | 43.125 | 47.500 | 90.625 |
| | 4 | 54.375 | 45.000 | 99.375 |
| Subtotals | | 206.250 | 168.125 | 374.375 |
| 280 | 1 | 51.071 | 38.571 | 89.642 |
| | 2 | 45.357 | 39.286 | 84.643 |
| | 3 | 50.000 | 41.071 | 91.071 |
| | 4 | 54.643 | 34.286 | 88.929 |
| Subtotals | | 201.071 | 153.214 | 354.285 |
| 400 | 1 | 46.000 | 39.250 | 85.250 |
| | 2 | 49.250 | 37.250 | 86.500 |
| | 3 | 43.500 | 38.750 | 82.250 |
| | 4 | 53.250 | 35.500 | 88.750 |
| Subtotals | | 192.000 | 150.750 | 342.750 |
| 520 | 1 | 45.385 | 28.654 | 74.039 |
| | 2 | 43.654 | 41.346 | 85.000 |
| | 3 | 41.538 | 38.654 | 80.192 |
| | 4 | 37.885 | 40.000 | 77.885 |
| Subtotals | | 168.462 | 148.654 | 317.116 |
| 640 | 1 | 36.875 | 33.438 | 70.313 |
| | 2 | 29.062 | 25.625 | 54.687 |
| | 3 | 31.875 | 33.594 | 65.469 |
| | 4 | 35.625 | 32.500 | 68.125 |
| Subtotals | | 133.437 | 125.157 | 258.594 |
| 960 | 1 | 23.830 | 22.553 | 46.383 |
| | 2 | 24.574 | 22.234 | 46.808 |
| | 3 | 16.596 | 18.830 | 35.426 |
| | 4 | 9.681 | 14.787 | 24.468 |
| Subtotals | | 74.681 | 78.404 | 153.085 |
| 1280 | 1 | 13.984 | 15.234 | 29.218 |
| | 2 | 15.078 | 12.500 | 27.578 |
| | 3 | 16.172 | 13.906 | 30.078 |
| | 4 | 13.438 | 13.516 | 26.954 |
| Subtotals | | 58.672 | 55.156 | 113.828 |
| Totals | | 1034.573 | 879.460 | 1914.033 |

$$SS_D = \frac{\Sigma_i \, Y_{i..}^2}{n_r n_s} - \frac{Y_{...}^2}{N} = d - (1)$$

$$= \frac{(374.375)^2 + (354.285)^2 + \cdots + (113.828)^2}{8} - (1)$$

$$= 73,372.1637 - (1) = 7952.1222 \tag{7.5}$$

$$SS_{R:D} = \frac{\Sigma_i \, \Sigma_j \, Y_{ij.}^2}{n_s} - \frac{\Sigma_i \, Y_{i..}^2}{n_r n_s} = dr - d$$

$$= \frac{(89.375)^2 + (95.000)^2 + \cdots + (26.954)^2}{2} - d$$

$$= 73,700.9610 - (d) = 328.7973 \tag{7.6}$$

$$\text{Split-plot SS} = \sum_i \sum_j \sum_k Y_{ijk}^2 - \frac{\Sigma_i \, \Sigma_j \, Y_{ij.}^2}{n_s} = drs - dr$$

$$= 74,840.7781 - 73,700.9610 = 1139.8171 \tag{7.7}$$

$$SS_S = \frac{\Sigma_k \, Y_{..k}^2}{n_d n_r} - \frac{Y_{...}^2}{N} = s - (1)$$

$$= \frac{(1034.573)^2 + (879.460)^2}{28} - (1)$$

$$= 65,849.6851 - (1) = 429.6436 \tag{7.8}$$

$$SS_{DS} = \frac{\Sigma_i \, \Sigma_k \, Y_{i.k}^2}{n_r} - \frac{\Sigma_i \, Y_{i..}^2}{n_r n_s} - \frac{\Sigma_k \, Y_{..k}^2}{n_d n_r} + \frac{Y_{...}^2}{N} = ds - d - s + (1)$$

$$= \frac{(206.250)^2 + (168.125)^2 + \cdots + (55.156)^2}{4} - d - s + (1)$$

$$= 74,113.7273 - d - s + (1) = 311.9200 \tag{7.9}$$

$$SS_{RS:D} = \sum_i \sum_j \sum_k Y_{ijk}^2 - \frac{\Sigma_i \, \Sigma_j \, Y_{ij.}^2}{n_s} - \frac{\Sigma_i \, \Sigma_k \, Y_{i.k}^2}{n_r} + \frac{\Sigma_i \, Y_{i..}^2}{n_r n_s}$$

$$= drs - dr - ds + d = 398.2535 \tag{7.10}$$

The outline of the analysis together with the degrees of freedom and the expectations of mean squares are shown in Table 7.4. The density levels and the sex effects are considered to be fixed effects and the replicates to be random effects.

From Table 7.5, which shows the completed analysis of variance, it can be seen that there is a highly significant effect of density on survival as well as a highly significant effect of sex. There is also a significant interaction between density and sex. Since the density treatments were applied in increasing levels, the appropriate partitioning of the treatment sum of squares is into linear, quadratic, and higher-degree components. Since the first three degrees of the polynomial account for most of the density sum of

**TABLE 7.4** SOURCES OF VARIATION, DEGREES OF FREEDOM, AND EXPECTATIONS OF MEAN SQUARES FOR DATA SHOWN IN TABLE 7.3

| Source | df | E(MS) |
|---|---|---|
| Whole plots | $n_d n_r - 1 = 27$ | |
| D | $n_d - 1 = 6$ | $\sigma_w^2 + 2\sigma_{r:d}^2 + 8\sigma_d^2$ |
| Error (a) (R:D) | $(n_r - 1)n_d = 21$ | $\sigma_w^2 + 2\sigma_{r:d}^2$ |
| Split plots | $(n_s - 1)n_d n_r = 28$ | |
| S | $n_s - 1 = 1$ | $\sigma_w^2 + \sigma_{rs:d}^2 + 28\sigma_s^2$ |
| DS | $(n_d - 1)(n_s - 1) = 6$ | $\sigma_w^2 + \sigma_{rs:d}^2 + 4\sigma_{ds}^2$ |
| Error (b) (RS:D) | $(n_r - 1)(n_s - 1)n_d = 21$ | $\sigma_w^2 + \sigma_{rs:d}^2$ |

$n_d = 7 \quad n_r = 4 \quad n_s = 2 \quad n_w = 1 \quad N = 56$

squares, it is not necessary to go beyond the cubic component. It can be seen that the linear trend or component accounts for the greatest portion of the density sum of squares, reflecting the decrease in survival with increasing density levels. The highly significant effect of the sexes reflects the fact that there was a greater mortality among the females. Since there was a significant interaction between density levels and sex, this was examined further by partitioning the interaction into the components of linear × linear, linear × quadratic, and linear × cubic effects of density × sex. The details of this type of partitioning are presented in Table 7.15. The highly significant effect of the linear × linear interaction emphasizes the difference in the straight line response to density levels between the two sexes. It is concluded that there is a highly significant difference between the linear regressions of survival on density for the two sexes. It should be noted that the tabulated orthogonal polynomial coefficients could not be used because the intervals between density levels were not equal. The necessary coefficients, calculated as shown in Section 5.9.3, are represented in Table 7.6.

**TABLE 7.5** ANALYSIS OF VARIANCE FOR COMPLETELY RANDOMIZED SPLIT-PLOT DESIGN

| Source | df | SS | MS | F |
|---|---|---|---|---|
| Whole plots | 27 | 8280.9195 | | |
| D | 6 | 7952.1222 | 1325.3537 | 84.65** |
|   Linear | 1 | 7700.3054 | 7700.3054 | 491.81** |
|   Quadratic | 1 | 0.1094 | 0.1094 | 0.01 |
|   Cubic | 1 | 218.8329 | 218.8329 | 13.98** |
|   Residual | 3 | 32.8745 | 10.9582 | 0.70 |
| R:D | 21 | 328.7973 | 15.6570 | |
| Split plots | 28 | 1139.8171 | | |
| S | 1 | 429.6346 | 429.6436 | 22.66** |
| DS | 6 | 311.9200 | 51.9867 | 2.74* |
|   Linear × linear | 1 | 229.2648 | 229.2648 | 12.09** |
|   Linear × quadratic | 1 | 27.0750 | 27.0750 | 1.43 |
|   Linear × cubic | 1 | 31.1368 | 31.1368 | 1.64 |
|   Residual | 3 | 24.4434 | 8.1478 | 0.43 |
| RS:D | 21 | 398.2535 | 18.9645 | |

**TABLE 7.6**  ORTHOGONAL POLYNOMIAL
              COEFFICIENTS FOR
              DENSITY LEVELS

| Linear | Quadratic | Cubic |
|--------|-----------|-------|
| −78 | 21.6370 | −622.4666 |
| −57 | 6.2065 | 239.0114 |
| −36 | −5.5394 | 524.9148 |
| −15 | −13.6007 | 397.2437 |
| 6 | −17.9774 | 17.9981 |
| 62 | −11.6350 | −991.2431 |
| 118 | 20.9090 | 434.5418 |

After completion of the initial analysis of variance, it may be desirable to investigate further the causes of significant differences among the main effects of the treatments by such procedures as single degree of freedom comparisons, trend analyses, or range tests. In any of these procedures, it is vital that the appropriate error mean square be employed. This creates no problem with the main effects of treatments since tests among the main effects in the whole plots use the whole-plot error (mean square) and tests among main effects in the split plots use the split-plot error term. However, a significant interaction between a whole-plot treatment and a split-plot treatment presents a different situation. A significant (and important) interaction requires an investigation of the simple effects among the subclasses (meaning differences between or among levels of $B$ at the same level of $A$ or differences between or among levels of $A$ at the same level of $B$) and different error terms are used depending on the tests conducted. This is a result of the fact that differences among simple effects are mixtures of main effects and interaction effects. In some cases, the main effects and the interaction effects have different error terms and thus a synthetic error term is necessary, a combination of the two error terms. This combination error term is constructed by using a weighted average of the two error terms, with the weights being the degrees of freedom for the respective error terms being combined, as will be shown below.

Let us consider a completely randomized design in which we have $A$ main effects in the whole-plot portion and $B$ main effects in the split-plot portion. The different types of comparison among the subclasses that are possible together with the error mean squares involved and used in the denominators of $F$ tests are shown in Table 7.7. While the concept of different error terms is presented in this table in terms of tests of simple effects, the same denominators are used when other procedures are used such as testing for the linear effects of $B$ within each level of $A$. Since simple effects are really nested effects, one can determine which error terms are involved in these tests by putting the comparisons into a nested form and recalling that *the sum of squares for a nested effect contains the sum of squares for the effect to the left of the colon, plus the sums of squares for all possible interactions of the effect to the left of the colon with all effects (considered singly and in combination) to the right of the colon.*

It can be seen that the last two comparisons in Table 7.7 are not nested effects but comparisons between $B$ at different levels of $A$ or $A$ at different levels of $B$. These

**TABLE 7.7** ERROR TERMS INVOLVED IN THE DENOMINATORS OF $F$ RATIOS FOR TESTING SIMPLE EFFECTS AND TESTS AMONG SUBCLASSES IN A COMPLETELY RANDOMIZED SPLIT-PLOT DESIGN

| Comparison | Nested equivalent | Effects involved | Error mean squares used in denominators of $F$ ratios |
|---|---|---|---|
| $\overline{AB}_{11} - \overline{AB}_{12}$ | $B{:}A$ | $B, AB$ | Error $(b)$ |
| $\overline{AB}_{21} - \overline{AB}_{22}$ | $B{:}A$ | $B, AB$ | Error $(b)$ |
| $\overline{AB}_{11} - \overline{AB}_{21}$ | $A{:}B$ | $A, AB$ | Error $(a)$ and error $(b)$ |
| $\overline{AB}_{12} - \overline{AB}_{22}$ | $A{:}B$ | $A, AB$ | Error $(a)$ and error $(b)$ |
| $\overline{AB}_{11} - \overline{AB}_{22}$ | None | $A, B, AB$ | Error $(a)$ and error $(b)$ |
| $\overline{AB}_{12} - \overline{AB}_{21}$ | None | $A, B, AB$ | Error $(a)$ and error $(b)$ |

are merely comparisons among subclasses in which $A$, $B$, and $AB$ effects are involved, since both $A$ and $B$ vary. In the first comparison only error $(b)$ is involved since both $B$ and $AB$ are tested by error $(b)$ mean square. In the third comparison both error $(a)$ and error $(b)$ are involved since $A$ is tested by error $(a)$ and $AB$ by error $(b)$.

As pointed out previously, the combination of error $(a)$ mean square and error $(b)$ mean square is a weighted average of the two error terms, the degrees of freedom of the respective mean squares being the weights. The appropriate error mean square for the present example, where error $(a) = R{:}D$ and error $(b) = RS{:}D$, would be found by pooling the error sums of squares and degrees of freedom as follows:

$$\frac{SS_{error(a)} + SS_{error(b)}}{(n_r - 1)n_d + (n_r - 1)(n_s - 1)n_d} = \frac{SS_{error(a)} + SS_{error(b)}}{(n_r - 1)n_d n_s}$$

$$= \frac{MS_{error(a)} + (n_s - 1)MS_{error(b)}}{n_s}$$

$$= \frac{(15.6570) + (1)(18.9645)}{2} = 17.3108 \qquad (7.11)$$

Once an error mean square is determined, it is necessary to decide on the appropriate degrees of freedom. When it is not necessary to synthesize a mean square, the degrees of freedom are those corresponding to the mean square used. In Section 3.17, we used Satterthwaite's approximation for this purpose, where mean squares were pooled and approximate degrees of freedom were developed for the pooled mean square. Using this same method, we can calculate the approximate degrees of freedom for the synthesized mean square shown in Equation (7.11) as

$$\frac{[MS_{error(a)} + (n_s - 1)MS_{error(b)}]^2}{MS^2_{error(a)}/(n_r - 1)n_d + [(n_s - 1)MS_{error(b)}]^2/(n_r - 1)(n_s - 1)n_d}$$

$$= \frac{[15.6570 + (1)(18.9645)]^2}{(15.6570)^2/(3)(7) + [(1)(18.9645)]^2/(3)(1)(7)} = 42 \qquad (7.12)$$

The value is rounded to the nearest integer.

The denominator of an $F$ ratio for a test involving a synthesized error term would be that shown in Equation (7.11). The numerator, of course, is the mean square for whatever effect is to be tested.

A test of the simple effects can also be carried out by a $t$ test, using a ratio of the mean difference over the standard error of the mean difference. If we wished to make a test of the difference between two levels of $D$ at the same level of $S$, we would have

$$t = \frac{\overline{DS}_{11} - \overline{DS}_{21}}{\sqrt{2[\text{MS}_{\text{error}(a)} + (n_r - 1)\text{MS}_{\text{error}(b)}]/n_r n_m n_s}} \tag{7.13}$$

where $n_r n_m = (4)(5) = 20$, the number of observations in a $DS$ subclass.

When a synthesized error mean square is used, the ratio of the mean difference and its standard error does not follow the $t$ distribution. Cochran and Cox (1957, p. 299) suggest the critical value of $t$ be calculated as

$$t = \frac{(n_s - 1)\text{MS}_{\text{error}(b)}t_b + \text{MS}_{\text{error}(a)}t_a}{(n_s - 1)\text{MS}_{\text{error}(b)} + \text{MS}_{\text{error}(a)}} \tag{7.14}$$

where $t_a$ and $t_b$ are the significance levels of $t$ corresponding to the degrees of freedom for $\text{MS}_{\text{error}(a)}$ and $\text{MS}_{\text{error}(b)}$, respectively.

As noted previously, it is possible in some experiments that the split plots can be divided, resulting in a split-split-plot design. Suppose that in the completely randomized split plot with $D$ main effects in the whole plots and $S$ main effects in the split plots, we now have $M$ main effects in the split-split plots. We would now have the added portion (or split-split-plot portion) shown in Table 7.8 added to our analysis. The expectations of mean squares in the table show that we now have two additional error terms, $RM{:}D$ = error $(c)$ and $RSM{:}D$ = error $(d)$. It can be seen that the main effects of $M$ and the interaction $DM$ are tested by error $(c)$, but the $SM$ and $DSM$ interactions are tested by error $(d)$.

If we find significant interactions in the split-split-plot portion of the analysis, we again may wish to look at the simple effects in this part of the analysis. We would then have simple effects involving main effects and interactions which are tested by the same error term as well as simple effects involving main effects and interactions which are tested by different error terms. In the latter case, we are once more faced

**TABLE 7.8**  DEGREES OF FREEDOM AND EXPECTATIONS OF MEAN SQUARES FOR THE SPLIT-SPLIT PORTION OF A COMPLETELY RANDOMIZED SPLIT-SPLIT-PLOT DESIGN

| Source | df | E(MS) |
|---|---|---|
| Split-split plots | $(n_m - 1)n_s n_d n_r = 224$ | |
| $M$ | $n_m - 1 = 4$ | $\sigma_w^2 + 2\sigma_{rm:d}^2 + 56\sigma_m^2$ |
| $DM$ | $(n_d - 1)(n_m - 1) = 24$ | $\sigma_w^2 + 2\sigma_{rm:d}^2 + 8\sigma_{dm}^2$ |
| $RM{:}D$ error $(c)$ | $(n_r - 1)(n_m - 1)n_d = 84$ | $\sigma_w^2 + 2\sigma_{rm:d}^2$ |
| $SM$ | $(n_s - 1)(n_m - 1) = 4$ | $\sigma_w^2 + \sigma_{rsm:d}^2 + 28\sigma_{sm}^2$ |
| $DSM$ | $(n_d - 1)(n_s - 1)(n_m - 1) = 24$ | $\sigma_w^2 + \sigma_{rsm:d}^2 + 4\sigma_{dsm}^2$ |
| $RSM{:}D$ error $(d)$ | $(n_r - 1)(n_s - 1)(n_m - 1)n_d = 84$ | $\sigma_w^2 + \sigma_{rsm:d}^2$ |
| $n_d = 7 \qquad n_r = 4 \qquad n_s = 2 \qquad n_m = 5 \qquad n_w = 1$ | | $N = 280$ |

with the necessity of synthesizing error terms as well as estimating degrees of freedom. For the purpose of generalization and ease of following the discussion, let us again consider that we have main effects $A$ in the whole plots, main effects $B$ in the split plots, and main effects $C$ in the split-split plots. Table 7.9 displays the simple effects involved in this situation, with the nested equivalents and the errors used in the denominators of the $F$ ratios. The $A$, $B$, and $C$ correspond with $D$, $S$, and $M$ in Table 7.8 and errors $(a)$, $(b)$, $(c)$, and $(d)$ correspond with $R{:}D$, $RS{:}D$, $RM{:}D$, and $RSM{:}D$. It should be pointed out that the errors $(c)$ and $(d)$, $RM{:}D$ and $RSM{:}D$, are frequently pooled into a single error.

When two error terms are to be combined, as seen in Table 7.9, the procedure is as shown in Equation (7.11). In one case, four error terms are combined and this can be accomplished as follows, where $E_a$ represents the error mean square $(a)$, and similarly for the other error terms, using the example we have been discussing:

$$\frac{SS_{E_a} + SS_{E_b} + SS_{E_c} + SS_{E_d}}{(n_r - 1)nd + (n_r - 1)(n_s - 1)nd + (n_r - 1)(n_m - 1)nd + (n_r - 1)(n_s - 1)(n_m - 1)n_d}$$

$$= \frac{SS_{E_a} + SS_{E_b} + SS_{E_c} + SS_{E_d}}{(n_r - 1)n_s n_m}$$

$$= \frac{MS_{E_a} + (n_s - 1)MS_{E_b} + (n_m - 1)MS_{E_c} + (n_s - 1)(n_m - 1)MS_{E_d}}{n_s n_m} \tag{7.15}$$

This, of course, is nothing more than pooling the corresponding sums of squares and degrees of freedom for the error terms involved. The denominator error mean squares resulting from combining the different pairs of error mean squares appearing in Table 7.9 are shown in Table 7.10. As shown in Equation (7.15), the same method is used when more than two error terms are combined.

Having found the denominator error mean squares for the pooled error terms in the split-split portion of the analysis, we must again calculate the approximate

**TABLE 7.9**  ERROR TERMS USED IN THE DENOMINATORS OF $F$ RATIOS FOR TESTING SIMPLE EFFECTS IN A COMPLETELY RANDOMIZED SPLIT-SPLIT-PLOT DESIGN

| Simple effect | Nested equivalent | Effects involved | Error mean squares used in denominators of $F$ ratios |
|---|---|---|---|
| $\overline{AB}_{11} - \overline{AB}_{21}$ | $A{:}B$ | $A$, $AB$ | Error $(a)$, error $(b)$ |
| $\overline{AC}_{11} - \overline{AC}_{21}$ | $A{:}C$ | $A$, $AC$ | Error $(a)$, error $(c)$ |
| $\overline{ABC}_{111} - \overline{ABC}_{211}$ | $A{:}BC$ | $A$, $AB$, $AC$, $ABC$ | Error $(a)$, error $(b)$, error $(c)$, error $(d)$ |
| $\overline{AB}_{11} - \overline{AB}_{12}$ | $B{:}A$ | $B$, $AB$ | Error $(b)$ |
| $\overline{BC}_{11} - \overline{BC}_{21}$ | $B{:}C$ | $B$, $BC$ | Error $(b)$, error $(d)$ |
| $\overline{ABC}_{111} - \overline{ABC}_{121}$ | $B{:}AC$ | $B$, $AB$, $BC$, $ABC$ | Error $(b)$, error $(d)$ |
| $\overline{AC}_{11} - \overline{AC}_{12}$ | $C{:}A$ | $C$, $AC$ | Error $(c)$ |
| $\overline{BC}_{11} - \overline{BC}_{12}$ | $C{:}B$ | $C$, $BC$ | Error $(c)$, error $(d)$ |
| $\overline{ABC}_{111} - \overline{ABC}_{112}$ | $C{:}AB$ | $C$, $AC$, $BC$, $ABC$ | Error $(c)$, error $(d)$ |

**TABLE 7.10   ERROR MEAN SQUARES FOR
DENOMINATORS WHEN POOLING
TWO ERROR MEAN SQUARES**

| Errors pooled | Error mean squares |
|---|---|
| Error ($a$) and error ($b$) | $[MS_{E_a} + (n_s - 1)MS_{E_b}]/n_s$ |
| Error ($a$) and error ($c$) | $[MS_{E_a} + (n_m - 1)MS_{E_c}]/n_m$ |
| Error ($b$) and error ($d$) | $[MS_{E_b} + (n_m - 1)MS_{E_d}]/n_m$ |
| Error ($c$) and error ($d$) | $[MS_{E_c} + (n_s - 1)MS_{E_d}]/n_s$ |

degrees of freedom to be used with these synthesized mean squares. The method shown in Equation (7.12) would be used and the formulas necessary for the combinations of two error terms are shown in Table 7.11. Error mean squares involving additional error terms would be formed in the same manner.

The procedures outlined in this discussion of randomized split-plot and split-split-plot designs can be followed in all the remaining split-plot designs discussed in this section. First, one can determine what the nested equivalents of the simple effects are and then decide what main effects and interaction effects are involved. That leads to a determination of which error terms are to be pooled to develop a synthetic error term. Once these errors are found, then the combination of error terms follows the procedures shown. Following the development of the appropriate denominator mean square, the approximate degrees of freedom for this mean square are calculated according to formula (7.12).

It should also be noted that the standard errors of the means of main effects use the error mean squares of those portions of the analysis in which the main effects are contained; that is, the standard errors for whole-plot main effects are calculated using the whole-plot error mean square. Standard errors for split-plot main effects use the split-plot error mean square and so on. It should also be noted, again, that when tests are merely among subclasses (e.g., tests between levels of $A$ at different levels of $B$, or between levels of $B$ at different levels of $A$) both of the main effects and the interaction are involved. While we have discussed tests among simple effects in terms of the difference between one level and another as a matter of expediency, it should be remembered that these same procedures lead to the appropriate error terms and degrees of freedom when conducting trend analyses or range tests among several levels of one effect at one level of another.

**TABLE 7.11   FORMULAS FOR DERIVING APPROXIMATE DEGREES OF FREEDOM
FOR ERROR MEAN SQUARES SHOWN IN TABLE 7.10**

| | |
|---|---|
| Error ($a$) and error ($b$) | $\dfrac{[MS_{E_a} + (n_s - 1)MS_{E_b}]^2}{MS_{E_a}^2/(n_r - 1)n_d + [(n_s - 1)MS_{E_b}]^2/(n_r - 1)(n_s - 1)n_d}$ |
| Error ($a$) and error ($c$) | $\dfrac{[MS_{E_a} + (n_m - 1)MS_{E_c}]}{MS_{E_a}^2/(n_r - 1)n_d + [(n_m - 1)MS_{E_c}]^2/(n_r - 1)(n_m - 1)n_d}$ |
| Error ($b$) and error ($d$) | $\dfrac{[MS_{E_b} + (n_m - 1)MS_{E_d}]^2}{MS_{E_b}^2/(n_r - 1)(n_s - 1)n_d + [(n_m - 1)MS_{E_d}]^2/(n_c - 1)(n_s - 1)(n_m - 1)n_d}$ |
| Error ($c$) and error ($d$) | $\dfrac{[MS_{E_c} + (n_m - 1)MS_{E_d}]^2}{MS_{E_c}^2/(n_r - 1)(n_m - 1)n_d + [(n_s - 1)MS_{E_d}]^2/(n_r - 1)(n_s - 1)(n_m - 1)n_d}$ |

## 7.3 FACTORIAL SPLIT-PLOT DESIGN

The example used to illustrate this design came from an experiment in bird mimicry conducted by C. J. Duncan and P. M. Sheppard in the Department of Zoology and Genetics, University of Liverpool, England (Duncan and Sheppard, 1965). The additional details of the experiment can be found in their paper. Six birds from each of three different types of breeding were used. Three birds from each group were offered six colored solutions of a liquid, with a gradient from very light to dark. At the darkest solution level in each shock level group the birds received a shock. There were no penalties at the remaining levels. The birds were offered each solution 10 times in a random sequence and the data were recorded as the number of times each solution was taken or accepted. The data are shown in Table 7.12.

The mathematical model used for this set of data is

$$Y_{ijkl} = \mu + g_i + s_j + gs_{ij} + b_{ijk} + c_l + gc_{il} + sc_{jl} + gsc_{ijl} + e_{ijkl} \qquad (7.16)$$

where the $g_i$ represent group effects, the $s_j$ represent shock effects, the $b_{ijk}$ represent birds nested within group $\times$ shock level subclass effects, and the $c_l$ represent solution effects.

In this analysis the bird is the whole plot and since an evaluation of each bird's performance is made over six solutions, the solutions become the split-plot effect. The split-plot sum of squares is then the sum of squares of solutions within birds. The effect of the birds nested within the group $\times$ shock level subclasses becomes the whole-plot error. The whole-plot sum of squares is the sum of squares among birds.

The necessary calculations follow.

$$\text{Total SS} = \sum_i \sum_j \sum_k \sum_l Y_{ijkl}^2 - \frac{Y_{....}^2}{N} = gsbc - (1)$$

$$= 4819 - \frac{(613)^2}{108} = 4819 - 3479.34 = 1339.66 \qquad (7.17)$$

**TABLE 7.12**  NUMBER OF TIMES DIFFERENT COLORED SOLUTIONS WERE ACCEPTED BY BIRDS FROM DIFFERENT GROUPS AT DIFFERENT SHOCK LEVELS

| | | Low-level shock | | | | | | | High-level shock | | | | | |
| | | Solution number | | | | | | | Solution number | | | | | |
| Group | Bird number | 1 | 2 | 3 | 4 | 5 | 6 | Bird number | 1 | 2 | 3 | 4 | 5 | 6 |
|---|---|---|---|---|---|---|---|---|---|---|---|---|---|---|
| 1 | 1 | 10 | 10 | 8 | 2 | 3 | 3 | 1 | 10 | 7 | 3 | 3 | 2 | 1 |
| | 2 | 10 | 10 | 10 | 10 | 6 | 2 | 2 | 10 | 10 | 6 | 3 | 3 | 3 |
| | 3 | 10 | 10 | 10 | 9 | 6 | 4 | 3 | 10 | 9 | 7 | 5 | 5 | 3 |
| 2 | 1 | 10 | 10 | 10 | 9 | 6 | 5 | 1 | 8 | 4 | 4 | 1 | 0 | 0 |
| | 2 | 10 | 10 | 10 | 10 | 5 | 4 | 2 | 8 | 5 | 3 | 2 | 1 | 0 |
| | 3 | 10 | 10 | 10 | 9 | 7 | 8 | 3 | 10 | 7 | 6 | 2 | 0 | 0 |
| 3 | 1 | 9 | 6 | 4 | 3 | 2 | 1 | 1 | 9 | 2 | 3 | 1 | 0 | 1 |
| | 2 | 9 | 7 | 3 | 3 | 1 | 0 | 2 | 8 | 5 | 6 | 3 | 2 | 1 |
| | 3 | 10 | 10 | 10 | 6 | 3 | 2 | 3 | 10 | 10 | 8 | 3 | 4 | 1 |

$$\text{Whole-plot SS} = \frac{\Sigma_i \, \Sigma_j \, \Sigma_k \, Y_{ijk\cdot}^2}{n_c} - \frac{Y_{\cdot\cdot\cdot\cdot}^2}{N} = gsb - (1)$$

$$= \frac{(36)^2 + (48)^2 + \cdots + (36)^2}{6} - (1)$$

$$= 3921.17 - (1) = 441.83 \tag{7.18}$$

$$\text{SS}_G = \frac{\Sigma_i \, Y_{i\cdot\cdot\cdot}^2}{n_s n_b n_c} - \frac{Y_{\cdot\cdot\cdot\cdot}^2}{N} = g - (1)$$

$$= \frac{(233)^2 + (214)^2 + (166)^2}{36} - (1)$$

$$= 3545.58 - (1) = 66.24 \tag{7.19}$$

$$\text{SS}_S = \frac{\Sigma_j \, Y_{\cdot j\cdot\cdot}^2}{n_g n_b n_c} - \frac{Y_{\cdot\cdot\cdot\cdot}^2}{N} = s - (1)$$

$$= \frac{(375)^2 + (238)^2}{54} - (1) = 3653.13 - (1) = 173.79 \tag{7.20}$$

$$\text{SS}_{GS} = \frac{\Sigma_i \, \Sigma_j \, Y_{ij\cdot\cdot}^2}{n_b n_c} - \frac{\Sigma_i \, Y_{i\cdot\cdot\cdot}^2}{n_s n_b n_c} - \frac{\Sigma_j \, Y_{\cdot j\cdot\cdot}^2}{n_g n_b n_c} + \frac{Y_{\cdot\cdot\cdot\cdot}^2}{N}$$

$$= gs - g - s + (1) = \frac{(133)^2 + (100)^2 + \cdots + (77)^2}{18} - g - s + (1)$$

$$= 3814.94 - g - s + (1) = 95.57 \tag{7.21}$$

$$\text{SS}_{B:GS} = \frac{\Sigma_i \, \Sigma_j \, \Sigma_k \, Y_{ijk\cdot}^2}{n_c} - \frac{\Sigma_i \, \Sigma_j \, Y_{ij\cdot\cdot}^2}{n_b n_c} = gsb - gs$$

$$= 3921.17 - 3814.94 = 106.23 \tag{7.22}$$

$$\text{Split-plot SS} = \sum_i \sum_j \sum_k \sum_l Y_{ijkl}^2 - \frac{\Sigma_i \, \Sigma_j \, \Sigma_k \, Y_{ijk\cdot}^2}{n_c} = gsbc - gsb$$

$$= 4819 - 3921.17 = 897.83 \tag{7.23}$$

$$\text{SS}_C = \frac{\Sigma_l \, Y_{\cdot\cdot\cdot l}^2}{n_g n_s n_b} - \frac{Y_{\cdot\cdot\cdot\cdot}^2}{N} = c - (1)$$

$$= \frac{(171)^2 + (142)^2 + \cdots + (39)^2}{18} - (1)$$

$$= 4208.83 - (1) = 729.49 \tag{7.24}$$

$$\text{SS}_{GC} = \frac{\Sigma_i \, \Sigma_l \, Y_{i\cdot\cdot l}^2}{n_s n_b} - \frac{\Sigma_i \, Y_{i\cdot\cdot\cdot}^2}{n_s n_b n_c} - \frac{\Sigma_l \, Y_{\cdot\cdot\cdot l}^2}{n_g n_s n_b} + \frac{Y_{\cdot\cdot\cdot\cdot}^2}{N}$$

$$= gc - g - c + (1) = \frac{(60)^2 + (56)^2 + \cdots + (6)^2}{6} - g - c + (1)$$

$$= 4289.83 - g - c + (1) = 14.76 \tag{7.25}$$

$$SS_{SC} = \frac{\Sigma_j \Sigma_l Y^2_{.j \cdot l}}{n_g n_b} - \frac{\Sigma_j Y^2_{.j \cdot \cdot}}{n_g n_b n_c} - \frac{\Sigma_l Y^2_{\cdot \cdot \cdot l}}{n_g n_s n_b} + \frac{Y^2_{\cdot \cdot \cdot \cdot}}{N}$$

$$= sc - s - c + (1) = \frac{(88)^2 + (83)^2 + \cdots + (10)^2}{9} - s - c + (1)$$

$$= 4416.11 - s - c + (1) = 33.49 \tag{7.26}$$

$$SS_{GSC} = \frac{\Sigma_i \Sigma_j \Sigma_l Y^2_{ij \cdot l}}{n_b} - \frac{\Sigma_i \Sigma_j Y^2_{ij \cdot \cdot}}{n_b n_c} - \frac{\Sigma_i \Sigma_l Y^2_{i \cdot \cdot l}}{n_s n_b}$$

$$- \frac{\Sigma_j \Sigma_l Y^2_{.j \cdot l}}{n_g n_b} + \frac{\Sigma_i Y^2_{i \cdot \cdot \cdot}}{n_s n_b n_c} + \frac{\Sigma_j Y^2_{.j \cdot \cdot}}{n_g n_b n_c} + \frac{\Sigma_l Y^2_{\cdot \cdot \cdot l}}{n_g n_s n_b} - \frac{Y^2_{\cdot \cdot \cdot \cdot}}{N}$$

$$= gsc - gs - gc - sc + g + s + c - (1)$$

$$= \frac{(30)^2 + (30)^2 + \cdots + (3)^2}{3} - gs - gc - sc + g + s + c - (1)$$

$$= 4616.33 - gs - gc - sc + g + s + c - (1) = 23.65 \tag{7.27}$$

$$SS_{BC:GS} = \sum_i \sum_j \sum_k \sum_l Y^2_{ijkl} - \frac{\Sigma_i \Sigma_j \Sigma_k Y^2_{ijk \cdot}}{n_c} - \frac{\Sigma_i \Sigma_j \Sigma_l Y^2_{ij \cdot l}}{n_b} + \frac{\Sigma_i \Sigma_j Y^2_{ij \cdot \cdot}}{n_b n_c}$$

$$= gsbc - gsb - gsc + gs = 96.44 \tag{7.28}$$

Assuming the breeding groups, shock levels, and solutions to be fixed effects and the birds to be random effects, the sources of variation, degrees of freedom, and expectations of mean squares are shown in Table 7.13.

Since the birds are the whole plots, there are one fewer degrees of freedom for whole plots than there are birds. The split plots are the solutions within birds which themselves are nested within breeding groups and shock levels. Thus, the split-plot

**TABLE 7.13**  SOURCE OF VARIATION, DEGREES OF FREEDOM, AND EXPECTATIONS OF MEAN SQUARES FOR DATA SHOWN IN TABLE 7.12

| Source | df | E(MS) |
|---|---|---|
| Whole plots | $n_g n_s n_b - 1 = 17$ | |
| G | $n_g - 1 = 2$ | $\sigma_w^2 + 6\sigma_{b:gs}^2 + 36\sigma_g^2$ |
| S | $n_s - 1 = 1$ | $\sigma_w^2 + 6\sigma_{b:gs}^2 + 54\sigma_s^2$ |
| GS | $(n_g - 1)(n_s - 1) = 2$ | $\sigma_w^2 + 6\sigma_{b:gs}^2 + 18\sigma_{gs}^2$ |
| B:GS | $(n_b - 1)n_g n_s = 12$ | $\sigma_w^2 + 6\sigma_{b:gs}^2$ |
| Split plots | $(n_c - 1)n_g n_s n_b = 90$ | |
| C | $n_c - 1 = 5$ | $\sigma_w^2 + \sigma_{bc:gs}^2 + 18\sigma_c^2$ |
| GC | $(n_g - 1)(n_c - 1) = 10$ | $\sigma_w^2 + \sigma_{bc:gs}^2 + 6\sigma_{gc}^2$ |
| SC | $(n_s - 1)(n_c - 1) = 5$ | $\sigma_w^2 + \sigma_{bc:gs}^2 + 9\sigma_{sc}^2$ |
| GSC | $(n_g - 1)(n_s - 1)(n_c - 1) = 10$ | $\sigma_w^2 + \sigma_{bc:gs}^2 + 3\sigma_{gsc}^2$ |
| BC:GS | $(n_b - 1)(n_c - 1)n_g n_s = 60$ | $\sigma_w^2 + \sigma_{bc:gs}^2$ |

| $n_g = 3$ | $n_s = 2$ | $n_b = 3$ | $n_c = 6$ | $n_w = 1$ | $N = 108$ |
|---|---|---|---|---|---|

**TABLE 7.14**  ANALYSIS OF VARIANCE FOR FACTORIAL
            SPLIT-PLOT DESIGN

| Source | df | SS | MS | F |
|---|---|---|---|---|
| Whole plots | 17 | 441.83 | | |
| G | 2 | 66.24 | 33.12 | 3.74 |
| S | 1 | 173.79 | 173.79 | 19.64** |
| GS | 2 | 95.57 | 47.78 | 5.40* |
| B:GS | 12 | 106.23 | 8.85 | |
| Split plots | 90 | 897.83 | | |
| C | 5 | 729.49 | 145.90 | 90.62** |
| GC | 10 | 14.76 | 1.48 | 0.92 |
| SC | 5 | 33.49 | 6.70 | 4.16** |
| GSC | 10 | 23.65 | 2.36 | 1.47 |
| BC:GS | 60 | 96.44 | 1.61 | |

degrees of freedom are calculated as for solutions within groups, shock levels, and bird subclasses. The resulting analysis is shown in Table 7.14. This analysis shows that there is a highly significant difference between shock levels, highly significant differences among solution levels, and a highly significant interaction between the two effects. In extracting further information from the analysis it is reasonable to partition the sum of squares among the solution levels into linear, quadratic, cubic, and so on components, assuming that equal differences in intensity of color existed from very light to dark. Furthermore, since there is a highly significant interaction between shock levels and solutions, it is informative to partition this interaction sum of squares into linear × linear, linear × quadratic, linear × cubic, and so on components. Since there is only one degree of freedom for shock levels, only the linear effect of this classification can be used. With more degrees of freedom available for shock levels, we could have had quadratic × linear, quadratic × quadratic, and so on.

The partitioning can be accomplished by setting up a table, as illustrated in Table 7.15, with the appropriate values and calculating the sums of squares in the usual

**TABLE 7.15**  ORTHOGONAL POLYNOMIAL COEFFICIENTS FOR SHOCK
            LEVEL × SOLUTION SUBCLASSES

| | | $SC_{11}$ | $SC_{12}$ | $SC_{13}$ | $SC_{14}$ | $SC_{15}$ | $SC_{16}$ | $SC_{21}$ | $SC_{22}$ | $SC_{23}$ | $SC_{24}$ | $SC_{25}$ | $SC_{26}$ |
|---|---|---|---|---|---|---|---|---|---|---|---|---|---|
| Subclass totals $T_j$ | | 88 | 83 | 75 | 61 | 39 | 29 | 83 | 59 | 46 | 23 | 17 | 10 |
| Shock level (linear) | $= Q_1$ | −1 | −1 | −1 | −1 | −1 | −1 | 1 | 1 | 1 | 1 | 1 | 1 |
| Solution (linear) | $= Q_2$ | −5 | −3 | −1 | 1 | 3 | 5 | −5 | −3 | −1 | 1 | 3 | 5 |
| Solution (quadratic) | $= Q_3$ | 5 | −1 | −4 | −4 | −1 | 5 | 5 | −1 | −4 | −4 | −1 | 5 |
| Solution (cubic) | $= Q_4$ | −5 | 7 | 4 | −4 | −7 | 5 | −5 | 7 | 4 | −4 | −7 | 5 |
| Solution (quartic) | $= Q_5$ | 1 | −3 | 2 | 2 | −3 | 1 | 1 | −3 | 2 | 2 | −3 | 1 |
| Solution (quintic) | $= Q_6$ | −1 | 5 | −10 | 10 | −5 | 1 | −1 | 5 | −10 | 10 | −5 | 1 |
| Linear × linear | $= Q_7$ | 5 | 3 | 1 | −1 | −3 | −5 | −5 | −3 | −1 | 1 | 3 | 5 |
| Linear × quadratic | $= Q_8$ | −5 | 1 | 4 | 4 | 1 | −5 | 5 | −1 | −4 | −4 | −1 | 5 |
| Linear × cubic | $= Q_9$ | 5 | −7 | −4 | 4 | 7 | −5 | −5 | 7 | 4 | −4 | −7 | 5 |
| Linear × quartic | $= Q_{10}$ | −1 | 3 | −2 | −2 | 3 | −1 | 1 | −3 | 2 | 2 | −3 | 1 |
| Linear × quintic | $= Q_{11}$ | 1 | −5 | 10 | −10 | 5 | −1 | −1 | 5 | −10 | 10 | −5 | 1 |

fashion. The orthogonal polynomials come from the tabulated values and are spread out over the subclasses (repeated) rather than over the overall totals so that the products of the rows can be used to obtain the partitioned interaction sum of squares. The coefficients for the linear effects of shock levels, the five degrees of polynomials for the solutions, and the products of these rows are displayed in the table.

The sums of squares for the various effects are calculated by the general formula

$$SS_{Q_i} = \frac{(\Sigma_j c_{ij} T_j)^2}{n \, \Sigma_j c_{ij}^2} = \frac{Q_i^2}{D_i}$$

where the $c_{ij}$ are the orthogonal polynomial coefficients, with $i$ representing the set of coefficients and running from 1 to 11 and $j$ representing the particular coefficient within a set and running from 1 to 12. $T_j$ represents a subclass total and $n$ is the number of observations in each subclass. The sums of squares for the 11 components are

$$SS_{Q_1} = \frac{Q_1^2}{D_1} = \frac{(-137)^2}{108} = 173.79^{**} \qquad SS_{Q_7} = \frac{(-73)^2}{1260} = 4.23^*$$

$$SS_{Q_2} = \frac{Q_2^2}{D_2} = \frac{(-955)^2}{1260} = 723.83^{**} \qquad SS_{Q_8} = \frac{(194)^2}{1512} = 24.89^{**}$$

$$SS_{Q_3} = \frac{Q_3^2}{D_3} = \frac{(32)^2}{1512} = 0.68 \qquad SS_{Q_9} = \frac{(-48)^2}{3240} = 0.71$$

$$SS_{Q_4} = \frac{Q_4^2}{D_4} = \frac{(90)^2}{3240} = 2.50 \qquad SS_{Q_{10}} = \frac{(-20)^2}{504} = 0.79$$

$$SS_{Q_5} = \frac{Q_5^2}{D_5} = \frac{(26)^2}{504} = 1.34 \qquad SS_{Q_{11}} = \frac{(-114)^2}{4536} = 2.87$$

$$SS_{Q_6} = \frac{Q_6^2}{D_6} = \frac{(-72)^2}{4536} = 1.14$$

These values could be entered into an analysis of variance table with all other effects but the table would be quite extensive even if we stopped after the quadratic tests. We have already seen that the linear effect of shock levels was highly significant. We now see that the linear effect of solutions is highly significant ($Q_2$ through $Q_{11}$ being tested by the split-plot error, $BC:GS$). It can also be seen that there is a significant interaction between the linear effect of shock levels and the linear effect of solutions ($Q_7$), indicating that there is a different linear regression of solutions at each of the shock levels. However, of particular interest in this analysis is the highly significant linear $\times$ quadratic effect. Since the quadratic effect of solutions was not significant, this comes as a bit of a surprise. Using the coefficients in Table 7.15, we can graph first the linear regression lines of the solutions at each shock level by methods presented earlier. For the linear regression coefficient for solutions for shock level 1, we use the polynomial coefficients shown in Table 7.15 for the first shock level only, in the line $Q_2$, and the same procedure for the second shock level. Thus,

$$b_l = \frac{\Sigma \, c_{2j} T_j}{n \, \Sigma \, c_{2j}^2} = \frac{-441}{(9)(70)} = -0.7000 \quad \text{for } S_1$$

$$b_l = \frac{\Sigma \, c_{2j} T_j}{n \, \Sigma \, c_{2j}^2} = \frac{-514}{(9)(70)} = -0.8159 \quad \text{for } S_2$$

For the quadratic regression, we use the coefficients for the quadratic regression of solutions separately for each shock level, yielding

$$b_q = \frac{\Sigma \, c_{3j} T_j}{n \, \Sigma \, c_{3j}^2} = \frac{-81}{(9)(84)} = -0.1071 \quad \text{for } S_1$$

$$b_q = \frac{\Sigma \, c_{3j} T_j}{n \, \Sigma \, c_{3j}^2} = \frac{113}{(9)(84)} = 0.1495 \quad \text{for } S_2$$

To complete the prediction equations, the mean of each shock level is required, and

$$\bar{y}_1 = \frac{Y_{\cdot 1 \cdot \cdot}}{n_g n_b n_c} = \frac{375}{54} = 6.9444 \quad \text{for } S_1$$

$$\bar{y}_2 = \frac{Y_{\cdot 2 \cdot \cdot}}{n_g n_b n_c} = \frac{238}{54} = 4.4074 \quad \text{for } S_2$$

We now have the prediction equations

$$\text{Linear} \quad \hat{Y} = 6.9444 - 0.7000 c_{2j} \quad \text{for } S_1$$

$$\text{Linear} \quad \hat{Y} = 4.4074 - 0.8159 c_{2j} \quad \text{for } S_2$$

$$\text{Quadratic} \quad \hat{Y} = 6.9444 - 0.7000 c_{2j} - 0.1071 c_{3j} \quad \text{for } S_1$$

$$\text{Quadratic} \quad \hat{Y} = 4.4074 - 0.8159 c_{2j} + 0.1495 c_{3j} \quad \text{for } S_2$$

Using these prediction equations, we can graph the linear and quadratic responses to the solutions at each level as shown in Figure 7.1.

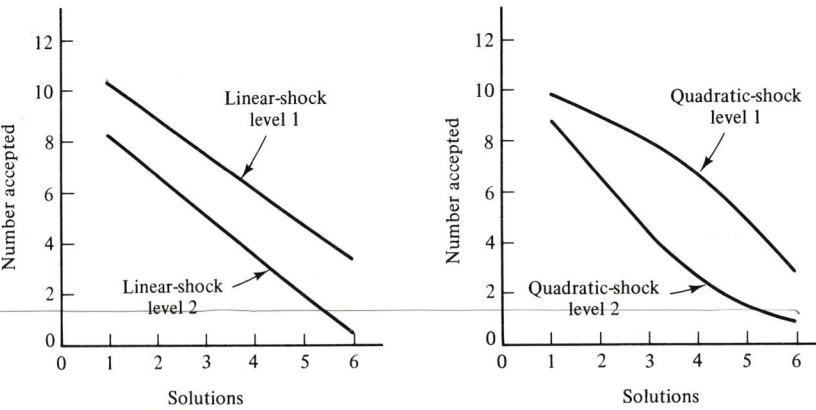

**Figure 7.1** Partitioning of interaction effects.

**TABLE 7.16** DEGREES OF FREEDOM AND EXPECTATIONS OF MEAN SQUARES FOR THE SPLIT-SPLIT PORTION OF A FACTORIAL SPLIT-SPLIT-PLOT DESIGN

| Source | df | E(MS) |
|---|---|---|
| Split-split plots | $(n_t - 1)n_g n_s n_b n_c = 432$ | |
| T | $n_t - 1 = 4$ | $\sigma_w^2 + 6\sigma_{bt:gs}^2 + 108\sigma_t^2$ |
| GT | $(n_g - 1)(n_t - 1) = 8$ | $\sigma_w^2 + 6\sigma_{bt:gs}^2 + 36\sigma_{gt}^2$ |
| ST | $(n_s - 1)(n_t - 1) = 4$ | $\sigma_w^2 + 6\sigma_{bt:gs}^2 + 54\sigma_{st}^2$ |
| GST | $(n_g - 1)(n_s - 1)(n_t - 1) = 8$ | $\sigma_w^2 + 6\sigma_{bt:gs}^2 + 18\sigma_{gst}^2$ |
| BT:GS | $(n_b - 1)(n_t - 1)n_g n_s = 48$ | $\sigma_w^2 + 6\sigma_{bt:gs}^2$ |
| CT | $(n_c - 1)(n_t - 1) = 20$ | $\sigma_w^2 + \sigma_{bct:gs}^2 + 18\sigma_{ct}^2$ |
| GCT | $(n_g - 1)(n_c - 1)(n_t - 1) = 40$ | $\sigma_w^2 + \sigma_{bct:gs}^2 + 6\sigma_{gct}^2$ |
| SCT | $(n_s - 1)(n_c - 1)(n_t - 1) = 20$ | $\sigma_w^2 + \sigma_{bct:gs}^2 + 9\sigma_{sct}^2$ |
| GSCT | $(n_g - 1)(n_s - 1)(n_c - 1)(n_t - 1) = 40$ | $\sigma_w^2 + \sigma_{bct:gs}^2 + 3\sigma_{gsct}^2$ |
| BCT:GS | $(n_b - 1)(n_c - 1)(n_t - 1)n_g n_s = 240$ | $\sigma_w^2 + \sigma_{bct:gs}^2$ |

$n_g = 3$    $n_s = 2$    $n_b = 3$    $n_c = 6$    $n_t = 5$    $n_w = 1$    $N = 540$

An examination of the first graph shows that while the difference is not glaring, the two linear regression lines do take on slightly different slopes. The second graph shows quite clearly why there is such a large interaction between linear and quadratic effects since the curve for the solutions for the first shock level follows a convex form and that for the second shock level follows a concave form. Averaging or pooling the two would result in a straight line, so the quadratic effect of solutions alone was not present.

It is also of interest to inspect the regression coefficients, where both linear coefficients are negative resulting from a decline in acceptance with darkening solutions. However, the quadratic coefficients differ in sign, the negative coefficient causing a convex curve and the positive coefficient causing a concave curve.

Again, it is possible to subdivide the split plots, leading to a split-split portion in the analysis. If we assume that the split plots are subdivided into five classes of the factor T, Table 7.16 shows the calculations necessary for the degrees of freedom together with the expectations of mean squares. The addition of a split-split-plot effect would result in some changes, of course, in the expectations of mean squares for the split-plot and whole-plot portions of the analysis. The treatments involved in the split plots or the split-split plots could be a factorial arrangement.

## 7.4 RANDOMIZED BLOCK, FACTORIAL SPLIT-PLOT DESIGN

This type of design is encountered quite frequently in experimental situations in many areas. The example we shall use resulted from an experiment conducted by Mary Ballew of the Department of Public Health at the University of Massachusetts. Nine blocks of four guinea pigs each were formed and each guinea pig in a block was subjected to one of four treatments. The treatments were composed of two diets, one containing a low level of vitamin C and one containing a high level, and two exposures to ozone, one of 0.0 parts per million and one of 0.3 parts per million, a factorial arrangement of treatments. Red blood cell counts were made on each guinea pig at one-half hour and three hours after exposure to the ozone, resulting in a split plot in time. The data are shown in Table 7.17, where $V_1$ represents a low level of vitamin C, $V_2$ represents a high level of vitamin C, $O_1$ represents no ozone, and $O_2$ represents ozone at 0.3 parts per million.

In writing out the mathematical model for this set of data, one could include the interactions of blocks with all other sources of variations involved. However, in this case, we shall consider each of the four combinations of vitamin and ozone as a treatment level and not attempt to consider all possible interactions. The whole-plot error term will therefore be a block × treatment interaction which is a pooling of the block × vitamin, block × ozone, and block × vitamin × ozone interactions. In the split-plot portion of the analysis all possible interactions with the blocks are also pooled. If it is suspected that the interactions of blocks with various effects are large, many of them could be tested by using a complete model. If the interactions are not significant at approximately the 20–25 percent level of probability, they may be pooled.

In this set of data, the interactions of blocks with vitamins and blocks with ozone were each tested for significance and neither exhibited significance at the 0.25 level, indicating that it was appropriate to pool these interactions with the error term.

The mathematical model would be

$$Y_{ijkl} = \mu + b_i + v_j + o_k + vo_{jk} + d_{ijk} + h_l + vh_{jl} + oh_{kl} + voh_{jkl} + e_{ijkl} \qquad (7.29)$$

where the $b_i$ represent block effects, the $v_j$ represent vitamin effects, the $o_k$ represent ozone effects, the $d_{ijk}$ stand for whole-plot errors, and the $h_l$ stand for hourly effects. The model could be written out including only the block, treatment, and hourly effects, later partitioning the $V$, $O$, and interaction effects. The necessary calculations follow.

$$\text{Total SS} = \sum_i \sum_j \sum_k \sum_l Y_{ijkl}^2 - \frac{Y_{....}^2}{N} = bvoh - (1)$$

$$= 2362.0312 - \frac{(408.74)^2}{72}$$

$$= 2362.0312 - 2320.3943 = 41.6369 \qquad (7.30)$$

$$\text{Whole-plot SS} = \frac{\sum_i \sum_j \sum_k Y_{ijk.}^2}{n_h} - \frac{Y_{....}^2}{N} = bvo - (1)$$

$$= \frac{(12.53)^2 + (11.39)^2 + \cdots + (8.62)^2}{2} - (1)$$

$$= 2355.7632 - (1) = 35.3689 \qquad (7.31)$$

$$\text{SS}_B = \frac{\sum_i Y_{i...}^2}{n_v n_o n_h} - \frac{Y_{....}^2}{N} = b - (1)$$

$$= \frac{(45.38)^2 + (48.30)^2 + \cdots + (36.31)^2}{8} - (1)$$

$$= 2335.1878 - (1) = 14.7935 \qquad (7.32)$$

**TABLE 7.17**  RED BLOOD CELL COUNTS OF GUINEA PIGS

| Block | $\frac{1}{2}$ Hour after exposure | | | | 3 Hours after exposure | | | | Totals |
|---|---|---|---|---|---|---|---|---|---|
| | $V_1O_1$ | $V_1O_2$ | $V_2O_1$ | $V_2O_2$ | $V_1O_1$ | $V_1O_2$ | $V_2O_1$ | $V_2O_2$ | |
| 1 | 6.59 | 5.49 | 4.84 | 5.50 | 5.94 | 5.90 | 5.25 | 5.77 | 45.28 |
| 2 | 6.42 | 6.39 | 6.00 | 5.75 | 6.39 | 6.52 | 6.36 | 4.47 | 48.30 |
| 3 | 5.38 | 6.18 | 6.92 | 6.45 | 5.38 | 6.36 | 6.33 | 5.31 | 48.31 |
| 4 | 5.58 | 5.64 | 6.51 | 6.15 | 5.09 | 6.25 | 6.29 | 5.04 | 46.55 |
| 5 | 5.47 | 5.51 | 6.02 | 5.23 | 4.38 | 4.93 | 5.63 | 5.49 | 42.66 |
| 6 | 6.30 | 6.36 | 6.11 | 5.79 | 6.39 | 5.90 | 5.71 | 5.45 | 48.01 |
| 7 | 6.07 | 5.43 | 6.49 | 5.99 | 5.49 | 5.50 | 6.22 | 5.55 | 46.74 |
| 8 | 6.33 | 5.85 | 6.02 | 6.56 | 5.06 | 5.16 | 5.77 | 5.83 | 46.58 |
| 9 | 3.81 | 3.98 | 6.77 | 4.49 | 3.35 | 3.53 | 6.25 | 4.13 | 36.31 |
| Totals | 51.95 | 50.83 | 55.68 | 51.91 | 47.47 | 50.05 | 53.81 | 47.04 | 408.74 |

$$SS_V = \frac{\Sigma_j\, Y^2_{.j..}}{n_b n_o n_h} - \frac{Y^2_{....}}{N} = v - (1)$$

$$= \frac{(200.30)^2 + (208.44)^2}{36} - (1)$$

$$= 2321.3145 - (1) = 0.9202 \tag{7.33}$$

$$SS_O = \frac{\Sigma_k\, Y^2_{..k.}}{n_b n_v n_h} - \frac{Y^2_{....}}{N} = o - (1)$$

$$= \frac{(208.91)^2 + (199.83)^2}{36} - (1)$$

$$= 2321.5394 - (1) = 1.1451 \tag{7.34}$$

$$SS_{VO} = \frac{\Sigma_j \Sigma_k\, Y^2_{.jk.}}{n_b n_h} - \frac{\Sigma_j\, Y^2_{.j..}}{n_b n_o n_h} - \frac{\Sigma_k\, Y^2_{..k.}}{n_b n_v n_h} + \frac{Y^2_{....}}{N}$$

$$= vo - v - o + (1)$$

$$= \frac{(99.42)^2 + (100.88)^2 + \cdots + (98.95)^2}{18} - v - o + (1)$$

$$= 2324.4596 - v - o + (1) = 2.0000 \tag{7.35}$$

$$SS_{\text{error}(a)} = \frac{\Sigma_i \Sigma_j \Sigma_k\, Y^2_{ijk.}}{n_h} - \frac{\Sigma_i\, Y^2_{i...}}{n_v n_o n_h} - \frac{\Sigma_j \Sigma_k\, Y^2_{.jk.}}{n_b n_h} + \frac{Y^2_{....}}{N}$$

$$= bvo - b - vo + (1)$$

$$= \text{Whole-plot SS} - SS_B - SS_V - SS_O - SS_{VO}$$

$$= 16.5101 \tag{7.36}$$

$$\text{Split-plot SS} = \sum_i \sum_j \sum_k \sum_l Y^2_{ijkl} - \frac{\Sigma_i \Sigma_j \Sigma_k\, Y^2_{ijk.}}{n_h} = bvoh - bvo$$

$$= 6.2680 \tag{7.37}$$

$$SS_H = \frac{\Sigma_l\, Y^2_{...l}}{n_b n_v n_o} - \frac{Y^2_{....}}{N} = h - (1)$$

$$= \frac{(210.37)^2 + (198.37)^2}{36} - (1)$$

$$= 2322.3943 - (1) = 2.0000 \qquad\qquad (7.38)$$

$$SS_{VH} = \frac{\Sigma_j\, \Sigma_l\, Y^2_{.j.l}}{n_b n_o} - \frac{\Sigma_j\, Y^2_{.j..}}{n_b n_o n_h} - \frac{\Sigma_l\, Y^2_{...l}}{n_b n_v n_o} + \frac{Y^2_{....}}{N}$$

$$= vh - v - h + (1)$$

$$= \frac{(102.78)^2 + (107.59)^2 + \cdots + (100.85)^2}{18} - v - h + (1)$$

$$= 2323.3450 - v - h + (1) = 0.0305 \qquad\qquad (7.39)$$

$$SS_{OH} = \frac{\Sigma_k\, \Sigma_l\, Y^2_{..kl}}{n_b n_v} - \frac{\Sigma_k\, Y^2_{..k.}}{n_b n_v n_h} - \frac{\Sigma_l\, Y^2_{...l}}{n_b n_v n_o} + \frac{Y^2_{....}}{N}$$

$$= oh - o - h + (1)$$

$$= \frac{(107.63)^2 + (102.74)^2 + \cdots + (97.09)^2}{18} - o - h + (1)$$

$$= 2323.5462 - o - h + (1) = 0.0068 \qquad\qquad (7.40)$$

$$SS_{VOH} = \frac{\Sigma_j\, \Sigma_k\, \Sigma_l\, Y^2_{.jkl}}{n_b} - \frac{\Sigma_j\, \Sigma_k\, Y^2_{.jk.}}{n_b n_h} - \frac{\Sigma_j\, \Sigma_l\, Y^2_{.j.l}}{n_b n_o} - \frac{\Sigma_k\, \Sigma_l\, Y^2_{..kl}}{n_b n_v}$$

$$+ \frac{\Sigma_j\, Y^2_{.j..}}{n_b n_o n_h} + \frac{\Sigma_k\, Y^2_{..k.}}{n_b n_v n_h} + \frac{\Sigma_l\, Y^2_{...l}}{n_b n_v n_o} - \frac{Y^2_{....}}{N}$$

$$= voh - vo - vh - oh + v + o + h - (1)$$

$$= \frac{(51.95)^2 + (50.83)^2 + \cdots + (47.04)^2}{9}$$

$$- vo - vh - oh + v + o + h - (1)$$

$$= 2327.1203 - vo - vh - oh + v + o + h - (1)$$

$$= 0.6234 \qquad\qquad (7.41)$$

$$SS_{error(b)} = \sum_i \sum_j \sum_k \sum_l Y^2_{ijkl} - \frac{\Sigma_i\, \Sigma_j\, \Sigma_k\, Y^2_{ijk.}}{n_h}$$

$$- \frac{\Sigma_j\, \Sigma_k\, \Sigma_l\, Y^2_{.jkl}}{n_b} + \frac{\Sigma_j\, \Sigma_k\, Y^2_{.jk.}}{n_b n_h}$$

$$= bvoh - bvo - voh + vo$$

$$= \text{Split-plot SS} - SS_H - SS_{VH} - SS_{OH} - SS_{VOH}$$

$$= 3.6073 \qquad\qquad (7.42)$$

**TABLE 7.18** DEGREES OF FREEDOM AND EXPECTATIONS OF MEAN SQUARES FOR DATA SHOWN IN TABLE 7.17

| Source | df | E(MS) |
|---|---|---|
| Whole plots | $n_b n_v n_o - 1 = 35$ | |
| B | $n_b - 1 = \ 8$ | $\sigma_e^2 + 2\sigma_d^2 + 8\sigma_b^2$ |
| V | $n_v - 1 = \ 1$ | $\sigma_e^2 + 2\sigma_d^2 + 36\sigma_v^2$ |
| O | $n_o - 1 = \ 1$ | $\sigma_e^2 + 2\sigma_d^2 + 36\sigma_o^2$ |
| VO | $(n_v - 1)(n_o - 1) = \ 1$ | $\sigma_e^2 + 2\sigma_d^2 + 18\sigma_{vo}^2$ |
| Error (a) | $(n_b - 1)(n_v n_o - 1) = 24$ | $\sigma_e^2 + 2\sigma_d^2$ |
| Split plots | $n_b n_v n_o(n_h - 1) = 36$ | |
| H | $n_h - 1 = \ 1$ | $\sigma_e^2 + 36\sigma_h^2$ |
| VH | $(n_v - 1)(n_h - 1) = \ 1$ | $\sigma_e^2 + 18\sigma_{vh}^2$ |
| OH | $(n_o - 1)(n_h - 1) = \ 1$ | $\sigma_e^2 + 18\sigma_{oh}^2$ |
| VOH | $(n_v - 1)(n_o - 1)(n_h - 1) = \ 1$ | $\sigma_e^2 + 9\sigma_{voh}^2$ |
| Error (b) | $(n_b - 1)(n_h - 1)n_v n_o = 32$ | $\sigma_e^2$ |

$n_b = 9 \qquad n_v = 2 \qquad n_o = 2 \qquad n_h = 2 \qquad n_w = 1 \qquad N = n_b n_v n_o n_h = 72$

While the interaction of blocks and the split-plot effects is frequently included in the model of a split-plot design in time, it was found to be very small in this set of data and so was included in the split-plot error.

The method of calculating the degrees of freedom together with the expectations of mean squares are shown in Table 7.18.

The split-plot error variance component has a coefficient of 1 and the whole-plot error variance component ($\sigma_d^2$) has a coefficient that is equal to the number of observations in each whole plot. All other coefficients are calculated as previously discussed. The degrees of freedom for the whole-plot error can be found, of course, by the difference between the degrees of freedom for the whole plots and the degrees of freedom for the other sources of variation in the whole plots. The degrees of freedom for the split-plot error can also be found in a similar fashion. The completed analysis of variance is shown in Table 7.19.

**TABLE 7.19** ANALYSIS OF VARIANCE OF RANDOMIZED BLOCK, FACTORIAL SPLIT-PLOT DESIGN

| Source | df | SS | MS | F |
|---|---|---|---|---|
| Whole plots | 35 | 35.3689 | | |
| B | 8 | 14.7935 | 1.8492 | 2.69* |
| V | 1 | 0.9202 | 0.9202 | 1.34 |
| O | 1 | 1.1451 | 1.1451 | 1.66 |
| VO | 1 | 2.0000 | 2.0000 | 2.91 |
| Error (a) | 24 | 16.5101 | 0.6879 | |
| Split plots | 36 | 6.2680 | | |
| H | 1 | 2.0000 | 2.0000 | 17.75** |
| VH | 1 | 0.0305 | 0.0305 | 0.27 |
| OH | 1 | 0.0068 | 0.0068 | 0.06 |
| VOH | 1 | 0.6234 | 0.6234 | 5.53* |
| Error (b) | 32 | 3.6073 | 0.1127 | |

The main conclusion from the analysis is the highly significant difference between the time of count of red blood cells after exposure to ozone ($H$). At one hour after exposure the mean red blood cell count was 5.84 and after three hours of exposure the mean count was 5.51. Note that the split-plot error mean square is considerably smaller than the whole-plot error mean square, as is usually the case. This type of design could include a split-split subdivision as noted for previous designs, with no complications in developing degrees of freedom, sums of squares, or expectations of mean squares.

## 7.5 RANDOMIZED BLOCK, SPLIT-PLOT DESIGN: MULTIPLE OBSERVATIONS

There are enough differences in the experiment with multiple observations within the whole plots and a single observation within each whole plot to make an example of this type of design worthwhile. The experiment that we shall discuss was conducted under the direction of Dr. Paul Jennings of the Department of Plant and Soil Science at the University of Massachusetts. The experiment was designed to measure the effect of a chilling treatment on the growth of corn. Thus, the treatments consisted of one control group and one group of plants that was chilled for a nine-hour period at night. The plants were grown in a growth chamber. Three blocks were used with three plants in each block × treatment subclass. The growth of the shoots was measured at daily intervals from 5 to 12 days. The data are displayed in Table 7.20.

The mathematical model for this experiment is

$$Y_{ijkl} = \mu + b_i + t_j + bt_{ij} + p_{ijk} + d_l + bd_{il} + td_{jl} + btd_{ijl} + e_{ijkl} \qquad (7.43)$$

where the $b_i$ represent block effects, the $t_j$ represent treatment effects, the $p_{ijk}$ stand for plants within the block × treatment subclasses, the $d_l$ stand for day effects, and $e_{ijkl} = pd_{ijkl} + w_{ijkl}$.

In this analysis, the plants are the whole plots and since they are evaluated over days, the days are the split-plot effects. That is, there are 18 plants in the block × treatment subclasses (three in each) and these 18 plants are evaluated on eight separate days. Each block × treatment × plant × day subclass (a single observation) is a split plot. The calculations necessary for the analysis follow.

$$\text{Total SS} = \sum_i \sum_j \sum_k \sum_l Y_{ijkl}^2 - \frac{Y_{\cdots\cdot}^2}{N} = btpd - (1)$$

$$= 53{,}692.16 - \frac{(2432.2)^2}{144} = 53{,}692.16 - 41{,}080.53$$

$$= 12{,}611.63 \qquad (7.44)$$

$$\text{Whole-plot SS} = \frac{\sum_i \sum_j \sum_k Y_{ijk\cdot}^2}{n_d} - \frac{Y_{\cdots\cdot}^2}{N} = btp - (1)$$

$$= \frac{(144.6)^2 + (167.2)^2 + \cdots + (115.4)^2}{8} - (1)$$

$$= 43{,}359.52 - (1) = 2278.99 \qquad (7.45)$$

**TABLE 7.20** LENGTHS OF SHOOTS OF CORN PLANTS (IN CENTIMETERS): $C$ = CONTROL, $T$ = TREATED

| | | Blocks | | | | | | |
|---|---|---|---|---|---|---|---|---|
| | | 1 | | 2 | | 3 | | |
| Day | Plant | C | T | C | T | C | T | Totals |
| 1 | 1 | 2.6 | 2.8 | 3.6 | 4.3 | 6.8 | 5.3 | 25.4 |
| | 2 | 4.4 | 4.3 | 3.9 | 4.6 | 5.1 | 5.3 | 27.6 |
| | 3 | 4.3 | 3.2 | 5.4 | 6.2 | 6.2 | 5.8 | 31.1 |
| 2 | 1 | 6.4 | 5.4 | 7.4 | 6.8 | 11.1 | 8.1 | 45.2 |
| | 2 | 9.1 | 7.1 | 8.0 | 7.3 | 9.7 | 7.8 | 49.0 |
| | 3 | 9.0 | 5.9 | 9.6 | 8.9 | 10.5 | 8.7 | 52.6 |
| 3 | 1 | 10.9 | 8.0 | 11.2 | 8.9 | 14.1 | 9.9 | 63.0 |
| | 2 | 13.6 | 9.6 | 11.9 | 9.5 | 13.1 | 9.9 | 67.6 |
| | 3 | 13.4 | 8.8 | 13.5 | 11.0 | 13.7 | 10.6 | 71.0 |
| 4 | 1 | 15.0 | 9.9 | 16.6 | 10.6 | 20.9 | 11.6 | 84.6 |
| | 2 | 19.3 | 11.4 | 17.4 | 11.2 | 18.0 | 11.5 | 88.8 |
| | 3 | 18.7 | 10.9 | 19.6 | 12.9 | 19.6 | 12.5 | 94.2 |
| 5 | 1 | 21.0 | 11.5 | 23.8 | 13.1 | 27.1 | 14.1 | 110.6 |
| | 2 | 25.4 | 13.8 | 23.6 | 13.9 | 24.8 | 13.7 | 115.2 |
| | 3 | 25.1 | 12.9 | 24.9 | 15.5 | 25.7 | 14.8 | 118.9 |
| 6 | 1 | 26.3 | 14.1 | 27.4 | 15.7 | 30.3 | 17.1 | 130.9 |
| | 2 | 29.3 | 16.7 | 27.2 | 16.5 | 28.5 | 16.7 | 134.9 |
| | 3 | 30.1 | 16.0 | 27.9 | 18.1 | 28.6 | 18.3 | 139.0 |
| 7 | 1 | 29.0 | 16.9 | 31.0 | 18.3 | 34.7 | 20.0 | 149.9 |
| | 2 | 31.1 | 19.4 | 31.8 | 19.3 | 31.9 | 19.5 | 153.0 |
| | 3 | 31.3 | 19.1 | 31.4 | 20.8 | 32.0 | 21.1 | 155.7 |
| 8 | 1 | 33.4 | 20.2 | 36.0 | 20.5 | 40.0 | 22.4 | 172.5 |
| | 2 | 35.0 | 24.2 | 35.3 | 21.5 | 37.5 | 22.3 | 175.8 |
| | 3 | 35.4 | 21.6 | 36.0 | 22.9 | 36.2 | 23.6 | 175.7 |
| Totals | | 479.1 | 293.7 | 484.4 | 318.3 | 526.1 | 330.6 | 2432.2 |

$$SS_B = \frac{\Sigma_i\, Y_{i\ldots}^2}{n_t n_p n_d} - \frac{Y_{\ldots}^2}{N} = b - (1)$$

$$= \frac{(772.8)^2 + (802.7)^2 + (856.7)^2}{48} - (1)$$

$$= 41{,}155.88 - (1) = 75.35 \qquad (7.46)$$

$$SS_T = \frac{\Sigma_j\, Y_{\cdot j\cdot\cdot}^2}{n_b n_p n_d} - \frac{Y_{\ldots}^2}{N} = t - (1)$$

$$= \frac{(1489.6)^2 + (942.6)^2}{72} - (1)$$

$$= 43{,}158.37 - (1) = 2077.84 \qquad (7.47)$$

$$SS_{BT} = \frac{\Sigma_i \, \Sigma_j \, Y_{ij..}^2}{n_p n_d} - \frac{\Sigma_i \, Y_{i...}^2}{n_t n_p n_d} - \frac{\Sigma_j \, Y_{.j..}^2}{n_b n_p n_d} + \frac{Y_{....}^2}{N}$$

$$= bt - b - t + (1)$$

$$= \frac{(479.1)^2 + (293.7)^2 + \cdots + (330.6)^2}{24} - b - t + (1)$$

$$= 43{,}243.01 - b - t + (1) = 9.29 \qquad (7.48)$$

$$SS_{P:BT} = \frac{\Sigma_i \, \Sigma_j \, \Sigma_k \, Y_{ijk.}^2}{n_d} - \frac{\Sigma_i \, \Sigma_j \, Y_{ij..}^2}{n_p n_d}$$

$$= btp - bt = 116.51 \qquad (7.49)$$

$$\text{Split-plot SS} = \sum_i \sum_j \sum_k \sum_l Y_{ijkl}^2 - \frac{\Sigma_i \, \Sigma_j \, \Sigma_k \, Y_{ijk.}^2}{n_d}$$

$$= btpd - btp = 10{,}332.64 \qquad (7.50)$$

$$SS_D = \frac{\Sigma_l \, Y_{...l}^2}{n_b n_t n_p} - \frac{Y_{....}^2}{N} = d - (1)$$

$$= \frac{(84.1)^2 + (146.8)^2 + \cdots + (524.0)^2}{18} - (1)$$

$$= 50{,}469.25 - (1) = 9388.72 \qquad (7.51)$$

$$SS_{BD} = \frac{\Sigma_i \, \Sigma_l \, Y_{i..l}^2}{n_t n_p} - \frac{\Sigma_i \, Y_{i...}^2}{n_t n_p n_d} - \frac{\Sigma_l \, Y_{...l}^2}{n_b n_t n_p} + \frac{Y_{....}^2}{N}$$

$$= bd - b - d + (1)$$

$$= \frac{(21.6)^2 + (42.9)^2 + \cdots + (182.0)^2}{6} - b - d + (1)$$

$$= 50{,}549.83 - b - d + (1) = 5.23 \qquad (7.52)$$

$$SS_{TD} = \frac{\Sigma_j \, \Sigma_l \, Y_{.j.l}^2}{n_b n_p} - \frac{\Sigma_j \, Y_{.j..}^2}{n_b n_p n_d} - \frac{\Sigma_l \, Y_{...l}^2}{n_b n_t n_p} + \frac{Y_{....}^2}{N}$$

$$= td - t - d + (1)$$

$$= \frac{(42.3)^2 + (80.8)^2 + \cdots + (199.2)^2}{9} - t - d + (1)$$

$$= 53{,}456.28 - t - d + (1) = 909.19 \qquad (7.53)$$

$$SS_{BTD} = \frac{\Sigma_i \, \Sigma_j \, \Sigma_l \, Y_{ij.l}^2}{n_p} - \frac{\Sigma_i \, \Sigma_j \, Y_{ij..}^2}{n_p n_d} - \frac{\Sigma_i \, \Sigma_l \, Y_{i..l}^2}{n_t n_p} - \frac{\Sigma_j \, \Sigma_l \, Y_{.j.l}^2}{n_b n_p}$$

$$+ \frac{\Sigma_i \, Y_{i...}^2}{n_t n_p n_d} + \frac{\Sigma_j \, Y_{.j..}^2}{n_b n_p n_d} + \frac{\Sigma_l \, Y_{...l}^2}{n_b n_t n_p} - \frac{Y_{....}^2}{N}$$

$$= btd - bt - bd - td + b + t + d - (1)$$

$$= \frac{(11.3)^2 + (24.5)^2 + \cdots + (68.3)^2}{3}$$

$$- bt - bd - td + b + t + d - (1)$$

$$= 53,554.98 - bt - bd - td + b + t + d - (1) = 8.83 \qquad (7.54)$$

$$\text{SS}_{PD:BT} = \sum_i \sum_j \sum_k \sum_l Y_{ijkl}^2 - \frac{\sum_i \sum_j \sum_k Y_{ijk\cdot}^2}{n_d} - \frac{\sum_i \sum_j \sum_l Y_{ij\cdot l}^2}{n_p}$$

$$+ \frac{\sum_i \sum_j Y_{ij\cdot\cdot}^2}{n_p n_d}$$

$$= btpd - btp - btd + bt = 20.67 \qquad (7.55)$$

Table 7.21 shows the sources of variation, degrees of freedom, and expectations of mean squares. The blocks and plants are considered to be random effects and the treatments and days are considered to be fixed effects.

One of the reasons for including this example is to demonstrate the different denominators used in $F$ tests in this analysis. It is customary to designate the last line in the whole-plot portion of the analysis as error ($a$) and the last line in the split-plot portion as error ($b$), which tends to imply that all effects in the whole plots are tested by error ($a$) and all effects in the split plots are tested by error ($b$). However, the expectations of mean squares show clearly that many different denominators are used for the appropriate $F$ tests. The completed analysis is shown in Table 7.22. Were a split-split-plot effect included, the extension would proceed in a fashion similar to that shown for the completely randomized and factorial split-plot designs presented in Sections 7.2 and 7.3.

The results of the analysis show that there is a highly significant effect of the two treatments, the nighttime chilling resulting in a much slower growth rate. There is

**TABLE 7.21** OUTLINE OF ANALYSIS OF VARIANCE FOR DATA SHOWN IN TABLE 7.20

| Source | df | E(MS) |
|---|---|---|
| Whole plots | $n_b n_t n_p - 1 = 17$ | |
| B | $n_b - 1 = 2$ | $\sigma_w^2 + 8\sigma_{p:bt}^2 + 48\sigma_b^2$ |
| T | $n_t - 1 = 1$ | $\sigma_w^2 + 8\sigma_{p:bt}^2 + 24\sigma_{bt}^2 + 72\sigma_t^2$ |
| BT | $(n_b - 1)(n_t - 1) = 2$ | $\sigma_w^2 + 8\sigma_{p:bt}^2 + 24\sigma_{bt}^2$ |
| P:BT | $(n_p - 1)n_b n_t = 12$ | $\sigma_w^2 + 8\sigma_{p:bt}^2$ |
| Split plots | $(n_d - 1)n_b n_t n_p = 126$ | |
| D | $n_d - 1 = 7$ | $\sigma_w^2 + \sigma_{pd:bt}^2 + 6\sigma_{bd}^2 + 18\sigma_d^2$ |
| BD | $(n_b - 1)(n_d - 1) = 14$ | $\sigma_w^2 + \sigma_{pd:bd}^2 + 6\sigma_{bd}^2$ |
| TD | $(n_t - 1)(n_d - 1) = 7$ | $\sigma_w^2 + \sigma_{pd:bt}^2 + 3\sigma_{btd}^2 + 9\sigma_{td}^2$ |
| BTD | $(n_b - 1)(n_t - 1)(n_d - 1) = 14$ | $\sigma_w^2 + \sigma_{pd:bt}^2 + 3\sigma_{btd}^2$ |
| PD:BT | $(n_p - 1)(n_d - 1)n_b n_t = 84$ | $\sigma_w^2 + \sigma_{pd:bt}^2$ |

$n_b = 3 \qquad n_t = 2 \qquad n_p = 3 \qquad n_d = 8 \qquad n_w = 1 \qquad N = 144$

**TABLE 7.22**　ANALYSIS OF VARIANCE DATA DISPLAYED IN TABLE 7.20

| Source | df | SS | MS | F |
|---|---|---|---|---|
| Whole plots | 17 | 2,278.99 | | |
| B | 2 | 75.35 | 37.68 | 37.68/9.71 =　3.88 |
| T | 1 | 2,077.84 | 2,077.84 | 2,077.84/4.64 =　447.81* |
| BT | 2 | 9.29 | 4.64 | 4.64/9.71 =　0.48 |
| P:BT | 12 | 116.51 | 9.71 | |
| Split plots | 126 | 10,332.64 | | |
| D | 7 | 9,388.72 | 1,341.25 | 1,341.25/0.37 = 3,625.00* |
| BD | 14 | 5.23 | 0.37 | 0.37/0.25 =　1.48 |
| TD | 7 | 909.19 | 129.88 | 129.88/0.65 =　206.16** |
| BDT | 14 | 8.83 | 0.63 | 0.63/0.25 =　2.52** |
| PD:BT | 84 | 20.67 | 0.25 | |

also a highly significant effect of days, as is certainly to be expected. In addition, there is a highly significant interaction between treatments and days. Owing to the remarkably low *PD:BT* mean square, a highly significant second-order interaction, *BTD,* is found; however, the sum of squares accounted for by this interaction is so small that the effect or importance of this interaction is obviously minor and can probably be ignored.

To explore the cause of the highly significant effect of treatments and the highly significant effect of the treatment × day interaction, we can set up Table 7.19 which contains the treatment × day subclass totals, the orthogonal polynomial coefficients for the linear effect of treatments, the linear through quintic effect of days, and the interaction of the linear effect of treatments with the linear through quintic effect of days. The coefficients for the interaction effects are formed by multiplying the linear through quintic coefficients of days in turn by the linear coefficients for treatments. The $c_{ij}$ in the table, of course, represent the orthogonal polynomial coefficients with $i$ running from 1 to 16, the number of treatment × day subclass totals.

The sum of squares for each degree of freedom is shown at the foot of Table 7.19, calculated by the usual formula

$$SS = \frac{(\sum c_{ij}T_j)^2}{n \sum c_{ij}^2}$$

where $n = 9$, the number of observations in each treatment × day subclass. Since the mean square is the same as the sum of squares for each single degree of freedom, division of these values by the appropriate mean square yields $F$ values, the results of which are shown in the form of asterisks or lack thereof on the sums of squares. The linear effects of treatments are tested by the *P:BT* line, the polynomial effects of days by the *BD* line, and the interaction components by the *BTD* line. Due to the particularly small mean square for *BTD,* all the interaction components are highly significant. However, in view of the much larger portion being accounted for by the linear × linear effect, this is the only sum of squares of importance.

The highly significant interaction of linear effect of treatments and linear effect of days indicates that the linear regression of growth over days differs greatly between the two treatments. We can develop the linear prediction equations for each treatment by calculating the linear regression coefficients for each treatment separately, using the first eight coefficients (Table 7.23) for the days for the first treatment and the last eight coefficients for the days for the second treatment,

**TABLE 7.23** ORTHOGONAL POLYNOMIAL COEFFICIENTS FOR TREATMENT × DAY SUBCLASS

| | | $T_1D_1$ | $T_1D_2$ | $T_1D_3$ | $T_1D_4$ | $T_1D_5$ | $T_1D_6$ | $T_1D_7$ | $T_1D_8$ | $T_2D_1$ | $T_2D_2$ | $T_2D_3$ | $T_2D_4$ | $T_2D_5$ | $T_2D_6$ | $T_2D_7$ | $T_2D_8$ |
|---|---|---|---|---|---|---|---|---|---|---|---|---|---|---|---|---|---|
| Totals | $C_{ij}$ | 42.3 | 80.8 | 115.4 | 165.1 | 221.4 | 255.6 | 284.2 | 324.8 | 41.8 | 66.0 | 86.2 | 102.5 | 123.3 | 149.2 | 174.4 | 199.2 |
| T (linear) | $c_{1j}$ | -1 | -1 | -1 | -1 | -1 | -1 | -1 | -1 | 1 | 1 | 1 | 1 | 1 | 1 | 1 | 1 |
| D (linear) | $c_{2j}$ | -7 | -5 | -3 | -1 | 1 | 3 | 5 | 7 | -7 | -5 | -3 | -1 | 1 | 3 | 5 | 5 |
| D (quadratic) | $c_{3j}$ | 7 | 1 | -3 | -5 | -5 | -3 | 1 | 7 | 7 | 1 | -3 | -5 | -5 | -3 | 1 | 7 |
| D (cubic) | $c_{4j}$ | -7 | 5 | 7 | 3 | -3 | -7 | -5 | 7 | -7 | 5 | 7 | 3 | -3 | -7 | -5 | 7 |
| D (quartic) | $c_{5j}$ | 7 | -13 | -3 | 9 | 9 | -3 | -13 | 7 | 7 | -13 | -3 | 9 | 9 | -3 | -13 | 7 |
| D (quintic) | $c_{6j}$ | -7 | 23 | -17 | -15 | 15 | 17 | -23 | 7 | -7 | 23 | -17 | -15 | 15 | 17 | -23 | 7 |
| T (linear) × D (linear) | $c_{7j}$ | 7 | 5 | 3 | 1 | -1 | -3 | -5 | -7 | -7 | -5 | -3 | -1 | 1 | 3 | 5 | 7 |
| T (linear) × D (quadratic) | $c_{8j}$ | -7 | -1 | 3 | 5 | 5 | 3 | -1 | -7 | 7 | 1 | -3 | -5 | -5 | -3 | 1 | 7 |
| T (linear) × D (cubie) | $c_{9j}$ | 7 | -5 | -7 | -3 | 3 | 7 | 5 | -7 | -7 | 5 | 7 | 3 | -3 | -7 | -5 | 7 |
| T (linear) × D (quartic) | $c_{10j}$ | -7 | 13 | 3 | -9 | -9 | 3 | 13 | -7 | 7 | -13 | -3 | 9 | 9 | -3 | -13 | 7 |
| T (linear) × D (quintic) | $c_{11j}$ | 7 | -23 | 17 | 15 | -15 | -17 | 23 | -7 | -7 | 23 | -17 | -15 | 15 | 17 | -23 | 7 |

**Days**

$SS_{lin} = 9376.86**$
$SS_{quad} = 0.11$
$SS_{cubic} = 3.74*$
$SS_{quar} = 0.55$
$SS_{quin} = 6.85**$

**Treatments × Day**

$SS_{lin \times lin} = 865.50**$
$SS_{lin \times quad} = 13.63**$
$SS_{lin \times cubic} = 12.76**$
$SS_{lin \times quar} = 8.25**$
$SS_{lin \times quin} = 7.30**$

**Treatments**

$SS_{lin} = 2077.84**$

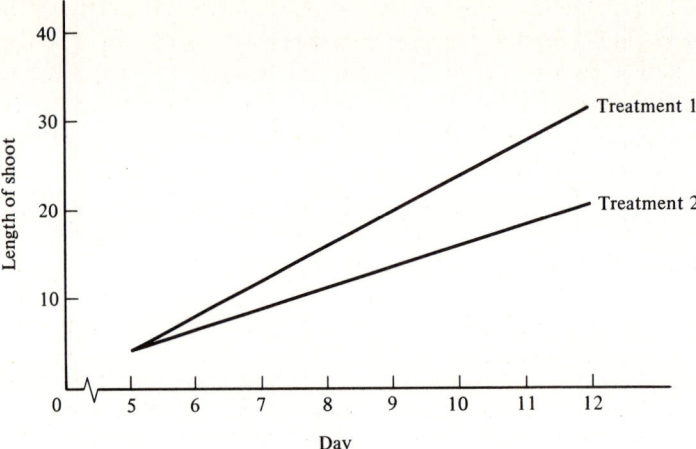

**Figure 7.2**  Linear regression lines for each treatment level.

$$b_l = \frac{\Sigma \, c_{2j}T_j}{n \, \Sigma \, c_{2j}^2} = \frac{3471.4}{(9)(168)} = 2.2959 \quad \text{for } T_1$$

$$j = 1, \ldots, 8$$

$$b_l = \frac{\Sigma \, c_{2j}T_j}{n \, \Sigma \, c_{2j}^2} = \frac{1853.6}{(9)(168)} = 1.2259 \quad \text{for } T_2$$

$$j = 9, \ldots, 16$$

and the means of the two treatments

$$\bar{y}_1 = \frac{Y_{.1..}}{n_b n_p n_d} = \frac{1489.6}{72} = 20.6889 \quad \text{for } T_1$$

$$\bar{y}_2 = \frac{Y_{.2..}}{n_b n_p n_d} = \frac{942.6}{72} = 13.0917 \quad \text{for } T_2$$

The prediction equations would be

$$\hat{y} = \bar{y}_1 + b_l c_{2j} = 20.69 + 2.30c_{2j} \quad \text{for } T_1$$

$$\hat{y} = \bar{y}_2 + b_l c_{2j} = 13.09 + 1.23c_{2j} \quad \text{for } T_2$$

The regression lines resulting from these equations are shown in Figure 7.2, where it can be seen that the slopes of the two lines do bear out the test of significance of the interaction.

It should be recognized that there could have been more than two levels of treatment, in which case we would have graphed a different regression line for each treatment level.

## 7.6  LATIN SQUARE SPLIT-PLOT DESIGN

As with the designs discussed previously in this chapter, split plots are frequently incorporated in the Latin square design. We earlier discussed a Latin square design

(Table 6.5) in which we ignored the split-plot aspect and analyzed only the whole-plot means. The complete set of data is shown in Table 7.24. For the last five days of each period of the experiment measurements were made on each cow. Thus, the period × cow cell is the whole plot on which repeated measures are made, the days becoming the split-plot factor. The split plots are the day × period × cow observations; the mathematical model may be written

$$Y_{ij(t)k} = \mu + c_i + p_j + t_{(t)} + d_{ij} + h_k + ch_{ik} + ph_{jk} + th_{(t)k} + e_{ijk} \qquad (7.56)$$

where the $c_i$ represent cow effects, the $p_j$ represent period effects, the $t_{(t)}$ represent treatment effects, the $d_{ij}$ stand for whole-plot errors, and the $h_k$ stand for day effects. The treatments are designated by the letters $A, B, C,$ and $D$ in the table. The calculations necessary for this analysis follow.

**TABLE 7.24**  TOTAL VOLATILE FATTY ACIDS MEASURED ON FOUR COWS, AT FOUR PERIODS, FOR FOUR RATIONS: MEASUREMENTS MADE ON EACH OF THE LAST FIVE DAYS OF EACH TRIAL PERIOD

| Period | Day | | 224 | | 225 | | 419 | | 426 | Totals |
|---|---|---|---|---|---|---|---|---|---|---|
| 1 | 1 | | 10.907 | | 10.720 | | 8.800 | | 8.560 | 38.987 |
| | 2 | | 11.040 | | 10.400 | | 9.734 | | 9.280 | 40.454 |
| | 3 | A | 9.280 | B | 9.920 | C | 8.747 | D | 8.267 | 36.214 |
| | 4 | | 9.280 | | 10.000 | | 8.960 | | 6.934 | 35.174 |
| | 5 | | 9.920 | | 9.974 | | 9.680 | | 6.587 | 36.161 |
| | | | 50.427 | | 51.014 | | 45.921 | | 39.628 | 186.990 |
| 2 | 1 | | 9.360 | | 9.654 | | 9.360 | | 9.600 | 37.974 |
| | 2 | | 9.280 | | 9.174 | | 8.800 | | 8.827 | 36.081 |
| | 3 | B | 9.840 | C | 8.960 | D | 8.987 | A | 10.400 | 38.187 |
| | 4 | | 9.494 | | 8.934 | | 8.694 | | 9.360 | 36.482 |
| | 5 | | 8.907 | | 9.227 | | 8.347 | | 9.254 | 35.735 |
| | | | 46.881 | | 45.949 | | 44.188 | | 47.441 | 184.459 |
| 3 | 1 | | 10.000 | | 10.107 | | 10.240 | | 9.014 | 39.361 |
| | 2 | | 9.414 | | 10.854 | | 10.400 | | 9.120 | 39.788 |
| | 3 | C | 9.094 | D | 9.707 | A | 10.374 | B | 8.080 | 37.255 |
| | 4 | | 10.160 | | 10.294 | | 11.360 | | 8.880 | 40.694 |
| | 5 | | 9.867 | | 10.507 | | 11.360 | | 8.427 | 40.161 |
| | | | 48.535 | | 51.469 | | 53.734 | | 43.521 | 197.259 |
| 4 | 1 | | 8.080 | | 10.227 | | 9.147 | | 8.840 | 36.294 |
| | 2 | | 8.627 | | 9.667 | | 10.854 | | 8.427 | 37.575 |
| | 3 | D | 8.707 | A | 9.654 | B | 10.467 | C | 8.320 | 37.148 |
| | 4 | | 8.320 | | 10.774 | | 10.947 | | 7.734 | 37.775 |
| | 5 | | 8.600 | | 9.600 | | 8.947 | | 8.254 | 35.401 |
| | | | 42.334 | | 49.922 | | 50.362 | | 41.575 | 184.193 |
| Totals | | | 188.177 | | 198.354 | | 194.205 | | 172.165 | 752.901 |
| | | $A$ = 201.524 | | $B$ = 191.778 | | $C$ = 181.980 | | $D$ = 177.619 | | |

$$\text{Total SS} = \sum_i \sum_j \sum_k Y_{ijk}^2 - \frac{Y^2_{...}}{N} = cph - (1)$$

$$= 7155.7288 - \frac{(752.901)^2}{80}$$

$$= 7155.7288 - 7085.7489 = 69.9799 \qquad (7.57)$$

$$\text{Whole-plot SS} = \frac{\sum_i \sum_j Y_{ij\cdot}^2}{n_h} - \frac{Y^2_{...}}{N} = cp - (1)$$

$$= \frac{(50.427)^2 + (51.014)^2 + \cdots + (41.575)^2}{5} - (1)$$

$$= 7134.4207 - (1) = 48.6718 \qquad (7.58)$$

$$\text{SS}_C = \frac{\sum_i \sum_j Y_{i\cdot\cdot}^2}{n_k n_h} - \frac{Y^2_{...}}{N} = c - (1)$$

$$= \frac{(188.177)^2 + (198.354)^2 + \cdots + (172.165)^2}{20} - (1)$$

$$= 7105.5631 - (1) = 19.8142 \qquad (7.59)$$

$$\text{SS}_P = \frac{\sum_j Y_{\cdot j\cdot}^2}{n_k n_h} - \frac{Y^2_{...}}{N} = p - (1)$$

$$= \frac{(186.990)^2 + (184.459)^2 + \cdots + (184.193)^2}{20} - (1)$$

$$= 7091.4279 - (1) = 5.6790 \qquad (7.60)$$

$$\text{SS}_T = \frac{\sum_t Y_{\cdot\cdot(t)\cdot}}{n_k n_h} - \frac{Y^2_{...}}{N} = t - (1)$$

$$= \frac{(201.524)^2 + (191.778)^2 + \cdots + (177.619)^2}{20} - (1)$$

$$= 7102.7977 - (1) = 17.0488 \qquad (7.61)$$

$$\text{Whole-plot error SS} = \text{SS}_{\text{whole plots}} - \text{SS}_C - \text{SS}_P - \text{SS}_T = 6.1298 \qquad (7.62)$$

$$\text{Split-plot SS} = \sum_i \sum_j \sum_k Y_{ijk}^2 - \frac{\sum_i \sum_j Y_{ij\cdot}^2}{n_h} = cph - cp = 21.3081 \qquad (7.63)$$

$$\text{SS}_H = \frac{\sum_k Y_{\cdot\cdot k}^2}{n_k^2} - \frac{Y^2_{...}}{N} = h - (1)$$

$$= \frac{(152.616)^2 + (153.898)^2 + \cdots + (147.458)^2}{16} - (1)$$

$$= 7087.5154 - (1) = 1.7665 \qquad (7.64)$$

$$\text{SS}_{CH} = \frac{\sum_i \sum_k Y_{i\cdot k}^2}{n_k} - \frac{\sum_i Y_{i\cdot\cdot}^2}{n_k n_h} - \frac{\sum_k Y_{\cdot\cdot k}^2}{n_k^2} + \frac{Y^2_{...}}{N} = ch - c - h + (1)$$

$$= \frac{(38.347)^2 + (38.361)^2 + \cdots + (32.522)^2}{4} - c - h + (1)$$

$$= 7110.5306 - c - h + (1) = 3.2010 \tag{7.65}$$

$$\text{SS}_{PH} = \frac{\Sigma_j \, \Sigma_k \, Y^2_{\cdot jk}}{n_k} - \frac{\Sigma_j \, Y^2_{\cdot j \cdot}}{n_k n_h} - \frac{\Sigma_k \, Y^2_{\cdot \cdot k}}{n_k^2} + \frac{Y^2_{\cdots}}{N} = ph - p - h + (1)$$

$$= \frac{(38.987)^2 + (40.454)^2 + \cdots + (35.401)^2}{4} - p - h + (1)$$

$$= 7100.3343 - p - h + (1) = 7.1399 \tag{7.66}$$

$$\text{SS}_{TH} = \frac{\Sigma_t \, \Sigma_k \, Y^2_{\cdot \cdot (t)k}}{n_k} - \frac{\Sigma_t \, Y^2_{\cdot \cdot (t) \cdot}}{n_k n_h} - \frac{\Sigma_k \, Y^2_{\cdot \cdot k}}{n_k^2} + \frac{Y^2_{\cdots}}{N} = th - t - h + (1)$$

$$= \frac{(40.974)^2 + (39.934)^2 + \cdots + (34.041)^2}{4} - t - h + (1)$$

$$= 7107.7765 - t - h + (1) = 3.2123 \tag{7.67}$$

Split-plot error SS = $\text{SS}_{\text{split plots}} - \text{SS}_H - \text{SS}_{CH} - \text{SS}_{PH} - \text{SS}_{TH} = 5.9884$ \quad (7.68)

where $N = n_k^2 n_h$ and $n_k$ is the size of the square.

The nature of this design does require a slight modification in the denominators when calculating the sums of squares. Summing over the rows and columns of a square for treatment effects is the same as summing over the square because of the fact that the treatment appears only once in each row and column. Therefore, instead of using $n_c n_p$ in the denominator, the value $n_k$ (size of the square) is used.

The outline of the analysis of variance for this set of data, including the calculation of degrees of freedom and expectations of mean squares, is shown in Table 7.25. Cow effects are considered to be random and period, treatment, and day effects are fixed. The calculation of the expectations of mean squares no longer follows all the rules presented earlier because the whole-plot and split-plot errors are really residuals or

**TABLE 7.25** OUTLINE OF ANALYSIS OF VARIANCE OF DATA SHOWN IN TABLE 7.24

| Source | df | E(MS) |
|---|---|---|
| Whole plots | $n_c n_p - 1 = 15$ | |
| C | $n_c - 1 = \phantom{0}3$ | $\sigma_e^2 + 5\sigma_d^2 + 20\sigma_c^2$ |
| P | $n_p - 1 = \phantom{0}3$ | $\sigma_e^2 + 5\sigma_d^2 + 20\sigma_p^2$ |
| T | $n_t - 1 = \phantom{0}3$ | $\sigma_e^2 + 5\sigma_d^2 + 20\sigma_t^2$ |
| Error (a) | $(n_c - 2)(n_p - 1) = \phantom{0}6$ | $\sigma_e^2 + 5\sigma_d^2$ |
| Split plots | $(n_h - 1)n_c n_p = 64$ | |
| H | $n_h - 1 = \phantom{0}4$ | $\sigma_e^2 + 4\sigma_{ch}^2 + 16\sigma_h^2$ |
| CH | $(n_c - 1)(n_h - 1) = 12$ | $\sigma_e^2 + 4\sigma_{ch}^2$ |
| PH | $(n_p - 1)(n_h - 1) = 12$ | $\sigma_e^2 + 4\sigma_{ph}^2$ |
| TH | $(n_t - 1)(n_h - 1) = 12$ | $\sigma_e^2 + 4\sigma_{th}^2$ |
| Error (b) | $(n_h - 1)(n_c - 2)(n_p - 1) = 24$ | $\sigma_e^2$ |

$n_c = 4$ \qquad $n_p = 4$ \qquad $n_t = 4$ \qquad $n_h = 5$ \qquad $N = n_c n_p n_h = 80$

remainders and there is no within, or $W$, term. While it is quite common for textbooks to use greek letters such as $\delta$ and $\varepsilon$ for whole-plot and split-plot errors, the authors have found it advantageous to use the letters $d$ and $e$ in keeping with the other models in this text. The coefficient for the variance component $\sigma_e^2$ is equal to the number of observations in any split plot, in this case, one, and the coefficient for the variance component $\sigma_d^2$ is equal to $n_h$, the number of observations in each whole plot. Coefficients for the remaining variance components are calculated by dividing $N$ by the product of the $n$'s corresponding to the subscripts in the variance components. Note that $n_t$ is omitted when calculating $N$ since any observation may be completely identified, ignoring the treatment classification. When dealing with a split-split-plot design, the coefficient for the split-split-plot error variance component would be the number of observations in any single split-split plot. The complete analysis of variance is shown in Table 7.26.

A common error in the analysis of this type of design is the failure to take into account the whole-plot error, resulting in a pooling of the whole-plot and split-plot error terms. Since the split-plot error term is normally expected to be smaller than the whole-plot error term, this can lead to a smaller error for testing whole-plot effects than is appropriate and a larger error for testing split-plot effects. The reverse could occur but is not as likely an event.

If a split-split-plot effect is involved in a Latin square design, the degrees of freedom, sums of squares, and expectations of mean squares are calculated as shown in the example for the crossover split-split-plot design discussed in Section 7.8.

The analysis of variance in Table 7.26 shows, of course, that the same results are obtained for the whole-plot analysis as were found in the previous analysis (see Table 6.7) when whole-plot means were analyzed. If the treatment sum of squares is partitioned into linear, quadratic, and cubic components, we find that there is a highly significant linear trend of treatments.

Although no effects were found to be significant in the split-plot portion of the analysis, a partitioning of the sum of squares for days ($H$) into linear, quadratic, cubic, and quartic components shows that the linear effect of days is very close to being significant at the 5 percent level. An even more interesting result occurs if we look more closely at the day $\times$ treatment interaction ($TH$). Although this interaction is

**TABLE 7.26**   ANALYSIS OF VARIANCE OF LATIN SQUARE,
SPLIT-PLOT DESIGN

| Source | df | SS | MS | $F$ |
|--------|----|----|----|-----|
| Whole plots | 15 | 48.6718 | | |
| $C$ | 3 | 19.8142 | 6.6047 | 6.47** |
| $P$ | 3 | 5.6790 | 1.8930 | 1.85 |
| $T$ | 3 | 17.0488 | 5.6829 | 5.56* |
| Error ($a$) | 6 | 6.1298 | 1.0216 | |
| Split plots | 64 | 21.3081 | | |
| $H$ | 4 | 1.7665 | 0.4416 | 1.66 |
| $CH$ | 12 | 3.2010 | 0.2668 | 1.07 |
| $PH$ | 12 | 7.1399 | 0.5950 | 2.38 |
| $TH$ | 12 | 3.2123 | 0.2677 | 1.07 |
| Error ($b$) | 24 | 5.9884 | 0.2495 | |

not significant, we can glean further information from the data by partitioning the sum of squares due to interaction. Just as we can partition the overall effect of treatments into linear, quadratic, and cubic components, we can also partition the treatment effects *for each day* into the same components. We can then test among linear, among quadratic, and among cubic components to see whether they differ among the days. This can be done by setting up Table 7.27 as shown by Winer (1971). Since we are

**TABLE 7.27**  PARTITIONING INTERACTION SUMS OF SQUARES INTO DIFFERENCES AMONG LINEAR, QUADRATIC, AND CUBIC EFFECTS OF TREATMENT AMONG DAYS

| Day | Treatment | | | | $Q_{i \cdot k} = \Sigma_j c_{ij} T_{jk}$ | | |
| | *A* | *B* | *C* | *D* | $Q_{1 \cdot k}$ | $Q_{2 \cdot k}$ | $Q_{3 \cdot k}$ |
|---|---|---|---|---|---|---|---|
| 1 | 40.974 | 38.241 | 37.294 | 36.107 | −15.548 | 1.546 | −2.026 |
| 2 | 39.934 | 39.654 | 36.749 | 37.561 | −10.024 | 1.092 | 6.342 |
| 3 | 39.708 | 38.307 | 35.121 | 35.668 | −15.306 | 1.948 | 5.518 |
| 4 | 40.774 | 39.321 | 35.788 | 34.242 | −23.129 | −0.093 | 4.067 |
| 5 | 40.134 | 36.255 | 37.028 | 34.041 | −17.506 | 0.892 | −8.412 |

Orthogonal polynomial coefficients $c_{ij}$

| | | | | | | |
|---|---|---|---|---|---|---|
| Linear | −3 | −1 | 1 | 3 | $\Sigma c_{1j}^2 = 20$ | $n_t = 4$ |
| Quadratic | 1 | −1 | −1 | 1 | $\Sigma c_{2j}^2 = 4$ | $n_d = 5$ |
| Cubic | −1 | 3 | −3 | 1 | $\Sigma c_{3j}^2 = 20$ | |

$T_{jk}$ = treatment × day subclass total  $\quad i$ = degree of polynomial
$n_t$ = number of treatments  $\quad\quad\quad\quad j$ = treatment number
$n_d$ = number of days  $\quad\quad\quad\quad\quad k$ = day number

$$\text{SS for differences among linear trends} = \frac{\Sigma_k Q_{1 \cdot k}^2}{n_t \Sigma_j c_{1j}^2} - \frac{(\Sigma_k Q_{1 \cdot k})^2}{n_t n_d \Sigma_j c_{1j}^2}$$

$$= \frac{1417.9052}{80} - \frac{(-81.513)^2}{400}$$

$$= 17.7238 - 16.6109 = 1.1129$$

$$\text{SS for differences among quadratic trends} = \frac{\Sigma_k Q_{2 \cdot k}^2}{n_t \Sigma_j c_{2j}^2} - \frac{(\Sigma_k Q_{2 \cdot k})^2}{n_t n_d \Sigma_j c_{2j}^2}$$

$$= \frac{8.1816}{16} - \frac{(5.385)^2}{80} = 0.5114 - 0.3625 = 0.1489$$

$$\text{SS for differences among cubic trends} = \frac{\Sigma_k Q_{3 \cdot k}^2}{n_t \Sigma_j c_{3j}^2} - \frac{(\Sigma_k Q_{3 \cdot k})^2}{n_t n_d \Sigma_j c_{3j}^2}$$

$$= \frac{162.0762}{80} - \frac{(5.489)^2}{400} = 2.0260 - 0.0753 = 1.9507$$

Overall treatment sum of squares partitioning

$$\text{SS}_{\text{lin}} = \frac{(\Sigma_k Q_{1 \cdot k})^2}{n_t n_d \Sigma_j c_{1j}^2} = \frac{(-81.513)^2}{400} = 16.6109$$

$$\text{SS}_{\text{quad}} = \frac{(\Sigma_k Q_{2 \cdot k})^2}{n_t n_d \Sigma_j c_{2j}^2} = \frac{(5.385)^2}{80} = 0.3625$$

$$\text{SS cubic} = \frac{(\Sigma_k Q_{3 \cdot k})^2}{n_t n_d \Sigma_j c_{3j}^2} = \frac{(5.489)^2}{400} = 0.0753$$

dealing with the treatment × day interaction, we set up a table of treatment × day subclass totals. The totals for each treatment × day subclass ($T_{jk}$) are multiplied by the orthogonal polynomial coefficients for the linear, quadratic, and cubic effects to obtain the $Q_{i.k}$ values. Thus,

$$Q_{1.1} = (-3)(40.974) + (-1)(38.241) + (1)(37.294) + (3)(36.107) = -15.548$$

In Table 7.27, $n_t$ represents the number of treatments and $n_d$ represents the number of days. It can be seen at the foot of the table that the correction terms used in calculating the sums of squares for trends among days give the overall sums of squares for the linear, quadratic, and cubic effects of treatments.

Since there are four degrees of freedom for the linear effects among the days as well as for the quadratic and cubic effects, the numerator mean squares for the tests among the linear, quadratic, and cubic effects are 0.2782, 0.0372, and 0.4877. Division of these mean squares by the mean square for error ($b$) of 0.2495 shows that none is significant. The linear, quadratic, and cubic effects of treatments are thus seen to be consistent from day to day. Since the original interaction was not significant, this is not surprising. However, it can be seen that with a significant interaction, or even one approaching significance a partitioning in this fashion can yield a great deal of additional information.

## 7.7 CROSSOVER SPLIT-PLOT DESIGN

We previously dealt with the crossover design with data from the Exercise Science Department at the University of Massachusetts (Table 6.24) in which we analyzed whole-plot means and ignored the split plots. Measurements were actually made prior to exercise and after exercise, which resulted in the data shown in Table 7.28 in which the letter preceding the dash in the body of the table presents the exercise treatment level. As noted previously, the two groups are made up of different sexes, males in group 1 and females in group 2. Time of measurement represents the pre- and post-exercise measurements. The similarity between the split-plot design and the nested design may be observed here. If the pre- and postexercise measurements were merely a pair of random measurements, they would be a nested rather than a main classification or effect. They would be nested within the whole plots and would make up a sampling error and there would be no split plots. It is vital in this design to recognize that the subject × day cells, and *not* the subjects themselves, are the whole plots. In crossover and Latin square designs, it is critical to recognize the row × column cell as being the whole plot. For example, in the data shown in Table 7.28, the subject one × day one cell in either group is the whole plot. The whole-plot sum of squares is then the sum of squares among these cells or subclasses. The split-plot effect is the time of measurement so the subject × day × time-of-measurement subclass is the split-plot. The split-plot sum of squares is then the sum of squares of the measurements within the whole plots. Since a large number of different classes and subclasses are involved in the calculations for the analysis of variance, several two-way tables are presented in Table 7.28. The mathematical model for the analysis of this set of data is

**TABLE 7.28** UNRESISTED ANKLE SIMPLE REACTION TIME FOR TWO EXERCISE TREATMENTS, TAKEN BEFORE AND AFTER EXERCISE (IN MILLISECONDS)

| Day | Time of measurement | Group 1 subjects 1 | 2 | 3 | 4 | 5 | 6 | 7 | 8 | Total |
|-----|------|------|------|------|------|------|------|------|------|------|
| 1 | 1 | A—228 | A—229 | B—198 | B—231 | A—245 | B—224 | B—187 | A—233 | 1775 |
|   | 2 | A—228 | A—212 | B—193 | B—219 | A—222 | B—224 | B—187 | A—218 | 1703 |
|   | Subtotal | 456 | 441 | 391 | 450 | 467 | 448 | 374 | 451 | 3478 |
| 2 | 1 | B—236 | B—236 | A—220 | A—220 | B—227 | A—223 | A—197 | B—226 | 1785 |
|   | 2 | B—225 | B—227 | A—208 | A—226 | B—203 | A—216 | A—190 | B—193 | 1688 |
|   | Subtotal | 461 | 463 | 428 | 446 | 430 | 439 | 387 | 419 | 3473 |
|   | Total | 917 | 904 | 819 | 896 | 897 | 887 | 761 | 870 | 6951 |

| Day | Time of measurement | Group 2 subjects 1 | 2 | 3 | 4 | 5 | 6 | 7 | 8 | Total |
|-----|------|------|------|------|------|------|------|------|------|------|
| 1 | 1 | A—236 | B—221 | B—194 | A—204 | A—222 | B—192 | B—190 | A—198 | 1657 |
|   | 2 | A—224 | B—225 | B—196 | A—171 | A—196 | B—187 | B—207 | A—202 | 1608 |
|   | Subtotal | 460 | 446 | 390 | 375 | 418 | 379 | 397 | 400 | 3265 |
| 2 | 1 | B—248 | A—237 | A—195 | B—189 | B—209 | A—202 | A—201 | B—213 | 1694 |
|   | 2 | B—248 | A—248 | A—212 | B—186 | B—203 | A—195 | A—191 | B—194 | 1677 |
|   | Subtotal | 496 | 485 | 407 | 375 | 412 | 397 | 392 | 407 | 3371 |
|   | Total | 956 | 931 | 797 | 750 | 830 | 776 | 789 | 807 | 6636 |

|       | $P_1$ | $P_2$ |       | $T_1$ | $T_2$ |       | $G_1$ | $G_2$ |       | $H_1$ | $H_2$ |
|-------|------|------|-------|------|------|-------|------|------|-------|------|------|
| $T_1$ | 3490 | 3359 | $G_1$ | 3515 | 3436 | $P_1$ | 3560 | 3351 | $P_1$ | 3432 | 3479 |
| $T_2$ | 3421 | 3317 | $G_2$ | 3334 | 3302 | $P_2$ | 3391 | 3285 | $P_2$ | 3311 | 3365 |
|       | 6911 | 6676 |       | 6849 | 6738 |       | 6951 | 6636 |       | 6743 | 6844 |

|       | $GS_{11}$ | $GS_{12}$ | $GS_{13}$ | $GS_{14}$ | $GS_{15}$ | $GS_{16}$ | $GS_{17}$ | $GS_{18}$ |
|-------|------|------|------|------|------|------|------|------|
| $P_1$ | 464 | 465 | 418 | 451 | 472 | 447 | 384 | 459 |
| $P_2$ | 453 | 439 | 401 | 445 | 425 | 440 | 377 | 411 |

|       | $GS_{21}$ | $GS_{22}$ | $GS_{23}$ | $GS_{24}$ | $GS_{25}$ | $GS_{26}$ | $GS_{27}$ | $GS_{28}$ |
|-------|------|------|------|------|------|------|------|------|
| $P_1$ | 484 | 458 | 389 | 393 | 431 | 394 | 391 | 411 |
| $P_2$ | 472 | 473 | 408 | 357 | 399 | 382 | 398 | 396 |

|       | $GT_{11}$ | $GT_{12}$ | $GT_{21}$ | $GT_{22}$ |
|-------|------|------|------|------|
| $P_1$ | 1795 | 1765 | 1695 | 1656 |
| $P_2$ | 1720 | 1671 | 1639 | 1646 |

$$Y_{ijk(t)l} = \mu + g_i + s_{ij} + h_k + t_{(t)} + gt_{i(t)} + d_{ijk} + p_l$$
$$+ gp_{il} + sp_{ijl} + hp_{kl} + tp_{(t)l} + gtp_{i(t)l} + e_{ijkl} \qquad (7.69)$$

where the $g_i$ represent groups, the $s_{ij}$ represent subjects within the groups, the $h_k$ represent days, the $t_{(t)}$ stand for treatments, the $d_{ijk}$ stand for whole-plot errors, the $p_l$

stand for times of measurement, and the $e_{ijkl}$ represent split-plot errors. The necessary calculations for the analysis follow.

$$\text{Total SS} = \sum_i \sum_j \sum_k \sum_l Y_{ijkl}^2 - \frac{Y_{....}^2}{N} = gshp - (1)$$

$$= 2{,}905{,}893.00 - \frac{(13{,}587)^2}{64} = 2{,}905{,}893.00 - 2{,}884{,}477.64$$

$$= 21{,}415.36 \tag{7.70}$$

$$\text{Whole-plot SS} = \frac{\sum_i \sum_j \sum_k Y_{ijk.}^2}{n_p} - \frac{Y_{....}^2}{N} = ghs - (1)$$

$$= \frac{(456)^2 + (441)^2 + \cdots + (407)^2}{2} - (1)$$

$$= 2{,}902{,}602.50 - (1) = 18{,}124.86 \tag{7.71}$$

$$SS_G = \frac{\sum_i Y_{i...}^2}{n_s n_h n_p} - \frac{Y_{....}^2}{N} = g - (1)$$

$$= \frac{(6951)^2 + (6636)^2}{32} - (1)$$

$$= 2{,}886{,}028.03 - (1) = 1550.39 \tag{7.72}$$

$$SS_{S:G} = \frac{\sum_i \sum_j Y_{ij..}^2}{n_h n_p} - \frac{\sum_i Y_{i...}^2}{n_s n_h n_p} = gs - g$$

$$= \frac{(917)^2 + (904)^2 + \cdots + (807)^2}{4} - g$$

$$= 2{,}900{,}583.25 - (g) = 14{,}555.22 \tag{7.73}$$

$$SS_H = \frac{\sum_k Y_{..k.}^2}{n_g n_s n_p} - \frac{Y_{....}^2}{N} = h - (1)$$

$$= \frac{(6743)^2 + (6844)^2}{32} - (1)$$

$$= 2{,}884{,}637.03 - (1) = 159.39 \tag{7.74}$$

$$SS_T = \frac{\sum_t Y_{...(t).}^2}{n_g n_s n_p} - \frac{Y_{....}^2}{N} = t - (1)$$

$$= \frac{(6849)^2 + (6738)^2}{32} - (1)$$

$$= 2{,}884{,}670.16 - (1) = 192.52 \tag{7.75}$$

$$SS_{GT} = \frac{\Sigma_i \, \Sigma_t \, Y_{i \cdot \cdot (t) \cdot}^2}{n_s n_p} - \frac{\Sigma_i \, Y_{i \cdot \cdot \cdot}^2}{n_s n_h n_p} - \frac{\Sigma_i \, Y_{\cdot \cdot \cdot (t) \cdot}^2}{n_g n_s n_p} + \frac{Y_{\cdot \cdot \cdot \cdot}^2}{N}$$

$$= gt - g - t + (1)$$

$$= \frac{(3515)^2 + (3436)^2 + \cdots + (3302)^2}{16}$$

$$- g - t + (1) = 2{,}886{,}255.06 - g - t + (1) = 34.51 \qquad (7.76)$$

Whole-plot error SS $= gsh - gs - h - gt + g + (1)$

$$= SS_{\text{whole plots}} - SS_G - SS_{S:G} - SS_H - SS_T - SS_{GT}$$

$$= 1632.83 \qquad (7.77)$$

Split-plot SS $= \displaystyle\sum_i \sum_j \sum_k \sum_l Y_{ijkl}^2 - \frac{\Sigma_i \, \Sigma_j \, \Sigma_k \, Y_{ijk \cdot}^2}{n_p} = gshp - gsh$

$$= 3290.50 \qquad (7.78)$$

$$SS_P = \frac{\Sigma_l \, Y_{\cdot \cdot \cdot l}^2}{n_g n_s n_h} - \frac{Y_{\cdot \cdot \cdot \cdot}^2}{N} = p - (1)$$

$$= \frac{(6911)^2 + (6676)^2}{32} - (1) = 2{,}885{,}340.53 - (1)$$

$$= 862.89 \qquad (7.79)$$

$$SS_{GP} = \frac{\Sigma_i \, \Sigma_l \, Y_{i \cdot \cdot l}^2}{n_s n_h} - \frac{\Sigma_i \, Y_{i \cdot \cdot \cdot}^2}{n_s n_h n_p} - \frac{\Sigma_l \, Y_{\cdot \cdot \cdot l}^2}{n_g n_s n_h} + \frac{Y_{\cdot \cdot \cdot \cdot}^2}{N}$$

$$= gp - g - p + (1)$$

$$= \frac{(3560)^2 + (3391)^2 + \cdots + (3285)^2}{16} - g - p + (1)$$

$$= 2{,}887{,}056.69 - g - p + (1) = 165.77 \qquad (7.80)$$

$$SS_{SP:G} = \frac{\Sigma_i \, \Sigma_j \, \Sigma_l \, Y_{ij \cdot l}^2}{n_h} - \frac{\Sigma_i \, \Sigma_j \, Y_{ij \cdot \cdot}^2}{n_h n_p} - \frac{\Sigma_i \, \Sigma_l \, Y_{i \cdot \cdot l}^2}{n_s n_h} + \frac{\Sigma_i \, Y_{i \cdot \cdot \cdot}^2}{n_s n_h n_p}$$

$$= gsp - gs - gp + g$$

$$= \frac{(464)^2 + (465)^2 + \cdots + (396)^2}{2} - gs - gp + g$$

$$= 2{,}902{,}883.50 - gs - gp + g = 1271.59 \qquad (7.81)$$

$$SS_{HP} = \frac{\Sigma_k \, \Sigma_l \, Y_{\cdot \cdot kl}^2}{n_g n_s} - \frac{\Sigma_k \, Y_{\cdot \cdot k \cdot}^2}{n_g n_s n_p} - \frac{\Sigma_l \, Y_{\cdot \cdot \cdot l}^2}{n_g n_s n_h} + \frac{Y_{\cdot \cdot \cdot \cdot}^2}{N}$$

$$= hp - h - p + (1)$$

$$= \frac{(3432)^2 + (3311)^2 + \cdots + (3365)^2}{16} - h - p + (1)$$

$$= 2{,}885{,}500.69 - h - p + (1) = 0.77 \tag{7.82}$$

$$\mathrm{SS}_{TP} = \frac{\Sigma_t\,\Sigma_l\,Y^2_{\ldots(t)l}}{n_g n_s} - \frac{\Sigma_t\,Y^2_{\ldots(t)\cdot}}{n_g n_s n_p} - \frac{\Sigma_l\,Y^2_{\ldots l}}{n_g n_s n_h} + \frac{Y^2_{\ldots\ldots}}{N}$$

$$= tp - t - p + (1)$$

$$= \frac{(3490)^2 + (3359)^2 + \cdots + (3317)^2}{16} - t - p + (1)$$

$$= 2{,}885{,}544.44 - t - p + (1) = 11.39 \tag{7.83}$$

$$\mathrm{SS}_{GTP} = \frac{\Sigma_i\,\Sigma_t\,\Sigma_l\,Y^2_{i\cdot\cdot(t)l}}{n_s} - \frac{\Sigma_i\,\Sigma_t\,Y^2_{i\cdot\cdot(t)\cdot}}{n_s n_p} - \frac{\Sigma_i\,\Sigma_l\,Y^2_{i\cdot\cdot l}}{n_s n_h} - \frac{\Sigma_t\,\Sigma_l\,Y^2_{\ldots(t)l}}{n_g n_s}$$

$$\quad + \frac{\Sigma_i\,Y^2_{i\ldots}}{n_s n_h n_p} + \frac{\Sigma_t\,Y^2_{\ldots(t)\cdot}}{n_g n_s n_p} + \frac{\Sigma_l\,Y^2_{\ldots l}}{n_g n_s n_h} - \frac{Y^2_{\ldots\ldots}}{N}$$

$$= gtp - gt - gp - tp + g + t + p - (1)$$

$$= \frac{(1795)^2 + (1720)^2 + \cdots + (1646)^2}{8} - gt - gp - tp + g + t + p - (1)$$

$$= 2{,}887{,}361.12 - gt - gp - tp + g + t + p - (1) = 66.01 \tag{7.84}$$

$$\text{Split-plot error SS} = (gsh - gs - h - gt + g + 1)(p - 1)$$

$$= gsph - gsp - hp - gtp + gp + p - \text{whole-plot error SS}$$

$$= \mathrm{SS}_{\text{split-plots}} - \mathrm{SS}_P \quad \mathrm{SS}_{GP} - \mathrm{SS}_{SP\cdot G} - \mathrm{SS}_{HP} - \mathrm{SS}_{TP} - \mathrm{SS}_{GTP}$$

$$= 912.08 \tag{7.85}$$

It can be seen that the notation used is somewhat awkward due to the fact that an observation is identified completely by the group, subject, day, and time. To distinguish the particular treatment or subclasses of treatments with other effects in the model, the notation differs slightly from our previous usage, both in the numerators and denominators of the formulas for the sums of squares where treatment effects are involved.

The outline of the analysis of variance together with the degrees of freedom and expectations of mean squares are shown in Table 7.29. The similarity to a nested design can be noted again here, particularly when calculating the degrees of freedom for the split plots. The formula is identical to that which would be used for observations nested within whole plots.

Again, the calculations of the expectations of mean squares no longer follow the rules laid down earlier. For a discussion of the changes involved, see the Latin square split-plot design explanation since the two situations are quite similar. Note that the expectations show that several different denominators are called for in the $F$ tests.

**TABLE 7.29** OUTLINE OF ANALYSIS OF VARIANCE FOR CROSSOVER SPLIT-PLOT DESIGN

| Source | df | E(MS) |
|--------|-----|-------|
| Whole plots | $n_g n_s n_h - 1 = 31$ | |
| G | $n_g - 1 = 1$ | $\sigma_e^2 + 2\sigma_d^2 + 4\sigma_{s:g}^2 + 32\sigma_g^2$ |
| S:G | $(n_s - 1)n_g = 14$ | $\sigma_e^2 + 2\sigma_d^2 + 4\sigma_{s:g}^2$ |
| H | $n_h - 1 = 1$ | $\sigma_e^2 + 2\sigma_d^2 + 32\sigma_h^2$ |
| T | $n_t - 1 = 1$ | $\sigma_e^2 + 2\sigma_d^2 + 32\sigma_t^2$ |
| GT | $(n_g - 1)(n_t - 1) = 1$ | $\sigma_e^2 + 2\sigma_d^2 + 16\sigma_{gt}^2$ |
| Error (a) | $(n_h - 1)(n_g n_s - 3) = 13$ | $\sigma_e^2 + 2\sigma_d^2$ |
| Split plots | $(n_p - 1)n_g n_s n_h = 32$ | |
| P | $n_p - 1 = 1$ | $\sigma_e^2 + 2\sigma_{sp:g}^2 + 32\sigma_p^2$ |
| GP | $(n_g - 1)(n_p - 1) = 1$ | $\sigma_e^2 + 2\sigma_{sp:g}^2 + 16\sigma_{gp}^2$ |
| SP:G | $(n_s - 1)(n_p - 1)n_g = 14$ | $\sigma_e^2 + 2\sigma_{sp:g}^2$ |
| HP | $(n_h - 1)(n_p - 1) = 1$ | $\sigma_e^2 + 16\sigma_{hp}^2$ |
| TP | $(n_t - 1)(n_p - 1) = 1$ | $\sigma_e^2 + 16\sigma_{tp}^2$ |
| GTP | $(n_g - 1)(n_t - 1)(n_p - 1) = 1$ | $\sigma_e^2 + 8\sigma_{gtp}^2$ |
| Error (b) | $(n_p - 1)(n_h - 1)(n_g n_s - 3) = 13$ | $\sigma_e^2$ |

$n_g = 2$  $n_s = 8$  $n_h = 2$  $n_t = 2$  $n_p = 2$  $N = n_g n_s n_h n_p = 64$

The subjects within the groups were considered to be a random effect and all others to be fixed effects. The completed analysis of variance is shown in Table 7.30.

It can be seen that, except for differences among subjects within groups (S:G), there were no significant differences in the whole-plot portion of the analysis. Pre- and postexercise treatments (P) were found to be highly significant in the split-plot portion of the analysis with the means for reaction time being 216 and 209 msec, respectively.

**TABLE 7.30** ANALYSIS OF VARIANCE OF CROSSOVER SPLIT-PLOT DESIGN

| Source | df | SS | MS | F |
|--------|-----|------|------|------|
| Whole plots | 31 | 18,124.86 | | |
| G | 1 | 1,550.39 | 1,550.39 | 1.49 |
| S:G | 14 | 14,555.22 | 1,039.55 | 8.28** |
| H | 1 | 159.39 | 159.39 | 1.27 |
| T | 1 | 192.52 | 192.52 | 1.53 |
| GT | 1 | 34.51 | 34.51 | 0.27 |
| Error (a) | 13 | 1,632.83 | 125.60 | |
| Split plots | 32 | 3,290.50 | | |
| P | 1 | 862.89 | 862.89 | 9.50** |
| GP | 1 | 165.77 | 165.77 | 1.83 |
| SP:G | 14 | 1,271.59 | 90.83 | 1.29 |
| HP | 1 | 0.77 | 0.77 | 0.01 |
| TP | 1 | 11.39 | 11.39 | 0.16 |
| GTP | 1 | 66.01 | 66.01 | 0.94 |
| Error (b) | 13 | 912.08 | 70.16 | |

It should be pointed out that a more appropriate analysis of the data given in Table 7.28 would be a covariance analysis where the reaction time taken prior to exercise would be used as the covariate. Analysis of covariance is the subject of Chapter 8.

## 7.8 CROSSOVER SPLIT-SPLIT-PLOT DESIGN

While the design discussed in this section is not one of the more common ones, it is used frequently enough to deserve attention. The example to be presented resulted from an experiment conducted by Douglas Lynch in the Department of Psychology at the University of Massachusetts. Two groups of subjects were used with 40 subjects in each group, thus making up the 80 columns of the crossover part of the design. Two different pairs or combinations of examinations were the treatment effects, each subject taking both pairs of examinations at different times, the time being the period or row effect. Each pair of examinations was separated as to contents, the contents being the split-plot effect. Finally, four questions were involved in the contents of each of the pairs of examinations, making up a split-split-plot effect. The variable analyzed was the proportion of correct answers. The proportions appeared to follow a normal distribution so no transformation of the data was deemed necessary.

Since there were 1280 observations involved in the study, the original observations are not included here. The mathematical model used was

$$Y_{ijk(t)lm} = \mu + g_i + s_{ij} + p_k + t_{(t)} + a_{ijk} + r_l + gr_{il} + sr_{ijl} + pr_{kl} + tr_{(t)l} + b_{ijkl}$$

$$+ q_m + gq_{im} + sq_{ijm} + pq_{km} + tq_{(t)m} + c_{ijkm} + rq_{lm} + grq_{ilm}$$

$$+ srq_{ijlm} + prq_{klm} + trq_{(t)lm} + e_{ijklm} \tag{7.86}$$

While this model looks rather formidable, a look at the sources of variation shown in Table 7.31 does help in following it. In the model, the $g_i$ represent group effects, the $s_{ij}$ subjects within groups, the $p_k$ period or time effects, the $t_{(t)}$ treatment effects, the $a_{ijk}$ the whole-plot errors or error $(a)$, the $r_l$ content effects, the $b_{ijkl}$ split-plot errors or error $(b)$, the $q_m$ question effects, the $c_{ijkm}$ error $(c)$, and the $e_{ijklm}$ error $(d)$.

Table 7.31 shows the formulas for calculating the degrees of freedom as well as the formulas for the sums of squares in symbolic notation. Calculations of the degrees of freedom in the split plots are merely the product of $(n_r - 1)$ with the degrees of freedom for all previous effects and the degrees of freedom in the split-split plots are the product of $(n_q - 1)$ with the degrees of freedom for all previous effects. Error $(a)$ degrees of freedom are the product of the number of columns less twice the number of rows minus one. Subtraction of 2 takes into account the degrees of freedom accounted for by the treatments.

The derivations of the formulas in symbolic notation for the sums of squares follow the same procedures as those for the degrees of freedom. However, when the formulas for the error terms are expanded, as they must be to calculate the sums of squares, they do become lengthy although they may be simplified. We shall look at each of the error terms separately.

**TABLE 7.31** CALCULATIONS FOR DEGREES OF FREEDOM AND SUMS OF SQUARES FOR CROSSOVER SPLIT-PLOT DESIGN

| Source | df | SS |
|---|---|---|
| Whole plots | $n_g n_s n_p - 1$ | $gsp - (1)$ |
| G | $n_g - 1$ | $g - (1)$ |
| S:G | $(n_s - 1)n_g$ | $gs - g$ |
| P | $n_p - 1$ | $p - (1)$ |
| T | $n_t - 1$ | $t - (1)$ |
| Error (a) | $(n_g n_s - 2)(n_p - 1)$ | $gsp - gs - p - t + (2)$ |
| Split plots | $(n_g n_s n_p)(n_r - 1)$ | $gspr - gsp$ |
| R | $n_r - 1$ | $r - (1)$ |
| GR | $(n_g - 1)(n_r - 1)$ | $gr - g - r + (1)$ |
| SR:G | $(n_s - 1)(n_r - 1)n_g$ | $sgr - gs - gr + g$ |
| PR | $(n_p - 1)(n_r - 1)$ | $pr - p - r + (1)$ |
| TR | $(n_t - 1)(n_r - 1)$ | $tr - t - r + (1)$ |
| Error (b) | $(n_g n_s - 2)(n_p - 1)(n_r - 1)$ | $(gsp - gs - p - t + 2)(r - 1)$ |
| Split-split plots | $(n_g n_s n_p n_r)(n_q - 1)$ | $gsprq - gspr$ |
| Q | $n_q - 1$ | $q - 1$ |
| GQ | $(n_g - 1)(n_q - 1)$ | $gq - g - q + (1)$ |
| SQ:G | $(n_s - 1)(n_q - 1)n_g$ | $gsq - sq - gq + g$ |
| PQ | $(n_p - 1)(n_q - 1)$ | $pq - p - q + (1)$ |
| TQ | $(n_t - 1)(n_q - 1)$ | $tq - t - q + (1)$ |
| Error (c) | $(n_g n_s - 2)(n_p - 1)(n_q - 1)$ | $(gsp - gs - p - t + 2)(q - 1)$ |
| RQ | $(n_r - 1)(n_q - 1)$ | $rq - r - q + (1)$ |
| GRQ | $(n_g - 1)(n_r - 1)(n_q - 1)$ | $grq - gr - gq - rq + g + r + q - (1)$ |
| SRQ:G | $(n_s - 1)(n_r - 1)(n_g - 1)n_g$ | $gsrq - grs - gsq - grq + gs + gr + gq - g$ |
| PRQ | $(n_p - 1)(n_r - 1)(n_q - 1)$ | $prq - pr - pq - rq + p + r - q - (1)$ |
| TRQ | $(n_t - 1)(n_r - 1)(n_q - 1)$ | $trq - tr - tq - rq + t + r + q - (1)$ |
| Error (d) | $(n_q n_s - 2)(n_p - 1)(n_r - 1)(n_q - 1)$ | $(gsp - gs - p - t + 2)(r - 1)(q - 1)$ |

Error ($a$) sum of squares is the difference between the whole-plot sum of squares and the sums of squares for all other whole-plot effects.

Error ($b$) sum of squares may be obtained as $gspr - gsr - pr - tr + 2r -$ error ($a$), or by the difference between the split-plot sum of squares and the sums of squares for all other split-plot effects. The split-plot sum of squares is the sum of squares among the $GSPR$ subclasses minus the $GSP$ subclass sum of squares.

Error ($c$) sum of squares may be obtained as $gspq - gsq - pq - tq + 2q -$ error ($a$), or a method similar to the one used for error ($b$). First, we can calculate the difference between the $GSPQ$ subclass sum of squares and the $GSP$ subclass sum of squares. Second, we can subtract the $SS_Q$, $SS_{GQ}$, $SS_{SP:Q}$, $SS_{PQ}$, and $S_{TQ}$ from this difference yielding error ($c$).

Error ($d$) sum of squares may be obtained as $gsprq - gsrq - prq - rtq + 2rq -$ error ($c$) $\times$ error ($b$) $\times$ error ($a$) or merely by the difference between the split-split-plot sum of squares and the sums of squares for all other effects in the split-split plots.

The completed analysis of variance is shown in Table 7.32, where it can be seen that there was a significant difference between the contents of the examinations ($R$). However, the highly significant difference among the questions ($Q$) appears to be of much greater importance—at least as far as accounting for variation in the study. In addition, many interactions involving the questions are significant or highly significant

**TABLE 7.32**   ANALYSIS OF VARIANCE AND EXPECTATIONS OF MEAN
SQUARES FOR CROSSOVER SPLIT-SPLIT-PLOT DESIGN

| Source | df | MS | F | E(MS) |
|--------|-----|--------|--------|-------|
| Whole plots | 159 | | | |
| $G$ | 1 | 0.30505 | 1.94 | $\sigma_e^2 + 2\sigma_c^2 + 4\sigma_b^2 + 8\sigma_a^2 + 16\sigma_{s:g}^2 + 640\sigma_g^2$ |
| $S{:}G$ | 78 | 0.15711 | 1.92 | $\sigma_e^2 + 2\sigma_c^2 + 4\sigma_b^2 + 8\sigma_a^2 + 16\sigma_{s:g}^2$ |
| $P$ | 1 | 0.00957 | 0.12 | $\sigma_e^2 + 2\sigma_c^2 + 4\sigma_b^2 + 8\sigma_a^2 + 640\sigma_p^2$ |
| $T$ | 1 | 0.09453 | 1.15 | $\sigma_e^2 + 2\sigma_c^2 + 4\sigma_b^2 + 8\sigma_a^2 + 640\sigma_t^2$ |
| Error ($a$) | 78 | 0.08201 | | $\sigma_e^2 + 2\sigma_c^2 + 4\sigma_b^2 + 8\sigma_a^2$ |
| Split plots | 160 | | | |
| $R$ | 1 | 0.26450 | 4.22* | $\sigma_e^2 + 4\sigma_b^2 + 8\sigma_{sr:g}^2 + 640\sigma_r^2$ |
| $GR$ | 1 | 0.00226 | 0.04 | $\sigma_e^2 + 4\sigma_b^2 + 8\sigma_{sr:g}^2 + 320\sigma_{gr}^2$ |
| $SR{:}G$ | 78 | 0.06275 | 1.18 | $\sigma_e^2 + 4\sigma_b^2 + 8\sigma_{sr:g}^2$ |
| $PR$ | 1 | 0.20910 | 3.93 | $\sigma_e^2 + 4\sigma_b^2 + 320\sigma_{pr}^2$ |
| $TR$ | 1 | 0.10549 | 1.98 | $\sigma_e^2 + 4\sigma_b^2 + 320\sigma_{tr}^2$ |
| Error ($b$) | 78 | 0.05324 | | $\sigma_e^2 + 4\sigma_b^2$ |
| Split-split plots | 960 | | | |
| $Q$ | 3 | 4.44272 | 57.13** | $\sigma_e^2 + 2\sigma_c^2 + 4\sigma_{sq:g}^2 + 320\sigma_q^2$ |
| $GQ$ | 3 | 0.55146 | 7.09** | $\sigma_e^2 + 2\sigma_c^2 + 4\sigma_{sq:g}^2 + 160\sigma_{gq}^2$ |
| $SQ{:}G$ | 234 | 0.07776 | 1.30 | $\sigma_e^2 + 2\sigma_c^2 + 4\sigma_{sq:g}^2$ |
| $PQ$ | 3 | 0.26046 | 4.36 | $\sigma_e^2 + 2\sigma_c^2 + 160\sigma_{pq}^2$ |
| $TQ$ | 3 | 0.29332 | 4.91** | $\sigma_e^2 + 2\sigma_c^2 + 160\sigma_{tq}^2$ |
| Error ($c$) | 234 | 0.05974 | | $\sigma_e^2 + 2\sigma_c^2$ |
| $RQ$ | 3 | 0.05607 | 0.97 | $\sigma_e^2 + 2\sigma_{srq:g}^2 + 160\sigma_{rq}^2$ |
| $GRQ$ | 3 | 0.17203 | 2.96* | $\sigma_e^2 + 2\sigma_{srq:g}^2 + 80\sigma_{grq}^2$ |
| $SRQ{:}G$ | 234 | 0.05803 | 0.99 | $\sigma_e^2 + 2\sigma_{srq:g}^2$ |
| $PRQ$ | 3 | 0.06820 | 1.16 | $\sigma_e^2 + 80\sigma_{prq}^2$ |
| $TRQ$ | 3 | 0.19810 | 3.37* | $\sigma_e^2 + 80\sigma_{trq}^2$ |
| Error ($d$) | 234 | 0.05885 | | $\sigma_e^2$ |

$$\Sigma_i\, \Sigma_j\, \Sigma_k\, \Sigma_l\, \Sigma_m\, Y_{ijklm}^2 = 827.99120 \qquad Y_{\ldots\ldots} = 961.10$$

so inspection of various two- and three-way tables involving the interacting effects
would yield information as to the nature of the interactions.

    While we have specifically noted four error terms, inspection of the expectations
of mean squares shown in Table 7.32 shows that there are several other denominators
involved in the various $F$ tests which serves to emphasize the vital importance of the
expectations of mean squares in analysis of variance.

## 7.9 ANALYSIS OF DATA FROM SPLIT-PLOT DESIGNS: DISPROPORTIONATE SUBCLASS FREQUENCIES

    Loss of data for one reason or another results in data with disproportionate subclass
frequencies. This often happens even in a carefully planned experiment and disproportionate frequencies will frequently occur in survey data. Therefore, when the un-

derlying model involves both fixed and random sets of effects, as in a split-plot design, the analysis should be completed using mixed-model methods such as those described for the two-way classification in Sections 3.21.1 and 3.21.2.

The model for the split-plot design can be written in general as follows:

$$Y_{ij} = \mu + a_i + e_i + b_j + e_j \tag{7.87}$$

where the $a_i$ represent all whole-plot effects, $e_i$ is an error deviation for the whole-plot effects, the $b_j$ represent all subplot effects including interactions of whole-plot effects with subplot effects, and $e_j$ is an error deviation for the subplot effects. The ANOVA desired is presented in Table 7.33 and with unequal subclass frequencies $k_3 \neq k_2$. Therefore, only approximate $F$ tests can be made for whole-plot effects.

The appropriate sums of squares for the ANOVA may be obtained by completing two separate least-squares analyses and then combining the results as follows:

1. Model: $Y_{ij} = \mu + a_i + b_j + e_{ij}$.
   (a) Set up the reduced set of equations after imposing the summation restrictions for $\mu$, the $a_i$, and the $b_j$.
   (b) This analysis will give the appropriate sums of squares for all whole-plot sets of effects (the $a_i$).
   (c) The remainder sum of squares will be

$$E_1 = \sum_i \sum_j Y_{ij}^2 - R(\mu, a, b)$$

2. Model: $Y_{ij} = \mu + a_i + e_i + b_j + e_j$.
   (a) Absorb equations for the $\mu + a_i + e_i$ subclass effects into the equations for the $b_j$ effects.
   (b) The appropriate sums of squares for all sets of subplot effects (the $b_j$) will be obtained from this analysis.
   (c) The remainder sum of squares will be

$$E_2 = \sum_i \sum_j Y_{ij}^2 - R(\mu, a, b, e_i)$$

   which is the appropriate sum of squares for the error term for the subplot sets of effects.

The sum of squares for the whole-plot error is now computed from

$$E_1 - E_2 = R(\mu, a, b, e_i) - R(\mu, a, b) \tag{7.88}$$

**TABLE 7.33**  ANALYSIS OF VARIANCE UNDER A SPLIT-PLOT MODEL

| Source | SS | $E(MS)$ |
|---|---|---|
| Whole-plot effects | $R(\mu, a, b) - R(\mu, b)$ | $\sigma_{e_j}^2 + k_3\sigma_{e_i}^2 + k_4\sigma_a^2$ |
| Whole-plot error | $R(\mu, a, b, e_i) - R(\mu, a, b)$ | $\sigma_{e_j}^2 + k_2\sigma_{e_i}^2$ |
| Subplot effects | $R(\mu, a, b, e_i) - R(\mu, a, e_i)$ | $\sigma_{e_j}^2 + k_1\sigma_b^2$ |
| Subplot error | $Y'Y - R(\mu, a, b, e_i)$ | $\sigma_{e_j}^2$ |

## 7.9.1 Completely Randomized Split-Plot Design

For the first illustration of how an analysis such as this is carried out, we shall use the data from the experiment conducted by Pedro Barbosa, T. Michael Peters, and N. C. Greenough of the University of Massachusetts (Table 7.3). However, to illustrate the analysis with disproportionate frequencies, we shall omit the observations presented in Table 7.34. The mathematical model for the analysis of these data is the same as that used for the balanced design analysis [Equation (7.2)].

The model for the first analysis is

$$Y_{ijk} = \mu + d_i + s_k + ds_{ik} + e_{ijk} \tag{7.89}$$

The reader will note that this is the model for the two-way classification with interaction, where both the $d_i$ and $s_k$ sets of effects are fixed. Therefore, the procedures previously described in Section 3.20.2 are followed to complete the least-squares analysis under this model. The ANOVA is presented in Table 7.35. As indicated in the table, the sum of squares obtained from this analysis for the whole-plot effects (densities, in this example) is the sum of squares desired for these effects; that is,

$$7009.86 = R(\mu, d, s, ds) - R(\mu, s, ds)$$
$$= \hat{\mathbf{B}}_d' \mathbf{Z}_d^{-1} \hat{\mathbf{B}}_d$$

Also, the sum of squares obtained for the orthogonal polynomials, which have been computed by weighted least-squares procedures as described in Section 5.17, are to be used in the final ANOVA. Likewise, the estimates obtained for the density effects from this analysis have been properly adjusted for unequal frequencies with respect to the sex and the density × sex interaction effects. These estimates were found to be as follows from the solution of the least-squares equations when the summation restrictions were imposed:

$$\hat{d}_1 = 12.7445 \qquad \hat{d}_5 = -1.7803$$

$$\hat{d}_2 = 10.0174 \qquad \hat{d}_6 = -15.4077$$

$$\hat{d}_3 = 8.7392 \qquad \hat{d}_7 = -\sum_{i=1}^{6} \hat{d}_i = -19.9443$$

$$\hat{d}_4 = 5.6312$$

To complete the analysis, the data are analyzed under the full model,

$$Y_{ijk} = \mu + d_i + r_{ij} + s_k + ds_{ik} + e_{ijk} \tag{7.90}$$

**TABLE 7.34**   OBSERVATIONS OMITTED

| Density | Replicate | Percent |
|---------|-----------|---------|
| 160 | 1 | 51.250 males |
| 280 | 2 | 39.286 females |
| 520 | 3 | 41.538 males |
| 960 | 2 | 22.234 females |
| 1280 | 2 | 15.078 males |

**TABLE 7.35**  ANALYSIS OF VARIANCE FROM
FIXED MODEL ANALYSIS

| Source | df | SS | MS |
|---|---|---|---|
| Densities (D) | 6 | 7009.8619 | 1168.3103 |
| Linear | 1 | 6763.7017 | 6763.7017 |
| Quadratic | 1 | 0.2238 | 0.2238 |
| Cubic | 1 | 211.6452 | 211.6452 |
| Residual | 3 | 34.2912 | 11.4304 |
| Sexes (S) | 1 | 411.4453 | 411.4453 |
| DS | 6 | 269.3060 | 44.8843 |
| Remainder | 37 | 715.7211 | 19.3438 |

and in this analysis the equations for $\mu + d_i + r_{ij}$ are absorbed into the equations for the $s_k$ and the $ds_{ik}$ effects. The computational procedures involved in the absorption of a set of equations was described in Section 3.21 under "Indirect Analysis." From this analysis the ANOVA presented in Table 7.36 is obtained. At this point it should be noted that the sums of squares for sexes and the density × sex interaction are quite different in the two analyses. These sums of squares would be identical in the two analyses if orthogonality existed. Also, with orthogonality the SS for $S$, $DS$, and Remainder would sum to the within $DR$ subclasses SS. The sum of squares for replicates within densities is computed as follows [Equation (7.88)]:

$$SS\ R{:}D = E_1 - E_2$$

$$= 715.7211 - 354.5825$$

$$= 361.1386$$

with $37 - 16 = 21$ degrees of freedom. The combined ANOVA from the two least-squares analyses is given in Table 7.37.

Several points should be noted concerning the results given in Table 7.37. Note that an $F$ value is given for $R{:}D$. This is a valid test for replicate within density effects only if one can assume that the $RS{:}D$ interaction is negligible (see Table 7.4). In any case, the mean squares for $R{:}D$ and $RS{:}D$ do not differ significantly from one another. Therefore, the analysis completed under the reduced model where replicate within density effects were omitted would be a satisfactory analysis for this particular set of data. However, since in most analyses of data of this type this may not be true, we shall proceed as though $R{:}D$ effects do exist and hence must be considered.

It should also be noted in Table 7.37 that the interaction of density with sex $(DS)$ is not significant at the 0.05 level of probability $(P = 0.08)$. However, to determine

**TABLE 7.36**  ANALYSIS OF VARIANCE UNDER FULL MODEL
AFTER ABSORPTION

| Source | df | SS | MS | F |
|---|---|---|---|---|
| Within DR subclasses | 23 | 1025.0361 | | |
| Sexes (S) | 1 | 343.4362 | 343.4362 | 15.50** |
| DS | 6 | 313.6395 | 52.2732 | 2.36 |
| Remainder | 16 | 354.5825 | 22.1614 | |

**TABLE 7.37**  ANALYSIS OF VARIANCE IN PERCENT LARVAL SURVIVAL
WITH DISPROPORTIONATE SUBCLASS FREQUENCIES

| Source | df | SS | MS | F |
|---|---|---|---|---|
| Whole plots | 27 | | | |
| D | 6 | 7009.8619 | 1168.3103 | 67.94** |
|    Linear | 1 | 6763.7017 | 6763.7017 | 393.31** |
|    Quadratic | 1 | 0.2238 | 0.2238 | 0.01 |
|    Cubic | 1 | 211.6452 | 211.6452 | 12.31** |
|    Residual | 3 | 34.2912 | 11.4304 | 0.66 |
| R:D | 21 | 361.1386 | 17.1971 | 0.78 |
| Split plots | 23 | | | |
| S | 1 | 343.4362 | 343.4362 | 15.50** |
| DS | 6 | 313.6395 | 52.2732 | 2.36 |
| RS:D | 16 | 354.5825 | 22.1614 | |

if the shape of the curve for densities is different for the two sexes, one must partition the sum of squares for the $DS$ interaction into linear $\times$ linear, quadratic $\times$ linear, and so on. To do this when unequal subclass frequencies exist, we partition the $DS$ subclass sum of squares as described in Section 5.18.

Before completing the surface fitting computational procedures, let us first examine the least-squares means obtained from the two analyses described above. These are given in Table 7.38 along with standard errors. The least-squares means for density classes were computed by using the estimate of $\mu$ when the overall mean is adjusted for density effects as estimated in the first analysis and for sex and the density $\times$ sex interaction effects as estimated from the second analysis where the $\mu + d_i + r_{ij}$ equations were absorbed. This estimate of $\mu$ (say $\tilde{\mu}$), which is given in Table 7.38, was also used to compute the least-squares means for sex and the density $\times$ sex subclasses. The approximate variances of the least-squares means were computed as follows:

**TABLE 7.38**  LEAST-SQUARES MEANS AND STANDARD ERRORS
FOR PERCENT LARVAL SURVIVAL

| Classification | Number of observations | Mean | SE | Classification | Number of observations | Mean | SE |
|---|---|---|---|---|---|---|---|
| Overall ($\tilde{\mu}$) | 51 | 36.10 | 0.67 | Density $\times$ sex | | | |
| Densities | | | | 160—M | 3 | 51.01 | 2.80 |
| 160 | 7 | 46.84 | 1.80 | 160—F | 4 | 42.67 | 2.45 |
| 280 | 7 | 44.11 | 1.80 | 280—M | 4 | 51.08 | 2.45 |
| 400 | 8 | 42.83 | 1.66 | 280—F | 3 | 37.15 | 2.80 |
| 520 | 7 | 39.73 | 1.80 | 400—M | 4 | 47.99 | 2.35 |
| 640 | 8 | 32.32 | 1.66 | 400—F | 4 | 37.68 | 2.35 |
| 960 | 7 | 18.69 | 1.80 | 520—M | 3 | 42.55 | 2.80 |
| 1280 | 7 | 14.15 | 1.80 | 520—F | 4 | 36.91 | 2.45 |
| Sexes | | | | 640—M | 4 | 33.35 | 2.35 |
| Males | 25 | 36.85 | 0.98 | 640—F | 4 | 31.28 | 2.35 |
| Females | 26 | 31.34 | 0.96 | 960—M | 4 | 17.68 | 2.45 |
| | | | | 960—F | 3 | 19.70 | 2.80 |
| | | | | 1280—M | 3 | 14.31 | 2.80 |
| | | | | 1280—F | 4 | 13.99 | 2.45 |

$$V(\tilde{\mu}) \simeq R^{\mu\mu}(\hat{\sigma}_e^2 + k\hat{\sigma}_{r:d}^2) \tag{7.91}$$

$$V(\tilde{\mu} + \hat{d}_i) \simeq (R^{\mu\mu} + R^{d_id_i} + 2R^{\mu d_i})(\hat{\sigma}_e^2 + k_i\hat{\sigma}_{r:d}^2) \tag{7.92}$$

$$V(\tilde{\mu} + \hat{s}_k) \simeq (R^{\mu\mu} + C_{s_ks_k} + 2R^{\mu s_k})\hat{\sigma}_e^2 + R^{\mu\mu}k\hat{\sigma}_{r:d}^2 \tag{7.93}$$

$$V(\tilde{\mu} + \hat{d}_i + \hat{s}_k + \hat{d}s_{ik}) \simeq (R^{\mu\mu} + R^{d_id_i} + C_{s_ks_k} + C_{ds_{ik}ds_{ik}} + 2R^{\mu d_i} + 2R^{\mu s_k}$$
$$+ 2R^{\mu ds_{ik}} + 2R^{d_is_k} + 2R^{d_ids_{ik}} + 2C_{s_kds_{ik}})(\hat{\sigma}_e^2 + k_{ik}\hat{\sigma}_{r:d}^2) \tag{7.94}$$

where $\mathbf{R}^{-1}$ is the inverse of the coefficient matrix from the first analysis, $\mathbf{C} = (\mathbf{S} - \mathbf{N'D^{-1}N})^{-1}$ as obtained after absorption of the $\mu + d_i + r_{ij}$ equations in the second analysis, and

$$k = \frac{\Sigma_i \Sigma_j n_{ij}^2 \cdot}{n \ldots} \tag{7.95}$$

$$k_i = \frac{\Sigma_j n_{ij}^2 \cdot}{n_i \cdot \cdot} \tag{7.96}$$

$$k_{ik} = \frac{\Sigma_j n_{ijk}^2}{n_{i \cdot k}} \tag{7.97}$$

The coefficient for $\sigma_{r:d}^2$ in the expected mean square for $R:D$ was computed as follows:

$$k_1 = \tfrac{1}{21}(51 - \operatorname{tr} \mathbf{R}^{-1}\mathbf{N}_s\mathbf{N}_s')$$
$$= \tfrac{1}{21}(51 - 14)$$
$$= 1.762$$

where $\mathbf{R}^{-1}$ is as defined above and $\mathbf{N}_s$ is the matrix that associates the $r_{ij}$ effects with $\mu$, $d_i$, $s_k$, and the $ds_{ik}$ sets of effects.

In our example problem, the estimate of $\sigma_{r:d}^2$ is negative ($-2.8176$). This variance component was assumed to be zero in the calculation of the standard errors of the least-squares means.

**Surface Fitting Procedures** The computer program used to partition the degrees of freedom among density $\times$ sex subclasses into single degree polynomials (LSMLMW— a modification of LSML76, Harvey, 1977b) first obtained the least-squares sum of squares for subclasses from $\hat{\mathbf{B}}_{DS}'\mathbf{Z}_{DS}^{-1}\hat{\mathbf{B}}_{DS}$. $\mathbf{Z}_{DS}$ was obtained by taking appropriate linear functions of elements in $\mathbf{R}^{-1}$ and $\mathbf{C}$ and elements in $\hat{\mathbf{B}}_{DS}$ were computed from

$$\hat{d}_i + \hat{s}_k + \hat{d}s_{ik} \tag{7.98}$$

A step-down weighted least-squares procedure, as described in Section 5.18, was then followed to obtain the results given in Table 7.39.

First, the reader should note that the sums of squares given for linear, quadratic, and cubic for densities are different in Tables 7.37 and 7.39. This will occur when unequal subclass frequencies exist if the sums of squares, as computed in Table 7.39, are not adjusted for the high-order polynomials in densities and in the density $\times$ sex interaction. Second, it should be noted that the cubic for density (a third degree polynomial) is highly significant. Hence, the best prediction equation for

**TABLE 7.39**  PARTITIONING OF THE DENSITY × SEX SUBCLASS
SUM OF SQUARES

| Source | Order eliminated | df | SS | MS | F |
|---|---|---|---|---|---|
| *DS* Subclasses | | 13 | 7838.9042 | | |
| Densities | | | | | |
|   Linear | 7 | 1 | 6912.0849 | 6912.0849 | 401.93** |
|   Quadratic | 4 | 1 | 0.3283 | 0.3283 | 0.02 |
|   Cubic | 3 | 1 | 217.9470 | 217.9470 | 12.67** |
| Sex | | | | | |
|   Linear | 6 | 1 | 325.0565 | 325.0565 | 14.67** |
| Density × sex | | | | | |
|   Linear × linear | 5 | 1 | 252.4551 | 252.4551 | 11.39** |
|   Quadratic × linear | 2 | 1 | 16.4718 | 16.4718 | 0.74 |
|   Cubic × linear | 1 | 1 | 48.7810 | 48.7810 | 2.20 |
| Residual | | 6 | 65.7796 | 10.9633 | 0.49 |
| *R:D* | | 21 | 361.1387 | 17.1971 | |
| *RS:D* | | 16 | 354.5825 | 22.1614 | |

plotting the surface should include linear, quadratic, and cubic for density, linear for sex, and the linear × linear interaction term. This equation, which was found to account for 98.3 percent of the variation among *DS* subclasses, is as follows:

$$\hat{Y} = 34.53 - 0.046426(D - \bar{D}) - 0.000017427(D - \bar{D})^2$$
$$+ 0.0000000607031(D - \bar{D})^3 - 5.34837(S - \bar{S})$$
$$+ 0.0120495(D - \bar{D})(S - \bar{S}) \tag{7.99}$$

where $\bar{D} = 605.7143$ and $\bar{S} = 1.5$, since sexes were coded 1 and 2 for males and females, respectively. Even though the variation among subclasses accounted for by these polynomials is large, the accuracy of prediction for densities that deviate widely from the average density will not be estimated accurately. For example, the estimate of percent larval survival for males in an initial density of 160 is 62.5 percent using this prediction equation which is 11.5 percent higher than that given by the least-squares mean (Table 7.38). The estimates of percent larval survival in a density of 1280 is negative for both sexes, which of course, is absurd. Hence, one must be cautious in attempting to predict with polynomial equations values for subclasses which lie outside the range included in the data set.

## 7.9.2 Completely Randomized Split-Plot Design: Multiple Observations

The data used in Section 3.21.1 (Table 3.62) that were used to illustrate least-squares computational procedures for a two-way classification mixed model will also serve as an example for this design. The nine sires in that data set were actually from three different lines of cattle with the first three sires being in line 1, the next two sires in line 2, and the last four sires in line 3. Therefore, the model to be considered here for the analysis of these data is as follows:

**TABLE 7.40**  ANALYSIS OF VARIANCE OF AVERAGE
DAILY GAIN OF 65 STEERS

| Source | df | SS |
|--------|-----|-----|
| Lines ($L$) | 2 | $R(\mu, l, a, la) - R(\mu, a, la)$ |
| Sires:lines ($S$:$L$) | 6 | $R(\mu, l, s, a, la) - R(\mu, l, a, la)$ |
| Age of dam ($A$) | 2 | $R(\mu, l, s, a, la) - R(\mu, l, s, la)$ |
| $LA$ | 4 | $R(\mu, l, s, a, la) - R(\mu, l, s, a)$ |
| $SA$:$L$ | 10 | $R(\mu, l, s, a, la, sa) - R(\mu, l, s, a, la)$ |
| Within | 40 | $\mathbf{Y'Y} - R(\mu, l, s, a, la, sa)$ |

$$Y_{ijkl} = \mu + l_i + s_{ij} + a_k + la_{ik} + sa_{ijk} + e_{ijkl}$$

$$i = 1, 2, 3$$

$$j = 1, 2, 3, 4$$

$$k = 1, 2, 3$$

$$l = 1, 2, \ldots, n_{ijk} \tag{7.100}$$

where the $l_i$ are the line effects, the $s_{ij}$ are the random sire effects within line, and the $a_k$ are the age-of-dam effects. With data such as these where there are multiple observations in the smallest subclasses, we are able to separate the "true" error ($\sigma_w^2$) from the highest-order interaction ($\sigma_{sa:l}^2$, in this example). The ANOVA desired for these data is presented in Table 7.40. When the sums of squares are computed as shown here, the expectation of the mean squares will be as presented in Table 7.41.

$$\sigma_{s':l}^2 = \frac{k_6}{k_7} \sigma_{sa:l}^2 + \sigma_{s:l}^2$$

is the variance that is tested for significance with the MS($S$:$L$)/MS(within). An approximate $F$ test for line effects is MS($L$)/MS($S$:$L$).

To complete this analysis by the indirect procedures described in Section 3.21.2 so that estimates of the variance components will be obtained and also least-squares means for the fixed effects, analyses must be completed under three models. The first analysis is completed under the reduced model

$$Y_{ikl} = \mu + l_i + a_k + la_{ik} + e_{ikl} \tag{7.101}$$

and constants are fitted only for the fixed effects. From this analysis, we obtain the $\mu$ equation which is used later to estimate the overall mean, the sum of squares for lines,

**TABLE 7.41**  EXPECTATIONS OF
MEAN SQUARES

| Source | $E$(MS) |
|--------|---------|
| Lines ($L$) | $\sigma_w^2 + k_8\sigma_{s':l}^2 + k_9\sigma_l^2$ |
| Sires:lines ($S$:$L$) | $\sigma_w^2 + k_6\sigma_{sa:l}^2 + k_7\sigma_{s:l}^2$ |
| Age of dam ($A$) | $\sigma_w^2 + k_4\sigma_{sa:l}^2 + k_5\sigma_a^2$ |
| $LA$ | $\sigma_w^2 + k_2\sigma_{sa:l}^2 + k_3\sigma_{la}^2$ |
| $SA$:$L$ | $\sigma_w^2 + k_1\sigma_{sa:l}^2$ |
| Within | $\sigma_w^2$ |

**TABLE 7.42**  LEAST-SQUARES REDUCED COEFFICIENT MATRIX UNDER THE FIXED MODEL AND THE $N_s$ MATRIX FOR SIRE WITHIN LINE EFFECTS

| Equation | $\hat{\mu}$ | $\hat{l}_1$ | $\hat{l}_2$ | $\hat{a}_1$ | $\hat{a}_2$ | $\hat{la}_{11}$ | $\hat{la}_{12}$ | $\hat{la}_{21}$ | $\hat{la}_{22}$ | $\hat{s}_{11}$ | $\hat{s}_{12}$ | $\hat{s}_{13}$ | $\hat{s}_{21}$ | $\hat{s}_{22}$ | $\hat{s}_{31}$ | $\hat{s}_{32}$ | $\hat{s}_{33}$ | $\hat{s}_{34}$ |
|---|---|---|---|---|---|---|---|---|---|---|---|---|---|---|---|---|---|---|
| $\mu$: | 65 | −8 | −14 | −25 | −21 | 0 | 1 | 5 | 5 | 8 | 8 | 5 | 8 | 7 | 6 | 8 | 7 | 8 |
| $l_1$: | | 50 | 29 | 0 | 1 | −20 | −17 | −10 | −9 | 8 | 8 | 5 | 0 | 0 | −6 | −8 | −7 | −8 |
| $l_2$: | | | 44 | 5 | 5 | −10 | −9 | −15 | −13 | 0 | 0 | 0 | 8 | 7 | −6 | −8 | −7 | −8 |
| $a_1$: | | | | 49 | 37 | −6 | −3 | −11 | −8 | −2 | −6 | −2 | −2 | −3 | −4 | −3 | 0 | −3 |
| $a_2$: | | | | | 53 | −3 | −5 | −8 | −11 | −2 | −4 | −2 | −2 | −2 | −2 | −4 | −2 | −1 |
| $la_{11}$: | Symmetric | | | | | 38 | 29 | 22 | 16 | −2 | −6 | −2 | 0 | 0 | 4 | 3 | 0 | 3 |
| $la_{12}$: | | | | | | | 41 | 16 | 23 | −2 | −4 | −2 | 0 | 0 | 2 | 4 | 2 | 1 |
| $la_{21}$: | | | | | | | | 33 | 24 | 0 | 0 | 0 | −2 | −3 | 4 | 3 | 0 | 3 |
| $la_{22}$: | | | | | | | | | 35 | 0 | 0 | 0 | −2 | −2 | 2 | 4 | 2 | 1 |

the remainder sum of squares, $\mathbf{Y'Y} - R(\mu, l, a, la)$, the number of $\sigma^2_{s:l}$ and $\sigma^2_{sa:l}$ components in $R(\mu, l, a, la)$, and the inverse elements involving $\mu$ and $l_i$ effects. The least-squares coefficient matrix for $\mu$, $l_i$, $a_k$, and $la_{ik}$ ($\mathbf{R}$) is given in Table 7.42 along with the matrix ($\mathbf{N}_s$) that associates the sire within line effects with the fixed sets of effects. The inverse of the coefficient matrix ($\mathbf{R}^{-1}$) is given in Table 7.43.

For the general case, as was pointed out in Section 3.21, the coefficient for a given variance component in the expectation of any reduction in sum of squares is

$$\text{tr } \mathbf{R}^{-1}\mathbf{N}_s\mathbf{N}'_s$$

Therefore, for the present example, to obtain the number of $\sigma^2_{s:l}$ and $\sigma^2_{sa:l}$ components in $R(\mu, l, a, la)$, we need $\mathbf{N}_s\mathbf{N}'_s$. This is found as follows:

$$\mathbf{N}_s\mathbf{N}'_s = \begin{bmatrix} 479 & -60 & -100 & -183 & -146 & -2 & 8 & 35 & 36 \\ & 366 & 213 & -2 & 8 & -146 & -124 & -72 & -66 \\ & & 326 & 35 & 36 & -72 & -66 & -109 & -96 \\ & & & 91 & 65 & 10 & 9 & -21 & -13 \\ \text{Symmetric} & & & & 57 & 9 & -1 & -13 & -17 \\ & & & & & 78 & 55 & 34 & 23 \\ & & & & & & 49 & 23 & 25 \\ & & & & & & & 47 & 33 \\ & & & & & & & & 33 \end{bmatrix} \quad (7.102)$$

**TABLE 7.43**  INVERSE OF THE REDUCED COEFFICIENT MATRIX FOR THE FIXED EFFECTS[a]

| Equation | $\hat{\mu}$ | $\hat{l}_1$ | $\hat{l}_2$ | $\hat{a}_1$ | $\hat{a}_2$ | $\hat{la}_{11}$ | $\hat{la}_{12}$ | $\hat{la}_{21}$ | $\hat{la}_{22}$ |
|---|---|---|---|---|---|---|---|---|---|
| $\mu$: | 20,872 | 1,730 | 5,363 | 9,992 | 1,086 | 4,443 | −1,466 | 810 | 457 |
| $l_1$: | | 43,474 | −27,965 | 4,443 | −1,466 | 24,427 | 706 | −15,245 | −77 |
| $l_2$: | | | 47,106 | 810 | 457 | −15,245 | −77 | 20,795 | 2,629 |
| $a_1$: | | | | 51,736 | −31,950 | 7,903 | −4,707 | 11,536 | −6,630 |
| $a_2$: | | | | | 42,829 | −4,707 | 1,995 | −6,630 | 11,183 |
| $la_{11}$: | Symmetric | | | | | 111,375 | −68,607 | −71,175 | 43,287 |
| $la_{12}$: | | | | | | | 87,654 | 43,287 | −56,007 |
| $la_{21}$: | | | | | | | | 115,008 | −70,530 |
| $la_{22}$: | | | | | | | | | 96,842 |

[a] Each element has been multiplied by $10^6$.

The reader can verify that tr $\mathbf{R}^{-1}\mathbf{N}_s\mathbf{N}'_s$, which equals $\Sigma_i \, \Sigma_j \, R^{ij}N_{ij}$ with $N_{ij}$ referring to the individual elements in the $\mathbf{N}_s\mathbf{N}'_s$ matrix, is 24.42683. This is the number of $\sigma^2_{s:l}$ components expected in $R(\mu, l, a, la)$. In most practical problems the number of random classes will be large and it will be essentially impossible to do these calculations with a hand calculator. The computer program of Harvey (LSML76) computes these numbers efficiently even though there are hundreds of random classes.

The right-hand members for the reduced set of equations under model (7.101) are as follows:

$$\begin{bmatrix} \mu: 156.74 & la_{11}: -3.17 \\ l_1: -17.69 & la_{12}: -1.95 \\ l_2: -29.96 & la_{21}: 7.85 \\ a_1: -58.50 & la_{22}: 6.86 \\ a_2: -48.73 & \end{bmatrix}$$

The solution of the least-squares equations for the first analysis is

$$\begin{bmatrix} \hat{\mu} \\ \hat{l}_1 \\ \hat{l}_2 \\ \hat{a}_1 \\ \hat{a}_2 \\ \hat{la}_{11} \\ \hat{la}_{12} \\ \hat{la}_{21} \\ \hat{la}_{22} \end{bmatrix} = \begin{bmatrix} 2.440962 \\ -0.047492 \\ 0.084038 \\ 0.022927 \\ 0.000134 \\ 0.006936 \\ -0.059604 \\ -0.017927 \\ -0.080134 \end{bmatrix} \qquad (7.103)$$

The remainder sum of squares from this analysis is

$$E_1 = \mathbf{Y}'\mathbf{Y} - R(\mu, l, a, la)$$

$$= 382.89980 - 378.97469$$

$$= 3.92511$$

The sum of squares obtained for lines in this analysis, as computed from $\hat{\mathbf{B}}'_L\mathbf{Z}_L^{-1}\hat{\mathbf{B}}_L$, equals

$$R(\mu, l, a, la) - R(\mu, a, la)$$

and this is the appropriate sum of squares for the final ANOVA.

$$\hat{\mathbf{B}}'_L\mathbf{Z}_L^{-1}\hat{\mathbf{B}}_L = [-0.047492 \quad 0.084038] \begin{bmatrix} 0.043474 & -0.027965 \\ -0.027965 & 0.047106 \end{bmatrix}^{-1} \begin{bmatrix} -0.047492 \\ 0.084038 \end{bmatrix}$$

$$= 0.150140$$

The second analysis is completed under the model

$$Y_{ijkl} = \mu + l_i + s_{ij} + a_k + la_{ik} + e_{ijkl} \qquad (7.104)$$

and the equations for $\mu + l_i + s_{ij}$ are absorbed into the equations for $a_k$ and $la_{ik}$. From this analysis, we shall obtain the appropriate sums of squares, inverse elements, and estimates for the $a_k$ and the $la_{ik}$ and the number of $\sigma^2_{sa:l}$ components in $R(\mu, l, s, a, la)$. Details of this analysis will not be given since they are essentially the same as described for the two-way classification mixed model with interaction (see Section 3.21.2). Estimates of the $a_k$ and $la_{ik}$ obtained from this analysis are as follows:

$$
\begin{bmatrix} \hat{a}_1 \\ \hat{a}_2 \\ \hat{la}_{11} \\ \hat{la}_{12} \\ \hat{la}_{21} \\ \hat{la}_{22} \end{bmatrix} = \begin{bmatrix} 0.036674 \\ -0.016041 \\ -0.080462 \\ -0.015322 \\ -0.016522 \\ -0.071534 \end{bmatrix} \tag{7.105}
$$

The reader will note that these differ somewhat from the estimates obtained for these effects in the first analysis since these are computed within sires and those from the first analysis were computed ignoring sire effects. The remainder sum of squares is

$$
\begin{aligned}
E_2 &= \mathbf{Y'Y} - R(\mu, l, s, a, la) \\
&= \mathbf{Y'Y} - R(\mu, l, s) - R(a, la \mid \mu, l, s) \\
&= 382.89980 - 379.26686 - 0.76783 \\
&= 3.63294 - 0.76783 \\
&= 2.86511
\end{aligned}
$$

As shown in Section 3.21.2, the number of $\sigma^2_{sa:l}$ components in $R(\mu, l, s, a, la)$ is computed from

$$
\begin{aligned}
\text{tr}(\mathbf{R}^{-1}\mathbf{N}_s\mathbf{N}'_s) &= \text{tr}(\mathbf{D}^{-1}\mathbf{PP'}) + \text{tr } \mathbf{CF} \\
&= 29.24762 + 14.15968 \\
&= 43.40730 \tag{7.106}
\end{aligned}
$$

The third analysis is completed under the full model

$$
Y_{ijkl} = \mu + l_i + s_{ij} + a_k + la_{ik} + sa_{ijk} + e_{ijkl} \tag{7.107}
$$

The only quantity obtained from this analysis is

$$
\begin{aligned}
E_3 &= \mathbf{Y'Y} - R(\mu, l, s, a, la, sa) \\
&= 382.89980 - 380.63251 \\
&= 2.26729
\end{aligned}
$$

If there were other sets of fixed effects to be fitted on a within $LRA$ subclass basis, these would be fitted in this analysis.

Combining the appropriate inverse elements from the first two analyses, we can construct the inverse matrix for the fixed effects. This inverse is given in Table 7.44. The coefficients of the variance components in the $S{:}L$ and the $SA{:}L$ mean squares are computed as follows:

$$
\begin{aligned}
k_1 &= \tfrac{1}{10}(65 - 43.40730) \\
&= \tfrac{1}{10}(21.59270)
\end{aligned}
$$

**TABLE 7.44**  FINAL INVERSE MATRIX FOR ALL FIXED EFFECTS[a]

| Equation | $\hat{\mu}$ | $\hat{l}_1$ | $\hat{l}_2$ | $\hat{a}_1$ | $\hat{a}_2$ | $\hat{la}_{11}$ | $\hat{la}_{12}$ | $\hat{la}_{21}$ | $\hat{la}_{22}$ |
|---|---|---|---|---|---|---|---|---|---|
| $\mu$: | 20,872 | 1,730 | 5,363 | 9,992 | 1,086 | 4,443 | −1,466 | 810 | 457 |
| $l_1$: | | 43,474 | −27,965 | 4,443 | −1,466 | 24,427 | 706 | −15,245 | −77 |
| $l_2$: | | | 47,106 | 810 | 457 | −15,245 | −77 | 20,795 | 2,629 |
| $a_1$: | | | | 56,160 | −34,337 | 9,833 | −4,720 | 8,234 | −4,804 |
| $a_2$: | | | | | 44,408 | −4,720 | 1,384 | −4,804 | 9,885 |
| $la_{11}$: | | | | | | 122,153 | −74,395 | −74,227 | 43,861 |
| $la_{12}$: | | | | | | | 90,199 | 43,861 | −55,677 |
| $la_{21}$: | | | | | | | | 120,554 | −73,479 |
| $la_{22}$: | | | | | | | | | 98,701 |

[a] Each element has been multiplied by $10^6$.

$$= 2.1593$$

$$k_6 = \tfrac{1}{6}(43.40730 - 24.42683)$$

$$= \tfrac{1}{6}(18.98047)$$

$$= 3.1634$$

$$k_7 = \tfrac{1}{6}(65 - 24.42683)$$

$$= \tfrac{1}{6}(40.57317)$$

$$= 6.7622$$

The sum of squares for sires within lines ($S{:}L$) is

$$R(\mu, l, s, a, la) - R(\mu, l, a, la) = E_1 - E_2$$

$$= 3.92511 - 2.86511$$

$$= 1.06000$$

and the sum of squares for the interaction of sire × age of dam within lines ($SA{:}L$) is

$$R(\mu, l, s, a, la, sa) - R(\mu, l, s, a, la) = E_2 - E_3$$

$$= 2.86511 - 2.26729$$

$$= 0.59782$$

The complete ANOVA is given in Table 7.45. Estimates of variance components are obtained as follows:

$$\hat{\sigma}_w^2 = 0.05668$$

$$\hat{\sigma}_{sa:l}^2 = \frac{0.05978 - 0.05668}{2.1593}$$

$$= 0.00144$$

$$\hat{\sigma}_{s:l}^2 = \frac{0.17667 - 0.05668 - (3.1634)(0.00144)}{6.7622}$$

$$= 0.01707$$

**TABLE 7.45** COMBINED ANOVA FOR GAINS OF STEERS

| Source | df | SS | MS | F | E(MS) |
|---|---|---|---|---|---|
| Lines | 2 | 0.15014 | 0.07507 | 0.42 | $\sigma_w^2 + k\sigma_{s:l}^2 + k\sigma_l^2$ |
| Sires:lines (S:L) | 6 | 1.06000 | 0.17667 | 3.12* | $\sigma_w^2 + 3.1634\sigma_{sa:l}^2 + 6.7622\sigma_{s:l}^2$ |
| Age of dam (A) | 2 | 0.02569 | 0.01284 | 0.22 | $\sigma_w^2 + k\sigma_{sa:l}^2 + k\sigma_a^2$ |
| Linear | 1 | 0.01989 | 0.01989 | 0.33 | |
| Quadratic | 1 | 0.00580 | 0.00580 | 0.10 | |
| LA | 4 | 0.64889 | 0.16222 | 2.71 | $\sigma_w^2 + k\sigma_{sa:l}^2 + k\sigma_{la}^2$ |
| SA:L | 10 | 0.59782 | 0.05978 | 1.06 | $\sigma_w^2 + 2.1593\sigma_{sa:l}^2$ |
| Within | 40 | 2.26729 | 0.05668 | | $\sigma_w^2$ |

These variance components may be used to compute two intraclass correlations. These are

$$r_1 = \frac{\hat{\sigma}_{s:l}^2 + \hat{\sigma}_{sa:l}^2}{\hat{\sigma}_{s:l}^2 + \hat{\sigma}_{sa:l}^2 + \hat{\sigma}_w^2}$$

$$= \frac{0.01707 + 0.00144}{0.01707 + 0.00144 + 0.05668}$$

$$= 0.246 \tag{7.108}$$

which estimates the correlation among paternal half sibs from the same age of dam, and

$$r_2 = \frac{\hat{\sigma}_{s:l}^2}{\hat{\sigma}_{s:l}^2 + \hat{\sigma}_{sa:l}^2 + \hat{\sigma}_w^2}$$

$$= \frac{0.01707}{0.01707 + 0.00144 + 0.05668}$$

$$- 0.227 \tag{7.109}$$

the correlation among half sibs who are from different ages of dam. Of course, the amount of data used here is too limited for these correlations to be of much value. They are calculated here to illustrate procedures that would be used when an adequate set of data is being analyzed. In fact, with the interaction of sires and age-of-dam within lines being nonsignificant ($P > 0.20$), the analysis should be rerun deleting this interaction from the model.

The estimates of the effects and the least-squares means for the fixed sets of effects along with standard errors are given in Table 7.46. The estimated variances of the least-squares means were computed as follows:

$$V(\tilde{\mu}) \simeq R^{\mu\mu}(\hat{\sigma}_w^2 + k_1\hat{\sigma}_{s:l}^2 + k_2\hat{\sigma}_{sa:l}^2) \tag{7.110}$$

with

$$k_1 = \frac{\Sigma_i \Sigma_j n_{ij.}^2}{n...} \quad \text{and} \quad k_2 = \frac{\Sigma_i \Sigma_j \Sigma_k n_{ijk}^2}{n...}$$

$$V(\tilde{\mu} + \hat{l}_i) \simeq (R^{\mu\mu} + R^{l_i l_i} + 2R^{\mu l_i})(\hat{\sigma}_w^2 + k_{1i}\hat{\sigma}_{s:l}^2 + k_{2i}\hat{\sigma}_{sa:l}^2) \tag{7.111}$$

with

$$k_{1i} = \frac{\Sigma_j n_{ij.}^2}{n_{i..}} \quad \text{and} \quad k_{2i} = \frac{\Sigma_j \Sigma_k n_{ijk}^2}{n_{i..}}$$

$$V(\tilde{\mu} + \hat{a}_k) \simeq (R^{\mu\mu} + C_{a_k a_k} + 2R^{\mu a_k})(\hat{\sigma}_w^2 + k_k\hat{\sigma}_{sa:l}^2) + R^{\mu\mu}k_1\hat{\sigma}_{s:l}^2 \tag{7.112}$$

**TABLE 7.46**  ESTIMATES OF FIXED EFFECTS AND LEAST-SQUARES MEANS WITH STANDARD ERRORS FOR STEER DATA

| Classification | Number of observations | Effects | | Means | |
|---|---|---|---|---|---|
| | | Estimate | SE | Estimate | SE |
| Overall ($\tilde{\mu}$) | 65 | 2.440 | 0.063 | 2.440 | 0.063 |
| **Lines** | | | | | |
| 1 | 21 | −0.047 | 0.090 | 2.393 | 0.112 |
| 2 | 15 | 0.084 | 0.095 | 2.524 | 0.122 |
| 3 | 29 | −0.037 | 0.080 | 2.403 | 0.088 |
| **Age of dam** | | | | | |
| 3 | 12 | 0.037 | 0.058 | 2.477 | 0.092 |
| 4 | 16 | −0.016 | 0.051 | 2.424 | 0.082 |
| 5-up | 37 | −0.021 | 0.045 | 2.419 | 0.067 |
| **LA Subclasses** | | | | | |
| 1  3 | 3 | −0.080 | 0.083 | 2.349 | 0.155 |
| 1  4 | 5 | −0.015 | 0.072 | 2.362 | 0.123 |
| 1  5-up | 13 | 0.095 | 0.061 | 2.468 | 0.089 |
| 2  3 | 3 | −0.017 | 0.083 | 2.544 | 0.154 |
| 2  4 | 4 | −0.072 | 0.075 | 2.436 | 0.136 |
| 2  5-up | 8 | 0.089 | 0.064 | 2.592 | 0.105 |
| 3  3 | 6 | 0.097 | 0.073 | 2.537 | 0.113 |
| 3  4 | 7 | 0.087 | 0.066 | 2.474 | 0.105 |
| 3  5-up | 16 | −0.184 | 0.055 | 2.198 | 0.075 |

with

$$k_k = \frac{\Sigma_i \, \Sigma_j \, n_{ijk}^2}{n_{..k}}$$

$$V(\tilde{\mu} + \hat{l}_i + \hat{a}_k + \hat{l}a_{ik}) \simeq \lambda_{ik}(\hat{\sigma}_w^2 + k_{ik}\hat{\sigma}_{sa:l}^2) + (R^{\mu\mu} + R^{l_il_i} + 2R^{\mu l_i})k_{ik}\hat{\sigma}_{s:l}^2 \qquad (7.113)$$

with $\lambda_{ik}$ being the appropriate linear function of the inverse elements and

$$k_{ik} = \frac{\Sigma_j \, n_{ijk}^2}{n_{i.k}} \qquad (7.114)$$

with $k_{ik}$ being estimated by

$$k_{ik} \simeq \frac{\Sigma_j \, n_{ij.}^2}{n_{i..}} \cdot \frac{1}{3}$$

$R^{-1}$ is the inverse matrix obtained in the first analysis and $C$ is the inverse obtained after absorption of the $\mu + l_i + s_{ij}$ equations in the second analysis. All necessary inverse elements for the calculation of the standard errors are given in Table 7.44. The $k$ values required for these calculations are computed by LSML76 as the equations are being set up during the third analysis. They are easily computed by hand from Table 3.62 where the data being analyzed here are listed.

The standard errors for the estimates of the effects are obtained from the above formulas by simply deleting that part of each formula which does not apply, that is,

$$V(\hat{l}_i) = R^{l_il_i}(\hat{\sigma}_w^2 + k_{1i}\hat{\sigma}_{s:l}^2 + k_{2i}\hat{\sigma}_{sa:l}^2) \qquad (7.115)$$

$$V(\hat{a}_k) = C_{a_ka_k}(\hat{\sigma}_w^2 + k_k\hat{\sigma}_{sa:l}^2) \qquad (7.116)$$

$$V(\hat{l}a_{ik}) = C_{la_{ik}la_{ik}}(\hat{\sigma}_w^2 + k_{ik}\hat{\sigma}_{sa:l}^2) \qquad (7.117)$$

It should be noted that the accuracy of differences among the $\hat{l}_i$, $\hat{a}_k$, and $l\hat{a}_{ik}$ are more directly proportional to the standard errors of the estimates of the effects than to the standard errors of the least-squares means.

### 7.9.3 Randomized Block, Split-Plot Design

To illustrate the least-squares analysis of data from this design when disproportionate frequencies exist, we shall use the data on red blood cell counts of guinea pigs from the experiment by Mary Ballew of the Department of Public Health at the University of Massachusetts (Table 7.17). However, we shall assume that 13 values are missing as shown in Table 7.47. Also, it will be noted that the four combinations of the two levels of vitamin C and the two levels of ozone are simply referred to as "treatments." Therefore, the mathematical model for the analysis of these data is

$$Y_{ijk} = \mu + b_i + t_j + bt_{ij} + h_k + th_{jk} + e_{ijk} \tag{7.118}$$

where the $b_i$ are random block effects, the $t_j$ are treatment effects, and the $h_k$ are hourly effects. With only one or no observation in each of the $BTH$ subclasses, the $BTH$ interaction is wholly confounded with the smallest within-subclass effects. If the $BH$ interaction is likely to exist, the analysis should be completed with this interaction included in the model. However, we shall assume there is good reason to believe the $BH$ interaction is negligible.

The error term for the whole-plot effects (blocks and treatments) is the mean square for the $BT$ interaction, which contains the variance due to the $BT$ interaction and the variance due to differences among animals, as well as the variance due to subplot error $\sigma_e^2$. Of course, the error term for the subplot effects (hour and treatment $\times$ hour interaction) is the mean square for remainder after accounting for the variation due to all other effects in the model, that is,

$$\text{Remainder SS} = \mathbf{Y}'\mathbf{Y} - R(\mu, b, t, bt, h, th) \tag{7.119}$$

The procedures that we want to use to compute the adjusted sums of squares for the different sources of variation must yield an ANOVA with the expected mean squares presented in Table 7.48. It should be noted that these are the same expected

**TABLE 7.47**  RED BLOOD CELL COUNTS OF GUINEA PIGS

| Block | $\frac{1}{2}$ Hour after exposure | | | | 3 Hours after exposure | | | |
|-------|-------|-------|-------|-------|-------|-------|-------|-------|
|       | $T_1$ | $T_2$ | $T_3$ | $T_4$ | $T_1$ | $T_2$ | $T_3$ | $T_4$ |
| 1 | 6.59 | — | — | 5.50 | 5.94 | 5.90 | — | 5.77 |
| 2 | — | 6.39 | 6.00 | 5.75 | 6.39 | 6.52 | 6.36 | — |
| 3 | 5.38 | 6.18 | 6.92 | 6.45 | — | 6.36 | 6.33 | 5.31 |
| 4 | 5.58 | 5.64 | 6.51 | 6.15 | 5.09 | — | 6.29 | 5.04 |
| 5 | 5.47 | 5.51 | 6.02 | 5.23 | 4.38 | 4.93 | 5.63 | — |
| 6 | 6.30 | 6.36 | — | 5.79 | 6.39 | 5.90 | 5.71 | 5.45 |
| 7 | 6.07 | 5.43 | 6.49 | 5.99 | — | — | 6.22 | 5.55 |
| 8 | — | 5.85 | 6.02 | 6.56 | 5.06 | 5.16 | 5.77 | 5.83 |
| 9 | 3.81 | 3.98 | — | 4.49 | 3.35 | 3.53 | 6.25 | 4.13 |

**TABLE 7.48** EXPECTED MEAN SQUARES DESIRED FROM LEAST-SQUARES ANALYSIS

| Source | df | E(MS) |
|---|---|---|
| Blocks (B) | 8 | $\sigma_e^2 + k_6\sigma_{bt}^2 + k_7\sigma_b^2$ |
| Treatments (T) | 3 | $\sigma_e^2 + k_4\sigma_{bt}^2 + k_5\sigma_t^2$ |
| BT | 23 | $\sigma_e^2 + k_3\sigma_{bt}^2$ |
| Hours (H) | 1 | $\sigma_e^2 + k_2\sigma_h^2$ |
| TH | 3 | $\sigma_e^2 + k_1\sigma_{th}^2$ |
| Remainder | 20 | $\sigma_e^2$ |

mean squares one would have if the data were balanced except for the coefficients of variance components. To obtain the adjusted sums of squares so that the E(MS) in Table 7.48 will apply, the sums of squares must be computed as in Table 7.49.

To complete this analysis by the indirect procedures so that estimates of the variance components and least-squares means for all sets of fixed effects will be obtained, analyses under three models must be completed. The first analysis is completed under the fixed model

$$Y_{ijl} = \mu + t_j + h_k + th_{jk} + e_{jkl} \tag{7.120}$$

The second analysis is completed under the reduced model

$$Y_{ijk} = \mu + b_i + t_j + h_k + th_{jk} + e_{ijk} \tag{7.121}$$

and the third analysis is completed under the full model [Equation (7.118)].

The first analysis is a standard least-squares analysis under a fixed model. The details involved in completing this analysis will not be given. However, from this analysis, we obtain the following:

1. tr $\mathbf{R}^{-1}\mathbf{N}_s\mathbf{N}_s'$, the number of $\sigma_b^2$ and $\sigma_{bt}^2$ components in $R(\mu, t, h, th)$.
2. The $\mu$ equation so that $\mu$ can be estimated later by adjusting for unequal frequencies with respect to all fixed effects.
3. Inverse elements for the $\mu$ row of the final inverse matrix for fixed sets of effects.
4. $E_1 = \mathbf{Y}'\mathbf{Y} - R(\mu, t, h, th)$.
5. Degrees of freedom for $E_1$ (the Remainder).

**TABLE 7.49** FORMULAS FOR COMPUTATION OF SUMS OF SQUARES

| Source | df | SS |
|---|---|---|
| Blocks (B) | 8 | $R(\mu, b, t, h, th) - R(\mu, t, h, th)$ |
| Treatments (T) | 3 | $R(\mu, b, t, h, th) - R(\mu, b, h, th)$ |
| BT | 23 | $R(\mu, b, t, bt, h, th) - R(\mu, b, t, h, th)$ |
| Hours (H) | 1 | $R(\mu, b, t, bt, h, th) - R(\mu, b, t, bt, th)$ |
| TH | 3 | $R(\mu, b, t, bt, h, th) - R(\mu, b, t, bt, h)$ |
| Remainder | 20 | $\mathbf{Y}'\mathbf{Y} - R(\mu, b, t, bt, h, th)$ |

In completing this first analysis these five quantities were found to be as follows:

1. tr $\mathbf{R}^{-1}\mathbf{N}_s\mathbf{N}_s' = 8.00000$.
2. $59\mu - 2\hat{t}_1 - \hat{t}_2 - 2\hat{t}_3 + \hat{h}_1 - 2\hat{th}_{11} - \hat{th}_{21} - 4\hat{th}_{31} = 334.95$.
3. $R^{\mu\mu} = \phantom{-}0.017175$,
   $R^{\mu t_1} = \phantom{-}0.000068$,
   $R^{\mu t_2} = -0.000434$,
   $R^{\mu t_3} = \phantom{-}0.001054$,
   $R^{\mu h_1} = -0.000124$,
   $R^{\mu th_{11}} = \phantom{-}0.000124$,
   $R^{\mu th_{21}} = -0.000992$,
   $R^{\mu th_{31}} = \phantom{-}0.002728$.
4. $E_1 = 1937.93470 - 1908.14875 = 29.78595$.
5. 51.

In the second analysis, the equations for $\mu + b_i$ are absorbed into the equations for $t_j$, $h_k$, and $th_{jk}$. From this analysis, we obtain the following:

1. tr $\mathbf{R}^{-1}\mathbf{N}_s\mathbf{N}_s' = $ tr $\mathbf{D}^{-1}\mathbf{PP}' + $ tr $\mathbf{CF}$, the number of $\sigma_{bt}^2$ components in $R(\mu, b, t, h, th)$.
2. $E_2 = \mathbf{Y}'\mathbf{Y} - R(\mu, b, t, h, th)$.
3. Degrees of freedom for $E_2$, the Remainder.
4. Sum of squares for treatments ($T$) as computed from $\hat{\mathbf{B}}_T'\mathbf{Z}_T^{-1}\hat{\mathbf{B}}_T$.
5. Degrees of freedom for treatments.
6. Estimates of $t_j$ effects.
7. Inverse elements involving $\hat{t}_j$.

When this second analysis was completed, these seven quantities were found to be as follows:

1. tr $\mathbf{D}^{-1}\mathbf{PP}' + $ tr $\mathbf{CF} = 16.27619 + 5.98277$
   $$= 22.25896.$$
2. $E_2 = \mathbf{Y}'\mathbf{Y} - R(\mu, b) - R(t, h, th \mid \mu, b)$
   $$= 1937.93470 - 1922.24661 - 5.23546$$
   $$= 10.45263.$$
3. 43.
4. 3.96868.
5. 3.
6. $\hat{t}_1 = -0.212913$, $\hat{t}_2 = -0.091779$, and $\hat{t}_3 = 0.472650$.
7.

|  | $\hat{t}_1$ | $\hat{t}_2$ | $\hat{t}_3$ | $\hat{h}_1$ | $\hat{th}_{11}$ | $\hat{th}_{21}$ | $\hat{th}_{31}$ |
|---|---|---|---|---|---|---|---|
| $\hat{t}_1$ | 0.055014 | −0.017759 | −0.021139 | −0.000398 | 0.000398 | 0.000542 | −0.003125 |
| $\hat{t}_2$ | −0.017759 | 0.051819 | −0.017940 | −0.001405 | 0.000474 | −0.002865 | −0.001585 |
| $\hat{t}_3$ | −0.021139 | −0.017940 | 0.057310 | 0.003915 | −0.003915 | −0.000716 | 0.005637 |

In the third analysis the equations for $\mu + b_i + t_j + bt_{ij}$ are absorbed into the equations for the $h_k$ and the $th_{jk}$. From this analysis we obtain the following:

1. $E_3 = \mathbf{Y'Y} - R(\mu, b, t, bt, h, th)$
   $= \mathbf{Y'Y} - R(\mu, b, t, bt) - R(h, th \mid \mu, b, t, bt).$
2. Degrees of freedom for $E_3$, the Remainder.
3. Sum of squares for hours $(H)$ and the treatment $\times$ hour $(TH)$ interaction as computed from $\hat{\mathbf{B}}_H' \mathbf{Z}_H^{-1} \hat{\mathbf{B}}_H$ and $\hat{\mathbf{B}}_{TH}' \mathbf{Z}_{TH}^{-1} \hat{\mathbf{B}}_{TH}$.
4. Degrees of freedom for $H$ and $TH$.
5. Estimates for $h_k$ and $th_{jk}$ effects.
6. Inverse elements involving the $\hat{h}_k$ and the $\hat{th}_{jk}$ effects.

The reader can verify that these six quantities are found to be as follows when this analysis is completed:

1. $E_3 = 1937.93470 - 1934.06660 - 2.18029$
   $= 1.68781.$
2. 20.
3. $H$: 1.91273 and for $TH$: 0.22816.
4. 1 and 3.
5. $\hat{h}_1 = \phantom{-}0.201042,$
   $\hat{th}_{11} = \phantom{-}0.058958,$
   $\hat{th}_{21} = -0.045208,$
   $\hat{th}_{31} = -0.087708.$
6.

|          | $\hat{h}_1$ | $\hat{th}_{11}$ | $\hat{th}_{21}$ | $\hat{th}_{31}$ |
|----------|-------------|-----------------|-----------------|-----------------|
| $\hat{h}_1$    | 0.021131 | 0.003869 | -0.000298 | -0.000298 |
| $\hat{th}_{11}$ |          | 0.071131 | -0.024702 | -0.024702 |
| $\hat{th}_{21}$ | Symmetric |         | 0.062798  | -0.020536 |
| $\hat{th}_{31}$ |          |          |           | 0.062798  |

We can now compute the following:

$$\text{SS Blocks} = E_1 - E_2$$

$$= 29.78595 - 10.45263$$

$$= 19.33333$$

$$\text{SS } BT = E_2 - E_3$$

$$= 10.45263 - 1.68781$$

$$= 8.76482$$

$$\text{df}(B) = \text{df}(E_1) - \text{df}(E_2)$$

$$= 51 - 43 = 8$$

$$\mathrm{df}(BT) = \mathrm{df}(E_2) - \mathrm{df}(E_3)$$

$$= 43 - 20 = 23$$

$$k_3 = \tfrac{1}{23}(59 - 22.25896)$$

$$= \tfrac{1}{23}(36.74104)$$

$$= 1.5974$$

$$k_6 = \tfrac{1}{8}(22.25896 - 8.00000)$$

$$= \tfrac{1}{8}(14.25896)$$

$$= 1.7824$$

$$k_7 = \tfrac{1}{8}(59 - 8.00000)$$

$$= \tfrac{1}{8}(51)$$

$$= 6.3750$$

The ANOVA obtained by combining the results from these three analyses is given in Table 7.50.

We now compute estimates of the variance components as follows:

$$\hat{\sigma}_e^2 = 0.08439$$

$$\hat{\sigma}_{bt}^2 = \frac{0.38108 - 0.08439}{1.5974}$$

$$= 0.18573$$

$$\hat{\sigma}_b^2 = \frac{2.41667 - 0.08439 - (1.7824)(0.18573)}{6.3750}$$

$$= 0.31392$$

The estimate of the correlation of red blood cell counts of Guinea pigs in the same block × treatment subclass (adjusted for hour effects) is

$$r_1 = \frac{\hat{\sigma}_b^2 + \hat{\sigma}_{bt}^2}{\hat{\sigma}_b^2 + \hat{\sigma}_{bt}^2 + \hat{\sigma}_e^2}$$

$$= \frac{0.31392 + 0.18573}{0.31392 + 0.18573 + 0.08439}$$

**TABLE 7.50**   COMBINED ANOVA FOR RED BLOOD CELL COUNTS OF GUINEA PIGS

| Source | df | SS | MS | F | E(MS) |
|---|---|---|---|---|---|
| Blocks (B) | 8 | 19.33333 | 2.41667 | 6.34** | $\sigma_e^2 + 1.7824\sigma_{bt}^2 + 6.3750\sigma_b^2$ |
| Treatments (T) | 3 | 3.96868 | 1.32289 | 3.47* | $\sigma_e^2 + k\sigma_{bt}^2 + k\sigma_t^2$ |
| BT | 23 | 8.76482 | 0.38108 | 4.52** | $\sigma_e^2 + 1.5974\sigma_{bt}^2$ |
| Hours (H) | 1 | 1.91273 | 1.91273 | 22.67** | $\sigma_e^2 + k\sigma_h^2$ |
| TH | 3 | 0.22816 | 0.07605 | 0.90 | $\sigma_e^2 + k\sigma_{th}^2$ |
| Remainder | 20 | 1.66781 | 0.08439 | | $\sigma_e^2$ |

$$= \frac{0.49965}{0.58404} = 0.856 \tag{7.122}$$

and the correlation among red blood cell counts in the same block, adjusted for treatment, hour, and treatment $\times$ hour interaction effects is

$$r_2 = \frac{\hat{\sigma}_b^2}{\hat{\sigma}_b^2 + \hat{\sigma}_{bt}^2 + \hat{\sigma}_e^2}$$

$$= \frac{0.31392}{0.58404} = 0.538 \tag{7.123}$$

The estimate of the overall mean adjusted for unequal frequencies with respect to all fixed effects is computed as follows:

$$\tilde{\mu} = \tfrac{1}{59}[334.95 + (2)(-0.212913) + (-0.091779) + (2)(0.472650)$$

$$- 0.201042 + (2)(0.058958) + (-0.045208) + (4)(-0.087708)]$$

$$= \tfrac{1}{59}(334.898529) = 5.6762 \tag{7.124}$$

The estimates of all fixed effects and the least-squares means with standard errors are given in Table 7.51. Again, it should be noted that the standard errors of the estimated

**TABLE 7.51** ESTIMATES OF FIXED EFFECTS AND LEAST-SQUARES MEANS WITH STANDARD ERRORS FOR GUINEA PIG DATA

| Classification | Number of observations | Effects Estimate | Effects SE | Means Estimate | Means SE |
|---|---|---|---|---|---|
| Overall ($\tilde{\mu}$) | 59 | 5.676 | 0.207 | 5.676 | 0.207 |
| Treatments | | a | | | |
| $V_1O_1$ | 14 | -0.213 | 0.149 | 5.463 | 0.256 |
| $V_1O_2$ | 15 | -0.092 | 0.147 | 5.584 | 0.253 |
| $V_2O_1$ | 14 | 0.473 | 0.157 | 6.149 | 0.262 |
| $V_2O_2$ | 16 | -0.168 | 0.148 | 5.508 | 0.253 |
| Hours | | b | | | |
| $\frac{1}{2}$ | 30 | 0.201 | 0.042 | 5.877 | 0.211 |
| 3 | 29 | -0.201 | 0.042 | 5.475 | 0.212 |
| TH Subclasses | | ns$^c$ | | | |
| $V_1O_1$ $\frac{1}{2}$ | 7 | 0.059 | 0.077 | 5.723 | 0.298 |
| $V_1O_1$ 3 | 7 | -0.059 | 0.077 | 5.203 | 0.298 |
| $V_1O_2$ $\frac{1}{2}$ | 8 | -0.045 | 0.073 | 5.740 | 0.274 |
| $V_1O_2$ 3 | 7 | 0.045 | 0.073 | 5.428 | 0.294 |
| $V_2O_1$ $\frac{1}{2}$ | 6 | -0.088 | 0.073 | 6.262 | 0.318 |
| $V_2O_1$ 3 | 8 | 0.088 | 0.073 | 6.036 | 0.273 |
| $V_2O_2$ $\frac{1}{2}$ | 9 | 0.074 | 0.069 | 5.783 | 0.255 |
| $V_2O_2$ 3 | 7 | -0.074 | 0.069 | 5.233 | 0.293 |

[a] $P < 0.05$.
[b] $P < 0.01$.
[c] ns, nonsignificantly different.

least-squares means are much larger than the standard errors of the estimates of the effects, since in the calculation of the former the sampling of the random effects must be considered.

Now, to complete the analysis, we need to make the linear comparisons of the treatment effects and the treatment × hour interaction effects as suggested by the nature of the treatments applied. The linear comparisons desired for treatment are as follows:

| | $\hat{t}_1$ $(V_1O_1)$ | $\hat{t}_2$ $(V_1O_2)$ | $\hat{t}_3$ $(V_2O_1)$ | $\hat{t}_4$ $(V_2O_2)$ |
|---|---|---|---|---|
| $V$ | 1 | 1 | −1 | −1 |
| $O$ | 1 | −1 | 1 | −1 |
| $VO$ | 1 | −1 | −1 | 1 |

Since the summation restrictions were used to obtain the $\hat{t}_i$, these contrasts may be expressed as linear functions of $\hat{t}_1$, $\hat{t}_2$, and $\hat{t}_3$ only; that is, $\hat{t}_4 = -\hat{t}_1 - \hat{t}_2 - \hat{t}_3$. To do this, we simply subtract the coefficients for $\hat{t}_4$ from the coefficients for the other $\hat{t}_i$. Hence,

$$V = V_1 - V_2 = 2\hat{t}_1 + 2\hat{t}_2$$

$$O = O_1 - O_2 = 2\hat{t}_1 + 2\hat{t}_3$$

$$VO = \text{Interaction} = -2\hat{t}_2 - 2\hat{t}_3$$

The variances of these linear contrasts are computed by making use of the appropriate inverse elements and the appropriate error term (MS for $BT$). For example,

$$V(V) = [(4)(0.055014) + (4)(0.051819) + (8)(-0.017759)](-0.38108)$$

$$= (0.285260)(0.38108)$$

$$= 0.108707$$

$$\text{SE}(V) = 0.329707$$

$$V = (2)(-0.212913) + (2)(-0.091779)$$

$$= -0.609384$$

$$t = \frac{-0.609384}{0.329707} = -1.848259$$

$$t^2 = F = \frac{\text{SS}}{\text{MS(Error)}}$$

or        $$\text{SS} = \text{MS(Error)}t^2$$

$$= (0.38108)(3.416062) = 1.3018$$

The contrasts desired among the $\hat{th}_{jk}$ are as follows:

|     | $\hat{th}_{11}$ | $\hat{th}_{21}$ | $\hat{th}_{31}$ | $\hat{th}_{41}$ | $\hat{th}_{12}$ | $\hat{th}_{22}$ | $\hat{th}_{32}$ | $\hat{th}_{42}$ |
|-----|-----|-----|-----|-----|-----|-----|-----|-----|
| V   | 1   | 1   | −1  | −1  | 1   | 1   | −1  | −1  |
| O   | 1   | −1  | 1   | −1  | 1   | −1  | 1   | −1  |
| H   | 1   | 1   | 1   | 1   | −1  | −1  | −1  | −1  |
| VH  | 1   | 1   | −1  | −1  | −1  | −1  | 1   | 1   |
| OH  | 1   | −1  | 1   | −1  | −1  | 1   | −1  | 1   |
| VOH | 1   | −1  | −1  | 1   | −1  | 1   | 1   | −1  |

The restrictions imposed on the interaction effects were $\Sigma_j \, \hat{th}_{jk} = \Sigma_k \, \hat{th}_{jk} = \Sigma_j \, \Sigma_k \, \hat{th}_{jk} = 0$. Therefore, the contrasts for $VH$, $OH$, and $VOH$ can be obtained from the estimates for $th_{11}$, $th_{21}$, and $th_{31}$. For example, the adjusted coefficients ($\lambda^A$'s) for $\hat{th}_{11}$, $\hat{th}_{21}$, and $th_{13}$ are computed as follows. For $VH$,

$$\lambda_{11}^A = \lambda_{11} - \lambda_{41} - \lambda_{12} + \lambda_{42} = 1 + 1 + 1 + 1 = 4$$

$$\lambda_{21}^A = \lambda_{21} - \lambda_{41} - \lambda_{22} + \lambda_{42} = 1 + 1 + 1 + 1 = 4$$

$$\lambda_{31}^A = \lambda_{31} - \lambda_{41} - \lambda_{32} + \lambda_{42} = -1 + 1 - 1 + 1 = 0$$

$$\therefore VH = 4\hat{th}_{11} + 4\hat{th}_{21}$$

The contrasts for $OH$ and $VOH$ are found to be

$$OH = 4\hat{th}_{11} + 4\hat{th}_{31}$$

$$VOH = -4\hat{th}_{21} - 4\hat{th}_{31}$$

The linear contrasts and sums of squares are now computed in the manner described above. A summary of the results from making these comparisons is given in Table 7.52. We can now see that the major reason for the significance of treatment effects (Table 7.50) is the vitamin C level × ozone level interaction. The difference between ozone levels changes sign from the low vitamin C level to the high vitamin C level (Table 7.51).

**TABLE 7.52**   TESTS OF SIGNIFICANCE FOR VITAMIN C
AND OZONE EFFECTS

| Comparison | SS = MS | F     | Difference in means | SE    |
|-----------|---------|-------|---------------------|-------|
| V         | 1.3018  | 3.42  | −0.305              | 0.165 |
| O         | 0.9631  | 2.53  | 0.260               | 0.163 |
| VO        | 1.9804  | 5.20* | −0.381              | 0.167 |
| VH        | 0.0022  | 0.03  | 0.014               | 0.084 |
| OH        | 0.0098  | 0.12  | −0.029              | 0.084 |
| VOH       | 0.2090  | 2.48  | 0.133               | 0.084 |

## EXERCISES

7.1. Dr. Donald L. Black of the Department of Veterinary and Animal Sciences at the University of Massachusetts conducted an experiment in which three groups of five rabbits each were included, each group being subjected to a different treatment, a completely randomized design. For each rabbit, a measurement was made on the right oviduct and a measurement made on the left oviduct, resulting in a split-plot design. The data are displayed in the table below, each observation representing the micrograms of norepinephrine per gram of tissue. Complete the appropriate analysis of variance for this set of data including the expectations of mean squares considering the treatment and oviduct effects to be fixed and the rabbit effect to be random.

| Treatment 1 | | | Treatment 2 | | | Treatment 3 | | |
|---|---|---|---|---|---|---|---|---|
| Rabbit | Left | Right | Rabbit | Left | Right | Rabbit | Left | Right |
| 1 | 1.28 | 1.05 | 1 | 1.22 | 0.63 | 1 | 1.73 | 1.92 |
| 2 | 8.58 | 7.85 | 2 | 4.33 | 4.26 | 2 | 6.71 | 6.88 |
| 3 | 2.86 | 2.74 | 3 | 1.84 | 3.02 | 3 | 4.08 | 3.68 |
| 4 | 1.72 | 1.97 | 4 | 1.69 | 2.12 | 4 | 3.12 | 2.92 |
| 5 | 2.90 | 2.27 | 5 | 2.27 | 2.51 | 5 | 8.16 | 3.47 |

7.2. Dr. Donald L. Anderson of the Department of Veterinary and Animal Sciences conducted an experiment with poultry which included two levels of calcium and two levels of lysine, a factorial experiment with six chickens per group. The weights of the birds were recorded at 10 and 20 weeks making this a split-plot design. Only a small portion of the data is shown here, the observations being the weights of the chickens in grams. Complete the appropriate analysis of variance for this set of data including the expectations of mean squares considering the calcium, lysine, and week effects to be fixed and the bird effects to be random.

| Low lysine | Low calcium | | | High calcium | | |
|---|---|---|---|---|---|---|
| | Bird | 10W | 20W | Bird | 10W | 20W |
| | 1 | 727 | 1300 | 1 | 852 | 1583 |
| | 2 | 738 | 1374 | 2 | 660 | 1505 |
| | 3 | 630 | 1513 | 3 | 784 | 1650 |
| | 4 | 575 | 1125 | 4 | 475 | 1204 |
| | 5 | 630 | 1243 | 5 | 680 | 1450 |
| | 6 | 735 | 1636 | 6 | 804 | 1591 |
| **High lysine** | Bird | 10W | 20W | Bird | 10W | 20W |
| | 1 | 730 | 1348 | 1 | 775 | 1293 |
| | 2 | 731 | 1657 | 2 | 712 | 1494 |
| | 3 | 556 | 1273 | 3 | 634 | 1572 |
| | 4 | 671 | 1348 | 4 | 508 | 1263 |
| | 5 | 673 | 1332 | 5 | 675 | 1347 |
| | 6 | 725 | 1542 | 6 | 777 | 1353 |

**7.3.** The data presented below, which resulted from a randomized block, split-split-plot design, were supplied by Dr. William M. Clapham formerly of the Plant and Soil Sciences Department at the University of Massachusetts. The data are the spring wheat yields in bushels per acre. Four blocks were used and four varieties were randomized across the four blocks. Each variety was split into two drill row spacings, 5.5 in. and 7.0 in. and each of the spacings (split plots) was divided into a portion to which 60 lb/acre of nitrogen was broadcast and one section with no nitrogen sidedressing. Complete the appropriate split-split-plot analysis of variance for this set of data, including the expectations of mean squares. Consider the blocks to be a random effect and the varieties, spacings, and nitrogen treatments to be fixed effects.

| Variety | N | Block 1 | | Block 2 | | Block 3 | | Block 4 | |
| | | 5.5 in. | 7.0 in. | 5.5 in. | 7.0 in. | 5.5 in. | 7.0 in. | 5.5 in. | 7.0 in. |
|---|---|---|---|---|---|---|---|---|---|
| 1 | + | 39 | 46 | 36 | 46 | 33 | 28 | 39 | 32 |
| | − | 34 | 49 | 31 | 38 | 32 | 34 | 42 | 31 |
| 2 | + | 39 | 37 | 45 | 47 | 56 | 38 | 53 | 44 |
| | − | 28 | 43 | 35 | 52 | 37 | 33 | 42 | 39 |
| 3 | + | 40 | 35 | 39 | 40 | 40 | 34 | 27 | 34 |
| | − | 37 | 28 | 36 | 30 | 37 | 39 | 43 | 30 |
| 4 | + | 45 | 50 | 42 | 49 | 45 | 44 | 55 | 40 |
| | − | 38 | 34 | 40 | 49 | 30 | 34 | 34 | 34 |

**7.4.** The data presented below were generated by Timothy C. Quick in conducting M.S. research at the Ohio State University. Three $3 \times 3$ Latin squares are included, the columns in the first square being three goats, the columns in the second square being three wool sheep, and the columns in the third square being three hair sheep. The rows are period effects. Each animal was evaluated on two separate days at each period, making the days a split-plot effect. The values above the broken line in each cell are the measurements for the first day and the values below the broken line are the measurements for the second day. The values represent fluid turnover in the rumen in times per day. Complete an analysis of variance for this set of data, including the expectations of mean squares. Note that the squares (a fixed effect) can interact with periods and treatments in the whole-plot portion of the analysis and with the days in the split-plot portion.

| | | I | | | | II | | | | III | | |
| | | 1 | 2 | 3 | | 1 | 2 | 3 | | 1 | 2 | 3 |
|---|---|---|---|---|---|---|---|---|---|---|---|---|
| 1 | | A 3.1 / 3.1 | B 2.2 / 1.8 | C 1.8 / 1.4 | 1 | A 2.5 / 1.5 | B 2.7 / 2.7 | C 2.5 / 1.6 | 1 | A 3.2 / 2.8 | B 3.2 / 1.8 | C 2.0 / 1.2 |
| 2 | | B 2.3 / 2.8 | C 1.7 / 2.1 | A 2.2 / 2.3 | 2 | B 2.2 / 2.6 | C 2.2 / 2.3 | A 3.3 / 3.5 | 2 | B 2.7 / 2.8 | C 2.7 / 2.2 | A 2.1 / 2.9 |
| 3 | | C 1.4 / 1.2 | A 1.3 / 0.7 | B 1.5 / 1.5 | 3 | C 2.3 / 2.1 | A 2.5 / 2.9 | B 2.8 / 2.8 | 3 | C 2.3 / 2.6 | A 3.9 / 2.1 | B 2.2 / 2.8 |

**7.5.** Scott Werme of the Department of Veterinary and Animal Sciences at the University of Massachusetts conducted a feeding trial which made use of a series of $2 \times 2$ Latin squares in a crossover split-plot design. A small portion of the data is reproduced here. Eight cows included in four $2 \times 2$ squares were alternated on two treatments ($T$) and four weekly ($W$) records were kept on each treatment for pounds of protein produced. Each cow's number is given in the table. Complete the appropriate analysis of variance for this set of data, including the expectations of mean squares. Consider the periods, treatments, and weeks to be fixed effects and the squares and animals within squares to be random effects. Do not include in the whole-plot portion of the analysis any interactions with squares.

| | | | I | | | | II | | | | III | | | | IV | | |
|---|---|---|---|---|---|---|---|---|---|---|---|---|---|---|---|---|---|
| P | W | T | #402 | T | #187 | T | #502 | T | #572 | T | #549 | T | #485 | T | #544 | T | #142 |
| 1 | 1 | 1 | 2.91 | 2 | 2.23 | 1 | 2.59 | 2 | 2.51 | 1 | 2.33 | 2 | 2.50 | 1 | 1.53 | 2 | 1.23 |
| | 2 | | 2.99 | | 2.53 | | 2.06 | | 2.14 | | 2.62 | | 2.22 | | 1.34 | | 1.21 |
| | 3 | | 2.81 | | 1.97 | | 2.40 | | 1.87 | | 2.41 | | 2.13 | | 1.49 | | 1.28 |
| | 4 | | 2.72 | | 2.02 | | 2.31 | | 2.02 | | 2.30 | | 1.97 | | 1.30 | | 1.24 |
| P | W | T | #402 | T | #187 | T | #502 | T | #572 | T | #549 | T | #485 | T | #544 | T | #142 |
| 2 | 1 | 2 | 2.44 | 1 | 1.70 | 2 | 1.97 | 1 | 1.76 | 2 | 2.02 | 1 | 1.88 | 2 | 1.08 | 1 | 1.10 |
| | 2 | | 1.97 | | 1.80 | | 1.78 | | 1.85 | | 0.96 | | 1.54 | | 1.22 | | 1.06 |
| | 3 | | 2.40 | | 1.52 | | 1.88 | | 2.11 | | 2.00 | | 1.97 | | 1.09 | | 1.12 |
| | 4 | | 2.55 | | 1.89 | | 1.75 | | 2.06 | | 1.88 | | 1.92 | | 0.91 | | 1.14 |

**7.6.** Using the data in Exercise 7.5, partition the sum of squares for weeks into linear, quadratic, and cubic components. Now, using the method shown in Table 7.11, partition the treatment $\times$ week interaction sum of squares into linear $\times$ linear, linear $\times$ quadratic, and linear $\times$ cubic components.

**7.7.** Again, using the data in Exercise 7.5, partition the sum of squares for the square $\times$ week interaction into: the sum of squares among the linear effects of weeks in each of the four squares; the sum of squares among the quadratic effects of weeks in each of the four squares; and the sum of squares among the cubic effects of weeks in each of the four squares using the method shown in Table 7.23. Test for the significance of the differences among each of these effects, remembering that each of the three sums of squares has three degrees of freedom.

**7.8.** The table below is a portion of the set of data collected by Jay Graves of the Exercise Science Department at the University of Massachusetts. The values in the table are serum creatine kinase levels before and at 6, 18, and 42 hours following three different isometric exercises. The design was that of a crossover split-plot. Complete the appropriate analysis of variance, writing out the expectations of mean squares. Consider the subjects to be random effects and the remaining classifications to be fixed effects.

| Subject | Day | Exercise | Hour 0 | Hour 6 | Hour 18 | Hour 42 |
|---------|-----|----------|--------|--------|---------|---------|
| 1 | 1 | A | 202 | 1768 | 1937 | 1142 |
| | 2 | B | 240 | 495 | 552 | 444 |
| | 3 | C | 255 | 742 | 1001 | 806 |
| 2 | 1 | B | 193 | 887 | 819 | 258 |
| | 2 | C | 202 | 864 | 825 | 486 |
| | 3 | A | 147 | 289 | 397 | 230 |

| Subject | Day | Exercise | Hour 0 | Hour 6 | Hour 18 | Hour 42 |
|---------|-----|----------|--------|--------|---------|---------|
| 3 | 1 | C | 172 | 533 | 523 | 287 |
|   | 2 | A | 121 | 183 | 206 | 200 |
|   | 3 | B | 154 | 231 | 297 | 198 |
| 4 | 1 | A | 242 | 572 | 515 | 139 |
|   | 2 | C | 162 | 320 | 360 | 421 |
|   | 3 | B | 193 | 247 | 284 | 239 |
| 5 | 1 | B | 1141 | 2297 | 2323 | 1488 |
|   | 2 | A | 565 | 1721 | 1890 | 1136 |
|   | 3 | C | 461 | 794 | 873 | 524 |
| 6 | 1 | C | 243 | 600 | 676 | 405 |
|   | 2 | B | 157 | 280 | 317 | 295 |
|   | 3 | A | 274 | 303 | 278 | 224 |

**7.9.** Eight bulls used in artificial insemination were equally divided into a high group and a low group based on percent of cows bred which did not return for a later breeding. Semen samples were collected over an eight-week period. The table below gives the percent normal sperm recorded for these semen samples. Since semen samples were not available in all weeks for all bulls, the data have unequal subclass numbers. Complete the analysis of these data using least-squares procedures as described in Section 7.9. Interpret the results.

| | Weeks | | | | | | | |
|-------|----|----|----|----|----|----|----|----|
| Bulls | 1 | 2 | 3 | 4 | 5 | 6 | 7 | 8 |
| High bulls | | | | | | | | |
| 1 | 79 | 83 | 81 | — | 76 | 81 | — | 79 |
| 2 | — | 78 | — | 77 | 70 | 84 | 82 | — |
| 3 | 77 | — | 86 | 90 | 75 | 73 | — | 80 |
| 4 | 77 | 81 | 88 | 79 | — | 87 | 75 | 80 |
| Low bulls | | | | | | | | |
| 5 | 32 | 58 | — | — | 66 | 58 | 62 | 67 |
| 6 | 33 | — | 35 | 24 | 33 | 29 | 28 | 36 |
| 7 | 61 | 65 | 66 | 67 | 68 | 68 | 73 | — |
| 8 | 28 | 37 | 29 | 46 | 40 | 45 | 30 | 54 |

**7.10.** The data given in the table below are the average daily gains (in pounds) of 87 steers from birth to weaning produced in a diallel cross of three lines of Herefords. The experiment was conducted over a four-year period and data were also available on heifers. However, to simplify computations only the data on steers obtained in one year of the experiment are given in the table. Complete the analysis of these data using least-squares procedures (Section 7.9) under the model that contains lines of sire, sires within lines or sire, lines of dam, interaction of lines of sire with lines of dam and interaction of sires within lines of sire with lines of dam. Obtain estimates of the variance components.

| Lines of dams | Lines of sire | | | | | | | | | | |
| --- | --- | --- | --- | --- | --- | --- | --- | --- | --- | --- | --- |
| | 1 | | | | 2 | | | 3 | | | |
| | 1 | 2 | 3 | 4 | 1 | 2 | 3 | 1 | 2 | 3 | 4 |
| 1 | 1.24 | 1.44 | 1.52 | 1.48 | 2.10 | 1.90 | 1.78 | 1.93 | 1.75 | 1.85 | 1.47 |
| | 1.66 | 1.54 | 1.58 | 1.52 | | | 1.65 | 1.71 | 1.50 | 2.10 | |
| | 1.54 | | 1.67 | 1.26 | | | 1.92 | | | 1.72 | |
| | 1.89 | | 1.64 | 1.96 | | | 1.85 | | | | |
| | 1.38 | | 1.84 | | | | | | | | |
| 2 | 1.90 | 1.93 | 1.86 | 1.85 | 1.71 | 1.65 | 1.88 | 1.64 | 1.76 | 1.84 | 1.78 |
| | 1.62 | 2.15 | 1.51 | 1.95 | 1.75 | 1.60 | 1.72 | 1.54 | | | 1.83 |
| | 1.76 | 1.86 | 1.86 | 1.85 | 1.80 | | 1.74 | 1.78 | | | 1.67 |
| | | | | 1.79 | | | 1.59 | | | | |
| 3 | 1.91 | 2.07 | 1.65 | 1.59 | 2.04 | 1.75 | 1.98 | 1.76 | 1.85 | 1.68 | 1.60 |
| | | 1.90 | | 1.80 | 1.90 | | 1.80 | 1.57 | 1.62 | 1.57 | 1.48 |
| | | | | 1.17 | | | | 1.46 | 1.60 | 1.56 | 1.52 |
| | | | | | | | | 1.70 | | | 1.36 |
| | | | | | | | | 1.44 | | | |

**7.11.** Complete the analysis of the data given in Exercise 7.10 under the reduced model where the interaction of sires within lines of sire with lines of dam is omitted. Compute estimates of the variance components and all least-squares means with standard errors. Obtain the sums of squares for the comparisons of pure-line progeny with cross-line progeny (1) over all the subclasses, (2) line 1 and line 2, (3) line 1 and line 3, and (4) line 2 and line 3.

# Chapter 8

# Analysis of Covariance

## 8.1 INTRODUCTION

The techniques of the analysis of variance and regression can be combined into what has been designated as the analysis of covariance. While curvilinear regression can be included in covariance analyses, they most commonly are limited to linear regression. Covariance can be used with all the common designs and is used to control experimental error and increase the precision of the experiment by reducing the error variance.

As with regression, we now deal with two observations per experimental subject rather than one. We have a continuous independent variable $X$ and a dependent variable $Y$. The extension from regression is that now the $X$ and $Y$ pairs are distributed into different treatments or combinations of treatments. While the analysis essentially adjusts the $Y$ values for the corresponding $X$ values and analyzes the adjusted $Y$'s, one would not use this procedure because it does not take into account the accuracy of making adjustments. The result of the analysis is that we now have treatment means adjusted for the regression of $Y$ on $X$ and we test for the significance of differences among the adjusted means. Thus, all means are usually adjusted to what it is estimated they would have been had each treatment had the same average $X$ value. The regression used for adjustment most frequently arises from a pooled regression, originating from the error line of the analysis of covariance. However, there are occasions when individual regressions for the separate classes or subclasses are fitted. Also, there are a number of problems where the regression or regressions must be estimated on an error line different from the lowest error.

The conducting of an analysis of covariance normally requires that the independent variable $X$ not be affected by the treatments. If $X$ is affected by the treatments, one may still use covariance but the interpretation is quite different. For example, different rations fed to pigs might affect appetite as well as gains. Thus, feed (when fed ad lib) intake will differ among rations. One might still like to know to what extent observed differences in gain can be explained by differences in feed intake.

If a pooled regression from the error line of the analysis of covariance is used, the assumption is made that the regression coefficients within the treatments do not differ, as well as the usual assumptions concerning the analysis of variance, namely, that the errors are normally and independently distributed with a mean of 0 and a variance of $\sigma_e^2$.

## 8.2 ONE-WAY CLASSIFICATION: ANALYSIS OF COVARIANCE

The data that will be used to illustrate this analysis are displayed in Table 8.1 and resulted from an experiment conducted by Wesley Autio in the Department of Plant and Soil Sciences at the University of Massachusetts. This is a portion of a much larger set of data and has been selected only for illustration of this method of analysis and does not necessarily reflect conclusions drawn from the complete experiment. The experiment was conducted to test the influence of four levels of treatments on the firmness of apples. The four levels of treatment consisted of one check and three different dates of application of the material AVG. Since it was known that the diameter of the fruit influenced the firmness, this was included as a covariate $X$.

The mathematical model for this set of data is

$$Y_{ij} = \mu + t_i + \beta(X_{ij} - \bar{x}.) + e_{ij}$$

$$i = 1, 2, \ldots, n_t \qquad j = 1, 2, \ldots, n_w \tag{8.1}$$

where the $t_i$ represent treatment effects and the $X_{ij}$ represent diameters of apples. The $n_w$ is the number of observations in each treatment. The calculations necessary to complete the analysis follow.

**TABLE 8.1**  FIRMNESS OF APPLES AS MEASURED BY PRESSURE, ($Y$ IN POUNDS/INCH$^2$) AND DIAMETER OF FRUIT ($X$ IN INCHES)

| | Treatment 1 | | Treatment 2 | | Treatment 3 | | Treatment 4 | |
|---|---|---|---|---|---|---|---|---|
| | X | Y | X | Y | X | Y | X | Y |
| | 1.80 | 22.0 | 2.27 | 19.5 | 2.70 | 17.0 | 2.04 | 22.5 |
| | 1.65 | 27.5 | 2.16 | 20.0 | 2.68 | 17.0 | 2.23 | 18.0 |
| | 1.89 | 22.5 | 2.25 | 19.0 | 2.36 | 17.5 | 2.61 | 18.5 |
| | 1.94 | 21.5 | 1.71 | 25.5 | 2.54 | 18.5 | 2.47 | 20.0 |
| | 1.96 | 23.0 | 2.52 | 20.0 | 2.00 | 19.0 | 2.38 | 19.5 |
| | 2.28 | 19.5 | 2.62 | 19.5 | 1.97 | 24.0 | 2.37 | 20.5 |
| | 2.10 | 19.0 | 2.45 | 18.5 | 2.07 | 20.0 | 2.37 | 21.0 |
| | 2.54 | 18.5 | 2.20 | 22.5 | 2.01 | 21.0 | 2.56 | 19.0 |
| | 2.40 | 19.0 | 1.83 | 29.0 | 2.30 | 20.0 | 2.74 | 18.0 |
| | 2.45 | 18.5 | 1.86 | 24.5 | 2.15 | 20.0 | 2.72 | 19.0 |
| $\Sigma X, \Sigma Y$ | 21.01 | 211.0 | 21.87 | 218.0 | 22.78 | 194.0 | 24.49 | 196.0 |
| $\Sigma X^2, \Sigma Y^2$ | 44.9623 | 4524.50 | 48.6709 | 4861.50 | 52.6120 | 3804.50 | 60.4053 | 3860.00 |
| $\Sigma XY$ | 436.565 | | 468.775 | | 437.640 | | 478.130 | |

The total sum of squares and cross-products are

$$\sum_i \sum_j (X_{ij} - \bar{x}.)^2 = \sum_i \sum_j X_{ij}^2 - \frac{X_{..}^2}{N} = \sum x^2$$

$$= 206.6505 - \frac{(90.15)^2}{40}$$

$$= 206.6505 - 203.1756 = 3.4749 \tag{8.2}$$

$$\sum_i \sum_j (X_{ij} - \bar{x}.)(Y_{ij} - \bar{y}.) = \sum_i \sum_j X_{ij}Y_{ij} - \frac{X_{..}Y_{..}}{N} = \sum xy$$

$$= 1821.1100 - \frac{(90.15)(819.0)}{40}$$

$$= 1821.1100 - 1845.8212 = -24.7112 \tag{8.3}$$

$$\sum_i \sum_j (Y_{ij} - \bar{y}.)^2 = \sum_i \sum_j Y_{ij}^2 - \frac{Y_{..}^2}{N} = \sum y^2$$

$$= 17,050.5000 - \frac{(819.0)^2}{40}$$

$$= 17,050.5000 - 16,769.0250 = 281.4750 \tag{8.4}$$

The among-treatment sums of squares and cross-products are

$$n_w \sum_i (\bar{x}_i - \bar{x}.)^2 = \sum_i \frac{X_{i.}^2}{n_w} - \frac{X_{..}^2}{N} = \sum x^2$$

$$= \frac{2038.4055}{10} - \frac{(90.15)^2}{40} = 203.8406 - 203.1756$$

$$= 0.6650 \tag{8.5}$$

$$n_w \sum_i (\bar{x}_i - \bar{x}.)(\bar{y}_i - \bar{y}.) = \sum_i \frac{X_{i.}Y_{i.}}{n_w} - \frac{X_{..}Y_{..}}{N} = \sum xy$$

$$= \frac{18,420.1300}{10} - \frac{(90.15)(819.0)}{40}$$

$$= 1842.0130 - 1845.8212 = -3.8082 \tag{8.6}$$

$$n_w \sum_i (\bar{y}_i - \bar{y}.)^2 = \sum_i \frac{Y_{i.}^2}{n_w} - \frac{Y_{..}^2}{N} = \sum y^2$$

$$= \frac{168,097.0000}{10} - \frac{(819.0)^2}{40}$$

$$= 16,809.7000 - 16,769.0250 = 40.6750 \tag{8.7}$$

The within-treatments sums of squares and cross-products found by differences between total and among-treatment values are

$$\sum_i \sum_j (X_{ij} - \bar{x}_i)^2 = \sum_i \sum_j X_{ij}^2 - \sum_i \frac{X_{i\cdot}^2}{n_w} = \sum x^2$$

$$= 206.6505 - 203.8406 = 2.8099 \qquad (8.8)$$

$$\sum_i \sum_j (X_{ij} - \bar{x}_i)(Y_{ij} - \bar{y}_i) = \sum_i \sum_j X_{ij} Y_{ij} - \sum_i \frac{X_{i\cdot} Y_{i\cdot}}{n_w} = \sum xy$$

$$= 1821.1100 - 1842.0130 = -20.9030 \qquad (8.9)$$

$$\sum_i \sum_j (Y_{ij} - \bar{y}_i)^2 = \sum_i \sum_j Y_{ij}^2 - \sum_i \frac{Y_{i\cdot}^2}{n_w} = \sum y^2$$

$$= 17,050.5000 - 16,809.7000 = 240.8000 \qquad (8.10)$$

The values calculated above are entered into Table 8.2, where a new section is developed to deal with the adjustment due to regression in the analysis of covariance, that labeled "Deviations from regression," where the sums of squares for the deviations from regression are calculated by the formula

$$\sum d_{y\cdot x}^2 = \sum y^2 - (\sum xy)^2 / \sum x^2 \qquad (8.11)$$

The appropriate $F$ ratio to test for the significance of the treatment effects is the mean square for adjusted means divided by the mean square for within, $F = 6.8146/2.4372 = 2.80$, with degrees of freedom of 3 and 35, a value which is very close to being significant at the 5 percent level.

Had the diameter of fruit not been taken into account, we would have had an analysis of variance using the values under the column headed $\sum y^2$ in Table 8.2. The result would have been an $F$ value of $13.5583/6.6889 = 2.03$, which is a very long way from being significant at the 5 percent level. Therefore, taking into account the diameter of the apples, we have removed the effect of some treatments having a larger average size than others, as well as removing the variation within the treatments due to diameter, thereby reducing the error term. It can be seen that by calculating the pooled regression coefficient

$$b = \sum xy / \sum x^2 = \frac{-20.9030}{2.8099} = -7.4391 \qquad (8.12)$$

that for an increase of 1 in. in diameter of the fruit there is a corresponding decrease of 7.4391 lb/in.$^2$ in the pressure measurement. The analysis of covariance allows for

**TABLE 8.2**   ANALYSIS OF COVARIANCE OF DATA DISPLAYED IN TABLE 8.1

| Source | df | Sums of squares and cross-products | | | Deviations from regression | | |
| | | $\sum x^2$ | $\sum xy$ | $\sum y^2$ | df | SS | MS |
|---|---|---|---|---|---|---|---|
| Total | 39 | 3.4749 | −24.7112 | 281.4750 | | | |
| $T$ | 3 | 0.6650 | −3.8082 | 40.6750 | | | |
| $W$ | 36 | 2.8099 | −20.9030 | 240.8000 | 35 | 85.3014 | 2.4372 |
| $T + W$ | 39 | 3.4749 | −24.7112 | 281.4750 | 38 | 105.7452 | |
| Among adjusted means | | | | | 3 | 20.4438 | 6.8146 |

**TABLE 8.3** SUMMARY TABLE OF ANALYSIS OF
COVARIANCE OF APPLE DATA

| Source | df | SS | MS | F |
|--------|-----|----------|----------|----------|
| Treatments | 3 | 20.4438 | 6.8146 | 2.80 |
| Regression | 1 | 155.4986 | 155.4986 | 63.80** |
| Error | 35 | 85.3014 | 2.4372 | |

the removal of this effect among treatment means and yields greater precision as seen by the reduction in the error sum of squares from 240.8000 to 85.3014.

Rather than putting the results of an analysis such as this in the large and awkward form of Table 8.2, we can summarize the results more simply, as shown in Table 8.3. The sum of squares due to regression

$$\text{SS}_b = (\sum xy)^2 / \sum x^2 = \frac{(-20.9030)^2}{2.8099} = 155.4986 \tag{8.13}$$

and the pooled regression coefficient shown in expression (8.12) are calculated from the error line of Table 8.2.

Since our test of significance is for differences among adjusted treatment means, we need to calculate these adjusted means to evaluate them properly. This adjustment makes use of the regression coefficient of $-7.4391$ which we have already calculated. The formula for calculating an adjusted mean is

$$\text{Adj } \bar{y}_i = \bar{y}_i - b(\bar{x}_i - \bar{x}.) \tag{8.14}$$

It is often desirable to present the means together with their standard errors to yield an estimate of the amount of variation involved with each mean. Formula (8.15) shows the method of calculating the standard errors,

$$s_{\text{Adj } y_i} = s \sqrt{\frac{1}{n_w} + \frac{(\bar{x}_i - \bar{x}.)}{\sum x^2}} \tag{8.15}$$

where $s$ is the square root of the error mean square, 2.4372, and $\sum x^2$ is obtained from the error line in the covariance table (Table 8.2), 2.8099. Table 8.4 presents the adjusted means accompanied by their standard errors. The overall mean is the average of the adjusted means, 20.475, since each treatment group having a larger than average mean of the $X$'s is adjusted upward because of the negative regression, while the treatment mean having a smaller than average mean of the $X$'s is adjusted downward. We are

**TABLE 8.4** ADJUSTED MEANS AND STANDARD ERRORS
FOR DATA DISPLAYED IN TABLE 8.1

| Adjusted mean | Standard error |
|---------------|----------------|
| Adj $\bar{y}_1 = 21.10 - (-7.4391)(2.101 - 2.2538) = 19.9633$ | 0.5138 |
| Adj $\bar{y}_2 = 21.80 - (-7.4391)(2.187 - 2.2538) = 21.3031$ | 0.4976 |
| Adj $\bar{y}_3 = 19.40 - (-7.4391)(2.278 - 2.2538) = 19.5800$ | 0.4942 |
| Adj $\bar{y}_4 = 19.60 - (-7.4391)(2.449 - 2.2538) = 21.0521$ | 0.5262 |

then estimating what the treatment means would have been if all the treatments had the same average $X$ values, thus removing the influence of the initial $X$ values from the conclusion as to the treatment effects.

Although treatments with the same number of observations in the one-way classification analysis of variance have the same standard error, this does not hold with the one-way (or multiway) analysis of covariance. As formula (8.15) demonstrates, the further that any treatment mean of the $X$'s is from the overall mean of the $X$'s, $\bar{x}_i - \bar{x}.$, the greater is the standard error.

While there may be slight or great differences among the average $X$'s for the treatments, the primary value of the covariance analysis is in the reduction in the error mean square. If there is a significant effect of regression, then a relatively large sum of squares will be removed from the error term that would otherwise be used in the analysis of variance. In the present example, Table 8.2 shows the error mean squares for the analysis of variance would be $240.8000/36 = 6.6889$ as compared to $2.4372$ in the analysis of covariance. In the analysis of variance the $F$ value would be $2.03$, a value that does not approach significance, as noted earlier. In spite of the smaller mean square among treatments due to the adjustment for regression, the $F$ value in the covariance analysis was extremely close to significance at the 5 percent level.

A particular feature of the calculations in the analysis of covariance is the addition of the treatment and error sums of squares and cross-products. If we had analyzed the data under the model

$$Y_i = \mu + \beta(X_i - \bar{x}.) + e_i \qquad (8.16)$$

our error sum of squares of deviations from regression would have been $105.7452$. However, with the model shown in Equation (8.1), which includes the treatment effects, the error sum of squares of deviations from regression is $85.3014$. The difference of $20.4438$ between these two error terms is the difference in reductions due to fitting the two models—the difference in reductions being the sum of squares due to treatments. The same situation exists in multiway classifications, the sums of squares for various classifications found by differences in reductions due to different models, reflected in the differences in error sums of squares.

The initial analysis of covariance, like the initial analysis of variance, does not complete the extraction of information from the data. If differences among treatments are significant, or approach significance, the usual mean separation procedures discussed previously, such as single degree of freedom comparisons, partitioning into linear, quadratic, cubic, and higher-degree components, or range tests, may be called upon. However, these procedures become more difficult and tedious with the introduction of the independent variable $X$, so the wisest course is to use readily available computer programs such as PROC GLM or PROC HARVEY in the Statistical Analysis System (SAS) or Harvey's "Least-Squares and Maximum Likelihood Mixed Model" program LSML76 (Harvey, 1977b). Options available in these programs allow for such procedures and remove the extended drudgery involved in the use of a desk or pocket calculator.

The appropriate expectations of mean squares for the analysis of covariance are those that would be obtained were the independent variable $X$ not included in the model except for the coefficients of the variance components other than the error component. While the example presented in this section has been one in which there were equal numbers in the different treatment groups, the same methods would be followed were the numbers not equal for the one-way classification.

## 8.3 COVARIANCE WITH CURVILINEAR REGRESSION

In the analysis of covariance it is possible to include the quadratic or cubic effect of the independent continuous variable. Degrees higher than cubic would probably be meaningless. It is a very time-consuming operation to do this by hand but the method will be demonstrated for the quadratic effect in the previous example for illustrative purposes. The model including the quadratic effect of $X$ is

$$Y_{ij} = \mu + t_i + \beta_l(X_{1ij} - \bar{x}_{1.}) + \beta_q(X_{2ij} - \bar{x}_{2.}) + e_{ij} \tag{8.17}$$

where $X_1 = X$ and $X_2 = X^2$ and the extra subscript is used to distinguish between $X_1$ and $X_2$.

As with the model including only the linear effect of $X$, each line of the covariance table must be adjusted for regression, in this case for both linear and quadratic regressions. The adjustment requires setting up normal equations for the linear and quadratic effects for each line. Thus, $\sum x_2^2$, $\sum x_1 x_2$, and $\sum x_2 y$ values must be calculated for the total, treatment, and error lines. The error values can be found by subtracting the treatment values from the total values. A new table of sums of squares and cross-products is now developed (Table 8.5), showing the previously calculated values as well as the new.

The normal equations needed to adjust the sums of squares for the linear and quadratic effects are

$$\sum x_1^2 b_l + \sum x_1 x_2 b_q = \sum x_1 y$$
$$\sum x_1 x_2 b_l + \sum x_2^2 b_q = \sum x_2 y \tag{8.18}$$

We can now solve these equations by inverting the coefficient matrix

$$\begin{bmatrix} \sum x_1^2 & \sum x_1 x_2 \\ \sum x_1 x_2 & \sum x_2^2 \end{bmatrix} \text{ to give } \begin{bmatrix} c^{11} & c^{12} \\ c^{21} & c^{22} \end{bmatrix}$$

where
$$c^{11} = \frac{\sum x_2^2}{\sum x_1^2 \sum x_2^2 - (\sum x_1 x_2)^2} \qquad c^{22} = \frac{\sum x_1^2}{\sum x_1^2 \sum x_2^2 - (\sum x_1 x_2)^2}$$

$$c^{12} = \frac{-\sum x_1 x_2}{\sum x_1^2 \sum x_2^2 - (\sum x_1 x_2)^2} \qquad c^{21} = c^{12}$$

**TABLE 8.5**  SUMS OF SQUARES AND CROSS-PRODUCTS FOR MODEL (8.17)

| Source | df | $\sum x_1^2$ | $\sum x_2^2$ | $\sum x_1 x_2$ | $\sum x_1 y$ | $\sum x_2 y$ | $\sum y^2$ |
|--------|-----|--------------|--------------|----------------|--------------|--------------|------------|
| Total | 39 | 3.4749 | 69.2494 | 15.4784 | −24.7112 | −107.3831 | 281.4750 |
| $T$ | 3 | 0.6650 | 13.1181 | 2.9532 | −3.8082 | −16.8222 | 40.6750 |
| $W$ | 36 | 2.8099 | 56.1313 | 12.5252 | −20.9030 | −90.5609 | 240.8000 |
| $T + W$ | 39 | 3.4749 | 69.2494 | 15.4784 | −24.7112 | −107.3831 | 281.4750 |

The regression coefficients are calculated as

$$b_l = c^{11} \sum x_1 y + c^{12} \sum x_2 y \quad \text{and} \quad b_q = c^{21} \sum x_1 y + c^{22} \sum x_2 y \qquad (8.19)$$

For the error line, we would have the equations

$$2.8099b_l + 12.5252b_q = -20.9030$$

$$12.5252b_l + 56.1313b_q = -90.5609$$

and inverting the coefficient matrix, we have

$$c^{11} = 66.60848893 \qquad c^{12} = -14.86309146 \qquad c^{22} = 3.33438194$$

Performing the multiplications shown in Equations (8.19), we find

$$b_l = -46.30230500 \quad \text{and} \quad b_q = 8.71857140$$

and the reduction in sum of squares due to the linear and quadratic regressions is

$$R(l, q) = b_l \sum x_1 y + b_q \sum x_2 y$$

$$= (-46.302305)(-20.9030) + (8.718571)(-90.5609) = 178.2954 \qquad (8.20)$$

which is subtracted from $\sum y^2$ in the error $(W)$ line, giving an adjusted sum of squares for error of

$$240.8000 - 178.2954 = 62.5046$$

The normal equations for the adjustment to the treatment plus error $(T + W)$ line are

$$3.4749b_l + 15.4784b_q = -24.7112$$

$$15.4784b_l + 69.2494b_q = -107.3831$$

and solving these equations, we find

$$b_l = -46.60635100 \quad \text{and} \quad b_q = 8.86662810$$

and the reduction due to the linear and quadratic regressions is

$$R(l, q) = (-46.60635100)(-24.7112) + (8.86662810)(-107.3831) = 199.5728$$

which is subtracted from $\sum y^2$ in the $T + W$ line giving an adjusted sum of squares of

$$281.4750 - 199.5728 = 81.9022$$

While the number of decimal places may seem excessive, they are necessary to prevent large rounding errors that would otherwise be involved in the calculations.

The adjusted sum of squares for treatments is the difference between the two adjusted sums of squares just obtained,

$$\text{Adj SS}_T = 81.9022 - 62.5046 = 19.3976$$

The sum of squares for linear regression has already been obtained and the sum of squares due to quadratic regression is found as the difference between the reduction due to fitting the linear and quadratic regressions and the reduction due to fitting the linear regression in the error lines,

**TABLE 8.6**　SUMMARY TABLE OF COVARIANCE ANALYSIS
OF MODEL (8.17)

| Source | df | SS | MS | F |
|---|---|---|---|---|
| Treatments | 3 | 19.3976 | 6.4659 | 3.52* |
| Linear regression | 1 | 155.4986 | 155.4986 | 84.58** |
| Quadratic regression | 1 | 22.7968 | 22.7968 | 12.40** |
| Error | 34 | 62.5046 | 1.8384 | |

$$SS_q = 178.2954 - 155.4986 = 22.7968$$

The completed analysis of covariance is shown in Table 8.6, where it can be seen that the quadratic effect of the diameter of the fruit did have a significant effect, resulting in a smaller error mean square than the previous analysis which included only the linear effect. The linear effect mean square, of course, remains the same, showing a highly significant effect, and the treatments now show significant differences.

Since the quadratic effect showed significance, it should be left in the model and new adjusted means would be necessary. The formula for calculating the adjusted mean is

$$\text{Adj } \bar{y}_i = \bar{y}_i - b_l(\bar{x}_{1i} - \bar{x}_1.) - b_q(\bar{x}_{2i} - \bar{x}_2.) \tag{8.21}$$

where the regression coefficients are those found earlier by Equations (8.19), $b_l = -46.302305$, and $b_q = 8.718571$. The overall mean for $X_2$ is the average of the $\Sigma X_{1ij}^2$, $206.6505/40 = 5.1663$, and the individual means for $X_2$ are the averages of the $X_{1ij}^2$ for each treatment.

If the standard errors of the mean are desired, they can be found by formula (8.22),

$S_{(\text{Adj } \bar{y}_i)}$

$$= s \sqrt{\frac{1}{n_w} + c^{11}(\bar{x}_{1i} - \bar{x}_1.)^2 + c^{22}(\bar{x}_{2i} - \bar{x}_2.)^2 + 2c^{12}(\bar{x}_{1i} - x_1.)(\bar{x}_{2i} - x_2.)} \tag{8.22}$$

where $s = \sqrt{1.8384}$, the error mean square from Table 8.6. The inverse elements are those from the error line, $c^{11} = 66.608489$, $c^{12} = -14.863091$, and $c^{22} = 3.334382$. Table 8.7 presents the adjusted means together with their standard errors. The overall mean is obtained as before as an average of the adjusted means, 20.473.

**TABLE 8.7**　ADJUSTED MEANS AND
STANDARD ERRORS FOR
DATA ANALYZED
UNDER MODEL (8.17)

| Adjusted mean | Standard error |
|---|---|
| $\bar{y}_1 = 19.867$ | 0.447 |
| $\bar{y}_2 = 21.316$ | 0.432 |
| $\bar{y}_3 = 19.693$ | 0.430 |
| $\bar{y}_4 = 21.016$ | 0.448 |

## 8.4 ONE-WAY CLASSIFICATION: SPLIT-PLOT DESIGN

The data used to demonstrate an analysis of covariance with a one-way classification, split-plot design were derived from an experiment conducted by Dr. Robert A. Coler of the Department of Environmental Sciences at the University of Massachusetts and are presented in Table 8.8. In the experiment, two locations in a river were sampled, with three sampling tubes at each location. While the sampling tubes were placed at the inner, middle, and outer portions of the locations, they were measuring essentially the same thing, with the position of the tube being of minor importance. Hence, the tubes were considered to be a random classification and the locations and days, or dates, as fixed classifications. The tubes were sampled at 10 different dates during the months of August and September. The experiment is a split-plot design with the tubes being the whole plots and the days being the split-plot effect.

The mathematical model for these data is

$$Y_{ijk} = \mu + l_i + t_{ij} + d_k + ld_{ik} + \beta(X_{ijk} - \bar{x}...) + e_{ijk} \qquad (8.23)$$

where the $l_i$ represent location effects, the $t_{ij}$ represent tubes within locations effects, the $d_k$ represent effects of the days, and $b$ represents the regression of milligrams of oxygen on rate of flow.

Since the calculations for total, location, and days follow the same form as those shown in Section 8.2, the details of those calculations are not shown. Also, the values for the error line can be found by subtracting the sums of squares and cross-products for all the effects in the model from the total sums of squares and cross-products. The formulas and calculations necessary for the sums of squares and cross-products for the tubes within locations and the interaction of locations and days are shown in formulas (8.24)–(8.29).

The sums of squares and cross-products for tubes within locations are as follows:

**TABLE 8.8** MEASUREMENTS (Y, IN MILLIGRAMS) OF OXYGEN GENERATED AT TWO LOCATIONS IN A RIVER AND FLOW RATE (X, IN MILLILITERS/SECOND)

| | | Location 1 | | | | | | Location 2 | | | | |
| | | Inner | | Middle | | Outer | | Inner | | Middle | | Outer | |
| Day | X | Y | X | Y | X | Y | X | Y | X | Y | X | Y |
|---|---|---|---|---|---|---|---|---|---|---|---|---|
| 1 | 4.75 | 1.3 | 3.13 | 1.5 | 2.06 | 2.2 | 3.17 | 2.3 | 5.56 | 1.6 | 5.56 | 1.5 |
| 3 | 3.45 | 0.7 | 3.33 | 0.3 | 3.85 | 1.1 | 2.78 | 1.0 | 4.76 | 1.0 | 4.55 | 1.1 |
| 4 | 3.13 | 0.4 | 3.45 | 0.3 | 4.76 | 0.3 | 2.86 | 1.3 | 5.88 | 0.5 | 4.55 | 0.7 |
| 9 | 7.69 | 0.3 | 4.00 | 0.3 | 6.67 | 0.2 | 4.54 | 1.4 | 7.69 | 0.8 | 5.56 | 0.9 |
| 10 | 8.33 | 0.4 | 4.35 | 0.8 | 9.09 | 0.6 | 5.26 | 1.6 | 8.33 | 1.2 | 6.67 | 1.5 |
| 11 | 5.88 | 1.1 | 4.55 | 1.3 | 6.25 | 1.1 | 4.35 | 3.0 | 5.88 | 2.1 | 4.76 | 2.2 |
| 13 | 4.55 | 1.0 | 3.23 | 1.0 | 3.33 | 1.1 | 3.03 | 1.9 | 3.85 | 1.8 | 4.55 | 1.1 |
| 15 | 5.26 | 1.3 | 2.70 | 1.9 | 4.17 | 1.8 | 3.03 | 2.8 | 3.45 | 2.2 | 3.45 | 2.4 |
| 16 | 4.55 | 1.3 | 5.26 | 1.1 | 4.55 | 1.1 | 2.63 | 3.1 | 2.86 | 3.3 | 3.56 | 1.9 |
| 20 | 6.67 | 0.9 | 6.25 | 0.8 | 4.76 | 0.9 | 4.76 | 2.5 | 5.88 | 1.9 | 6.67 | 1.2 |
| $\Sigma X, \Sigma Y$ | 54.26 | 8.7 | 40.25 | 9.3 | 49.49 | 10.4 | 36.41 | 20.9 | 54.14 | 16.4 | 49.88 | 14.5 |
| $\Sigma XY$ | 45.255 | | 36.470 | | 44.316 | | 75.905 | | 80.222 | | 70.064 | |

$$\sum x^2 = \frac{\Sigma_i \, \Sigma_j \, X_{ij\cdot}^2}{n_d} - \frac{\Sigma_i \, X_{i\cdot\cdot}^2}{n_t n_d}$$

$$= \frac{(54.26)^2 + (40.25)^2 + \cdots + (49.88)^2}{10} - \frac{(144.00)^2 + (140.43)^2}{30}$$

$$= 1375.8312 - 1348.5528 = 27.2784 \tag{8.24}$$

$$\sum xy = \frac{\Sigma_i \, \Sigma_j \, X_{ij\cdot} \cdot Y_{ij\cdot}}{n_d} - \frac{\Sigma_i \, X_{i\cdot\cdot} \cdot Y_{i\cdot\cdot}}{n_t n_d}$$

$$= \frac{(54.26)(8.7) + (40.25)(9.3) + \cdots + (49.88)(14.5)}{10}$$

$$\quad - \frac{(144.00)(28.4) + (140.43)(51.8)}{30}$$

$$= 373.3208 - 378.7958 = -5.4570 \tag{8.25}$$

$$\sum y^2 = \frac{\Sigma_i \, \Sigma_j \, Y_{ij\cdot}^2}{n_d} - \frac{\Sigma_i \, Y_{i\cdot\cdot}^2}{n_t n_d}$$

$$= \frac{(8.7)^2 + (9.3)^2 + \cdots + (14.5)^2}{10} - \frac{(28.4)^2 + (51.8)^2}{30}$$

$$= 118.6360 - 116.3267 = 2.3093 \tag{8.26}$$

The sums of squares and cross-products for the location $\times$ day interaction are:

$$\sum x^2 = \frac{\Sigma_i \, \Sigma_k \, X_{i\cdot k}^2}{n_t} - \frac{\Sigma_i \, X_{i\cdot\cdot}^2}{n_t n_d} - \frac{\Sigma_k \, X_{\cdot\cdot k}^2}{n_l n_t} + \frac{X_{\cdot\cdot\cdot}^2}{N}$$

$$= 1436.1462 - 1348.5528 - 1425.5473 + 1348.3404 = 10.3865 \tag{8.27}$$

$$\sum xy = \frac{\Sigma_i \, \Sigma_k \, X_{i\cdot k} Y_{i\cdot k}}{n_t} - \frac{\Sigma_i \, X_{i\cdot\cdot} \cdot Y_{i\cdot\cdot}}{n_t n_d} - \frac{\Sigma_k \, X_{\cdot\cdot k} Y_{\cdot\cdot k}}{n_l n_t} + \frac{X_{\cdot\cdot\cdot} \cdot Y_{\cdot\cdot\cdot}}{N}$$

$$= 362.2990 - 378.7958 - 368.4185 + 380.1881 = -4.7272 \tag{8.28}$$

$$\sum y^2 = \frac{\Sigma_i \, \Sigma_k \, Y_{i\cdot k}^2}{n_t} - \frac{\Sigma_i \, Y_{i\cdot\cdot}^2}{n_t n_d} - \frac{\Sigma_k \, Y_{\cdot\cdot k}^2}{n_l n_t} + \frac{Y_{\cdot\cdot\cdot}^2}{N}$$

$$= 134.8733 - 116.3267 - 123.1933 + 107.2007 = 2.5540 \tag{8.29}$$

The values calculated above, as well as those noted earlier but not shown, are entered in Table 8.9, the initial analysis of covariance table. The same procedure as in the one-way classification without a split-plot effect is followed to obtain the adjusted sum of squares for each line in the analysis of covariance table. The sums of squares and cross-products for each line are added in turn to the error sums of squares and cross-products, from which the deviations from regression are found. The subtraction of the deviation sum of squares due to the combined values from the error line yields the appropriate adjusted sum of squares.

**TABLE 8.9**  ANALYSIS OF COVARIANCE OF OXYGEN GENERATED DATA

| Source | df | Sums of squares and cross-products | | | Deviations from regression | | |
|--------|-----|----------|----------|----------|-----|---------|--------|
|        |    | $\Sigma\, x^2$ | $\Sigma\, xy$ | $\Sigma\, y^2$ | df | SS | MS |
| Total | 59 | 147.1549 | −27.9561 | 32.8793 | | | |
| *L* | 1 | 0.2124 | −1.3923 | 9.1260 | | | |
| *T:L* | 4 | 27.2784 | −5.4750 | 2.3093 | | | |
| *D* | 9 | 77.2069 | −11.7696 | 15.9926 | | | |
| *LD* | 9 | 10.3865 | −4.7272 | 2.5540 | | | |
| Error (*E*) | 36 | 32.0707 | −4.5920 | 2.8974 | 35 | 2.2399 | 0.0640 |
| *L + E* | 37 | 32.2831 | −5.9843 | 12.0234 | 36 | 10.9141 | |
| *L* adjusted | | | | | 1 | 8.6742 | 8.6742 |
| *T:L + E* | 40 | 59.3491 | −10.0670 | 5.2067 | 39 | 3.4991 | |
| *T:L* adjusted | | | | | 4 | 1.2592 | 0.3148 |
| *D + E* | 45 | 109.2776 | −16.3616 | 18.8900 | 44 | 16.4403 | |
| *D* adjusted | | | | | 9 | 14.2004 | 1.5778 |
| *LD + E* | 45 | 42.4572 | −9.3192 | 5.4514 | 44 | 3.4059 | |
| *LD* adjusted | | | | | 9 | 1.1660 | 0.1296 |

The summary of the analysis of covariance is shown in Table 8.10. The sum of squares due to regression shown in the summary table is calculated from the error line of Table 8.9 as

$$SS_b = \frac{(\Sigma\, xy)^2}{\Sigma\, x^2} = \frac{(-4.5920)^2}{32.0707} = 0.6575$$

The analysis shows that there is a highly significant difference between the two locations in the river, highly significant differences among days, and a highly significant regression of oxygen generated on flow rate. One is normally interested in the trend over time in such an experiment, so it would be logical to partition the sum of squares for days into at least linear, quadratic, and cubic components. This could be done most conveniently by a computer program designed to perform this type of partitioning, such as Harvey's mixed-model least-squares computer program, LSML76 (Harvey, 1977b).

The adjusted means and standard errors for all classes and subclasses can be calculated as shown by formulas (8.14) and (8.15). Only the adjusted means for lo-

**TABLE 8.10**  ANALYSIS OF COVARIANCE OF OXYGEN
GENERATED DATA: SUMMARY TABLE

| Source | df | SS | MS | *F* |
|--------|-----|--------|--------|----------|
| Whole plots | 5 | | | |
| *L* | 1 | 8.6742 | 8.6742 | 27.55** |
| *T:L* | 4 | 1.2592 | 0.3148 | |
| Split plots | 54 | | | |
| *D* | 9 | 14.2004 | 1.5778 | 24.65** |
| *LD* | 9 | 1.1660 | 0.1296 | 2.02 |
| Regression | 1 | 0.6575 | 0.6575 | 10.27** |
| Error | 35 | 2.2399 | 0.0640 | |

cations (together with the standard errors) have been calculated and they are shown in Table 8.11.

As with the example of covariance in the one-way classification, the quadratic or cubic effect of the continuous independent variable might be included in the analysis if found to be significant. The methods shown for this in the one-way classification would apply here.

The analysis that we have completed has made the assumption that the regression in the whole-plot portion of the analysis does not differ from the regression in the split-plot portion. This would normally be true unless the treatments have an effect on the covariate.

In the preceding example, the whole-plot regression coefficient would be calculated from the $T{:}L$ (or whole-plot error) line as

$$b_{wp} = \frac{\Sigma \ xy}{\Sigma \ x^2}$$

$$= \frac{-5.4750}{27.2784} = -0.2007 \tag{8.30}$$

and the split-plot regression coefficient would be calculated from the split-plot error line as

$$b_{sp} = \frac{\Sigma \ xy}{\Sigma \ x^2}$$

$$= \frac{-4.5920}{32.0707} = -0.1432 \tag{8.31}$$

While there would seem to be no reason to suspect that these coefficients are significantly different, a test of this difference can be made (Winer, 1971) as

$$t' = \frac{b_{wp} - b_{sp}}{\sqrt{s_1^2 + s_2^2}} \tag{8.32}$$

where $s_1^2$ and $s_2^2$ are the error variances for $b_{wp}$ and $b_{sp}$, respectively.

Now,

$$s_1^2 = \frac{s_{wp}^2}{\Sigma \ x^2} \tag{8.33}$$

$$s_{wp}^2 = \frac{\Sigma \ d_{y \cdot x}^2}{(n_t - 1)n_l - 1} \tag{8.34}$$

$$\Sigma \ d_{y \cdot x}^2 = \Sigma \ y^2 - \frac{(\Sigma \ xy)^2}{\Sigma \ x^2} \tag{8.35}$$

**TABLE 8.11**  ADJUSTED MEANS AND STANDARD ERRORS FOR LOCATIONS

|  | Adjusted mean |  | Standard error |
|---|---|---|---|
| Adj $\bar{y}_1..$ | = 0.9467 − (−0.1432)(0.0595) | = 0.9552 | 0.462 |
| Adj $\bar{y}_2..$ | = 1.7267 − (−0.1432)(−0.0595) | = 1.7182 | 0.462 |

Solving for these values, we have

$$\Sigma\, d_{y \cdot x}^2 = \Sigma\, y^2 - \frac{(\Sigma\, xy)^2}{\Sigma\, x^2} = 2.3093 - \frac{(-5.4750)^2}{27.2784}$$

$$= 2.3093 - 1.0989 = 1.2104$$

$$s_{wp}^2 = \frac{\Sigma\, d_{y \cdot x}^2}{(n_t - 1)n_l - 1} = \frac{1.2104}{3} = 0.4035$$

$$s_1^2 = \frac{s_{wp}^2}{\Sigma\, x^2} = \frac{0.4035}{27.2784} = 0.0148$$

Similarly, using values from the split-plot error line, we have

$$s_2^2 = \frac{s_{sp}^2}{\Sigma\, x^2} \tag{8.36}$$

$$s_{sp}^2 = \frac{\Sigma\, d_{y \cdot x}^2}{(n_t - 1)(n_d - 1)n_l - 1} \tag{8.37}$$

$$\Sigma\, d_{y \cdot x}^2 = \Sigma\, y^2 - \frac{(\Sigma\, xy)}{\Sigma\, x^2} \tag{8.38}$$

Solving for these values we have

$$\Sigma\, d_{y \cdot x}^2 = \Sigma\, y^2 - \frac{(\Sigma\, xy)^2}{\Sigma\, x^2} = 2.8974 - \frac{(-4.5920)^2}{32.0707}$$

$$= 2.8974 - 0.6575 = 2.2399$$

$$s_{sp}^2 = \frac{\Sigma\, d_{y \cdot x}^2}{(n_t - 1)(n_d - 1)n_l - 1} = \frac{2.2399}{35} = 0.0640$$

$$s_2^2 = \frac{s_{sp}^2}{\Sigma\, x^2} = \frac{0.0640}{32.0707} = 0.0020$$

The $t'$ test would now be

$$t' = \frac{b_{wp} - b_{sp}}{\sqrt{s_1^2 + s_2^2}} = \frac{-0.2007 + 0.1432}{\sqrt{0.0148 + 0.0020}} = \frac{-0.0575}{0.1296} = 0.44$$

The sampling distribution of $t'$ is not that of the usual $t$ but may be approximated by the usual $t$ distribution with degrees of freedom of $f$, where

$$f = \frac{(s_1^2 + s_2^2)^2}{s_1^4/f_{wp} + s_2^4/f_{sp}} \tag{8.39}$$

where $f_{wp}$ and $f_{sp}$ are the degrees of freedom for $s_{wp}^2$ and $s_{sp}^2$, respectively. In our example,

$$f = \frac{(0.0148 + 0.0020)^2}{0.00007301 + 0.00000011} = \frac{0.000282}{0.000073} = 3.86$$

The $t'$ statistic is obviously not significant, but if it had been then the whole-plot effects should have been tested by an error term adjusted for the whole-plot regression and the whole-plot means adjusted by the whole-plot regression. If the regressions are not different then the split-plot regression would be used throughout the analysis.

## 8.5  ONE-WAY CLASSIFICATION: SPLIT-PLOT DESIGN AND SINGLE COVARIATE PER WHOLE PLOT

In the example discussed in Section 8.4, each $Y$ value had a corresponding $X$ value. However, there are occasions where several $Y$ values are associated with the same $X$ value. Suppose that in a one-way or multiway classification, a measurement such as weight, height, intelligence quotient, or speed of movement is made on each individual (the experimental unit). If the individuals are subjected to different treatments and their performances measured over time, each individual would have only one covariate measurement but several observations over time. In such a situation only the between-individuals (whole-plot) comparisons are adjusted for the covariate. The within individual or split-plot comparisons will have no adjustment for the covariate.

As an example of this type of design, let us make use of the data shown in Table 8.8, only this time we assume that only one $X$ value, or flow rate measurement, was made for each tube. We can take the average of the flow rate for each tube to be the single assumed $X$ value. This would result in data as shown in Table 8.12.

The mathematical model for this design is

$$Y_{ijk} = \mu + l_i + \beta(X_{ij} - \bar{x}..) + h_{ij} + d_k + ld_{ik} + e_{ijk} \qquad (8.40)$$

where $h_{ij}$ is the whole-plot error and $e_{ijk}$ is the split-plot error.

Since we now have only six different $X$ values, the total sums of squares for $X$

**TABLE 8.12**  MEASUREMENTS OF OXYGEN ($Y$, IN MILLIGRAMS) GENERATED AT TWO LOCATIONS IN A RIVER AND FLOW RATE ($X$, IN MILLILITERS/SECOND)

| | | Location 1 | | | | | | Location 2 | | | | |
| | | Inner | | Middle | | Outer | | Inner | | Middle | | Outer | |
| Day | X | Y | X | Y | X | Y | X | Y | X | Y | X | Y |
|---|---|---|---|---|---|---|---|---|---|---|---|---|
| 1 | 5.426 | 1.3 | 4.025 | 1.5 | 4.949 | 2.2 | 3.641 | 2.3 | 5.414 | 1.6 | 4.988 | 1.5 |
| 3 | 5.426 | 0.7 | 4.025 | 0.3 | 4.949 | 1.1 | 3.641 | 1.0 | 5.414 | 1.0 | 4.988 | 1.1 |
| 4 | 5.426 | 0.4 | 4.025 | 0.3 | 4.949 | 0.3 | 3.641 | 1.3 | 5.414 | 0.5 | 4.988 | 0.7 |
| 9 | 5.426 | 0.3 | 4.025 | 0.3 | 4.949 | 0.2 | 3.641 | 1.4 | 5.414 | 0.8 | 4.988 | 0.9 |
| 10 | 5.426 | 0.4 | 4.025 | 0.8 | 4.949 | 0.6 | 3.641 | 1.6 | 5.414 | 1.2 | 4.988 | 1.5 |
| 11 | 5.426 | 1.1 | 4.025 | 1.3 | 4.949 | 1.1 | 3.641 | 3.0 | 5.414 | 2.1 | 4.988 | 2.2 |
| 13 | 5.426 | 1.0 | 4.025 | 1.0 | 4.949 | 1.1 | 3.641 | 1.9 | 5.414 | 1.8 | 4.988 | 1.1 |
| 15 | 5.426 | 1.3 | 4.025 | 1.9 | 4.494 | 1.8 | 3.641 | 2.8 | 5.414 | 2.2 | 4.988 | 2.4 |
| 16 | 5.426 | 1.3 | 4.025 | 1.1 | 4.949 | 1.1 | 3.641 | 3.1 | 5.414 | 3.3 | 4.988 | 1.9 |
| 20 | 5.426 | 0.9 | 4.025 | 0.8 | 4.949 | 0.9 | 3.641 | 2.5 | 5.414 | 1.9 | 4.988 | 1.2 |
| $\Sigma X, \Sigma Y$ | 54.260 | 8.7 | 40.250 | 9.3 | 49.490 | 10.4 | 36.410 | 20.9 | 54.140 | 16.4 | 49.880 | 14.5 |
| $\Sigma XY$ | 47.2062 | | 37.4325 | | 51.4696 | | 76.0969 | | 88.7896 | | 72.3260 | |

and the sum of the cross-products of $X$ and $Y$ change from the previous analysis. We calculate the values for the total sums of squares and cross-products as

$$\sum x^2 = \sum_i \sum_j \sum_k X_{ijk}^2 - \frac{X_{\cdot\cdot\cdot}^2}{N}$$

$$= 1375.8312 - \frac{(284.43)^2}{60} = 27.4908 \tag{8.41}$$

$$\sum xy = \sum_i \sum_j \sum_k X_{ijk} Y_{ijk} - \frac{X_{\cdot\cdot\cdot} Y_{\cdot\cdot\cdot}}{N}$$

$$= 373.3208 - \frac{(284.43)(80.2)}{60} = -6.8673 \tag{8.42}$$

$$\sum y^2 = \sum_i \sum_j \sum_k Y_{ijk}^2 - \frac{Y_{\cdot\cdot\cdot}^2}{N}$$

$$= 140.08 - \frac{(80.2)^2}{60} = 32.8793 \tag{8.43}$$

The sums of squares and cross-products for the locations and tubes within locations are exactly as before but there are drastic changes for the days, location $\times$ day interaction, and error. The sums of squares and cross-products for the days are

$$\sum x^2 = \frac{\sum_k X_{\cdot\cdot k}^2}{n_l n_t} - \frac{X_{\cdot\cdot\cdot}^2}{N}$$

$$= \frac{(28.443)^2 + (28.443)^2 + \cdots + (28.443)^2}{6} - \frac{(284.43)^2}{60}$$

$$= 1348.3404 - 1348.3404 = 0 \tag{8.44}$$

$$\sum xy = \frac{\sum_k Y_{\cdot\cdot k} Y_{\cdot\cdot k}}{n_l n_t} - \frac{X_{\cdot\cdot\cdot} Y_{\cdot\cdot\cdot}}{N}$$

$$= \frac{(28.443)(10.4) + (28.443)(5.2) + \cdots + (28.443)(8.2)}{6}$$

$$- \frac{(284.43)(80.2)}{60}$$

$$= 380.1881 - 380.1881 = 0 \tag{8.45}$$

$$\sum y^2 = \frac{\sum_k Y_{\cdot\cdot k}^2}{n_l n_t} - \frac{Y_{\cdot\cdot\cdot}^2}{N}$$

$$= \frac{(10.4)^2 + (5.2)^2 + \cdots + (8.2)^2}{6} - \frac{(80.2)^2}{60}$$

$$= 123.1933 - 107.2007 = 15.9926 \tag{8.46}$$

**TABLE 8.13**  ANALYSIS OF COVARIANCE OF OXYGEN GENERATED DATA: SINGLE $X$ PER TUBE

| Source | df | $\Sigma x^2$ | $\Sigma xy$ | $\Sigma y^2$ | df | SS | MS |
|--------|-----|----------|----------|----------|-----|--------|--------|
| | | Sums of squares and cross-products | | | Deviations from regression | | |
| Total | 59 | 27.4908 | −6.8673 | 32.8793 | | | |
| $L$ | 1 | 0.2124 | −1.3923 | 9.1260 | | | |
| $T{:}L$ | 4 | 27.2784 | −5.4750 | 2.3093 | 3 | 1.2104 | 0.4035 |
| $L + T{:}L$ | 5 | 27.4908 | −6.8673 | 11.4353 | 4 | 9.7198 | |
| Adjusted $L$ | | | | | 1 | 8.5094 | 8.5094 |
| $D$ | 9 | 0 | 0 | 15.9926 | | | |
| $LD$ | 9 | 0 | 0 | 2.5540 | | | |
| $TD{:}L$ | 36 | 0 | 0 | 2.8974 | | | |

The sums of squares and cross-products for the location $\times$ day interaction are

$$\sum x^2 = \frac{\Sigma_i \Sigma_k X_{i \cdot k}^2}{n_t} - \frac{\Sigma_i X_{i \cdot \cdot}^2}{n_t n_d} - \frac{\Sigma_k X_{\cdot \cdot k}^2}{n_l n_t} + \frac{X_{\cdot \cdot \cdot}^2}{N}$$

$$= 1348.5528 - 1348.5528 - 1348.3404 + 1348.3404 = 0 \qquad (8.47)$$

$$\sum xy = \frac{\Sigma_i \Sigma_k X_{i \cdot k} Y_{i \cdot k}}{n_t} - \frac{\Sigma_i X_{i \cdot \cdot} Y_{i \cdot \cdot}}{n_t n_d} - \frac{\Sigma_k Y_{\cdot \cdot k} Y_{\cdot \cdot k}}{n_l n_t} + \frac{X_{\cdot \cdot \cdot} Y_{\cdot \cdot \cdot}}{N}$$

$$= 378.7958 - 378.7958 - 380.1881 + 380.1881 = 0 \qquad (8.48)$$

$$\sum y^2 = \frac{\Sigma_i \Sigma_k Y_{i \cdot k}^2}{n_t} - \frac{\Sigma_i Y_{i \cdot \cdot}^2}{n_t n_d} - \frac{\Sigma_k Y_{\cdot \cdot k}^2}{n_l n_t} + \frac{Y_{\cdot \cdot \cdot}^2}{N}$$

$$= 134.8733 - 116.1933 - 123.1933 + 107.2007 = 2.5540 \qquad (8.49)$$

The error sums of squares and cross-products are found as before by subtracting the $L$, $T{:}L$, $D$, and $LD$ values from the totals.

Using the values calculated here together with the values for $L$ and $T{:}L$ from Table 8.9, we can set up a preliminary analysis of covariance table, Table 8.13, this time adjusting only the whole-plot effects.

**TABLE 8.14**  ANALYSIS OF COVARIANCE OF OXYGEN GENERATION DATA: SUMMARY TABLE

| Source | df | SS | MS | F |
|--------|-----|--------|--------|---------|
| Whole plots | 5 | | | |
| $L$ | 1 | 8.5094 | 8.5094 | 21.09* |
| Regression | 1 | 1.0989 | 1.0989 | 2.72 |
| Error ($a$) | 3 | 1.2104 | 0.4035 | |
| Split plots | 54 | | | |
| $D$ | 9 | 15.9926 | 1.7770 | 22.07** |
| $LD$ | 9 | 2.5540 | 0.2838 | 3.53** |
| Error ($b$) | 36 | 2.8974 | 0.0805 | |

The 0's in the $\Sigma\, x^2$ and $\Sigma\, xy$ for the split-plot effects show that no adjustment will be made in the split plots. Since adjustments are to be made only in the whole plots, the summary table (Table 8.14) takes on a little different form than the previous summary table, Table 8.10. One degree of freedom is lost from the whole-plot error and the sum of squares for regression is now calculated in the whole-plot portion of the analysis. Since the degrees of freedom for error in the whole-plot analysis were only four to start with, losing one degree of freedom from the error is serious. In designing such an experiment it would be advisable to have a greater number of degrees of freedom for error if at all possible.

## 8.6  COVARIANCE WITH LATIN SQUARE AND CROSSOVER DESIGNS

When each variate is accompanied by a covariate (every $Y$ value has a corresponding $X$ value), the analysis of covariance can be used in Latin square and crossover designs. The procedures followed in the analysis of these types of design are the same as shown for the examples in Sections 8.2–8.5. However, when only one $X$ value is available for each subject or unit making up a column, in a Latin square or crossover design, covariance cannot be used if columns are also included in the model since there would be a confounding between the regression effect and the column effect. As an example of this situation in a Latin square, let us look at the data shown in Table 8.15, a set of hypothetical data where $X$ represents some type of baseline measurement made on each subject.

We have a choice in the analysis of these data as to whether we include the columns as a source in our model or the covariate $X$; we cannot include both. Analyzing the data by both models leads to the results shown in Table 8.16, where it can be seen that the linear, quadratic, and cubic regression components account for the three degrees of freedom available among the columns together with the sum of squares among columns. The analysis of covariance can be merely a partitioning of the column sum of squares into linear, quadratic, and cubic components. Both column effects and polynomial regression effects of a covariate cannot be included in the same model since they are accounting for the same sum of squares. However, if the relationship of $X$ and $Y$ is known to be linear, one could adjust the sum of squares for columns for the linear effects of $X$.

This same type of confounding of column effect with regression effects exists with a crossover design. As an example, let us look at the figures shown in Table 8.17,

**TABLE 8.15**  LATIN SQUARE DESIGN WITH SINGLE $X$
              FOR EACH SUBJECT

| Row | Subject 1<br>$X = 10$ | Subject 2<br>$X = 12$ | Subject 3<br>$X = 18$ | Subject 4<br>$X = 21$ |
|-----|-----------|-----------|-----------|-----------|
| 1 | $A = 2$ | $B = 7$ | $C = 20$ | $D = 20$ |
| 2 | $D = 40$ | $A = 3$ | $B = 8$ | $C = 23$ |
| 3 | $C = 15$ | $D = 32$ | $A = 2$ | $B = 12$ |
| 4 | $B = 9$ | $C = 18$ | $D = 18$ | $A = 5$ |

**TABLE 8.16**    ANALYSIS OF VARIANCE AND ANALYSIS OF COVARIANCE
OF DATA SHOWN IN TABLE 8.15

| Analysis of variance | | | Analysis of covariance | | |
|---|---|---|---|---|---|
| Source | df | SS | Source | df | SS |
| Total | 15 | 1783.75 | Total | 15 | 1783.75 |
| Rows | 3 | 102.25 | Rows | 3 | 102.25 |
| Columns | 3 | 42.75 | Treatments | 3 | 1406.75 |
| Treatments | 3 | 1406.75 | Regression | 3 | 42.75 |
| Error | 6 | 232.00 | Linear | 1 | 13.21 |
| | | | Quadratic | 1 | 26.13 |
| | | | Cubic | 1 | 3.41 |
| | | | Error | 6 | 232.00 |

which represent a crossover design with six subjects or columns with a single $X$ value for each subject. We can again analyze these data using two different models. One analysis will include the columns, or subjects, and omit the covariate while the second analysis will omit the columns and include the covariate through the quintic degree polynomial. We need to go to the quintic degree to include the five degrees of freedom among the six columns. The results of the two analyses are shown in Table 8.18. It can again be seen that the regressions due to the covariate account for the same sum of squares as do the columns, so both cannot be included in the same analysis, unless one is willing to assume a low-order relationship between $X$ and $Y$. Even then, only differences among subjects would be adjusted for the effects of $X$ on $Y$ since treatments and rows are orthogonal with the $X$ effects.

## 8.7  MULTIPLE COVARIANCE

The previous examples of covariance have included only one continuous, independent variable. However, it is possible to have two or more such variables. While the methods are quite similar to previous examples, it nevertheless seems worthwhile to look at such a design. The data used for this example are displayed in Table 8.19 and have been selected from a set of data used in the Harvey publication "Least-Squares Analysis of Data with Unequal Subclass Numbers" (1979). The data have been rearranged slightly for this example. Since it is known that both initial weight and initial age could have an influence on the rate of gain of steers on a feeding trial, each of these

**TABLE 8.17**    CROSSOVER DESIGN WITH A SINGLE $X$ FOR EACH COLUMN

| Row | Subject 1 $X = 5$ | Subject 2 $X = 7$ | Subject 3 $X = 3$ | Subject 4 $X = 9$ | Subject 5 $X = 4$ | Subject 6 $X = 1$ |
|---|---|---|---|---|---|---|
| 1 | $A = 1$ | $B = 3$ | $C = 7$ | $A = 2$ | $B = 6$ | $C = 12$ |
| 2 | $B = 4$ | $C = 12$ | $A = 3$ | $C = 9$ | $A = 4$ | $B = 7$ |
| 3 | $C = 8$ | $A = 3$ | $B = 6$ | $B = 8$ | $C = 14$ | $A = 1$ |

**TABLE 8.18**   ANALYSIS OF VARIANCE AND ANALYSIS OF COVARIANCE
OF DATA SHOWN IN TABLE 8.17

| Analysis of variance | | | Analysis of covariance | | |
|---|---|---|---|---|---|
| Source | df | SS | Source | df | SS |
| Total | 17 | 255.778 | Total | 17 | 255.778 |
| Rows | 2 | 8.111 | Rows | 2 | 8.111 |
| Columns | 5 | 23.111 | Treatments | 2 | 193.778 |
| Treatments | 2 | 193.778 | Regression | 5 | 23.111 |
| Error | 10 | 30.778 | Linear | 1 | 0.262 |
| | | | Quadratic | 1 | 1.387 |
| | | | Cubic | 1 | 0.310 |
| | | | Quartic | 1 | 0.007 |
| | | | Quintic | 1 | 21.145 |
| | | | Error | 10 | 30.778 |

factors was included as an independent, continuous variable. The model for this analysis
is

$$Y_{ijk} = \mu + l_i + s_{ij} + \beta_1(X_{1ij} - \bar{x}_1..) + \beta_2(X_{2ij} - \bar{x}_2..) + e_{ijk} \qquad (8.50)$$

where the $l_i$ represent line effects, the $s_{ij}$ represent sire effects (within lines), $X_1$ stands
for initial age, and $X_2$ stands for initial weight.

Again each line of the analysis of covariance table must be adjusted, in this case,
the adjustment being for the linear regression of $X_1$ and the linear regression of $X_2$.
This requires setting up normal equations in a fashion identical to that in Section 8.3,
where quadratic regression in covariance was discussed. Of course, quadratic or even
cubic components can be included in multiple covariance.

**TABLE 8.19**   AGE AT WEANING ($X_1$), INITIAL WEIGHT AT BEGINNING OF FEEDING
TRIAL ($X_2$), AND AVERAGE DAILY GAIN ($Y$) OF 36 HEREFORD
STEERS, FOUR STEERS FROM THREE DIFFERENT SIRES OF
THREE DIFFERENT LINES

| | | Line 1 | | | | | Line 2 | | | | | Line 3 | | |
|---|---|---|---|---|---|---|---|---|---|---|---|---|---|---|
| Sire | Steer | $X_1$ | $X_2$ | $Y$ | Sire | Steer | $X_1$ | $X_2$ | $Y$ | Sire | Steer | $X_1$ | $X_2$ | $Y$ |
| 1 | 1 | 192 | 390 | 2.24 | 1 | 1 | 169 | 443 | 2.94 | 1 | 1 | 184 | 411 | 3.00 |
| | 2 | 154 | 403 | 2.65 | | 2 | 158 | 381 | 2.50 | | 2 | 184 | 420 | 2.49 |
| | 3 | 185 | 432 | 2.41 | | 3 | 158 | 365 | 2.44 | | 3 | 187 | 427 | 2.25 |
| | 4 | 183 | 457 | 2.25 | | 4 | 169 | 386 | 2.44 | | 4 | 184 | 409 | 2.49 |
| 2 | 1 | 188 | 439 | 2.29 | 2 | 1 | 144 | 339 | 2.15 | 2 | 1 | 205 | 472 | 2.57 |
| | 2 | 178 | 407 | 2.26 | | 2 | 159 | 419 | 2.54 | | 2 | 193 | 430 | 2.37 |
| | 3 | 198 | 498 | 1.97 | | 3 | 152 | 469 | 2.74 | | 3 | 162 | 375 | 2.64 |
| | 4 | 193 | 459 | 2.14 | | 4 | 149 | 376 | 2.50 | | 4 | 206 | 451 | 2.37 |
| 3 | 1 | 154 | 389 | 2.38 | 3 | 1 | 189 | 395 | 2.65 | 3 | 1 | 200 | 466 | 2.16 |
| | 2 | 184 | 414 | 2.46 | | 2 | 187 | 447 | 2.52 | | 2 | 184 | 356 | 2.33 |
| | 3 | 174 | 483 | 2.29 | | 3 | 165 | 430 | 2.67 | | 3 | 175 | 449 | 2.52 |
| | 4 | 170 | 430 | 2.30 | | 4 | 181 | 453 | 2.79 | | 4 | 178 | 360 | 2.45 |

**TABLE 8.20**    SUMS OF SQUARES AND CROSS-PRODUCTS FOR MODEL (8.33)

| Source | df | $\Sigma\, x_1^2$ | $\Sigma\, x_2^2$ | $\Sigma\, x_1 x_2$ | $\Sigma\, x_1 y$ | $\Sigma\, x_2 y$ | $\Sigma\, y^2$ |
|--------|-----|-------------|--------------|--------------|-----------|-----------|----------|
| Total | 35 | 9,320.7500 | 52,267.2222 | 11,294.1667 | −34.5267 | −23.8089 | 1.7158 |
| L | 2 | 2,958.1667 | 3,737.7222 | 1,894.5833 | −19.6500 | −40.7789 | 0.4455 |
| S:L | 6 | 2,593.3333 | 6,330.0000 | 3,197.3334 | 3.2658 | 0.0325 | 0.2541 |
| W | 27 | 3,769.2500 | 42,199.5000 | 6,202.2500 | −18.1425 | 16.9375 | 1.0162 |

All the sums of squares and cross-products needed in this analysis are shown in Table 8.20. The normal equations necessary for the regression adjustments in each line are written as shown in Equations (8.18) and, of course, solved similarly. The regression coefficients in this case are those for the linear regressions of $X_1$ and $X_2$. The reductions due to regression are also calculated as shown previously in Equation (8.20). For the error line, we would have the equations

$$3769.25b_1 + 6202.25b_1 = -18.1425$$

$$6,202.25b_1 + 42,199.50b_1 = 16.9375 \tag{8.51}$$

Inverting the $2 \times 2$ matrix of sums of squares and cross-products, we find $c^{11} = 0.00034993$, $c^{12} = -0.00005143$, and $c^{22} = 0.00003126$ and $b_1 = -0.00721970$ and $b_2 = 0.00146254$.

The reduction in sum of squares is then

$$R(b_1, b_2) = (-0.00721970)(-18.1425) + (0.00146254)(16.9375) = 0.1558 \tag{8.52}$$

which is subtracted from $\Sigma\, y^2$ in the error line giving an adjusted error sum of squares of

$$1.0162 - 0.1558 = 0.8604$$

Pooling the $L$ and $W$ sums of squares and cross-products, we have the equations

$$6727.4166b_1 + 8096.8333b_2 = -37.7925$$

$$8,096.8333b_1 + 45,937.2221b_2 = -23.8414 \tag{8.53}$$

and inverting the $2 \times 2$ matrix of sums of squares and cross-products, we obtain the inverse elements of $c^{11} = 0.00018867$, $c^{12} = -0.00003325$, and $c^{22} = 0.00002763$, which result in coefficients of $b_1 = -0.00633758$ and $b_2 = 0.00059786$. The reduction in the sum of squares due to the two regressions is

$$R(b_1, b_2) = (-0.00633758)(-37.7925) + (0.00059786)(-23.8414)$$

$$= 0.2553 \tag{8.54}$$

This reduction in sum of squares is subtracted from the pooled $\Sigma\, y^2$ of the $L + W$ lines $0.4455 + 1.0162 = 1.4617$, giving a sum of $1.2064$. Now subtracting the adjusted error sum of squares of $0.8604$ from $1.2064$, we have the sum of squares for $L$ adjusted for both regressions, a value of $0.3460$ which is shown in Table 8.21. We proceed in the same fashion for the $S:L$ classification, and the pooled sums of squares and cross-products lead to the equations

**TABLE 8.21**  REGRESSION COEFFICIENTS, REDUCTIONS IN SUMS OF SQUARES,
AND ADJUSTED SUMS OF SQUARES

| Source | $b_1$ | $b_2$ | $R(b_1, b_2)$ | df | SS |
|---|---|---|---|---|---|
| Error | −0.00721970 | 0.00146254 | 0.1558 | 25 | 0.8604 |
| $L + W$ | −0.00633758 | 0.00059786 | 0.2553 | 27 | 1.2064 |
| $L$ adjusted | | | | 2 | 0.3460 |
| $S{:}L + W$ | −0.00399885 | 0.00112426 | 0.0786 | 31 | 1.1917 |
| $S{:}L$ adjusted | | | | 6 | 0.3313 |

$$6{,}362.5833b_1 + 9{,}399.5834b_2 = -14.8767$$
$$9{,}339.5834b_1 + 48{,}529.5000b_2 = 16.9700 \tag{8.55}$$

and the inversion of the coefficient matrix yields the inverse elements of $c^{11} = 0.00022016$, $c^{12} = -0.00004264$, and $c^{22} = 0.00002887$. We then find $b_1 = -0.00399885$ and $b_2 = 0.00112426$. The reduction in sum of squares of

$$R(b_1, b_2) = (0.000399885)(-14.8767) + (0.00112426)(16.9700) = 0.0786 \tag{8.56}$$

is subtracted from the pooled $\Sigma\, y^2$ from the $S{:}L$ and $W$ lines in Table 8.20 of 0.2541 + 1.0162 = 1.2703, giving 1.2703 − 0.0786 = 1.1917. We now subtract the adjusted sum of squares for $W$ of 0.8604 from this value giving an adjusted sum of squares of 0.3313 for $S{:}L$, shown in Table 8.21.

The sum of squares for the linear regression of $Y$ on $X_1$ is found by

$$\frac{b_1^2}{c^{11}} = \frac{-0.00721970}{0.00034993} = 0.1490 \tag{8.57}$$

and $Y$ on $X_2$ by

$$\frac{b_2^2}{c^{22}} = \frac{0.00146254}{0.00003126} = 0.0684 \tag{8.58}$$

These values come from the calculations in the $W$ or error line.

The completed analysis of covariance is shown in Table 8.22. It can be seen that the only effect showing significance is the linear regression of age at weaning on rate of gain in this sample of data.

The formula for calculating the adjusted line mean is

$$\text{Adj } \bar{y}_{i.} = \bar{y}_{i.} - b_1(\bar{x}_{1i.} - \bar{x}_{1..}) - b_2(\bar{x}_{2i.} - \bar{x}_{2..}) \tag{8.59}$$

**TABLE 8.22**  SUMMARY TABLE OF ANALYSIS OF
MULTIPLE COVARIANCE

| Source | df | SS | MS | F |
|---|---|---|---|---|
| $L$ | 2 | 0.3460 | 0.1730 | 3.13 |
| $S{:}L$ | 6 | 0.3313 | 0.0552 | 1.60 |
| Linear regression, $X_1$ | 1 | 0.1490 | 0.1490 | 4.33** |
| Linear regression, $X_2$ | 1 | 0.0684 | 0.0684 | 1.99 |
| Error | 25 | 0.8604 | 0.0344 | |

**TABLE 8.23** ADJUSTED MEANS AND THEIR STANDARD ERRORS FOR DATA ANALYZED UNDER MODEL (8.33)

| Adjusted mean | Standard error |
|---|---|
| $\bar{y}_{1.} = 2.3010$ | 0.0549 |
| $\bar{y}_{2.} = 2.5032$ | 0.0654 |
| $\bar{y}_{3.} = 2.5425$ | 0.0638 |

where the regression coefficients are those shown earlier,

$$b_1 = -0.007220 \quad \text{and} \quad b_2 = 0.001463$$

If the standard errors of the means are desired, they are determined by formula (8.60), where $s = \sqrt{0.0552}$.

$$S_{(\text{Adj } \bar{y}_{i.})}$$
$$= s\sqrt{(1/n_w) + c^{11}(\bar{x}_{1i.} - \bar{x}_1..)^2 + c^{22}(\bar{x}_{2i.} - \bar{x}_2..)^2 + 2c^{12}(\bar{x}_{1i.} - \bar{x}_1..)(\bar{x}_{2i.} - \bar{x}_2..)}$$
$$(8.60)$$

The adjusted means for the lines and their standard errors are shown in Table 8.23. The $s$ value and the inverse elements are derived from the error line.

## 8.8 COVARIANCE ANALYSIS WITH DISPROPORTIONATE SUBCLASS FREQUENCIES

Covariance analysis of data from a one-way classification can be accomplished by standard methods even though unequal numbers exist from class to class, as pointed out at the end of Section 8.2. However, identical results may be obtained with general least-squares procedures, which involve matrix arithmetic. These general procedures are applicable to a wide range of problems. If mean separation procedures or a set of orthogonal comparisons are desired from the least-squares or adjusted treatment means, it is usually best to obtain the inverse of the least-squares coefficient matrix.

We shall first consider a covariance analysis for a one-way classification with unequal numbers. The underlying mathematical model for this analysis may be written in two ways. If the continuous independent variable $(X)$ is to be taken as a deviation from the mean of $X$ observed in the sample $(\bar{x}.)$, then the model is

$$Y_{ij} = \mu + a_i + b(X_{ij} - \bar{x}.) + e_{ij}$$

$$i = 1, \ldots, n_a$$

$$j = 1, \ldots, n_i \qquad (8.61)$$

where $Y_{ij}$ is the $j$th observation in the $i$th $A$ class, $\mu$ is the overall mean with equal frequencies and with $X_{ij} = \bar{x}.$, $a_i$ is the effect of the $i$th $A$ class, $b$ is the regression of

**TABLE 8.24**  LEAST-SQUARES EQUATIONS
UNDER MODEL (8.62)

|        | $\hat{\alpha}$ | $\hat{a}_i$ |   | $\hat{b}$ | RHM |
|--------|------|------|---|------|-----|
| $\alpha$: | $n.$ | $n_i$ | 0 | $X.$ | $Y.$ |
| $a_i$: | $n_i$ | 0 | $n_i$ | $X_i$ | $Y_i$ |
| $b$: | $X.$ |   | $X_i$ | $\sum_i \sum_j X_{ij}^2$ | $\sum_i \sum_j X_{ij} Y_{ij}$ |

$Y_{ij}$ on $X_{ij}$ on a within-$A$ class basis, and the $e_{ij}$ are random errors. The $X_{ij}$ are assumed to be fixed and measured without error.

In practice, it is more convenient to work with the values of $X_{ij}$ instead of the deviations $X_{ij} - \bar{x}.$. When this is done the underlying model is

$$Y_{ij} = \alpha + a_i + bX_{ij} + e_{ij} \tag{8.62}$$

where the definitions of all common terms for models (8.61) and (8.62) remain the same. The new symbol $\alpha$ is the population mean when $X_{ij} = 0$. Using this model, we must obtain the estimate of $\mu$ from

$$\hat{\mu} = \hat{\alpha} + \hat{b}\bar{x}. \tag{8.63}$$

The least-squares equations under model (8.62) are as follows:

$$n.\hat{\alpha} + n_1\hat{a}_1 + n_2\hat{a}_2 + \cdots + n_a\hat{a}_a + X.\hat{b} = Y.$$

$$n_1\hat{\alpha} + n_1\hat{a}_1 + 0\hat{a}_2 + \cdots + 0\hat{a}_a + X_1\hat{b} = Y_1$$

$$n_2\hat{\alpha} + 0\hat{a}_1 + n_2\hat{a}_2 + \cdots + 0\hat{a}_a + X_2\hat{b} = Y_2$$

$$\vdots \qquad \vdots \qquad \vdots \qquad \qquad \vdots \qquad \vdots \qquad \vdots$$

$$n_a\hat{\alpha} + 0\hat{a}_1 + 0\hat{a}_2 + \cdots + n_a\hat{a}_a + X_a\hat{b} = Y_a$$

$$X.\hat{\alpha} + X_1\hat{a}_1 + X_2\hat{a}_2 + \cdots + X_a\hat{a}_a + \sum_i \sum_j X_{ij}^2\hat{b} = \sum_i \sum_j X_{ij} Y_{ij}$$

The least-squares equations are listed in tabular form in Table 8.24. The 0's in the $a_i$, $\hat{a}_i$ section of the table signify that the off-diagonal elements in this section are all zero. When the restriction is imposed that $\sum_i \hat{a}_i = 0$, the equations are those listed in Table 8.25.

**TABLE 8.25**  LEAST-SQUARES EQUATIONS AFTER
SUMMATION RESTRICTION

|        | $\hat{\alpha}$ | $\hat{a}_i$ | $\hat{b}$ | RHM |
|--------|------|------|------|-----|
| $\alpha$: | $n.$ | $n_i - n_a$ | $X.$ | $Y.$ |
|        |      | $n_a$ |      |     |
| $a_i$: | $n_i - n_a$ | $n_i + n_a$ | $X_i - X_a$ | $Y_i - Y_a$ |
|        |      | $n_a$ |      |     |
| $b$: | $X.$ | $X_i - X_a$ | $\sum_i \sum_j X_{ij}^2$ | $\sum_i \sum_j X_{ij} Y_{ij}$ |

**TABLE 8.26** LEAST-SQUARES EQUATIONS UNDER MODEL (8.61) WITH RESTRICTIONS

|  | $\hat{\mu}$ | $\hat{a}_i$ | $\hat{b}$ | RHM |
|---|---|---|---|---|
| $\mu$: | $n.$ | $n_i - \dfrac{n_a}{n_a}$ | $0$ | $Y.$ |
| $a_i$: | $n_i - n_a$ | $n_i + \dfrac{n_a}{n_a}$ | $\sum_j(X_{ij} - \bar{x}.) - \sum_j(X_{aj} - \bar{x}.)$ | $Y_i - Y_a$ |
| $b$: | $0$ | $\sum_j(X_{ij} - \bar{x}.) - \sum_j(X_{aj} - \bar{x}.)$ | $\sum_i \sum_j(X_{ij} - \bar{x}.)^2$ | $\sum_i \sum_j(X_{ij} - \bar{x}.)Y_{ij}$ |

The least-squares equations under model (8.61) when the restriction is imposed that $\sum_i \hat{a}_i = 0$ are listed in Table 8.26. It should now be noted that the elements in the equation for $b$ under model (8.61) are easily computed from the elements in the equation for $b$ under model (8.62). Note that

$$0 = X. - \frac{n.X.}{n.}$$

$$\sum_j(X_{ij} - \bar{x}.) - \sum_j(X_{aj} - \bar{x}.) = X_i - X_a - \frac{(n_i - n_a)X.}{n.}$$

$$\sum_i \sum_j(X_{ij} - \bar{x}.)^2 = \sum_i \sum_j X_{ij}^2 - \frac{X.^2}{n.}$$

$$\sum_i \sum_j(X_{ij} - \bar{x}.)Y_{ij} = \sum_i \sum_j X_{ij}Y_{ij} - \frac{X.Y.}{n.}$$

Hence, one can construct the set of least-squares equations under model (8.62) and then adjust the elements in the $b$ row and column to obtain the equations under model

**TABLE 8.27** HYPOTHETICAL DATA SET

|  | Ration number | | | | | |
|---|---|---|---|---|---|---|
|  | 1 | | 2 | | 3 | |
| Pig number | Initial weight | Gain | Initial weight | Gain | Initial weight | Gain |
| 1 | 5 | 3 | 4 | 5 | 8 | 7 |
| 2 | 9 | 5 | 7 | 6 | 7 | 6 |
| 3 | 11 | 6 | 0 | 2 | 3 | 4 |
| 4 | 3 | 2 | 8 | 7 | 2 | 3 |
| 5 |  |  | 10 | 8 | 8 | 6 |
| 6 |  |  | 2 | 3 | 2 | 4 |
| 7 |  |  | 12 | 9 |  |  |
| 8 |  |  | 5 | 8 |  |  |
| Totals | 28 | 16 | 48 | 48 | 30 | 30 |
| Means | 7 | 4 | 6 | 6 | 5 | 5 |

(8.61). This is desirable for two reasons: (1) an estimate of $\mu$ is obtained directly from solving the equations and (2) the least-squares coefficient matrix is better conditioned and fewer rounding errors will occur from the inversion process.

Standard procedures as outlined in Section 2.9.1 are used to obtain the inverse of the reduced coefficient matrix, the estimates of $\mu$, $a_i$, and $b_j$ and the sums of squares for the analysis of variance. The computational procedures will now be illustrated using a hypothetical set of data from Harvey (1979). These data are given in Table 8.27.

The model underlying the analysis of the data in Table 8.27 is

$$Y_{ij} = \alpha + r_i + bW_{ij} + e_{ij} \tag{8.64}$$

where $Y_{ij}$ is the gain for the $j$th pig in the $i$th ration, $r_i$ is the effect of the $i$th ration, $b$ is the regression of gain on initial weight, $W_{ij}$ is the initial weight, and the $e_{ij}$ are random errors. The $\mathbf{X}_r$ matrix (see Section 3.20) with the restriction imposed that $\Sigma_i\,\hat{r}_i = 0$ and the $\mathbf{Y}$ column vector are given in Table 8.28. The set of least-squares equations are as follows:

$$\mathbf{X}_r'\mathbf{X}_r\hat{\mathbf{B}} = \mathbf{X}_r'\mathbf{Y}$$

$$\underbrace{\begin{bmatrix} 18 & -2 & 2 & 106 \\ -2 & 10 & 6 & -2 \\ 2 & 6 & 14 & 18 \\ 106 & -2 & 18 & 832 \end{bmatrix}}_{\mathbf{X}_r'\mathbf{X}_r} \underbrace{\begin{bmatrix} \hat{\alpha} \\ \hat{r}_1 \\ \hat{r}_2 \\ \hat{b} \end{bmatrix}}_{\hat{\mathbf{B}}} = \underbrace{\begin{bmatrix} 94 \\ -14 \\ 18 \\ 656 \end{bmatrix}}_{\mathbf{X}_r'\mathbf{Y}}$$

**TABLE 8.28**   REDUCED **X** MATRIX AND **Y** COLUMN VECTOR
FOR DATA IN TABLE 8.27

| Ration | Pig number | $\alpha$ | $r_1$ | $r_2$ | $b$ | $Y$ |
|--------|-----------|----------|-------|-------|-----|-----|
| 1 | 1 | 1 | 1 | 0 | 5 | 3 |
|   | 2 | 1 | 1 | 0 | 9 | 5 |
|   | 3 | 1 | 1 | 0 | 11 | 6 |
|   | 4 | 1 | 1 | 0 | 3 | 2 |
| 2 | 1 | 1 | 0 | 1 | 4 | 5 |
|   | 2 | 1 | 0 | 1 | 7 | 6 |
|   | 3 | 1 | 0 | 1 | 0 | 2 |
|   | 4 | 1 | 0 | 1 | 8 | 7 |
|   | 5 | 1 | 0 | 1 | 10 | 8 |
|   | 6 | 1 | 0 | 1 | 2 | 3 |
|   | 7 | 1 | 0 | 1 | 12 | 9 |
|   | 8 | 1 | 0 | 1 | 5 | 8 |
| 3 | 1 | 1 | -1 | -1 | 8 | 7 |
|   | 2 | 1 | -1 | -1 | 7 | 6 |
|   | 3 | 1 | -1 | -1 | 3 | 4 |
|   | 4 | 1 | -1 | -1 | 2 | 3 |
|   | 5 | 1 | -1 | -1 | 8 | 6 |
|   | 6 | 1 | -1 | -1 | 2 | 4 |

To obtain an estimate of the overall mean expected with equal numbers with $W_{ij} = \bar{w}$. and to reduce rounding errors, the elements in the equation for $\hat{b}$ are adjusted for the mean of $W$ to yield the following equations:

$$
\begin{bmatrix}
18 & -2 & 2 & 0 \\
-2 & 10 & 6 & 9.77778 \\
2 & 6 & 14 & 6.22222 \\
0 & 9.77778 & 6.22222 & 207.77778
\end{bmatrix}
\begin{bmatrix}
\hat{\mu} \\
\hat{r}_1 \\
\hat{r}_2 \\
\hat{b}
\end{bmatrix}
=
\begin{bmatrix}
94 \\
-14 \\
18 \\
102.44444
\end{bmatrix}
$$

where

$$0 = 106 - \frac{(18)(106)}{18}$$

$$9.77778 = -2 - \frac{(-2)(106)}{18}$$

$$6.22222 = 18 - \frac{(2)(106)}{18}$$

$$207.77778 = 832 - \frac{(106)^2}{18}$$

$$102.44444 = 656 - \frac{(106)(94)}{18}$$

The inverse of the coefficient matrix is

$$
(\mathbf{X}'_r\mathbf{X}_r)^{-1} =
\begin{bmatrix}
0.060248 & 0.023709 & -0.018520 & 0.000000 \\
0.023709 & 0.148569 & -0.064810 & -0.005050 \\
-0.018520 & -0.064810 & 0.101852 & 0.000000 \\
0.000000 & -0.005050 & 0.000000 & 0.005051
\end{bmatrix}
$$

and the estimates of the constants are as follows:

$$\hat{\mu} = 4.94052 \qquad \hat{r}_3 = -(\hat{r}_1 + \hat{r}_2)$$
$$\hat{r}_1 = -1.53540 \qquad \qquad = 0.53540$$
$$\hat{r}_2 = 1.00000 \qquad \hat{b} = 0.53535$$

The adjusted means for rations (least-squares means) are

$$\hat{\mu} + \hat{a}_1 = 4.94 - 1.54 = 3.40$$
$$\hat{\mu} + \hat{a}_2 = 4.94 + 1.00 = 5.94$$
$$\hat{\mu} + \hat{a}_3 = 4.94 + 0.54 = 5.48$$

The total reduction due to fitting all constants is

$$R(\mu, r, b) = \hat{\mathbf{B}}'\mathbf{X}'_r\mathbf{Y} = 558.747$$

The sum of squares for rations is

**TABLE 8.29** ANALYSIS OF VARIANCE FOR DATA
IN TABLE 8.27

| Source | df | SS | MS | F |
|--------|-----|--------|--------|---------|
| Rations | 2 | 17.350 | 8.675 | 13.12** |
| Regression | 1 | 56.741 | 56.741 | 85.84** |
| Error | 14 | 9.253 | 0.661 | |

$$\hat{\mathbf{B}}_r'\mathbf{Z}_r^{-1}\hat{\mathbf{B}}_r = [-1.53540 \quad 1]\begin{bmatrix} 0.148569 & -0.064810 \\ -0.064810 & 0.101852 \end{bmatrix}^{-1}\begin{bmatrix} -1.53540 \\ 1 \end{bmatrix}$$

$$= [-1.53540 \quad 1]\begin{bmatrix} 9.31711 & 5.92862 \\ 5.92862 & 13.59064 \end{bmatrix}\begin{bmatrix} -1.53540 \\ 1 \end{bmatrix}$$

$$= [-8.37687 \quad 4.48784]\begin{bmatrix} -1.53540 \\ 1 \end{bmatrix}$$

$$= 17.350$$

The sum of squares for the regression of gain on initial weight is

$$\hat{\mathbf{B}}_b'\mathbf{Z}_b^{-1}\hat{\mathbf{B}}_b = \frac{(0.53535)^2}{0.005051} = 56.741$$

The error sum of squares is

$$\mathbf{Y}'\mathbf{Y} - R(\mu, r, b) = 568 - 558.747 = 9.253$$

The analysis of variance table is set up in Table 8.29.

Linear comparisons among ration means may be tested for significance in the usual manner by making use of the inverse elements and the error variance. For example,

$$\hat{r}_1 - \hat{r}_2 = -1.5354 - 1 = -2.5354$$

$$V(\hat{r}_1 - \hat{r}_2) = [0.148569 + 0.101852 - (2)(-0.064810)](0.661)$$

$$= 0.25121$$

$$s_{\hat{r}_1-\hat{r}_2} = 0.5012$$

Hence, the test for significance of this difference is

$$t = \frac{-2.5354}{0.5012} = -5.06**$$

with 14 degrees of freedom.

## 8.9 INDIVIDUAL CLASS AND/OR SUBCLASS REGRESSIONS

A question often asked by many investigators is whether the regression of a dependent variable on an independent variable is the same for different classes or subclasses. In effect, this is asking whether an interaction exists between a discrete variate and a continuous variate. Methods for computing the separate regression coefficients and for making tests of significance are given in many textbooks on statistics. The purpose of this section is to describe a general least-squares method of simultaneously computing the separate partial regression coefficients, appropriate estimates of other constants being fitted, and the analysis of variance when unequal numbers exist.

For the case of the one-way classification with individual class regressions a convenient model is as follows:

$$Y_{ij} = \mu + a_i + b(X_{ij} - \bar{x}.) + b_i(X_{ij} - \bar{x}.) + e_{ij} \tag{8.65}$$

Under this model, when the restrictions are imposed that $\Sigma_i \, \hat{a}_i = \Sigma_i \, \hat{b}_i = 0$, $\hat{\mu}$ will estimate the mean if equal numbers exist and when $X_{ij} = \bar{x}.$; $\hat{a}_i$ will estimate the deviation of the $i$th $A$ class mean from the overall mean when adjusted to $X_{ij} = \hat{x}.$ with the $i$th individual class regression; $\hat{b}$ will estimate the average of the individual class regressions and $\hat{b}_i$ will estimate the deviation of the $i$th individual class regression from the average regression.

The data given in Table 8.27 with two changes in initial weights will be used to illustrate the least-squares analysis under model (8.65). The reduced **X** matrix and the **Y** column vector for these data are given in Table 8.30. The least-squares equations obtained from computing sums of squares and sums of cross-products among the columns given in Table 8.30 are as follows:

$$
\begin{bmatrix}
18 & -2 & 2 & 0 & 8 & 4 \\
-2 & 10 & 6 & 8 & 0 & -4 \\
2 & 6 & 14 & 4 & -4 & -4 \\
0 & 8 & 4 & 206 & -4 & 66 \\
8 & 0 & -4 & -4 & 92 & 48 \\
4 & -4 & -4 & 66 & 48 & 162
\end{bmatrix}
\begin{bmatrix}
\hat{\mu} \\
\hat{r}_1 \\
\hat{r}_2 \\
\hat{b} \\
\hat{b}_1 \\
\hat{b}_2
\end{bmatrix}
=
\begin{bmatrix}
94 \\
-14 \\
18 \\
102 \\
34 \\
62
\end{bmatrix}
$$

**TABLE 8.30**  REDUCED **X** MATRIX AND **Y** COLUMN VECTOR FOR DATA IN TABLE 8.27 UNDER MODEL (8.65)

| Ration | Pig number | $\mu$ | $r_1$ | $r_2$ | $b$ | $b_1$ | $b_2$ | $Y$ |
|---|---|---|---|---|---|---|---|---|
| 1 | 1 | 1 | 1 | 0 | −1 | −1 | 0 | 3 |
|   | 2 | 1 | 1 | 0 | 3 | 3 | 0 | 5 |
|   | 3 | 1 | 1 | 0 | 5 | 5 | 0 | 6 |
|   | 4 | 1 | 1 | 0 | −3 | −3 | 0 | 2 |
| 2 | 1 | 1 | 0 | 1 | −2 | 0 | −2 | 5 |
|   | 2 | 1 | 0 | 1 | 1 | 0 | 1 | 6 |
|   | 3 | 1 | 0 | 1 | −6 | 0 | −6 | 2 |
|   | 4 | 1 | 0 | 1 | 2 | 0 | 2 | 7 |
|   | 5 | 1 | 0 | 1 | 4 | 0 | 4 | 8 |
|   | 6 | 1 | 0 | 1 | −4 | 0 | −4 | 3 |
|   | 7 | 1 | 0 | 1 | 6 | 0 | 6 | 9 |
|   | 8 | 1 | 0 | 1 | −1 | 0 | −1 | 8 |
| 3 | 1 | 1 | −1 | −1 | 2 | −2 | −2 | 7 |
|   | 2 | 1 | −1 | −1 | 1 | −1 | −1 | 6 |
|   | 3 | 1 | −1 | −1 | −3 | 3 | 3 | 4 |
|   | 4 | 1 | −1 | −1 | −4 | 4 | 4 | 3 |
|   | $5^a$ | 1 | −1 | −1 | 3 | −3 | −3 | 6 |
|   | $6^a$ | 1 | −1 | −1 | −3 | 3 | 3 | 4 |

[a] Initial weight was increased by 1 for each of these two observations to simplify the arithmetic.

The inverse of the coefficient matrix is

$$(\mathbf{X}'_r\mathbf{X}_r)^{-1}$$

$$=
\begin{bmatrix}
0.064052 & 0.027614 & -0.022390 & -0.001140 & -0.007190 & -0.001144 \\
 & 0.155719 & -0.069280 & -0.007190 & -0.009480 & 0.007190 \\
 & & 0.105719 & 0.001144 & 0.007190 & -0.001140 \\
 & \textit{Symmetric} & & 0.006203 & 0.002130 & -0.003280 \\
 & & & & 0.014537 & -0.005050 \\
 & & & & & 0.009127
\end{bmatrix}$$

and the estimates of the constants are as follows:

$$\hat{\mu} = 4.94118 \qquad \hat{b} = 0.51557$$

$$\hat{r}_1 = -1.44120 \qquad \hat{b}_1 = -0.01557$$

$$\hat{r}_2 = 1.05882 \qquad \hat{b}_2 = 0.04584$$

$$\hat{r}_3 = -(\hat{r}_1 + \hat{r}_2) \qquad \hat{b}_3 = -(\hat{b}_1 + \hat{b}_2)$$

$$= 0.38238 \qquad = -0.03027$$

The least-squares means for rations and the individual ration regression coefficients are

$$\hat{\mu} + \hat{r}_1 = 3.5000 \qquad \hat{b} + \hat{b}_1 = 0.5000$$

$$\hat{\mu} + \hat{r}_2 = 6.0000 \qquad \hat{b} + \hat{b}_2 = 0.5614$$

$$\hat{\mu} + \hat{r}_3 = 5.3236 \qquad \hat{b} + \hat{b}_3 = 0.4854$$

The total reduction due to fitting all constants is

$$R(\mu, r, b, b_i) = \hat{\mathbf{B}}'\mathbf{X}'_r\mathbf{Y} = 558.606$$

and the error sum of squares is

$$\mathbf{Y}'\mathbf{Y} - R(\mathbf{B}) = 568 - 558.606 = 9.394$$

The sum of squares for differences among rations when each ration mean is adjusted to $\bar{w}.$ using the respective regression coefficient is

$$\hat{\mathbf{B}}'_r\mathbf{Z}_r^{-1}\hat{\mathbf{B}}_r = [-1.44120 \quad 1.05882]\begin{bmatrix} 0.155719 & -0.069280 \\ -0.069280 & 0.105719 \end{bmatrix}^{-1}\begin{bmatrix} -1.44120 \\ 1.05882 \end{bmatrix}$$

$$= [-1.44120 \quad 1.05882]\begin{bmatrix} 9.064661 & 5.940286 \\ 5.940286 & 13.351838 \end{bmatrix}\begin{bmatrix} -1.44120 \\ 1.05882 \end{bmatrix}$$

$$= [-6.774296 \quad 5.576053]\begin{bmatrix} -1.44120 \\ 1.05882 \end{bmatrix}$$

$$= 15.667$$

The sum of squares for the average regression $b$ is

$$\hat{\mathbf{B}}'_b\mathbf{Z}_b^{-1}\hat{\mathbf{B}}_b = \frac{(0.51557)^2}{0.006203} = 42.852$$

and the sum of squares for differences among the individual class regression coefficients is

$$\hat{\mathbf{B}}'_{bi}\mathbf{Z}^{-1}_{bi}\hat{\mathbf{B}}_{bi} = [-0.01557 \quad 0.04584]\begin{bmatrix} 0.014537 & -0.005050 \\ -0.005050 & 0.009127 \end{bmatrix}^{-1}\begin{bmatrix} -0.01557 \\ 0.04584 \end{bmatrix}$$

$$= [-0.01557 \quad 0.04584]\begin{bmatrix} 85.1584 & 47.1185 \\ 47.1185 & 135.6358 \end{bmatrix}\begin{bmatrix} -0.01557 \\ 0.04584 \end{bmatrix}$$

$$= [0.833996 \quad 5.483910]\begin{bmatrix} -0.01557 \\ 0.04584 \end{bmatrix}$$

$$= 0.238$$

The analysis of variance table is set up in Table 8.31.

It is clear from this ANOVA that the regression coefficients do not differ among rations. Hence, the ANOVA previously completed, where the pooled regression was fitted, is much preferred over this analysis because of loss of efficiency in estimating the common regression and differences among adjusted ration means. For example, the standard error of the pooled regression coefficient is

$$s_b = [(0.005051)(0.661)]^{1/2} = 0.058$$

whereas the standard error of the average regression coefficient is

$$s_b = [(0.006203)(0.783)]^{1/2} = 0.070$$

Also, the standard error of the difference between the least-squares means for rations 1 and 2 for the pooled covariate analysis is

$$s_{\hat{r}_1-\hat{r}_2} = \{[0.148569 + 0.101852 - (2)(-0.064810)](0.661)\}^{1/2}$$

$$= 0.501$$

and when the individual class regressions are fitted this standard error is

$$s_{\hat{r}_1-\hat{r}_2} = \{[0.155719 + 0.105719 - (2)(-0.069280)](0.783)\}^{1/2}$$

$$= 0.560$$

This additional "cost" of adjusting the ration means exists in the second analysis whether or not differences exist among the individual regressions. Therefore, a step-down procedure should always be followed when sources of variation are found to be insignificant. A suggested significance level to use in this step-down procedure is 0.20.

**TABLE 8.31**   ANALYSIS OF VARIANCE WITH
INDIVIDUAL REGRESSIONS

| Source | df | SS | MS | F |
|--------|----|----|----|----|
| Rations | 2 | 15.667 | 7.833 | 10.00** |
| Regressions | | | | |
| Average | 1 | 42.852 | 42.852 | 54.73** |
| Rations | 2 | 0.238 | 0.119 | 0.15 |
| Error | 12 | 9.394 | 0.783 | |

The least-squares procedures described in this section are easily extended to analyses of data under more complex models. With multiple covariance analyses where the partial regressions are fitted on an error line basis, one simply adds additional columns to the reduced **X** matrix (Table 8.28). Of course, one may also desire to determine if the regressions for each independent continuous variable differ among classes or among subclasses. This can easily be done by extending model (8.65) to include the additional regression coefficients to be fitted. Additional columns would then be added to the reduced **X** matrix in Table 8.30.

## EXERCISES

**8.1.** The data presented below have been extracted from the results of a steer-feeding trial conducted by Dr. Joseph P. Tritschler at the University of Florida. Three rations were tested, with eight animals per ration being used for this example. $X$ represents the final weight of a steer in pounds and $Y$ represents the rib eye area in square inches. Conduct an analysis of covariance on this set of data and calculate the adjusted treatment means. Next, carry out an analysis of variance for these data noting the difference in conclusions regarding the significance of the treatment effects.

| Ration 1 | | Ration 2 | | Ration 3 | |
|---|---|---|---|---|---|
| $X$ | $Y$ | $X$ | $Y$ | $X$ | $Y$ |
| 855 | 8.7 | 870 | 10.1 | 1030 | 11.3 |
| 900 | 10.1 | 890 | 9.5 | 950 | 11.0 |
| 825 | 9.5 | 880 | 9.7 | 920 | 9.5 |
| 805 | 8.7 | 1035 | 10.3 | 855 | 10.5 |
| 880 | 9.3 | 875 | 9.1 | 1025 | 11.1 |
| 950 | 10.1 | 985 | 9.6 | 1000 | 9.2 |
| 955 | 9.9 | 1130 | 10.8 | 1040 | 10.9 |
| 900 | 8.9 | 925 | 10.4 | 1115 | 12.0 |

**8.2.** The data presented below have been selected from an experiment conducted by Solandt et al. (1943) which compares the effects of different drugs in delaying the atrophy of denervated muscles. A number of rats were put randomly into four groups and a certain muscle was deprived of its nerve supply by severing the appropriate nerves. Each of the four groups was assigned to treatment by one of the drugs. Four days after treatment was begun four rats were selected at random from each of the four groups and measures of atrophy were obtained from them. This procedure was repeated after 8 and 12 days. Only three treatments and the 4- and 12-day measurements are included in this exercise. For a detailed analysis and discussion of this experiment see Delury (1948). $X_1$ represents initial body weight, $X_2$ represents final body weight, and $Y$ represents the weight of the denervated muscle. Using only $X_2$, conduct an analysis of covariance for this set of data, a two-way classification. Next, calculate an analysis of variance and compare results. Compute the adjusted means for the days and treatments.

| Day | Treatment | $X_1$ | $X_2$ | $Y$ | Day | Treatment | $X_1$ | $X_2$ | $Y$ |
|---|---|---|---|---|---|---|---|---|---|
| 4 | $A$ | 217 | 196 | 0.94 | 12 | $A$ | 198 | 165 | 0.34 |
| | | 246 | 218 | 1.16 | | | 175 | 150 | 0.43 |
| | | 256 | 216 | 1.26 | | | 199 | 159 | 0.41 |
| | | 200 | 165 | 0.85 | | | 224 | 163 | 0.48 |

| Day | Treatment | $X_1$ | $X_2$ | Y | Day | Treatment | $X_1$ | $X_2$ | Y |
|-----|-----------|-------|-------|------|-----|-----------|-------|-------|------|
| 4 | B | 198 | 202 | 1.19 | 12 | B | 233 | 242 | 0.41 |
|   |   | 248 | 231 | 1.15 |   |   | 250 | 226 | 0.87 |
|   |   | 180 | 187 | 0.86 |   |   | 289 | 300 | 0.91 |
|   |   | 218 | 230 | 1.21 |   |   | 255 | 252 | 0.87 |
|   | C | 264 | 231 | 1.22 |   | C | 204 | 181 | 0.57 |
|   |   | 200 | 170 | 0.90 |   |   | 234 | 181 | 0.80 |
|   |   | 210 | 189 | 1.00 |   |   | 211 | 180 | 0.69 |
|   |   | 192 | 185 | 1.00 |   |   | 214 | 200 | 0.84 |

**8.3.** Using both $X_1$ and $X_2$ in the data shown in Exercise 8.2, complete an analysis of multiple covariance, comparing the results with those found in the two analyses of Exercise 8.2. Calculate the adjusted means for the days and treatments.

**8.4.** Given the following set of data, which conforms to a one-way classification, split-plot (or repeated measures) design, complete an analysis of covariance. Eight steers are allotted at random to two rations and the amounts of gain in pounds for two periods are recorded. Age in days is the covariate.

| | | Period 1 | | Period 2 | |
|--------|-------|-------------|-----------|-------------|-----------|
| Ration | Steer | Initial age | Gain (lb) | Initial age | Gain (lb) |
| 1 | 1 | 176 | 75 | 176 | 80 |
|   | 2 | 167 | 78 | 167 | 83 |
|   | 3 | 165 | 84 | 165 | 76 |
|   | 4 | 162 | 73 | 162 | 75 |
| 2 | 1 | 196 | 72 | 196 | 69 |
|   | 2 | 194 | 69 | 194 | 65 |
|   | 3 | 180 | 56 | 180 | 60 |
|   | 4 | 171 | 81 | 171 | 77 |

**8.5.** Barrows and gilts from three breeds of swine were slaughtered at approximately 100 kg. The carcass weight (in kilograms) and backfat thickness (in centimeters) measurements obtained on 97 pigs that were randomly selected from the 173 pigs included in this experiment are given in the table below. Complete the following ANOVA for backfat thickness using least-squares procedures:

| | | Breeds | | | | | |
|-----|-----|-----|-----|-----|-----|-----|-----|
| | | 1 | | 2 | | 3 | |
| Sex | Carcass weight | Backfat | Carcass weight | Backfat | Carcass weight | Backfat |
| Barrows (1) | 65.9 | 3.63 | 75.7 | 3.89 | 68.2 | 3.23 |
|   | 73.9 | 3.38 | 78.3 | 3.23 | 71.7 | 2.29 |
|   | 68.5 | 2.79 | 69.9 | 3.63 | 62.4 | 2.72 |
|   | 65.4 | 3.30 | 71.7 | 2.97 | 71.2 | 2.87 |
|   | 69.0 | 3.30 | 83.2 | 3.23 | 63.7 | 2.79 |
|   | 62.8 | 3.12 | 72.1 | 3.56 | 63.3 | 2.54 |
|   | 67.2 | 3.38 | 72.9 | 2.72 | 69.0 | 2.87 |
|   | 72.5 | 3.38 | 72.1 | 3.38 | 65.0 | 2.62 |
|   | 70.3 | 3.30 | 67.2 | 2.79 | 76.9 | 3.56 |

| | Breeds | | | | | |
| | 1 | | 2 | | 3 | |
| Sex | Carcass weight | Backfat | Carcass weight | Backfat | Carcass weight | Backfat |
|---|---|---|---|---|---|---|
| Barrows (1) | 65.0 | 2.79 | 73.4 | 3.38 | 84.0 | 3.30 |
| | 68.5 | 3.12 | 80.5 | 3.38 | 74.8 | 2.72 |
| | 63.7 | 3.05 | 68.5 | 3.38 | 68.2 | 2.54 |
| | 64.5 | 3.63 | 65.0 | 2.72 | 67.7 | 3.12 |
| | 64.2 | 3.12 | 64.2 | 2.97 | 65.0 | 2.97 |
| | 69.4 | 3.23 | 65.0 | 3.05 | | |
| | 68.5 | 3.89 | 63.7 | 3.38 | | |
| | | | 67.7 | 2.72 | | |
| | | | 71.7 | 2.87 | | |
| | | | 73.9 | 3.81 | | |
| Gilts (2) | 75.2 | 4.06 | 85.8 | 2.79 | 62.8 | 3.89 |
| | 65.9 | 3.81 | 80.5 | 3.73 | 72.9 | 3.73 |
| | 70.3 | 3.05 | 85.8 | 3.63 | 67.2 | 3.30 |
| | 69.9 | 3.30 | 75.7 | 3.81 | 68.5 | 3.30 |
| | 63.3 | 2.79 | 79.6 | 4.24 | 72.1 | 3.12 |
| | 64.2 | 2.79 | 72.9 | 3.63 | 66.8 | 3.30 |
| | 73.4 | 3.56 | 68.5 | 3.12 | 67.2 | 3.73 |
| | 63.7 | 3.56 | 90.2 | 2.87 | 65.9 | 3.38 |
| | 79.6 | 3.89 | 64.2 | 3.12 | 63.3 | 3.23 |
| | 64.2 | 2.79 | 61.5 | 3.56 | 70.8 | 3.63 |
| | 64.5 | 3.63 | 90.2 | 2.87 | 60.5 | 2.97 |
| | 64.5 | 3.38 | 72.5 | 3.73 | 70.3 | 3.05 |
| | 69.0 | 2.79 | 64.5 | 2.87 | 65.0 | 3.30 |
| | 66.8 | 2.62 | 86.7 | 2.97 | | |
| | 64.2 | 2.79 | 67.2 | 3.63 | | |
| | 65.0 | 3.12 | | | | |
| | 90.2 | 4.83 | | | | |
| | 78.7 | 3.73 | | | | |
| | 62.4 | 2.97 | | | | |
| | 63.7 | 3.56 | | | | |

Prepare a table giving the least-squares means with standard errors for breeds, sexes, and breed × sex subclasses. Also, give the quadratic prediction equations for each breed × sex subclass. Interpret the results.

| Source | df |
|---|---|
| Breeds ($B$) | 2 |
| Sexes ($S$) | 1 |
| $BS$ | 2 |
| Regression ($CWT$) | |
|   Linear—average | 1 |
|   Among breeds—linear | 2 |
|   Quadratic—average | 1 |
|   Among breeds—quadratic | 2 |
| Remainder | 85 |

# Chapter 9

# Matrix Algebra

## 9.1 INTRODUCTION

Many of the analytical techniques used throughout this text rely on the application of matrix algebra. The use of this mathematical tool has increased greatly in the field of statistics and it is of considerable advantage for a researcher dealing with large amounts of data to have a working knowledge of some of the more commonly used matrix operations. Extremely large groups of numbers can be condensed quite simply in matrix notation, allowing the designation of complex operations by a few symbols. Only the more common matrix operations are considered here. More detailed information on the subject can be found in texts by Searle (1966), Hohn (1958), and Graybill (1969).

## 9.2 DEFINITION

A matrix is defined as a rectangular or square array of numbers, or elements, composed of one or more rows and one or more columns. Each row has the same number of elements as does each column. A matrix is usually said to have $r$ rows and $c$ columns (or $m$ rows and $n$ columns) and is designated by a capital letter. The numbers of rows and columns are designated as $m$ and $n$ in the following matrix $\mathbf{C}$.

$$\begin{bmatrix} c_{11} & c_{12} & \cdots & c_{1n} \\ c_{21} & c_{22} & \cdots & c_{2n} \\ \vdots & \vdots & \cdots & \vdots \\ c_{m1} & c_{m2} & & c_{mn} \end{bmatrix} = \mathbf{C} = [c_{ij}]$$

## 9.3 ORDER

The order of a matrix defines the size of the matrix, that is, the number of rows and columns included in the matrix. If the matrix **C** has $r$ rows and $c$ columns, it is of the order $r \times c$ and can be written $\mathbf{C}_{r \times c}$. When the number of rows of a matrix is equal to the number of columns, the matrix is described as a square matrix and may be designated as having order $r$.

## 9.4 COEFFICIENT MATRIX

In dealing with a system of linear equations such as

$$4x + 12y = 14$$

$$9x - 10y = -18$$

the matrix $\begin{bmatrix} 4 & 12 \\ 9 & -10 \end{bmatrix}$ is called the coefficient matrix, the order of this matrix being $2 \times 2$.

## 9.5 VECTOR

The values 14 and $-18$ in the above set of equations can also be put into matrix form. A column such as $\begin{bmatrix} 14 \\ -18 \end{bmatrix}$ is known as a column vector. Similarly, $\begin{bmatrix} 14 & -18 \end{bmatrix}$ is called a row vector.

## 9.6 ADDITION OF MATRICES

Two matrices **A** and **B** of the same order are added by adding corresponding elements. For example,

$$\mathbf{A} = \begin{bmatrix} 7 & 32 \\ 24 & 9 \end{bmatrix} \quad \mathbf{B} = \begin{bmatrix} -4 & -6 \\ -8 & 12 \end{bmatrix} \quad \mathbf{A} + \mathbf{B} = \begin{bmatrix} 7 - 4 & 32 - 6 \\ 24 - 8 & 9 + 12 \end{bmatrix} = \begin{bmatrix} 3 & 26 \\ 16 & 21 \end{bmatrix}$$

## 9.7 SUBTRACTION OF MATRICES

The difference between two matrices of the same order can be found by subtracting corresponding elements. For example,

$$\mathbf{A} = \begin{bmatrix} 6 & 9 \\ 4 & 15 \\ 8 & 11 \end{bmatrix} \quad \mathbf{B} = \begin{bmatrix} 4 & 2 \\ 6 & 9 \\ 7 & 2 \end{bmatrix} \quad \mathbf{A} - \mathbf{B} = \begin{bmatrix} 6 - 4 & 9 - 2 \\ 4 - 6 & 15 - 9 \\ 8 - 7 & 11 - 2 \end{bmatrix} = \begin{bmatrix} 2 & 7 \\ -2 & 6 \\ 1 & 9 \end{bmatrix}$$

## 9.8 SCALAR MULTIPLICATION

A *scalar* is a value expressed as one number. In order to multiply a matrix $\mathbf{C}$ by a scalar $\alpha$, every element of $\mathbf{C}$ is multiplied by $\alpha$, the product written $[(\alpha C_{ij})]$, $\alpha\mathbf{C}$, or $\mathbf{C}\alpha$. For example,

$$(2)\begin{bmatrix} a & b \\ c & d \end{bmatrix} = \begin{bmatrix} 2a & 2b \\ 2c & 2d \end{bmatrix}$$

In the same fashion, a scalar can be factored out of a matrix by reversing the above procedure.

## 9.9 MULTIPLICATION

The product of two matrices is obtained by multiplying the elements of each column of the second matrix by the elements of each row of the first matrix in turn, summing after each row by column multiplication. The value of the $i$th row by the $j$th column multiplication is known as the *inner product* of the $i$th row and $j$th column and assumes the $ij$ position in the product matrix. For example,

$$\overset{\mathbf{A}}{\begin{bmatrix} 2 & 6 \\ 4 & -2 \\ 5 & 7 \end{bmatrix}} \overset{\mathbf{B}}{\begin{bmatrix} a & b \\ c & d \end{bmatrix}} = \overset{\mathbf{AB}}{\begin{bmatrix} (2a+6c) & (2b+6d) \\ (4a-2c) & (4b-2d) \\ (5a+7c) & (5b+7d) \end{bmatrix}}$$

To perform the multiplication, the number of columns in $\mathbf{A}$ must be the same as the number of rows in $\mathbf{B}$. The product $\mathbf{AB}$ has the same number of rows as does the matrix $\mathbf{A}$ and the same number of columns as the matrix $\mathbf{B}$. $\mathbf{B}$ is said to be premultiplied by $\mathbf{A}$ and $\mathbf{A}$ is said to be postmultiplied by $\mathbf{B}$. Note that the products $\mathbf{AB}$ and $\mathbf{BA}$ are not necessarily equal, that is, $\mathbf{AB}$ does not necessarily equal $\mathbf{BA}$, although this is possible.

$$\overset{\mathbf{A}}{\begin{bmatrix} 0 & 2 & 4 & 6 \\ 6 & 4 & 2 & 0 \end{bmatrix}} \overset{\mathbf{B}}{\begin{bmatrix} 0 & 3 \\ 1 & 2 \\ 2 & 1 \\ 3 & 0 \end{bmatrix}} = \overset{\mathbf{AB}}{\begin{bmatrix} 28 & 8 \\ 8 & 28 \end{bmatrix}}$$

$$\overset{\mathbf{B}}{\begin{bmatrix} 0 & 3 \\ 1 & 2 \\ 2 & 1 \\ 3 & 0 \end{bmatrix}} \overset{\mathbf{A}}{\begin{bmatrix} 0 & 2 & 4 & 6 \\ 6 & 4 & 2 & 0 \end{bmatrix}} = \overset{\mathbf{BA}}{\begin{bmatrix} 18 & 12 & 6 & 0 \\ 12 & 10 & 8 & 6 \\ 6 & 8 & 10 & 12 \\ 0 & 6 & 12 & 18 \end{bmatrix}}$$

## 9.10 ELEMENT BY ELEMENT MULTIPLICATION

Sometimes referred to as term by term multiplication, element by element multiplication is performed by multiplying the corresponding elements of two matrices of the same order. For example,

$$
\overset{\mathbf{A}}{\begin{bmatrix} 4 & 9 \\ 7 & 4 \\ 3 & 8 \end{bmatrix}} \overset{\mathbf{B}}{\begin{bmatrix} 7 & 3 \\ 2 & 1 \\ 3 & 2 \end{bmatrix}} = \overset{\mathbf{AB}}{\begin{bmatrix} (4 \times 7) & (9 \times 3) \\ (7 \times 2) & (4 \times 1) \\ (3 \times 3) & (8 \times 2) \end{bmatrix}} = \overset{\mathbf{AB}}{\begin{bmatrix} 28 & 27 \\ 14 & 4 \\ 9 & 16 \end{bmatrix}}
$$

## 9.11 TRANSPOSE OF A MATRIX

The transpose of a matrix $\mathbf{A}$ is obtained by interchanging rows and columns of the matrix $\mathbf{A}$ and is designated as $\mathbf{A}^T$ or $\mathbf{A}'$. For example,

$$
\overset{\mathbf{A}}{\begin{bmatrix} 97 & 63 & 29 \\ 104 & 79 & 62 \end{bmatrix}} \overset{\mathbf{A'}}{\begin{bmatrix} 97 & 104 \\ 63 & 79 \\ 29 & 62 \end{bmatrix}}
$$

It is of interest to note that the transpose of a product matrix $(\mathbf{AB})'$ is equal to the product of the transposed matrices $\mathbf{B}'$ and $\mathbf{A}'$, that is, $(\mathbf{AB})' = \mathbf{B}'\mathbf{A}'$. Also, $(\mathbf{ABC})' = \mathbf{C}'\mathbf{B}'\mathbf{A}'$.

## 9.12 TRACE

The elements of a square matrix that have the same row and column subscripts, such as $a_{11}$, $a_{22}$, and so forth, are called *diagonal elements*. The sum of all the diagonal elements is called the *trace* of the matrix, and has considerable utility in many statistical analyses. The $a_{11}$ element is called the *leading term* or *leading element* of the matrix and all elements other than the diagonals are referred to as off-diagonal elements. For example,

$$
\overset{\mathbf{A}}{\begin{bmatrix} \underline{12} & 2 & 4 \\ 2 & \underline{8} & 3 \\ 4 & 3 & \underline{9} \end{bmatrix}}
\qquad
\begin{aligned}
&\text{Trace } \mathbf{A} = 12 + 8 + 9 = 29 \\
&\text{Leading element} = 12
\end{aligned}
$$

Referring to the product matrices $\mathbf{AB}$ and $\mathbf{BA}$ in Section 9.9, we can see that the trace of each is 56. It is true of any product matrix that trace$(\mathbf{AB})$ = trace$(\mathbf{BA})$. It can also be shown that trace$(\mathbf{ABC})$ = trace$(\mathbf{BCA})$ = trace$(\mathbf{CAB})$.

## 9.13 SYMMETRIC MATRIX

A symmetric matrix is a square matrix whose transpose is identical with the original matrix. Thus, $a_{ij} = a_{ji}$ for each pair of off-diagonal elements. For example,

$$\mathbf{A} \begin{bmatrix} 8 & 6 & 4 \\ 6 & 7 & 2 \\ 4 & 2 & 1 \end{bmatrix} \qquad \begin{array}{l} a_{12} = a_{21} = 6 \\ a_{13} = a_{31} = 4 \\ a_{23} = a_{32} = 2 \end{array}$$

## 9.14 IDENTITY MATRIX

An identity or unit matrix is a square matrix having all diagonal elements of 1 and all off-diagonal elements of 0 and is denoted by $\mathbf{I}$.

## 9.15 MATRIX INVERSION

The inverse of a matrix $\mathbf{A}$ is designated as $\mathbf{A}^{-1}$. It is customary to call a square matrix $\mathbf{A}$ whose determinant $\neq 0$ a *nonsingular* matrix, whereas if determinant $\mathbf{A} = 0$, $\mathbf{A}$ is called a *singular matrix*. Thus, matrix $\mathbf{A}$ has an inverse if and only if $\mathbf{A}$ is nonsingular. The determinant of matrix $\mathbf{A}$ is a scalar value denoted by "det $\mathbf{A}$."

The inverse of a $1 \times 1$ matrix is defined as the reciprocal of the element in that matrix. For example,

$$\mathbf{A} = [10] \qquad \mathbf{A}^{-1} = [\tfrac{1}{10}]$$

The inversion of matrices of order $2 \times 2$ and $3 \times 3$ by the *method of determinants* is performed by the following set of operations.

Step (i). Replace each element of $\mathbf{A}$ by its cofactor $A_{ij}$.

Step (ii). Divide each element of this matrix of cofactors by determinant $\mathbf{A}$.

Step (iii). Transpose the results.

The cofactor $A_{ij}$ of $a_{ij}$ is $(-1)^{i+j}$ times the determinant of the submatrix of order $n - 1$ obtained by deleting the $i$th row and $j$th column from $\mathbf{A}$. The determinant of a $1 \times 1$ matrix $[a]$ is defined as $[a]$.

The determinant of any square *submatrix* is called a *minor*, $[M_{ij}]$. An example of inversion of a $2 \times 2$ matrix is the following:

$$\mathbf{A} = \begin{bmatrix} a_{11} & a_{12} \\ a_{21} & a_{22} \end{bmatrix}$$

Cofactor $A_{11}$ of $a_{11} = (-1)^{1+1} a_{22} = a_{22}$

Cofactor $A_{12}$ of $a_{12} = (-1)^{1+2} a_{21} = -a_{21}$

Cofactor $A_{21}$ of $a_{21} = (-1)^{2+1} a_{12} = -a_{12}$

Cofactor $A_{22}$ of $a_{22} = (-1)^{2+2} a_{11} = a_{11}$

To proceed with step (ii) we must define the determinant of the $2 \times 2$ matrix. This determinant is defined as

$$\det \begin{bmatrix} a_{11} & a_{12} \\ a_{21} & a_{22} \end{bmatrix} = a_{11}a_{22} - a_{12}a_{21}$$

Dividing each of the cofactors by the determinant, we now have

$$\begin{bmatrix} \dfrac{a_{22}}{a_{11}a_{22} - a_{12}a_{21}} & -\dfrac{a_{21}}{a_{11}a_{22} - a_{12}a_{21}} \\ -\dfrac{a_{12}}{a_{11}a_{22} - a_{12}a_{21}} & \dfrac{a_{11}}{a_{11}a_{22} - a_{12}a_{21}} \end{bmatrix} = \left( \dfrac{1}{a_{11}a_{22} - a_{12}a_{21}} \right) \begin{bmatrix} a_{22} & -a_{21} \\ -a_{12} & a_{11} \end{bmatrix}$$

Now proceeding with step (iii), which is transposing the results, we have

$$\left( \dfrac{1}{a_{11}a_{22} - a_{12}a_{21}} \right) \begin{bmatrix} a_{22} & -a_{12} \\ -a_{21} & a_{11} \end{bmatrix}$$

which is the inverse of the original matrix.

To be certain that we do have the correct inverse, we can make use of the fact that $AA^{-1} = I$ and multiply the inverse by the original matrix.

$$\overset{A}{\begin{bmatrix} a_{11} & a_{12} \\ a_{21} & a_{22} \end{bmatrix}} \overset{A^{-1}}{\left( \dfrac{1}{a_{11}a_{22} - a_{12}a_{21}} \right) \begin{bmatrix} a_{22} & -a_{12} \\ -a_{21} & a_{11} \end{bmatrix}}$$

$$= \left( \dfrac{1}{a_{11}a_{22} - a_{12}a_{21}} \right) \overset{AA^{-1}}{\begin{bmatrix} a_{11}a_{22} - a_{12}a_{21} & -a_{11}a_{12} + a_{12}a_{11} \\ a_{21}a_{22} - a_{22}a_{21} & -a_{21}a_{12} + a_{22}a_{11} \end{bmatrix}} = \begin{bmatrix} 1 & 0 \\ 0 & 1 \end{bmatrix}$$

A numerical example is

$$\overset{A}{\begin{bmatrix} 4 & 12 \\ 9 & -10 \end{bmatrix}}$$

Cofactor of $a_{11} = (-1)^{1+1}(-10) = -10$

Cofactor of $a_{12} = (-1)^{1+2}(9) = -9$

Cofactor of $a_{21} = (-1)^{2+1}(12) = -12$

Cofactor of $a_{22} = (-1)^{2+2}(4) = 4$

$$\text{Matrix of cofactors} = \begin{bmatrix} -10 & -9 \\ -12 & 4 \end{bmatrix}$$

The determinant of $A$ can be found as $a_{11}a_{22} - a_{12}a_{21}$ as shown earlier or by multiplying elements of the corresponding rows or the corresponding columns of the matrix $A$ and the matrix of cofactors. Thus,

$$\text{Determinant } A = |A| = (4)(-10) + (12)(-9) = -148 \quad \text{or}$$

$$(9)(-12) + (-10)(4) = -148 \quad \text{or}$$

$$(4)(-10) + (9)(-12) = -148$$

Now, dividing the matrix of cofactors by the determinant,

$$\begin{bmatrix} \dfrac{-10}{-148} & \dfrac{-9}{-148} \\[2mm] \dfrac{-12}{-148} & \dfrac{4}{-148} \end{bmatrix}$$

and transposing the results

$$\mathbf{A}^{-1} = \begin{bmatrix} \dfrac{10}{148} & \dfrac{12}{148} \\[2mm] \dfrac{9}{148} & -\dfrac{4}{148} \end{bmatrix}$$

and

$$\overset{\mathbf{A}}{\begin{bmatrix} 4 & 12 \\ 9 & -10 \end{bmatrix}} \overset{\mathbf{A}^{-1}}{\begin{bmatrix} \dfrac{10}{148} & \dfrac{12}{148} \\[2mm] \dfrac{9}{148} & -\dfrac{4}{148} \end{bmatrix}} = \overset{\mathbf{I}}{\begin{bmatrix} 1 & 0 \\ 0 & 1 \end{bmatrix}}$$

This procedure can be stated more succinctly for inverting a $2 \times 2$ matrix:

1. Calculate the determinant.
2. Interchange the diagonal elements.
3. Change the signs of the off-diagonal elements.
4. Divide all elements by the determinant.

A numerical example of the inversion of a $3 \times 3$ matrix is

$$\mathbf{A} = \begin{bmatrix} 4 & -1 & 2 \\ 3 & 4 & 1 \\ -2 & -2 & 4 \end{bmatrix}$$

Cofactor of $a_{11} = (-1)^{1+1}(18) = 18$

Cofactor of $a_{12} = (-1)^{1+2}(14) = -14$

Cofactor of $a_{13} = (-1)^{1+3}(2) = 2$

Cofactor of $a_{21} = (-1)^{2+1}(0) = 0$

Cofactor of $a_{22} = (-1)^{2+2}(20) = 20$

Cofactor of $a_{23} = (-1)^{2+3}(-10) = 10$

Cofactor of $a_{31} = (-1)^{3+1}(-9) = -9$

Cofactor of $a_{32} = (-1)^{3+2}(-2) = 2$

Cofactor of $a_{33} = (-1)^{3+3}(19) = 19$

In calculating the cofactor of $a_{11}$, for example, we would delete the first row and column, leaving the matrix

$$\begin{bmatrix} 4 & 1 \\ -2 & 4 \end{bmatrix}$$

The determinant of this submatrix is $(4)(4) - (1)(-2) = 18$. The matrix of cofactors is now

$$\begin{bmatrix} 18 & -14 & 2 \\ 0 & 20 & 10 \\ -9 & 2 & 19 \end{bmatrix}$$

The determinant of the matrix $\mathbf{A}$ is defined as

$$a_{11}a_{22}a_{33} + a_{12}a_{23}a_{31} + a_{13}a_{32}a_{21} - a_{13}a_{22}a_{31} - a_{12}a_{21}a_{33} - a_{11}a_{32}a_{23}$$

$$= (4)(4)(4) + (-1)(1)(-2) + (2)(-2)(3) - (2)(4)(-2) - (-1)(3)(4) - (4)(-2)(1) = 90$$

The determinant can also be found by the product of corresponding rows or columns of the matrix $\mathbf{A}$ and its matrix of cofactors. Multiplying the first rows of each and the first columns of each we find

$$|\mathbf{A}| = (4)(18) + (-1)(-14) + (2)(2) = 90$$

$$|\mathbf{A}| = (4)(18) + (3)(0) + (-2)(-9) = 90$$

Dividing by the determinant and transposing, we have

$$\mathbf{A}^{-1} \begin{bmatrix} \dfrac{18}{90} & \dfrac{0}{90} & -\dfrac{9}{90} \\[2ex] \dfrac{-14}{90} & \dfrac{20}{90} & \dfrac{2}{90} \\[2ex] \dfrac{2}{90} & \dfrac{10}{90} & \dfrac{19}{90} \end{bmatrix}$$

and

$$\mathbf{A}\mathbf{A}^{-1} \begin{bmatrix} \dfrac{72 + 14 + 4}{90} & \dfrac{-20 + 20}{90} & \dfrac{-36 - 2 + 38}{90} \\[2ex] \dfrac{54 - 56 + 2}{90} & \dfrac{80 + 10}{90} & \dfrac{-27 + 8 + 19}{90} \\[2ex] \dfrac{-36 + 28 + 8}{90} & \dfrac{-40 + 40}{90} & \dfrac{18 - 4 + 76}{90} \end{bmatrix} = \begin{bmatrix} 1 & 0 & 0 \\ 0 & 1 & 0 \\ 0 & 0 & 1 \end{bmatrix} = \mathbf{I}$$

A second method of inversion, presented by Crout (1951), is frequently used for the inversion of matrices of order $3 \times 3$ and larger. While the method described here is used only for symmetric matrices, it can be adapted readily to the inversion of non-symmetric matrices. The inversion process follows the pattern shown, where $\mathbf{A}$ is the matrix to be inverted and $\mathbf{C}$ is the matrix inverse, $\mathbf{A}^{-1}$.

$$\mathbf{A}$$

$$\begin{bmatrix} a_{11} & a_{12} & a_{13} \\ a_{21} & a_{22} & a_{23} \\ a_{31} & a_{32} & a_{33} \end{bmatrix}$$

$$
\mathbf{I}
$$
$$
\begin{bmatrix} 1 & 0 & 0 \\ 0 & 1 & 0 \\ 0 & 0 & 1 \end{bmatrix}
$$
$$
[S_1 \quad S_2 \quad S_3]
$$

$$
\mathbf{B}
$$
$$
\begin{bmatrix} x_{11} & x_{12} & x_{13} \\ x_{21} & x_{22} & x_{23} \\ x_{31} & x_{32} & x_{33} \end{bmatrix}
\begin{bmatrix} b_{11} & & \\ b_{21} & b_{22} & \\ b_{31} & b_{32} & b_{33} \end{bmatrix}
\begin{bmatrix} R_1 \\ R_2 \\ R_3 \end{bmatrix}
$$

$$
\mathbf{N} \qquad\qquad \mathbf{C}
$$
$$
\begin{bmatrix} 1 & x_{42} & x_{43} \\ & 1 & x_{53} \\ & & 1 \end{bmatrix}
\begin{bmatrix} c^{11} & c^{12} & c^{13} \\ c^{21} & c^{22} & c^{23} \\ c^{31} & c^{32} & c^{33} \end{bmatrix}
$$
$$
[T_1 \quad T_2 \quad T_3]
$$

The capital $T$'s are used as column checks and the sum of all the entries from the diagonal (including the diagonal) to the bottom of the column should agree within rounding error with the $T$'s calculated as shown below.

The steps in the inversion process are as follows:

1. Write the identity or unit matrix below the matrix to be inverted.
2. Add the elements in each column, including those in the identity matrix, to develop the $S$ values.
3. Copy the elements in the first column of the matrix to be inverted, the identity matrix, and $S_1$, that is, $T_1 = S_1$.
4. Divide all entries below the diagonal element by the diagonal element: $x_{21}/x_{11} = x_{12}$, $x_{31}/x_{11} = x_{13}$, $1/x_{11} = b_{11}$, and $T_1/x_{11} = R_1$. $R_1$ is the sum of all elements to the right of the diagonal plus 1, a row check.
5. Calculate the column and row in which $x_{22}$ appears: $x_{22} = a_{22} - x_{21}x_{12}$, $x_{32} = a_{32} - x_{31}x_{12}$, $x_{42} = 0 - (1)(x_{12})$, and $T_2 = S_2 - T_1x_{12}$. $T_2$ is the sum of all values in that column from the diagonal down (including the diagonal). $x_{23} = x_{32}/x_{22}$, $b_{21} = x_{42}/x_{22}$, $b_{22} = 1/x_{22}$, and $R_2 = T_2/x_{22}$.
6. Calculate the column and row in which $x_{33}$ appears: $x_{33} = a_{33} - x_{31}x_{13} - x_{32}x_{23}$, $x_{43} = 0 - (1)(x_{13}) - x_{42}x_{23}$, $x_{53} = 0 - (1)(x_{23})$, $T_3 = S_3 - T_1x_{13} - T_2x_{23}$, $b_{31} = x_{43}/x_{33}$, $b_{32} = x_{53}/x_{33}$, $b_{33} = 1/x_{33}$, and $R_3 = T_3/x_{33}$.
7. $\mathbf{C} = \mathbf{NB}$.

Following the pattern seen in the above calculations, one can invert much larger matrices with this procedure. For example,

$$
\mathbf{A}
$$
$$
\begin{bmatrix} 12 & 2 & 1 \\ 2 & 8 & 3 \\ 1 & 3 & 7 \end{bmatrix}
$$

$$
\mathbf{I} \quad
\begin{bmatrix}
1 & 0 & 0 \\
0 & 1 & 0 \\
0 & 0 & 1
\end{bmatrix}
$$

$$[16 \quad 14 \quad 12]$$

$$
\begin{bmatrix}
12 & \dfrac{1}{6} & \dfrac{1}{12} \\[6pt]
2 & \dfrac{23}{3} & \dfrac{17}{46} \\[6pt]
1 & \dfrac{17}{6} & \dfrac{270}{46}
\end{bmatrix}
\overset{\mathbf{B}}{
\begin{bmatrix}
\dfrac{1}{12} & & \\[6pt]
\dfrac{-1}{46} & \dfrac{3}{23} & \\[6pt]
\dfrac{-1}{270} & \dfrac{-17}{270} & \dfrac{46}{270}
\end{bmatrix}}
\begin{bmatrix}
\dfrac{4}{3} \\[6pt]
\dfrac{34}{23} \\[6pt]
\dfrac{149}{135}
\end{bmatrix}
$$

$$
\overset{\mathbf{N}}{
\begin{bmatrix}
1 & \dfrac{-1}{6} & \dfrac{-1}{46} \\[6pt]
0 & 1 & \dfrac{-17}{46} \\[6pt]
0 & 0 & 1
\end{bmatrix}}
\overset{\mathbf{C}}{
\begin{bmatrix}
47 & -11 & -2 \\
-11 & 83 & -34 \\
-2 & -34 & 92
\end{bmatrix}}
\left(\dfrac{1}{540}\right)
$$

$$
\begin{bmatrix}
16 & \dfrac{34}{3} & \dfrac{298}{46}
\end{bmatrix}
$$

## 9.16 SOLUTION OF EQUATIONS

Among the many uses of a matrix inverse, the most common one in the field of statistical analysis is in the solution of normal or least-squares equations. Additionally, the elements of the matrix inverse serve many purposes. We can see the use of the inverse in solving equations by considering the following set of equations, presented earlier in Section 9.4:

$$4x + 12y = 14$$

$$9x - 10y = -18$$

Given this set of equations, we want to find values of $x$ and $y$ that are said to satisfy the equations. With such a simple set as this, we can find these values readily by the method of elimination. To do this, we multiply the first equation by 9 and the second equation by 4. By subtracting the second equation from the first, we eliminate the $x$ values and we can solve for $y$ as

$$(1) \qquad 36x + 108y = 126$$

$$(2) \qquad 36x - 40y = -72$$

$$(1) - (2) \qquad 148y = 198$$

$$y = \frac{99}{74}$$

Substituting this value of $y$ in the first equation we can now solve for $x$ as

$$4x + 12\left(\frac{99}{74}\right) = 14$$

$$4x = 14 - \frac{1188}{74}$$

$$x = -\frac{38}{74}$$

Now substituting the values we have found for $x$ and $y$ in the two original equations, we find

$$(4)\left(\frac{-38}{74}\right) + (12)\left(\frac{99}{74}\right) = \frac{1036}{74} = 14$$

$$(9)\left(-\frac{38}{74}\right) - (10)\left(\frac{99}{74}\right) = -\frac{1332}{74} = -18$$

Thus, we can see that the values we have found for $x$ and $y$ satisfy the equations.

However, with a large number of equations the work becomes prohibitive in solving equations by this method. In addition, the inverse elements, needed frequently in statistical analysis, are not obtained. Instead, we can solve these equations by making use of the matrix procedures discussed in this chapter. We can put the coefficients of $x$ and $y$ into matrix form and the values of 14 and $-18$ into a column vector and express the equations in matrix form as

$$\underset{\mathbf{C}}{\begin{bmatrix} 4 & 12 \\ 9 & -10 \end{bmatrix}} \underset{\mathbf{B}}{\begin{bmatrix} x \\ y \end{bmatrix}} = \underset{\mathbf{Y}}{\begin{bmatrix} 14 \\ -18 \end{bmatrix}}$$

We can express this relationship as

$$\mathbf{CB} = \mathbf{Y}$$

and if we could divide both sides of the equation by $\mathbf{C}$ we could solve for $\mathbf{B}$, the matrix of unknown values of $x$ and $y$. However, such an operation does not exist in matrix algebra. Instead, we take advantage of the fact that $\mathbf{CC^{-1}} = \mathbf{I}$ and multiply both sides of the equation by $\mathbf{C^{-1}}$, giving

$$\mathbf{C^{-1}CB} = \mathbf{C^{-1}Y}$$

and since $\mathbf{CC^{-1}} = \mathbf{I}$ and $\mathbf{IB} = \mathbf{B}$ we have

$$\mathbf{B} = \mathbf{C^{-1}Y}$$

We have previously seen that

$$\mathbf{C}^{-1} = \begin{bmatrix} \dfrac{10}{148} & \dfrac{12}{148} \\[2ex] \dfrac{12}{148} & -\dfrac{4}{148} \end{bmatrix} \quad \text{and} \quad \mathbf{Y} = \begin{bmatrix} 14 \\ -18 \end{bmatrix}$$

leading to

$$\mathbf{C}^{-1}\mathbf{Y} = \begin{bmatrix} -\dfrac{38}{74} = x \\[2ex] \dfrac{99}{74} = y \end{bmatrix}$$

and yielding the same values of $x$ and $y$ found by the method of elimination. The procedure has found great utility in statistical analysis and is used repeatedly throughout this text.

## 9.17 NECESSITY FOR RESTRICTIONS OR CONSTRAINTS

Very frequently, the coefficient matrices developed in statistical procedures are singular; that is, their determinants are zero. To obtain an inverse to the matrix, and thus a solution to the equations, some restriction or constraint must be applied to the coefficient matrix, which is frequently referred to as the left-hand members (LHM). The same restrictions must be applied to the values to the right of the equalities, referred to as the right-hand members (RHM) when using the matrix to solve equations.

Let us look at the following matrix which can represent the coefficient matrix (LHM) in a one-way analysis of variance with two levels of $A$ with five observations in each level. $\hat{\mu}$ represents an estimate of the population mean and the $\hat{a}_i$ represent estimates of treatment deviations from the population mean:

$$\begin{array}{cc} & \begin{array}{ccc} \hat{\mu} & \hat{a}_1 & \hat{a}_2 \end{array} \\ \begin{array}{c} \mu \\ a_1 \\ a_2 \end{array} & \begin{bmatrix} 10 & 5 & 5 \\ 5 & 5 & \\ 5 & & 5 \end{bmatrix} \end{array}$$

If we were to attempt to invert this matrix we would find that its determinant is 0. To invert this matrix it must be reduced to the appropriate rank or number of degrees of freedom that it represents. The row (or column) *rank* of the matrix is defined as the number of linearly independent rows (or columns) therein (Searle, 1966). Since there is one degree of freedom for the mean and one degree of freedom between the two levels of $A$, this $3 \times 3$ matrix must be reduced to a $2 \times 2$. To reduce the matrix to the size equal to its rank certain restrictions or constraints must be imposed.

One constraint that may be imposed, and the type most commonly used, is that $\Sigma_i \, \hat{a}_i = 0$; that is, the sum of the deviations must equal 0. This constraint is imposed by subtracting the last (or any) equation for $\hat{a}_i$ from all other equations for $\hat{a}_i$ by column and row (or row and column) in turn. Thus, subtraction by column yields

$$\begin{array}{cc} & \hat{\mu} \quad \hat{a}_1 \\ \begin{matrix} \mu \\ a_1 \\ a_2 \end{matrix} & \begin{bmatrix} 10 & 0 \\ 5 & 5 \\ 5 & -5 \end{bmatrix} \end{array}$$

and subtraction by row yields

$$\begin{array}{cc} & \hat{\mu} \quad \hat{a}_1 \\ \begin{matrix} \mu \\ a_1 \end{matrix} & \begin{bmatrix} 10 & 0 \\ 0 & 10 \end{bmatrix} \end{array}$$

Since this matrix does have a nonzero determinant, its inverse can be found and then used in the solution of equations. The same constraint must be imposed on the RHMs, which in this case would mean subtracting the sum for the second level of $A$ from the first (or all others if there are more than two levels of $A$).

A second constraint that is sometimes used is to consider any one of the $a_i$ (usually the last) to be zero. To impose this constraint we merely delete the last row and column, giving

$$\begin{array}{cc} & \hat{\mu} \quad \hat{a}_1 \\ \begin{matrix} \mu \\ a_1 \end{matrix} & \begin{bmatrix} 10 & 5 \\ 5 & 10 \end{bmatrix} \end{array}$$

a matrix that does have an inverse. The same constraint, of course, must be imposed on the RHMs.

## 9.18  INVERSION OF SPECIAL MATRICES

The inversion of a diagonal matrix, one whose off-diagonal elements are all zero, is obtained readily since it is made up of the reciprocals of the diagonals. For example,

$$\mathbf{A} = \begin{bmatrix} 5 & 0 & 0 \\ 0 & 7 & 0 \\ 0 & 0 & 12 \end{bmatrix} \qquad \mathbf{A}^{-1} = \begin{bmatrix} \frac{1}{5} & 0 & 0 \\ 0 & \frac{1}{7} & 0 \\ 0 & 0 & \frac{1}{12} \end{bmatrix}$$

Sometimes matrices can be broken up into separate submatrices, each independent of one another. In such cases, each submatrix can be inverted separately. The following matrix can be partitioned into three submatrices and each inverted separately since there are no ties to the other submatrices. For example,

$$A = \begin{bmatrix} 45 & 0 & 0 & 0 & 0 \\ 0 & 30 & 15 & 0 & 0 \\ 0 & 15 & 30 & 0 & 0 \\ 0 & 0 & 0 & 30 & 15 \\ 0 & 0 & 0 & 15 & 30 \end{bmatrix} \qquad A^{-1} = \begin{bmatrix} \dfrac{1}{45} & 0 & 0 & 0 & 0 \\ 0 & \dfrac{2}{45} & \dfrac{-1}{45} & 0 & 0 \\ 0 & \dfrac{-1}{45} & \dfrac{2}{45} & 0 & 0 \\ 0 & 0 & 0 & \dfrac{2}{45} & \dfrac{-1}{45} \\ 0 & 0 & 0 & \dfrac{-1}{45} & \dfrac{2}{45} \end{bmatrix}$$

## 9.19 INVERSION BY PARTITIONING

On occasion it is useful to partition a matrix into submatrices and complete the inversion as described below as shown by Searle (1966). One major advantage of this procedure is that the inversion of smaller matrices is involved. The procedure is shown first for a symmetric matrix.

Consider the following matrix, which is partitioned as shown:

$$L = \begin{bmatrix} 20 & 6 & 4 & 2 \\ 6 & 10 & 2 & 2 \\ 4 & 2 & 8 & 2 \\ 2 & 2 & 2 & 4 \end{bmatrix}$$

We then have

$$L = \begin{bmatrix} D & N \\ N' & S \end{bmatrix} \quad \text{and} \quad L^{-1} = \begin{bmatrix} D & N \\ N' & S \end{bmatrix}^{-1} = \begin{bmatrix} A & G \\ G' & C \end{bmatrix}$$

Premultiplying $L^{-1}$ by $L$ we have

$$\begin{bmatrix} D & N \\ N' & S \end{bmatrix}\begin{bmatrix} A & G \\ G' & C \end{bmatrix} = I = \begin{bmatrix} I & 0 \\ 0 & I \end{bmatrix}$$

From this we can see that

$$DA + NG' = I$$

$$DG + NC = 0$$

$$N'A + SG' = 0$$

$$N'G + SC = I$$

We can now solve sequentially for **A**, **G**, and **C** (with **G'** = transpose of **G**):

$$DA + NG' = I \qquad\qquad DG + NC = 0$$

$$DA = I - NG' \qquad\qquad DG = -NC$$

$$A = D^{-1}(I - NG') \qquad\qquad G = -D^{-1}NC$$

$$N'G + SC = I$$

$$-N'D^{-1}NC + SC = I$$

$$(S - N'D^{-1}N)C = I$$

$$C = (S - N'D^{-1}N)^{-1}$$

As a first step in the inversion of the given matrix **L**, we can first find $D^{-1}$,

$$D = \begin{bmatrix} 20 & 6 \\ 6 & 10 \end{bmatrix} \qquad D^{-1} = \begin{bmatrix} \dfrac{5}{82} & -\dfrac{3}{82} \\[2mm] -\dfrac{3}{82} & \dfrac{10}{82} \end{bmatrix}$$

Next, perform the multiplication $N'D^{-1}N$,

$$\underset{N'}{\begin{bmatrix} 4 & 2 \\ 2 & 2 \end{bmatrix}} \underset{D^{-1}}{\begin{bmatrix} \dfrac{5}{82} & -\dfrac{3}{82} \\[2mm] -\dfrac{3}{82} & \dfrac{10}{82} \end{bmatrix}} = \underset{N'D^{-1}}{\begin{bmatrix} \dfrac{14}{82} & \dfrac{8}{82} \\[2mm] \dfrac{4}{82} & \dfrac{14}{82} \end{bmatrix}} \quad \text{and} \quad N'D^{-1}N = \begin{bmatrix} \dfrac{36}{41} & \dfrac{22}{41} \\[2mm] \dfrac{22}{41} & \dfrac{18}{41} \end{bmatrix}$$

Now, perform the subtraction $S - N'D^{-1}N$,

$$\underset{S}{\begin{bmatrix} 8 & 2 \\ 2 & 4 \end{bmatrix}} - \underset{N'D^{-1}N}{\begin{bmatrix} \dfrac{36}{41} & \dfrac{22}{41} \\[2mm] \dfrac{22}{41} & \dfrac{18}{41} \end{bmatrix}} = \underset{S-N'D^{-1}N}{\begin{bmatrix} \dfrac{292}{41} & \dfrac{60}{41} \\[2mm] \dfrac{60}{41} & \dfrac{146}{41} \end{bmatrix}}$$

We can now obtain $C = (S - N'D^{-1}N)^{-1}$,

$$C = \begin{bmatrix} \dfrac{73}{476} & -\dfrac{30}{476} \\[2mm] -\dfrac{30}{476} & \dfrac{146}{476} \end{bmatrix}$$

Now, calculate $G = -D^{-1}NC$,

$$\underset{-D^{-1}N}{\begin{bmatrix} -\dfrac{7}{41} & -\dfrac{2}{41} \\[2mm] -\dfrac{4}{41} & -\dfrac{7}{41} \end{bmatrix}} \underset{C}{\begin{bmatrix} \dfrac{73}{476} & -\dfrac{30}{476} \\[2mm] -\dfrac{30}{476} & \dfrac{146}{476} \end{bmatrix}} = \underset{-D^{-1}NC=G}{\begin{bmatrix} -\dfrac{11}{476} & -\dfrac{2}{476} \\[2mm] -\dfrac{2}{476} & -\dfrac{22}{476} \end{bmatrix}}$$

Last, calculate $\mathbf{A} = \mathbf{D}^{-1}(\mathbf{I} - \mathbf{NG}') = \mathbf{D}^{-1} - \mathbf{D}^{-1}\mathbf{NG}'$,

$$
\underset{\mathbf{D}^{-1}\mathbf{N}}{\begin{bmatrix} \dfrac{7}{41} & \dfrac{2}{41} \\[2mm] \dfrac{4}{41} & \dfrac{7}{41} \end{bmatrix}}
\underset{\mathbf{G}'}{\begin{bmatrix} -\dfrac{11}{476} & -\dfrac{2}{476} \\[2mm] -\dfrac{2}{476} & -\dfrac{22}{476} \end{bmatrix}}
=
\underset{\mathbf{D}^{-1}\mathbf{NG}'}{\begin{bmatrix} -\dfrac{81}{19,516} & -\dfrac{58}{19,516} \\[2mm] -\dfrac{58}{19,516} & -\dfrac{162}{19,516} \end{bmatrix}}
$$

$$
\underset{\mathbf{A}=\mathbf{D}^{-1}-\mathbf{D}^{-1}\mathbf{NG}'}{\begin{bmatrix} \dfrac{31}{476} & -\dfrac{16}{476} \\[2mm] -\dfrac{16}{476} & \dfrac{62}{476} \end{bmatrix}}
$$

The completed inverse is then

$$
\mathbf{L}^{-1} = \left(\frac{1}{476}\right) \begin{bmatrix} 31 & -16 & -11 & -2 \\ -16 & 62 & -2 & -22 \\ -11 & -2 & 73 & -30 \\ -2 & -22 & -30 & 146 \end{bmatrix}
$$

For the inversion of a nonsymmetric matrix by partitioning, let us consider the matrix $\mathbf{L}$, partitioned as

$$
\mathbf{L} = \begin{bmatrix} \mathbf{D} & \mathbf{N} \\ \mathbf{M} & \mathbf{S} \end{bmatrix} \qquad \mathbf{L}^{-1} = \begin{bmatrix} \mathbf{A} & \mathbf{G} \\ \mathbf{F} & \mathbf{C} \end{bmatrix}
$$

Now,

$$
\begin{bmatrix} \mathbf{D} & \mathbf{N} \\ \mathbf{M} & \mathbf{S} \end{bmatrix}\begin{bmatrix} \mathbf{A} & \mathbf{G} \\ \mathbf{F} & \mathbf{C} \end{bmatrix} = \mathbf{I} = \begin{bmatrix} \mathbf{I} & \mathbf{0} \\ \mathbf{0} & \mathbf{I} \end{bmatrix}
$$

resulting in

$$
\mathbf{DA} + \mathbf{NF} = \mathbf{I} \qquad\qquad \mathbf{DG} + \mathbf{NC} = \mathbf{0}
$$
$$
\mathbf{DA} = \mathbf{I} - \mathbf{NF} \qquad\qquad \mathbf{DG} = -\mathbf{NC}
$$
$$
\mathbf{A} = \mathbf{D}^{-1}(\mathbf{I} - \mathbf{NF}) \qquad\qquad \mathbf{G} = -\mathbf{D}^{-1}\mathbf{NC}
$$
$$
\mathbf{MG} + \mathbf{SC} = \mathbf{I}
$$
$$
-\mathbf{MD}^{-1}\mathbf{NC} + \mathbf{SC} = \mathbf{I}
$$
$$
(\mathbf{S} - \mathbf{MD}^{-1}\mathbf{N})\mathbf{C} = \mathbf{I}
$$
$$
\mathbf{C} = (\mathbf{S} - \mathbf{MD}^{-1}\mathbf{N})^{-1}
$$

Since it is also true that

$$
\begin{bmatrix} \mathbf{A} & \mathbf{G} \\ \mathbf{F} & \mathbf{C} \end{bmatrix}\begin{bmatrix} \mathbf{D} & \mathbf{N} \\ \mathbf{M} & \mathbf{S} \end{bmatrix} = \mathbf{I} = \begin{bmatrix} \mathbf{I} & \mathbf{0} \\ \mathbf{0} & \mathbf{I} \end{bmatrix}
$$

we find

$$FD + CM = 0$$

$$F = -CMD^{-1}$$

which completes the inverse.

## EXERCISES

**9.1.** Given the matrices

$$A = \begin{bmatrix} 4 & 7 & 9 \\ 3 & 4 & 12 \\ 2 & 8 & 14 \end{bmatrix} \quad B = \begin{bmatrix} 9 & 12 & 14 \\ 6 & 5 & 9 \\ 4 & 7 & 12 \end{bmatrix} \quad C = \begin{bmatrix} 10 & 20 & 22 \\ 19 & 22 & 15 \\ 20 & 16 & 12 \end{bmatrix}$$

perform the following operations:
(a) **A + B.**
(b) **B − C.**
(c) **B + C − A.**
(d) Show that **A′ + B′ = (A + B)′.**

**9.2.** Given the matrices

$$A = \begin{bmatrix} 2 & 4 \\ 6 & 8 \\ 10 & 12 \end{bmatrix} \quad B = \begin{bmatrix} 6 & 8 & 10 \\ 12 & 14 & 16 \end{bmatrix}$$

perform the following operations:
(a) **AB.**
(b) **BA.**
(c) **BB′.**
(d) **B′B.**
Show that
(e) trace **BB′** = trace **B′B.**
(f) **(AB)′ = B′A′.**

**9.3.** Given the matrices

$$A = \begin{bmatrix} 8 & 12 & 14 \\ 7 & 4 & 2 \\ 3 & 9 & 6 \end{bmatrix} \quad B = \begin{bmatrix} 7 & 4 & 6 \\ 3 & 2 & 1 \\ 1 & 2 & 3 \end{bmatrix}$$

(a) Calculate **AA′, AB′, BA′,** and **BB′** and show that

$$(A + B)(A + B)' = AA' + AB' + BA' + BB'.$$

(b) Show that (i) **AB′ = (BA′)′** and (ii) **A(B + B′) = AB + AB′.**

**9.4.** Given the matrices

$$A = \begin{bmatrix} 6 & 8 & 10 \\ 12 & 14 & 16 \\ 2 & 4 & 6 \end{bmatrix} \quad B = \begin{bmatrix} 3 & 6 & 9 \\ 12 & 15 & 18 \\ 21 & 24 & 27 \end{bmatrix}$$

(a) Perform the multiplication $\frac{1}{3}$**A** times $\frac{1}{2}$**B.**
(b) Find the term by term product of **AB.**

**9.5.** Invert the following matrices, using the method of determinants. In each case, check to be certain that the matrix times its inverse is equal to the identity matrix.

$$A = \begin{bmatrix} 4 & 7 \\ 2 & 6 \end{bmatrix} \qquad B = \begin{bmatrix} 12 & -11 \\ -4 & 12 \end{bmatrix} \qquad C = \begin{bmatrix} 4 & -7 \\ -6 & 8 \end{bmatrix}$$

$$X = \begin{bmatrix} 12 & 7 & 8 \\ 7 & 14 & 6 \\ 2 & 5 & 8 \end{bmatrix} \qquad Y = \begin{bmatrix} 12 & 2 & 1 \\ 2 & 8 & 3 \\ 1 & 3 & 7 \end{bmatrix} \qquad Z = \begin{bmatrix} 18 & 4 & 2 \\ 4 & 12 & 4 \\ 2 & 4 & 10 \end{bmatrix}$$

**9.6.** Invert the following matrices using the Crout procedure.

$$M = \begin{bmatrix} 15 & 2 & 1 \\ 2 & 10 & 4 \\ 1 & 4 & 9 \end{bmatrix} \qquad N = \begin{bmatrix} 6 & 2 & 1 \\ 2 & 4 & 1 \\ 1 & 1 & 3 \end{bmatrix}$$

**9.7.** Solve the following sets of equations by using the matrix inverse, following the equation $B = C^{-1}Y$, where $B$ is the column vector of unknowns, $C^{-1}$ is the inverse of the coefficient matrix (LHMs), and $Y$ is the vector of right-hand members of the equations.

(a) $2x + 6y = 12$
    $5x - 3y = 8.$
(b) $4a + 3c = 15$
    $7a + 2c = 20.$
(c) $x + y = 5$
    $3x + 4y = 16.$

**9.8.** Invert the following matrix by the method of partitioning, following the steps outlined in Section 9.19. You may prefer to work with decimals rather than fractions.

$$L = \begin{bmatrix} 27 & 2 & 4 & 1 \\ & D & & N & \\ 2 & 12 & 5 & 5 \\ \hline 4 & 5 & 14 & 5 \\ & N' & & S & \\ 1 & 5 & 5 & 11 \end{bmatrix}$$

# Appendix

**TABLE A.1**  TEN THOUSAND RANDOM DIGITS

|    | 00–04 | 05–09 | 10–14 | 15–19 | 20–24 | 25–29 | 30–34 | 35–39 | 40–44 | 45–49 |
|----|-------|-------|-------|-------|-------|-------|-------|-------|-------|-------|
| 00 | 22808 | 04391 | 45529 | 53968 | 57136 | 98228 | 85485 | 13801 | 68194 | 56382 |
| 01 | 49305 | 36965 | 44849 | 64987 | 59501 | 35141 | 50159 | 57369 | 76913 | 75739 |
| 02 | 81934 | 19920 | 73316 | 69243 | 69605 | 17022 | 53264 | 83417 | 55193 | 92929 |
| 03 | 10840 | 13508 | 48120 | 22467 | 54505 | 70536 | 91206 | 81038 | 22418 | 34800 |
| 04 | 99555 | 73289 | 59605 | 37105 | 24621 | 44100 | 72832 | 12268 | 97089 | 68112 |
| 05 | 32677 | 45709 | 62337 | 35132 | 45128 | 96761 | 08745 | 53388 | 98353 | 46724 |
| 06 | 09401 | 75407 | 27704 | 11569 | 52842 | 83543 | 44750 | 03177 | 50511 | 15301 |
| 07 | 73424 | 31711 | 65519 | 74869 | 56744 | 40864 | 75315 | 89866 | 96563 | 75142 |
| 08 | 37075 | 81378 | 59472 | 71858 | 86903 | 66860 | 03757 | 32723 | 54273 | 45477 |
| 09 | 02060 | 37158 | 55244 | 44812 | 45369 | 78939 | 08048 | 28036 | 40946 | 03898 |
| 10 | 94719 | 43565 | 40028 | 79866 | 43137 | 28063 | 52513 | 66405 | 71511 | 66135 |
| 11 | 70234 | 48272 | 59621 | 88778 | 16536 | 36505 | 41724 | 24776 | 63971 | 01685 |
| 12 | 07972 | 71752 | 92745 | 86465 | 01845 | 27416 | 50519 | 48458 | 68460 | 63113 |
| 13 | 58521 | 64882 | 26993 | 48104 | 61307 | 73933 | 17214 | 44827 | 88306 | 78177 |
| 14 | 32580 | 45202 | 21148 | 09684 | 39411 | 04892 | 02055 | 75276 | 51831 | 85686 |
| 15 | 88796 | 30829 | 35009 | 22695 | 23694 | 11220 | 71006 | 26720 | 39476 | 60538 |
| 16 | 31525 | 82746 | 78935 | 82980 | 61236 | 28940 | 96341 | 13790 | 66247 | 33839 |
| 17 | 02747 | 35989 | 70387 | 89571 | 34570 | 17002 | 79223 | 96817 | 31681 | 15207 |
| 18 | 46651 | 28987 | 20625 | 61347 | 63981 | 41085 | 67412 | 29053 | 00724 | 14841 |
| 19 | 43598 | 14436 | 33521 | 55637 | 39789 | 26560 | 66404 | 71802 | 18763 | 80560 |
| 20 | 30596 | 92319 | 11474 | 64546 | 60030 | 73795 | 60809 | 24016 | 29166 | 36059 |
| 21 | 56198 | 64370 | 85771 | 62633 | 78240 | 05766 | 32419 | 35769 | 14057 | 80674 |
| 22 | 68266 | 67544 | 06464 | 84956 | 18431 | 04015 | 89049 | 15098 | 12018 | 89338 |
| 23 | 31107 | 28597 | 65102 | 75599 | 17496 | 87590 | 68848 | 33021 | 69855 | 54015 |
| 24 | 37555 | 05069 | 38680 | 87274 | 55152 | 21792 | 77219 | 48732 | 03377 | 01160 |

| | 00–04 | 05–09 | 10–14 | 15–19 | 20–24 | 25–29 | 30–34 | 35–39 | 40–44 | 45–49 |
|---|---|---|---|---|---|---|---|---|---|---|
| 25 | 90463 | 27249 | 43845 | 94391 | 12145 | 36882 | 48906 | 52336 | 00780 | 74407 |
| 26 | 99189 | 88731 | 93531 | 52638 | 54989 | 04237 | 32978 | 59902 | 05463 | 09245 |
| 27 | 37631 | 74016 | 89072 | 59598 | 55356 | 27346 | 80856 | 80875 | 52850 | 36548 |
| 28 | 73829 | 21651 | 50141 | 76142 | 72303 | 06694 | 61697 | 76662 | 23745 | 96282 |
| 29 | 15634 | 89428 | 47090 | 12094 | 42134 | 62381 | 87236 | 90118 | 53463 | 46969 |
| 30 | 00571 | 45172 | 78532 | 63863 | 98597 | 15742 | 41967 | 11821 | 91389 | 07476 |
| 31 | 83374 | 10184 | 56384 | 27050 | 77700 | 13875 | 96607 | 76479 | 80535 | 17454 |
| 32 | 78666 | 85645 | 13181 | 08700 | 08289 | 62956 | 64439 | 39150 | 95690 | 18555 |
| 33 | 47890 | 88197 | 21368 | 65254 | 35917 | 54035 | 83028 | 84636 | 38186 | 50581 |
| 34 | 56238 | 13559 | 79344 | 83198 | 94642 | 35165 | 40188 | 21456 | 67024 | 62771 |
| 35 | 36369 | 32234 | 38129 | 59963 | 99237 | 72648 | 66504 | 99065 | 61161 | 16186 |
| 36 | 42934 | 34578 | 28968 | 74028 | 42164 | 56647 | 76806 | 61023 | 33099 | 48293 |
| 37 | 09010 | 15226 | 43474 | 30174 | 26727 | 39317 | 48508 | 55438 | 85336 | 40762 |
| 38 | 83897 | 90073 | 72941 | 85613 | 85569 | 24183 | 08247 | 15946 | 02957 | 68504 |
| 39 | 82206 | 01230 | 93252 | 89045 | 25141 | 91943 | 75531 | 87420 | 99012 | 80751 |
| 40 | 14175 | 32992 | 49046 | 41272 | 94040 | 44929 | 98531 | 27712 | 05106 | 35242 |
| 41 | 58968 | 88367 | 70927 | 74765 | 18635 | 85122 | 27722 | 95388 | 61523 | 91745 |
| 42 | 62601 | 04595 | 76926 | 11007 | 67631 | 64641 | 07994 | 04639 | 39314 | 83126 |
| 43 | 97030 | 71165 | 47032 | 85021 | 65554 | 66774 | 21560 | 04121 | 57297 | 85415 |
| 44 | 89074 | 31587 | 21360 | 41673 | 71192 | 85795 | 82757 | 52928 | 62586 | 02179 |
| 45 | 07806 | 81312 | 81215 | 99858 | 26762 | 28993 | 74951 | 64680 | 50934 | 32011 |
| 46 | 91540 | 86466 | 13229 | 76624 | 44092 | 96604 | 08590 | 89705 | 03424 | 48033 |
| 47 | 99279 | 27334 | 33804 | 77988 | 93592 | 90708 | 56780 | 70097 | 39907 | 51006 |
| 48 | 63224 | 05074 | 83941 | 25034 | 43516 | 22840 | 35230 | 66048 | 80754 | 46302 |
| 49 | 98361 | 97513 | 27529 | 66419 | 35328 | 19738 | 82366 | 38573 | 50967 | 72754 |

| | 50–54 | 55–59 | 60–64 | 65–69 | 70–74 | 75–79 | 80–84 | 85–89 | 90–94 | 95–99 |
|---|---|---|---|---|---|---|---|---|---|---|
| 00 | 53330 | 26487 | 85005 | 06384 | 13822 | 83736 | 95876 | 71355 | 31226 | 56063 |
| 01 | 96990 | 62825 | 97110 | 73006 | 32661 | 63408 | 03893 | 10333 | 41902 | 69175 |
| 02 | 30385 | 16588 | 63609 | 09132 | 53081 | 14478 | 50813 | 22887 | 03746 | 10289 |
| 03 | 75252 | 66905 | 60536 | 13408 | 25158 | 35825 | 10447 | 47375 | 89249 | 91238 |
| 04 | 52615 | 66504 | 78496 | 90443 | 84414 | 31981 | 88768 | 49629 | 15174 | 99795 |
| 05 | 39992 | 51082 | 74547 | 31022 | 71980 | 40900 | 84729 | 34286 | 96944 | 49502 |
| 06 | 51788 | 87155 | 13272 | 92461 | 06466 | 25392 | 22330 | 17336 | 42528 | 78628 |
| 07 | 88569 | 35645 | 50602 | 94043 | 35316 | 66344 | 78064 | 89651 | 89025 | 12722 |
| 08 | 14513 | 34794 | 44976 | 71244 | 60548 | 03041 | 03300 | 46389 | 25340 | 23804 |
| 09 | 50257 | 53477 | 24546 | 01377 | 20292 | 85097 | 00660 | 39561 | 62367 | 61424 |
| 10 | 35170 | 69025 | 46214 | 27085 | 83416 | 48597 | 19494 | 49380 | 28469 | 77549 |
| 11 | 22225 | 83437 | 43912 | 30337 | 75784 | 77689 | 60425 | 85588 | 93438 | 61343 |
| 12 | 90103 | 12542 | 97828 | 85859 | 85859 | 64101 | 00924 | 89012 | 17889 | 01154 |
| 13 | 68240 | 89649 | 85705 | 18937 | 30114 | 89827 | 89460 | 01998 | 81745 | 31281 |
| 14 | 01589 | 18335 | 24024 | 39498 | 82052 | 07868 | 49486 | 25155 | 61730 | 08946 |
| 15 | 36375 | 61694 | 90654 | 16475 | 92703 | 59561 | 45517 | 90922 | 93357 | 00207 |
| 16 | 11237 | 60921 | 51162 | 74153 | 94774 | 84150 | 39274 | 10089 | 45020 | 09624 |
| 17 | 48667 | 68353 | 40567 | 79819 | 48551 | 26789 | 07281 | 14669 | 00576 | 17435 |
| 18 | 99286 | 42806 | 02956 | 73762 | 04419 | 21676 | 67533 | 50553 | 21115 | 26742 |
| 19 | 44651 | 48349 | 13003 | 39656 | 99757 | 74964 | 00141 | 21387 | 66777 | 68533 |
| 20 | 83251 | 70164 | 05732 | 66842 | 77717 | 25305 | 36218 | 85600 | 23736 | 06629 |
| 21 | 41551 | 54630 | 88759 | 10085 | 48806 | 08724 | 50685 | 95638 | 20829 | 37264 |

**TABLE A.1**  (*Continued*)

| | 50–54 | 55–59 | 60–64 | 65–69 | 70–74 | 75–79 | 80–84 | 85–89 | 90–94 | 95–99 |
|---|---|---|---|---|---|---|---|---|---|---|
| 22 | 68990 | 51280 | 51368 | 73661 | 21764 | 71552 | 69654 | 17776 | 51935 | 53169 |
| 23 | 63393 | 76820 | 33106 | 23322 | 16783 | 35630 | 50938 | 90047 | 97577 | 27699 |
| 24 | 93317 | 87564 | 32371 | 04190 | 27608 | 40658 | 11517 | 19646 | 82335 | 60088 |
| 25 | 48546 | 41090 | 69890 | 58014 | 04093 | 39286 | 12253 | 55859 | 83853 | 15023 |
| 26 | 31435 | 57566 | 99741 | 77250 | 43165 | 31150 | 20735 | 57406 | 85891 | 04806 |
| 27 | 56405 | 29392 | 76998 | 66849 | 29175 | 11641 | 85284 | 89978 | 73169 | 62140 |
| 28 | 70102 | 50882 | 85960 | 85955 | 03828 | 69417 | 55854 | 63173 | 60485 | 00327 |
| 29 | 92746 | 32004 | 52242 | 94763 | 32955 | 39848 | 09724 | 30029 | 45196 | 67606 |
| 30 | 67737 | 34389 | 57920 | 47081 | 60714 | 04935 | 48278 | 90687 | 99290 | 18554 |
| 31 | 35606 | 76646 | 14813 | 51114 | 52492 | 46778 | 08156 | 22372 | 59999 | 43938 |
| 32 | 64836 | 28649 | 45759 | 45788 | 43183 | 25275 | 25300 | 21548 | 33941 | 66314 |
| 33 | 86319 | 92367 | 37873 | 48993 | 71443 | 22768 | 69124 | 65611 | 79267 | 49709 |
| 34 | 90632 | 32314 | 24446 | 60301 | 31376 | 13575 | 99663 | 81929 | 39343 | 17648 |
| 35 | 83752 | 51966 | 43895 | 03129 | 37539 | 72989 | 52393 | 45542 | 70344 | 96712 |
| 36 | 56755 | 21142 | 86355 | 33569 | 63096 | 66780 | 97539 | 75150 | 25718 | 33724 |
| 37 | 14100 | 28857 | 60648 | 86304 | 97397 | 97210 | 74842 | 87483 | 51558 | 52883 |
| 38 | 69227 | 24872 | 48057 | 29318 | 74385 | 02097 | 63266 | 26950 | 73173 | 53025 |
| 39 | 77718 | 56967 | 36560 | 87155 | 26021 | 70903 | 32086 | 11722 | 32053 | 63723 |
| 40 | 09550 | 38799 | 88929 | 80877 | 87779 | 99905 | 17122 | 25985 | 16866 | 76005 |
| 41 | 12404 | 42453 | 88609 | 89149 | 85892 | 96045 | 10310 | 45021 | 62023 | 70061 |
| 42 | 07985 | 27418 | 92734 | 80000 | 58969 | 99011 | 73815 | 49705 | 68076 | 69605 |
| 43 | 58124 | 53830 | 08705 | 20916 | 46048 | 30342 | 86530 | 72608 | 93074 | 80937 |
| 44 | 46173 | 77223 | 75661 | 57691 | 24055 | 27568 | 41227 | 58542 | 73196 | 44886 |
| 45 | 13476 | 72301 | 85793 | 80516 | 59479 | 66985 | 24801 | 84009 | 71317 | 87321 |
| 46 | 82472 | 98647 | 17053 | 94591 | 36790 | 42275 | 51154 | 77765 | 01115 | 09331 |
| 47 | 55370 | 63433 | 80653 | 30739 | 68821 | 46854 | 41939 | 38962 | 20703 | 69424 |
| 48 | 89274 | 74795 | 82231 | 69384 | 53605 | 67860 | 01309 | 27273 | 76316 | 54253 |
| 49 | 55242 | 74511 | 62992 | 17981 | 17323 | 79325 | 35238 | 21393 | 13114 | 70084 |

| | 00–04 | 05–09 | 10–14 | 15–19 | 20–24 | 25–29 | 30–34 | 35–39 | 40–44 | 45–49 |
|---|---|---|---|---|---|---|---|---|---|---|
| 50 | 27791 | 82504 | 33523 | 27623 | 16597 | 32089 | 81596 | 78429 | 14111 | 68245 |
| 51 | 33147 | 46058 | 92388 | 10150 | 63224 | 26003 | 56427 | 29945 | 44546 | 50233 |
| 52 | 67243 | 10454 | 40269 | 44324 | 46013 | 00061 | 21622 | 68213 | 47749 | 76398 |
| 53 | 78176 | 70368 | 95523 | 09134 | 31178 | 33857 | 26171 | 07063 | 41984 | 99310 |
| 54 | 70199 | 70547 | 94431 | 45423 | 48695 | 01370 | 68065 | 61982 | 20200 | 27066 |
| 55 | 19840 | 01143 | 18606 | 07622 | 77282 | 68422 | 70767 | 33026 | 15135 | 91212 |
| 56 | 32970 | 28267 | 17695 | 20571 | 50227 | 69447 | 45535 | 16845 | 68283 | 15919 |
| 57 | 43233 | 53872 | 68520 | 70013 | 31395 | 60361 | 39034 | 59444 | 17066 | 07418 |
| 58 | 08514 | 23921 | 16685 | 89184 | 71512 | 82239 | 72947 | 69523 | 75618 | 79826 |
| 59 | 28595 | 51196 | 96108 | 84384 | 80359 | 02346 | 60581 | 01488 | 63177 | 47496 |
| 60 | 83334 | 81552 | 88223 | 29934 | 68663 | 23726 | 18429 | 84855 | 26897 | 94782 |
| 61 | 66112 | 95787 | 84997 | 91207 | 67576 | 27496 | 01603 | 22395 | 41546 | 68178 |
| 62 | 25245 | 14749 | 30653 | 42355 | 88625 | 37412 | 87384 | 09392 | 11273 | 28116 |
| 63 | 21861 | 22185 | 41576 | 15238 | 92294 | 50643 | 69848 | 48020 | 19785 | 41518 |
| 64 | 74506 | 40569 | 90770 | 40812 | 57730 | 84150 | 91500 | 53850 | 52104 | 37988 |
| 65 | 23271 | 39549 | 33042 | 10661 | 37312 | 50914 | 73027 | 21010 | 76788 | 64037 |
| 66 | 08548 | 16021 | 64715 | 08275 | 50987 | 67327 | 11431 | 31492 | 86970 | 47335 |
| 67 | 14236 | 80869 | 90798 | 85659 | 10079 | 28535 | 35938 | 10710 | 67046 | 74021 |

**TABLE A.1**  (*Continued*)

|    | 00–04 | 05–09 | 10–14 | 15–19 | 20–24 | 25–29 | 30–34 | 35–39 | 40–44 | 45–49 |
|----|-------|-------|-------|-------|-------|-------|-------|-------|-------|-------|
| 68 | 55270 | 49583 | 86467 | 40633 | 27952 | 27187 | 35058 | 66628 | 94372 | 75665 |
| 69 | 02301 | 05524 | 91801 | 23647 | 51330 | 35677 | 05972 | 90729 | 26650 | 81684 |
| 70 | 72843 | 03767 | 62590 | 92077 | 91552 | 76853 | 45812 | 15503 | 93138 | 87788 |
| 71 | 49248 | 43346 | 29503 | 22494 | 08051 | 09035 | 75802 | 63967 | 74257 | 00046 |
| 72 | 62598 | 99092 | 87806 | 42727 | 30659 | 10118 | 83000 | 96198 | 47155 | 00361 |
| 73 | 27510 | 69457 | 98616 | 62172 | 07056 | 61015 | 22159 | 65590 | 51082 | 34912 |
| 74 | 84167 | 66640 | 69100 | 22944 | 19833 | 23961 | 80834 | 37418 | 42284 | 12951 |
| 75 | 14722 | 88488 | 54999 | 55244 | 03301 | 37344 | 01053 | 79305 | 94771 | 95215 |
| 76 | 46696 | 05477 | 32442 | 18738 | 43021 | 72933 | 14995 | 30408 | 64043 | 67834 |
| 77 | 13938 | 09867 | 28949 | 94761 | 38419 | 38695 | 90165 | 82841 | 75399 | 09932 |
| 78 | 48778 | 56434 | 42495 | 07050 | 35250 | 09660 | 56192 | 34793 | 36146 | 96806 |
| 79 | 00571 | 71281 | 01563 | 66448 | 94560 | 55920 | 31580 | 26640 | 91262 | 30863 |
| 80 | 96050 | 57641 | 21798 | 14917 | 21836 | 15053 | 33566 | 51177 | 91786 | 12610 |
| 81 | 30870 | 81575 | 14019 | 07831 | 81840 | 25506 | 29358 | 88668 | 42742 | 62048 |
| 82 | 59153 | 29135 | 00712 | 73025 | 14263 | 17253 | 95662 | 75535 | 26170 | 95240 |
| 83 | 78283 | 70379 | 54969 | 05821 | 26485 | 28990 | 40207 | 00434 | 38863 | 61892 |
| 84 | 12175 | 95800 | 41106 | 93962 | 06245 | 00883 | 65337 | 75506 | 66294 | 62241 |
| 85 | 14192 | 39242 | 17691 | 29448 | 84078 | 14545 | 39417 | 83649 | 26495 | 41672 |
| 86 | 69060 | 38669 | 00849 | 24991 | 84252 | 41611 | 62773 | 63024 | 57079 | 59283 |
| 87 | 46154 | 11705 | 29355 | 71523 | 21377 | 36745 | 00766 | 21549 | 51796 | 81340 |
| 88 | 93419 | 54353 | 41269 | 07014 | 28352 | 77594 | 57293 | 59219 | 26098 | 63041 |
| 89 | 13201 | 04017 | 68889 | 81388 | 60829 | 46231 | 46161 | 01360 | 25839 | 52380 |
| 90 | 62264 | 99963 | 98226 | 29972 | 95169 | 07546 | 01574 | 94986 | 06123 | 52804 |
| 91 | 58030 | 30054 | 27479 | 70354 | 12351 | 33761 | 94357 | 81081 | 74418 | 74297 |
| 92 | 81242 | 26739 | 92304 | 81425 | 29052 | 37708 | 49370 | 46749 | 59613 | 50749 |
| 93 | 16372 | 70531 | 92036 | 54496 | 50521 | 83872 | 30064 | 67555 | 40354 | 23671 |
| 94 | 54191 | 04574 | 58634 | 91370 | 40041 | 77649 | 42030 | 42547 | 47593 | 07435 |
| 95 | 15933 | 92602 | 19496 | 18703 | 63380 | 58017 | 14665 | 88867 | 84807 | 44672 |
| 96 | 21518 | 77770 | 53826 | 97114 | 82062 | 34592 | 87400 | 64938 | 75540 | 54751 |
| 97 | 34524 | 64627 | 92997 | 21198 | 14976 | 07071 | 91566 | 44335 | 83237 | 24335 |
| 98 | 46557 | 67780 | 59432 | 23250 | 63352 | 43890 | 07109 | 07911 | 85956 | 62699 |
| 99 | 31929 | 13996 | 05126 | 83561 | 03244 | 33635 | 26952 | 01638 | 22788 | 26393 |

|    | 50–54 | 55–59 | 60–64 | 65–69 | 70–74 | 75–79 | 80–84 | 85–89 | 90–94 | 95–99 |
|----|-------|-------|-------|-------|-------|-------|-------|-------|-------|-------|
| 50 | 03674 | 36059 | 46810 | 58367 | 82676 | 15051 | 57977 | 49410 | 02971 | 05797 |
| 51 | 26136 | 80623 | 96505 | 91089 | 02309 | 54743 | 15831 | 45538 | 96456 | 87272 |
| 52 | 61716 | 80405 | 84735 | 12997 | 86386 | 61606 | 75091 | 84996 | 76070 | 54923 |
| 53 | 67051 | 63246 | 99547 | 81223 | 52485 | 90333 | 24697 | 06266 | 07388 | 70389 |
| 54 | 17284 | 60347 | 87314 | 30218 | 87983 | 45426 | 84153 | 10569 | 64042 | 95618 |
| 55 | 12543 | 23999 | 95777 | 28105 | 66073 | 35174 | 67706 | 05181 | 35176 | 85558 |
| 56 | 45494 | 93037 | 29209 | 70724 | 86438 | 65354 | 71209 | 27969 | 85321 | 10216 |
| 57 | 39262 | 15415 | 93940 | 41615 | 43605 | 95675 | 53916 | 29580 | 07048 | 95838 |
| 58 | 29094 | 58703 | 92144 | 14287 | 50165 | 85661 | 95740 | 61118 | 36668 | 96852 |
| 59 | 77988 | 03222 | 57805 | 00725 | 91543 | 80021 | 16442 | 63360 | 33620 | 39324 |
| 60 | 02758 | 86823 | 52423 | 32355 | 96707 | 47448 | 06453 | 59430 | 43952 | 16775 |
| 61 | 46702 | 37467 | 66803 | 49344 | 59519 | 92717 | 97110 | 82087 | 36785 | 00880 |
| 62 | 61759 | 95153 | 80090 | 60626 | 55917 | 92812 | 63544 | 82295 | 50729 | 20116 |
| 63 | 82316 | 11402 | 28078 | 75325 | 43963 | 63105 | 09294 | 30285 | 61473 | 53613 |
| 64 | 92754 | 74241 | 14315 | 49697 | 61970 | 66711 | 61707 | 81589 | 53936 | 82115 |

**TABLE A.1** (*Continued*)

| | 50–54 | 55–59 | 60–64 | 65–69 | 70–74 | 75–79 | 80–84 | 85–89 | 90–94 | 95–99 |
|---|---|---|---|---|---|---|---|---|---|---|
| 65 | 37907 | 24080 | 31741 | 86653 | 81460 | 32304 | 99590 | 56644 | 41521 | 91172 |
| 66 | 16619 | 75264 | 12279 | 18996 | 16716 | 81959 | 65722 | 10058 | 91522 | 65410 |
| 67 | 66640 | 06195 | 84416 | 32836 | 53178 | 93810 | 36766 | 59778 | 26612 | 69017 |
| 68 | 45208 | 58525 | 07714 | 77126 | 67986 | 73140 | 12026 | 75550 | 84912 | 64691 |
| 69 | 00910 | 40237 | 91035 | 29125 | 03534 | 47246 | 64698 | 00608 | 39537 | 71755 |
| 70 | 19965 | 46945 | 59357 | 15551 | 20335 | 03145 | 21519 | 37882 | 99146 | 70161 |
| 71 | 37538 | 05747 | 54982 | 00494 | 51866 | 86172 | 82679 | 04152 | 56369 | 20356 |
| 72 | 38571 | 69663 | 03287 | 28101 | 46753 | 55715 | 93527 | 30508 | 19722 | 02072 |
| 73 | 76711 | 02864 | 00880 | 85518 | 25834 | 52317 | 48070 | 51582 | 03374 | 19540 |
| 74 | 07128 | 44400 | 48015 | 41449 | 21109 | 38948 | 21816 | 52089 | 64529 | 21510 |
| 75 | 00882 | 89357 | 80906 | 76476 | 58420 | 95793 | 34043 | 00991 | 38937 | 39859 |
| 76 | 96160 | 18580 | 40549 | 46562 | 45106 | 53768 | 76097 | 60504 | 85273 | 63076 |
| 77 | 13443 | 22235 | 46210 | 47755 | 05802 | 00311 | 15171 | 23818 | 89870 | 47578 |
| 78 | 99494 | 35395 | 71411 | 48281 | 92151 | 84465 | 63651 | 15969 | 61345 | 13324 |
| 79 | 90647 | 11809 | 96365 | 52409 | 17977 | 05971 | 33835 | 03889 | 43733 | 66100 |
| 80 | 33050 | 48785 | 92200 | 59319 | 36977 | 41111 | 28002 | 51580 | 10573 | 21763 |
| 81 | 21257 | 15066 | 72630 | 23206 | 03106 | 53140 | 50292 | 64012 | 83184 | 81304 |
| 82 | 45362 | 94234 | 81800 | 83980 | 97244 | 09691 | 08435 | 66723 | 06150 | 54972 |
| 83 | 93322 | 58684 | 95695 | 19096 | 98108 | 47678 | 98061 | 87193 | 99992 | 82870 |
| 84 | 20374 | 61803 | 62508 | 83696 | 54449 | 53649 | 86447 | 66115 | 90857 | 69114 |
| 85 | 00715 | 13209 | 17080 | 06890 | 38022 | 76469 | 27696 | 30778 | 31836 | 96676 |
| 86 | 85519 | 93677 | 90186 | 09579 | 98760 | 50320 | 98077 | 46048 | 79700 | 81431 |
| 87 | 71948 | 15871 | 84502 | 41330 | 46675 | 51342 | 93431 | 55566 | 90819 | 68923 |
| 88 | 43427 | 95500 | 02004 | 51802 | 59668 | 17806 | 87605 | 33010 | 20991 | 76269 |
| 89 | 64854 | 28815 | 74959 | 03531 | 77051 | 51807 | 89005 | 18898 | 23716 | 45862 |
| 90 | 62195 | 29095 | 23982 | 75883 | 41561 | 25897 | 43595 | 92703 | 86676 | 32038 |
| 91 | 61186 | 54041 | 60984 | 61602 | 18482 | 57941 | 59657 | 35924 | 21738 | 30646 |
| 92 | 88585 | 40218 | 69965 | 74354 | 62274 | 38948 | 44813 | 31558 | 40625 | 22477 |
| 93 | 15598 | 21389 | 79016 | 92151 | 21926 | 49901 | 16835 | 88055 | 30545 | 60306 |
| 94 | 27097 | 89653 | 21558 | 72731 | 66694 | 36703 | 92172 | 46129 | 32660 | 91356 |
| 95 | 40537 | 85697 | 78182 | 39711 | 59270 | 21934 | 78647 | 94801 | 78832 | 37287 |
| 96 | 74828 | 06544 | 13078 | 59528 | 31100 | 11132 | 91256 | 85899 | 72492 | 18200 |
| 97 | 43297 | 83195 | 66218 | 65838 | 63255 | 72093 | 38976 | 44892 | 96861 | 97848 |
| 98 | 32663 | 58127 | 73258 | 09220 | 49701 | 92357 | 43700 | 37214 | 56844 | 02048 |
| 99 | 45551 | 31330 | 08152 | 23712 | 23963 | 58274 | 94583 | 03761 | 73429 | 47328 |

*Source:* This table is reproduced from Jerrold H. Zar, *Biostatistical Analysis,* 2nd ed., © 1984, pp. 653–656. Reprinted by permission of Prentice-Hall, Inc., Englewood Cliffs, NJ.

**TABLE A.2**  DISTRIBUTION OF *t*: TWO-TAILED TESTS

| df | \multicolumn{9}{c}{Probability of a larger value of *t*, sign ignored} |
| --- | --- | --- | --- | --- | --- | --- | --- | --- | --- |
| df | 0.500 | 0.400 | 0.300 | 0.200 | 0.100 | 0.050 | 0.020 | 0.010 | 0.001 |
| 1 | 1.000 | 1.376 | 1.963 | 3.078 | 6.314 | 12.706 | 31.821 | 63.657 | 636.619 |
| 2 | 0.816 | 1.061 | 1.386 | 1.886 | 2.920 | 4.303 | 6.965 | 9.925 | 31.598 |
| 3 | 0.765 | 0.978 | 1.250 | 1.638 | 2.353 | 3.182 | 3.541 | 5.841 | 12.941 |
| 4 | 0.741 | 0.941 | 1.190 | 1.533 | 2.132 | 2.776 | 3.747 | 4.604 | 8.610 |
| 5 | 0.727 | 0.920 | 1.156 | 1.476 | 2.015 | 2.571 | 3.365 | 4.032 | 6.859 |
| 6 | 0.718 | 0.906 | 1.134 | 1.440 | 1.943 | 2.447 | 3.143 | 3.707 | 5.959 |
| 7 | 0.711 | 0.896 | 1.119 | 1.415 | 1.895 | 2.365 | 2.998 | 3.499 | 5.405 |
| 8 | 0.706 | 0.889 | 1.108 | 1.397 | 1.860 | 2.306 | 2.896 | 3.355 | 5.041 |
| 9 | 0.703 | 0.883 | 1.100 | 1.383 | 1.833 | 2.262 | 2.821 | 3.250 | 4.781 |
| 10 | 0.700 | 0.879 | 1.093 | 1.372 | 1.812 | 2.228 | 2.764 | 3.169 | 4.587 |
| 11 | 0.697 | 0.876 | 1.088 | 1.363 | 1.796 | 2.201 | 2.718 | 3.106 | 4.437 |
| 12 | 0.695 | 0.873 | 1.083 | 1.356 | 1.782 | 2.179 | 2.681 | 3.055 | 4.318 |
| 13 | 0.694 | 0.870 | 1.079 | 1.350 | 1.771 | 2.160 | 2.650 | 3.012 | 4.221 |
| 14 | 0.692 | 0.868 | 1.076 | 1.345 | 1.761 | 2.145 | 2.624 | 2.977 | 4.140 |
| 15 | 0.691 | 0.866 | 1.074 | 1.341 | 1.753 | 2.131 | 2.602 | 2.947 | 4.073 |
| 16 | 0.690 | 0.865 | 1.071 | 1.337 | 1.746 | 2.120 | 2.583 | 2.921 | 4.015 |
| 17 | 0.689 | 0.863 | 1.069 | 1.333 | 1.740 | 2.110 | 2.567 | 2.898 | 3.965 |
| 18 | 0.688 | 0.862 | 1.067 | 1.330 | 1.734 | 2.101 | 2.552 | 2.878 | 3.922 |
| 19 | 0.688 | 0.861 | 1.066 | 1.328 | 1.729 | 2.093 | 2.539 | 2.861 | 3.883 |
| 20 | 0.687 | 0.860 | 1.064 | 1.325 | 1.725 | 2.086 | 2.528 | 2.845 | 3.850 |
| 21 | 0.686 | 0.859 | 1.063 | 1.323 | 1.721 | 2.080 | 2.518 | 2.831 | 3.819 |
| 22 | 0.686 | 0.858 | 1.061 | 1.321 | 1.717 | 2.074 | 2.408 | 2.819 | 3.792 |
| 23 | 0.685 | 0.858 | 1.060 | 1.319 | 1.714 | 2.069 | 2.500 | 2.807 | 3.767 |
| 24 | 0.685 | 0.857 | 1.059 | 1.318 | 1.711 | 2.064 | 2.492 | 2.797 | 3.745 |
| 25 | 0.684 | 0.856 | 1.058 | 1.316 | 1.708 | 2.060 | 2.485 | 2.787 | 3.725 |
| 26 | 0.684 | 0.856 | 1.058 | 1.315 | 1.706 | 2.056 | 2.479 | 2.779 | 3.707 |
| 27 | 0.684 | 0.855 | 1.057 | 1.314 | 1.703 | 2.052 | 2.473 | 2.771 | 3.690 |
| 28 | 0.683 | 0.855 | 1.056 | 1.313 | 1.701 | 2.048 | 2.467 | 2.763 | 3.674 |
| 29 | 0.683 | 0.854 | 1.055 | 1.311 | 1.699 | 2.045 | 2.462 | 2.756 | 3.659 |
| 30 | 0.683 | 0.854 | 1.055 | 1.310 | 1.697 | 2.042 | 2.457 | 2.750 | 3.646 |
| 40 | 0.681 | 0.851 | 1.050 | 1.303 | 1.684 | 2.021 | 2.423 | 2.704 | 3.551 |
| 60 | 0.679 | 0.848 | 1.046 | 1.296 | 1.671 | 2.000 | 2.390 | 2.600 | 3.460 |
| 120 | 0.674 | 0.845 | 1.041 | 1.289 | 1.658 | 1.980 | 2.358 | 2.617 | 3.373 |
| | 0.674 | 0.842 | 1.036 | 1.282 | 1.645 | 1.960 | 2.326 | 2.576 | 3.291 |
| df | 0.250 | 0.200 | 0.150 | 0.100 | 0.050 | 0.025 | 0.010 | 0.005 | 0.0005 |

Probability of a larger positive value of *t*, one-tailed test

*Source:* This table is taken from Table III of Fisher and Yates, *Statistical Tables for Biological, Agricultural, and Medical Research,* published by Longman Group, Ltd., London (previously published by Oliver and Boyd, Ltd., Edinburgh) and by permission of the authors and publishers.

**TABLE A.3** VALUES OF THE *F* DISTRIBUTION

| Denomi-nator df | Probability of a larger F | Numerator df | | | | | | | | |
|---|---|---|---|---|---|---|---|---|---|---|
| | | 1 | 2 | 3 | 4 | 5 | 6 | 7 | 8 | 9 |
| 1 | 0.100 | 39.86 | 49.50 | 53.59 | 55.83 | 57.24 | 58.20 | 58.91 | 59.44 | 59.86 |
| | 0.050 | 161.4 | 199.5 | 215.7 | 224.6 | 230.2 | 234.0 | 236.8 | 238.9 | 240.5 |
| | 0.025 | 647.9 | 799.5 | 864.2 | 899.6 | 921.8 | 937.1 | 948.2 | 956.7 | 963.3 |
| | 0.010 | 4,052 | 4,999.5 | 5,403 | 5,625 | 5,764 | 5,859 | 5,928 | 5,982 | 6,022 |
| | 0.005 | 16,211 | 20,000 | 21,615 | 22,500 | 23,056 | 23,437 | 23,715 | 23,925 | 24,091 |
| 2 | 0.100 | 8.53 | 9.00 | 9.16 | 9.24 | 9.29 | 9.33 | 9.35 | 9.37 | 9.38 |
| | 0.050 | 18.51 | 19.00 | 19.16 | 19.25 | 19.30 | 19.33 | 19.35 | 19.37 | 19.38 |
| | 0.025 | 38.51 | 39.00 | 39.17 | 39.25 | 39.30 | 39.33 | 39.36 | 39.37 | 39.39 |
| | 0.010 | 98.50 | 99.00 | 99.17 | 99.25 | 99.30 | 99.33 | 99.36 | 99.37 | 99.39 |
| | 0.005 | 198.5 | 199.0 | 199.2 | 199.2 | 199.3 | 199.3 | 199.4 | 199.4 | 199.4 |
| 3 | 0.100 | 5.54 | 5.46 | 5.39 | 5.34 | 5.31 | 5.28 | 5.27 | 5.25 | 5.24 |
| | 0.050 | 10.13 | 9.55 | 9.28 | 9.12 | 9.01 | 8.94 | 8.89 | 8.85 | 8.81 |
| | 0.025 | 17.44 | 16.04 | 15.44 | 15.10 | 14.88 | 14.73 | 14.62 | 14.54 | 14.47 |
| | 0.010 | 34.12 | 30.82 | 29.46 | 28.71 | 28.24 | 27.91 | 27.67 | 27.49 | 27.35 |
| | 0.005 | 55.55 | 49.80 | 47.47 | 46.19 | 45.39 | 44.84 | 44.43 | 44.13 | 43.88 |
| 4 | 0.100 | 4.54 | 4.32 | 4.19 | 4.11 | 4.05 | 4.01 | 3.98 | 3.95 | 3.94 |
| | 0.050 | 7.71 | 6.94 | 6.59 | 6.39 | 6.26 | 6.16 | 6.09 | 6.04 | 6.00 |
| | 0.025 | 12.22 | 10.65 | 9.98 | 9.60 | 9.36 | 9.20 | 9.07 | 8.98 | 8.90 |
| | 0.010 | 21.20 | 18.00 | 16.69 | 15.98 | 15.52 | 15.21 | 14.98 | 14.80 | 14.66 |
| | 0.005 | 31.33 | 26.28 | 24.26 | 23.15 | 22.46 | 21.97 | 21.62 | 21.35 | 21.14 |
| 5 | 0.100 | 4.06 | 3.78 | 3.62 | 3.52 | 3.45 | 3.40 | 3.37 | 3.34 | 3.32 |
| | 0.050 | 6.61 | 5.79 | 5.41 | 5.19 | 5.05 | 4.95 | 4.88 | 4.82 | 4.77 |
| | 0.025 | 10.01 | 8.43 | 7.76 | 7.39 | 7.15 | 6.98 | 6.85 | 6.76 | 6.68 |
| | 0.010 | 16.26 | 13.27 | 12.06 | 11.39 | 10.97 | 10.67 | 10.46 | 10.29 | 10.16 |
| | 0.005 | 22.78 | 18.31 | 16.53 | 15.56 | 14.94 | 14.51 | 14.20 | 13.96 | 13.77 |
| 6 | 0.100 | 3.78 | 3.46 | 3.29 | 3.18 | 3.11 | 3.05 | 3.01 | 2.98 | 2.96 |
| | 0.050 | 5.99 | 5.14 | 4.76 | 4.53 | 4.39 | 4.28 | 4.21 | 4.15 | 4.10 |
| | 0.025 | 8.81 | 7.26 | 6.60 | 6.23 | 5.99 | 5.82 | 5.70 | 5.60 | 5.52 |
| | 0.010 | 13.75 | 10.92 | 9.78 | 9.15 | 8.75 | 8.47 | 8.26 | 8.10 | 7.98 |
| | 0.005 | 18.63 | 14.54 | 12.92 | 12.03 | 11.46 | 11.07 | 10.79 | 10.57 | 10.39 |
| 7 | 0.100 | 3.59 | 3.26 | 3.07 | 2.96 | 2.88 | 2.83 | 2.78 | 2.75 | 2.72 |
| | 0.050 | 5.59 | 4.74 | 4.35 | 4.12 | 3.97 | 3.87 | 3.79 | 3.73 | 3.68 |
| | 0.025 | 8.07 | 6.54 | 5.89 | 5.52 | 5.29 | 5.12 | 4.99 | 4.90 | 4.82 |
| | 0.010 | 12.25 | 9.55 | 8.45 | 7.85 | 7.46 | 7.19 | 6.99 | 6.84 | 6.72 |
| | 0.005 | 16.24 | 12.40 | 10.88 | 10.05 | 9.52 | 9.16 | 8.89 | 8.68 | 8.51 |
| 8 | 0.100 | 3.46 | 3.11 | 2.92 | 2.81 | 2.73 | 2.67 | 2.62 | 2.59 | 2.56 |
| | 0.050 | 5.32 | 4.46 | 4.07 | 3.84 | 3.69 | 3.58 | 3.50 | 3.44 | 3.39 |
| | 0.025 | 7.57 | 6.06 | 5.42 | 5.05 | 4.82 | 4.65 | 4.53 | 4.43 | 4.36 |
| | 0.010 | 11.26 | 8.65 | 7.59 | 7.01 | 6.63 | 6.37 | 6.18 | 6.03 | 5.91 |
| | 0.005 | 14.69 | 11.04 | 9.60 | 8.81 | 8.30 | 7.95 | 7.69 | 7.50 | 7.34 |
| 9 | 0.100 | 3.36 | 3.01 | 2.81 | 2.69 | 2.61 | 2.55 | 2.51 | 2.47 | 2.44 |
| | 0.050 | 5.12 | 4.26 | 3.86 | 3.63 | 3.48 | 3.37 | 3.29 | 3.23 | 3.18 |
| | 0.025 | 7.21 | 5.71 | 5.08 | 4.72 | 4.48 | 4.32 | 4.20 | 4.10 | 4.03 |
| | 0.010 | 10.56 | 8.02 | 6.99 | 6.42 | 6.06 | 5.80 | 5.61 | 5.47 | 5.35 |
| | 0.005 | 13.61 | 10.11 | 8.72 | 7.96 | 7.47 | 7.13 | 6.88 | 6.69 | 6.54 |
| 10 | 0.100 | 3.29 | 2.92 | 2.73 | 2.61 | 2.52 | 2.46 | 2.41 | 2.38 | 2.35 |
| | 0.050 | 4.96 | 4.10 | 3.71 | 3.48 | 3.33 | 3.22 | 3.14 | 3.07 | 3.02 |
| | 0.025 | 6.94 | 5.46 | 4.83 | 4.47 | 4.24 | 4.07 | 3.95 | 3.85 | 3.78 |
| | 0.010 | 10.04 | 7.56 | 6.55 | 5.99 | 5.64 | 5.39 | 5.20 | 5.06 | 4.94 |
| | 0.005 | 12.83 | 9.43 | 8.08 | 7.34 | 6.87 | 6.54 | 6.30 | 6.12 | 5.97 |

| Numerator df | | | | | | | | | | | |
|---|---|---|---|---|---|---|---|---|---|---|---|
| 10 | 12 | 15 | 20 | 24 | 30 | 40 | 60 | 120 | ∞ | P | df |
| 60.19 | 60.71 | 61.22 | 61.74 | 62.00 | 62.26 | 62.53 | 62.79 | 63.06 | 63.33 | 0.100 | 1 |
| 241.9 | 243.9 | 245.9 | 248.0 | 249.1 | 250.1 | 251.1 | 252.2 | 253.3 | 254.3 | 0.050 | |
| 968.6 | 976.7 | 984.9 | 993.1 | 997.2 | 1,001 | 1,006 | 1,010 | 1,014 | 1,018 | 0.025 | |
| 6,056 | 6,106 | 6,157 | 6,209 | 6,235 | 6,261 | 6,287 | 6,313 | 6,339 | 6,366 | 0.010 | |
| 24,224 | 24,426 | 24,630 | 24,836 | 24,940 | 25,044 | 25,148 | 25,253 | 25,359 | 25,465 | 0.005 | |
| 9.39 | 9.41 | 9.42 | 9.44 | 9.45 | 9.46 | 9.47 | 9.47 | 9.48 | 9.49 | 0.100 | 2 |
| 19.40 | 19.41 | 19.43 | 19.45 | 19.45 | 19.46 | 19.47 | 19.48 | 19.49 | 19.50 | 0.050 | |
| 39.40 | 39.41 | 39.43 | 39.45 | 39.46 | 39.46 | 39.47 | 39.48 | 39.49 | 39.50 | 0.025 | |
| 99.40 | 99.42 | 99.43 | 99.45 | 99.46 | 99.47 | 99.47 | 99.48 | 99.49 | 99.50 | 0.010 | |
| 199.4 | 199.4 | 199.4 | 199.4 | 199.5 | 199.5 | 199.5 | 199.5 | 199.5 | 199.5 | 0.005 | |
| 5.23 | 5.22 | 5.20 | 5.18 | 5.18 | 5.17 | 5.16 | 5.15 | 5.14 | 5.13 | 0.100 | 3 |
| 8.79 | 8.74 | 8.70 | 8.66 | 8.64 | 8.62 | 8.59 | 8.57 | 8.55 | 8.53 | 0.050 | |
| 14.42 | 14.34 | 14.25 | 14.17 | 14.12 | 14.08 | 14.04 | 13.99 | 13.95 | 13.90 | 0.025 | |
| 27.23 | 27.05 | 26.87 | 26.69 | 26.60 | 26.50 | 26.41 | 26.32 | 26.22 | 26.13 | 0.010 | |
| 43.69 | 43.39 | 43.08 | 42.78 | 42.62 | 42.47 | 42.31 | 42.15 | 41.99 | 41.83 | 0.005 | |
| 3.92 | 3.90 | 3.87 | 3.84 | 3.83 | 3.82 | 3.80 | 3.79 | 3.78 | 3.76 | 0.100 | 4 |
| 5.96 | 5.91 | 5.86 | 5.80 | 5.77 | 5.75 | 5.72 | 5.69 | 5.66 | 5.63 | 0.050 | |
| 8.84 | 8.75 | 8.66 | 8.56 | 8.51 | 8.46 | 8.41 | 8.36 | 8.31 | 8.26 | 0.025 | |
| 14.55 | 14.37 | 14.20 | 14.02 | 13.93 | 13.84 | 13.75 | 13.65 | 13.56 | 13.46 | 0.010 | |
| 20.97 | 20.70 | 20.44 | 20.17 | 20.03 | 19.89 | 19.75 | 19.61 | 19.47 | 19.32 | 0.005 | |
| 3.30 | 3.27 | 3.24 | 3.21 | 3.19 | 3.17 | 3.16 | 3.14 | 3.12 | 3.10 | 0.100 | 5 |
| 4.74 | 4.68 | 4.62 | 4.56 | 4.53 | 4.50 | 4.46 | 4.43 | 4.40 | 4.36 | 0.050 | |
| 6.62 | 6.52 | 6.43 | 6.33 | 6.28 | 6.23 | 6.18 | 6.12 | 6.07 | 6.02 | 0.025 | |
| 10.05 | 9.89 | 9.72 | 9.55 | 9.47 | 9.38 | 9.29 | 9.20 | 9.11 | 9.02 | 0.010 | |
| 13.62 | 13.38 | 13.15 | 12.90 | 12.78 | 12.66 | 12.53 | 12.40 | 12.27 | 12.14 | 0.005 | |
| 2.94 | 2.90 | 2.87 | 2.84 | 2.82 | 2.80 | 2.78 | 2.76 | 2.74 | 2.72 | 0.100 | 6 |
| 4.06 | 4.00 | 3.94 | 3.87 | 3.84 | 3.81 | 3.77 | 3.74 | 3.70 | 3.67 | 0.050 | |
| 5.46 | 5.37 | 5.27 | 5.17 | 5.12 | 5.07 | 5.01 | 4.96 | 4.90 | 4.85 | 0.025 | |
| 7.87 | 7.72 | 7.56 | 7.40 | 7.31 | 7.23 | 7.14 | 7.06 | 6.97 | 6.88 | 0.010 | |
| 10.25 | 10.03 | 9.81 | 9.59 | 9.47 | 9.36 | 9.24 | 9.12 | 9.00 | 8.88 | 0.005 | |
| 2.70 | 2.67 | 2.63 | 2.59 | 2.58 | 2.56 | 2.54 | 2.51 | 2.49 | 2.47 | 0.100 | 7 |
| 3.64 | 3.57 | 3.51 | 3.44 | 3.41 | 3.38 | 3.34 | 3.30 | 3.27 | 3.23 | 0.050 | |
| 4.76 | 4.67 | 4.57 | 4.47 | 4.42 | 4.36 | 4.31 | 4.25 | 4.20 | 4.14 | 0.025 | |
| 6.62 | 6.47 | 6.31 | 6.16 | 6.07 | 5.99 | 5.91 | 5.82 | 5.74 | 5.65 | 0.010 | |
| 8.38 | 8.18 | 7.97 | 7.75 | 7.65 | 7.53 | 7.42 | 7.31 | 7.19 | 7.08 | 0.005 | |
| 2.54 | 2.50 | 2.46 | 2.42 | 2.40 | 2.38 | 2.36 | 2.34 | 2.32 | 2.29 | 0.100 | 8 |
| 3.35 | 3.28 | 3.22 | 3.15 | 3.12 | 3.08 | 3.04 | 3.01 | 2.97 | 2.93 | 0.050 | |
| 4.30 | 4.20 | 4.10 | 4.00 | 3.95 | 3.89 | 3.84 | 3.78 | 3.73 | 3.67 | 0.025 | |
| 5.81 | 5.67 | 5.52 | 5.36 | 5.28 | 5.20 | 5.12 | 5.03 | 4.95 | 4.86 | 0.010 | |
| 7.21 | 7.01 | 6.81 | 6.61 | 6.50 | 6.40 | 6.29 | 6.18 | 6.06 | 5.95 | 0.005 | |
| 2.42 | 2.38 | 2.34 | 2.30 | 2.28 | 2.25 | 2.23 | 2.21 | 2.18 | 2.16 | 0.100 | 9 |
| 3.14 | 3.07 | 3.01 | 2.94 | 2.90 | 2.86 | 2.83 | 2.79 | 2.75 | 2.71 | 0.050 | |
| 3.96 | 3.87 | 3.77 | 3.67 | 3.61 | 3.56 | 3.51 | 3.45 | 3.39 | 3.33 | 0.025 | |
| 5.26 | 5.11 | 4.96 | 4.81 | 4.73 | 4.65 | 4.57 | 4.48 | 4.40 | 4.31 | 0.010 | |
| 6.42 | 6.23 | 6.03 | 5.83 | 5.73 | 5.62 | 5.52 | 5.41 | 5.30 | 5.19 | 0.005 | |
| 2.32 | 2.28 | 2.24 | 2.20 | 2.18 | 2.16 | 2.13 | 2.11 | 2.08 | 2.06 | 0.100 | 10 |
| 2.98 | 2.91 | 2.85 | 2.77 | 2.74 | 2.70 | 2.66 | 2.62 | 2.58 | 2.54 | 0.050 | |
| 3.72 | 3.62 | 3.52 | 3.42 | 3.37 | 3.31 | 3.26 | 3.20 | 3.14 | 3.08 | 0.025 | |
| 4.85 | 4.71 | 4.56 | 4.41 | 4.33 | 4.25 | 4.17 | 4.08 | 4.00 | 3.91 | 0.010 | |
| 5.85 | 5.66 | 5.47 | 5.27 | 5.17 | 5.07 | 4.97 | 4.86 | 4.75 | 4.64 | 0.005 | |

| Denominator df | Probability of a larger F | Numerator df | | | | | | | | |
|---|---|---|---|---|---|---|---|---|---|---|
| | | 1 | 2 | 3 | 4 | 5 | 6 | 7 | 8 | 9 |
| 11 | 0.100 | 3.23 | 2.86 | 2.66 | 2.54 | 2.45 | 2.39 | 2.34 | 2.30 | 2.27 |
| | 0.050 | 4.84 | 3.98 | 3.59 | 3.36 | 3.20 | 3.09 | 3.01 | 2.95 | 2.90 |
| | 0.025 | 6.72 | 5.26 | 4.63 | 4.28 | 4.04 | 3.88 | 3.76 | 3.66 | 3.59 |
| | 0.010 | 9.65 | 7.21 | 6.22 | 5.67 | 5.32 | 5.07 | 4.89 | 4.74 | 4.63 |
| | 0.005 | 12.23 | 8.91 | 7.60 | 6.88 | 6.42 | 6.10 | 5.86 | 5.68 | 5.54 |
| 12 | 0.100 | 3.18 | 2.81 | 2.61 | 2.48 | 2.39 | 2.33 | 2.28 | 2.24 | 2.21 |
| | 0.050 | 4.75 | 3.89 | 3.49 | 3.26 | 3.11 | 3.00 | 2.91 | 2.85 | 2.80 |
| | 0.025 | 6.55 | 5.10 | 4.47 | 4.12 | 3.89 | 3.73 | 3.61 | 3.51 | 3.44 |
| | 0.010 | 9.33 | 6.93 | 5.95 | 5.41 | 5.06 | 4.82 | 4.64 | 4.50 | 4.39 |
| | 0.005 | 11.75 | 8.51 | 7.23 | 6.52 | 6.07 | 5.76 | 5.52 | 5.35 | 5.20 |
| 13 | 0.100 | 3.14 | 2.76 | 2.56 | 2.43 | 2.35 | 2.28 | 2.23 | 2.20 | 2.16 |
| | 0.050 | 4.67 | 3.81 | 3.41 | 3.18 | 3.03 | 2.92 | 2.83 | 2.77 | 2.71 |
| | 0.025 | 6.41 | 4.97 | 4.35 | 4.00 | 3.77 | 3.60 | 3.48 | 3.39 | 3.31 |
| | 0.010 | 9.07 | 6.70 | 5.74 | 5.21 | 4.86 | 4.62 | 4.44 | 4.30 | 4.19 |
| | 0.005 | 11.37 | 8.19 | 6.93 | 6.23 | 5.79 | 5.48 | 5.25 | 5.08 | 4.94 |
| 14 | 0.100 | 3.10 | 2.73 | 2.52 | 2.39 | 2.31 | 2.24 | 2.19 | 2.15 | 2.12 |
| | 0.050 | 4.60 | 3.74 | 3.34 | 3.11 | 2.96 | 2.85 | 2.76 | 2.70 | 2.65 |
| | 0.025 | 6.30 | 4.86 | 4.24 | 3.89 | 3.66 | 3.50 | 3.38 | 3.29 | 3.21 |
| | 0.010 | 8.86 | 6.51 | 5.56 | 5.04 | 4.69 | 4.46 | 4.28 | 4.14 | 4.03 |
| | 0.005 | 11.06 | 7.92 | 6.68 | 6.00 | 5.56 | 5.26 | 5.03 | 4.86 | 4.72 |
| 15 | 0.100 | 3.07 | 2.70 | 2.49 | 2.36 | 2.27 | 2.21 | 2.16 | 2.12 | 2.09 |
| | 0.050 | 4.54 | 3.68 | 3.29 | 3.06 | 2.90 | 2.79 | 2.71 | 2.64 | 2.59 |
| | 0.025 | 6.20 | 4.77 | 4.15 | 3.80 | 3.58 | 3.41 | 3.29 | 3.20 | 3.12 |
| | 0.010 | 8.68 | 6.36 | 5.42 | 4.89 | 4.56 | 4.32 | 4.14 | 4.00 | 3.89 |
| | 0.005 | 10.80 | 7.70 | 6.48 | 5.80 | 5.37 | 5.07 | 4.85 | 4.67 | 4.54 |
| 16 | 0.100 | 3.05 | 2.67 | 2.46 | 2.33 | 2.24 | 2.18 | 2.13 | 2.09 | 2.06 |
| | 0.050 | 4.49 | 3.63 | 3.24 | 3.01 | 2.85 | 2.74 | 2.66 | 2.59 | 2.54 |
| | 0.025 | 6.12 | 4.69 | 4.08 | 3.73 | 3.50 | 3.34 | 3.22 | 3.12 | 3.05 |
| | 0.010 | 8.53 | 6.23 | 5.29 | 4.77 | 4.44 | 4.20 | 4.03 | 3.89 | 3.78 |
| | 0.005 | 10.58 | 7.51 | 6.30 | 5.64 | 5.21 | 4.91 | 4.69 | 4.52 | 4.38 |
| 17 | 0.100 | 3.03 | 2.64 | 2.44 | 2.31 | 2.22 | 2.15 | 2.10 | 2.06 | 2.03 |
| | 0.050 | 4.45 | 3.59 | 3.20 | 2.96 | 2.81 | 2.70 | 2.61 | 2.55 | 2.49 |
| | 0.025 | 6.04 | 4.62 | 4.01 | 3.66 | 3.44 | 3.28 | 3.16 | 3.06 | 2.98 |
| | 0.010 | 8.40 | 6.11 | 5.18 | 4.67 | 4.34 | 4.10 | 3.93 | 3.79 | 3.68 |
| | 0.005 | 10.38 | 7.35 | 6.16 | 5.50 | 5.07 | 4.78 | 4.56 | 4.39 | 4.25 |
| 18 | 0.100 | 3.01 | 2.62 | 2.42 | 2.29 | 2.20 | 2.13 | 2.08 | 2.04 | 2.00 |
| | 0.050 | 4.41 | 3.55 | 3.16 | 2.93 | 2.77 | 2.66 | 2.58 | 2.51 | 2.46 |
| | 0.025 | 5.98 | 4.56 | 3.95 | 3.61 | 3.38 | 3.22 | 3.10 | 3.01 | 2.93 |
| | 0.010 | 8.29 | 6.01 | 5.09 | 4.58 | 4.25 | 4.01 | 3.84 | 3.71 | 3.60 |
| | 0.005 | 10.22 | 7.21 | 6.03 | 5.37 | 4.96 | 4.66 | 4.44 | 4.28 | 4.14 |
| 19 | 0.100 | 2.99 | 2.61 | 2.40 | 2.27 | 2.18 | 2.11 | 2.06 | 2.02 | 1.98 |
| | 0.050 | 4.38 | 3.52 | 3.13 | 2.90 | 2.74 | 2.63 | 2.54 | 2.48 | 2.42 |
| | 0.025 | 5.92 | 4.51 | 3.90 | 3.56 | 3.33 | 3.17 | 3.05 | 2.96 | 2.88 |
| | 0.010 | 8.18 | 5.93 | 5.01 | 4.50 | 4.17 | 3.94 | 3.77 | 3.63 | 3.52 |
| | 0.005 | 10.07 | 7.09 | 5.92 | 5.27 | 4.85 | 4.56 | 4.34 | 4.18 | 4.04 |
| 20 | 0.100 | 2.97 | 2.59 | 2.38 | 2.25 | 2.16 | 2.09 | 2.04 | 2.00 | 1.96 |
| | 0.050 | 4.35 | 3.49 | 3.10 | 2.87 | 2.71 | 2.60 | 2.51 | 2.45 | 2.39 |
| | 0.025 | 5.87 | 4.46 | 3.86 | 3.51 | 3.29 | 3.13 | 3.01 | 2.91 | 2.84 |
| | 0.010 | 8.10 | 5.85 | 4.94 | 4.43 | 4.10 | 3.87 | 3.70 | 3.56 | 3.46 |
| | 0.005 | 9.94 | 6.99 | 5.82 | 5.17 | 4.76 | 4.47 | 4.26 | 4.09 | 3.96 |

| 10 | 12 | 15 | 20 | 24 | 30 | 40 | 60 | 120 | ∞ | P | df |
|----|----|----|----|----|----|----|----|-----|---|---|----|
| 2.25 | 2.21 | 2.17 | 2.12 | 2.10 | 2.08 | 2.05 | 2.03 | 2.00 | 1.97 | 0.100 | 11 |
| 2.85 | 2.79 | 2.72 | 2.65 | 2.61 | 2.57 | 2.53 | 2.49 | 2.45 | 2.40 | 0.050 | |
| 3.53 | 3.43 | 3.33 | 3.23 | 3.17 | 3.12 | 3.06 | 3.00 | 2.94 | 2.88 | 0.025 | |
| 4.54 | 4.40 | 4.25 | 4.10 | 4.02 | 3.94 | 3.86 | 3.78 | 3.69 | 3.60 | 0.010 | |
| 5.42 | 5.24 | 5.05 | 4.86 | 4.76 | 4.65 | 4.55 | 4.44 | 4.34 | 4.23 | 0.005 | |
| 2.19 | 2.15 | 2.10 | 2.06 | 2.04 | 2.01 | 1.99 | 1.96 | 1.93 | 1.90 | 0.100 | 12 |
| 2.75 | 2.69 | 2.62 | 2.54 | 2.51 | 2.47 | 2.43 | 2.38 | 2.34 | 2.30 | 0.050 | |
| 3.37 | 3.28 | 3.18 | 3.07 | 3.02 | 2.96 | 2.91 | 2.85 | 2.79 | 2.72 | 0.025 | |
| 4.30 | 4.16 | 4.01 | 3.86 | 3.78 | 3.70 | 3.62 | 3.54 | 3.45 | 3.36 | 0.010 | |
| 5.09 | 4.91 | 4.72 | 4.53 | 4.43 | 4.33 | 4.23 | 4.12 | 4.01 | 3.90 | 0.005 | |
| 2.14 | 2.10 | 2.05 | 2.01 | 1.98 | 1.96 | 1.93 | 1.90 | 1.88 | 1.85 | 0.100 | 13 |
| 2.67 | 2.60 | 2.53 | 2.46 | 2.42 | 2.38 | 2.34 | 2.30 | 2.25 | 2.21 | 0.050 | |
| 3.25 | 3.15 | 3.05 | 2.95 | 2.89 | 2.84 | 2.78 | 2.72 | 2.66 | 2.60 | 0.025 | |
| 4.10 | 3.96 | 3.82 | 3.66 | 3.59 | 3.51 | 3.43 | 3.34 | 3.25 | 3.17 | 0.010 | |
| 4.82 | 4.64 | 4.46 | 4.27 | 4.17 | 4.07 | 3.97 | 3.87 | 3.76 | 3.65 | 0.005 | |
| 2.10 | 2.05 | 2.01 | 1.96 | 1.94 | 1.91 | 1.89 | 1.86 | 1.83 | 1.80 | 0.100 | 14 |
| 2.60 | 2.53 | 2.46 | 2.39 | 2.35 | 2.31 | 2.27 | 2.22 | 2.18 | 2.13 | 0.050 | |
| 3.15 | 3.05 | 2.95 | 2.84 | 2.79 | 2.73 | 2.67 | 2.61 | 2.55 | 2.49 | 0.025 | |
| 3.94 | 3.80 | 3.66 | 3.51 | 3.43 | 3.35 | 3.27 | 3.18 | 3.09 | 3.00 | 0.010 | |
| 4.60 | 4.43 | 4.25 | 4.06 | 3.96 | 3.86 | 3.76 | 3.66 | 3.55 | 3.44 | 0.005 | |
| 2.06 | 2.02 | 1.97 | 1.92 | 1.90 | 1.87 | 1.85 | 1.82 | 1.79 | 1.76 | 0.100 | 15 |
| 2.54 | 2.48 | 2.40 | 2.33 | 2.29 | 2.25 | 2.20 | 2.16 | 2.11 | 2.07 | 0.050 | |
| 3.06 | 2.96 | 2.86 | 2.76 | 2.70 | 2.64 | 2.59 | 2.52 | 2.46 | 2.40 | 0.025 | |
| 3.80 | 3.67 | 3.52 | 3.37 | 3.29 | 3.21 | 3.13 | 3.05 | 2.96 | 2.87 | 0.010 | |
| 4.42 | 4.25 | 4.07 | 3.88 | 3.79 | 3.69 | 3.58 | 3.48 | 3.37 | 3.26 | 0.005 | |
| 2.03 | 1.99 | 1.94 | 1.89 | 1.87 | 1.84 | 1.81 | 1.78 | 1.75 | 1.72 | 0.100 | 16 |
| 2.49 | 2.42 | 2.35 | 2.28 | 2.24 | 2.19 | 2.15 | 2.11 | 2.06 | 2.01 | 0.050 | |
| 2.99 | 2.89 | 2.79 | 2.68 | 2.63 | 2.57 | 2.51 | 2.45 | 2.38 | 2.32 | 0.025 | |
| 3.69 | 3.55 | 3.41 | 3.26 | 3.18 | 3.10 | 3.02 | 2.93 | 2.84 | 2.75 | 0.010 | |
| 4.27 | 4.10 | 3.92 | 3.73 | 3.64 | 3.54 | 3.44 | 3.33 | 3.22 | 3.11 | 0.005 | |
| 2.00 | 1.96 | 1.91 | 1.86 | 1.84 | 1.81 | 1.78 | 1.75 | 1.72 | 1.69 | 0.100 | 17 |
| 2.45 | 2.38 | 2.31 | 2.23 | 2.19 | 2.15 | 2.10 | 2.06 | 2.01 | 1.96 | 0.050 | |
| 2.92 | 2.82 | 2.72 | 2.62 | 2.56 | 2.50 | 2.44 | 2.38 | 2.32 | 2.25 | 0.025 | |
| 3.59 | 3.46 | 3.31 | 3.16 | 3.08 | 3.00 | 2.92 | 2.83 | 2.75 | 2.65 | 0.010 | |
| 4.14 | 3.97 | 3.79 | 3.61 | 3.51 | 3.41 | 3.31 | 3.21 | 3.10 | 2.98 | 0.005 | |
| 1.98 | 1.93 | 1.89 | 1.84 | 1.81 | 1.78 | 1.75 | 1.72 | 1.69 | 1.66 | 0.100 | 18 |
| 2.41 | 2.34 | 2.27 | 2.19 | 2.15 | 2.11 | 2.06 | 2.02 | 1.97 | 1.92 | 0.050 | |
| 2.87 | 2.77 | 2.67 | 2.56 | 2.50 | 2.44 | 2.38 | 2.32 | 2.26 | 2.19 | 0.025 | |
| 3.51 | 3.37 | 3.23 | 3.08 | 3.00 | 2.92 | 2.84 | 2.75 | 2.66 | 2.57 | 0.010 | |
| 4.03 | 3.86 | 3.68 | 3.50 | 3.40 | 3.30 | 3.20 | 3.10 | 2.99 | 2.87 | 0.005 | |
| 1.96 | 1.91 | 1.86 | 1.81 | 1.79 | 1.76 | 1.73 | 1.70 | 1.67 | 1.63 | 0.100 | 19 |
| 2.38 | 2.31 | 2.23 | 2.16 | 2.11 | 2.07 | 2.03 | 1.98 | 1.93 | 1.88 | 0.050 | |
| 2.82 | 2.72 | 2.62 | 2.51 | 2.45 | 2.39 | 2.33 | 2.27 | 2.20 | 2.13 | 0.025 | |
| 3.43 | 3.30 | 3.15 | 3.00 | 2.92 | 2.84 | 2.76 | 2.67 | 2.58 | 2.49 | 0.010 | |
| 3.93 | 3.76 | 3.59 | 3.40 | 3.31 | 3.21 | 3.11 | 3.00 | 2.89 | 2.78 | 0.005 | |
| 1.94 | 1.89 | 1.84 | 1.79 | 1.77 | 1.74 | 1.71 | 1.68 | 1.64 | 1.61 | 0.100 | 20 |
| 2.35 | 2.28 | 2.20 | 2.12 | 2.08 | 2.04 | 1.99 | 1.95 | 1.90 | 1.84 | 0.050 | |
| 2.77 | 2.68 | 2.57 | 2.46 | 2.41 | 2.35 | 2.29 | 2.22 | 2.16 | 2.09 | 0.025 | |
| 3.37 | 3.23 | 3.09 | 2.94 | 2.86 | 2.78 | 2.69 | 2.61 | 2.52 | 2.42 | 0.010 | |
| 3.85 | 3.68 | 3.50 | 3.32 | 3.22 | 3.12 | 3.02 | 2.92 | 2.81 | 2.69 | 0.005 | |

**TABLE A.3** *(Continued)*

| Denomi-<br>nator<br>df | Probability<br>of a larger<br>F | Numerator df | | | | | | | | |
|---|---|---|---|---|---|---|---|---|---|---|
| | | 1 | 2 | 3 | 4 | 5 | 6 | 7 | 8 | 9 |
| 21 | 0.100 | 2.96 | 2.57 | 2.36 | 2.23 | 2.14 | 2.08 | 2.02 | 1.98 | 1.95 |
| | 0.050 | 4.32 | 3.47 | 3.07 | 2.84 | 2.68 | 2.57 | 2.49 | 2.42 | 2.37 |
| | 0.025 | 5.83 | 4.42 | 3.82 | 3.48 | 3.25 | 3.09 | 2.97 | 2.87 | 2.80 |
| | 0.010 | 8.02 | 5.78 | 4.87 | 4.37 | 4.04 | 3.81 | 3.64 | 3.51 | 3.40 |
| | 0.005 | 9.83 | 6.89 | 5.73 | 5.09 | 4.68 | 4.39 | 4.18 | 4.01 | 3.88 |
| 22 | 0.100 | 2.95 | 2.56 | 2.35 | 2.22 | 2.13 | 2.06 | 2.01 | 1.97 | 1.93 |
| | 0.050 | 4.30 | 3.44 | 3.05 | 2.82 | 2.66 | 2.55 | 2.46 | 2.40 | 2.34 |
| | 0.025 | 5.79 | 4.38 | 3.78 | 3.44 | 3.22 | 3.05 | 2.93 | 2.84 | 2.76 |
| | 0.010 | 7.95 | 5.72 | 4.82 | 4.31 | 3.99 | 3.76 | 3.59 | 3.45 | 3.35 |
| | 0.005 | 9.73 | 6.81 | 5.65 | 5.02 | 4.61 | 4.32 | 4.11 | 3.94 | 3.81 |
| 23 | 0.100 | 2.94 | 2.55 | 2.34 | 2.21 | 2.11 | 2.05 | 1.99 | 1.95 | 1.92 |
| | 0.050 | 4.28 | 3.42 | 3.03 | 2.80 | 2.64 | 2.53 | 2.44 | 2.37 | 2.32 |
| | 0.025 | 5.75 | 4.35 | 3.75 | 3.41 | 3.18 | 3.02 | 2.90 | 2.81 | 2.73 |
| | 0.010 | 7.88 | 5.66 | 4.76 | 4.26 | 3.94 | 3.71 | 3.54 | 3.41 | 3.30 |
| | 0.005 | 9.63 | 6.73 | 5.58 | 4.95 | 4.54 | 4.26 | 4.05 | 3.88 | 3.75 |
| 24 | 0.100 | 2.93 | 2.54 | 2.33 | 2.19 | 2.10 | 2.04 | 1.98 | 1.94 | 1.91 |
| | 0.050 | 4.26 | 3.40 | 3.01 | 2.78 | 2.62 | 2.51 | 2.42 | 2.36 | 2.30 |
| | 0.025 | 5.72 | 4.32 | 3.72 | 3.38 | 3.15 | 2.99 | 2.87 | 2.78 | 2.70 |
| | 0.010 | 7.82 | 5.61 | 4.72 | 4.22 | 3.90 | 3.67 | 3.50 | 3.36 | 3.26 |
| | 0.005 | 9.55 | 6.66 | 5.52 | 4.89 | 4.49 | 4.20 | 3.99 | 3.83 | 3.69 |
| 25 | 0.100 | 2.92 | 2.53 | 2.32 | 2.18 | 2.09 | 2.02 | 1.97 | 1.93 | 1.89 |
| | 0.050 | 4.24 | 3.39 | 2.99 | 2.76 | 2.60 | 2.49 | 2.40 | 2.34 | 2.28 |
| | 0.025 | 5.69 | 4.29 | 3.69 | 3.35 | 3.13 | 2.97 | 2.85 | 2.75 | 2.68 |
| | 0.010 | 7.77 | 5.57 | 4.68 | 4.18 | 3.85 | 3.63 | 3.46 | 3.32 | 3.22 |
| | 0.005 | 9.48 | 6.60 | 5.46 | 4.84 | 4.43 | 4.15 | 3.94 | 3.78 | 3.64 |
| 26 | 0.100 | 2.91 | 2.52 | 2.31 | 2.17 | 2.08 | 2.01 | 1.96 | 1.92 | 1.88 |
| | 0.050 | 4.23 | 3.37 | 2.98 | 2.74 | 2.59 | 2.47 | 2.39 | 2.32 | 2.27 |
| | 0.025 | 5.66 | 4.27 | 3.67 | 3.33 | 3.10 | 2.94 | 2.82 | 2.73 | 2.65 |
| | 0.010 | 7.72 | 5.53 | 4.64 | 4.14 | 3.82 | 3.59 | 3.42 | 3.29 | 3.18 |
| | 0.005 | 9.41 | 6.54 | 5.41 | 4.79 | 4.38 | 4.10 | 3.89 | 3.73 | 3.60 |
| 27 | 0.100 | 2.90 | 2.51 | 2.30 | 2.17 | 2.07 | 2.00 | 1.95 | 1.91 | 1.87 |
| | 0.050 | 4.21 | 3.35 | 2.96 | 2.73 | 2.57 | 2.46 | 2.37 | 2.31 | 2.25 |
| | 0.025 | 5.63 | 4.24 | 3.65 | 3.31 | 3.08 | 2.92 | 2.80 | 2.71 | 2.63 |
| | 0.010 | 7.68 | 5.49 | 4.60 | 4.11 | 3.78 | 3.56 | 3.39 | 3.26 | 3.15 |
| | 0.005 | 9.34 | 6.49 | 5.36 | 4.74 | 4.34 | 4.06 | 3.85 | 3.69 | 3.56 |
| 28 | 0.100 | 2.89 | 2.50 | 2.29 | 2.16 | 2.06 | 2.00 | 1.94 | 1.90 | 1.87 |
| | 0.050 | 4.20 | 3.34 | 2.95 | 2.71 | 2.56 | 2.45 | 2.36 | 2.29 | 2.24 |
| | 0.025 | 5.61 | 4.22 | 3.63 | 3.29 | 3.06 | 2.90 | 2.78 | 2.69 | 2.61 |
| | 0.010 | 7.64 | 5.45 | 4.57 | 4.07 | 3.75 | 3.53 | 3.36 | 3.23 | 3.12 |
| | 0.005 | 9.28 | 6.44 | 5.32 | 4.70 | 4.30 | 4.02 | 3.81 | 3.65 | 3.52 |
| 29 | 0.100 | 2.89 | 2.50 | 2.28 | 2.15 | 2.06 | 1.99 | 1.93 | 1.89 | 1.86 |
| | 0.050 | 4.18 | 3.33 | 2.93 | 2.70 | 2.55 | 2.43 | 2.35 | 2.28 | 2.22 |
| | 0.025 | 5.59 | 4.20 | 3.61 | 3.27 | 3.04 | 2.88 | 2.76 | 2.67 | 2.59 |
| | 0.010 | 7.60 | 5.42 | 4.54 | 4.04 | 3.73 | 3.50 | 3.33 | 3.20 | 3.09 |
| | 0.005 | 9.23 | 6.40 | 5.28 | 4.66 | 4.26 | 3.98 | 3.77 | 3.61 | 3.48 |
| 30 | 0.100 | 2.88 | 2.49 | 2.28 | 2.14 | 2.05 | 1.98 | 1.93 | 1.88 | 1.85 |
| | 0.050 | 4.17 | 3.32 | 2.92 | 2.69 | 2.53 | 2.42 | 2.33 | 2.27 | 2.21 |
| | 0.025 | 5.57 | 4.18 | 3.59 | 3.25 | 3.03 | 2.87 | 2.75 | 2.65 | 2.57 |
| | 0.010 | 7.56 | 5.39 | 4.51 | 4.02 | 3.70 | 3.47 | 3.30 | 3.17 | 3.07 |
| | 0.005 | 9.18 | 6.35 | 5.24 | 4.62 | 4.23 | 3.95 | 3.74 | 3.58 | 3.45 |

| Numerator df | | | | | | | | | | | |
|---|---|---|---|---|---|---|---|---|---|---|---|
| 10 | 12 | 15 | 20 | 24 | 30 | 40 | 60 | 120 | ∞ | P | df |
| 1.92 | 1.87 | 1.83 | 1.78 | 1.75 | 1.72 | 1.69 | 1.66 | 1.62 | 1.59 | 0.100 | 21 |
| 2.32 | 2.25 | 2.18 | 2.10 | 2.05 | 2.01 | 1.96 | 1.92 | 1.87 | 1.81 | 0.050 | |
| 2.73 | 2.64 | 2.53 | 2.42 | 2.37 | 2.31 | 2.25 | 2.18 | 2.11 | 2.04 | 0.025 | |
| 3.31 | 3.17 | 3.03 | 2.88 | 2.80 | 2.72 | 2.64 | 2.55 | 2.46 | 2.36 | 0.010 | |
| 3.77 | 3.60 | 3.43 | 3.24 | 3.15 | 3.05 | 2.95 | 2.84 | 2.73 | 2.61 | 0.005 | |
| 1.90 | 1.86 | 1.81 | 1.76 | 1.73 | 1.70 | 1.67 | 1.64 | 1.60 | 1.57 | 0.100 | 22 |
| 2.30 | 2.23 | 2.15 | 2.07 | 2.03 | 1.98 | 1.94 | 1.89 | 1.84 | 1.78 | 0.050 | |
| 2.70 | 2.60 | 2.50 | 2.39 | 2.33 | 2.27 | 2.21 | 2.14 | 2.08 | 2.00 | 0.025 | |
| 3.26 | 3.12 | 2.98 | 2.83 | 2.75 | 2.67 | 2.58 | 2.50 | 2.40 | 2.31 | 0.010 | |
| 3.70 | 3.54 | 3.36 | 3.18 | 3.08 | 2.98 | 2.88 | 2.77 | 2.66 | 2.55 | 0.005 | |
| 1.89 | 1.84 | 1.80 | 1.74 | 1.72 | 1.69 | 1.66 | 1.62 | 1.59 | 1.55 | 0.100 | 23 |
| 2.27 | 2.20 | 2.13 | 2.05 | 2.01 | 1.96 | 1.91 | 1.86 | 1.81 | 1.76 | 0.050 | |
| 2.67 | 2.57 | 2.47 | 2.36 | 2.30 | 2.24 | 2.18 | 2.11 | 2.04 | 1.97 | 0.025 | |
| 3.21 | 3.07 | 2.93 | 2.78 | 2.70 | 2.62 | 2.54 | 2.45 | 2.35 | 2.26 | 0.010 | |
| 3.64 | 3.47 | 3.30 | 3.12 | 3.02 | 2.92 | 2.82 | 2.71 | 2.60 | 2.48 | 0.005 | |
| 1.88 | 1.83 | 1.78 | 1.73 | 1.70 | 1.67 | 1.64 | 1.61 | 1.57 | 1.53 | 0.100 | 24 |
| 2.25 | 2.18 | 2.11 | 2.03 | 1.98 | 1.94 | 1.89 | 1.84 | 1.79 | 1.73 | 0.050 | |
| 2.64 | 2.54 | 2.44 | 2.33 | 2.27 | 2.21 | 2.15 | 2.08 | 2.01 | 1.94 | 0.025 | |
| 3.17 | 3.03 | 2.89 | 2.74 | 2.66 | 2.58 | 2.49 | 2.40 | 2.31 | 2.21 | 0.010 | |
| 3.59 | 3.42 | 3.25 | 3.06 | 2.97 | 2.87 | 2.77 | 2.66 | 2.55 | 2.43 | 0.005 | |
| 1.87 | 1.82 | 1.77 | 1.72 | 1.69 | 1.66 | 1.63 | 1.59 | 1.56 | 1.52 | 0.100 | 25 |
| 2.24 | 2.16 | 2.09 | 2.01 | 1.96 | 1.92 | 1.87 | 1.82 | 1.77 | 1.71 | 0.050 | |
| 2.61 | 2.51 | 2.41 | 2.30 | 2.24 | 2.18 | 2.12 | 2.05 | 1.98 | 1.91 | 0.025 | |
| 3.13 | 2.99 | 2.85 | 2.70 | 2.62 | 2.54 | 2.45 | 2.36 | 2.27 | 2.17 | 0.010 | |
| 3.54 | 3.37 | 3.20 | 3.01 | 2.92 | 2.82 | 2.72 | 2.61 | 2.50 | 2.38 | 0.005 | |
| 1.86 | 1.81 | 1.76 | 1.71 | 1.68 | 1.65 | 1.61 | 1.58 | 1.54 | 1.50 | 0.100 | 26 |
| 2.22 | 2.15 | 2.07 | 1.99 | 1.95 | 1.90 | 1.85 | 1.80 | 1.75 | 1.69 | 0.050 | |
| 2.59 | 2.49 | 2.39 | 2.28 | 2.22 | 2.16 | 2.09 | 2.03 | 1.95 | 1.88 | 0.025 | |
| 3.09 | 2.96 | 2.81 | 2.66 | 2.58 | 2.50 | 2.42 | 2.33 | 2.23 | 2.13 | 0.010 | |
| 3.49 | 3.33 | 3.15 | 2.97 | 2.87 | 2.77 | 2.67 | 2.56 | 2.45 | 2.33 | 0.005 | |
| 1.85 | 1.80 | 1.75 | 1.70 | 1.67 | 1.64 | 1.60 | 1.57 | 1.53 | 1.49 | 0.100 | 27 |
| 2.20 | 2.13 | 2.06 | 1.97 | 1.93 | 1.88 | 1.84 | 1.79 | 1.73 | 1.67 | 0.050 | |
| 2.57 | 2.47 | 2.36 | 2.25 | 2.19 | 2.13 | 2.07 | 2.00 | 1.93 | 1.85 | 0.025 | |
| 3.06 | 2.93 | 2.78 | 2.63 | 2.55 | 2.47 | 2.38 | 2.29 | 2.20 | 2.10 | 0.010 | |
| 3.45 | 3.28 | 3.11 | 2.93 | 2.83 | 2.73 | 2.63 | 2.52 | 2.41 | 2.29 | 0.005 | |
| 1.84 | 1.79 | 1.74 | 1.69 | 1.66 | 1.63 | 1.59 | 1.56 | 1.52 | 1.48 | 0.100 | 28 |
| 2.19 | 2.12 | 2.04 | 1.96 | 1.91 | 1.87 | 1.82 | 1.77 | 1.71 | 1.65 | 0.050 | |
| 2.55 | 2.45 | 2.34 | 2.23 | 2.17 | 2.11 | 2.05 | 1.98 | 1.91 | 1.83 | 0.025 | |
| 3.03 | 2.90 | 2.75 | 2.60 | 2.52 | 2.44 | 2.35 | 2.26 | 2.17 | 2.06 | 0.010 | |
| 3.41 | 3.25 | 3.07 | 2.89 | 2.79 | 2.69 | 2.59 | 2.48 | 2.37 | 2.25 | 0.005 | |
| 1.83 | 1.78 | 1.73 | 1.68 | 1.65 | 1.62 | 1.58 | 1.55 | 1.51 | 1.47 | 0.100 | 29 |
| 2.18 | 2.10 | 2.03 | 1.94 | 1.90 | 1.85 | 1.81 | 1.75 | 1.70 | 1.64 | 0.050 | |
| 2.53 | 2.43 | 2.32 | 2.21 | 2.15 | 2.09 | 2.03 | 1.96 | 1.89 | 1.81 | 0.025 | |
| 3.00 | 2.87 | 2.73 | 2.57 | 2.49 | 2.41 | 2.33 | 2.23 | 2.14 | 2.03 | 0.010 | |
| 3.38 | 3.21 | 3.04 | 2.86 | 2.76 | 2.66 | 2.56 | 2.45 | 2.33 | 2.21 | 0.005 | |
| 1.82 | 1.77 | 1.72 | 1.67 | 1.64 | 1.61 | 1.57 | 1.54 | 1.50 | 1.46 | 0.100 | 30 |
| 2.16 | 2.09 | 2.01 | 1.93 | 1.89 | 1.84 | 1.79 | 1.74 | 1.68 | 1.62 | 0.050 | |
| 2.51 | 2.41 | 2.31 | 2.20 | 2.14 | 2.07 | 2.01 | 1.94 | 1.87 | 1.79 | 0.025 | |
| 2.98 | 2.84 | 2.70 | 2.55 | 2.47 | 2.39 | 2.30 | 2.21 | 2.11 | 2.01 | 0.010 | |
| 3.34 | 3.18 | 3.01 | 2.82 | 2.73 | 2.63 | 2.52 | 2.42 | 2.30 | 2.18 | 0.005 | |

| Denomi- nator df | Probability of a larger F | Numerator df | | | | | | | | |
|---|---|---|---|---|---|---|---|---|---|---|
| | | 1 | 2 | 3 | 4 | 5 | 6 | 7 | 8 | 9 |
| 40 | 0.100 | 2.84 | 2.44 | 2.23 | 2.09 | 2.00 | 1.93 | 1.87 | 1.83 | 1.79 |
| | 0.050 | 4.08 | 3.23 | 2.84 | 2.61 | 2.45 | 2.34 | 2.25 | 2.18 | 2.12 |
| | 0.025 | 5.42 | 4.05 | 3.46 | 3.13 | 2.90 | 2.74 | 2.62 | 2.53 | 2.45 |
| | 0.010 | 7.31 | 5.18 | 4.31 | 3.83 | 3.51 | 3.29 | 3.12 | 2.99 | 2.89 |
| | 0.005 | 8.83 | 6.07 | 4.98 | 4.37 | 3.99 | 3.71 | 3.51 | 3.35 | 3.22 |
| 60 | 0.100 | 2.79 | 2.39 | 2.18 | 2.04 | 1.95 | 1.87 | 1.82 | 1.77 | 1.74 |
| | 0.050 | 4.00 | 3.15 | 2.76 | 2.53 | 2.37 | 2.25 | 2.17 | 2.10 | 2.04 |
| | 0.025 | 5.29 | 3.93 | 3.34 | 3.01 | 2.79 | 2.63 | 2.51 | 2.41 | 2.33 |
| | 0.010 | 7.08 | 4.98 | 4.13 | 3.65 | 3.34 | 3.12 | 2.95 | 2.82 | 2.72 |
| | 0.005 | 8.49 | 5.79 | 4.73 | 4.14 | 3.76 | 3.49 | 3.29 | 3.13 | 3.01 |
| 120 | 0.100 | 2.75 | 2.35 | 2.13 | 1.99 | 1.90 | 1.82 | 1.77 | 1.72 | 1.68 |
| | 0.050 | 3.92 | 3.07 | 2.68 | 2.45 | 2.29 | 2.17 | 2.09 | 2.02 | 1.96 |
| | 0.025 | 5.15 | 3.80 | 3.23 | 2.89 | 2.67 | 2.52 | 2.39 | 2.30 | 2.22 |
| | 0.010 | 6.85 | 4.79 | 3.95 | 3.48 | 3.17 | 2.96 | 2.79 | 2.66 | 2.56 |
| | 0.005 | 8.18 | 5.54 | 4.50 | 3.92 | 3.55 | 3.28 | 3.09 | 2.93 | 2.81 |
| ∞ | 0.100 | 2.71 | 2.30 | 2.08 | 1.94 | 1.85 | 1.77 | 1.72 | 1.67 | 1.63 |
| | 0.050 | 3.84 | 3.00 | 2.60 | 2.37 | 2.21 | 2.10 | 2.01 | 1.94 | 1.88 |
| | 0.025 | 5.02 | 3.69 | 3.12 | 2.79 | 2.57 | 2.41 | 2.29 | 2.19 | 2.11 |
| | 0.010 | 6.63 | 4.61 | 3.78 | 3.32 | 3.02 | 2.80 | 2.64 | 2.51 | 2.41 |
| | 0.005 | 7.88 | 5.30 | 4.28 | 3.72 | 3.35 | 3.09 | 2.90 | 2.74 | 2.62 |

*Source:* This table is condensed from M. Merrington and C. M. Thompson, "Tables of Percentage Points of the Inverted Beta (*F*) Distribution," *Biometrika*, Vol. 33, 73 (1943), and is published with the permission of the Biometrika Trustees.

| | | | | | Numerator df | | | | | | | |
|---|---|---|---|---|---|---|---|---|---|---|---|---|
| 10 | 12 | 15 | 20 | 24 | 30 | 40 | 60 | 120 | $\infty$ | P | df |
| 1.76 | 1.71 | 1.66 | 1.61 | 1.57 | 1.54 | 1.51 | 1.47 | 1.42 | 1.38 | 0.100 | 40 |
| 2.08 | 2.00 | 1.92 | 1.84 | 1.79 | 1.74 | 1.69 | 1.64 | 1.58 | 1.51 | 0.050 | |
| 2.39 | 2.29 | 2.18 | 2.07 | 2.01 | 1.94 | 1.88 | 1.80 | 1.72 | 1.64 | 0.025 | |
| 2.80 | 2.66 | 2.52 | 2.37 | 2.29 | 2.20 | 2.11 | 2.02 | 1.92 | 1.80 | 0.010 | |
| 3.12 | 2.95 | 2.78 | 2.60 | 2.50 | 2.40 | 2.30 | 2.18 | 2.06 | 1.93 | 0.005 | |
| 1.71 | 1.66 | 1.60 | 1.54 | 1.51 | 1.48 | 1.44 | 1.40 | 1.35 | 1.29 | 0.100 | 60 |
| 1.99 | 1.92 | 1.84 | 1.75 | 1.70 | 1.65 | 1.59 | 1.53 | 1.47 | 1.39 | 0.050 | |
| 2.27 | 2.17 | 2.06 | 1.94 | 1.88 | 1.82 | 1.74 | 1.67 | 1.58 | 1.48 | 0.025 | |
| 2.63 | 2.50 | 2.35 | 2.20 | 2.12 | 2.03 | 1.94 | 1.84 | 1.73 | 1.60 | 0.010 | |
| 2.90 | 2.74 | 2.57 | 2.39 | 2.29 | 2.19 | 2.08 | 1.96 | 1.83 | 1.69 | 0.005 | |
| 1.65 | 1.60 | 1.55 | 1.48 | 1.45 | 1.41 | 1.37 | 1.32 | 1.26 | 1.19 | 0.100 | 120 |
| 1.91 | 1.83 | 1.75 | 1.66 | 1.61 | 1.55 | 1.50 | 1.43 | 1.35 | 1.25 | 0.050 | |
| 2.16 | 2.05 | 1.94 | 1.82 | 1.76 | 1.69 | 1.61 | 1.53 | 1.43 | 1.31 | 0.025 | |
| 2.47 | 2.34 | 2.19 | 2.03 | 1.95 | 1.86 | 1.76 | 1.66 | 1.53 | 1.38 | 0.010 | |
| 2.71 | 2.54 | 2.37 | 2.19 | 2.09 | 1.98 | 1.87 | 1.75 | 1.61 | 1.43 | 0.005 | |
| 1.60 | 1.55 | 1.49 | 1.42 | 1.38 | 1.34 | 1.30 | 1.24 | 1.17 | 1.00 | 0.100 | $\infty$ |
| 1.83 | 1.75 | 1.67 | 1.57 | 1.52 | 1.46 | 1.39 | 1.32 | 1.22 | 1.00 | 0.050 | |
| 2.05 | 1.94 | 1.83 | 1.71 | 1.64 | 1.57 | 1.48 | 1.39 | 1.27 | 1.09 | 0.025 | |
| 2.32 | 2.18 | 2.04 | 1.88 | 1.79 | 1.70 | 1.59 | 1.47 | 1.32 | 1.00 | 0.010 | |
| 2.52 | 2.36 | 2.19 | 2.00 | 1.90 | 1.79 | 1.67 | 1.53 | 1.36 | 1.00 | 0.005 | |

**TABLE A.4** CUMULATIVE DISTRIBUTION OF CHI-SQUARE

Probability of a greater value

| Degrees of freedom | 0.005 | 0.010 | 0.025 | 0.050 | 0.100 | 0.250 | 0.500 | 0.750 | 0.900 | 0.950 | 0.975 | 0.990 | 0.995 |
|---|---|---|---|---|---|---|---|---|---|---|---|---|---|
| 1 | 7.88 | 6.63 | 5.02 | 3.84 | 2.71 | 1.32 | 0.45 | 0.10 | 0.02 | .... | .... | .... | .... |
| 2 | 10.60 | 9.21 | 7.38 | 5.99 | 4.61 | 2.77 | 1.39 | 0.58 | 0.21 | 0.10 | 0.05 | 0.02 | 0.01 |
| 3 | 12.84 | 11.34 | 9.35 | 7.81 | 6.25 | 4.11 | 2.37 | 1.21 | 0.58 | 0.35 | 0.22 | 0.11 | 0.07 |
| 4 | 14.86 | 13.28 | 11.14 | 9.49 | 7.78 | 5.39 | 3.36 | 1.92 | 1.06 | 0.71 | 0.48 | 0.30 | 0.21 |
| 5 | 16.75 | 15.09 | 12.83 | 11.07 | 9.24 | 6.63 | 4.35 | 2.67 | 1.61 | 1.15 | 0.83 | 0.55 | 0.41 |
| 6 | 18.55 | 16.81 | 14.45 | 12.59 | 10.64 | 7.84 | 5.35 | 3.45 | 2.20 | 1.64 | 1.24 | 0.87 | 0.68 |
| 7 | 20.28 | 18.48 | 16.01 | 14.07 | 12.02 | 9.04 | 6.35 | 4.25 | 2.83 | 2.17 | 1.69 | 1.24 | 0.99 |
| 8 | 21.96 | 20.09 | 17.53 | 15.51 | 13.36 | 10.22 | 7.34 | 5.07 | 3.49 | 2.73 | 2.18 | 1.65 | 1.34 |
| 9 | 23.59 | 21.67 | 19.02 | 16.92 | 14.68 | 11.39 | 8.34 | 5.90 | 4.17 | 3.33 | 2.70 | 2.09 | 1.73 |
| 10 | 25.19 | 23.21 | 20.48 | 18.31 | 15.99 | 12.55 | 9.34 | 6.74 | 4.87 | 3.94 | 3.25 | 2.56 | 2.16 |
| 11 | 26.76 | 24.72 | 21.92 | 19.68 | 17.28 | 13.70 | 10.34 | 7.58 | 5.58 | 4.57 | 3.82 | 3.05 | 2.60 |
| 12 | 28.30 | 26.22 | 23.34 | 21.03 | 18.55 | 14.85 | 11.34 | 8.44 | 6.30 | 5.23 | 4.40 | 3.57 | 3.07 |
| 13 | 29.82 | 27.69 | 24.74 | 22.36 | 19.81 | 15.98 | 12.34 | 9.30 | 7.04 | 5.89 | 5.01 | 4.11 | 3.57 |
| 14 | 31.32 | 29.14 | 26.12 | 23.68 | 21.06 | 17.12 | 13.34 | 10.17 | 7.79 | 6.57 | 5.63 | 4.66 | 4.07 |
| 15 | 32.80 | 30.58 | 27.49 | 25.00 | 22.31 | 18.25 | 14.34 | 11.04 | 8.55 | 7.26 | 6.27 | 5.23 | 4.60 |
| 16 | 34.27 | 32.00 | 28.85 | 26.30 | 23.54 | 19.37 | 15.34 | 11.91 | 9.31 | 7.96 | 6.91 | 5.81 | 5.14 |
| 17 | 35.72 | 33.41 | 30.19 | 27.59 | 24.77 | 20.49 | 16.34 | 12.79 | 10.09 | 8.67 | 7.56 | 6.41 | 5.70 |
| 18 | 37.16 | 34.81 | 31.53 | 28.87 | 25.99 | 21.60 | 17.34 | 13.68 | 10.86 | 9.39 | 8.23 | 7.01 | 6.26 |
| 19 | 38.58 | 36.19 | 32.85 | 30.14 | 27.20 | 22.72 | 18.34 | 14.56 | 11.65 | 10.12 | 8.91 | 7.63 | 6.84 |
| 20 | 40.00 | 37.57 | 34.17 | 31.41 | 28.41 | 23.83 | 19.34 | 15.45 | 12.44 | 10.85 | 9.59 | 8.26 | 7.43 |
| 21 | 41.40 | 38.93 | 35.48 | 32.67 | 29.62 | 24.93 | 20.34 | 16.34 | 13.24 | 11.59 | 10.28 | 8.90 | 8.03 |
| 22 | 42.80 | 40.29 | 36.78 | 33.92 | 30.81 | 26.04 | 21.34 | 17.24 | 14.04 | 12.34 | 10.98 | 9.54 | 8.64 |
| 23 | 44.18 | 41.64 | 38.08 | 35.17 | 32.01 | 27.14 | 22.34 | 18.14 | 14.85 | 13.09 | 11.69 | 10.20 | 9.26 |
| 24 | 45.56 | 42.98 | 39.36 | 36.42 | 33.20 | 28.24 | 23.34 | 19.04 | 15.66 | 13.85 | 12.40 | 10.86 | 9.89 |
| 25 | 46.93 | 44.31 | 40.65 | 37.65 | 34.38 | 29.34 | 24.34 | 19.94 | 16.47 | 14.61 | 13.12 | 11.52 | 10.52 |
| 26 | 48.29 | 45.64 | 41.92 | 38.89 | 35.56 | 30.43 | 25.34 | 20.84 | 17.29 | 15.38 | 13.84 | 12.20 | 11.16 |
| 27 | 49.64 | 46.96 | 43.19 | 40.11 | 36.74 | 31.53 | 26.34 | 21.75 | 18.11 | 16.15 | 14.57 | 12.88 | 11.81 |
| 28 | 50.99 | 48.28 | 44.46 | 41.34 | 37.92 | 32.62 | 27.34 | 22.66 | 18.94 | 16.93 | 15.31 | 13.56 | 12.46 |
| 29 | 52.34 | 49.59 | 45.72 | 42.56 | 39.09 | 33.71 | 28.34 | 23.57 | 19.77 | 17.71 | 16.05 | 14.26 | 13.12 |
| 30 | 53.67 | 50.89 | 46.98 | 43.77 | 40.26 | 34.80 | 29.34 | 24.48 | 20.60 | 18.49 | 16.79 | 14.95 | 13.79 |
| 40 | 66.77 | 63.69 | 59.34 | 55.76 | 51.80 | 45.62 | 39.34 | 33.66 | 29.05 | 26.51 | 24.43 | 22.16 | 20.71 |
| 50 | 79.49 | 76.15 | 71.42 | 67.50 | 63.17 | 56.33 | 49.33 | 42.94 | 37.69 | 34.76 | 32.36 | 29.71 | 27.99 |
| 60 | 91.95 | 88.38 | 83.30 | 79.08 | 74.40 | 66.98 | 59.33 | 52.29 | 46.46 | 43.19 | 40.48 | 37.48 | 35.53 |
| 70 | 104.22 | 100.42 | 95.02 | 90.53 | 85.53 | 77.58 | 69.33 | 61.70 | 55.33 | 51.74 | 48.76 | 45.44 | 43.28 |
| 80 | 116.32 | 112.33 | 106.63 | 101.88 | 96.58 | 88.13 | 79.33 | 71.14 | 64.28 | 60.39 | 57.15 | 53.54 | 51.17 |
| 90 | 128.30 | 124.12 | 118.14 | 113.14 | 107.56 | 98.64 | 89.33 | 80.62 | 73.29 | 69.13 | 65.65 | 61.75 | 59.20 |
| 100 | 140.17 | 135.81 | 129.56 | 124.34 | 118.50 | 109.14 | 99.33 | 90.13 | 82.36 | 77.93 | 74.22 | 70.06 | 67.33 |

*Source:* This table is condensed from Catherine M. Thompson, "Table of Percentage Points of the $X^2$ Distribution," *Biometrika*, Vol. 32, 187 (1941–42), and is published with the permission of the Biometrika Trustees.

**TABLE A.5** SIGNIFICANT STUDENTIZED RANGES FOR 5 PERCENT AND 1 PERCENT LEVEL NEW MULTIPLE-RANGE TEST

| Error df | Significance level | \(p\) = Number of means for range being tested | | | | | | | | | | | | | |
|---|---|---|---|---|---|---|---|---|---|---|---|---|---|---|---|
| | | 2 | 3 | 4 | 5 | 6 | 7 | 8 | 9 | 10 | 12 | 14 | 16 | 18 | 20 |
| 1 | 0.05 | 18.0 | 18.0 | 18.0 | 18.0 | 18.0 | 18.0 | 18.0 | 18.0 | 18.0 | 18.0 | 18.0 | 18.0 | 18.0 | 18.0 |
| | 0.01 | 90.0 | 90.0 | 90.0 | 90.0 | 90.0 | 90.0 | 90.0 | 90.0 | 90.0 | 90.0 | 90.0 | 90.0 | 90.0 | 90.0 |
| 2 | 0.05 | 6.09 | 6.09 | 6.09 | 6.09 | 6.09 | 6.09 | 6.09 | 6.09 | 6.09 | 6.09 | 6.09 | 6.09 | 6.09 | 6.09 |
| | 0.01 | 14.0 | 14.0 | 14.0 | 14.0 | 14.0 | 14.0 | 14.0 | 14.0 | 14.0 | 14.0 | 14.0 | 14.0 | 14.0 | 14.0 |
| 3 | 0.05 | 4.50 | 4.50 | 4.50 | 4.50 | 4.50 | 4.50 | 4.50 | 4.50 | 4.50 | 4.50 | 4.50 | 4.50 | 4.50 | 4.50 |
| | 0.01 | 8.26 | 8.5 | 8.6 | 8.7 | 8.8 | 8.9 | 8.9 | 9.0 | 9.0 | 9.0 | 9.1 | 9.2 | 9.3 | 9.3 |
| 4 | 0.05 | 3.93 | 4.01 | 4.02 | 4.02 | 4.02 | 4.02 | 4.02 | 4.02 | 4.02 | 4.02 | 4.02 | 4.02 | 4.02 | 4.02 |
| | 0.01 | 6.51 | 6.8 | 6.9 | 7.0 | 7.1 | 7.1 | 7.2 | 7.2 | 7.3 | 7.3 | 7.4 | 7.4 | 7.5 | 7.5 |
| 5 | 0.05 | 3.64 | 3.74 | 3.79 | 3.83 | 3.83 | 3.83 | 3.83 | 3.83 | 3.83 | 3.83 | 3.83 | 3.83 | 3.83 | 3.83 |
| | 0.01 | 5.70 | 5.96 | 6.11 | 6.18 | 6.26 | 6.33 | 6.40 | 6.44 | 6.5 | 6.6 | 6.6 | 6.7 | 6.7 | 6.8 |
| 6 | 0.05 | 3.46 | 3.58 | 3.64 | 3.68 | 3.68 | 3.68 | 3.68 | 3.68 | 3.68 | 3.68 | 3.68 | 3.68 | 3.68 | 3.68 |
| | 0.01 | 5.24 | 5.51 | 5.65 | 5.73 | 5.81 | 5.88 | 5.95 | 6.00 | 6.0 | 6.1 | 6.2 | 6.2 | 6.3 | 6.3 |
| 7 | 0.05 | 3.35 | 3.47 | 3.54 | 3.58 | 3.60 | 3.61 | 3.61 | 3.61 | 3.61 | 3.61 | 3.61 | 3.61 | 3.61 | 3.61 |
| | 0.01 | 4.95 | 5.22 | 5.37 | 5.45 | 5.53 | 5.61 | 5.69 | 5.73 | 5.8 | 5.8 | 5.9 | 5.9 | 6.0 | 6.0 |
| 8 | 0.05 | 3.26 | 3.39 | 3.47 | 3.52 | 3.55 | 3.56 | 3.56 | 3.56 | 3.56 | 3.56 | 3.56 | 3.56 | 3.56 | 3.56 |
| | 0.01 | 4.74 | 5.00 | 5.14 | 5.23 | 5.32 | 5.40 | 5.47 | 5.51 | 5.5 | 5.6 | 5.7 | 5.7 | 5.8 | 5.8 |
| 9 | 0.05 | 3.20 | 3.34 | 3.41 | 3.47 | 3.50 | 3.52 | 3.52 | 3.52 | 3.52 | 3.52 | 3.52 | 3.52 | 3.52 | 3.52 |
| | 0.01 | 4.60 | 4.86 | 4.99 | 5.08 | 5.17 | 5.25 | 5.32 | 5.36 | 5.4 | 5.5 | 5.5 | 5.6 | 5.7 | 5.7 |
| 10 | 0.05 | 3.15 | 3.30 | 3.37 | 3.43 | 3.46 | 3.47 | 3.47 | 3.47 | 3.47 | 3.47 | 3.47 | 3.47 | 3.47 | 3.48 |
| | 0.01 | 4.48 | 4.73 | 4.88 | 4.96 | 5.06 | 5.13 | 5.20 | 5.24 | 5.28 | 5.36 | 5.42 | 5.48 | 5.54 | 5.55 |
| 11 | 0.05 | 3.11 | 3.27 | 3.35 | 3.39 | 3.43 | 3.44 | 3.45 | 3.46 | 3.46 | 3.46 | 3.46 | 3.46 | 3.47 | 3.48 |
| | 0.01 | 4.39 | 4.63 | 4.77 | 4.86 | 4.94 | 5.01 | 5.06 | 5.12 | 5.15 | 5.24 | 5.28 | 5.34 | 5.38 | 5.39 |
| 12 | 0.05 | 3.08 | 3.23 | 3.33 | 3.36 | 3.40 | 3.42 | 3.44 | 3.44 | 3.46 | 3.46 | 3.46 | 3.46 | 3.47 | 3.48 |
| | 0.01 | 4.32 | 4.55 | 4.68 | 4.76 | 4.84 | 4.92 | 4.96 | 5.02 | 5.07 | 5.13 | 5.17 | 5.22 | 5.24 | 5.26 |
| 13 | 0.05 | 3.06 | 3.21 | 3.30 | 3.35 | 3.38 | 3.41 | 3.42 | 3.44 | 3.45 | 3.45 | 3.46 | 3.46 | 3.47 | 3.47 |
| | 0.01 | 4.26 | 4.48 | 4.62 | 4.69 | 4.74 | 4.84 | 4.88 | 4.94 | 4.98 | 5.04 | 5.08 | 5.13 | 5.14 | 5.15 |
| 14 | 0.05 | 3.03 | 3.18 | 3.27 | 3.33 | 3.37 | 3.39 | 3.41 | 3.42 | 3.44 | 3.45 | 3.46 | 3.46 | 3.47 | 3.47 |
| | 0.01 | 4.21 | 4.42 | 4.55 | 4.63 | 4.70 | 4.78 | 4.83 | 4.87 | 4.91 | 4.96 | 5.00 | 5.04 | 5.06 | 5.07 |

| df | α | | | | | | | | | | | | | | |
|---|---|---|---|---|---|---|---|---|---|---|---|---|---|---|---|
| 15 | 0.05 | 3.01 | 3.16 | 3.25 | 3.31 | 3.36 | 3.38 | 3.40 | 3.42 | 3.43 | 3.44 | 3.45 | 3.46 | 3.47 | 3.47 |
|  | 0.01 | 4.17 | 4.37 | 4.50 | 4.58 | 4.64 | 4.72 | 4.77 | 4.81 | 4.84 | 4.90 | 4.94 | 4.97 | 4.99 | 5.00 |
| 16 | 0.05 | 3.00 | 3.15 | 3.23 | 3.30 | 3.34 | 3.37 | 3.39 | 3.41 | 3.43 | 3.44 | 3.45 | 3.46 | 3.47 | 3.47 |
|  | 0.01 | 4.13 | 4.34 | 4.45 | 4.54 | 4.60 | 4.67 | 4.72 | 4.76 | 4.79 | 4.84 | 4.88 | 4.91 | 4.93 | 4.94 |
| 17 | 0.05 | 2.98 | 3.13 | 3.22 | 3.28 | 3.33 | 3.36 | 3.38 | 3.40 | 3.42 | 3.44 | 3.45 | 3.46 | 3.47 | 3.47 |
|  | 0.01 | 4.10 | 4.30 | 4.41 | 4.50 | 4.56 | 4.63 | 4.68 | 4.72 | 4.75 | 4.80 | 4.83 | 4.86 | 4.88 | 4.89 |
| 18 | 0.05 | 2.97 | 3.12 | 3.21 | 3.27 | 3.32 | 3.35 | 3.37 | 3.39 | 3.41 | 3.43 | 3.45 | 3.46 | 3.47 | 3.47 |
|  | 0.01 | 4.07 | 4.27 | 4.38 | 4.46 | 4.53 | 4.59 | 4.64 | 4.68 | 4.71 | 4.76 | 4.79 | 4.82 | 4.84 | 4.85 |
| 19 | 0.05 | 2.96 | 3.11 | 3.19 | 3.26 | 3.31 | 3.35 | 3.37 | 3.39 | 3.41 | 3.43 | 3.44 | 3.46 | 3.47 | 3.47 |
|  | 0.01 | 4.05 | 4.24 | 4.35 | 4.43 | 4.50 | 4.56 | 4.61 | 4.64 | 4.67 | 4.72 | 4.76 | 4.79 | 4.81 | 4.82 |
| 20 | 0.05 | 2.95 | 3.10 | 3.18 | 3.25 | 3.30 | 3.34 | 3.36 | 3.38 | 3.40 | 3.43 | 3.44 | 3.46 | 3.46 | 3.47 |
|  | 0.01 | 4.02 | 4.22 | 4.33 | 4.40 | 4.47 | 4.53 | 4.58 | 4.61 | 4.65 | 4.69 | 4.73 | 4.76 | 4.78 | 4.79 |
| 22 | 0.05 | 2.93 | 3.08 | 3.17 | 3.24 | 3.29 | 3.32 | 3.35 | 3.37 | 3.39 | 3.42 | 3.44 | 3.45 | 3.46 | 3.47 |
|  | 0.01 | 3.99 | 4.17 | 4.28 | 4.36 | 4.42 | 4.48 | 4.53 | 4.57 | 4.60 | 4.65 | 4.68 | 4.71 | 4.74 | 4.75 |
| 24 | 0.05 | 2.92 | 3.07 | 3.15 | 3.22 | 3.28 | 3.31 | 3.34 | 3.37 | 3.38 | 3.41 | 3.44 | 3.45 | 3.46 | 3.47 |
|  | 0.01 | 3.96 | 4.14 | 4.24 | 4.33 | 4.39 | 4.44 | 4.49 | 4.53 | 4.57 | 4.62 | 4.64 | 4.67 | 4.70 | 4.72 |
| 26 | 0.05 | 2.91 | 3.06 | 3.14 | 3.21 | 3.27 | 3.30 | 3.34 | 3.36 | 3.38 | 3.41 | 3.43 | 3.45 | 3.46 | 3.47 |
|  | 0.01 | 3.93 | 4.11 | 4.21 | 4.30 | 4.36 | 4.41 | 4.46 | 4.50 | 4.53 | 4.58 | 4.62 | 4.65 | 4.65 | 4.69 |
| 28 | 0.05 | 2.90 | 3.04 | 3.13 | 3.20 | 3.26 | 3.30 | 3.33 | 3.35 | 3.37 | 3.40 | 3.43 | 3.45 | 3.46 | 3.47 |
|  | 0.01 | 3.91 | 4.08 | 4.18 | 4.28 | 4.34 | 4.39 | 4.43 | 4.47 | 4.51 | 4.56 | 4.60 | 4.62 | 4.65 | 4.67 |
| 30 | 0.05 | 2.89 | 3.04 | 3.12 | 3.20 | 3.25 | 3.29 | 3.32 | 3.35 | 3.37 | 3.40 | 3.43 | 3.44 | 3.46 | 3.47 |
|  | 0.01 | 3.89 | 4.06 | 4.16 | 4.22 | 4.32 | 4.36 | 4.41 | 4.45 | 4.48 | 4.54 | 4.58 | 4.61 | 4.63 | 4.65 |
| 40 | 0.05 | 2.86 | 3.01 | 3.10 | 3.17 | 3.22 | 3.27 | 3.30 | 3.33 | 3.35 | 3.39 | 3.42 | 3.44 | 3.46 | 3.47 |
|  | 0.01 | 3.82 | 3.99 | 4.10 | 4.17 | 4.24 | 4.30 | 4.34 | 4.37 | 4.41 | 4.46 | 4.51 | 4.54 | 4.57 | 4.59 |
| 60 | 0.05 | 2.83 | 2.98 | 3.08 | 3.14 | 3.20 | 3.24 | 3.28 | 3.31 | 3.33 | 3.37 | 3.40 | 3.43 | 3.45 | 3.47 |
|  | 0.01 | 3.76 | 3.92 | 4.03 | 4.12 | 4.17 | 4.23 | 4.27 | 4.31 | 4.34 | 4.39 | 4.44 | 4.47 | 4.50 | 4.53 |
| 100 | 0.05 | 2.80 | 2.95 | 3.05 | 3.12 | 3.18 | 3.22 | 3.26 | 3.29 | 3.32 | 3.36 | 3.40 | 3.42 | 3.45 | 3.47 |
|  | 0.01 | 3.71 | 3.86 | 3.98 | 4.06 | 4.11 | 4.17 | 4.21 | 4.25 | 4.29 | 4.35 | 4.38 | 4.42 | 4.45 | 4.48 |
| ∞ | 0.05 | 2.77 | 2.92 | 3.02 | 3.09 | 3.15 | 3.19 | 3.23 | 3.26 | 3.29 | 3.34 | 3.38 | 3.41 | 3.44 | 3.47 |
|  | 0.01 | 3.64 | 3.80 | 3.90 | 3.98 | 4.04 | 4.09 | 4.14 | 4.17 | 4.20 | 4.26 | 4.31 | 4.34 | 4.38 | 4.41 |

*Source:* This table is reproduced from D. B. Duncan, "Multiple Range and Multiple $F$ Tests," *Biometrics*, Vol. 11, 1–42 (1955), with permission from The Biometric Society.

**TABLE A.6** FIVE AND ONE PERCENT POINTS FOR *r* AND *R*

| Degrees of freedom | P | Number of variables 2 | 3 | 4 | 5 | Degrees of freedom | P | Number of variables 2 | 3 | 4 | 5 |
|---|---|---|---|---|---|---|---|---|---|---|---|
| 1 | 0.05 | 0.997 | 0.999 | 0.999 | 0.999 | 24 | 0.05 | 0.388 | 0.470 | 0.523 | 0.562 |
|   | 0.01 | 1.000 | 1.000 | 1.000 | 1.000 |   | 0.01 | 0.496 | 0.565 | 0.609 | 0.642 |
| 2 | 0.05 | 0.950 | 0.975 | 0.983 | 0.987 | 25 | 0.05 | 0.381 | 0.462 | 0.514 | 0.553 |
|   | 0.01 | 0.990 | 0.995 | 0.997 | 0.998 |   | 0.01 | 0.487 | 0.555 | 0.600 | 0.633 |
| 3 | 0.05 | 0.878 | 0.930 | 0.950 | 0.961 | 26 | 0.05 | 0.374 | 0.454 | 0.506 | 0.545 |
|   | 0.01 | 0.959 | 0.976 | 0.983 | 0.987 |   | 0.01 | 0.478 | 0.546 | 0.590 | 0.624 |
| 4 | 0.05 | 0.811 | 0.881 | 0.912 | 0.930 | 27 | 0.05 | 0.367 | 0.446 | 0.498 | 0.536 |
|   | 0.01 | 0.917 | 0.949 | 0.962 | 0.970 |   | 0.01 | 0.470 | 0.538 | 0.582 | 0.615 |
| 5 | 0.05 | 0.754 | 0.836 | 0.874 | 0.898 | 28 | 0.05 | 0.361 | 0.439 | 0.490 | 0.529 |
|   | 0.01 | 0.874 | 0.917 | 0.937 | 0.947 |   | 0.01 | 0.463 | 0.530 | 0.573 | 0.606 |
| 6 | 0.05 | 0.707 | 0.795 | 0.839 | 0.867 | 29 | 0.05 | 0.355 | 0.432 | 0.482 | 0.521 |
|   | 0.01 | 0.834 | 0.886 | 0.911 | 0.927 |   | 0.01 | 0.456 | 0.522 | 0.565 | 0.598 |
| 7 | 0.05 | 0.666 | 0.758 | 0.807 | 0.838 | 30 | 0.05 | 0.349 | 0.426 | 0.476 | 0.514 |
|   | 0.01 | 0.798 | 0.855 | 0.885 | 0.904 |   | 0.01 | 0.449 | 0.514 | 0.558 | 0.591 |
| 8 | 0.05 | 0.632 | 0.726 | 0.777 | 0.811 | 35 | 0.05 | 0.325 | 0.397 | 0.445 | 0.482 |
|   | 0.01 | 0.765 | 0.827 | 0.860 | 0.822 |   | 0.01 | 0.418 | 0.481 | 0.523 | 0.556 |
| 9 | 0.05 | 0.602 | 0.697 | 0.750 | 0.786 | 40 | 0.05 | 0.304 | 0.373 | 0.419 | 0.455 |
|   | 0.01 | 0.735 | 0.800 | 0.836 | 0.861 |   | 0.01 | 0.393 | 0.454 | 0.494 | 0.526 |
| 10 | 0.05 | 0.576 | 0.671 | 0.726 | 0.763 | 45 | 0.05 | 0.288 | 0.353 | 0.397 | 0.432 |
|   | 0.01 | 0.708 | 0.776 | 0.814 | 0.840 |   | 0.01 | 0.372 | 0.430 | 0.470 | 0.501 |
| 11 | 0.05 | 0.553 | 0.648 | 0.703 | 0.741 | 50 | 0.05 | 0.273 | 0.336 | 0.379 | 0.412 |
|   | 0.01 | 0.684 | 0.753 | 0.793 | 0.821 |   | 0.01 | 0.354 | 0.410 | 0.449 | 0.479 |
| 12 | 0.05 | 0.532 | 0.627 | 0.683 | 0.722 | 60 | 0.05 | 0.250 | 0.308 | 0.348 | 0.380 |
|   | 0.01 | 0.661 | 0.732 | 0.773 | 0.802 |   | 0.01 | 0.325 | 0.377 | 0.414 | 0.442 |
| 13 | 0.05 | 0.514 | 0.608 | 0.664 | 0.703 | 70 | 0.05 | 0.232 | 0.286 | 0.324 | 0.354 |
|   | 0.01 | 0.641 | 0.712 | 0.755 | 0.785 |   | 0.01 | 0.302 | 0.351 | 0.386 | 0.413 |
| 14 | 0.05 | 0.497 | 0.590 | 0.646 | 0.686 | 80 | 0.05 | 0.217 | 0.269 | 0.304 | 0.332 |
|   | 0.01 | 0.623 | 0.694 | 0.737 | 0.768 |   | 0.01 | 0.283 | 0.330 | 0.362 | 0.389 |
| 15 | 0.05 | 0.482 | 0.574 | 0.630 | 0.670 | 90 | 0.05 | 0.205 | 0.254 | 0.288 | 0.315 |
|   | 0.01 | 0.606 | 0.677 | 0.721 | 0.752 |   | 0.01 | 0.267 | 0.312 | 0.343 | 0.368 |
| 16 | 0.05 | 0.468 | 0.559 | 0.615 | 0.655 | 100 | 0.05 | 0.195 | 0.241 | 0.274 | 0.300 |
|   | 0.01 | 0.590 | 0.662 | 0.706 | 0.738 |   | 0.01 | 0.254 | 0.297 | 0.327 | 0.351 |
| 17 | 0.05 | 0.456 | 0.545 | 0.601 | 0.641 | 125 | 0.05 | 0.174 | 0.216 | 0.246 | 0.269 |
|   | 0.01 | 0.575 | 0.647 | 0.691 | 0.724 |   | 0.01 | 0.228 | 0.266 | 0.294 | 0.316 |
| 18 | 0.05 | 0.444 | 0.532 | 0.587 | 0.628 | 150 | 0.05 | 0.159 | 0.198 | 0.225 | 0.247 |
|   | 0.01 | 0.561 | 0.633 | 0.678 | 0.710 |   | 0.01 | 0.208 | 0.244 | 0.270 | 0.290 |
| 19 | 0.05 | 0.433 | 0.520 | 0.575 | 0.615 | 200 | 0.05 | 0.138 | 0.172 | 0.196 | 0.215 |
|   | 0.01 | 0.549 | 0.620 | 0.665 | 0.698 |   | 0.01 | 0.181 | 0.212 | 0.234 | 0.253 |
| 20 | 0.05 | 0.423 | 0.509 | 0.563 | 0.604 | 300 | 0.05 | 0.113 | 0.141 | 0.160 | 0.176 |
|   | 0.01 | 0.537 | 0.608 | 0.652 | 0.685 |   | 0.01 | 0.148 | 0.174 | 0.192 | 0.208 |
| 21 | 0.05 | 0.413 | 0.498 | 0.552 | 0.592 | 400 | 0.05 | 0.098 | 0.122 | 0.139 | 0.153 |
|   | 0.01 | 0.526 | 0.596 | 0.641 | 0.674 |   | 0.01 | 0.128 | 0.151 | 0.167 | 0.180 |
| 22 | 0.05 | 0.404 | 0.488 | 0.542 | 0.582 | 500 | 0.05 | 0.088 | 0.109 | 0.124 | 0.137 |
|   | 0.01 | 0.515 | 0.585 | 0.630 | 0.663 |   | 0.01 | 0.115 | 0.135 | 0.150 | 0.162 |
| 23 | 0.05 | 0.396 | 0.479 | 0.532 | 0.572 | 1000 | 0.05 | 0.062 | 0.077 | 0.088 | 0.097 |
|   | 0.01 | 0.505 | 0.574 | 0.619 | 0.652 |   | 0.01 | 0.081 | 0.096 | 0.106 | 0.115 |

*Source:* This table is reproduced from George W. Snedecor, *Statistical Methods,* 4th ed., 1946, with the permission of Iowa State University Press.

**TABLE A.7**  TABLE OF $Z = \frac{1}{2} \ln(1 + r)/(1 - r)$ TO TRANSFORM THE CORRELATION COEFFICIENT

| $r$ | 0.00 | 0.01 | 0.02 | 0.03 | 0.04 | 0.05 | 0.06 | 0.07 | 0.08 | 0.09 |
|-----|------|------|------|------|------|------|------|------|------|------|
| 0.0 | 0.000 | 0.010 | 0.020 | 0.030 | 0.040 | 0.050 | 0.060 | 0.070 | 0.080 | 0.090 |
| 0.1 | 0.100 | 0.110 | 0.121 | 0.131 | 0.141 | 0.151 | 0.161 | 0.172 | 0.182 | 0.192 |
| 0.2 | 0.203 | 0.213 | 0.224 | 0.234 | 0.245 | 0.255 | 0.266 | 0.277 | 0.288 | 0.299 |
| 0.3 | 0.310 | 0.321 | 0.332 | 0.343 | 0.354 | 0.365 | 0.377 | 0.388 | 0.400 | 0.412 |
| 0.4 | 0.424 | 0.436 | 0.448 | 0.460 | 0.472 | 0.485 | 0.497 | 0.510 | 0.523 | 0.536 |
| 0.5 | 0.549 | 0.563 | 0.576 | 0.590 | 0.604 | 0.618 | 0.633 | 0.648 | 0.662 | 0.678 |
| 0.6 | 0.693 | 0.709 | 0.725 | 0.741 | 0.758 | 0.775 | 0.793 | 0.811 | 0.829 | 0.848 |
| 0.7 | 0.867 | 0.887 | 0.908 | 0.929 | 0.950 | 0.973 | 0.996 | 1.020 | 1.045 | 1.071 |
| 0.8 | 1.099 | 1.127 | 1.157 | 1.188 | 1.221 | 1.256 | 1.293 | 1.333 | 1.376 | 1.422 |

| $r$ | 0.000 | 0.001 | 0.002 | 0.003 | 0.004 | 0.005 | 0.006 | 0.007 | 0.008 | 0.009 |
|-----|-------|-------|-------|-------|-------|-------|-------|-------|-------|-------|
| 0.90 | 1.472 | 1.478 | 1.483 | 1.488 | 1.494 | 1.499 | 1.505 | 1.510 | 1.516 | 1.522 |
| 0.91 | 1.528 | 1.533 | 1.539 | 1.545 | 1.551 | 1.557 | 1.564 | 1.570 | 1.576 | 1.583 |
| 0.92 | 1.589 | 1.596 | 1.602 | 1.609 | 1.616 | 1.623 | 1.630 | 1.637 | 1.644 | 1.651 |
| 0.93 | 1.658 | 1.666 | 1.673 | 1.681 | 1.689 | 1.697 | 1.705 | 1.713 | 1.721 | 1.730 |
| 0.94 | 1.738 | 1.747 | 1.756 | 1.764 | 1.774 | 1.783 | 1.792 | 1.802 | 1.812 | 1.822 |
| 0.95 | 1.832 | 1.842 | 1.853 | 1.863 | 1.874 | 1.886 | 1.897 | 1.909 | 1.921 | 1.933 |
| 0.96 | 1.946 | 1.959 | 1.972 | 1.986 | 2.000 | 2.014 | 2.029 | 2.044 | 2.060 | 2.076 |
| 0.97 | 2.092 | 2.109 | 2.127 | 2.146 | 2.165 | 2.185 | 2.205 | 2.227 | 2.249 | 2.273 |
| 0.98 | 2.298 | 2.323 | 2.351 | 2.380 | 2.410 | 2.443 | 2.477 | 2.515 | 2.555 | 2.599 |
| 0.99 | 2.646 | 2.700 | 2.759 | 2.826 | 2.903 | 2.994 | 3.106 | 3.250 | 3.453 | 3.800 |

*Source:* This table is reproduced from G. W. Snedecor and W. G. Cochran, *Statistical Methods,* 6th ed., 1967, with permission of Iowa State University Press.

**TABLE A.8** TABLE OF $r$ IN TERMS OF $z^a$

| z | 0.00 | 0.01 | 0.02 | 0.03 | 0.04 | 0.05 | 0.06 | 0.07 | 0.08 | 0.09 |
|-----|-------|-------|-------|-------|-------|-------|-------|-------|-------|-------|
| 0.0 | 0.000 | 0.010 | 0.020 | 0.030 | 0.040 | 0.050 | 0.060 | 0.070 | 0.080 | 0.090 |
| 0.1 | 0.100 | 0.110 | 0.119 | 0.129 | 0.139 | 0.149 | 0.159 | 0.168 | 0.178 | 0.187 |
| 0.2 | 0.197 | 0.207 | 0.216 | 0.226 | 0.236 | 0.245 | 0.254 | 0.264 | 0.273 | 0.282 |
| 0.3 | 0.291 | 0.300 | 0.310 | 0.319 | 0.327 | 0.336 | 0.345 | 0.354 | 0.363 | 0.371 |
| 0.4 | 0.380 | 0.389 | 0.397 | 0.405 | 0.414 | 0.422 | 0.430 | 0.438 | 0.446 | 0.454 |
| 0.5 | 0.462 | 0.470 | 0.478 | 0.485 | 0.493 | 0.500 | 0.508 | 0.515 | 0.523 | 0.530 |
| 0.6 | 0.537 | 0.544 | 0.551 | 0.558 | 0.565 | 0.572 | 0.578 | 0.585 | 0.592 | 0.598 |
| 0.7 | 0.604 | 0.611 | 0.617 | 0.623 | 0.629 | 0.635 | 0.641 | 0.647 | 0.653 | 0.658 |
| 0.8 | 0.664 | 0.670 | 0.675 | 0.680 | 0.686 | 0.691 | 0.696 | 0.701 | 0.706 | 0.711 |
| 0.9 | 0.716 | 0.721 | 0.726 | 0.731 | 0.735 | 0.740 | 0.744 | 0.749 | 0.753 | 0.757 |
| 1.0 | 0.762 | 0.766 | 0.770 | 0.774 | 0.778 | 0.782 | 0.786 | 0.790 | 0.793 | 0.797 |
| 1.1 | 0.800 | 0.804 | 0.808 | 0.811 | 0.814 | 0.818 | 0.821 | 0.824 | 0.828 | 0.831 |
| 1.2 | 0.834 | 0.837 | 0.840 | 0.843 | 0.846 | 0.848 | 0.851 | 0.854 | 0.856 | 0.859 |
| 1.3 | 0.862 | 0.864 | 0.867 | 0.869 | 0.872 | 0.874 | 0.876 | 0.879 | 0.881 | 0.883 |
| 1.4 | 0.885 | 0.888 | 0.890 | 0.892 | 0.894 | 0.896 | 0.898 | 0.900 | 0.902 | 0.903 |
| 1.5 | 0.905 | 0.907 | 0.909 | 0.910 | 0.912 | 0.914 | 0.915 | 0.917 | 0.919 | 0.920 |
| 1.6 | 0.922 | 0.923 | 0.925 | 0.926 | 0.928 | 0.929 | 0.930 | 0.932 | 0.933 | 0.934 |
| 1.7 | 0.935 | 0.937 | 0.938 | 0.939 | 0.940 | 0.941 | 0.942 | 0.944 | 0.945 | 0.946 |
| 1.8 | 0.947 | 0.948 | 0.949 | 0.950 | 0.951 | 0.952 | 0.953 | 0.954 | 0.954 | 0.955 |
| 1.9 | 0.956 | 0.957 | 0.958 | 0.959 | 0.960 | 0.960 | 0.961 | 0.962 | 0.963 | 0.963 |
| 2.0 | 0.964 | 0.965 | 0.965 | 0.966 | 0.967 | 0.967 | 0.968 | 0.969 | 0.969 | 0.970 |
| 2.1 | 0.970 | 0.971 | 0.972 | 0.972 | 0.973 | 0.973 | 0.974 | 0.974 | 0.975 | 0.975 |
| 2.2 | 0.976 | 0.976 | 0.977 | 0.977 | 0.978 | 0.978 | 0.978 | 0.979 | 0.979 | 0.980 |
| 2.3 | 0.980 | 0.980 | 0.981 | 0.981 | 0.982 | 0.982 | 0.982 | 0.983 | 0.983 | 0.983 |
| 2.4 | 0.984 | 0.984 | 0.984 | 0.985 | 0.985 | 0.985 | 0.986 | 0.986 | 0.986 | 0.986 |
| 2.5 | 0.987 | 0.987 | 0.987 | 0.987 | 0.988 | 0.988 | 0.988 | 0.988 | 0.989 | 0.989 |
| 2.6 | 0.989 | 0.989 | 0.989 | 0.990 | 0.990 | 0.990 | 0.990 | 0.990 | 0.991 | 0.991 |
| 2.7 | 0.991 | 0.991 | 0.991 | 0.992 | 0.992 | 0.992 | 0.992 | 0.992 | 0.992 | 0.992 |
| 2.8 | 0.993 | 0.993 | 0.993 | 0.993 | 0.993 | 0.993 | 0.994 | 0.994 | 0.994 | 0.994 |
| 2.9 | 0.994 | 0.994 | 0.994 | 0.994 | 0.994 | 0.995 | 0.995 | 0.995 | 0.995 | 0.995 |

[a] $r = (e^{2z} - 1)(e^{2z} + 1)$.

*Source:* This table is reproduced from G. W. Snedecor and W. G. Cochran, *Statistical Methods,* 6th ed., 1967, with permission of Iowa State University Press.

**TABLE A.9**  THE ANGULAR TRANSFORMATION

| p | 0 | 1 | 2 | 3 | 4 | 5 | 6 | 7 | 8 | 9 | p |
|---|---|---|---|---|---|---|---|---|---|---|---|
| 0.000 | 0.00 | 0.57 | 0.81 | 0.99 | 1.15 | 1.28 | 1.40 | 1.52 | 1.62 | 1.72 | 0.000 |
| 0.001 | 1.81 | 1.90 | 1.99 | 2.07 | 2.14 | 2.22 | 2.29 | 2.36 | 2.43 | 2.50 | 0.001 |
| 0.002 | 2.56 | 2.63 | 2.69 | 2.75 | 2.81 | 2.87 | 2.92 | 2.98 | 3.03 | 3.09 | 0.002 |
| 0.003 | 3.14 | 3.19 | 3.24 | 3.29 | 3.34 | 3.39 | 3.44 | 3.49 | 3.53 | 3.58 | 0.003 |
| 0.004 | 3.63 | 3.67 | 3.72 | 3.76 | 3.80 | 3.85 | 3.89 | 3.93 | 3.97 | 4.01 | 0.004 |
| 0.005 | 4.05 | 4.10 | 4.14 | 4.17 | 4.21 | 4.25 | 4.29 | 4.33 | 4.37 | 4.41 | 0.005 |
| 0.006 | 4.44 | 4.48 | 4.52 | 4.55 | 4.59 | 4.62 | 4.66 | 4.70 | 4.73 | 4.76 | 0.006 |
| 0.007 | 4.80 | 4.83 | 4.87 | 4.90 | 4.93 | 4.97 | 5.00 | 5.03 | 5.07 | 5.10 | 0.007 |
| 0.008 | 5.13 | 5.16 | 5.20 | 5.23 | 5.26 | 5.29 | 5.32 | 5.35 | 5.38 | 5.41 | 0.008 |
| 0.009 | 5.44 | 5.47 | 5.50 | 5.53 | 5.56 | 5.59 | 5.62 | 5.65 | 5.68 | 5.71 | 0.009 |
| 0.01 | 5.74 | 6.02 | 6.29 | 6.55 | 6.80 | 7.03 | 7.27 | 7.49 | 7.71 | 7.92 | 0.01 |
| 0.02 | 8.13 | 8.33 | 8.53 | 8.72 | 8.91 | 9.10 | 9.28 | 9.46 | 9.63 | 9.80 | 0.02 |
| 0.03 | 9.97 | 10.14 | 10.30 | 10.47 | 10.63 | 10.78 | 10.94 | 11.09 | 11.24 | 11.39 | 0.03 |
| 0.04 | 11.54 | 11.68 | 11.83 | 11.97 | 12.11 | 12.25 | 12.38 | 12.52 | 12.66 | 12.79 | 0.04 |
| 0.05 | 12.92 | 13.05 | 13.18 | 13.31 | 13.44 | 13.56 | 13.69 | 13.81 | 13.94 | 14.06 | 0.05 |
| 0.06 | 14.18 | 14.30 | 14.42 | 14.54 | 14.65 | 14.77 | 14.89 | 15.00 | 15.12 | 15.23 | 0.06 |
| 0.07 | 15.34 | 15.45 | 15.56 | 15.68 | 15.79 | 15.89 | 16.00 | 16.11 | 16.22 | 16.32 | 0.07 |
| 0.08 | 16.43 | 16.54 | 16.64 | 16.74 | 16.85 | 16.95 | 17.05 | 17.15 | 17.26 | 17.36 | 0.08 |
| 0.09 | 17.46 | 17.56 | 17.66 | 17.76 | 17.85 | 17.95 | 18.05 | 18.15 | 18.24 | 18.34 | 0.09 |
| 0.10 | 18.43 | 18.53 | 18.63 | 18.72 | 18.81 | 18.91 | 19.00 | 19.09 | 19.19 | 19.28 | 0.10 |
| 0.11 | 19.37 | 19.46 | 19.55 | 19.64 | 19.73 | 19.82 | 19.91 | 20.00 | 20.09 | 20.18 | 0.11 |
| 0.12 | 20.27 | 20.36 | 20.44 | 20.53 | 20.62 | 20.70 | 20.79 | 20.88 | 20.96 | 21.05 | 0.12 |
| 0.13 | 21.13 | 21.22 | 21.30 | 21.39 | 21.47 | 21.56 | 21.64 | 21.72 | 21.81 | 21.89 | 0.13 |
| 0.14 | 21.97 | 22.06 | 22.14 | 22.22 | 22.30 | 22.38 | 22.46 | 22.54 | 22.63 | 22.71 | 0.14 |
| 0.15 | 22.79 | 22.87 | 22.95 | 23.03 | 23.11 | 23.18 | 23.26 | 23.34 | 23.42 | 23.50 | 0.15 |
| 0.16 | 23.58 | 23.66 | 23.73 | 23.81 | 23.89 | 23.97 | 24.04 | 24.12 | 24.20 | 24.27 | 0.16 |
| 0.17 | 24.35 | 24.43 | 24.50 | 24.58 | 24.65 | 24.73 | 24.80 | 24.88 | 24.95 | 25.03 | 0.17 |
| 0.18 | 25.10 | 25.18 | 25.25 | 25.33 | 25.40 | 25.47 | 25.55 | 25.62 | 25.70 | 25.77 | 0.18 |
| 0.19 | 25.84 | 25.91 | 25.99 | 26.06 | 26.13 | 26.21 | 26.28 | 26.35 | 26.42 | 26.49 | 0.19 |
| 0.20 | 26.57 | 26.64 | 26.71 | 26.78 | 26.85 | 26.92 | 26.99 | 27.06 | 27.13 | 27.20 | 0.20 |
| 0.21 | 27.27 | 27.35 | 27.42 | 27.49 | 27.56 | 27.62 | 27.69 | 27.76 | 27.83 | 27.90 | 0.21 |
| 0.22 | 27.97 | 28.04 | 28.11 | 28.18 | 28.25 | 28.32 | 28.39 | 28.45 | 28.52 | 28.59 | 0.22 |
| 0.23 | 28.66 | 28.73 | 28.79 | 28.86 | 28.93 | 29.00 | 29.06 | 29.13 | 29.20 | 29.27 | 0.23 |
| 0.24 | 29.33 | 29.40 | 29.47 | 29.53 | 29.60 | 29.67 | 29.73 | 29.80 | 29.87 | 29.93 | 0.24 |
| 0.25 | 30.00 | 30.07 | 30.13 | 30.20 | 30.26 | 30.33 | 30.40 | 30.46 | 30.53 | 30.59 | 0.25 |
| 0.26 | 30.66 | 30.72 | 30.79 | 30.85 | 30.92 | 30.98 | 31.05 | 31.11 | 31.18 | 31.24 | 0.26 |
| 0.27 | 31.31 | 31.37 | 31.44 | 31.50 | 31.56 | 31.63 | 31.69 | 31.76 | 31.82 | 31.88 | 0.27 |
| 0.28 | 31.95 | 32.01 | 32.08 | 32.14 | 32.20 | 32.27 | 32.33 | 32.39 | 32.46 | 32.52 | 0.28 |
| 0.29 | 32.58 | 32.65 | 32.71 | 32.77 | 32.83 | 32.90 | 32.96 | 33.02 | 33.09 | 33.15 | 0.29 |
| 0.30 | 33.21 | 33.27 | 33.34 | 33.40 | 33.46 | 33.52 | 33.58 | 33.65 | 33.71 | 33.77 | 0.30 |
| 0.31 | 33.83 | 33.90 | 33.96 | 34.02 | 34.08 | 34.14 | 34.20 | 34.27 | 34.33 | 34.39 | 0.31 |
| 0.32 | 34.45 | 34.51 | 34.57 | 34.63 | 34.70 | 34.76 | 34.82 | 34.88 | 34.94 | 35.00 | 0.32 |
| 0.33 | 35.06 | 35.12 | 35.18 | 35.24 | 35.30 | 35.37 | 35.43 | 35.49 | 35.55 | 35.61 | 0.33 |
| 0.34 | 35.67 | 35.73 | 35.79 | 35.85 | 35.91 | 35.97 | 36.03 | 36.09 | 36.15 | 36.21 | 0.34 |
| 0.35 | 36.27 | 36.33 | 36.39 | 36.45 | 36.51 | 36.57 | 36.63 | 36.69 | 36.75 | 36.81 | 0.35 |
| 0.36 | 36.87 | 36.93 | 36.99 | 37.05 | 37.11 | 37.17 | 37.23 | 37.29 | 37.35 | 37.41 | 0.36 |
| 0.37 | 37.46 | 37.52 | 37.58 | 37.64 | 37.70 | 37.76 | 37.82 | 37.88 | 37.94 | 38.00 | 0.37 |
| 0.38 | 38.06 | 38.12 | 38.17 | 38.23 | 38.29 | 38.35 | 38.41 | 38.47 | 38.53 | 38.59 | 0.38 |
| 0.39 | 38.65 | 38.70 | 38.76 | 38.82 | 38.88 | 38.94 | 39.00 | 39.06 | 39.11 | 39.17 | 0.39 |

| p | 0 | 1 | 2 | 3 | 4 | 5 | 6 | 7 | 8 | 9 | p |
|---|---|---|---|---|---|---|---|---|---|---|---|
| 0.40 | 39.23 | 39.29 | 39.35 | 39.41 | 39.47 | 39.52 | 39.58 | 39.64 | 39.70 | 39.76 | 0.40 |
| 0.41 | 39.82 | 39.87 | 39.93 | 39.99 | 40.05 | 40.11 | 40.16 | 40.22 | 40.28 | 40.34 | 0.41 |
| 0.42 | 40.40 | 40.45 | 40.51 | 40.57 | 40.63 | 40.69 | 40.74 | 40.80 | 40.86 | 40.92 | 0.42 |
| 0.43 | 40.98 | 41.03 | 41.09 | 41.15 | 41.21 | 41.27 | 41.32 | 41.38 | 41.44 | 41.50 | 0.43 |
| 0.44 | 41.55 | 41.61 | 41.67 | 41.73 | 41.78 | 41.84 | 41.90 | 41.96 | 42.02 | 42.07 | 0.44 |
| 0.45 | 42.13 | 42.19 | 42.25 | 42.30 | 42.36 | 42.42 | 42.48 | 42.53 | 42.59 | 42.65 | 0.45 |
| 0.46 | 42.71 | 42.76 | 42.82 | 42.88 | 42.94 | 42.99 | 43.05 | 43.11 | 43.17 | 43.22 | 0.46 |
| 0.47 | 43.28 | 43.34 | 43.39 | 43.45 | 43.51 | 43.57 | 43.62 | 43.68 | 43.74 | 43.80 | 0.47 |
| 0.48 | 43.85 | 43.91 | 43.97 | 44.03 | 44.08 | 44.14 | 44.20 | 44.26 | 44.31 | 44.37 | 0.48 |
| 0.49 | 44.43 | 44.48 | 44.54 | 44.60 | 44.66 | 44.71 | 44.77 | 44.83 | 44.89 | 44.94 | 0.49 |
| 0.50 | 45.00 | 45.06 | 45.11 | 45.17 | 45.23 | 45.29 | 45.34 | 45.40 | 45.46 | 45.52 | 0.50 |
| 0.51 | 45.57 | 45.63 | 45.69 | 45.74 | 45.80 | 45.86 | 45.92 | 45.97 | 46.03 | 46.09 | 0.51 |
| 0.52 | 46.15 | 46.20 | 46.26 | 46.32 | 46.38 | 46.43 | 46.49 | 46.55 | 46.61 | 46.66 | 0.52 |
| 0.53 | 46.72 | 46.78 | 46.83 | 46.89 | 46.95 | 47.01 | 47.06 | 47.12 | 47.18 | 47.24 | 0.53 |
| 0.54 | 47.29 | 47.35 | 47.41 | 47.47 | 47.52 | 47.58 | 47.64 | 47.70 | 47.75 | 47.81 | 0.54 |
| 0.55 | 47.87 | 47.93 | 47.98 | 48.04 | 48.10 | 48.16 | 48.22 | 48.27 | 48.33 | 48.39 | 0.55 |
| 0.56 | 48.45 | 48.50 | 48.56 | 48.62 | 48.68 | 48.73 | 48.79 | 48.85 | 48.91 | 48.97 | 0.56 |
| 0.57 | 49.02 | 49.08 | 49.14 | 49.20 | 49.26 | 49.31 | 49.37 | 49.43 | 49.49 | 49.55 | 0.57 |
| 0.58 | 49.60 | 49.66 | 49.72 | 49.78 | 49.84 | 49.89 | 49.95 | 50.01 | 50.07 | 50.13 | 0.58 |
| 0.59 | 50.18 | 50.24 | 50.30 | 50.36 | 50.42 | 50.48 | 50.53 | 50.59 | 50.65 | 50.71 | 0.59 |
| 0.60 | 50.77 | 50.83 | 50.89 | 50.94 | 51.00 | 51.06 | 51.12 | 51.18 | 51.24 | 51.30 | 0.60 |
| 0.61 | 51.35 | 51.41 | 51.47 | 51.53 | 51.59 | 51.65 | 51.71 | 51.77 | 51.83 | 51.88 | 0.61 |
| 0.62 | 51.94 | 52.00 | 52.06 | 52.12 | 52.18 | 52.24 | 52.30 | 52.36 | 52.42 | 52.48 | 0.62 |
| 0.63 | 52.54 | 52.59 | 52.65 | 52.71 | 52.77 | 52.83 | 52.89 | 52.95 | 53.01 | 53.07 | 0.63 |
| 0.64 | 53.13 | 53.19 | 53.25 | 53.31 | 53.37 | 53.43 | 53.49 | 53.55 | 53.61 | 53.67 | 0.64 |
| 0.65 | 53.73 | 53.79 | 53.85 | 53.91 | 53.97 | 54.03 | 54.09 | 54.15 | 54.21 | 54.27 | 0.65 |
| 0.66 | 54.33 | 54.39 | 54.45 | 54.51 | 54.57 | 54.63 | 54.70 | 54.76 | 54.82 | 54.88 | 0.66 |
| 0.67 | 54.94 | 55.00 | 55.06 | 55.12 | 55.18 | 55.24 | 55.30 | 55.37 | 55.43 | 55.49 | 0.67 |
| 0.68 | 55.55 | 55.61 | 55.67 | 55.73 | 55.80 | 55.86 | 55.92 | 55.98 | 56.04 | 56.10 | 0.68 |
| 0.69 | 56.17 | 56.23 | 56.29 | 56.35 | 56.42 | 56.48 | 56.54 | 56.60 | 56.66 | 56.73 | 0.69 |
| 0.70 | 56.79 | 56.85 | 56.91 | 56.98 | 57.04 | 57.10 | 57.17 | 57.23 | 57.29 | 57.35 | 0.70 |
| 0.71 | 57.42 | 57.48 | 57.54 | 57.61 | 57.67 | 57.73 | 57.80 | 57.86 | 57.92 | 57.99 | 0.71 |
| 0.72 | 58.05 | 58.12 | 58.18 | 58.24 | 58.31 | 58.37 | 58.44 | 58.50 | 58.56 | 58.63 | 0.72 |
| 0.73 | 58.69 | 58.76 | 58.82 | 58.89 | 58.95 | 59.02 | 59.08 | 59.15 | 59.21 | 59.28 | 0.73 |
| 0.74 | 59.34 | 59.41 | 59.47 | 59.54 | 59.60 | 59.67 | 59.74 | 59.80 | 59.87 | 59.93 | 0.74 |
| 0.75 | 60.00 | 60.07 | 60.13 | 60.20 | 60.27 | 60.33 | 60.40 | 60.47 | 60.53 | 60.60 | 0.75 |
| 0.76 | 60.67 | 60.73 | 60.80 | 60.87 | 60.94 | 61.00 | 61.07 | 61.14 | 61.21 | 61.27 | 0.76 |
| 0.77 | 61.34 | 61.41 | 61.48 | 61.55 | 61.61 | 61.68 | 61.75 | 61.82 | 61.89 | 61.96 | 0.77 |
| 0.78 | 62.03 | 62.10 | 62.17 | 62.24 | 62.31 | 62.38 | 62.44 | 62.51 | 62.58 | 62.65 | 0.78 |
| 0.79 | 62.73 | 62.80 | 62.87 | 62.94 | 63.01 | 63.08 | 63.15 | 63.22 | 63.29 | 63.36 | 0.79 |
| 0.80 | 63.43 | 63.51 | 63.58 | 63.65 | 63.72 | 63.79 | 63.87 | 63.94 | 64.01 | 64.09 | 0.80 |
| 0.81 | 64.16 | 64.23 | 64.30 | 64.38 | 64.45 | 64.53 | 64.60 | 64.67 | 64.75 | 64.82 | 0.81 |
| 0.82 | 64.90 | 64.97 | 65.05 | 65.12 | 65.20 | 65.27 | 65.35 | 65.42 | 65.50 | 65.57 | 0.82 |
| 0.83 | 65.65 | 65.73 | 65.80 | 65.88 | 65.96 | 66.03 | 66.11 | 66.19 | 66.27 | 66.34 | 0.83 |
| 0.84 | 66.42 | 66.50 | 66.58 | 66.66 | 66.74 | 66.82 | 66.89 | 66.97 | 67.05 | 67.13 | 0.84 |

**TABLE A.9** (Continued)

| p | 0 | 1 | 2 | 3 | 4 | 5 | 6 | 7 | 8 | 9 | p |
|---|---|---|---|---|---|---|---|---|---|---|---|
| 0.85 | 67.21 | 67.29 | 67.37 | 67.46 | 67.54 | 67.62 | 67.70 | 67.78 | 67.86 | 67.94 | 0.85 |
| 0.86 | 68.03 | 68.11 | 68.19 | 68.28 | 68.36 | 68.44 | 68.53 | 68.61 | 68.70 | 68.78 | 0.86 |
| 0.87 | 68.87 | 68.95 | 69.04 | 69.12 | 69.21 | 69.30 | 69.38 | 69.47 | 69.56 | 69.64 | 0.87 |
| 0.88 | 69.73 | 69.82 | 69.91 | 70.00 | 70.09 | 70.18 | 70.27 | 70.36 | 70.45 | 70.54 | 0.88 |
| 0.89 | 70.63 | 70.72 | 70.81 | 70.91 | 71.00 | 71.09 | 71.19 | 71.28 | 71.37 | 71.47 | 0.89 |
| 0.90 | 71.57 | 71.66 | 71.76 | 71.85 | 71.95 | 72.05 | 72.15 | 72.24 | 72.34 | 72.44 | 0.90 |
| 0.91 | 72.54 | 72.64 | 72.74 | 72.85 | 72.95 | 73.05 | 73.15 | 73.26 | 73.36 | 73.46 | 0.91 |
| 0.92 | 73.57 | 73.68 | 73.78 | 73.89 | 74.00 | 74.11 | 74.21 | 74.32 | 74.44 | 74.55 | 0.92 |
| 0.93 | 74.66 | 74.77 | 74.88 | 75.00 | 75.11 | 75.23 | 75.35 | 75.46 | 75.58 | 75.70 | 0.93 |
| 0.94 | 75.82 | 75.94 | 76.06 | 76.19 | 76.31 | 76.44 | 76.56 | 76.69 | 76.82 | 76.95 | 0.94 |
| 0.95 | 77.08 | 77.21 | 77.34 | 77.48 | 77.62 | 77.75 | 77.89 | 78.03 | 78.17 | 78.32 | 0.95 |
| 0.96 | 78.46 | 78.61 | 78.76 | 78.91 | 79.06 | 79.22 | 79.37 | 79.53 | 79.70 | 79.86 | 0.96 |
| 0.97 | 80.03 | 80.20 | 80.37 | 80.54 | 80.72 | 80.90 | 81.09 | 81.28 | 81.47 | 81.67 | 0.97 |
| 0.98 | 81.87 | 82.08 | 82.29 | 82.51 | 82.73 | 82.97 | 83.20 | 83.45 | 83.71 | 83.98 | 0.98 |
| 0.990 | 84.26 | 84.29 | 84.32 | 84.35 | 84.38 | 84.41 | 84.44 | 84.47 | 84.50 | 84.53 | 0.990 |
| 0.991 | 84.56 | 84.59 | 84.62 | 84.65 | 84.68 | 84.71 | 84.74 | 84.77 | 84.80 | 84.84 | 0.991 |
| 0.992 | 84.87 | 84.90 | 84.93 | 84.97 | 85.00 | 85.03 | 85.07 | 85.10 | 85.13 | 85.17 | 0.992 |
| 0.993 | 85.20 | 85.24 | 85.27 | 85.30 | 85.34 | 85.38 | 85.41 | 85.45 | 85.48 | 85.52 | 0.993 |
| 0.994 | 85.56 | 85.59 | 85.63 | 85.67 | 85.71 | 85.75 | 85.79 | 85.83 | 85.86 | 85.90 | 0.994 |
| 0.995 | 85.95 | 85.99 | 86.03 | 86.07 | 86.11 | 86.15 | 86.20 | 86.24 | 86.28 | 86.33 | 0.995 |
| 0.996 | 86.37 | 86.42 | 86.47 | 86.51 | 86.56 | 86.61 | 86.66 | 86.71 | 86.76 | 86.81 | 0.996 |
| 0.997 | 86.86 | 86.91 | 86.97 | 87.02 | 87.08 | 87.13 | 87.19 | 87.25 | 87.31 | 87.37 | 0.997 |
| 0.998 | 87.44 | 87.50 | 87.57 | 87.64 | 87.71 | 87.78 | 87.86 | 87.93 | 88.01 | 88.10 | 0.998 |
| 0.999 | 88.19 | 88.28 | 88.38 | 88.48 | 88.60 | 88.72 | 88.85 | 89.01 | 89.19 | 89.43 | 0.999 |
| 1.000 | 90.00 | | | | | | | | | | |

*Source:* This table is reproduced from F. James Rohlf and Robert R. Sokal, *Statistical Tables,* 2nd ed., W. H. Freeman and Company, copyright © 1981.

**TABLE A.10** PERCENTAGE POINTS OF THE STUDENTIZED RANGE,
$q_x = (\bar{y}_{max} - \bar{y}_{min})/s_{\bar{y}}$

| ν \ n | 2 | 3 | 4 | 5 | 6 | 7 | 8 | 9 | 10 |
|---|---|---|---|---|---|---|---|---|---|
| | | | | Upper 5% points | | | | | |
| 1 | 17.97 | 26.98 | 32.82 | 37.08 | 40.41 | 43.12 | 45.40 | 47.36 | 49.07 |
| 2 | 6.08 | 8.33 | 9.80 | 10.88 | 11.74 | 12.44 | 13.03 | 13.54 | 13.99 |
| 3 | 4.50 | 5.91 | 6.82 | 7.50 | 8.04 | 8.48 | 8.85 | 9.18 | 9.46 |
| 4 | 3.93 | 5.04 | 5.76 | 6.29 | 6.71 | 7.05 | 7.35 | 7.60 | 7.83 |
| 5 | 3.64 | 4.60 | 5.22 | 5.67 | 6.03 | 6.33 | 6.58 | 6.80 | 6.99 |
| 6 | 3.46 | 4.34 | 4.90 | 5.30 | 5.63 | 5.90 | 6.12 | 6.32 | 6.49 |
| 7 | 3.34 | 4.16 | 4.68 | 5.06 | 5.36 | 5.61 | 5.82 | 6.00 | 6.16 |
| 8 | 3.26 | 4.04 | 4.53 | 4.89 | 5.17 | 5.40 | 5.60 | 5.77 | 5.92 |
| 9 | 3.20 | 3.95 | 4.41 | 4.76 | 5.02 | 5.24 | 5.43 | 5.59 | 5.74 |
| 10 | 3.15 | 3.88 | 4.33 | 4.65 | 4.91 | 5.12 | 5.30 | 5.46 | 5.60 |
| 11 | 3.11 | 3.82 | 4.26 | 4.57 | 4.82 | 5.03 | 5.20 | 5.35 | 5.49 |
| 12 | 3.08 | 3.77 | 4.20 | 4.51 | 4.75 | 4.95 | 5.12 | 5.27 | 5.39 |
| 13 | 3.06 | 3.73 | 4.15 | 4.45 | 4.69 | 4.88 | 5.05 | 5.19 | 5.32 |
| 14 | 3.03 | 3.70 | 4.11 | 4.41 | 4.64 | 4.83 | 4.99 | 5.13 | 5.25 |
| 15 | 3.01 | 3.67 | 4.08 | 4.37 | 4.59 | 4.78 | 4.94 | 5.08 | 5.20 |
| 16 | 3.00 | 3.65 | 4.05 | 4.33 | 4.56 | 4.74 | 4.90 | 5.03 | 5.15 |
| 17 | 2.98 | 3.63 | 4.02 | 4.30 | 4.52 | 4.70 | 4.86 | 4.99 | 5.11 |
| 18 | 2.97 | 3.61 | 4.00 | 4.28 | 4.49 | 4.67 | 4.82 | 4.96 | 5.07 |
| 19 | 2.96 | 3.59 | 3.98 | 4.25 | 4.47 | 4.65 | 4.79 | 4.92 | 5.04 |
| 20 | 2.95 | 3.58 | 3.96 | 4.23 | 4.45 | 4.62 | 4.77 | 4.90 | 5.01 |
| 24 | 2.92 | 3.53 | 3.90 | 4.17 | 4.37 | 4.54 | 4.68 | 4.81 | 4.92 |
| 30 | 2.89 | 3.49 | 3.85 | 4.10 | 4.30 | 4.46 | 4.60 | 4.72 | 4.82 |
| 40 | 2.86 | 3.44 | 3.79 | 4.04 | 4.23 | 4.39 | 4.52 | 4.63 | 4.73 |
| 60 | 2.83 | 3.40 | 3.74 | 3.98 | 4.16 | 4.31 | 4.44 | 4.55 | 4.65 |
| 120 | 2.80 | 3.36 | 3.68 | 3.92 | 4.10 | 4.24 | 4.36 | 4.47 | 4.56 |
| ∞ | 2.77 | 3.31 | 3.63 | 3.86 | 4.03 | 4.17 | 4.29 | 4.39 | 4.47 |
| | | | | Upper 1% points | | | | | |
| 1 | 90.03 | 135.0 | 164.3 | 185.6 | 202.2 | 215.8 | 227.2 | 237.0 | 245.6 |
| 2 | 14.04 | 19.02 | 22.29 | 24.72 | 26.63 | 28.20 | 29.53 | 30.68 | 31.69 |
| 3 | 8.26 | 10.62 | 12.17 | 13.33 | 14.24 | 15.00 | 15.64 | 16.20 | 16.69 |
| 4 | 6.51 | 8.12 | 9.17 | 9.96 | 10.58 | 11.10 | 11.55 | 11.93 | 12.27 |
| 5 | 5.70 | 6.98 | 7.80 | 8.42 | 8.91 | 9.32 | 9.67 | 9.97 | 10.24 |
| 6 | 5.24 | 6.33 | 7.03 | 7.56 | 7.97 | 8.32 | 8.61 | 8.87 | 9.10 |
| 7 | 4.95 | 5.92 | 6.54 | 7.01 | 7.37 | 7.68 | 7.94 | 8.17 | 8.37 |
| 8 | 4.75 | 5.64 | 6.20 | 6.62 | 6.96 | 7.24 | 7.47 | 7.68 | 7.86 |
| 9 | 4.60 | 5.43 | 5.96 | 6.35 | 6.66 | 6.91 | 7.13 | 7.33 | 7.49 |
| 10 | 4.48 | 5.27 | 5.77 | 6.14 | 6.43 | 6.67 | 6.87 | 7.05 | 7.21 |
| 11 | 4.39 | 5.15 | 5.62 | 5.97 | 6.25 | 6.48 | 6.67 | 6.84 | 6.99 |
| 12 | 4.32 | 5.05 | 5.50 | 5.84 | 6.10 | 6.32 | 6.51 | 6.67 | 6.81 |
| 13 | 4.26 | 4.96 | 5.40 | 5.73 | 5.98 | 6.19 | 6.37 | 6.53 | 6.67 |
| 14 | 4.21 | 4.89 | 5.32 | 5.63 | 5.88 | 6.08 | 6.26 | 6.41 | 6.54 |
| 15 | 4.17 | 4.84 | 5.25 | 5.56 | 5.80 | 5.99 | 6.16 | 6.31 | 6.44 |
| 16 | 4.13 | 4.79 | 5.19 | 5.49 | 5.72 | 5.92 | 6.08 | 6.22 | 6.35 |
| 17 | 4.10 | 4.74 | 5.14 | 5.43 | 5.66 | 5.85 | 6.01 | 6.15 | 6.27 |
| 18 | 4.07 | 4.70 | 5.09 | 5.38 | 5.60 | 5.79 | 5.94 | 6.08 | 6.20 |
| 19 | 4.05 | 4.67 | 5.05 | 5.33 | 5.55 | 5.73 | 5.89 | 6.02 | 6.14 |
| 20 | 4.02 | 4.64 | 5.02 | 5.29 | 5.51 | 5.69 | 5.84 | 5.97 | 6.09 |
| 24 | 3.96 | 4.55 | 4.91 | 5.17 | 5.37 | 5.54 | 5.69 | 5.81 | 5.92 |
| 30 | 3.89 | 4.45 | 4.80 | 5.05 | 5.24 | 5.40 | 5.54 | 5.65 | 5.76 |
| 40 | 3.82 | 4.37 | 4.70 | 4.93 | 5.11 | 5.26 | 5.39 | 5.50 | 5.60 |
| 60 | 3.76 | 4.28 | 4.59 | 4.82 | 4.99 | 5.13 | 5.25 | 5.36 | 5.45 |
| 120 | 3.70 | 4.20 | 4.50 | 4.71 | 4.87 | 5.01 | 5.12 | 5.21 | 5.30 |
| ∞ | 3.64 | 4.12 | 4.40 | 4.60 | 4.76 | 4.88 | 4.99 | 5.08 | 5.16 |

| 11 | 12 | 13 | 14 | 15 | 16 | 17 | 18 | 19 | 20 |
|---|---|---|---|---|---|---|---|---|---|
| | | | | Upper 5% points | | | | | |
| 50.59 | 51.96 | 53.20 | 54.33 | 55.36 | 56.32 | 57.22 | 58.04 | 58.83 | 59.56 |
| 14.39 | 14.75 | 15.08 | 15.38 | 15.65 | 15.91 | 16.14 | 16.37 | 16.57 | 16.77 |
| 9.72 | 9.95 | 10.15 | 10.35 | 10.52 | 10.69 | 10.84 | 10.98 | 11.11 | 11.24 |
| 8.03 | 8.21 | 8.37 | 8.52 | 8.66 | 8.79 | 8.91 | 9.03 | 9.13 | 9.23 |
| 7.17 | 7.32 | 7.47 | 7.60 | 7.72 | 7.83 | 7.93 | 8.03 | 8.12 | 8.21 |
| 6.65 | 6.79 | 6.92 | 7.03 | 7.14 | 7.24 | 7.34 | 7.43 | 7.51 | 7.59 |
| 6.30 | 6.43 | 6.55 | 6.66 | 6.76 | 6.85 | 6.94 | 7.02 | 7.10 | 7.17 |
| 6.05 | 6.18 | 6.29 | 6.39 | 6.48 | 6.57 | 6.65 | 6.73 | 6.80 | 6.87 |
| 5.87 | 5.98 | 6.09 | 6.19 | 6.28 | 6.36 | 6.44 | 6.51 | 6.58 | 6.64 |
| 5.72 | 5.83 | 5.93 | 6.03 | 6.11 | 6.19 | 6.27 | 6.34 | 6.40 | 6.47 |
| 5.61 | 5.71 | 5.81 | 5.90 | 5.98 | 6.06 | 6.13 | 6.20 | 6.27 | 6.33 |
| 5.51 | 5.61 | 5.71 | 5.80 | 5.88 | 5.95 | 6.02 | 6.09 | 6.15 | 6.21 |
| 5.43 | 5.53 | 5.63 | 5.71 | 5.79 | 5.86 | 5.93 | 5.99 | 6.05 | 6.11 |
| 5.36 | 5.46 | 5.55 | 5.64 | 5.71 | 5.79 | 5.85 | 5.91 | 5.97 | 6.03 |
| 5.31 | 5.40 | 5.49 | 5.57 | 5.65 | 5.72 | 5.78 | 5.85 | 5.90 | 5.96 |
| 5.26 | 5.35 | 5.44 | 5.52 | 5.59 | 5.66 | 5.73 | 5.79 | 5.84 | 5.90 |
| 5.21 | 5.31 | 5.39 | 5.47 | 5.54 | 5.61 | 5.67 | 5.73 | 5.79 | 5.84 |
| 5.17 | 5.27 | 5.35 | 5.43 | 5.50 | 5.57 | 5.63 | 5.69 | 5.74 | 5.79 |
| 5.14 | 5.23 | 5.31 | 5.39 | 5.46 | 5.53 | 5.59 | 5.65 | 5.70 | 5.75 |
| 5.11 | 5.20 | 5.28 | 5.36 | 5.43 | 5.49 | 5.55 | 5.61 | 5.66 | 5.71 |
| 5.01 | 5.10 | 5.18 | 5.25 | 5.32 | 5.38 | 5.44 | 5.49 | 5.55 | 5.59 |
| 4.92 | 5.00 | 5.08 | 5.15 | 5.21 | 5.27 | 5.33 | 5.38 | 5.43 | 5.47 |
| 4.82 | 4.90 | 4.98 | 5.04 | 5.11 | 5.16 | 5.22 | 5.27 | 5.31 | 5.36 |
| 4.73 | 4.81 | 4.88 | 4.94 | 5.00 | 5.06 | 5.11 | 5.15 | 5.20 | 5.24 |
| 4.64 | 4.71 | 4.78 | 4.84 | 4.90 | 4.95 | 5.00 | 5.04 | 5.09 | 5.13 |
| 4.55 | 4.62 | 4.68 | 4.74 | 4.80 | 4.85 | 4.89 | 4.93 | 4.97 | 5.01 |
| | | | | Upper 1% points | | | | | |
| 253.2 | 260.0 | 266.2 | 271.8 | 277.0 | 281.8 | 286.3 | 290.4 | 294.3 | 298.0 |
| 32.59 | 33.40 | 34.13 | 34.81 | 35.43 | 36.00 | 36.53 | 37.03 | 37.50 | 37.95 |
| 17.13 | 17.53 | 17.89 | 18.22 | 18.52 | 18.81 | 19.07 | 19.32 | 19.55 | 19.77 |
| 12.57 | 12.84 | 13.09 | 13.32 | 13.53 | 13.73 | 13.91 | 14.08 | 14.24 | 14.40 |
| 10.48 | 10.70 | 10.89 | 11.08 | 11.24 | 11.40 | 11.55 | 11.68 | 11.81 | 11.93 |
| 9.30 | 9.48 | 9.65 | 9.81 | 9.95 | 10.08 | 10.21 | 10.32 | 10.43 | 10.54 |
| 8.55 | 8.71 | 8.86 | 9.00 | 9.12 | 9.24 | 9.35 | 9.46 | 9.55 | 9.65 |
| 8.03 | 8.18 | 8.31 | 8.44 | 8.55 | 8.66 | 8.76 | 8.85 | 8.94 | 9.03 |
| 7.65 | 7.78 | 7.91 | 8.03 | 8.13 | 8.23 | 8.33 | 8.41 | 8.49 | 8.57 |
| 7.36 | 7.49 | 7.60 | 7.71 | 7.81 | 7.91 | 7.99 | 8.08 | 8.15 | 8.23 |
| 7.13 | 7.25 | 7.36 | 7.46 | 7.56 | 7.65 | 7.73 | 7.81 | 7.88 | 7.95 |
| 6.94 | 7.06 | 7.17 | 7.26 | 7.36 | 7.44 | 7.52 | 7.59 | 7.66 | 7.73 |
| 6.79 | 6.90 | 7.01 | 7.10 | 7.19 | 7.27 | 7.35 | 7.42 | 7.48 | 7.55 |
| 6.66 | 6.77 | 6.87 | 6.96 | 7.05 | 7.13 | 7.20 | 7.27 | 7.33 | 7.39 |
| 6.55 | 6.66 | 6.76 | 6.84 | 6.93 | 7.00 | 7.07 | 7.14 | 7.20 | 7.26 |
| 6.46 | 6.56 | 6.66 | 6.74 | 6.82 | 6.90 | 6.97 | 7.03 | 7.09 | 7.15 |
| 6.38 | 6.48 | 6.57 | 6.66 | 6.73 | 6.81 | 6.87 | 6.94 | 7.00 | 7.05 |
| 6.31 | 6.41 | 6.50 | 6.58 | 6.65 | 6.73 | 6.79 | 6.85 | 6.91 | 6.97 |
| 6.25 | 6.34 | 6.43 | 6.51 | 6.58 | 6.65 | 6.72 | 6.78 | 6.84 | 6.89 |
| 6.19 | 6.28 | 6.37 | 6.45 | 6.52 | 6.59 | 6.65 | 6.71 | 6.77 | 6.82 |
| 6.02 | 6.11 | 6.19 | 6.26 | 6.33 | 6.39 | 6.45 | 6.51 | 6.56 | 6.61 |
| 5.85 | 5.93 | 6.01 | 6.08 | 6.14 | 6.20 | 6.26 | 6.31 | 6.36 | 6.41 |
| 5.69 | 5.76 | 5.83 | 5.90 | 5.96 | 6.02 | 6.07 | 6.12 | 6.16 | 6.21 |
| 5.53 | 5.60 | 5.67 | 5.73 | 5.78 | 5.84 | 5.89 | 5.93 | 5.97 | 6.01 |
| 5.37 | 5.44 | 5.50 | 5.56 | 5.61 | 5.66 | 5.71 | 5.75 | 5.79 | 5.83 |
| 5.23 | 5.29 | 5.35 | 5.40 | 5.45 | 5.49 | 5.54 | 5.57 | 5.61 | 5.65 |

$n$: size of sample from which range obtained. $v$: degrees of freedom of independent $s_v$.

*Source:* This table is reproduced from "Percentage Points of the Studentized Range, $q_x = (\chi_n - \chi_1)/s_y$," from *Biometrika Tables for Statisticians,* Vol. 1, 3rd ed., 1966, with the permission of the Biometrika Trustees.

**TABLE A.11A**  CRITICAL VALUES OF $q'$ FOR THE ONE-TAILED DUNNETT'S TEST

| $v$ | $p = 2$ | 3 | 4 | 5 | 6 | 7 | 8 | 9 | 10 |
|---|---|---|---|---|---|---|---|---|---|
| | | | | $\alpha = 0.05$ | | | | | |
| 5 | 2.02 | 2.44 | 2.68 | 2.85 | 2.98 | 3.08 | 3.16 | 3.24 | 3.30 |
| 6 | 1.94 | 2.34 | 2.56 | 2.71 | 2.83 | 2.92 | 3.00 | 3.07 | 3.12 |
| 7 | 1.89 | 2.27 | 2.48 | 2.62 | 2.73 | 2.82 | 2.89 | 2.95 | 3.01 |
| 8 | 1.86 | 2.22 | 2.42 | 2.55 | 2.66 | 2.74 | 2.81 | 2.87 | 2.92 |
| 9 | 1.83 | 2.18 | 2.37 | 2.50 | 2.60 | 2.68 | 2.75 | 2.81 | 2.86 |
| 10 | 1.81 | 2.15 | 2.34 | 2.47 | 2.56 | 2.64 | 2.70 | 2.76 | 2.81 |
| 11 | 1.80 | 2.13 | 2.31 | 2.44 | 2.53 | 2.60 | 2.67 | 2.72 | 2.77 |
| 12 | 1.78 | 2.11 | 2.29 | 2.41 | 2.50 | 2.58 | 2.64 | 2.69 | 2.74 |
| 13 | 1.77 | 2.09 | 2.27 | 2.39 | 2.48 | 2.55 | 2.61 | 2.66 | 2.71 |
| 14 | 1.76 | 2.08 | 2.25 | 2.37 | 2.46 | 2.53 | 2.59 | 2.64 | 2.69 |
| 15 | 1.75 | 2.07 | 2.24 | 2.36 | 2.44 | 2.51 | 2.57 | 2.62 | 2.67 |
| 16 | 1.75 | 2.06 | 2.23 | 2.34 | 2.43 | 2.50 | 2.56 | 2.61 | 2.65 |
| 17 | 1.74 | 2.05 | 2.22 | 2.33 | 2.42 | 2.49 | 2.54 | 2.59 | 2.64 |
| 18 | 1.73 | 2.04 | 2.21 | 2.32 | 2.41 | 2.48 | 2.53 | 2.58 | 2.62 |
| 19 | 1.73 | 2.03 | 2.20 | 2.31 | 2.40 | 2.47 | 2.52 | 2.57 | 2.61 |
| 20 | 1.72 | 2.03 | 2.19 | 2.30 | 2.39 | 2.46 | 2.51 | 2.56 | 2.60 |
| 24 | 1.71 | 2.01 | 2.17 | 2.28 | 2.36 | 2.43 | 2.48 | 2.53 | 2.57 |
| 30 | 1.70 | 1.99 | 2.15 | 2.25 | 2.33 | 2.40 | 2.45 | 2.50 | 2.54 |
| 40 | 1.68 | 1.97 | 2.13 | 2.23 | 2.31 | 2.37 | 2.42 | 2.47 | 2.51 |
| 60 | 1.67 | 1.95 | 2.10 | 2.21 | 2.28 | 2.35 | 2.39 | 2.44 | 2.48 |
| 120 | 1.66 | 1.93 | 2.08 | 2.18 | 2.26 | 2.32 | 2.37 | 2.41 | 2.45 |
| $\infty$ | 1.64 | 1.92 | 2.06 | 2.16 | 2.23 | 2.29 | 2.34 | 2.38 | 2.42 |
| | | | | $\alpha = 0.01$ | | | | | |
| 5 | 3.37 | 3.90 | 4.21 | 4.43 | 4.60 | 4.73 | 4.85 | 4.94 | 5.03 |
| 6 | 3.14 | 3.61 | 3.88 | 4.07 | 4.21 | 4.33 | 4.43 | 4.51 | 4.59 |
| 7 | 3.00 | 3.42 | 3.66 | 3.83 | 3.96 | 4.07 | 4.15 | 4.23 | 4.30 |
| 8 | 2.90 | 3.29 | 3.51 | 3.67 | 3.79 | 3.88 | 3.96 | 4.03 | 4.09 |
| 9 | 2.82 | 3.19 | 3.40 | 3.55 | 3.66 | 3.75 | 3.82 | 3.89 | 3.94 |
| 10 | 2.76 | 3.11 | 3.31 | 3.45 | 3.56 | 3.64 | 3.71 | 3.78 | 3.83 |
| 11 | 2.72 | 3.06 | 3.25 | 3.38 | 3.48 | 3.56 | 3.63 | 3.69 | 3.74 |
| 12 | 2.68 | 3.01 | 3.19 | 3.32 | 3.42 | 3.50 | 3.56 | 3.62 | 3.67 |
| 13 | 2.65 | 2.97 | 3.15 | 3.27 | 3.37 | 3.44 | 3.51 | 3.56 | 3.61 |
| 14 | 2.62 | 2.94 | 3.11 | 3.23 | 3.32 | 3.40 | 3.46 | 3.51 | 3.56 |
| 15 | 2.60 | 2.91 | 3.08 | 3.20 | 3.29 | 3.36 | 3.42 | 3.47 | 3.52 |
| 16 | 2.58 | 2.88 | 3.05 | 3.17 | 3.26 | 3.33 | 3.39 | 3.44 | 3.48 |
| 17 | 2.57 | 2.86 | 3.03 | 3.14 | 3.23 | 3.30 | 3.36 | 3.41 | 3.45 |
| 18 | 2.55 | 2.84 | 3.01 | 3.12 | 3.21 | 3.27 | 3.33 | 3.38 | 3.42 |
| 19 | 2.54 | 2.83 | 2.99 | 3.10 | 3.18 | 3.25 | 3.31 | 3.36 | 3.40 |
| 20 | 2.53 | 2.81 | 2.97 | 3.08 | 3.17 | 3.23 | 3.29 | 3.34 | 3.38 |
| 24 | 2.49 | 2.77 | 2.92 | 3.03 | 3.11 | 3.17 | 3.22 | 3.27 | 3.31 |
| 30 | 2.46 | 2.72 | 2.87 | 2.97 | 3.05 | 3.11 | 3.16 | 3.21 | 3.24 |
| 40 | 2.42 | 2.68 | 2.82 | 2.92 | 2.99 | 3.05 | 3.10 | 3.14 | 3.18 |
| 60 | 2.39 | 2.64 | 2.78 | 2.87 | 2.94 | 3.00 | 3.04 | 3.08 | 3.12 |
| 120 | 2.36 | 2.60 | 2.73 | 2.82 | 2.89 | 2.94 | 2.99 | 3.03 | 3.06 |
| $\infty$ | 2.33 | 2.56 | 2.68 | 2.77 | 2.84 | 2.89 | 2.93 | 2.97 | 3.00 |

*Source:* This table is reproduced from Charles W. Dunnett, "Comparing Several Treatments with a Control," *Am. Stat. Assoc.,* Vol. 50, 1096–1121 (1955), with permission of the Board of Directors of the American Statistical Association.

**TABLE A.11B** CRITICAL VALUES OF $q'$ FOR THE TWO-TAILED DUNNETT'S TEST

| $v$ | $p = 2$ | 3 | 4 | 5 | 6 | 7 | 8 | 9 | 10 | 11 | 12 | 13 | 16 | 21 |
|---|---|---|---|---|---|---|---|---|---|---|---|---|---|---|
| | | | | | | $\alpha = 0.05$ | | | | | | | | |
| 5 | 2.57 | 3.03 | 3.29 | 3.48 | 3.62 | 3.73 | 3.82 | 3.90 | 3.97 | 4.03 | 4.09 | 4.14 | 4.26 | 4.42 |
| 6 | 2.45 | 2.86 | 3.10 | 3.26 | 3.39 | 3.49 | 3.57 | 3.64 | 3.71 | 3.76 | 3.81 | 3.86 | 3.97 | 4.11 |
| 7 | 2.36 | 2.75 | 2.97 | 3.12 | 3.24 | 3.33 | 3.41 | 3.47 | 3.53 | 3.58 | 3.63 | 3.67 | 3.78 | 3.91 |
| 8 | 2.31 | 2.67 | 2.88 | 3.02 | 3.13 | 3.22 | 3.29 | 3.35 | 3.41 | 3.46 | 3.50 | 3.54 | 3.64 | 3.76 |
| 9 | 2.26 | 2.61 | 2.81 | 2.95 | 3.05 | 3.14 | 3.20 | 3.26 | 3.32 | 3.36 | 3.40 | 3.44 | 3.53 | 3.65 |
| 10 | 2.23 | 2.57 | 2.76 | 2.89 | 2.99 | 3.07 | 3.14 | 3.19 | 3.24 | 3.29 | 3.33 | 3.36 | 3.45 | 3.57 |
| 11 | 2.20 | 2.53 | 2.72 | 2.84 | 2.94 | 3.02 | 3.08 | 3.14 | 3.19 | 3.23 | 3.27 | 3.30 | 3.39 | 3.50 |
| 12 | 2.18 | 2.50 | 2.68 | 2.81 | 2.90 | 2.98 | 3.04 | 3.09 | 3.14 | 3.18 | 3.22 | 3.25 | 3.34 | 3.45 |
| 13 | 2.16 | 2.48 | 2.65 | 2.78 | 2.87 | 2.94 | 3.00 | 3.06 | 3.10 | 3.14 | 3.18 | 3.21 | 3.28 | 3.40 |
| 14 | 2.14 | 2.46 | 2.63 | 2.75 | 2.84 | 2.91 | 2.97 | 3.02 | 3.07 | 3.11 | 3.14 | 3.18 | 3.26 | 3.36 |
| 15 | 2.13 | 2.44 | 2.61 | 2.73 | 2.82 | 2.89 | 2.95 | 3.00 | 3.04 | 3.08 | 3.12 | 3.15 | 3.23 | 3.33 |
| 16 | 2.12 | 2.42 | 2.59 | 2.71 | 2.80 | 2.87 | 2.92 | 2.97 | 3.02 | 3.06 | 3.09 | 3.12 | 3.20 | 3.30 |
| 17 | 2.11 | 2.41 | 2.58 | 2.69 | 2.78 | 2.85 | 2.90 | 2.95 | 3.00 | 3.03 | 3.07 | 3.10 | 3.18 | 3.27 |
| 18 | 2.10 | 2.40 | 2.56 | 2.68 | 2.76 | 2.83 | 2.89 | 2.94 | 2.98 | 3.01 | 3.05 | 3.08 | 3.16 | 3.25 |
| 19 | 2.09 | 2.39 | 2.55 | 2.66 | 2.75 | 2.81 | 2.87 | 2.92 | 2.96 | 3.00 | 3.03 | 3.06 | 3.14 | 3.23 |
| 20 | 2.09 | 2.38 | 2.54 | 2.65 | 2.73 | 2.80 | 2.86 | 2.90 | 2.95 | 2.98 | 3.02 | 3.05 | 3.12 | 3.22 |
| 24 | 2.06 | 2.35 | 2.51 | 2.61 | 2.70 | 2.76 | 2.81 | 2.86 | 2.90 | 2.94 | 2.97 | 3.00 | 3.07 | 3.16 |
| 30 | 2.04 | 2.32 | 2.47 | 2.58 | 2.66 | 2.72 | 2.77 | 2.82 | 2.86 | 2.89 | 2.92 | 2.95 | 3.02 | 3.11 |
| 40 | 2.02 | 2.29 | 2.44 | 2.54 | 2.62 | 2.68 | 2.73 | 2.77 | 2.81 | 2.85 | 2.87 | 2.90 | 2.97 | 3.06 |
| 60 | 2.00 | 2.27 | 2.41 | 2.51 | 2.58 | 2.64 | 2.69 | 2.73 | 2.77 | 2.80 | 2.83 | 2.85 | 2.92 | 3.00 |
| 120 | 1.98 | 2.24 | 2.38 | 2.47 | 2.55 | 2.60 | 2.65 | 2.69 | 2.73 | 2.76 | 2.79 | 2.81 | 2.87 | 2.95 |
| $\infty$ | 1.96 | 2.21 | 2.35 | 2.44 | 2.51 | 2.57 | 2.61 | 2.65 | 2.69 | 2.72 | 2.74 | 2.77 | 2.83 | 2.91 |
| | | | | | | $\alpha = 0.01$ | | | | | | | | |
| 5 | 4.03 | 4.63 | 4.98 | 5.22 | 5.41 | 5.56 | 5.69 | 5.80 | 5.89 | 5.98 | 6.05 | 6.12 | 6.30 | 6.52 |
| 6 | 3.71 | 4.21 | 4.51 | 4.71 | 4.87 | 5.00 | 5.10 | 5.20 | 5.28 | 5.35 | 5.41 | 5.47 | 5.62 | 5.81 |
| 7 | 3.50 | 3.95 | 4.21 | 4.39 | 4.53 | 4.64 | 4.74 | 4.82 | 4.89 | 4.95 | 5.01 | 5.06 | 5.19 | 5.36 |
| 8 | 3.36 | 3.77 | 4.00 | 4.17 | 4.29 | 4.40 | 4.48 | 4.56 | 4.62 | 4.68 | 4.73 | 4.78 | 4.90 | 5.05 |
| 9 | 3.25 | 3.63 | 3.85 | 4.01 | 4.12 | 4.22 | 4.30 | 4.37 | 4.43 | 4.48 | 4.53 | 4.57 | 4.68 | 4.82 |
| 10 | 3.17 | 3.53 | 3.74 | 3.88 | 3.99 | 4.08 | 4.16 | 4.22 | 4.28 | 4.33 | 4.37 | 4.42 | 4.52 | 4.65 |
| 11 | 3.11 | 3.45 | 3.65 | 3.79 | 3.89 | 3.98 | 4.05 | 4.11 | 4.16 | 4.21 | 4.25 | 4.29 | 4.30 | 4.52 |
| 12 | 3.05 | 3.39 | 3.58 | 3.71 | 3.81 | 3.89 | 3.96 | 4.02 | 4.07 | 4.12 | 4.16 | 4.19 | 4.29 | 4.41 |
| 13 | 3.01 | 3.33 | 3.52 | 3.65 | 3.74 | 3.82 | 3.89 | 3.94 | 3.99 | 4.04 | 4.08 | 4.11 | 4.20 | 4.32 |
| 14 | 2.98 | 3.29 | 3.47 | 3.59 | 3.69 | 3.76 | 3.83 | 3.88 | 3.93 | 3.97 | 4.01 | 4.05 | 4.13 | 4.24 |
| 15 | 2.95 | 3.25 | 3.43 | 3.55 | 3.64 | 3.71 | 3.78 | 3.83 | 3.88 | 3.92 | 3.95 | 3.99 | 4.07 | 4.18 |
| 16 | 2.92 | 3.22 | 3.39 | 3.51 | 3.60 | 3.67 | 3.73 | 3.78 | 3.83 | 3.87 | 3.91 | 3.94 | 4.02 | 4.13 |
| 17 | 2.90 | 3.19 | 3.36 | 3.47 | 3.56 | 3.63 | 3.69 | 3.74 | 3.79 | 3.83 | 3.86 | 3.90 | 3.98 | 4.08 |
| 18 | 2.88 | 3.17 | 3.33 | 3.44 | 3.53 | 3.60 | 3.66 | 3.71 | 3.75 | 3.79 | 3.83 | 3.86 | 3.94 | 4.04 |
| 19 | 2.86 | 3.15 | 3.31 | 3.42 | 3.50 | 3.57 | 3.63 | 3.68 | 3.72 | 3.76 | 3.79 | 3.83 | 3.90 | 4.00 |
| 20 | 2.85 | 3.13 | 3.29 | 3.40 | 3.48 | 3.55 | 3.60 | 3.65 | 3.69 | 3.73 | 3.77 | 3.80 | 3.87 | 3.97 |
| 24 | 2.80 | 3.07 | 3.22 | 3.32 | 3.40 | 3.47 | 3.52 | 3.57 | 3.61 | 3.64 | 3.68 | 3.70 | 3.78 | 3.87 |
| 30 | 2.75 | 3.01 | 3.15 | 3.25 | 3.33 | 3.39 | 3.44 | 3.49 | 3.52 | 3.56 | 3.59 | 3.62 | 3.69 | 3.78 |
| 40 | 2.70 | 2.95 | 3.09 | 3.19 | 3.26 | 3.32 | 3.37 | 3.41 | 3.44 | 3.48 | 3.51 | 3.53 | 3.60 | 3.68 |
| 60 | 2.66 | 2.90 | 3.03 | 3.12 | 3.19 | 3.25 | 3.29 | 3.33 | 3.37 | 3.40 | 3.42 | 3.45 | 3.51 | 3.59 |
| 120 | 2.62 | 2.85 | 2.97 | 3.06 | 3.12 | 3.18 | 3.22 | 3.26 | 3.29 | 3.32 | 3.35 | 3.37 | 3.43 | 3.51 |
| $\infty$ | 2.58 | 2.79 | 2.92 | 3.00 | 3.06 | 3.11 | 3.15 | 3.19 | 3.22 | 3.25 | 3.27 | 3.29 | 3.35 | 2.42 |

*Source:* This table is reproduced from Charles W. Dunnett, "Comparing Several Treatments with a Control," *Am. Stat. Assoc.*, Vol. 50, 1096–1121 (1955), with permission of the Board of Directors of the American Statistical Association.

**TABLE A.12**  ORTHOGONAL POLYNOMIAL COEFFICIENTS

| | $n = 8$ | | | | | $n = 9$ | | | | | $n = 10$ | | | |
|---|---|---|---|---|---|---|---|---|---|---|---|---|---|---|
| $c_{1j}$ | $c_{2j}$ | $c_{3j}$ | $c_{4j}$ | $c_{5j}$ | $c_{1j}$ | $c_{2j}$ | $c_{3j}$ | $c_{4j}$ | $c_{5j}$ | $c_{1j}$ | $c_{2j}$ | $c_{3j}$ | $c_{4j}$ | $c_{5j}$ |
| −7 | +7 | −7 | 7 | −7 | 0 | −20 | 0 | +18 | 0 | +1 | −4 | −12 | +18 | +6 |
| −5 | +1 | +5 | −13 | +23 | +1 | −17 | −9 | +9 | +9 | +3 | −3 | −31 | +3 | +11 |
| −3 | −3 | +7 | −3 | −17 | +2 | −8 | −13 | −11 | +4 | +5 | −1 | −35 | −17 | +1 |
| −1 | −5 | +3 | 9 | −15 | +3 | +7 | −7 | −21 | −11 | +7 | +2 | −14 | −22 | −14 |
| +1 | −5 | −3 | 9 | +15 | +4 | +28 | 14 | +14 | +4 | +9 | +6 | +42 | +18 | +6 |
| +3 | −3 | −7 | −3 | +17 | | | | | | | | | | |
| +5 | +1 | −5 | −13 | −23 | | | | | | | | | | |
| +7 | +7 | +7 | 7 | +7 | | | | | | | | | | |
| $D$  168 | 168 | 264 | 616 | 2,184 | 60 | 2,772 | 990 | 2,002 | 468 | 330 | 132 | 8,580 | 2,860 | 780 |
| $\lambda$  2 | 1 | $\frac{2}{3}$ | $\frac{7}{12}$ | $\frac{7}{10}$ | 1 | 3 | $\frac{5}{6}$ | $\frac{7}{12}$ | $\frac{3}{20}$ | 2 | $\frac{1}{2}$ | $\frac{5}{3}$ | $\frac{5}{12}$ | $\frac{1}{10}$ |

| | $n = 11$ | | | | | $n = 12$ | | | | | $n = 13$ | | | |
|---|---|---|---|---|---|---|---|---|---|---|---|---|---|---|
| $c_{1j}$ | $c_{2j}$ | $c_{3j}$ | $c_{4j}$ | $c_{5j}$ | $c_{1j}$ | $c_{2j}$ | $c_{3j}$ | $c_{4j}$ | $c_{5j}$ | $c_{1j}$ | $c_{2j}$ | $c_{3j}$ | $c_{4j}$ | $c_{5j}$ |
| 0 | −10 | 0 | +6 | 0 | +1 | −35 | −8 | +28 | +20 | 0 | 14 | 0 | +84 | 0 |
| +1 | −9 | −14 | +4 | +4 | +3 | −29 | −19 | +12 | +44 | +1 | −13 | −4 | +64 | +20 |
| +2 | −6 | −23 | −1 | +4 | +5 | −17 | −25 | −13 | +29 | +2 | −10 | −7 | +11 | +26 |
| +3 | −1 | −22 | −6 | −1 | +7 | +1 | −21 | −33 | −21 | +3 | −5 | −8 | −54 | +11 |
| +4 | +6 | −6 | −6 | −6 | +9 | +25 | −3 | −27 | −57 | +4 | +2 | −6 | −96 | −18 |
| +5 | +15 | +30 | +6 | +3 | +11 | +55 | +33 | +33 | +33 | +5 | +11 | 0 | −66 | −33 |
| | | | | | | | | | | +6 | +22 | +11 | +99 | +22 |
| $D$  110 | 858 | 4,290 | 286 | 156 | 572 | 12,012 | 5,148 | 8,008 | 15,912 | 182 | 2,002 | 572 | 68,068 | 6,188 |
| $\lambda$  1 | 1 | $\frac{5}{6}$ | $\frac{1}{12}$ | $\frac{1}{40}$ | 2 | 3 | $\frac{2}{3}$ | $\frac{7}{24}$ | $\frac{3}{20}$ | 1 | 1 | $\frac{1}{6}$ | $\frac{7}{12}$ | $\frac{7}{120}$ |

| | $n = 14$ | | | | | $n = 15$ | | | |
|---|---|---|---|---|---|---|---|---|---|
| $c_{1j}$ | $c_{2j}$ | $c_{3j}$ | $c_{4j}$ | $c_{5j}$ | $c_{1j}$ | $c_{2j}$ | $c_{3j}$ | $c_{4j}$ | $c_{5j}$ |
| +1 | −8 | −24 | +108 | +60 | 0 | −56 | 0 | +756 | 0 |
| +3 | −7 | −67 | +63 | +145 | +1 | −53 | −27 | +621 | +675 |
| +5 | −5 | −95 | −13 | +139 | +2 | −44 | −49 | +251 | +1,000 |
| +7 | −2 | −98 | −92 | +28 | +3 | −29 | −61 | −249 | +751 |
| +9 | +2 | −66 | −132 | −132 | +4 | −8 | −58 | −704 | −44 |
| +11 | +7 | +11 | −77 | −187 | +5 | +19 | −35 | −869 | −979 |
| +13 | +13 | +143 | +143 | +143 | +6 | +52 | +13 | −429 | −1,144 |
| | | | | | +7 | +91 | +91 | +1,001 | +1,001 |
| $D$  910 | 728 | 97,240 | 136,136 | 235,144 | 280 | 37,128 | 39,780 | 6,466,460 | 10,581,480 |
| $\lambda$  2 | $\frac{1}{2}$ | $\frac{5}{3}$ | $\frac{7}{12}$ | $\frac{7}{30}$ | 1 | 3 | $\frac{5}{6}$ | $\frac{35}{12}$ | $\frac{21}{20}$ |

The $D$ values are the sums of squares of the polynomial coefficients, and $\lambda$ is the multiplier necessary to yield integral values. The leading coefficients for $n = 9$ through $n = 15$ have been omitted in order to save space. They may be constructed readily, however, since these values are those printed, read in reverse order. For the linear, cubic, and quintic values, the signs are also reversed. The central values are printed where the number of levels is odd.

# Solutions

## CHAPTER 1

**1.1.** Strains: qualitative, independent, and discrete.
Yield: quantitative, dependent, and continuous.

**1.2.** Groups: qualitative, independent, and discrete.
Yield: quantitative, dependent, and discrete.

**1.3.** $n = 10$, $Y_6 = 11.6$, $Y_8 = 10.1$, $Y_{i-7} = Y_2 = 9.4$, $Y_9 - 1.0 = 11.2 - 1.0 = 10.2$, $\bar{y} = 11.0$.

**1.4.** $\bar{y} = Y./n = 240/6 = 40.0$

$$y_i = Y_i - \bar{y} = -8, -16, +16, +8, -4, +4$$

$$\sum_i (Y_i - \bar{y}) = 0$$

$$s^2 = \frac{\sum_i (Y_i - \bar{y})^2}{n-1} = \frac{672}{5} = 134.4$$

$$s^2 = \frac{(\sum_i Y_i^2 - Y.^2/n)}{n-1} = \frac{10{,}272 - (240)^2/6}{5} = 134.4$$

$$s = \sqrt{134.4} = 11.59$$

$$s_{\bar{y}} = \frac{s}{\sqrt{n}} = \sqrt{\frac{s^2}{n}} = \sqrt{\frac{134.4}{6}} = 4.73$$

**1.5.**
$$\sum_i (Y_i - \bar{y})^2 = \sum_i (Y_i^2 - 2Y_i\bar{y} + \bar{y})^2$$

$$= \sum_i Y_i^2 - 2 \sum_i Y_i\bar{y} + n\bar{y}^2$$

$$= \sum_i Y_i^2 - 2n\bar{y}^2 + n\bar{y}^2$$

$$= \sum_i Y_i^2 - n\bar{y}^2$$

$$= \sum_i Y_i^2 - \frac{Y_{\cdot}^2}{n}$$

**1.6.** $\bar{y} = \dfrac{Y_{\cdot}}{n} = \dfrac{720}{8} = 90$

$$s^2 = \frac{(\sum_i Y_i^2 - Y_{\cdot}^2/n)}{n-1} = \frac{64{,}894 - (720)^2/8}{7} = 13.4286$$

$$s = \sqrt{13.4286} = 3.6645$$

$$s_{\bar{y}} = \sqrt{\frac{s^2}{n}} = \sqrt{\frac{13.4286}{8}} = 1.2956$$

$$-t_{0.05}s_{\bar{y}} = (-2.365)(1.2950) = -3.064$$

$$t_{0.05}s_{\bar{y}} = (2.365)(1.2950) = 3.064$$

$$\text{CI} = 90 \pm 3.064 \quad \text{or} \quad 86.936 \text{ to } 93.064$$

$$\text{CV} = \frac{(100)s}{\bar{y}} = \frac{(100)(3.6645)}{90} = 4.072$$

**1.7.** $\bar{y} = 81.00$; $s^2 = 200$; $s = 14.1421$; $s_{\bar{y}} = 4.0825$; CI = 68.32 to 93.68; CV = 17.4594.

## CHAPTER 2

**2.1.**

| Source | df | SS | MS | F |
|---|---|---|---|---|
| T | 2 | 5988.0225 | 2994.0112 | 31.19** |
| W | 21 | 2015.8775 | 95.9942 | |

Highly significant differences among treatments.
$\hat{t}_i = -11.0000$; $\hat{t}_2 = -11.3375$; $\hat{t}_3 = 22.3375$; $\sum_i n_w \hat{t}_i^2 = 5988.0224$.

**2.2.**

| Source | df | SS | MS | F |
|---|---|---|---|---|
| T | 2 | 31.4121 | 15.7060 | 24.93** |
| W | 13 | 8.1882 | 0.6299 | |

Highly significant differences among treatments.
$e_{11}$ through $e_{34}$: 1.3129, 0.1429, −0.0571, −1.0971, −1.3271, 1.1729, −0.1471; −0.23, 0.29, −0.23, −0.18, 0.35; 0.735, 0.575, −0.705, −0.605. $\sum_i \sum_j (Y_{ij} - \bar{y}_i)^2 = \sum_i \sum_j e_{ij}^2 = 8.1882$.

**2.3.**

| Source | df | SS | MS | F |
|---|---|---|---|---|
| B | 1 | 7411.25 | 7411.25 | 28.07** |
| W | 18 | 4753.30 | 264.0722 | |

$\bar{d} = 38.5$; $s_{\bar{d}} = 7.2674$; $t = 38.5/7.2674 = 5.298$**; $t^2 = 28.07 = F$.

**2.4.**

| Source | df | SS | MS | F |
|---|---|---|---|---|
| B | 1 | 6.0953 | 6.0953 | 2.10(ns) |
| W | 12 | 34.8333 | 2.9028 | |

$\bar{d} = 1.3333$; $s_{\bar{d}} = 0.9201$; $t = 1.449$(ns); $t^2 = 2.10 = F$.

**2.5.**

| Source | df | SS | MS | F |
|---|---|---|---|---|
| B | 1 | 19,323.375 | 19,323.375 | 10.62* |
| S:B | 4 | 7,280.500 | 1,820.125 | 7.71** |
| W | 18 | 4,249.750 | 236.097 | |

**2.6.**

| Source | df | SS | MS | F |
|---|---|---|---|---|
| A | 1 | 68,201.051 | 68,201.051 | 38.69** |
| F:A | 5 | 8,813.733 | 1,762.747 | 40.16** |
| W | 22 | 965.600 | 43.891 | |

**2.7.** The least-squares equations are

$$\begin{bmatrix} 24 & 8 & 8 & 8 \\ 8 & 8 & 0 & 0 \\ 8 & 0 & 8 & 0 \\ 8 & 0 & 0 & 8 \end{bmatrix} \begin{bmatrix} \hat{\mu} \\ \hat{t}_1 \\ \hat{t}_2 \\ \hat{t}_3 \end{bmatrix} = \begin{bmatrix} 1417.2 \\ 384.4 \\ 381.7 \\ 651.1 \end{bmatrix}$$

After imposing restrictions that $\Sigma_i \hat{t}_i = 0$, the equations are

$$\begin{bmatrix} 24 & 0 & 0 \\ 0 & 16 & 8 \\ 0 & 8 & 16 \end{bmatrix} \begin{bmatrix} \hat{\mu} \\ \hat{t}_1 \\ \hat{t}_2 \end{bmatrix} = \begin{bmatrix} 1417.2 \\ -266.7 \\ -269.4 \end{bmatrix}$$

The solutions of the equations are

$$\hat{\mu} = 59.05 \qquad \hat{t}_1 = -\frac{2112.0}{192} = -11.0000$$

$$\hat{t}_2 = -\frac{2176.8}{192} = -11.3375 \qquad \hat{t}_3 = -\hat{t}_1 - \hat{t}_2 = 11.0000 + 11.3375 = 22.3375$$

The total reduction in SS is

$$R(\mu, t_i) = (59.05)(1417.2) + (11)(266.7) + (11.3375)(269.4)$$

$$= 89,673.6825$$

The reduction due to the overall mean (correction term) is

$$R(\mu) = (59.05)^2(24) \doteq 83,685.66 = \frac{(1417.2)^2}{24}$$

Total uncorrected SS = 91,689.56

$$\text{SS Treatments} = R(\mu, t_i) - R(\mu)$$

$$= 89,673.6825 - 83,685.66$$

$$= 5,988.0225$$

$$\text{SS Within} = \sum_i \sum_j Y_{ij}^2 - R(\mu, t_i)$$

$$= 91{,}689.56 - 89{,}673.6825$$

$$= 2{,}015.8775$$

**2.8.** The least-squares equations are

$$\begin{bmatrix} 16 & 7 & 5 & 4 \\ 7 & 7 & 0 & 0 \\ 5 & 0 & 5 & 0 \\ 4 & 0 & 0 & 4 \end{bmatrix} \begin{bmatrix} \hat{\mu} \\ \hat{t}_1 \\ \hat{t}_2 \\ \hat{t}_3 \end{bmatrix} = \begin{bmatrix} 37.24 \\ 24.34 \\ 1.40 \\ 11.50 \end{bmatrix}$$

After imposing restrictions that $\sum_i \hat{t}_i = 0$, the equations are

$$\begin{bmatrix} 16 & 3 & 1 \\ 3 & 11 & 4 \\ 1 & 4 & 9 \end{bmatrix} \begin{bmatrix} \hat{\mu} \\ \hat{t}_1 \\ \hat{t}_2 \end{bmatrix} = \begin{bmatrix} 37.24 \\ 12.84 \\ -10.10 \end{bmatrix}$$

The solutions of the equations are

$$\begin{bmatrix} \hat{\mu} \\ \hat{t}_1 \\ \hat{t}_2 \end{bmatrix} = \begin{bmatrix} 16 & 3 & 1 \\ 3 & 11 & 4 \\ 1 & 4 & 9 \end{bmatrix}^{-1} \begin{bmatrix} 37.24 \\ 12.84 \\ -10.10 \end{bmatrix}$$

$$= \frac{1}{1260} \begin{bmatrix} 83 & -23 & 1 \\ -23 & 143 & -61 \\ 1 & -61 & 167 \end{bmatrix} \begin{bmatrix} 37.24 \\ 12.84 \\ -10.10 \end{bmatrix}$$

$$= \begin{bmatrix} 2.210714 \\ 1.266429 \\ -1.930714 \end{bmatrix}$$

$$\hat{t}_3 = -1.266429 + 1.930714 = 0.664285$$

The total reduction in SS is

$$R(\mu, t_i) = 118.088149$$

The reduction due to the overall mean is

$$R(\mu) = \frac{(37.24)^2}{16} = 86.676100$$

Total uncorrected SS $= 126.276400$

SS Treatments $= R(\mu, t_i) - R(\mu)$

$$= 118.088149 - 86.676100$$

$$= 31.412049$$

This sum of squares may be computed by the direct method as follows:

$$\text{SS}_T = \hat{\mathbf{B}}_T' \mathbf{Z}_T^{-1} \hat{\mathbf{B}}_T$$

$$= [1.266429 \ \ -1.930714] \left[ \frac{1}{1260} \begin{pmatrix} 143 & -61 \\ -61 & 167 \end{pmatrix} \right]^{-1} \begin{bmatrix} 1.266429 \\ -1.930714 \end{bmatrix}$$

$$= [1.266429 \quad -1.930714] \begin{bmatrix} 10.437500 & 3.812500 \\ 3.812500 & 8.937500 \end{bmatrix} \begin{bmatrix} 1.266429 \\ -1.930714 \end{bmatrix}$$

$$= [5.857506 \quad -12.427496] \begin{bmatrix} 1.266429 \\ -1.930714 \end{bmatrix}$$

$$= 31.412056$$

$$SS_W = \sum_i \sum_j Y_{ij}^2 - R(\mu, t_i)$$

$$= 126.2764 - 118.088149$$

$$= 8.188251$$

| Source | df | SS | MS | F |
|--------|-----|---------|---------|---------|
| T | 2 | 31.4121 | 15.7060 | 24.93** |
| W | 13 | 8.1883 | 0.6299 | |

# CHAPTER 3

**3.1.**

| Source | df | SS | MS | F |
|--------|-----|----------|----------|------|
| B | 8 | 0.001084 | 0.001355 | |
| T | 1 | 0.000044 | 0.000044 | 0.58 |
| BT | 8 | 0.000610 | 0.000076 | |

**3.2.** $\bar{d} = 003111$; $s_{\bar{d}} = 0.004118$; $t = 0.76$; $t^2 = 0.58 = F$.

**3.3.**

| Source | df | SS | MS | F |
|--------|-----|----------|----------|---------|
| S | 1 | 6.1256 | 6.1256 | 0.72 |
| N | 1 | 178.8906 | 178.8906 | 21.11** |
| SN | 1 | 1.2657 | 1.2657 | 0.15 |
| W | 12 | 101.7075 | 8.4756 | |

$\hat{sn}_{11} = -0.2815$ $\quad$ $\hat{sn}_{12} = 0.2815$ $\quad$ $\sum_i \sum_j \hat{sn}_{ij}^2 = 1.2656$

$\hat{sn}_{21} = 0.2815$ $\quad$ $\hat{sn}_{22} = -0.2815$

**3.4.**

| Source | df | SS | MS |
|--------|-----|---------|--------|
| S | 1 | 9.1260 | 9.1260 |
| T:S | 4 | 2.3093 | 0.5773 |
| D | 9 | 15.9926 | 1.7770 |
| SD | 9 | 2.5540 | 0.2838 |
| TD:S | 36 | 2.8974 | 0.0805 |

**3.5. (a)**

| Source | df | SS | E(MS) | F |
|---|---|---|---|---|
| $A$ | 3 | $a-1$ | $\sigma_w^2 + 2\sigma_{c:ab}^2 + 30\sigma_a^2$ | $A/C{:}AB$ |
| $B$ | 2 | $b-1$ | $\sigma_w^2 + 2\sigma_{c:ab}^2 + 40\sigma_b^2$ | $B/C{:}AB$ |
| $AB$ | 6 | $ab-a-b+1$ | $\sigma_w^2 + 2\sigma_{c:ab}^2 + 10\sigma_{ab}^2$ | $AB/C{:}AB$ |
| $C{:}AB$ | 48 | $abc-ab$ | $\sigma_w^2 + 2\sigma_{c:ab}^2$ | $C{:}AB/W$ |
| $W$ | 60 | $\sum_i \sum_j \sum_k Y_{ijk}^2 - abc$ | $\sigma_w^2$ | |

**(b)**

| Source | df | SS | E(MS) | F |
|---|---|---|---|---|
| $A$ | 4 | $a-1$ | $\sigma_w^2 + 21\sigma_{b:a}^2 + 84\sigma_a^2$ | $A/B{:}A$ |
| $B{:}A$ | 15 | $ab-a$ | $\sigma_w^2 + 21\sigma_{b:a}^2$ | $B{:}A/W$ |
| $C$ | 2 | $c-1$ | $\sigma_w^2 + 7\sigma_{bc:a}^2 + 140\sigma_c^2$ | $C/BC{:}A$ |
| $BC{:}A$ | 30 | $abc-ab-ac+a$ | $\sigma_w^2 + 7\sigma_{bc:a}^2$ | $BC{:}A/W$ |
| $W$ | 360 | $\sum_i \sum_j \sum_k \sum_l Y_{ijkl}^2 - abc$ | $\sigma_w^2$ | |

**(c)**

| Source | df | SS | E(MS) | F |
|---|---|---|---|---|
| $A$ | 2 | $a-1$ | $\sigma_w^2 + \sigma_{d:abc}^2 + 2\sigma_{c:ab}^2 + 40\sigma_a^2$ | $A/C{:}AB$ |
| $B$ | 4 | $b-1$ | $\sigma_w^2 + \sigma_{d:abc}^2 + 2\sigma_{c:ab}^2 + 24\sigma_b^2$ | $B/C{:}AB$ |
| $AB$ | 8 | $ab-a-b+1$ | $\sigma_w^2 + \sigma_{d:abc}^2 + 2\sigma_{c:ab}^2 + 8\sigma_{ab}^2$ | $AB/C{:}AB$ |
| $C{:}AB$ | 45 | $abc-ab$ | $\sigma_w^2 + \sigma_{d:abc}^2 + 2\sigma_{c:ab}^2$ | $C{:}AB/D{:}ABC$ |
| $D{:}ABC$ | 60 | $abcd-abc$ | $\sigma_w^2 + \sigma_{d:abc}^2$ | |

**(d)**

| Source | df | SS | E(MS) | F |
|---|---|---|---|---|
| $A$ | 3 | $a-1$ | $\sigma_w^2 + 6\sigma_{bd:ac}^2 + 18\sigma_{ad:c}^2 + 60\sigma_{b:a}^2 + 180\sigma_a^2$ | |
| $B{:}A$ | 8 | $ab-a$ | $\sigma_w^2 + 6\sigma_{bd:ac}^2 + 60\sigma_{b:a}^2$ | $B{:}A/BD{:}AC$ |
| $C$ | 4 | $c-1$ | $\sigma_w^2 + 6\sigma_{bd:ac}^2 + 72\sigma_{d:c}^2 + 12\sigma_{bc:a}^2 + 144\sigma_c^2$ | |
| $AC$ | 12 | $ac-a-c+1$ | $\sigma_w^2 + 6\sigma_{bd:ac}^2 + 18\sigma_{ad:c}^2 + 12\sigma_{bc:a}^2 + 36\sigma_{ac}^2$ | |
| $BC{:}A$ | 32 | $abc-ab-ac+a$ | $\sigma_w^2 + 6\sigma_{bd:ac}^2 + 12\sigma_{bc:a}^2$ | $BC{:}A/BD{:}AC$ |
| $D{:}C$ | 5 | $cd-c$ | $\sigma_w^2 + 6\sigma_{bd:ac}^2 + 72\sigma_{d:c}^2$ | $D{:}C/BD{:}AC$ |
| $AD{:}C$ | 15 | $acd-ac-cd+c$ | $\sigma_w^2 + 6\sigma_{bd:ac}^2 + 18\sigma_{ad:c}^2$ | $AD{:}C/BD{:}AC$ |
| $BD{:}AC$ | 40 | $abcd-abc-acd+ac$ | $\sigma_w^2 + 6\sigma_{bd:ac}^2$ | $BD{:}AC/W$ |
| $W$ | 600 | $\sum_i \sum_j \sum_k \sum_l Y_{ijkl}^2 - abcd$ | $\sigma_w^2$ | |

**(e)**

| Source | df | SS | E(MS) | F |
|---|---|---|---|---|
| $A$ | 4 | $a-1$ | $\sigma_w^2 + 24\sigma_{bc:a}^2 + 96\sigma_{c:a}^2 + 72\sigma_{b:a}^2 + 288\sigma_a^2$ | |
| $B{:}A$ | 15 | $ab-a$ | $\sigma_w^2 + 24\sigma_{bc:a}^2 + 72\sigma_{b:a}^2$ | $B{:}A/AB{:}A$ |
| $C{:}A$ | 10 | $ac-a$ | $\sigma_w^2 + 24\sigma_{bc:a}^2 + 96\sigma_{c:a}^2$ | $C{:}A/BC{:}A$ |
| $D$ | 5 | $d-1$ | $\sigma_w^2 + 4\sigma_{bcd:a}^2 + 16\sigma_{cd:a}^2 + 12\sigma_{bd:a}^2 + 240\sigma_d^2$ | |

| | | | |
|---|---|---|---|
| $AD$ | 20 | $ad - a - d + 1$ | $\sigma_w^2 + 4\sigma_{bcd:a}^2 + 16\sigma_{cd:a}^2$ |
| | | | $\quad + 12\sigma_{bd:a}^2 + 48\sigma_{ad}^2$ |
| $BC:A$ | 30 | $abc - ab - ac + a$ | $\sigma_w^2 + 24\sigma_{bc:a}^2$ | $BC:A/W$ |
| $BD:A$ | 75 | $abd - ab - ad + a$ | $\sigma_w^2 + 4\sigma_{bcd:a}^2 + 12\sigma_{bd:a}^2$ | $BD:A/BCD:A$ |
| $CD:A$ | 50 | $acd - ac - ad + a$ | $\sigma_w^2 + 4\sigma_{bcd:a}^2 + 16\sigma_{cd:a}^2$ | $CD:A/BCD:A$ |
| $BCD:A$ | 150 | $abcd - abc - abd - acd$ | $\sigma_w^2 + 4\sigma_{bcd:a}^2$ | $BCD:A/W$ |
| | | $\quad + ab + ac + ad - a$ |
| $W$ | 1080 | $\Sigma_i \Sigma_j \Sigma_k \Sigma_l \Sigma_m Y_{ijklm}^2$ | $\sigma_w^2$ |
| | | $\quad - abcd$ |

**3.6.**

| Source | df | SS | MS | F |
|---|---|---|---|---|
| $C$ | 1 | 4,2666 | 4.2666 | 23.20** |
| $W$ | 58 | 10,6667 | 10.6667 | |

**3.7.** $s_1^2 = 0.0746$; $s_2^2 = 30,122$; $F = 40.38**$.

**3.8.** Corrected $X^2 = 43.40**$.

**3.9.**

| Source | df | SS | MS | F |
|---|---|---|---|---|
| $S$ | 1 | 1,734.0000 | 1,734.0000 | 48.59** |
| $N$ | 2 | 14,648.8525 | 7,324.4262 | 205.05** |
| $SN$ | 2 | 696.1825 | 348.0910 | 9.75 |
| Nonadditivity | 1 | 552.7442 | 552.7442 | 3.85 |
| Residual | 1 | 143.4383 | 143.4383 | |
| $W$ | 18 | 642.3250 | 35.6847 | |

**3.10.**

| Source | df | SS | MS | F |
|---|---|---|---|---|
| $S$ | 1 | 0.1094 | 0.1094 | 21.04** |
| $N$ | 2 | 1.7659 | 0.8830 | 169.81** |
| $SN$ | 2 | 0.0327 | 0.0164 | 3.15 |
| $W$ | 18 | 0.0945 | 0.0052 | |

Interaction is no longer significant.

**3.11.**

| Source | df | SS | MS | F |
|---|---|---|---|---|
| $S$ | 3 | 235.3333 | 78.4444 | |
| $M$ | 2 | 917.1667 | 458.5834 | 226.15** |
| $SM$ | 6 | 12.1667 | 2.0278 | |

Using arcsines, we have the following:

| Source | df | SS | MS | F |
|---|---|---|---|---|
| $S$ | 3 | 312.5898 | 104.1966 | |
| $M$ | 2 | 699.7156 | 349.8578 | 38.79** |
| $SM$ | 6 | 54.1121 | 9.0187 | |

**3.12.**

| Source | df | SS | MS | F |
|--------|----|----|----|----|
| B | 3 | 566.1875 | 188.7292 | |
| T | 3 | 17,372.1875 | 5,790.7292 | 28.98** |
| BT | 9 | 1,798.5625 | 199.8403 | |

Using square roots, we have the following:

| Source | df | SS | MS | F |
|--------|----|----|----|----|
| B | 3 | 4.0676 | 1.3559 | |
| T | 3 | 96.6656 | 32.2219 | 40.36** |
| BT | 9 | 7.1855 | 0.7984 | |

**3.13.**

| Source | df | E(MS) |
|--------|----|-------|
| S | 1 | $\sigma_w^2 + \sigma_{td:s}^2 + 3\sigma_{sd}^2 + 10\sigma_{t:s}^2 + 30\sigma_s^2$ |
| T:S | 4 | $\sigma_w^2 + \sigma_{td:s}^2 + 10\sigma_{t:s}^2$ |
| D | 9 | $\sigma_w^2 + \sigma_{td:s}^2 + 6\sigma_d^2$ |
| SD | 9 | $\sigma_w^2 + \sigma_{td:s}^2 + 3\sigma_{sd}^2$ |
| TD:S | 36 | $\sigma_w^2 + \sigma_{td:s}^2$ |

$$S + TD:S = 2\sigma_w^2 + 2\sigma_{td:s}^2 + 3\sigma_{sd}^2 + 10\sigma_{t:s}^2 + 30\sigma_s^2$$

$$SD + T:S = 2\sigma_w^2 + 2\sigma_{td:s}^2 + 3\sigma_{sd}^2 + 10\sigma_{t:s}^2$$

**3.14.** For $A$,     $$F' = \frac{\text{MS}_a + \text{MS}_{BC:A}}{\text{MS}_{B:A} + \text{MS}_{C:A}}$$

For $D$,     $$F' = \frac{\text{MS}_D + \text{MS}_{BLD:A}}{\text{MS}_{BD:A} + \text{MS}_{CD:A}}$$

For $AD$,    $$F' = \frac{\text{MS}_{AD} + \text{MS}_{BCD:A}}{\text{MS}_{BD:A} + \text{MS}_{CD:A}}$$

**3.15.**

| Source | df | SS | MS | F |
|--------|----|----|----|----|
| Sires | 5 | 13,624 | 2,725 | 2.99* |
| Within | 31 | 28,203 | 910 | |

$k_0 = 6.0216$; $\sigma_s^2 = 301$; $\sigma_e^2 = 910$; $r = 0.25$; $\bar{y} = 450.2$; $s_{\bar{y}} = 9.0$.

**3.16.**

| Source | df | SS | MS | F | E(MS) |
|--------|----|----|----|----|----|
| Families (F) | 3 | 35,726.2 | 11,908.7 | 2.10 | $\sigma_e^2 + 3.8174\sigma_{p:f}^2 + 21.1216\sigma_f^2$ |
| Plots:F | 20 | 113,269.0 | 5,663.4 | 1.39 | $\sigma_e^2 + 3.4827\sigma_{p:f}^2$ |
| Trees | 61 | 249,243.1 | 4,086.0 | | $\sigma_e^2$ |

$\sigma_e^2 = 4,086$; $\sigma_{p:f}^2 = 452.9$; $\sigma_f^2 = 288.5$; $r_1 = 741.0/4827.4 = 0.15$; $r_2 = 288.5/4827.4 = 0.06$; $\bar{y} = 232.9$; $s_{\bar{y}} = 11.9$.

**3.17.** The least-squares equations are

$$
\begin{bmatrix}
14 & 2 & 0 & -1 & -2 & 1 \\
2 & 14 & -2 & 1 & 0 & -1 \\
0 & -2 & 10 & 5 & 0 & 1 \\
-1 & 1 & 5 & 9 & 1 & 3 \\
-2 & 0 & 0 & 1 & 10 & 5 \\
1 & -1 & 1 & 3 & 5 & 9
\end{bmatrix}
\begin{bmatrix}
\hat{\mu} \\
\hat{n}_1 \\
\hat{l}_1 \\
\hat{l}_2 \\
\hat{nl}_{11} \\
\hat{nl}_{12}
\end{bmatrix}
=
\begin{bmatrix}
550.7 \\
-22.1 \\
-228.1 \\
-158.1 \\
17.3 \\
90.5
\end{bmatrix}
$$

The solutions of the equations are

$$
\begin{bmatrix}
\hat{\mu} \\
\hat{n}_1 \\
\hat{l}_1 \\
\hat{l}_2 \\
\hat{nl}_{11} \\
\hat{nl}_{12}
\end{bmatrix}
=
\begin{bmatrix}
0.083333 & -0.01852 & -0.01389 & 0.027778 & 0.032407 & -0.03704 \\
 & 0.083333 & 0.032407 & -0.03704 & -0.01389 & 0.027778 \\
 & & 0.152778 & -0.09722 & -0.00463 & 0.023148 \\
 & \text{Symmetric} & & 0.194444 & 0.023148 & -0.07407 \\
 & & & & 0.152778 & -0.09722 \\
 & & & & & 0.194444
\end{bmatrix}
$$

$$
*
\begin{bmatrix}
550.7 \\
-22.1 \\
-228.1 \\
-158.1 \\
17.3 \\
90.5
\end{bmatrix}
$$

$$
=
\begin{bmatrix}
42.2861 \\
-11.303 \\
-25.828 \\
1.2472 \\
9.3944 \\
1.3361
\end{bmatrix}
\qquad
\begin{aligned}
\hat{n}_2 &= 11.303 \\
\hat{l}_3 &= 24.5808 \\
\hat{nl}_{21} &= -9.3944 \\
\hat{nl}_{22} &= -1.3361 \\
\hat{nl}_{13} &= -10.7305 \\
\hat{nl}_{23} &= 10.7305
\end{aligned}
$$

$R(\mu, n, l, nl) = 29{,}514.3762$

$\sum_i \sum_j \sum_k Y_{ijk}^2 = 29{,}823.97$

SS Within = 309.59

$$SS(N) = \frac{(-11.303)^2}{0.083333} = 1533.0998$$

$$SS(L) = [-25.828 \quad 1.2472]
\begin{bmatrix}
0.152778 & -0.09722 \\
-0.09722 & 0.194444
\end{bmatrix}^{-1}
\begin{bmatrix}
-25.828 \\
1.2472
\end{bmatrix}$$

$$= 6106.5131$$

$$SS(NL) = [9.3944 \quad 1.3361]
\begin{bmatrix}
0.152778 & -0.09722 \\
-0.09722 & 0.194444
\end{bmatrix}^{-1}
\begin{bmatrix}
9.3944 \\
1.3361
\end{bmatrix}$$

$$= 981.2089$$

| ANOVA | | | | |
|---|---|---|---|---|
| Source | df | SS | MS | F |
| Nitrogen source ($N$) | 1 | 1533.10 | 1533.10 | 39.61** |
| Levels ($L$) | 2 | 6106.51 | 3053.26 | 78.90** |
| $NL$ | 2 | 981.21 | 490.60 | 12.68** |
| Within | 8 | 309.59 | 38.70 | |

The least-squares means and standard errors are as follows:

| Source | | LS Mean | SE | n |
|---|---|---|---|---|
| Overall | | 42.29 | 1.80 | 14 |
| Nitrogen sources | | | | |
|   Ammonium sulfate | | 30.98 | 2.24 | 8 |
|   Nitrate | | 53.59 | 2.81 | 6 |
| Levels | | | | |
|   1 | | 16.46 | 2.84 | 5 |
|   2 | | 43.53 | 3.59 | 4 |
|   3 | | 66.87 | 2.84 | 5 |
| $NL$ Subclasses | | | | |
|   Ammonium sulfate | 1 | 14.55 | 4.40 | 2 |
| | 2 | 33.57 | 3.59 | 3 |
| | 3 | 44.83 | 3.59 | 3 |
|   Nitrate | 1 | 18.37 | 3.59 | 3 |
| | 2 | 53.50 | 6.22 | 1 |
| | 3 | 88.90 | 4.40 | 2 |

# CHAPTER 4

**4.1.**

| Source | df | SS | MS | F |
|---|---|---|---|---|
| $T$ | 4 | 0.001777 | 0.000444 | 4.04* |
| $Q_1$ | 1 | 0.000333 | 0.000333 | 3.03 |
| $Q_2$ | 1 | 0.001121 | 0.001121 | 10.19** |
| $Q_3$ | 1 | 0.000193 | 0.000193 | 1.75 |
| $Q_4$ | 1 | 0.000131 | 0.000131 | 1.19 |
| $W$ | 25 | 0.002760 | 0.000110 | |

**4.2.**

| Source | df | SS | MS | F |
|---|---|---|---|---|
| $B$ | 7 | 16.9544 | 2.4221 | 0.71 |
| $T$ | 7 | 153.8894 | 21.9842 | 6.49** |
| $P_1$ | 1 | 29.4306 | 29.4306 | 8.68** |
| $P_2$ | 3 | 107.4363 | 35.8121 | 10.57** |
| $P_3$ | 3 | 17.0225 | 5.6742 | 1.67 |
| $BT$ | 49 | 166.0606 | 3.3890 | |

**4.3.** $t_1 = \dfrac{0.0334}{0.0195} = 1.1713 \qquad t^2 = 2.93$

$t_2 = \dfrac{0.0274}{0.00872} = 3.142** \qquad t^2 = 9.87$

$$t_3 = \frac{0.0114}{0.00872} = 1.307 \qquad t^2 = 1.71$$

$$t_4 = \frac{0.0094}{0.00872} = 1.078 \qquad t^2 = 1.16$$

**4.4.** $t_1 = 4.18^{**}$; $t_2 = 0.75$; $t_3 = 1.24$; $t_4 = 3.43^{**}$; df = 22.

**4.5.** $Q = Q_1 - Q_2 = 29.83333 - 60.8333 = -31$; $s_Q = 10.1543$; $t = 3.053^*$; df = 12.

**4.6.**

| Treatment | Mean | LSD | Duncan's | SND | Tukey's | Scheffé's |
|---|---|---|---|---|---|---|
| 3 | 58.520 | *a* | *a* | *a* | *a* | *a* |
| 5 | 51.560 | *a* | *a* | *a* | *ab* | *ab* |
| 2 | 38.200 | *b* | *b* | *b* | *bc* | *bc* |
| 1 | 30.320 | *bc* | *bc* | *bc* | *cd* | *cd* |
| 4 | 24.880 | *c* | *c* | *c* | *d* | *d* |

**4.7.**

| Source | df | SS | MS | F |
|---|---|---|---|---|
| B | 7 | 6.9211 | 0.9887 | |
| T | 4 | 22.5028 | 5.6257 | 4.26* |
| BT | 28 | 37.0117 | 1.3218 | |

| Strain | Mean | |
|---|---|---|
| Nonspur | 2.824 | |
| Spur 1 | 2.109 | + |
| Spur 3 | 1.944 | + |
| Spur 4 | 1.855 | + |
| Spur 2 | 0.512 | − |

+, Does not differ from control; −, differs from control.

**4.8.**

| Source | df | SS | MS | F |
|---|---|---|---|---|
| Years | 3 | 115,359 | 38,453 | 1.44 |
| Ages | 2 | 123,460 | 61,730 | 2.31 |
| Remainder | 664 | 17,769,807 | 26,762 | |

The estimates of constants, least-squares means, and standard errors follow:

| Classification | Number | Constant | LS Mean | SE |
|---|---|---|---|---|
| Overall | 670 | 1,420.47 | 1,420.47 | 6.51 |
| Years | | | | |
| 1969 | 162 | 14.12 | 1,434.59 | 12.90 |
| 1970 | 112 | −9.14 | 1,411.33 | 15.58 |
| 1971 | 213 | −16.23 | 1,404.25 | 11.26 |
| 1972 | 183 | 11.25 | 1,431.72 | 13.06 |
| Ages | | | | |
| 2 | 202 | −20.60 | 1,399.88 | 12.06 |
| 3 | 222 | 13.55 | 1,434.03 | 11.05 |
| 4–up | 246 | 7.05 | 1,427.52 | 11.13 |

The results of $t$ tests for age follow:

| Comparison | $\hat{a}_i - \hat{a}_j$ | $\sqrt{\dfrac{1}{c^{ii} + c^{jj} - 2c^{ij}}}$ | Product | $t_{0.05}\hat{\sigma}_e$ |
|---|---|---|---|---|
| $\hat{a}_1 - \hat{a}_2$ | −34.1499 | 9.9522 | −339.87* | 323.91 |
| $\hat{a}_1 - \hat{a}_3$ | −27.6447 | 9.6832 | −267.69 | 323.91 |
| $\hat{a}_2 - \hat{a}_3$ | 6.5052 | 10.6124 | 69.04 | 323.91 |

# CHAPTER 5

**5.1.** (a) $b = 3.2242$; (b) $\hat{Y} = 35.5333 - 3.2242X$; (c) $SS_b = 857.6485$; (d) $\Sigma\ d_{y \cdot x}^2 = 117.9515$; (e) $s_{y \cdot x}^2 = 14.7439$; (f) $s_b = 0.4227$; (g) $t = 7.63**$; (h) $F = 58.17**$; (i) CI $= -4.1989$ to $-2.2495$.

**5.2.** $Y - \hat{Y} = 4.6911, -5.0847, 2.1395, -3.6363, -1.4121, -1.1879, 5.0363, -0.7395, 3.4847, -3.2911.\ \Sigma(Y - \hat{Y}) = 0;\ \Sigma(Y - \hat{Y})^2 = 117.9515.$

**5.3.** $t_{0.05}s_{\hat{y}}$ for $X = 1$ through 10: 5.2, 4.4, 3.7, 3.2, 2.8, 2.8, 3.2, 3.7, 4.4, 5.2.

**5.4.** $t = 2.4000/0.8950 = 2.68*$; df = 16.

**5.5.** $t = a_1 - a_2/s_{a_1 - a_2} = 8.6000/3.0025 = 2.864**$; df = 17.

**5.6.** $F = 167.2455/27.6803 = 6.04$; df = 2, 24.

**5.7.** (a) $r_1 = -0.3918$; (b) $t = 0.3918/0.3253 = 1.20$ (ns); (c) CI $= -0.819$ to $0.316$; (d) $r_2 = 0.0424$; (e) $t = 0.457/0.5345 = 0.86$ (ns); (f) $r_3 = -0.3883$; (g) $\chi^2 = 0.960$ (ns); (h) average $r = 0.254$.

**5.8.** Linear: $t = 0.0173/0.0107 = 1.617$; df = 7; $F = 0.0180/0.0069 = 2.61$; df = 1, 7. Quadratic: $t = 0.01004/0.00307 = 3.27*$; df = 6; $F = 0.0311/0.0029 = 10.72*$; df = 1, 6; $\hat{Y} = 1.1701 + 0.0831X_1 - 0.0100X_2$. Linear: $t = 0.1018/0.0452 = 2.25$; df = 7; $F = 0.6222/0.1227 = 5.07$; df = 1, 7. Quadratic: $t = 0.0183/0.0202 = 0.906$; df = 6; $F = 0.1027/0.1260 = 0.82$; df = 1, 6. Cubic: $t = 0.0194/0.0056 = 3.46*$; df = 5; $F = 0.5338/0.0444 = 12.02*$; df = 1, 5; $\hat{Y} = 17.8808 + 1.3038X_1 - 0.3085X_2 + 0.0194X_3$.

**5.9.** Linear:    $Q_1 = -1.04$    $D_1 = 60$    $b_l = -0.0173$    $SS_l = 0.0180$
Quadratic:  $Q_2 = -9.18$   $D_2 = 2772$   $b_q = -0.003348$   $SS_q = 0.0311$

$$Y = \bar{y} + c_{1j}b_l + c_{2j}b_q = 1.2689 - 0.0173c_{1j} - 0.003348c_{2j}$$

| $X$ | 1 | 2 | 3 | 4 | 5 | 6 | 7 | 8 | 9 |
|---|---|---|---|---|---|---|---|---|---|
| $\hat{Y}$ | 1.244 | 1.297 | 1.330 | 1.343 | 1.336 | 1.308 | 1.261 | 1.195 | 1.106 |

Linear:    $Q_1 = -6.11$    $D_1 = 60$    $b_l = -0.1018$    $SS_l = 0.6222$
Quadratic:  $Q_2 = -16.87$   $D_2 = 2772$   $b_q = -0.0061$   $SS_q = 0.027$
Cubic:     $Q_3 = 22.90$    $D_3 = 990$    $b_c = 0.0232$    $SS_c = 0.5339$

| $X$ | 1 | 2 | 3 | 4 | 5 | 6 | 7 | 8 | 9 |
|---|---|---|---|---|---|---|---|---|---|
| $\hat{Y}$ | 18.91 | 19.42 | 19.55 | 19.41 | 19.12 | 18.79 | 18.54 | 18.49 | 19.74 |

**5.10** (a) $\hat{Y} = \bar{y} + bc_{1j} = 1.2689 - 0.0173c_{1j}$

(b) $\hat{Y} = a + bX = 1.3554 - 0.0173X$

(c) $\hat{Y} = 1.2689 - (0.0173)(X + a_1)(\lambda/d)$   $\lambda = 1$   $d = 1$   $n = 9$   $a = -5$
$= 1.2689 - (0.0173)(X - 5)(1/1)$
$= 1.3554 - 0.0173X$

**5.11.**

| | | | | |
|---|---|---|---|---|
| $c_{1j}$ | $-7$ | $-3$ | $3$ | $7$ |
| $c_{2j}$ | $29$ | $-29$ | $-29$ | $29$ |
| $c_{3j}$ | $-3$ | $7$ | $-7$ | $3$ |

**5.12.** $b_{y1 \cdot 23} = 0.4908$; $b_{y2 \cdot 13} = 2.9306$; $b_{y3 \cdot 12} = 1.0132$.

| Source | df | SS | MS | F |
|---|---|---|---|---|
| Regressions | 3 | 103.1999 | 34.4000 | 2.43 |
| Residual | 8 | 113.2901 | 14.1613 | |

$s_{by1 \cdot 23} = 0.4481$   $t = \dfrac{0.4908}{0.4481} = 1.095$   $t^2 = 1.200$

$s_{by2 \cdot 13} = 1.7749$   $t = \dfrac{2.9306}{1.7749} = 1.651$   $t^2 = 2.726$

$s_{by3 \cdot 12} = 0.5761$   $t = \dfrac{1.0132}{0.5761} = 1.759$   $t^2 = 3.094$

| Source | df | SS | MS | F |
|---|---|---|---|---|
| $X_1$ | 1 | 16.9899 | 16.9899 | 1.20 |
| $X_2$ | 1 | 38.6073 | 38.6073 | 2.73 |
| $X_3$ | 1 | 43.7984 | 43.7984 | 3.09 |
| Residual | 8 | 113.2901 | 14.1613 | |

$$R(X_1, X_2, X_3) - R(X_1, X_2) = 103.1999 - 59.4015 = 43.7984$$

$$b'_{y3 \cdot 12} = 0.5475 \qquad b'_{y2 \cdot 13} = 0.5134 \qquad b'_{y1 \cdot 23} = 0.3787$$

$$R^2 = \frac{103.1999}{216.49} = 0.4767$$

**5.13.**   $r_{12} = 0.8278$   $r_{13} = -0.3918$   $r_{14} = 0.4704$   $r_{23} = -0.5676$

$r_{24} = 0.5154$   $r_{34} = -0.8661$

$t_{12} = 4.17**$   $t_{13} = 1.20$   $t_{14} = 1.51$   $t_{23} = 1.95$

$t_{24} = 1.70$   $t_{34} = 4.90**$

$r_{12 \cdot 3} = 0.7992$   $r_{14 \cdot 3} = 0.2851$   $r_{24 \cdot 3} = 0.0578$   $r_{12 \cdot 34} = 0.8180$

$t_{12 \cdot 3} = 3.52**$   $t_{14 \cdot 3} = 0.79$   $t_{24 \cdot 3} = 0.15$   $t_{12 \cdot 34} = 3.48*$

$r_{12 \cdot 3} = \dfrac{2.918964}{3.652750} = 0.7991$   $r_{13 \cdot 2} = \dfrac{0.376146}{2.227938} = 0.1688$

$r_{23 \cdot 1} = \dfrac{-1.173283}{2.489743} = -0.4712$

$r_{12 \cdot 34} = 0.8179$   $r_{13 \cdot 24} = 0.4188$   $r_{14 \cdot 23} = 0.3978$   $r_{23 \cdot 14} = -0.4905$

$r_{24 \cdot 13} = -0.2949$   $r_{34 \cdot 12} = -0.8439$

**5.14.** $R^2 = 103.1999/216.49 = 0.4767; R = 0.6904.$

**5.15. (a)**

| Source | df | SS | MS | F |
|--------|-----|----------|----------|---------|
| $R$ | 1 | 0.000204 | 0.000204 | 0.48 |
| $D$ | 2 | 0.038933 | 0.019366 | 45.91** |
| $RD$ | 2 | 0.000433 | 0.000216 | |
| $W$ | 18 | 0.007625 | 0.000424 | |

**(b)** $Q_1 = 0.07; Q_2 = 0.08; Q_3 = -1.36; Q_4 = -0.08; Q_5 = 0.04.$

| Source | df | MS | F |
|--------|-----|----------|---------|
| $R_L$ | 1 | 0.000204 | 0.48 |
| $D_L$ | 1 | 0.000400 | 0.94 |
| $D_Q$ | 1 | 0.038533 | 90.88** |
| $R_L D_L$ | 1 | 0.000400 | 0.94 |
| $R_L D_Q$ | 1 | 0.000033 | 0.08 |
| $W$ | 18 | 0.000424 | |

**5.16.** The following table presents the ANOVA with SS for polynomials:

| Source | df | SS | MS | F |
|--------|-----|---------------|----------|---------|
| Years | 3 | 49,041.60 | 16,347 | 0.65 |
| Linear | 1 | 47,871.45 | 47,871 | 1.90 |
| Quadratic | 1 | 465.69 | 466 | 0.02 |
| Cubic | 1 | 704.46 | 704 | 0.03 |
| Ages | 2 | 216,809.12 | 108,405 | 4.24* |
| Linear | 1 | 42,446.84 | 42,447 | 1.68 |
| Quadratic | 1 | 174,362.28 | 174,362 | 6.91** |
| Years × ages | 6 | 1,161,229.51 | 193,538 | 7.67** |
| Within | 658 | 16,608,578.00 | 25,241 | |

The following table presents the partitioning of the subclass sum of squares:

| Source | df | Order of elimination | SS | MS | F |
|--------|-----|------------|--------------|---------|---------|
| Subclasses | 11 | — | 1,472,569.6 | 133,870 | 5.30** |
| Years ($T$) | | | | | |
| Linear | 1 | 10 | 6,356.6 | 6,357 | 0.25 |
| Quadratic | 1 | 7 | 1.8 | 2 | 0.00 |
| Cubic | 1 | 4 | 7.4 | 7 | 0.00 |
| Ages ($A$) | | | | | |
| Linear | 1 | 11 | 119,815.9 | 119,816 | 4.75* |
| Quadratic | 1 | 8 | 233,950.4 | 233,950 | 9.27** |
| Years × ages | | | | | |
| $TL \times AL$ | 1 | 9 | 698,692.2 | 698,692 | 27.68** |
| $TL \times AQ$ | 1 | 5 | 7,018.6 | 7,019 | 0.28 |
| $TQ \times AL$ | 1 | 6 | 115,615.2 | 115,615 | 4.58* |
| $TQ \times AQ$ | 1 | 2 | 11,098.7 | 11,099 | 0.44 |
| $TC \times AL$ | 1 | 3 | 229,315.4 | 229,315 | 9.09** |
| $TC \times AQ$ | 1 | 1 | 50,697.4 | 50,697 | 2.01 |
| Error | 658 | — | 16,608,578.0 | 25,241 | |

The prediction equation is

$$\hat{Y}_{kl} = 1438.267 - 14.1134(T - 2.5) - 2.0871(T - 2.5)^2 + 2.7191(T - 2.5)^3$$
$$+ 50.7210(A - 3) - 36.4228(A - 3)^2 - 37.8226(T - 2.5)(A - 3)$$
$$- 2.1323(T - 2.5)(A - 3)^2 - 21.9930(T - 2.5)^2(A - 3)$$
$$+ 38.6979(T - 2.5)^3(A - 3)$$

where $\hat{\beta}_0 = 1438.267 = 1411.376 + \frac{1}{4}(2.0871)(5) + \frac{1}{3}(36.4228)(2)$

$$R^2 = 0.9580 \qquad R = 0.98$$

The estimates of constants, least-squares means, and standard errors follow:

| Classification | | Number | Constant | LS Mean | SE |
|---|---|---|---|---|---|
| Overall | | 670 | 1411.38 | 1411.38 | 8.17 |
| Years | | | | | |
| 1969 | | 162 | 13.24 | 1424.62 | 12.66 |
| 1970 | | 112 | 7.67 | 1419.05 | 15.79 |
| 1971 | | 213 | −5.44 | 1405.94 | 11.03 |
| 1972 | | 183 | −15.47 | 1395.91 | 23.14 |
| Ages | | | | | |
| 2 | | 202 | −35.51 | 1375.87 | 18.42 |
| 3 | | 222 | 27.05 | 1438.43 | 10.85 |
| 4-up | | 246 | 8.46 | 1419.84 | 11.96 |
| Years × ages | | | | | |
| 1969 | 2 | 61 | 100.92 | 1490.03 | 20.34 |
| | 3 | 59 | −14.60 | 1437.07 | 20.68 |
| | 4-up | 42 | −86.32 | 1346.76 | 24.51 |
| 1970 | 2 | 49 | −44.05 | 1339.49 | 22.70 |
| | 3 | 40 | 25.86 | 1471.96 | 25.12 |
| | 4-up | 23 | 18.19 | 1445.70 | 33.13 |
| 1971 | 2 | 86 | −4.31 | 1366.12 | 17.13 |
| | 3 | 58 | −10.46 | 1422.53 | 20.86 |
| | 4-up | 69 | 14.78 | 1429.17 | 19.13 |
| 1972 | 2 | 6 | −52.56 | 1307.84 | 64.86 |
| | 3 | 65 | −0.80 | 1422.16 | 19.71 |
| | 4-up | 112 | 53.36 | 1457.73 | 15.01 |

## CHAPTER 6

**6.1.**

| Source | df | SS | MS | F |
|---|---|---|---|---|
| $T$ | 3 | 14,486.20 | 4,828.73 | 5.04** |
| $W$ | 26 | 34,495.03 | 958.20 | |

**6.2.**

| Source | df | SS | MS | F | E(MS) |
|---|---|---|---|---|---|
| $B$ | 3 | 54,362 | 18,121 | | $\sigma_w^2 + 4\sigma_b^2$ |
| $T$ | 3 | 206,394 | 68,798 | 9.87** | $\sigma_w^2 + \sigma_{bt}^2 + 4\sigma_t^2$ |
| $BT$ | 9 | 62,700 | 6,967 | | $\sigma_w^2 + \sigma_{bt}^2$ |

**6.3.** 126.94 percent.

**6.4.**

| Source | df | SS | MS | F | E(MS) |
|---|---|---|---|---|---|
| S | 3 | 15.4389 | 5.1463 | 2.81 | $\sigma_e^2 + 4\sigma_s^2$ |
| P | 3 | 24.4702 | 8.1567 | 4.45 | $\sigma_e^2 + 4\sigma_p^2$ |
| T | 3 | 314.6126 | 104.8709 | 57.25** | $\sigma_e^2 + 4\sigma_t^2$ |
| Error | 6 | 10.9901 | 1.8317 | | $\sigma_e^2$ |

**6.5.**

| Source | df | SS | MS | F | E(MS) |
|---|---|---|---|---|---|
| S | 1 | 15.7173 | 15.7173 | 11.56* | $\sigma_e^2 + 3\sigma_{g:s}^2 + 9\sigma_s^2$ |
| G:S | 4 | 6.4408 | 1.6102 | 0.12 | $\sigma_e^2 + 3\sigma_{g:s}^2$ |
| P | 2 | 3.9138 | 1.9569 | 0.17 | $\sigma_e^2 + 6\sigma_p^2$ |
| T | 2 | 2.9719 | 1.4860 | 0.13 | $\sigma_e^2 + 6\sigma_t^2$ |
| SP | 2 | 28.2058 | 14.1029 | 1.21 | $\sigma_e^2 + 3\sigma_{sp}^2$ |
| ST | 2 | 2.0887 | 1.0444 | 0.09 | $\sigma_e^2 + 3\sigma_{st}^2$ |
| Error | 4 | 45.6453 | 11.4113 | | $\sigma_e^2$ |

**6.6.**

| Source | df | SS | MS | F | E(MS) |
|---|---|---|---|---|---|
| G | 1 | 242.00 | 242.00 | 0.25 | $\sigma_e^2 + 3\sigma_{s:g}^2 + 9\sigma_g^2$ |
| S:G | 4 | 3837.78 | 959.44 | 2.71 | $\sigma_e^2 + 3\sigma_{s:g}^2$ |
| P | 2 | 112.11 | 56.06 | 0.16 | $\sigma_e^2 + 6\sigma_p^2$ |
| T | 2 | 694.78 | 347.39 | 0.98 | $\sigma_e^2 + 6\sigma_t^2$ |
| L | 2 | 7535.44 | 3767.72 | 10.62* | $\sigma_e^2 + 6\sigma_l^2$ |
| Error | 6 | 2127.67 | 354.61 | | $\sigma_e^2$ |

**6.7.**

| Source | df | SS | MS | F | E(MS) |
|---|---|---|---|---|---|
| S | 3 | 23.5249 | 7.8416 | 8.98* | $\sigma_e^2 + 2\sigma_{c:s}^2 + 4\sigma_s^2$ |
| C:S | 4 | 3.4945 | 0.8736 | 1.85 | $\sigma_e^2 + 2\sigma_{c:s}^2$ |
| P | 1 | 8.6583 | 8.6583 | 18.34** | $\sigma_e^2 + 8\sigma_p^2$ |
| T | 1 | 0.0637 | 0.0637 | 0.13 | $\sigma_e^2 + 8\sigma_t^2$ |
| Error | 6 | 2.8333 | 0.4722 | | $\upsilon_e^2$ |

**6.8.**

| Source | df | SS | MS | F | E(MS) |
|---|---|---|---|---|---|
| S | 1 | 2,667.10 | 2,667.10 | 75.47** | $\sigma_w^2 + 24\sigma_s^2$ |
| L | 2 | 18,616.51 | 9,308.26 | 263.39** | $\sigma_w^2 + 16\sigma_l^2$ |
| P | 1 | 68.64 | 68.64 | 1.94 | $\sigma_w^2 + 24\sigma_p^2$ |
| SL | 2 | 2,266.25 | 1,133.12 | 32.06** | $\sigma_w^2 + 8\sigma_{sl}^2$ |
| SP | 1 | 93.52 | 93.52 | 93.52 | $\sigma_w^2 + 12\sigma_{sp}^2$ |
| LP | 2 | 129.21 | 64.60 | 1.83 | $\sigma_w^2 + 8\sigma_{lp}^2$ |
| SLP | 2 | 19.63 | 9.82 | 0.28 | $\sigma_w^2 + 4\sigma_{slp}^2$ |
| W | 36 | 1,272.12 | 35.34 | | $\sigma_w^2$ |

## CHAPTER 7

**7.1.**

| Source | df | SS | MS | F | E(MS) |
|---|---|---|---|---|---|
| Whole plots | 14 | 125.7518 | | | |
| T | 2 | 17.6346 | 8.8173 | 0.98 | $\sigma_w^2 + 2\sigma_{r:t}^2 + 10\sigma_t^2$ |
| R:T | 12 | 108.1172 | 9.0098 | | $\sigma_w^2 + 2\sigma_{r:t}^2$ |
| Split plots | 15 | 12.6543 | | | |
| O | 1 | 0.9013 | 0.9013 | 1.05 | $\sigma_w^2 + \sigma_{ro:t}^2 + 15\sigma_o^2$ |
| TO | 2 | 1.8840 | 0.7420 | 0.87 | $\sigma_w^2 + \sigma_{ro:t}^2 + 5\sigma_{to}^2$ |
| RO:T | 12 | 9.8690 | 0.8558 | | $\sigma_w^2 + \sigma_{ro:t}^2$ |

**7.2.**

| Source | df | SS | MS | F | E(MS) |
|---|---|---|---|---|---|
| Whole plots | 23 | 562,205 | | | |
| C | 1 | 14,317 | 14,317 | 0.56 | $\sigma_w^2 + 2\sigma_{b:cl}^2 + 24\sigma_c^2$ |
| L | 1 | 4,700 | 4,700 | 0.18 | $\sigma_w^2 + 2\sigma_{b:cl}^2 + 24\sigma_l^2$ |
| CL | 1 | 29,751 | 29,751 | 1.16 | $\sigma_w^2 + 2\sigma_{b:cl}^2 + 12\sigma_{cl}^2$ |
| B:CL | 20 | 513,437 | 25,672 | | $\sigma_w^2 + 2\sigma_{b:cl}^2$ |
| Split plots | 24 | 6,585,651 | | | |
| P | 1 | 6,408,677 | 6,408,677 | 796.11 | $\sigma_w^2 + \sigma_{bp:cl}^2 + 24\sigma_p^2$ |
| CP | 1 | 3,317 | 3,317 | 0.41 | $\sigma_w^2 + \sigma_{bp:cl}^2 + 12\sigma_{cp}^2$ |
| LP | 1 | 1,093 | 1,093 | 0.14 | $\sigma_w^2 + \sigma_{bp:cl}^2 + 12\sigma_{lp}^2$ |
| CLP | 1 | 11,563 | 11,563 | 1.43 | $\sigma_w^2 + \sigma_{bp:cl}^2 + 6\sigma_{clp}^2$ |
| BP:CL | 20 | 161,001 | 8,050 | | $\sigma_w^2 + \sigma_{bp:cl}^2$ |

**7.3.**

| Source | df | SS | MS | F | E(MS) |
|---|---|---|---|---|---|
| Whole plots | 15 | 990.44 | | | |
| B | 3 | 117.57 | 39.19 | | $\sigma_w^2 + 16\sigma_b^2$ |
| V | 3 | 476.82 | 158.94 | 3.61 | $\sigma_w^2 + 4\sigma_{bv}^2 + 16\sigma_v^2$ |
| BV | 9 | 396.05 | 44.01 | | $\sigma_w^2 + 4\sigma_{bv}^2$ |
| Split plots | 16 | 757.00 | | | |
| S | 1 | 1.00 | 1.00 | 0.01 | $\sigma_w^2 + 8\sigma_{bs}^2 + 32\sigma_s^2$ |
| BS | 3 | 372.12 | 124.04 | | $\sigma_w^2 + 8\sigma_{bs}^2$ |
| VS | 3 | 73.62 | 24.54 | 0.71 | $\sigma_w^2 + 2\sigma_{bvs}^2 + 8\sigma_{vs}^2$ |
| BVS | 9 | 310.26 | 34.47 | | $\sigma_w^2 + 2\sigma_{bvs}^2$ |
| Split-split plots | 32 | 1284.00 | | | |
| N | 1 | 324.00 | 324.00 | 141.48** | $\sigma_w^2 + 8\sigma_{bn}^2 + 16\sigma_n^2$ |
| BN | 3 | 6.87 | 2.29 | | $\sigma_w^2 + 8\sigma_{bn}^2$ |
| VN | 3 | 211.87 | 70.62 | 2.09 | $\sigma_w^2 + 2\sigma_{bvn}^2 + 8\sigma_{vn}^2$ |
| BVN | 9 | 304.26 | 33.81 | | $\sigma_w^2 + 2\sigma_{bvn}^2$ |
| SN | 1 | 39.06 | 39.06 | 2.56 | $\sigma_w^2 + 4\sigma_{bsn}^2 + 16\sigma_{sn}^2$ |
| BSN | 3 | 45.82 | 15.27 | | $\sigma_w^2 + 4\sigma_{bsn}^2$ |
| VSN | 3 | 177.57 | 59.19 | 3.05 | $\sigma_w^2 + \sigma_{bvsn}^2 + 4\sigma_{vsn}^2$ |
| BVSN | 9 | 174.55 | 19.39 | | $\sigma_w^2 + \sigma_{bvsn}^2$ |

**7.4.**

| Source | df | SS | MS | F | E(MS) |
|---|---|---|---|---|---|
| Whole plots | 26 | 17.1259 | | | |
| S | 2 | 4.3670 | 2.1835 | 3.78 | $\sigma_e^2 + 2\sigma_d^2 + 6\sigma_{a:s}^2 + 18\sigma_s^2$ |
| A:S | 6 | 3.4656 | 0.5776 | 2.44 | $\sigma_e^2 + 2\sigma_d^2 + 6\sigma_{a:s}^2$ |
| R | 2 | 1.0237 | 0.5118 | 2.16 | $\sigma_e^2 + 2\sigma_d^2 + 18\sigma_r^2$ |
| T | 2 | 3.2070 | 1.6035 | 6.77* | $\sigma_e^2 + 2\sigma_d^2 + 18\sigma_t^2$ |
| SR | 4 | 3.5719 | 0.8930 | 3.77 | $\sigma_e^2 + 2\sigma_d^2 + 6\sigma_{sr}^2$ |
| ST | 4 | 0.0686 | 0.0172 | 0.07 | $\sigma_e^2 + 2\sigma_d^2 + 6\sigma_{st}^2$ |
| Error (a) | 6 | 1.4221 | 0.2370 | | $\sigma_e^2 + 2\sigma_d^2$ |
| Split plots | 27 | 5.3550 | | | |
| P | 1 | 0.4090 | 0.4090 | 1.20 | $\sigma_e^2 + 3\sigma_{ap:s}^2 + 27\sigma_p^2$ |
| SP | 2 | 0.2005 | 0.1002 | 0.29 | $\sigma_e^2 + 3\sigma_{ap:s}^2 + 9\sigma_{sp}^2$ |
| AP:S | 6 | 2.0521 | 0.3420 | 2.90 | $\sigma_e^2 + 3\sigma_{ap:s}^2$ |
| RP | 2 | 1.5215 | 0.7608 | 6.44* | $\sigma_e^2 + 9\sigma_{rp}^2$ |
| TP | 2 | 0.1560 | 0.0780 | 0.66 | $\sigma_e^2 + 9\sigma_{tp}^2$ |
| SRP | 4 | 0.2273 | 0.0568 | 0.48 | $\sigma_e^2 + 3\sigma_{srp}^2$ |
| STP | 4 | 0.0795 | 0.0199 | 0.17 | $\sigma_e^2 + 3\sigma_{stp}^2$ |
| Error (b) | 6 | 0.7091 | 0.1181 | | $\sigma_e^2$ |

**7.5.**

| Source | df | SS | MS | F | E(MS) |
|---|---|---|---|---|---|
| Whole plots | 15 | 15.4314 | | | |
| S | 3 | 10.6913 | 3.5638 | 8.45* | $\sigma_e^2 + 4\sigma_d^2 + 8\sigma_{a:s}^2 + 16\sigma_s^2$ |
| A:S | 4 | 1.6864 | 0.4216 | 9.31* | $\sigma_e^2 + 4\sigma_d^2 + 8\sigma_{a:s}^2$ |
| P | 1 | 2.5840 | 2.5840 | 57.04** | $\sigma_e^2 + 4\sigma_d^2 + 32\sigma_p^2$ |
| T | 1 | 0.1980 | 0.1980 | 4.37 | $\sigma_e^2 + 4\sigma_d^2 + 16\sigma_t^2$ |
| Error (a) | 6 | 0.2717 | 0.0453 | | $\sigma_e^2 + 4\sigma_d^2$ |
| Split plots | 48 | 2.1780 | | | |
| W | 3 | 0.2070 | 0.0690 | 2.16 | $\sigma_e^2 + 2\sigma_{aw:s}^2 + 8\sigma_{sw}^2 + 16\sigma_w^2$ |
| SW | 9 | 0.2876 | 0.0320 | 1.39 | $\sigma_e^2 + 2\sigma_{aw:s}^2 + 8\sigma_{sw}^2$ |
| AW:S | 12 | 0.2775 | 0.0231 | 0.45 | $\sigma_e^2 + 2\sigma_{aw:s}^2$ |
| PW | 3 | 0.3960 | 0.1320 | 2.60 | $\sigma_e^2 + 8\sigma_{pw}^2$ |
| TW | 3 | 0.0957 | 0.0319 | 0.63 | $\sigma_e^2 + 8\sigma_{tw}^2$ |
| Error (b) | 18 | 0.9142 | 0.0508 | | $\sigma_e^2$ |

**7.6.** Linear = 0.0562; quadratic = 0.0638; cubic = 0.0871; linear $\times$ linear = 0.0611; linear $\times$ quadratic = 0.0272; linear $\times$ cubic = 0.0074.

**7.7.**

| Source | df | SS | MS | F |
|---|---|---|---|---|
| Among linear trends | 2 | 0.0056 | 0.0028 | 0.12 |
| Among quadratic trends | 2 | 0.0597 | 0.0298 | 1.29 |
| Among cubic trends | 2 | 0.2221 | 0.1110 | 4.81* |
| Error | 12 | 0.2775 | 0.0231 | |

**7.8.**

| Source | df | SS | MS | F |
|---|---|---|---|---|
| Whole plots | 17 | 13,820,949 | | |
| Subjects ($S$) | 5 | 9,232,162 | 1,846,432 | 6.34** |
| Days ($D$) | 2 | 2,157,203 | 1,078,602 | 3.70 |
| Treatments ($T$) | 2 | 101,242 | 50,621 | <1 |
| Error ($a$) | 8 | 2,330,342 | 291,293 | |
| Split plots | 54 | 5,607,413 | | |
| Hours ($H$) | 3 | 2,826,997 | 942,332 | 10.27** |
| $SH$ | 15 | 1,375,786 | 91,719 | 3.71** |
| $DH$ | 6 | 713,728 | 118,953 | 4.81** |
| $TH$ | 6 | 97,600 | 16,267 | <1 |
| Error ($b$) | 24 | 593,300 | 24,721 | |

| Source | df | E(MS) |
|---|---|---|
| Whole plots | 17 | |
| $S$ | 5 | $\sigma_e^2 + 4\sigma_d^2 + 12\sigma_s^2$ |
| $D$ | 2 | $\sigma_e^2 + 4\sigma_d^2 + 24\sigma_d^2$ |
| $T$ | 2 | $\sigma_e^2 + 4\sigma_d^2 + 24\sigma_t^2$ |
| Error ($a$) | 8 | $\sigma_e^2 + 4\sigma_d^2$ |
| Split plots | 54 | |
| $H$ | 3 | $\sigma_e^2 + 3\sigma_h^2$ |
| $SH$ | 15 | $\sigma_e^2 + 3\sigma_{sh}^2$ |
| $DH$ | 6 | $\sigma_e^2 + 6\sigma_{dh}^2$ |
| $TH$ | 6 | $\sigma_e^2 + 6\sigma_{th}^2$ |
| Error ($b$) | 24 | $\sigma_e^2$ |

**7.9.** The following table presents the ANOVA with SS for polynomials.

| Source | df | SS | MS | F |
|---|---|---|---|---|
| $H$ vs. $L$ ($G$) | 1 | 12,816.90 | 12,816.90 | 13.22** |
| Bulls:$G$ ($B$:$G$) | 6 | 5,818.74 | 969.79 | 25.47** |
| Weeks ($W$) | 7 | 459.41 | 65.63 | 1.72 |
| Linear | 1 | 229.54 | 229.54 | 6.03* |
| Quadratic | 1 | 11.64 | 11.64 | 0.31 |
| Cubic | 1 | 162.00 | 162.00 | 4.25* |
| Quartic | 1 | 3.46 | 3.46 | 0.09 |
| Quintic | 1 | 10.85 | 10.85 | 0.28 |
| Residual | 2 | 41.91 | 20.95 | 0.55 |
| $GW$ | 7 | 502.02 | 71.72 | 1.88 |
| $BW$:$G$ | 30 | 1,142.34 | 38.08 | |

The variance component estimates are

$$\hat{\sigma}_e^2 = 38.08$$

$$\hat{\sigma}_{b:g}^2 = \frac{969.79 - 38.08}{6} = 155.29$$

The following table presents the partitioning of the bull group by weeks subclass sum of squares:

| Source | df | Order of elimination | SS | MS | F |
|---|---|---|---|---|---|
| Subclass | 15 | | 14,414.67 | 960.98 | 25.24** |
| H vs. L (G) | | | | | |
|   Linear | 1 | 7 | 13,412.12 | 13,412.12 | 13.83** |
| Weeks (W) | | | | | |
|   Linear | 1 | 6 | 252.94 | 252.94 | 6.64* |
|   Quadratic | 1 | 4 | 20.92 | 20.92 | 0.55 |
|   Cubic | 1 | 3 | 174.65 | 174.65 | 4.59* |
| GW | | | | | |
|   Linear × linear | 1 | 5 | 274.79 | 274.79 | 7.22* |
|   Linear × quadratic | 1 | 2 | 0.46 | 0.46 | 0.01 |
|   Linear × cubic | 1 | 1 | 8.29 | 8.29 | 0.22 |
| Residual | 8 | | 270.50 | 33.81 | 0.89 |
| Bulls:G | 6 | | 5,818.74 | 969.79 | |
| BW:G | 30 | | 1,142.34 | 38.08 | |

The prediction equation is

$$\hat{Y}_{ik} = 64.6502 - 32.0484(G - 1.5) - 1.1131(W - 4.5) - 0.1402(W - 4.5)^2$$
$$+ 0.2136(W - 4.5)^3 + 2.0240(G - 1.5)(W - 4.5)$$
$$R^2 = 0.981 \quad \text{and} \quad R = 0.99$$

Estimates of constants, least-squares means, and standard errors follow:

| Classification | Number | Constant | LS Mean | SE |
|---|---|---|---|---|
| Overall ($\hat{\mu}$) | 52 | 63.91 | 63.91 | 4.60 |
| Bull groups (G) | | | | |
|   High | 24 | 15.95 | 79.87 | 6.50 |
|   Low | 28 | −15.95 | 47.96 | 6.44 |
| Weeks (W) | | | | |
|   1 | 7 | −6.31 | 57.60 | 5.11 |
|   2 | 6 | −0.07 | 63.84 | 5.20 |
|   3 | 6 | 1.11 | 65.02 | 5.20 |
|   4 | 6 | 1.03 | 64.94 | 5.20 |
|   5 | 7 | −1.43 | 62.49 | 5.11 |
|   6 | 8 | 1.42 | 65.34 | 5.02 |
|   7 | 6 | −0.63 | 63.28 | 5.29 |
|   8 | 6 | 4.88 | 68.80 | 5.20 |
| G × W Subclasses | | | | |
|   High  1 | 3 | 3.84 | 77.39 | 7.38 |
|         2 | 3 | 1.13 | 80.93 | 7.38 |
|         3 | 3 | 3.75 | 84.72 | 7.38 |
|         4 | 3 | 1.33 | 82.22 | 7.38 |
|         5 | 3 | −4.53 | 73.91 | 7.43 |
|         6 | 4 | 0.07 | 81.36 | 6.26 |
|         7 | 2 | −0.23 | 79.00 | 9.27 |
|         8 | 3 | −5.36 | 79.39 | 7.38 |

| Classification | | Number | Constant | LS Mean | SE |
|---|---|---|---|---|---|
| Low | 1 | 4 | −3.84 | 37.81 | 6.63 |
| | 2 | 3 | −1.13 | 46.75 | 7.80 |
| | 3 | 3 | −3.75 | 45.32 | 7.76 |
| | 4 | 3 | −1.33 | 47.66 | 7.76 |
| | 5 | 4 | 4.53 | 51.06 | 6.63 |
| | 6 | 4 | −0.07 | 49.31 | 6.63 |
| | 7 | 4 | 0.23 | 47.56 | 6.63 |
| | 8 | 3 | 5.36 | 58.20 | 7.80 |

**7.10.** The following table presents the combined ANOVA:

| Source | df | SS | MS | F |
|---|---|---|---|---|
| Lines sire (LS) | 2 | 0.227882 | 0.11394 | 3.13 |
| Sires: LS | 8 | 0.291639 | 0.03646 | 1.39 |
| Lines dam (LD) | 2 | 0.022222 | 0.01111 | 0.31 |
| LS × LD | 4 | 0.639430 | 0.15986 | 4.44* |
| Sires × LD:LS | 16 | 0.575679 | 0.03598 | 1.37 |
| Within | 54 | 1.417992 | 0.02626 | |

| Source | E(MS) |
|---|---|
| Sires: LS | $\sigma_w^2 + 2.9009\sigma_{sd:ls}^2 + 7.5586\sigma_{s:ls}^2$ |
| Sires × LD:LS | $\sigma_w^2 + 2.3288\sigma_{sd:ls}^2$ |
| Within | $\sigma_w^2$ |

The variance component estimates are

$$\hat{\sigma}_w^2 = 0.02626$$
$$\hat{\sigma}_{sd:ls}^2 = 0.00417$$
$$\hat{\sigma}_{s:ls}^2 = -0.00025$$

**7.11.** The following table presents the combined ANOVA:

| Source | df | SS | MS | F |
|---|---|---|---|---|
| Lines sire (LS) | 2 | 0.227882 | 0.11394 | 3.13 |
| Sires:LS | 8 | 0.291639 | 0.03646 | 1.28 |
| Lines dam (LD) | 2 | 0.022222 | 0.01111 | 0.39 |
| LS × LD | 4 | 0.639430 | 0.15986 | 5.61** |
| Heterosis | | | | |
|   Overall | 1 | 0.584506 | 0.58451 | 20.52** |
|   Lines 1 and 2 | 1 | 0.410484 | 0.41048 | 14.41** |
|   Lines 1 and 3 | 1 | 0.190887 | 0.19089 | 6.70* |
|   Lines 2 and 3 | 1 | 0.217617 | 0.21762 | 7.64** |
| Remainder | 70 | 1.993671 | 0.02848 | |

| Source | $E(MS)$ |
|---|---|
| Sires:$LS$ | $\sigma_e^2 + 7.5586\sigma_{s:ls}^2$ |
| Remainder | $\sigma_e^2$ |

The variance component estimates are

$$\hat{\sigma}_e^2 = 0.02848$$
$$\hat{\sigma}_{s:ls}^2 = 0.00105$$

Estimates of constants, least-squares means, and standard errors follow:

| Classification | | Number | Constant | LS Mean | SE |
|---|---|---|---|---|---|
| Overall ($\hat{\mu}$) | | 87 | 1.742 | 1.742 | 0.022 |
| Line of sire ($LS$) | | | | | |
| 1 | | 36 | −0.030 | 1.712 | 0.035 |
| 2 | | 20 | 0.083 | 1.825 | 0.044 |
| 3 | | 31 | −0.053 | 1.689 | 0.036 |
| Line of dam ($LD$) | | | | | |
| 1 | | 30 | −0.013 | 1.729 | 0.036 |
| 2 | | 30 | 0.024 | 1.766 | 0.034 |
| 3 | | 27 | −0.011 | 1.731 | 0.038 |
| $LS \times LD$ Subclasses | | | | | |
| 1 | 1 | 16 | −0.120 | 1.579 | 0.046 |
| | 2 | 13 | 0.099 | 1.835 | 0.049 |
| | 3 | 7 | 0.021 | 1.722 | 0.068 |
| 2 | 1 | 6 | 0.067 | 1.879 | 0.073 |
| | 2 | 9 | −0.136 | 1.713 | 0.059 |
| | 3 | 5 | 0.069 | 1.883 | 0.080 |
| 3 | 1 | 8 | 0.053 | 1.729 | 0.064 |
| | 2 | 8 | 0.037 | 1.750 | 0.064 |
| | 3 | 15 | −0.090 | 1.588 | 0.046 |

# CHAPTER 8

**8.1.**

| Source | df | SS | MS | $F$ |
|---|---|---|---|---|
| $T$ | 2 | 1.5895 | 0.7948 | 2.32 |
| Regression | 1 | 3.8770 | 3.8770 | 11.32** |
| Error | 20 | 6.8505 | 0.3425 | |

| Source | df | SS | MS | $F$ |
|---|---|---|---|---|
| $T$ | 2 | 6.6908 | 3.3454 | 6.55** |
| $W$ | 21 | 10.7275 | 0.5108 | |

**8.2.**

| Source | df | SS | MS | F |
|---|---|---|---|---|
| $D$ | 1 | 1.0483 | 1.0483 | 65.15** |
| $T$ | 2 | 0.0683 | 0.0342 | 2.12 |
| $DT$ | 2 | 0.0704 | 0.0352 | 2.19 |
| Regression | 1 | 0.1931 | 0.1931 | 11.99** |
| Error | 17 | 0.2736 | 0.0161 | |

| Source | df | SS | MS | F |
|---|---|---|---|---|
| $D$ | 1 | 1.0923 | 1.0923 | 42.17** |
| $T$ | 2 | 0.1702 | 0.0851 | 3.28 |
| $DT$ | 2 | 0.1344 | 0.0672 | 2.59 |
| $W$ | 18 | 0.4666 | 0.0259 | |

Adj $\bar{y}_{d_1} = 1.0576$    Adj $\bar{y}_{d_2} = 0.6391$    Adj $\bar{y}_{t_1} = 0.8362$

Adj $\bar{y}_{t_2} = 0.7788$    Adj $\bar{y}_{t_3} = 0.9300$

**8.3.**

| Source | df | SS | MS | F |
|---|---|---|---|---|
| $D$ | 1 | 1.1036 | 1.1036 | 81.15** |
| $T$ | 2 | 0.0611 | 0.0306 | 2.25 |
| $DT$ | 2 | 0.0681 | 0.0340 | 2.50 |
| Regression of $X_1$ | 1 | 0.0565 | 0.0565 | 4.15 |
| Regression of $X_2$ | 1 | 0.0021 | 0.0021 | 0.15 |
| Error | 16 | 0.2170 | 0.0136 | |

Adj $\bar{y}_{d_1} = 1.0706$    Adj $\bar{y}_{d_2} = 0.6261$    Adj $\bar{y}_{t_1} = 0.7817$

Adj $\bar{y}_{t_2} = 0.8542$    Adj $\bar{y}_{t_3} = 0.9090$

**8.4.**

| Source | df | SS | MS | F |
|---|---|---|---|---|
| Whole plots | 7 | | | |
| $\quad R$ | 1 | 86.75 | 86.75 | 4.14 |
| $\quad$ Regression | 1 | 20.95 | 20.95 | 0.22 |
| $\quad$ Error | 5 | 485.43 | 97.09 | |
| Split plots | 8 | | | |
| $\quad P$ | 1 | 0.56 | 0.56 | 0.04 |
| $\quad PR$ | 1 | 7.56 | 7.56 | 0.57 |
| $\quad$ Error | 6 | 79.38 | 13.23 | |

**8.5.** The following table presents the ANOVA:

| Source | df | SS | MS | F |
|---|---|---|---|---|
| Breeds ($B$) | 2 | 0.82101 | 0.4105 | 3.70* |
| Sexes ($S$) | 1 | 1.73044 | 1.7304 | 15.60** |
| $BS$ | 2 | 1.45574 | 0.7279 | 6.56** |

| Source | df | SS | MS | F |
|---|---|---|---|---|
| Regression (*CWT*) | | | | |
| Linear—average | 1 | 2.19894 | 2.1989 | 19.82** |
| Among breeds—linear | 2 | 0.12349 | 0.0617 | 0.56 |
| Quadratic—average | 1 | 0.01139 | 0.01139 | 0.10 |
| Among breeds—quadratic | 2 | 1.84810 | 0.9241 | 8.33** |
| Remainder | 85 | 9.42910 | 0.1109 | |

The prediction equations are as follows:

For breed 1—sex 1:

$$\hat{Y} = 3.347 + 0.03890(W - 70.154) + 0.00199(W - 70.154)^2$$

For breed 1—sex 2:

$$\hat{Y} = 3.297 + 0.03890(W - 70.154) + 0.00199(W - 70.154)^2$$

For breed 2—sex 1:

$$\hat{Y} = 3.277 + 0.03826(W - 70.154) + 0.00371(W - 70.154)^2$$

For breed 2—sex 2:

$$\hat{Y} = 3.618 + 0.03826(W - 70.154) + 0.00371(W - 70.154)^2$$

For breed 3—sex 1:

$$\hat{Y} = 2.849 + 0.02161(W - 70.154) + 0.00102(W - 70.154)^2$$

For breed 3—sex 2:

$$\hat{Y} = 3.422 + 0.02161(W - 70.154) + 0.00102(W - 70.154)^2$$

Estimates of constants, least-squares means, and standard errors follow:

| Classification | | Number | Constant | LS Mean | SE |
|---|---|---|---|---|---|
| Overall | | 97 | 3.302 | 3.302 | 0.046 |
| Breeds (*B*) | | | | | |
| 1 | | 36 | 0.020 | 3.323 | 0.079 |
| 2 | | 34 | 0.146 | 3.448 | 0.076 |
| 3 | | 27 | −0.166 | 3.136 | 0.086 |
| Sexes (*S*) | | | | | |
| 1 (Male) | | 49 | −0.144 | 3.158 | 0.055 |
| 2 (Female) | | 48 | 0.144 | 3.446 | 0.063 |
| *B* × *S* Subclasses | | | | | |
| 1 | 1 | 16 | 0.169 | 3.347 | 0.096 |
| | 2 | 20 | −0.169 | 3.297 | 0.100 |
| 2 | 1 | 19 | −0.027 | 3.277 | 0.079 |
| | 2 | 15 | 0.027 | 3.619 | 0.117 |
| 3 | 1 | 14 | −0.142 | 2.850 | 0.108 |
| | 2 | 13 | 0.142 | 3.422 | 0.110 |

| Classification | Number | Constant | LS Mean | SE |
|---|---|---|---|---|
| Regressions (*CWT*) | | | | |
| Linear—average | — | 0.03292 | 0.03292 | 0.00739 |
| Breed 1 | — | 0.00597 | 0.03889 | 0.01235 |
| 2 | — | 0.00534 | 0.03826 | 0.01258 |
| 3 | — | −0.01131 | 0.02161 | 0.01347 |
| Quadratic—average | — | −0.00024 | −0.00024 | 0.00074 |
| Breed 1 | — | 0.00223 | 0.00199 | 0.00110 |
| 2 | — | −0.00347 | −0.00371 | 0.00097 |
| 3 | — | 0.00126 | 0.00102 | 0.00168 |

# CHAPTER 9

**9.1.**

**(a)** $A + B = \begin{bmatrix} 13 & 19 & 23 \\ 9 & 9 & 21 \\ 6 & 15 & 26 \end{bmatrix}$  **(b)** $B - C = \begin{bmatrix} -1 & -8 & -8 \\ -13 & -17 & -6 \\ -16 & -9 & 0 \end{bmatrix}$

**(c)** $B + C - A = \begin{bmatrix} 15 & 25 & 27 \\ 22 & 23 & 12 \\ 22 & 15 & 10 \end{bmatrix}$

**(d)** $A' + B' = \begin{bmatrix} 13 & 9 & 6 \\ 19 & 9 & 15 \\ 23 & 21 & 26 \end{bmatrix} = (A + B)' = \begin{bmatrix} 13 & 9 & 6 \\ 19 & 9 & 15 \\ 23 & 21 & 26 \end{bmatrix}$

**9.2.**

**(a)** $AB = \begin{bmatrix} 60 & 72 & 84 \\ 132 & 160 & 188 \\ 204 & 248 & 292 \end{bmatrix}$  **(b)** $BA = \begin{bmatrix} 160 & 208 \\ 268 & 352 \end{bmatrix}$

**(c)** $BB' = \begin{bmatrix} 200 & 344 \\ 344 & 596 \end{bmatrix}$  **(d)** $B'B = \begin{bmatrix} 180 & 216 & 252 \\ 216 & 260 & 304 \\ 252 & 304 & 356 \end{bmatrix}$

**(e)** Trace $BB' = 796 = $ trace $B'B$.

**(f)** $(AB)' = \begin{bmatrix} 60 & 132 & 204 \\ 72 & 160 & 248 \\ 84 & 188 & 292 \end{bmatrix} = B'A' = \begin{bmatrix} 60 & 132 & 204 \\ 72 & 160 & 248 \\ 84 & 188 & 292 \end{bmatrix}$

**9.3.**

**(a)** $AA' = \begin{bmatrix} 404 & 132 & 216 \\ 132 & 69 & 69 \\ 216 & 69 & 126 \end{bmatrix}$  $AB' = \begin{bmatrix} 188 & 62 & 74 \\ 77 & 31 & 21 \\ 93 & 33 & 39 \end{bmatrix}$

$BA' = \begin{bmatrix} 188 & 77 & 93 \\ 62 & 31 & 33 \\ 74 & 21 & 39 \end{bmatrix}$  $BB' = \begin{bmatrix} 101 & 35 & 33 \\ 35 & 14 & 10 \\ 33 & 10 & 14 \end{bmatrix}$

$(A + B)(A + B)' = \begin{bmatrix} 881 & 306 & 416 \\ 306 & 145 & 133 \\ 416 & 133 & 218 \end{bmatrix} = AA' + AB' + BA' + BB'$

**(b)** (i) $AB' = \begin{bmatrix} 188 & 62 & 74 \\ 77 & 31 & 21 \\ 93 & 33 & 39 \end{bmatrix} = (BA')'$

(ii) $A(B + B') = \begin{bmatrix} 294 & 146 & 176 \\ 140 & 71 & 73 \\ 147 & 75 & 84 \end{bmatrix} = AB + AB'$

**9.4.**

**(a)** $(\tfrac{1}{3}A)(\tfrac{1}{2}B) = \begin{bmatrix} 323 & 66 & 1341 \\ 90 & 111 & 132 \\ 30 & 36 & 42 \end{bmatrix}$

**(b)** $AB = \begin{bmatrix} 18 & 48 & 90 \\ 144 & 210 & 288 \\ 42 & 96 & 162 \end{bmatrix}$

**9.5.** $A^{-1} = \begin{bmatrix} 0.6 & -0.7 \\ -0.2 & 0.4 \end{bmatrix}$ $\qquad B^{-1} = \begin{bmatrix} 0.12 & 0.11 \\ 0.04 & 0.12 \end{bmatrix}$ $\qquad C^{-1} = \begin{bmatrix} -0.8 & -0.7 \\ -0.6 & -0.4 \end{bmatrix}$

$X^{-1} = \left(\dfrac{1}{732}\right) \begin{bmatrix} 82 & -16 & -70 \\ -44 & 80 & -16 \\ 7 & -46 & 119 \end{bmatrix}$ $\qquad Y^{-1} = \left(\dfrac{1}{540}\right) \begin{bmatrix} 47 & -11 & -2 \\ -11 & 83 & -34 \\ -2 & -34 & 92 \end{bmatrix}$

$Z^{-1} = \left(\dfrac{1}{1728}\right) \begin{bmatrix} 104 & -32 & -8 \\ -32 & 176 & -64 \\ -8 & -64 & 200 \end{bmatrix}$

**9.6.** $M^{-1} = \left(\dfrac{1}{540}\right) \begin{bmatrix} 37 & -7 & -1 \\ -7 & 67 & -29 \\ -1 & -29 & 73 \end{bmatrix}$ $\qquad N^{-1} = \left(\dfrac{1}{54}\right) \begin{bmatrix} 11 & -5 & -2 \\ -5 & 17 & -4 \\ -2 & -4 & 20 \end{bmatrix}$

**9.7. (a)** $x = 7/3$ $\qquad$ **(b)** $x = 30/13$ $\qquad$ **(c)** $x = 4$

$\qquad\quad y = 11/9$ $\qquad\quad y = 25/13$ $\qquad\quad y = 1$

**9.8.** $L^{-1} = \left(\dfrac{1}{30,240}\right) \begin{bmatrix} 1173 & -93 & -333 & 87 \\ -93 & 3333 & -747 & -1167 \\ -333 & -747 & 2853 & -927 \\ 87 & -1167 & -927 & 3693 \end{bmatrix}$

# References

Anderson, R. L. "Missing-Plot Techniques." *Biometrics,* Vol. 2, No. 3, 41–47 (1946).

Anderson, R. L. and E. E. Houseman. "Tables of Orthogonal Values Extended to $N = 104$." *Iowa State Exp. Sta. Res. Bull.* No. 297 (Apr. 1942).

Anderson, V. L. and R. A. McLean. *Design of Experiments: A Realistic Approach.* Dekker, New York (1974).

Aspin, A. A. "Tables for Use in Comparisons Whose Accuracy Involves Two Variances Separately Estimated." *Biometrika,* Vol. 36, 290–293 (1949).

Bancroft, T. A. *Topics in Intermediate Statistical Methods,* Vol. 1. Iowa State University Press, Ames (1968).

Bartlett, M. S. "Properties of Sufficiency and Statistical Tests." *Proc. R. Soc.,* Vol. A160, 268–282 (1937a).

Bartlett, M. S. "Some Examples of Statistical Methods of Research in Agriculture and Applied Biology." *J. R. Stat. Soc. Suppl.,* Vol. 4, 137–183 (1937b).

Beyer, William H., Editor. *Handbook of Tables for Probability Statistics.* CRC Press, Boca Raton, FL (1979).

Box, G. E. P. "Problems in the Analysis of Growth and Wear Curves." *Biometrics,* Vol. 6, No. 4, 362–389 (1950).

Bozivich, Helen, T. A. Bancroft, and H. O. Hartley. "Power of Analysis of Variance Test Procedures for Certain Incompletely Specified Models. I." *Ann. Math. Stat.,* Vol. 27, No. 4, 1017–1043 (1956).

Carmer, S. G. and R. D. Seif. "Calculation of Orthogonal Coefficients When Treatments Are Unequally Replicated and/or Unequally Spaced." *Agron. J.,* Vol. 55, 387–389 (1963).

Carmer, S. G. and M. R. Swanson. "Evaluation of Ten Pairwise Multiple Comparison Procedures by Monte Carlo Methods." *J. Am. Stat. Assoc.,* Vol. 68, 66–74 (1973).

Cochran, William G. "Testing a Linear Relation Among Variances." *Biometrics,* Vol. 7, No. 1, 17–32 (1951).

Cochran, William G. and Gertrude M. Cox. *Experimental Designs.* Wiley, New York (1957).

Cole, J. W. L. and J. E. Grizzle. "Applications of Multivariate Analysis of Variance to Repeated Measurements." *Biometrics,* Vol. 22, No. 4, 810–828 (1966).

Cooley, William W. and Paul R. Lohnes. *Multivariate Data Analysis.* Wiley, New York (1971).

Crout, Prescott D. "A Short Method for Evaluating Determinants and Solving Systems of Linear Equations with Real or Complex Coefficients." *Trans. Am. Inst. Electr. Eng.,* Vol. 60, 1–7 (1951).

Danford, M. B., Harry M. Hughes, and R. C. McNee. "On the Analysis of Repeated Measurements Experiments." *Biometrics,* Vol. 16, No. 4, 547–565 (1960).

Delury, D. B. "The Analysis of Covariance." *Biometrics,* Vol. 4, No. 3, 153–170 (1948).

Dixon, W. J. "Analysis of Extreme Values." *Ann. Math. Stat.,* Vol. 21, 488 (1950).

Dixon, Wilfred J. and Frank J. Massey, Jr. *Introduction to Statistical Analysis,* 3rd ed. McGraw-Hill, New York (1969).

Duncan, D. B. "Multiple Range and *F*-tests." *Biometrics,* Vol. 11, No. 1, 1–42 (1955).

Duncan, D. B. "Multiple Range Tests for Correlated and Heteroscedastic Means." *Biometrics,* Vol. 13, No. 2, 164–176 (1957).

Duncan, C. J. and P. M. Sheppard. "Sensory Evolution and Its Role in the Evolution of Batesian Mimicry." *Behavior,* 269–282 (1965).

Dunn, O. J. and V. Clark. "Comparison of Tests of the Equality of Dependent Correlation Coefficients." *J. Am. Stat. Assoc.,* Vol. 66, 904–908 (1971).

Dunnett, C. W. "A Multiple Comparison Procedure for Comparing Several Treatments with a Control." *J. Am. Stat. Assoc.,* Vol. 50, 1096–1121 (1955).

Federer, Walter T. *Experimental Design.* Macmillan, New York (1955).

Fisher, R. A. "Applications of 'Student's' Distribution." *Metron,* Vol. 5, 90–104 (1926).

Fisher, R. A. "The Use of Multiple Measurements in Taxonomic Problems." *Ann. Eugen.,* Vol. 7, 179–188 (1936).

Fisher, R. A. *The Design of Experiments,* six editions. Oliver and Boyd, London (1935–1951).

Fisher, R. A. and F. Yates. *Statistical Tables,* 5th ed. Oliver and Boyd, Edinburgh (1957).

Gill, John L. *Design and Analysis of Experiments,* Vols. I and II. Iowa State University Press, Ames (1978).

Grandage, A. "Queries and Notes." *Biometrics,* Vol. 14, No. 2, 287–289 (1958).

Graybill, Franklin A. *Introduction to Matrix Algebra with Applications in Statistics.* Wadsworth, Belmont, CA (1969).

Harvey, W. R. "Least Squares Analysis of Data with Unequal Subclass Numbers." USDA, ARS 20-8 (1960).

Harvey, W. R. "Estimation of Variance and Covariance Components in the Mixed Model." *Biometrics,* Vol. 26, No. 3, 485–505 (1970).

Harvey, W. R. "Missing Subclass Problems Using Method 3 of Henderson in Mixed Models." *Proc. Stat. Assoc.,* 22–26 (1977a).

Harvey, W. R. "User's Guide for LSML76." The Ohio State University, Columbus (1977b).

Harvey, W. R. "Least-Squares Analysis of Data with Unequal Subclass Numbers." Science and Education Administration, Agricultural Research (1979).

Harvey, W. R. "Estimation of Fixed Effects in Mixed Models with Unbalanced Data." In *Proceedings of the 6th Annual SAS User's Group International Conference,* Orlando, FL, Feb. 8–11, pp. 78–84. SAS Institute, Cary, NC (1981).

Harvey, W. R. "Least-Squares Analysis of Discrete Data." *J. Anim. Sci.,* Vol. 54, No. 5, 1067–1071 (1982a).

Harvey, W. R. "Mixed Model Capabilities of LSML76." *J. Anim. Sci.,* Vol. 54, No. 6, 1279–1285 (1982b).

Harvey, W. R., M. E. Hourihan, and C. E. Terrill. "Estimation of the Mean Fiber Length of Individual Grease Fleeces." *J. Anim. Sci.,* Vol. 27, No. 5, 1224–1228 (1968).

Harvey, W. R., R. H. Ross, and D. L. Fourt. "The Importance of Judges, Heifers and Ages in Causing Variation in Type Ratings of Young Dairy Heifers." In *Proceedings of the 34th Annual Meeting of the Western Division of the American Dairy Association,* pp. 29–49. American Dairy Science Association, Champaign, IL (1953).

Harvey, W. R. and L. A. Swiger. "Orthogonal Polynomial Fitting with Arbitrary Spacing and Correlated Means." In Proceedings of the 3rd Annual SAS User's Group International Conference, Las Vegas, NV, Jan. 30–Feb. 1, pp. 108–112. SAS Institute, Cary, NC (1978).

Henderson, C. R. "Estimation of Variance Components and Covariance Components." *Biometrics,* Vol. 9, 226–252 (1953).

Henderson, C. R. "Design and Analysis of Animal Husbandry Experiments." *Am. Soc. Anim. Prod.,* pp. 1–55 (1959) [2nd ed., 1969].

Hohn, Franz E. *Elementary Matrix Algebra.* Macmillan, New York (1958).

Keuls, M. "The Use of the 'Studentized Range' in Connection with an Analysis of Variance." *Euphytica,* Vol. 1, 112–122 (1952).

Kramer, C. Y. "Extension of Multiple Range Tests to Group Means with Unequal Number of Replications." *Biometrics,* Vol. 12, No. 3, 307–310 (1956).

Kramer, C. Y. "Extension of Multiple Range Tests to Group Correlated Adjusted Means." *Biometrics,* Vol. 13, No. 1, 13–18 (1957).

Kroll, W. P., W. L. Kilmer, L. L. Bultman, and J. Boucher. "Prediction of Male and Female Isometric Arm Strength by Anthropometric Measures." *Human Biology* (to be published) (1983).

Lachenbruck, Peter A. *Discriminant Analysis.* Hafner Press, New York (1975).

Li, Jerome C. R. *Introduction to Statistical Theory.* Edwards Brothers, Ann Arbor, MI (1957).

Lindeman, Richard H., Peter F. Merenda, and Ruth Z. Bold. *Introduction to Bivariate and Multivariate Analysis.* Scott, Foresman, Chicago (1980).

Mills, H. A., A. V. Barker, and D. N. Maynard. "Nitrate Accumulation in Radish as Affected by Nitrapyrin." *Agron. J.,* Vol. 68, 13–17 (1976).

Newman, D. "The Distribution of the Range in Samples from a Normal Population Expressed in Terms of an Independent Estimate of Standard Deviation." *Biometrika,* Vol. 31, 20–30 (1939).

Nie, Norman H., C. Hadlai Hull, Jean G. Jenkins, Karin Steinbrenner, and Dale H. Bent. *Statistical Package for the Social Sciences.* McGraw-Hill, New York (1975).

Ochoa, P. G., W. L. Mangus, J. S. Brinks, and A. H. Denham. "Effect of Creep Feeding Bull Calves on Dam Most Probable Producing Ability Values." *J. Anim. Sci.,* Vol. 53, No. 3, 567–574 (1981).

*SAS User's Guide: Statistics.* SAS Institute, Cary, NC (1982).

Satterthwaite, F. E. "An Approximate Distribution of Estimates of Variance Components." *Biometrics,* Vol. 2, No. 6, 110–114 (1946).

Scheffé, H. "A Method for Judging All Contrasts in the Analysis of Variance." *Biometrika,* Vol. 40, 87–104 (1953).

Searle, S. R. *Matrix Algebra for the Biological Sciences.* Wiley, New York (1966).

Shelby, C. E., W. R. Harvey, R. T. Clark, J. R. Quesenberry, and R. R. Woodward. "Estimates of Phenotypic and Genetic Parameters in Ten Years of Miles City R.O.P. Steer Data." *J. Anim. Sci.,* Vol. 22, No. 2, 346–353 (1963).

Snedecor, George W. *Analysis of Variance and Covariance.* Collegiate Press, Ames, IA (1934).

Snedecor, George W. and William G. Cochran. *Statistical Methods,* 7th ed. Iowa State University Press, Ames (1980).

Snedecor, G. W. and G. M. Cox. *Iowa Agric. Exp. Sta. Res. Bull.* 180 (1935).

Sokal, Robert R. and F. James Rohlf. *Biometry.* W. H. Freeman, San Francisco (1969).

Solandt, D. Y., D. B. DeLury, and John Hunter. "The Effect of Atropine and Quinidine Sulphate on Atrophy and Fibrillation in Denervated Skeletal Muscle." *Am. J. Physiology,* Vol. 140, No. 2, 247–255 (1943).

Steel, Robert B. D. and James H. Torrie. *Principles and Procedures of Statistics,* 2nd ed. McGraw-Hill, New York (1980).

Student (W. S. Gossett). "The Probable Error of a Mean." *Biometrika,* Vol. 6, No. 1, 1–25 (1908).

*SUGI Supplemental Library Users Guide.* SAS Institute, Cary, NC (1983).

Tukey, J. W. "One Degree of Freedom for Additivity." *Biometrics,* Vol. 5, 232–242 (1949).

Tukey, J. W. "The Problem of Multiple Comparisons." Unpublished Ditto Notes, Princeton University, 396 pp. (1953).

Welch, B. L. "The Generalization of Student's Problem when Several Different Population Variances Are Involved." *Biometrika,* Vol. 34, 28–35 (1947).

Williams, E. J. "Experimental Designs Balanced for the Estimation of Residual Effects of Treatments." *Australian J. Sci. Res. A,* Vol. 2, 149–168 (1949).

Winer, B. J. *Statistical Principles in Experimental Design,* 2nd ed. McGraw-Hill, New York (1971).

Yates, F. "The Analysis of Replicated Experiments When the Field Results Are Incomplete." *Empire J. Exper. Agr.,* Vol. 1, 129–142 (1933).

Yates, F. "The Analysis of Multiple Classifications with Unequal Numbers in the Different Subclasses." *J. Am. Stat. Assoc.,* Vol. 29, 51–66 (1934).

# Index